ENGINEERING MATERIALS

ENGINEERING MATERIALS

S L Kakani

Former Executive Director

Institute of Technology and Management, Bhilwara

(*Affiliated to Rajasthan Technical University, Kota*)

Amit Kakani

Publishing Globally

NEW AGE INTERNATIONAL (P) LIMITED, PUBLISHERS

New Delhi • Bangalore • Chennai • Cochin • Guwahati

Hyderabad • Kolkata • Lucknow • Mumbai

Visit us at www.newagepublishers.com

Published by New Age International (P) Ltd., Publishers
First Edition: 2014

BRANCHES

- **Bangalore** 37/10, 8th Cross (Near Hanuman Temple), Azad Nagar, Chamarajpet, Bangalore-560 018
  Tel.: (080) 26756823, Telefax: 26756820, **E-mail: bangalore@newagepublishers.com**
- **Chennai** 26, Damodaran Street, T. Nagar, Chennai-600 017
  Tel.: (044) 24353401, Telefax: 24351463, **E-mail: chennai@newagepublishers.com**
- **Cochin** CC-39/1016, Carrier Station Road, Ernakulam South, Cochin-682 016
  Tel.: (0484) 2377004, Telefax: 4051303, **E-mail: cochin@newagepublishers.com**
- **Guwahati** Hemsen Complex, Mohd. Shah Road, Paltan Bazar, Near Starline Hotel, Guwahati-781 008
  Tel.: (0361) 2513881, Telefax: 2543669, **E-mail: guwahati@newagepublishers.com**
- **Hyderabad** 105, 1st Floor, Madhiray Kaveri Tower, 3-2-19, Azam Jahi Road, Nimboliadda, Hyderabad-500 027
  Tel.: (040) 24652456, Telefax: 24652457, **E-mail: hyderabad@newagepublishers.com**
- **Kolkata** RDB Chambers (Formerly Lotus Cinema) 106A, 1st Floor, S.N. Banerjee Road, Kolkata-700 014
  Tel.: (033) 22273773, Telefax: 22275247, **E-mail: kolkata@newagepublishers.com**
- **Lucknow** 16-A, Jopling Road, Lucknow-226 001
  Tel.: (0522) 2209578, 4045297, Telefax: 2204098, **E-mail: lucknow@newagepublishers.com**
- **Mumbai** 142C, Victor House, Ground Floor, N.M. Joshi Marg, Lower Parel, Mumbai-400 013
  Tel.: (022) 24927869, Telefax: 24915415, **E-mail: mumbai@newagepublishers.com**
- **New Delhi** 22, Golden House, Daryaganj, New Delhi-110 002
  Tel.: (011) 23262368, 23262370, Telefax: 43551305, **E-mail: sales@newagepublishers.com**

ISBN: 978-81-224-3621-1
₹ 299.00
C-13-06-6975

Printed in India at Ajit Printers, Delhi.
Typeset at Pegitek Graphics, Delhi.

PUBLISHING FOR ONE WORLD
NEW AGE INTERNATIONAL (P) LIMITED, PUBLISHERS
7/30 A, Daryaganj, New Delhi-110002
Visit us at **www.newagepublishers.com**

Preface

Materials have always been the necessity of mankind. The early historical ages of human civilization *e.g.* Stone age, Bronze age and Iron age clearly reveals the importance of materials in human life. Materials are the building blocks for all technological developments. During the 20th century, tremendous progress have been made in understanding and analysing the structure—property correlation for almost all types of materials—from metals, ceramics, composites, semiconductors, polymers, nanomaterials, smart materials, metamaterials, nonlinear optical materials, etc. This has made essential for all engineering graduates and diploma holders to undergo a basic course in engineering materials in order to understand their nature, properties and applications which are influenced to a great extent by the structure of materials.

Organized in ten chapters, this book fully covers the syllabi of the subject prescribed for degree and diploma students in engineering and M.Sc. (Materials), and preparing for GATE, and other entrance examinations.

A reasonably wide coverage in sufficient depth has been attempted, giving the importance to the basic principles, structure, standards and specifications, experimental details, theory and applications. All efforts have been made to cover latest developments in the field, *e.g.* special topics on nano-materials smart materials, non-linear optical materials, polymers, ceramics, composites are dealt in sufficient depth.

To make text more useful, good number of worked out problems, summary, review questions, exercises, short-question answers, objective questions are provided with each chapter.

We are thankful to M/S New Age International (P) Limited Publishers for their untiring efforts in bringing out the book with excellent printing and nice get up within the shortest possible time period.

Suggestions for the improvement of the book are most welcome.

S.L. Kakani
<slkakani28@gmail.com>
Amit Kakani

Contents

Contents

CHAPTER

1

Introduction: Classification and Properties of Materials

1.1 INTRODUCTION

Materials have always been an integral part of human civilization and social development, *e.g.* we designate periods in the past as the *stone, Bronze* and *Iron* ages. Recent advances in technologies rely on sophisticated materials—all of them used devices, products, and systems that consist of materials. With the rapid advances in computer technology, design engineering have become quite sophisticated. Nowadays, mathematical models have been developed to study stress and process kinetics; computers are used to make detailed drawings and machines are controlled by minicomputers.

Materials that are used by us are in one form or another—some in their pure elemental form, *e.g.* copper, silver, nickel, etc.; some in the form of alloys and compounds, *e.g.* steel, brass etc. and some as composites, *e.g.* fibreglass, wood, etc. The selection of materials and most appropriate manufacturing process depends on several factors, but the most important considerations are shape complexity and properties of material. However, the properties of materials are ultimately linked with the *microstructure* and *processing*.

Many times, a materials problem is one of selecting the right material from many thousands that are available. There are several criteria on which the final decision is normally based. First of all, the in-service conditions must be characterized, for these will dictate the properties required of the material. However, rarely does a material possess the maximum or ideal combination of properties. Clearly, it may be necessary to trade off one characteristic for another. For example, a material having a high strength will have only a limited ductility. This means, one will have to make a reasonable compromise between two or more properties may be necessary.

The second consideration in the selection of a material is any deterioration of material properties that may occur during service operation, *e.g.* significant reductions in mechanical strength may result from exposure to elevated temperatures or corrosive environments.

Finally, probably the overriding consideration is that of cost. A material may have the ideal set of properties but is prohibitively expensive. Obviously, some compromise will have to be made.

An engineer must have a good knowledge of the various characteristics and structure—property relationships, as well as processing techniques of materials. This helps him to make judicious materials choices based on these criteria.

1.2 CLASSIFICATION OF MATERIALS

Based on chemical make up and atomic structure, solid materials have been conveniently grouped into three basic categories: *metals, ceramics* and *polymers*. Most materials fall into one distinct grouping or another, although there are also some intermediates. In addition to these, there are also three other groups of important engineering materials: *composites, semiconductors* and *biomaterials*. There are also *advanced materials* utilized in high-technology (or high-tech). Recently, a group of new and state-of-the art materials called as *smart* (or *intelligent*) materials being developed. Very recently, scientists have developed *nano-engineering materials*. A brief description of the material types and representative characteristics are:

(i) Metals

Normally metallic materials are combinations of metallic elements. Metallic materials have large number of *nonlocalized electrons*, *i.e.* electrons are not bound to particular atoms. Many properties of metals are directly attributable to these electrons. All metals are characterized by metallic properties, *e.g.* luster, opacity, malleability, ductility and electrical conductivity. Although metals compose about three fourth of the known elements but few find service in their pure form. The desired properties for engineering purposes are often found in *alloys*. Typical examples of metallic materials are iron, aluminium, copper, zinc, etc. and their alloys. They can be used either in bulk or powder form. Metals are extremely good conductors of electricity and heat are not transparent to visible light; a polished metal surface has a lustrous appearance. Moreover, metals are quite strong, yet deformable, which accounts for their extensive use in structural applications.

Metallic materials are always crystalline in nature. Scientists have developed *amorphous* (non-crystalline) alloys by very rapid cooling of a melt or by very high-energy mechanical miling.

Recently, scientists have developed materials through rapid solidification called as *quasicrystals*. These are neither crystalline nor amorphous, but form an ordered structure somewhere between two known structures. These materials are expected to exhibit far reaching electrical properties.

(ii) Ceramics

These are crystalline compounds between metallic and non-metallic elements. They are most frequently oxides, nitrides and carbides. Nowadays graphite is also categorized in ceramics. The wide range of materials which falls within this classification include ceramics that are composed of clay minerals, cement and glass. Glass is grouped with this class because it has similar properties but most glasses are *amorphous*. Ceramics are characterised by high hardness, abrasion resistance, brittleness and chemical inertness. Ceramics are typically insulative to the passage of electricity and heat, and are more resistant to high temperatures and harsh environments than metals and polymers. With regard to mechanical behaviour, these materials are hard but very brittle. These materials are widely categorized into oxide and non-oxide ceramics.

(iii) Polymers

Many of these are organic substances and derivatives of carbon and hydrogen. Polymers include the familiar plastic and rubber materials. Usually polymers are classified into three categories: *thermoplastic* polymers, *thermosetting* polymers and elastomers, better called as rubbers. Polymers have very large molecular structures. Most plastic polymers are light in weight and are soft in comparison to metals. Polymer materials have typically low densities and may be extremely flexible and widely used as insulators, both thermal and electrical. Few examples of polymers are polyesters, phenolics, polyethylene, nylon and rubber. The overriding consideration of the selection of a given *polymer* is whether or not the material can be processed into the required article easily and economically.

(iv) Composites

A composite is a composition of two or more materials in the first three categories, *e.g.* metals, ceramics and polymers, that has properties from its constituents. Large number of composite materials have been engineered. Few typical examples of composite materials are wood, clad metals, fibre glass, reinforced plastics, cemented carbides, etc. Fibre glass is a most familiar composite material, in which glass fibres are embedded within a polymeric material. A composite is designed to display a combination of the best characteristics of each of the component materials. Fibre glass acquires strength from the glass and the flexibility from the polymer. A true composite structure should show matrix material completely surrounding its reinforcing material in which the two phases act together to exhibit desired characteristics. These materials as a class of engineering material provide almost an unlimited potential for higher strength, stiffness, and corrosion resistance over the 'pure' material systems of metals, ceramics and polymers. Many of the recent developments of materials have involved composite materials. Probably, the composites will be the steels of this century.

Nowadays, the rapidly expanding field of *nano composites* is generating many exciting new materials with novel properties. The general class of nano composite organic or inorganic material is a fast growing field of research. Significant efforts are going on to obtain control of nano composite materials depend not only on the properties of their individual parents but also on their morphology and interfacial characteristics. The lamellar class of intercalated organic/inorganic nano composites and namely those systems that exhibit electronic properties in atleast one of the composites offers the possibility of obtaining well ordered systems some of which may lead to unusual electrical and mechanical properties. Polymer-based nano composites are also being developed for electronic applications such as thin-film capacitors in integrated circuits and solid polymer electrolytes for batteries. No doubt, the field of nano composites is of broad scientific interest with extremely impressive technological promise.

The properties of these four varieties of materials have been contrasted in Table 1.1. Figure 1.1 shows the evolution of various types of composites along with the classification of the engineering materials.

Table 1.1 Four varieties of materials and the contrast between their properties

Properties	*Metals*	*Ceramics*	*Polymers*
• Strength	Moderately high	High (hard)	Low
• Resistance to loss of strength with rise in temperature	Moderate	High	Poor
• Ease of fabrication	Good (ductile)	Poor (brittle)	Very good (ductile)
• Conduction (Thermal and electrical)	Good conductor	Insulator	Insulator
• Resistance to chemical attack	Poor	Inert	Inert
• Dimensional stability (stiffness)	High	High	Poor
• Specific gravity	Very high	Low (light)	Low (light)
• Lusture	Excellent	Poor	Poor

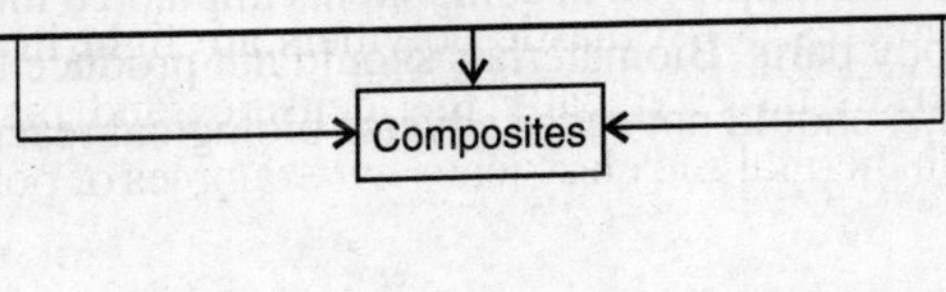

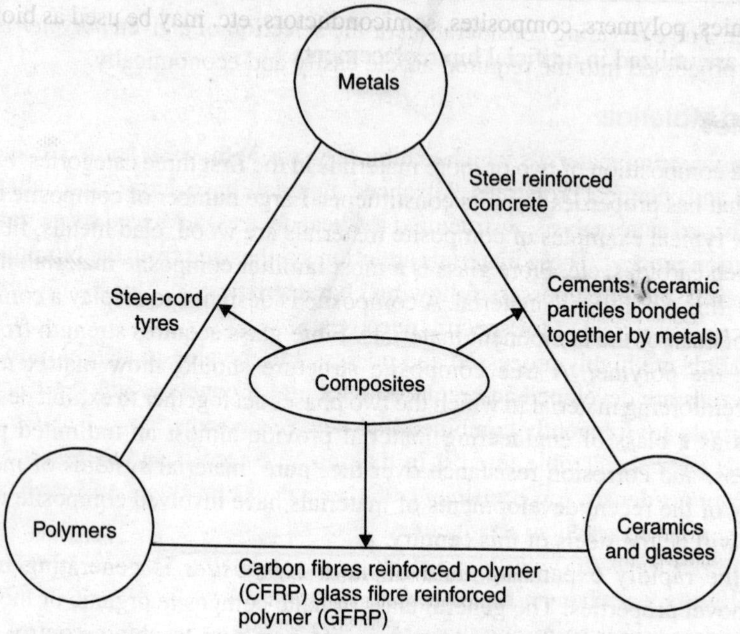

Fig. 1.1 Evolution of various types of composites. Figure also shows the classification of engineering materials.

(v) Semiconductors

These materials have electrical properties that are intermediate between electrical conductors and insulators. Moreover, the electrical characteristics of semiconducting materials are extremely sensitive to the presence of minute concentrations of impurity atoms; these concentrations may be controlled over very small spatial regions. Silicon, Germanium and compounds listed in Table 1.2 form the vast majority of semiconducting crystals.

Table 1.2 Semiconducting compounds

Column in periodic table	*Compounds*
IV – IV	SiC
III – V	AlP, AlAs, AlSb, GaP, GaAs, GaSb, InP, InAs, InSb
II – VI	ZnO, ZnS, CdS, CdSe, CdTe
I – VII	CuCl, AgI

These semiconducting materials are used in a number of solid state devices, *e.g.* diodes, transistors, photoelectric devices, solar batteries, radiation detectors, thermistors and lasers. The semiconductors have made possible the advent of *integrated circuitary* that has completely revolutionized the electronics and computer industries.

(vi) Biomaterials

These materials are employed in components implanted into the human body for replacement of diseased or damaged body parts. Biomaterials should not produce toxic substances and must be compatible with body tissues, *i.e.* should not cause adverse biological reactions. We may note all of the above materials, *i.e.* metals, ceramics, polymers, composites, semiconductors, etc. may be used as biomaterials. Some of the biomaterials are utilized in artificial hip replacements.

(vii) Advanced Materials

These are new engineering materials which exhibit high strength, great hardness, and superior thermal, electrical, optical and chemical properties. Advanced materials have dramatically altered communication technologies, reshaped data analysis, restructured medical devices, advanced space travel and transformed industrial production process. These materials are often synthesized from the biproducts of conventional commodity materials and often possess following characteristics:

- These materials are created for specific purposes,
- These materials are highly processed and possess a high value-to weight ratio,
- These materials are developed and replaced with high frequency, and
- These materials are frequently combined into new composites.

Nowadays, there is considerable interest in making advanced materials that are usually graded by chemical composition, density or coefficient of thermal expansion of material or based on micro-structural features, *e.g.* a particular arrangement of second-phase particles or fibres in a matrix. Such materials are referred as *functionally graded materials.* Instead of having a step function, one may strive to achieve a gradual change. Such gradual change will reduce the chances of mechanical and thermal stresses, generally present otherwise. We may note that the concept of a functionally graded material is applicable to any material metal, polymer or ceramic. A lot of research work is going on these materials.

(viii) Materials of the Future

(a) Smart Materials

Smart or intelligent materials form a group of new and state of art materials now being developed that will have a significant influence on many of present-day technologies. The adjective 'smart' implies that these materials are able to sense changes in their environments and then respond to these changes in predetermined manners—traits that are also found in living organisms. In addition, the concept of smart materials is being extended to rather sophisticated systems that consist of both smart and traditional materials.

The field of smart materials attempts to combine the *sensor* (that detects an input signal), *actuator* (that performs a responsive and adaptive function) and the control circuit or as one integrated unit. Acutators may be called upon to change shape, position, natural frequency, or mechanical characteristics in response to changes in temperature, electric fields, and or magnetic fields.

Usually, four types of materials are commonly used for actuators: *shape memory alloys, piezo-electric ceramics, magnetostrictive materials,* and *electrorheological/magnetorheological fluids*. Shape memory alloys are metals that, after having been deformed, revert back to their original shapes when temperature is changed. Piezoelectric ceramics expand and contract in response to an applied electric field (or voltage); conversely these materials also generate an electric field when their dimensions are altered. The behaviour of magnetostrictive materials is analogous to that of the piezoelectric ceramic materials, except that they are responsive to magnetic fields. Also, electrorheological and magnetorheological fluids are liquids that experience dramatic changes in viscocity upon application of electric and magnetic fields, respectively.

The combined system of sensor, actuator and control circuit or as one IC unit, emulates a biological system (Fig. 1.2).

These are known as smart sensors, microsystem technology (MST) or micromechanical systems (MEMS). Materials/devices employed as sensors include optical fibres, piezoelectric materials (including some polymers) and MEMS.

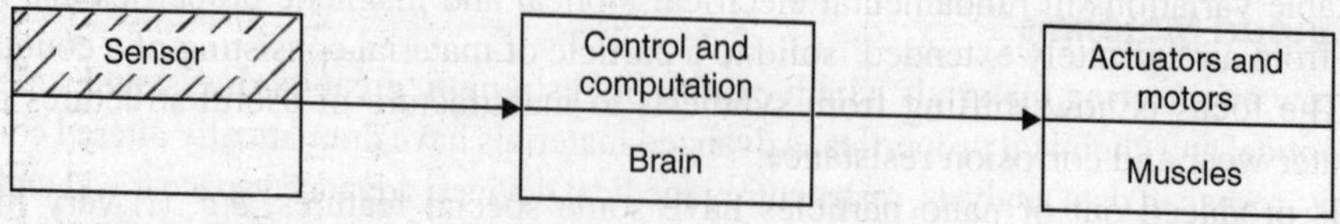

Fig. 1.2 Integrated sensor—actuator systems with controller are analogous to biological systems.

For example, one type of smart system is used in helicopters to reduce aero-dynamic cockpit noise that is created by the rotating rotor blades. Piezoelectric sensors inserted into the blades, monitor blade stresses and deformations; feedback signals from these sensors are fed into a computer controlled adaptive device, which generates noise cancelling antidose.

MEMS devices are small in size, light weight, low cost, reliable with large batch fabrication technology. They generally consist of sensors that gather environmental information such as pressure, temperature, acceleration, etc., integrated electronics to process the data collected and actuators to influence and control the environment in the desired manner.

The MEMS technology involves a large number of materials. Silicon forms the backbone of these systems also due to its excellent mechanical properties as well as mature micro-fabrication technology including lithography, etching, and bonding. Other materials having piezoelectric, piezoresistive, ferroelectric and other properties are widely used for sensing and actuating functions in conjunction with silicon.

(*b*) *Nano Materials*

Nano-structured (NS) materials are defined as solids having microstructural features in the range of 1–100 nm (= (1–100) × 10^{-9}m) in at least in one dimension. These materials have outstanding mechanical and physical properties due to their extremely fine grain size and high grain boundary volume fraction. Usually, the clusters of atoms consisting of typically hundreds to thousands on the nanometer scale are called as nanoclusters. These small group of atoms, in general, go by different names such as nano particles, nanocrystals, quantum dots and quantum boxes. Significant work in being carried out in the domain of nano-structured materials and nano tubes since they were found to have potential for high technology engineering applications. Nano-structured materials exhibit properties which are quite different from their bulk properties. These materials contain a controlled morphology with atleast one nano scale dimension. Nano crystals, nano wires and nano tubes of a large number of inorganic materials have been synthesized and characterized in the last few years. Some of the nano materials exhibit properties of potential technological value. This is particularly true for nano-structures of semiconducting materials such as metal chalcogenides and nitrides. The mixing of nano-particles with polymers to form composite materials has been practiced for decades. For example, the clay reinforced resin known as Bakelite is the first mass-produced polymer-nanoparticle composites and fundamentally transformed the nature of practical household materials. Even before bakelite, nano composites were finding applications in the form of nano particle-toughened automobile tires prepared by blending carbon black, zinc oxide, and/or magnesium sulfate particles with vulcanized rubber. Despite these early successes, the broad scientific community was not galvanized by nano composites until the early 1990s, when reports revealed that adding mica to nylon produced a five-fold increase in the yield and tensile strength of the material. Subsequent developments have further contributed to the surging interest in polymer–nano particle composites.

Significant progress has been made in various aspects of synthesis of nano-structured materials. The explosion of both academic and industrial interest in these materials over the past decade arises from the remarkable variations in fundamental electrical, optical and magnetic properties that occur as one progresses from an 'infinitely extended' solid to a particle of material consisting of a countable number of atoms. The focus is now shifting from synthesis to manufacture of useful structures and coatings having greater wear and corrosion resistance.

Materials produced out of nano particles have some special features, *e.g.* (*i*) very high ductility (*ii*) very high hardness ~4 to 5 times more than usual conventional materials (*iii*) transparent ceramics achievable (*iv*) manipulation of colour (*v*) extremely high coercivity magnets (*vi*) developing conducting inks and polymers.

1.3 MODERN MATERIALS' NEEDS

Material science has expanded from the traditional metallurgy and ceramics into new areas such as electronic polymers, complex fluids, intelligent materials, organic composites, structural composites, biomedical materials (for implants and other applications), biomimetics, artificial tissues, biocompatible materials, "auxetic" materials (which grow fatter when stretched), elastomers, dielectric ceramics (which yield thinner dielectric layers for more compact electronics), ferroelectric films (for non-volatile memories), more efficient photovoltaic converters, ceramic superconductors, improved battery technologies, self-assembling materials, fuel cell materials, optoelectronics, artificial diamonds, improved sensors (based on metal oxides, or conducting polymers), grated light values, ceramic coatings in air (by plasma deposition), electrostrictive polymers, chemical—mechanical polishing, alkali-metal thermoelectric converters, luminescent silicon, planar optical displays without phosphors, MEMS, and super molecular materials. However, there still remain technological challenges, including the development of even more sophisticated and specialized materials, as well as consideration of the environmental impact of materials production. Material scientists are interested in *green* approaches, by entering the field of environmental—biological science, by developing environmentally friendly processing techniques and by inventing more recyclable materials.

Nano materials is another emerging field called as *Nano materials by severe plastic deformation* (SPD) which involves the application of very high strains and flow stresses to work pieces. The respective new processes yields microstructural features and properties in materials (notably metals and alloys) that differ features and properties in materials (notably metals and alloys) that differ from those known for conventional cold worked materials. Specifically, pore-free grain refinements down to nanometer dimensions, and dislocation accumulations upto the limiting density of 10^{16} m^{-2} are observed. SPD yields an increase in tensile ductility without a substantial loss in strength and fatigue behaviour. Furthermore, unusual phase transformations leading to highly metastable states have been reported and are associated with a formation of supersaturated solid solutions, disordering, amorphization, and a high thermal stability. Moreover, superplastic elongations in alloys that are generally not superplastic can be achieved. This affords a superplastic flow at strain rates significantly faster than in conventional alloys, enabling the rapid fabrication of complex parts. Finally, the magnetic properties of severely plastic deformed materials are different from their conventional counterparts. In particular, one observes an enhanced zemanence in hard magnetic materials, a decrease of coercivity, (*i.e.* energy loss) in soft magnetic materials, and an induced magnetic anisotropy.

Clearly, the fields of materials is extending into new territory, and this trend is expected to continue. Figure 1.3 displays the development of essential material properties during the 20th century.

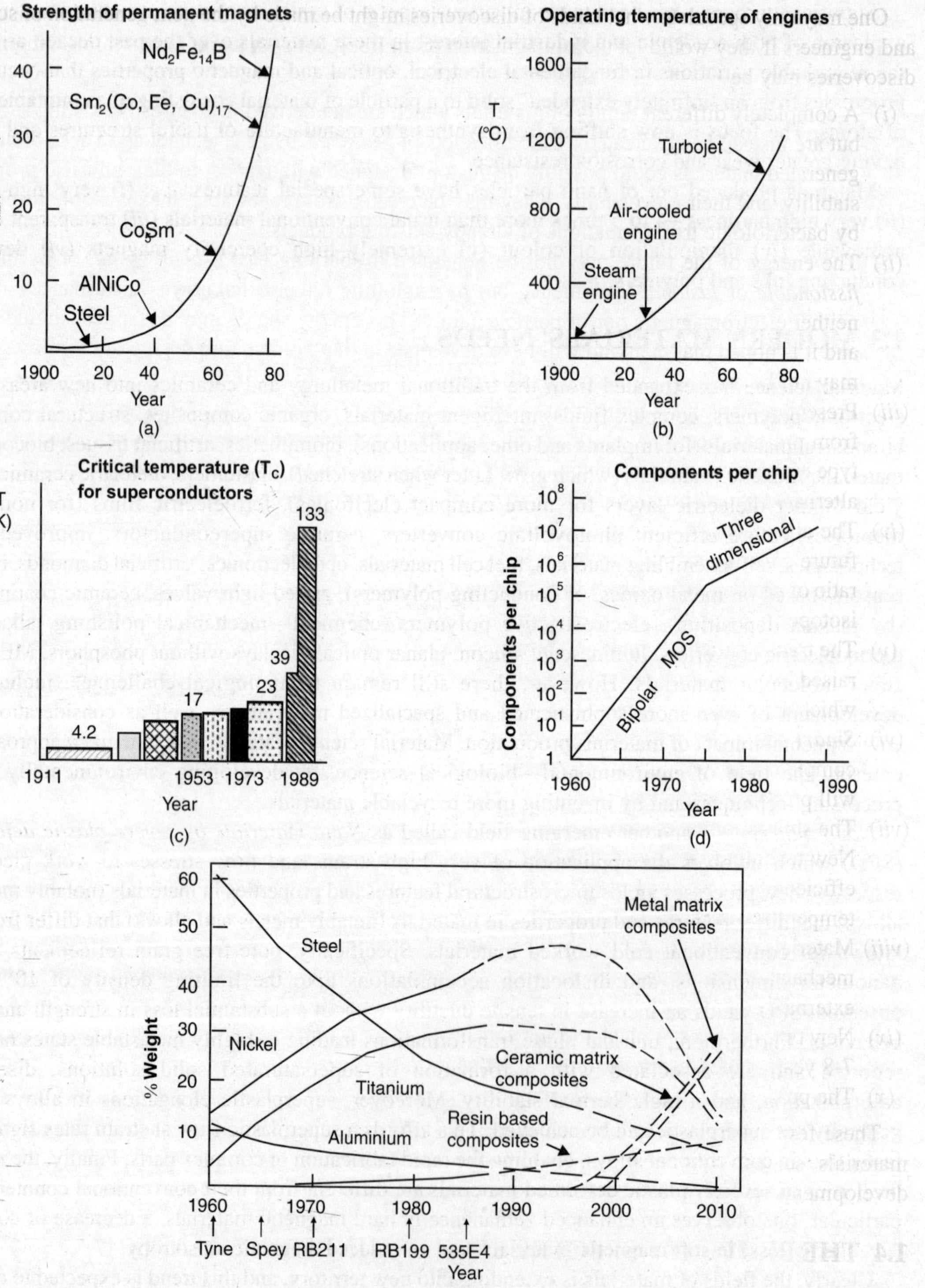

Fig. 1.3 Improvement of materials properties during the 20th century.

One may only speculate what kinds of discoveries might be made by the next generation of scientists and engineers if they would let their imagination roam freely into yet unexplored realms. Among these discoveries may be

(*i*) A completely different family of materials which are not derived from already existing substances but are instead newly created by modification of *genes, i.e.* by gene technology. These biologically generated materials could possibly be custom-designed with respect to their physical properties, stability, and their recyclability. They may be created from renewable, inexhaustible resources or by bacteriologic transformations of already existing products.

(*ii*) The energy of the future may not be generated by burning *wood, coal,* or *oil,* or by involving *fissionable* or *fusionable* elements, but by exploiting hitherto unknown "disturbances" that are neither electromagnetic nor of particle nature. This energy source may be tapped, should it exist, and it is hoped that mankind will have developed at that point a high degree of morality so that it may not be misused for destructive purposes.

(*iii*) Presently, mankind is quite fixated on the concept that matter consists exclusively of atoms built from protons, neutrons, electrons and a handful of other particles. It is not possible that another type of matter does exist which is built of different particles beyond our imagination. May be this alternate form of matter will be discovered in this century, should it exist.

(*iv*) The abundance of *radioactive waste* produced from reactors, etc. surely will be a challenge to future generations. New techniques will have to be found which are capable of manipulating the ratio of protons and neutrons in radioactive elements which will transform them into nonradioactive isotopes.

(*v*) The transition temperature (T_c) at which *superconduction* commences may again be substantially raised by employing new materials which have a striking similarity to fibres spun by animals or which are otherwise created in the body of animals or humans such as in nerve cells.

(*vi*) *Smart* acoustic materials may be discovered which compensate incoming sound with a complementary sound, thus eliminating any noise. In particular, the acoustical properties of materials will probably be studied more intensely in the future than they have been in the past.

(*vii*) The storage of energy in batteries, etc. is at present most inefficient and requires bulky devices. New techniques will probably be discovered that raise the energy to mass ratio and increase the efficiency by involving a *plasma* technology, which is harnessed in containers consisting of high-temperature resistant materials.

(*viii*) Materials may be found which, when weakened by fatigue or cracking, will activate a self-healing mechanism that returns the material, while in use, to its originally intended properties without external intervention.

(*ix*) New types of trees or plant species may be genetically engineered which can be harvested in about 7-8 years rather than in the present 30-year time interval.

(*x*) The process of photosynthesis may be copied and used to create new materials and energy.

These few instances may serve as stimuli for research scientists and engineers and thus may lead materials science into new dimensions for the betterment of mankind. We hope that future developments will definitely be linked to societal issues.

1.4 THE PROPERTY SPECTRUM OF ENGINEERING MATERIALS

Properties of materials play an important role in their selection for specific purposes. The engineer has to decide the properties required of a material for a part under design and then he has to weigh the properties of candidate materials. There are literally speaking hundreds of properties that are measured in laboratories for the purpose of comparing various materials, but one has to concentrate on the more important ones.

The major properties to be considered in material selection can be put under four categories: chemical, physical, mechanical and dimensional properties. Figure 1.4 gives a brief account of these properties.

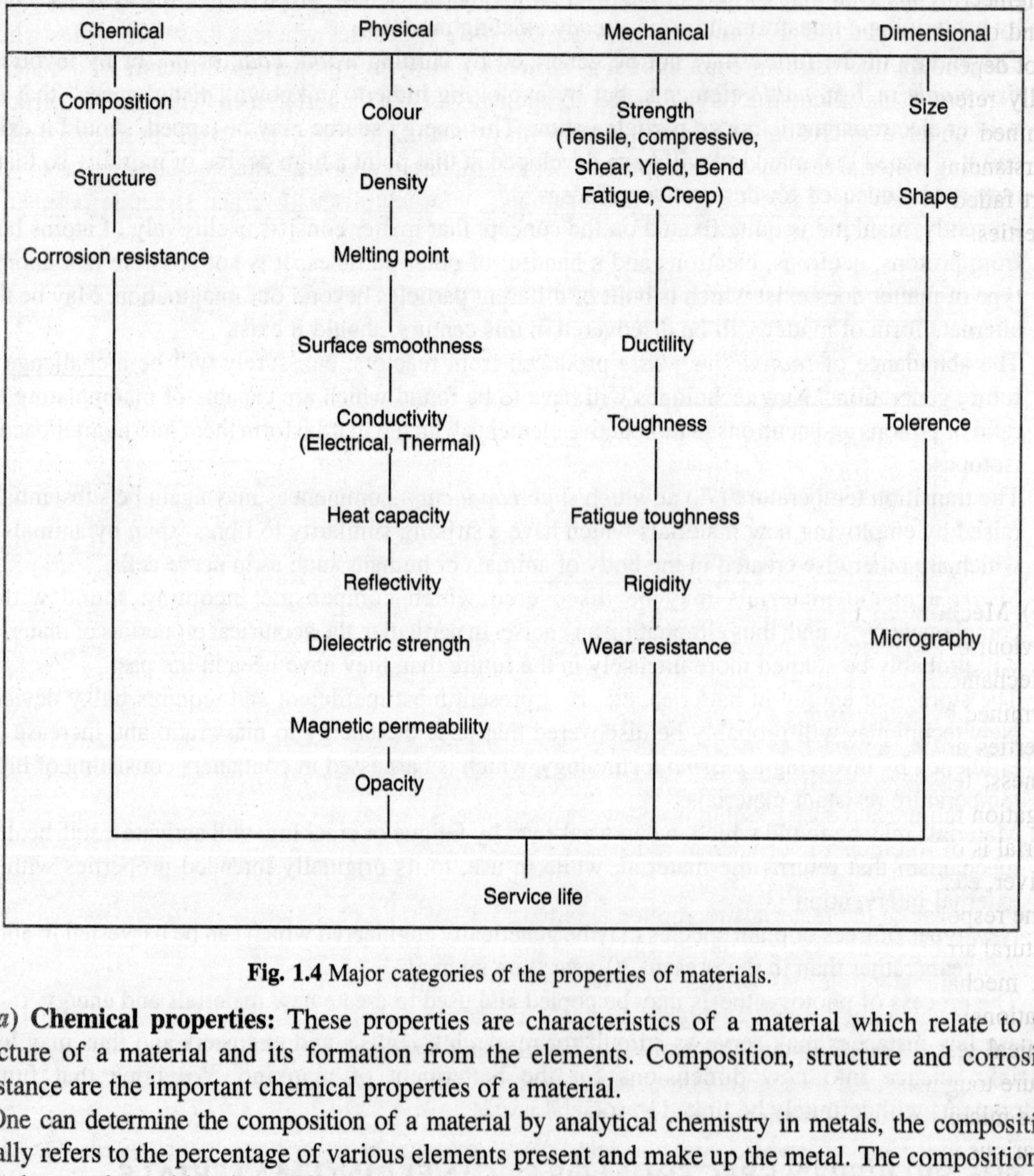

Fig. 1.4 Major categories of the properties of materials.

(*a*) **Chemical properties:** These properties are characteristics of a material which relate to the structure of a material and its formation from the elements. Composition, structure and corrosion resistance are the important chemical properties of a material.

One can determine the composition of a material by analytical chemistry in metals, the composition usually refers to the percentage of various elements present and make up the metal. The composition is a fundamental consideration while selecting a material. The engineer should have atleast some idea about the composition of the material.

The chemical properties describe the combining tendencies, corrosion characteristics, reactivity, solubilities, etc., of a material. Few of the chemical properties are:

(*i*) **Corrosion:** This refers the deterioration of a material by chemical reaction with the environment. Corrosion degrades material properties and thereby reduces the economic value and utility of the material. Corrosion attracts both metals and non-metals. In soils, corrosion of concrete by sulphates is a common problem.

(*ii*) *Chemical composition*, and (*iii*) *Alkalinity or acidity*.

For metallurgist, studies of the structure of the materials is one of the most useful tool. Properties of an engineering material may or may not depend on its microstructure. Properties of the material which depend on microstructure are known as *structure-sensitive* properties whereas those properties which do not depend on microstructure are known as *structure-insensitive* properties. Structure of a material usually refers to its microstructure, *i.e.* the components of the material seen when the material is examined under a microscope. Microstructures play invaluable roles in solving problems and in understanding material responses to treatments. Microstructure properties help one to analyze as to why a part failed in service. Table 1.3 gives few examples of structure-sensitive and structure-insensitive properties.

Table 1.3 Structure-sensitive and Structure-insensitive properties of materials

Structure-sensitive properties of materials	*Structure-insensitive properties of materials*
• Strength (yield, creep, fracture)	Density Melting point
• Electrical conductivity	Elastic constants Heat capacity
• Magnetic properties	Coefficient of thermal expansion

(*b*) **Mechanical properties:** These include those characteristics of material that describe the behaviour of the material under the action of external forces.

Mechanically properties usually relate to the elastic and plastic behaviour of the material and can be determined by conducting experimental tests on the specimen of the material. Most of the mechanical properties are determined by destructive tests. Important mechanical properties of the materials are hardness, tensile strength, modulus of elasticity, yield strength, impact, reduction in area, percent elongation fatigue and creep strengths and wear resistance. The knowledge of mechanical properties of material is of foremost importance in order to construct a mechanically sound structure, *e.g.* bridge on the river, etc.

The response of the materials to applied external forces will depend on the type of the bonding and structural arrangement of atoms or molecules and the type and number of imperfections present. This is why, mechanical properties of the materials are quite sensitive to manufacturing processes and operations, which may result in highly variable characteristics even in the materials of the same chemical tests. The important mechanical properties are strength, hardness, ductility, stiffness and fracture toughness. All mechanical properties, except modulus of rigidity *i.e.* elastic constants, depend on the microstructure of the material and hence they are called as *structure-sensitive* properties.

Various mechanical properties are: elasticity, plasticity, toughness, resilience, tensile strength, yield strength, impact strength, ductility, malleability, brittleness, hardness, fatigue, creep, wear resistance, etc. Mechanical properties of materials are affected due to: alloy contents, grain size, crystal imperfections, cold working, manufacturing, etc.

(*c*) **Physical properties:** These are the characteristics of materials that are determined by nature. There are many physical properties of materials and it is not possible for anyone to describe briefly all of them. However, important physical properties of materials include colour, density, melting point, electrical conductivity, thermal expansion and magnetic properties.

Some mechanical properties of materials are given in Table 1.4.

Table 1.4 Some mechanical properties of materials

Material	*Modulus of elasticity, E [GPa]*	*Yield strength σ_y [MPa]*	*Tensile strength σ_r [MPa]*
Diamond	1,000	50,000	same
SiC	450	10,000	same
W	406	1000	1510
Cast irons	170–190	230–1030	400–1200
Low carbon steel, hot rolled	196	180–260	325–485
Carbon steels, water-quenched and tempered	~200	260–1300	500–1800
Fe	196	50	200
Cu	124	60	400
Si	107	–	–
10% Sn bronze	100	190	–
SiO_2 (silica glass)	94	7200	about the same
Au	82	40	220
Al	69	40	200
Soda glass	69	3600	about the same
Concrete	50	25*	–
Wood to grain	9–16	–	33–50*; 73–121
Pb	14	11	14
Spider drag line	2.8–4.7	–	870–1420
Nylon	3	49–87	60–100
Wood⁺ to grain	0.6–1	5*	3–10*; 2–8
Rubbers	0.01–0.1	–	30
PVC	0.003–0.01	45	–

* Compression; ⁺tensile

Note: The data listed here are average values.

(*d*) **Dimensional properties:** These properties form an important selection factor but usually not listed in handbooks. In selecting a material, the available size, shape and tolerances on materials are important. Surface roughness, waviness and micrography are other dimensional properties. The microscopic characteristics of a surface are termed microtopography and it includes various surface characteristics, *e.g.* roughness, surface layout, height, roughness width, waviness width, waviness height, etc. Using a *profilometer,* total surface profile, which is the net of the surface roughness and waviness is usually measured. A profilometer is a device that electronically measures surface texture with a styles and yield contour maps, single-line surface profiles and roughness data.

1.5 MATERIALS SELECTION

Selection of a material for a specific application is a major challenge usually encountered by a material engineer. While selecting materials, one may keep the following considerations in mind:

(*i*) Based on the preliminary considerations, select the engineering materials, *i.e.* metals, or ceramics or polymers, etc.
(*ii*) Select such materials which can be developed with the required specific properties.
(*iii*) Select only those materials which can be easily manufactured or processed into desired shape, size and dimensions.
(*iv*) Calculate the cost of the materials carefully and select only those materials which can be processed or manufactured most economically. While calculating the cost of materials, one must include both direct and indirect cost. While deciding cost, reliability has to keep in mind. For example, in aerospace, satellite applications, chemical and petrochemical industries, etc. reliability is one major requirement. In such cases high cost can not be deciding factor.

Now-a-days, lots of data for various materials under varying service conditions are available. This has made the task of selecting the materials for specific purposes somewhat easier.

SUGGESTED READINGS

1. S.L. Kakani and A. Kakani, 'Material Science', New Age International, New Delhi (2nd edition, 2010).
2. B.S. Mitchell, 'An Introduction to Materials Engineering and Science', Wiley Inter Science (2004).
3. R.E. Hummel, 'Understanding Materials Science', Springer (2nd edition, 2004).
4. L.H. Van Vlack, 'Elements of Material Science and Engineering', Pearson (6th edition, 2005).
5. W.D. Callister, Jr., 'Materials Science and Engineering, John Wiley (2003).
6. M. Philips and W. Bolton, 'Technology of Engineering Materials', Butterworth-Heinemann, Oxford (2002).
7. J.F. Schackelford, 'Introduction to Materials Science for Engineers', Prentice Hall, N.J. (5th edition, 2000).
8. J. Taylor, 'Materials', Cambridge University Press (2002).

REVIEW QUESTIONS

1. What do you understand by engineering materials? How these materials are classified?
2. What is a metal? Mention its most important property. Give few examples of metals.
3. What is a ceramic? Mention few examples of ceramics used in industry.
4. What are plastics? Why they are quite popular?
5. What is a composite material? Give few examples of these materials.
6. What are the important physical properties of a material? Mention few of them.
7. What do you understand by chemical properties of materials? What role they play in selecting a material for specific engineering purpose.
8. What are mechanical properties of a material. Explain their role in selecting a material.
9. What are advanced materials? Explain their importance.
10. What are nano-structured materials? How they are going to affect technology?
11. How the development of materials may take place in future?

SHORT ANSWER QUESTIONS

1. Give few examples of metallic properties.

Ans. High electrical and thermal conductivities, good ductility and malleability, opacity and bright lusture.

2. What do you understand by metallic properties?

Ans. Metals and their alloys are bonded by metallic bond. A metallic bond is characterized by the presence of free electrons. These free electrons are mainly responsible for the various properties of the materials.

3. What are metalloids? Give few examples of metalloids.

Ans. There are some elements which are neither metals nor non-metals. These are called metalloids, *e.g.* boron, silicon, germanium, antimony and tellurium.

4. What are the important properties of solid materials?

Ans. Mechanical, electrical, thermal, magnetic, optical and deteriorative.

5. Justify the statement, 'the development of many technologies that make our existence so comfortable has been intimately associated with the accessibility of suitable materials'.

Ans. An advancement in the understanding of a material type is often the forerunner to the stepwise progression to a technology, *e.g.* automobiles would not have been possible without the availability of inexpensive steel or some other comparable substitute. Sophisticated electronic devices rely on components that are made from semiconducting materials.

6. What are ceramics? Enlist salient characteristics of these materials.

Ans. These are compounds between metallic and non-metallic elements; they are most frequently oxides, nitrides and carbides. The wide range of materials that falls within this category includes ceramics that are composed of clay minerals, cement and glass. These materials are strong and hard but brittle in nature. These materials are typically insulative to the passage of electricity and heat, and are more resistant to high temperatures and harsh environment than metals and polymers.

7. Cite few applications of ceramics.

Ans. Household pottery, fire bricks, table ware, decorative idols, building materials, such as port land cement, concrete, bricks, cutting tools, etc. Optical and electrical application include integrated circuits (ICs) optical fibre systems and variety of sensing devices.

8. What are polymers? Mention their important characteristics.

Ans. These include the familiar plastic and rubber materials. Many of the polymers are organic compounds that are chemically based on carbon, hydrogen, and other non-metallic elements. Moreover, they have very large molecular structures. These materials are poor conductors of heat and electricity, have low densities, low strength levels and low melting temperatures. They may be extremely flexible and have high degree of plasticity.

9. What are composites? Why are they considered to be superior in comparison to other materials?

Ans. A composite is designed to consist of more than one material type and display a combination of the best characteristics of each of the component materials, *e.g.* fibre glass. In fibre glass, glass fibres are embedded within a polymeric material. Obviously, fibre glass acquires strength from the glass and flexibility from the material. Wood—the most widely material–is a natural composite and has a merit of being replenishable. Polymers have greater strength and toughness when they are part of a reinforced composite. Many of the recent material developments have involved composite materials. Advanced aircraft and aerospace vehicles rely heavily on composites.

10. Give example of a natural composite.

Ans. Wood

11. Which metal is most abundant in the earth's crust?

Ans. Aluminium, iron and calcium follow it.

12. How metals are classified?

Ans. Metals are usually classified as ferrous and non-ferrous metals. Iron falls under ferrous metal and all other metals are non-ferrous metals. Metals are classified into following groups based on their salient features:

(*i*) light metals, (*ii*) nobel metals, (*iii*) alkali-earth metals, (*iv*) rare-earth metals, (*v*) nuclear metals, (*vi*) refractory metals, (*vii*) low melting metals, etc.

13. Does alkali-earth metals amongst the various classes of metals possess lowest density?

Ans. Yes, Lithium has a relative density of 0.534.

14. What is the major role of semiconductor materials in electronics and computer industries?

Ans. The semiconductors have made possible the advent of *integrated circuitry* that has totally revolutionized the electronics and computer industries.

15. What are biomaterials?

Ans. The boundary between the disciplines of biology and materials engineering is becoming increasingly blurred. Molecules that were once though of as being biological 'macromolecules', such as DNA are now being used as templates to produce molecules with desired gene sequences, and live tissue cultures are being used to reconstruct human body parts much as an auto body is repaired with fiber glass and epoxy. These materials must not produce toxic substances and must be compatible with body tissues. Metals, ceramics, polymers, composites, and semiconductors may be used as biological materials.

16. What are *biomimetic* structures?

Ans. Materials developed as a result of adaptation of structural elements found in nature are called biomemetic structures–an area that is receiving a considerable attention today.

17. What are advanced materials?

Ans. Materials that are utilized in high-technology or high-tech applications, *e.g.* VCRS, CD players, computers, fibre optic systems, aircraft, spacecraft, military rocketry are sometimes termed as advanced materials. These are typically either traditional materials whose properties have been enhanced or newly developed, high-performance materials. Furthermore, they may be all material types, *e.g.* metals, ceramics, polymers and are *normally* relatively expensive.

18. What are *smart* or intelligent materials?

Ans. There are a group of new and state-of-art-materials now being developed that will have a significant influence on many of our technologies. The adjective 'smart' implies that these materials are able to sense changes in their environments and then respond to these changes in predetermined manners—traits that are also found in living organisms. The smart materials and structural systems are based on the concept to incorporate sensors, actuators, etc. in the materials themselves and to integrate the materials with the structures, thereby imparting such functions as to: (*i*) self-evaluate the occurrance and deterioration of strains and damages, (*ii*) self-control vibrations and noises, (*iii*) self-change their own shape depending on conditions and (*iv*) self-control the propagation of damages and repair them as the need arises.

Sensor detects an input signal and an actuator performs a responsive and adaptive functions. Actuators may be called upon to change shape, position, natural frequency or mechanical characteristics in response to changes in temperature, electric fields, and/or magnetic fields. Shape memory alloys, piezoelectric ceramics, magneto-restrictive materials, and electro-rheological/magneto-rheological fluids are commonly used for actuators. Optical fibers, piezoelectric materials (including some polymers), and microelectromechanical devices (MEMS) are used as sensors.

19. What are precious metals?

Ans. Silver, platinum, gold, palladium, rhodium, and iridium are termed as precious metals. Their corrosion resistance is excellent and these metals are used for dental fillings, electrical contacts, resistance thermometers and thermo-couples.

20. Does graphite belong to ceramic materials?

Ans. Yes, this is a crystalline form of carbon having a typical hexagonal layered structure.

21. Is tungsten carbide a ceramic?

Ans. Yes, this is an inorganic compound.

22. Write the names of three metals which are widely used as alloying elements but not as base metals.

Ans. Manganese, chromium and vanadium are basically used as alloying elements.

23. How a material selection is of prime importance in engineering design?

Ans. Selection of material for material design ultimately determines the success or failure of design to be adopted. Obviously, proper selection of material is of prime importance.

24. What are Whiskers?

Ans. These are very thin, single crystal of high perfection that has an extremely large length-diameter ratio.

CHAPTER 2

Bonding and Structure of Materials

2.1 INTRODUCTION

One can obtain a wealth of information by looking at the structure of a material. Though there are many levels of structure, *e.g. atomic structure, crystal structure, microstructure, macrostructure,* many physical properties of a material can be related directly to the arrangement and types of bonds that make up the material.

The *atom* is the basic unit of internal structure of materials. Different materials possess different structures and due to this there are differences in the properties of various materials. The structure of a material exhibits material's internal and surface details. One can examine these details and expressed in different orders of magnification varying from a few times to several thousands. One can express the structure of a solid material in order of decreasing magnification as:

(*i*) Atomic structure,
(*ii*) Crystal structure,
(*iii*) Microstructure,
(*iv*) Macrostructure.

The atom consists of nucleus containing almost entire mass (due to neutrons and protons) at the center, around which electrons revolve round in various orbits. The number of electrons present in an atom is equal to the number of protons, *i.e.* atom is a neutral entity. Electrons belonging to an atom may be classified as the outer and inner electrons. The inner electrons are more firmly bound to the nucleus of the atom and they are termed as *core* electrons. The outer electrons are least firmly bound to the atomic nucleus and revolve round the nucleus in outer orbits. They are called as *valence electrons.* Valence electrons can be easily removed from the atom. When an atom looses a valence electron, it becomes a *positive ion.*

Each electron in an atom performs two types of motion simultaneously: (*i*) one motion is revolution around the nucleus in definite orbits and (*ii*) the other is *spinning* on its own axis. We may note that electrons revolving around the nucleus never fall into the nucleus due to the fact that electrostatic force of attraction between electrons and nucleus is balanced by the kinetic energy of rotation and the potential energy of the electron.

Atomic number is at the top, the symbol and atomic weights of the element are in the next line and below it is given the electronic configuration.

Periods / Groups	s^1 and s^2		d^1 to d^{10}										p^1 and p^6					
	IA	IIA	IIIB	IVB	VB	VIB	VIIB	VIIIB			IB	IIB	IIIA	IVA	VA	VIA	VIIA	ZERO
1.	1 H 1.008 $1s^1$																	2 He 4.003 Is^2
2.	3 Li 6.936 (He) $2s^1$	4 Be 9.012 (He) $2s^2$											5 B 10.81 (He) $2s^2 2p^1$	6 C 12.01 (He) $2s^2 2p^2$	7 N 14.01 (He) $2s^2 2p^3$	8 0 16.00 (He) $2s^2 2p^4$	9 F 19.00 (He) $2s^2 2p^5$	10 Ne 20.18 (He) $2s^2 2p^6$
3.	11 Na 22.99 (Ne) $3s^1$	12 Mg 24.31 (Ne) $3s^2$											13 Al 26.98 (Ne) $3s^2 3p^1$	14 Si 28.09 (Ne) $3s^2 3p^2$	15 P 30.97 (Ne) $3s^2 3p^3$	16 S 32.06 (Ne) $3s^2 3p^4$	17 Cl 35.45 (Ne) $3s^2 3p^5$	18 Ar 39.95 (Ne) $3s^2 3p^6$
4.	19 K 39.10 (Ar) $4s^1$	20 Ca 40.08 (Ar) $4s^2$	21 Sc 44.96 (Ar) $3d^1 4s^2$	22 Ti 47.90 (Ar) $3d^2 4s^2$	23 V 50.94 (Ar) $3d^3 4s^2$	24 Cr 52.00 (Ar) $3d^5 4s^1$	25 Mn 54.94 (Ar) $3d^5 4s^2$	26 Fe 55.85 (Ar) $3d^6 4s^2$	27 Co 58.93 (Ar) $3d^7 4s^2$	28 Ni 58.71 (Ar) $3d^8 4s^2$	29 Cu 63.54 (Ar) $3d^{10} 4s^1$	30 Zn 65.37 (Ar) $3d^{10} 4s^2$	31 Ga 69.72 (Ar)$3d^{10} 4s^2$ $4p^1$	32 Ge 72.59 (Ar)$3d^{10} 4s^2$ $4p^2$	33 As 74.92 (Ar)$3d^{10} 4s^2$ $4p^3$	34 Se 78.96 (Ar)$3d^{10} 4s^2$ $4p^4$	35 Br 79.91 (Ar)$3d^{10} 4s^2$ $4p^5$	36 Kr 83.80 (Ar)$3d^{10} 4s^2$ $4p^6$
5.	37 Rb 85.47 (Kr)$5s^1$	38 Sr 87.62 (Kr)$5s^2$	39 Y 88.91 (Kr)$4d^1 5s^2$	40 Zr 91.22 (Kr)$4d^2 5s^2$	41 Nb 92.91 (Kr)$4d^4 5s^1$	42 Mo 95.94 (Kr)$4d^5 5s^1$	43 Tc 99.0 (Kr)$4d^6 5s^1$	44 Ru 101.1 (Kr)$4d^7 5s^1$	45 Rh 102.9 (Kr)$4d^8 5s^1$	46 Pd 106.4 (Kr)$4d^{10}$	47 Ag 107.9 (Kr)$4d^{10} 5s^1$	48 Cd 112.4 (Kr)$4d^{10} 5s^2$	49 In 114.8 (Kr)$4d^{10} 5s^2$ $5p^1$	50 Sn 118.7 (Kr)$4d^{10} 5s^2$ $5p^2$	51 Sb 121.7 (Kr)$4d^{10} 5s^5$ $5p^3$	52 Te 127.6 (Kr)$4d^{10} 5s^2$ $5p^4$	53 I 126.9 (Kr)$4d^{10} 5s^2$ $5p^5$	54 Xe 131.3 (Kr)$4d^{10} 5s^2$ $5p^6$
6.	55 Cs 132.9 (Xe) $6s^5$	56 Ba 137.3 (Xe) $6s^2$	57 La 138.9 (Xe) $5d^1 6s^2$	72 Hf 178.5 (Xe) $4f^{14} 5d^2$ $6s^2$	73 Ta 180.9 (Xe) $4f^{14} 5d^3$ $6s^2$	74 W 183.9 (Xe) $4f^{14} 5d^4$ $6s^2$	75 Re 186.2 (Xe) $4f^{14} 5d^5$ $6s^2$	76 Os 190.2 (Xe) $4f^{14} 5d^6$ $6s^2$	77 Ir 192.2 (Xe) $4f^{14} 5d^9$	78 Pt 195.1 (Xe) $4f^{14} 5d^9$ $6s^1$	79 Au 197.0 (Xe) $4f^{14} 5d^{10}$ $6s^1$	80 Hg 200.6 (Xe) $4f^{14} 5d^{10}$ $6s^2$	81 Tl 204.4 (Xe) $4f^{14} 5d^{10}$ $6s^2 6p^1$	82 Pb 207.2 (Xe) $4f^{14} 5d^{10}$ $6s^2 6p^2$	83 Bi 209.0 (Xe) $4f^{14} 5d^{10}$ $6s^2 6p^3$	84 Po 210.0 (Xe) $4f^{14} 5d^{10}$ $6s^2 6p^4$	85 At 211 (Xe) $4f^{14} 5d^{10}$ $6s^2 6p^5$	86 Rn 222 (Xe) $4f^{14} 5d^{10}$ $6s^2 6p^6$
7.	87 Fr 223 (Rn)$7s^2$	88 Ra 226 (Rn)$7s^2$	89 Ac 227 (Rn)$6d^1 7s^2$															

Lanthanides ($4f^1$ to $4f^{14}$)

58	59	60	61	62	63	64	65	66	67	68	69	70	71
Ce 140.1	Pr 140.9	Nd 144.3	Pm 147.0	Sm 150.3	Eu 152.0	Gd 157.3	Tb 158.9	Dy 162.5	Ho 164.9	Er 167.3	Tm 168.9	Yb 173.0	Lu 175.0
(Xe) $4f^2 6s^2$	(Xe) $4f^3 6s^2$	(Xe) $4f^4 6s^2$	(Xe) $4f^5 6s^2$	(Xe) $4f^6 6s^2$	(Xe) $4f^7 6s^2$	(Xe) $4f^7 5d^1$ $6s^2$	(Xe) $4f^8 5d^1$ $6s^2$	(Xe) $4f^{10} 6s^2$	(Xe) $4f^{11} 6s^2$	(Xe) $4f^{12} 6s^2$	(Xe) $4f^{13} 6s^2$	(Xe) $4f^{14} 6s^2$	(Xe) $4f^{14} 5d^1$ $6s^2$

Actinides ($5f^1$ to $5f^{14}$)

90	91	92	93	94	95	96	97	98	99	100	101	102	103
Th 232	Pa 231	U 238	Np 237	Pu 244	Am 243	Cm 247	Bk 247	Cf 251	Es 254	Fm 257	Md 256	No 254	Lw 257
(Rn) $6d^2 7s^2$	(Rn) $5f^2 6d^1$ $7s^2$	(Rn) $5f^3 6d^1$ $7s^2$	(Rn) $5f^5 7s^2$	(Rn) $5f^6 7s^2$	(Rn) $5f^7 7s^2$	(Rn) $5f^7 6d^1$ $7s^2$	(Rn) $5f^8 6d^1$ $7s^2$	(Rn) $5f^{10} 7s^2$	(Rn) $5f^{11} 7s^2$	(Rn) $5f^{12} 7s^2$	(Rn) $5f^{13} 7s^2$	(Rn) $5f^{14} 7s^2$	(Rn) $5f^{14} 6d^1$ $7s^2$

Fig. 2.1 Periodic table of elements.

The number of protons in an atom of an element denotes the *atomic number* of the element. The periodic table (Fig. 2.1) of elements displays elements arranged with ascending atomic number. Every element is denoted by its symbol and atomic number. A table of the relative abundance of elements (Table 2.1) shows that most of our universe is made up of hydrogen and helium. Table 2.2 shows the relative abundance of selected elements in the earth's crest.

Table 2.1 Relative Abundance of Elements in the Universe

Element	*Relative abundance (Si = 1)*
Hydrogen (H)	12,000
Helium (He)	2,800
Oxygen (O)	16
Nitrogen (N)	8
Carbon (C)	3
Iron (Fe)	2.6
Silicon (Si)	1
Magnesium (Mg)	0.89
Sulfur (S)	0.33
Nickel (Ni)	0.21
Aluminum (Al)	0.09
Calcium (Ca)	0.07
Sodium (Na)	0.045
Chlorine (Cl)	0.025

Table 2.2 Relative Abundance of Selected Elements in the Earth's Crust

Element	*Relative abundance (ppm)*	*Element*	*Relative abundance (ppm)*
Oxygen (O)	466,000	Fluorine (F)	300
Silicon (Si)	277,200	Strontium (Sr)	300
Aluminum (AI)	81,300	Barium (Ba)	250
Iron (Fe)	50,000	Zirconium (Zr)	220
Calcium (Ca)	36,300	Chromium (Cr)	200
Sodium (Na)	28,300	Vanadium (V)	150
Potassium (K)	25,900	Zinc (Zn)	132
Magnesium (Mg)	20,900	Nickel (Ni)	80
Titanium (Ti)	4,400	Molybdenum (Mo)	15
Hydrogen (H)	1,400	Uranium (U)	4
Phosphorus (P)	1,180	Mercury (Hg)	0.5
Manganese (Mn)	1,000	Silver (Ag)	0.1
Sulfur(S)	520	Platinum (Pt)	0.005
Carbon (C)	320	Gold (Au)	0.005
Chlorine (Cl)	314	Helium (He)	0.003

The number of neutrons present in an atom of an element can vary, *i.e.* each atom of an element contains the same number of protons, but may have different number of neutrons. Such atoms are termed as *isotopes*. Many elements can exist in several isotopic forms. Thus atoms of the same elements have two or more different masses, which are called *isotopes.* The *atomic weight* of an element corresponds to the weighted average of the atomic masses of the atom's naturally occurring isotopes. The *atomic mass unit* (*amu*) may be used for computations of atomic weight. A scale has been established whereby 1 amu is defined as 1/12 of the atomic mass of the most common isotope of carbon 12(^{12}C) (A = 12.00000). Within this scheme the masses of protons and neutrons are slightly greater than unity, and

$$A = Z + N$$

The atomic weight of an element or the molecular weight of a compound may be specified on the basis of amu per atom (molecule) or mass per mole of the material. In one mole of a substance there are 6.023×10^{23} (Avogadro's number, N) of atoms or molecules. These two atomic weight schemes are related through

$$1 \text{ amu/atom (or molecule)} = 1 \text{ g/m}$$

For example, the atomic weight of iron is 55.85 amu/atom, or 55.85 g/mol.

Many elements in the periodic table are solids at room temperature (Fig. 2.2*a* white boxes). As we heat to 1000°C, many of the IIIA–VA elements melt (light shaded, Fig. 2.2*a*) and many of the alkali metals vaporised (dark shaded, Fig. 2.2*a*), but most of the transition elements still remains in the solid form. At 2000°C, the alkali earths are molten, and many of the transition elements begun to melt too. The *highest melting point* element is carbon (Fig. 2.2*d*). We may note that this is in an inert atmosphere. Tungsten, platinum, molybdenum, have exceptional high temperature properties.

In addition, there are many elements, in and of themselves, materials of construction. Aluminium and copper are just few examples of elements that are used extensively for fabricating mechanical parts. These elements have special electrical characteristics, too. Silver and gold, in addition to their use for jewellery, also widely used for variety of electrical components.

2.2 TRENDS IN PERIODIC TABLE

A closer look at the periodic table reveal some interesting trends. Table 2.3 shows that there is a general trend to decrease *melting point* as we go down a column for alkali metals and alkali earth elements (column I A and II A), but that the column trends for the transition metals appears to different. There are some trends that are more uniform, however, and are related to the electronic structure of the element.

2.2.1 First Ionization Energy (IE or V) or Ionization Potential

It is the energy required to remove the most weakly bound (usually outermost) electron from an isolated gaseous atom

$$\text{Atom (ground state } g) + \text{IE} \rightarrow \text{Positive ion } (g) + e^- \qquad ...(1)$$

One can calculate I E or V (in eV) using the energy of the outermost electron as given by the Bohr model of the atom and Schrödinger's wave equation (in eV):

$$V = \frac{13.6\ Z^2}{n^2}\ \text{eV} \qquad ...(2)$$

where Z is the effective nuclear charge and *n* is the principal quantum number.

Figure 2.3(*a*) shows the general trend in the periodic table for the ionization energy to increase from bottom to right. The quantity related to the ionization energy or potential is the *work function* (W). The work function is the energy necessary to remove an electron from the metal surface in *thermoelectric* or *photoelectric emission.*

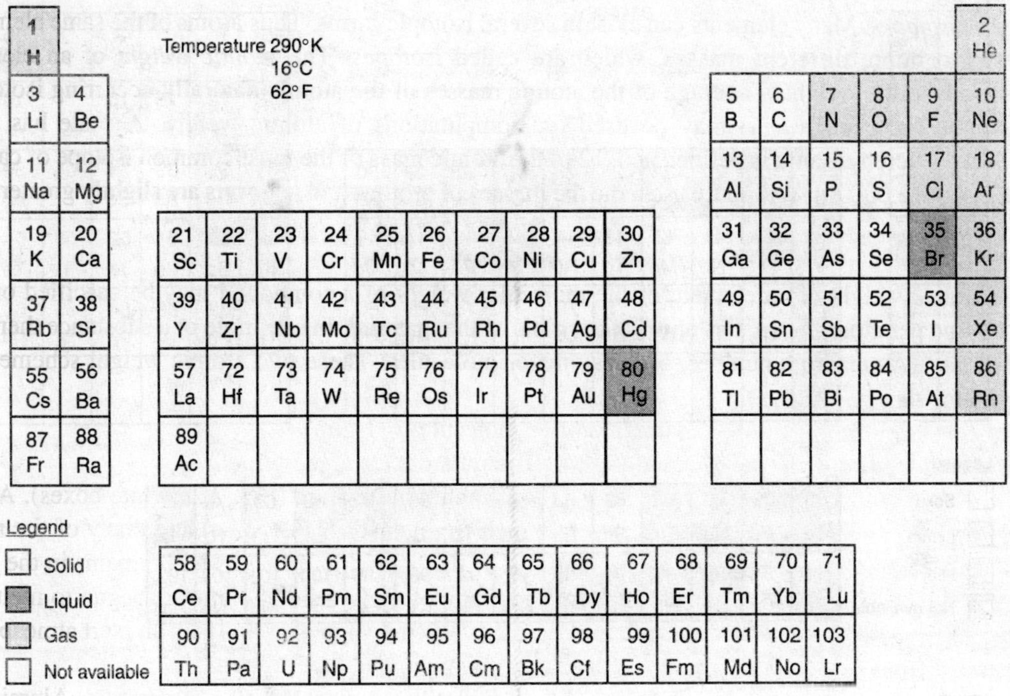

(a)

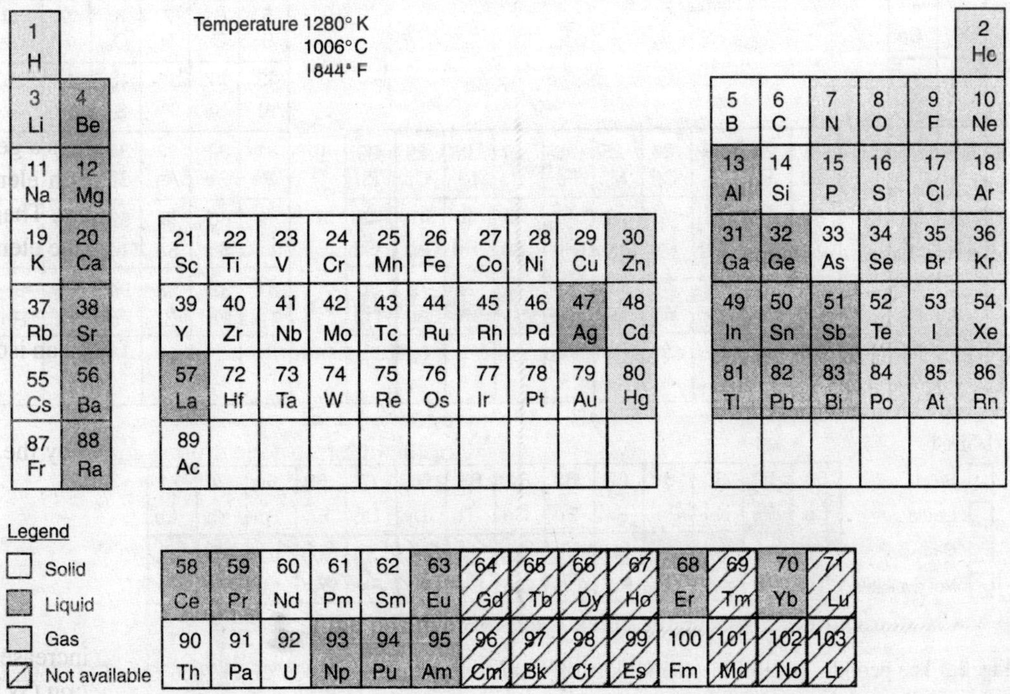

(b)

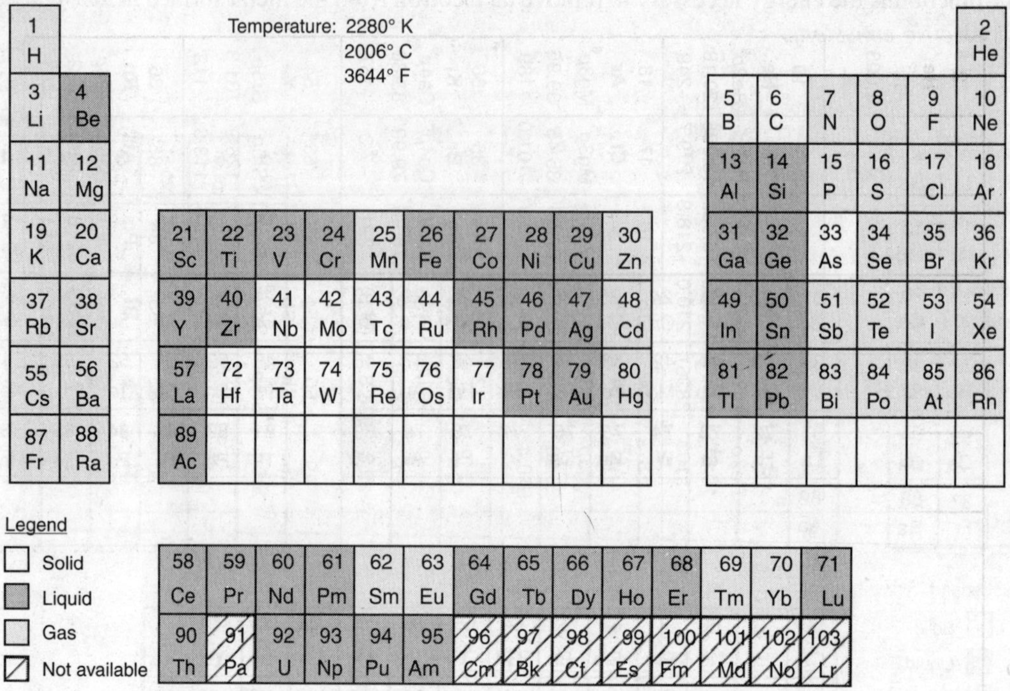

(c)

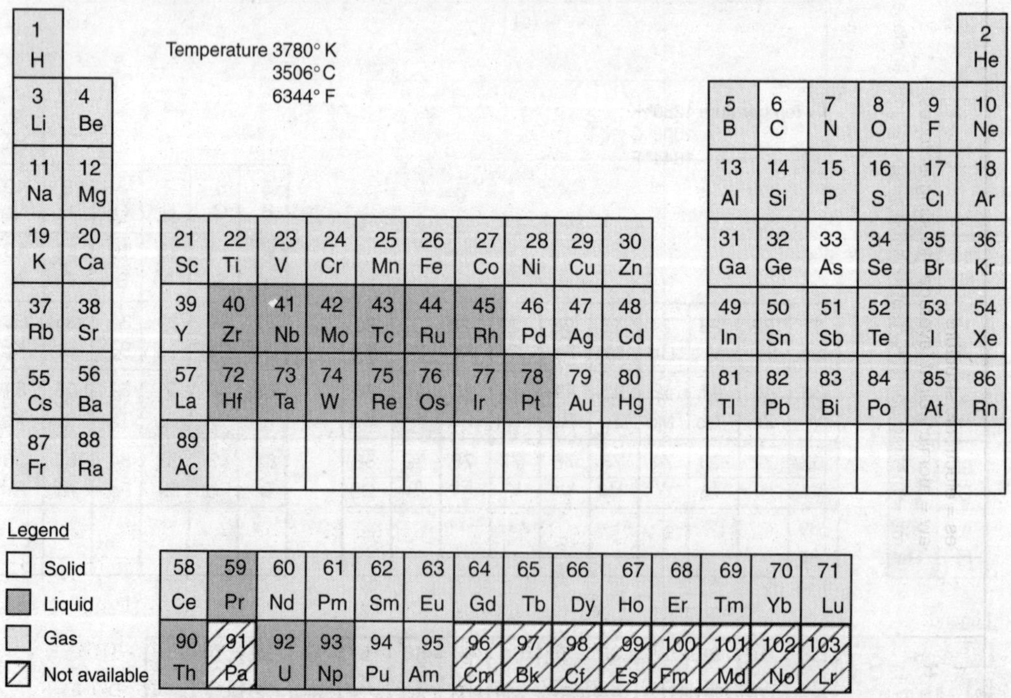

(d)

Fig. 2.2 The periodic table of the elements at (*a*) room temperature, (*b*) 1006°C (*c*) 2006°C and (*d*) 3506°C.

Table 2.3 Electronic Structure and Melting Points of the Elements

	1	es = electronic structure aw = atomic weight (average including isotopes) mp = melting point, °C (sublimation temperatures enclosed in parentheses)																2
	H																	**He**
es	1s^1																	1s^2
aw	1.008																	4.003
mp	–259																	—
	3	4											5	6	7	8	9	10
	Li	**Be**											**B**	**C**	**N**	**O**	**F**	**Ne**
es	He2s^1	He2s^2											Be2p^1	Be2p^2	Be2p^3	Be2p^4	Be2p^5	Be2p^6
aw	6.94	9.012											10.81	12.01	14.006	15.999	18.998	20.18
mp	180.5	1289											~2103	(3836)	–210.0	–218.8	–219.7	–249
	11	12											13	14	15	16	17	18
	Na	**Mg**											**Al**	**Si**	**P**	**S**	**Cl**	**Ar**
es	Ne3s^1	Ne3s^2											Mg3p^1	Mg3p^2	Mg3p^3	Mg3p^4	Mg3p^5	Mg3p^6
aw	22.99	24.30											26.98	28.09	30.974	32.06	35.45	39.95
mp	97.8	649											660.4	1414	44.1	112.8	–101.0	–189
	19	20	21	22	23	24	25	26	27	28	29	30	31	32	33	34	35	36
	K	**Ca**	**Sc**	**Ti**	**V**	**Cr**	**Mn**	**Fe**	**Co**	**Ni**	**Cu**	**Zn**	**Ga**	**Ge**	**As**	**Se**	**Br**	**Kr**
es	Ar4s^1	Ar4s^2	Ca3d^1	Ca3d^2	Ca3d^3	K3d^5	Ca3d^5	Ca3d^6	Ca3d^7	Ca3d^8	K3d^{10}	Ca3d^{10}	Ca4p^1	Ca4p^2	Ca4p^3	Ca4p^4	Ca4p^5	Ca4p^6
aw	39.10	40.08	44.96	47.9	50.94	51.9	54.93	55.85	58.93	58.71	63.55	65.37	69.72	72.59	74.92	78.96	79.90	83.80
mp	63.2	840	1541	1672	1929	1863	1246	1538	1494	1455	1084.5	419.6	29.8	938.3	(603)	221	–7.2	–157
	37	38	39	40	41	42	43	44	45	46	47	48	49	50	51	52	53	54
	Rb	**Sr**	**Y**	**Zr**	**Nb**	**Mo**	**Tc**	**Ru**	**Rh**	**Pd**	**Ag**	**Cd**	**In**	**Sn**	**Sb**	**Te**	**I**	**Xe**
es	Kr5s^1	Kr5s^2	Sr4d^1	Sr4d^2	Rb4d^4	Rb4d^5	Rb4d^6	Rb4d^7	Rb4d^8	Kr4d^{10}	Rb4d^{10}	Sr4d^{10}	Sr5p^1	Sr5p^2	Sr5p^3	Sr5p^4	Sr5p^5	Sr5p^6
aw	85.47	87.62	88.91	91.22	92.91	95.94	98.91	101.7	102.9	106.4	107.87	112.4	114.8	118.7	121.8	127.6	126.9	131.3
mp	39.5	769	1528	1865	2471	2623	2204	2254	1963	1554	961.9	321.1	156.6	232.0	630.7	449.6	113.6	–112
	55	56	57	72	73	74	75	76	77	78	79	80	81	82	83	84	85	86
	Cs	**Ba**	**La**	**Hf**	**Ta**	**W**	**Re**	**Os**	**Ir**	**Pt**	**Au**	**Hg**	**Tl**	**Pb**	**Bi**	**Po**	**At**	**Rn**
es	Xe6s^1	Xe6s^2	Ba5d^1	Ba5d^2	Ba5d^3	Ba5d^4	Ba5d^5	Ba5d^6	Xe5d^9	Cs5d^9	Cs5d^{10}	Ba5d^{10}	Ba6p^1	Ba6p^2	Ba6p^3	Ba6p^4	Ba6p^5	Ba6p^6
aw	132.9	137.3	138.9	178.5	180.9	183.9	186.2	190.2	192.2	1951	196.97	200.6	204.4	207.2	208.9	210	210	222
mp	28.4	729	921	2231	3020	3387	3186	3033	2447	1772	1064.4	–38.9	304	327.5	271.4	254	—	–71

The first ionization energies of atoms of elements belonging to two periods in periodic table is shown in Fig. 2.3(e). The energies are expressed in eV.

$$1 \text{ eV} = 1.602 \times 10^{-19}\text{C} = 1.602 \times 10^{-19} \text{ J/mole} = 96.49 \text{ kJ/mole.}$$

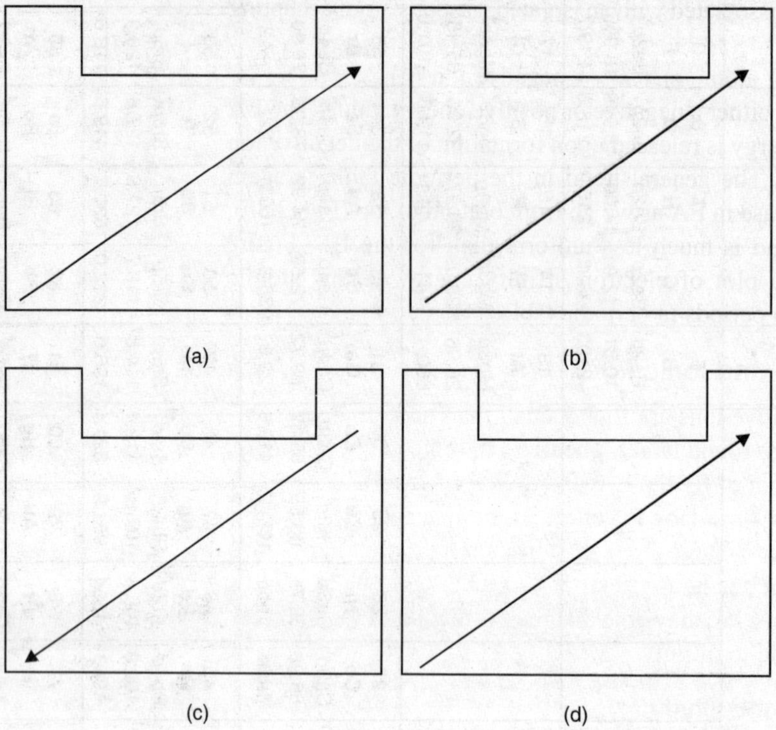

Fig. 2.3 Some important trends in the periodic table for (a) ionization energy, (b) electron affinity, (c) atomic and ionic radii and (d) electronegativity. Increasing values are in the direction of the arrow.

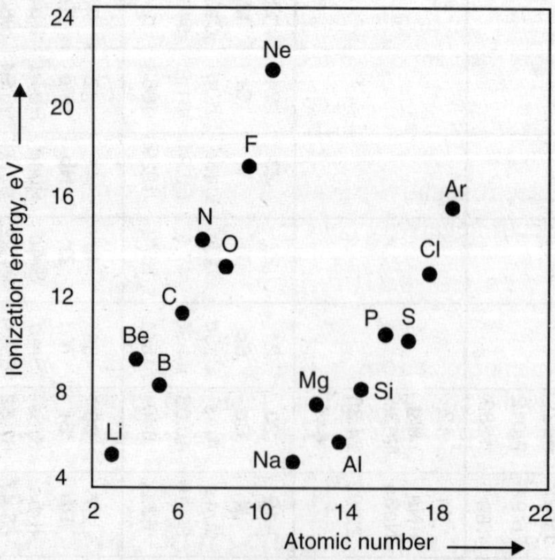

Fig. 2.3 (e) The first ionization energies of two periods in periodic table as a function of the atomic number of the elements.

2.2.2 Electron Affinity (EA)

This is the reverse process to the ionization energy. It is the energy change (in eV) associated with an isolated gaseous atom accepting one electron

$$\text{atom } (g) + e^- \rightarrow \text{negative ion } (g)$$

EA can have either a negative or positive energy value. The EA is positive if energy is released upon formation of the negative ion and vice versa. The general trend in the periodic table is again toward an increase in EA as we go from bottom to top (Fig. 2.3*b*), though this trend is much less uniform than for the IE. Figure 2.3(*f*) shows a plot of electron affinity energy versus atomic number for two periods in periodic table.

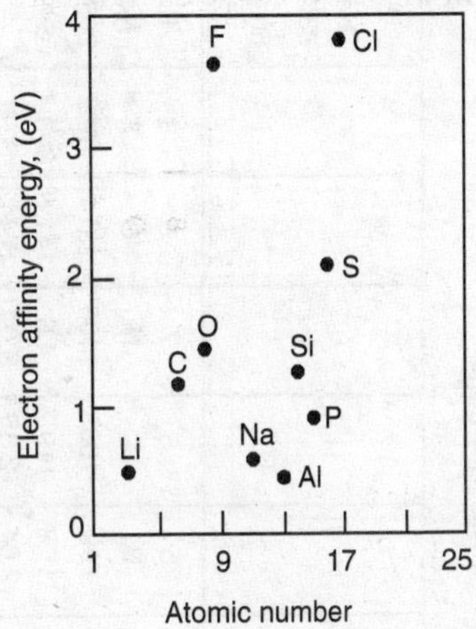

Fig. 2.3 (*f*) Electron affinity of elements of two periods in periodic table with increase in atomic number of elements.

2.2.3 Atomic and Ionic Radii

In general, positive ions are found small than neutral atoms, why negative ions are found larger. The trend in ionic and atomic radii is opposite to that of IE and EA (Fig. 2.3*c*). In general, there is an increase in radius from top to bottom, right to left. In this case, the effective nuclear charge increases from left to right, the inner electrons cannot shield as effectively, and the outer electrons are drawn close to the nucleus, reducing the atomic radius. We may note that the estimation of radii are only approximations because the orbitals, in theory, extend to infinity.

2.2.4 Electronegativity (χ)

IE and EA are characteristics of isolated atoms and they say very little about how two atoms will interact with each other, *i.e.* about chemical bonding. Electronegativity (χ) is such a quantity, whose values are always relative to one another for a given method of calculation, and values from one method should not be used with values from another method.

According to Mulliken, χ is the average of the IE and EA, *i.e.*

$$\chi = \frac{\text{IE} + \text{EA}}{2} \quad ...(3)$$

There are other types of electronegativity scales as well, the most widely utilized of which is the one from the developer of the electronegativity concept due to Pauling:

$$\chi = \frac{0.31(n + 1 \pm c)}{r} + 0.5 \quad ...(4)$$

where n is the number of valence electrons, c is any formal valence change on the atom and the sign corresponding to it, and r is the covalent radius. Table 2.4 gives the typical electronegativity values along with the values of IE and EA. The concept of electronegativity is quite useful in the study of chemical bonding between atoms.

2.3 ELECTRON ENERGY LEVELS AND QUANTUM NUMBERS

The number of electrons in a given orbit and the motion they execute in an atom are fixed. According to *Pauli exclusion principle only two electrons can occupy the same level and that these two must be of*

Table 2.4 Ionization Energies, Electron Affinities and Electronegativities of the Elements[a]

IE EA χ	1 **H** 1310 67.4 2.20																	2 **He** 2372 –60.2 —
IE EA χ	3 **Li** 519 77.0 0.98	4 **Be** 900 –18.4 1.57											5 **B** 799 31.8 2.04	6 **C** 1088 119.7 2.55	7 **N** 1406 4.6 3.04	8 **O** 1314 141.8 3.44	9 **F** 1682 349.4 3.98	10 **Ne** 2080 –54.8 —
IE EA χ	11 **Na** 498 117.2 0.93	12 **Mg** 736 0 1.31											13 **Al** 577 50.2 1.61	14 **Si** 787 138.1 1.90	15 **P** 1063 75.3 2.19	16 **S** 1000 199.6 2.58	17 **Cl** 1255 356.1 3.16	18 **Ar** 2372 –60.2 —
IE EA χ	19 **K** 1310 67.4 2.20	20 **Ca** 590 — 1.00	21 **Sc** 632 — 1.36	22 **Ti** 661 — 1.54	23 **V** 653 — 1.63	24 **Cr** 653 — 1.66	25 **Mn** 715 — 1.55	26 **Fe** 761 — 1.8	27 **Co** 757 — 1.88	28 **Ni** 736 — 1.91	29 **Cu** 745 — 1.90	30 **Zn** 904 — 1.65	31 **Ga** 577 — 1.81	32 **Ge** 782 — 2.01	33 **As** 966 — 2.18	34 **Se** 941 — 2.55	35 **Br** 1142 333.0 2.96	36 **Kr** 1351 — —
IE EA χ	37 **Rb** 402 — 0.82	38 **Sr** 548 — 0.95	39 **Y** 636 — 1.22	40 **Zr** 669 — 1.33	41 **Nb** 653 — 1.6	42 **Mo** 695 — 2.16	43 **Tc** 699 — 1.9	44 **Ru** 724 — 2.28	45 **Rh** 745 — 2.2	46 **Pd** 803 — 2.20	47 **Ag** 732 — 1.93	48 **Cd** 866 — 1.69	49 **In** 556 — 1.78	50 **Sn** 707 — 1.96	51 **Sb** 833 — 2.05	52 **Te** 870 — 2.1	53 **I** 1008 304.2 2.66	54 **Xe** 1172 — —
IE EA χ	55 **Cs** 377 — 0.79	56 **Ba** 502 — 0.89	57 **La** 540 — 1.10	72 **Hf** 531 — 1.3	73 **Ta** 577 — 1.5	74 **W** 770 — 2.36	75 **Re** 761 — 1.9	76 **Os** 841 — 2.2	77 **Ir** 887 — 2.2	78 **Pt** 866 — 2.38	79 **Au** 891 — 2.54	80 **Hg** 1008 — 2.00	81 **Tl** 590 — 2.04	82 **Pb** 715 — 2.33	83 **Bi** 774 — 2.02	84 **Po** — — 2.0	85 **At** — — 2.2	86 **Rn** 1038 — —

opposite magnetic spin. Obviously, in all elements whose atomic numbers are greater than 2, there exist more than one orbit.

According to de-Broglie, a moving particle has wave properties associated with it. The wavelength associated with a wavelength is given by

$$\lambda = \frac{h}{p} = \frac{h}{mv} \qquad ...(5)$$

where λ is the wavelength of the moving particle whose momentum is $p(= mv)$. h is Planck's constant. v is the velocity of the moving particle.

Einstein found that in an atomic transition process a definite packet of energy, called quantum of energy ($h\nu$) is involved, governed by

$$E_2 - E_1 = h\nu = h\frac{c}{\lambda} \qquad ...(6)$$

where E_1 and E_2 are energies of atomic levels 1 and 2, ν is the frequency of radiation and λ is the wavelength of the radiation. c is the velocity of light.

As stated earlier, the electrons in an atom are in continuous motion around the nucleus, one can express the motion of electron (negatively charged particle) in the force field around the force field of positively charged nucleus by Schrödinger wave equation

$$\nabla^2\psi + \frac{2m}{\hbar^2}(E - V)\psi = 0 \qquad ...(7)$$

where ψ is complex wave function, whose complex conjugate is ψ^*, where $\psi\psi^*$ is proportional to the electron density (if $\psi = a + ib$, $\psi^* = a - ib$, $i = \sqrt{-1}$, a and b are real functions). m is the mass of the electron, $\hbar = h/2\pi$, E is the total energy and V is the potential energy and function of position, *i.e.* $V = V(x, y, z)$. In the case of hydrogen atom $V = -e^2/r$, where e is the charge on the electron and r is its radial distance from the nucleus. One can determine the potential energy of the other atoms by the positively charged nucleus and other negatively charged electrons. The solution of Eqn. (7) give spatial distribution of electron density around the atom. We may note that Eqn. (7) has many solutions, but the physically meaningful and acceptable solutions are obtained only when the electrons are permitted to have specific values of energy. This means that energy of electrons are quantized. These allowed values of energy or the allowed electron states in an atom are specified by the solutions in terms of four quantum states n, l, m_l and m_s. n is the *principal quantum number* and it is a positive integer, which is a measure of the energy of the main shell. l, the second quantum number is a measure of *angular momentum of the electron* and can have values between zero and $(n - 1)$. For a hydrogen like atom, the orbital angular momentum is $\hbar\sqrt{l(l+1)}$. The third quantum number m_l, which is a measure of the component of the angular momentum in a particular direction gives the component of the orbital angular momentum in Z-direction. Of course, speaking of such an axis in an atom is meaningless. Therefore, z-axis is taken as that parallel to a weak magnetic field. m_l can have any values from $-l$ to $+l$. For a hydrogen like atom the component of orbital angular momentum is $\hbar m_l \left(= \frac{h}{2\pi} m_l\right)$. We know that the orbital angular momentum representing the motion of a charged particle in an electric field is associated with a magnetic moment and the component of magnetic moment along the z-axis is given by

$$\left(\frac{e}{2m_e}\right)\left(\frac{h}{2\pi}m_l\right)=\left(\frac{eh}{4\pi m_e}\right)m_l \quad \text{...(8)}$$

The quantity $eh/4\pi m_e$ is a fundamental quantity, called as *Bohr-magneton* and is assigned the symbol μ_B. The Bohr-magneton has a value of 9.27×10^{-24} A-m^2.

The fourth quantum number, m_s determines the direction of the spinning of the electron and produces the angular momentum due to spin. Spin magnetic moments may be only in an "up" direction or in an antiparallel "down" direction. The spin angular momentum has a numerical value $\frac{1}{2}\left(\frac{h}{2\pi}\right)$, *i.e.*

$$p_s = s(h/2\pi) \quad \text{...(9)}$$

All the four quantum numbers constitute the *quantum state* of an atom. Table 2.5 shows the various quantum states of the electrons in the first three *K*, *L* and *M* shells of the atom.

Possible values of four quantum numbers n, l, m_l and m_s which specify the electron state are given below:

n : 1, 2, 3, 4, ...

l : 0, 1, 2, 3, ...$(n-1)$

m_l : $-l$, ..., 0, ..., $+l$

m_s : +1/2, −1/2

Table 2.5 Various Quantum states of the Electrons in the First Three Shells *K*, *L* and *M* of the Atom

S.No.	*Shell*	*Quantum numbers* n	l	m_l	m_2	*Number of states*	*Designation*
1.	First, '*K*' shell	1	0	0	+1/2, −1/2	2	$1s^2$
2.	Second, '*L*' shell	2	0	0	+1/2, −1/2	2	$2s^2$
			1	+1	+1/2, −1/2		
				0	+1/2, −1/2	6	$2p^6$
				−1	+1/2, −1/2		
3.	Third, '*M*' shell	3	0	0	+1/2, −1/2	2	$3s^2$
			1	+1	+1/2, −1/2		
				0	+1/2, −1/2	6	$3p^6$
				−1	+1/2, −1/2		
			2	+2	+1/2, −1/2		
				+1	+1/2, −1/2		
				0	+1/2, −1/2	10	$3d^{10}$
				−1	+1/2, −1/2		
				−	+1/2, −1/2		

The electrons in an atom are arranged according to Pauli exclusion principle and in accordance with the four quantum numbers in different shells, designated as *K*, *L*, *M*,..., etc. shells. These shells are further divided into subshells depending upon the number of electrons in each shell. Each subshell (orbit) in an atom is at a definite distance and the nucleus exerts a definite attractive force on the electrons in this subshell. This force on the electron due to the nucleus is known as the energy of that subshell (orbit). Each subshell which can be occupied by the electrons is termed as *energy level* in that atom. Obviously, each electron of the atom belongs to a particular subshell (orbit) and hence occupies a definite energy level in the atom. The total energy of an atom is the sum of the different energy levels occupied by different electrons. We may note that the arrangement of electrons in different energy levels (orbits) is such that it produces a minimum total energy in the atom. Clearly, this arrangement of electrons becomes the most *stable state* of the atom. When an electron occupies a high energy level (orbit), it is called an *excite state* of the atom.

Table 2.6 shows the various shells, subshells and the maximum number of electrons which can occupy a given subshell.

Table 2.6 Possible Shells, Subshells and Orbital Electrons which can Occupy them in the Stable State of the Atoms

Sl. No.	*Energy levels*	*Shell number*	*Various subshells*	*Possible electrons per subshell*
1.	4	*P*	6*p*	6
			4*f*	14
			5*d*	10
			6*s*	2
2.	3	*O*	5*p*	6
			4*d*	10
			5*s*	2
3.	3	*N*	4*p*	6
			3*d*	10
			4*s*	2
4.	2	*M*	3*p*	6
			3*s*	2
5.	2	*L*	2*p*	6
			2*s*	2
6.	1	*K*	1*s*	2

The energy of an electron in $n = \infty$ electron state is taken as zero. Obviously, small values of n correspond to large negative values of energy for the electron state. However, the relative energies of electron states are functions of the atomic number of the element. A schematic illustration of the variation of the relative energies of the electron energy states with increase in atomic number of elements is shown in Fig. 2.4. Electrons always try to occupy the lowest, possible energy states in accordance with Pauli exclusion principle.

One can visualise the manner in which the electron in a given energy level positions itself around the nucleus of the atom by plotting the probability ($\psi^*\psi$ or $|\psi^2|$) of finding the electron at various locations around the nucleus of the atom. Figure 2.5 shows an approximate representation of the probability of finding an electron in the 1*s* and 2*p* states at locations around a nucleus of an atom.

Figure 2.5(*a*) reveals that *s*-states are spherically symmetrical and the electron in this state is all around the nucleus of the atom. Figure 2.5(*b*) depicts the 2*p* states with $m_l = 0$ and $m_l = \pm 1$. We can see that in the three substates of the 2*p* shell, the electron density tends to concentrate in three distinctly separate direction, which are perpendicular to each other. Clearly, electrons in these states, *i.e.* $m_l = 0$ and $m_l = \pm 1$ possess directional characteristics. However, when the 2*p* shell is full with its maximum capacity of 6 electrons, the electron distribution becomes spherically symmetrical. We may remember that the complete sub-group of states np^6, nd^{10} and nf^{14} give rise to spherically symmetrical probability patterns.

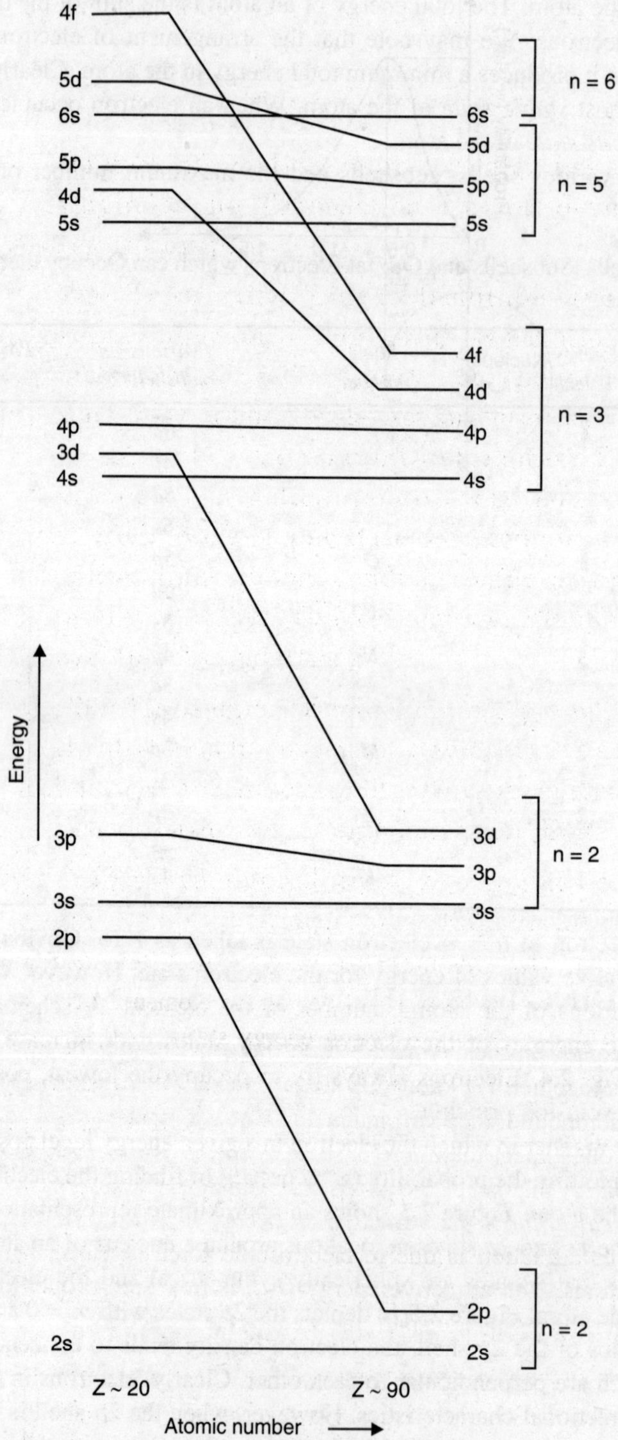

Fig. 2.4 Energies of electrons in various states representing the relative order of the shells with atomic number (Z) of the elements.

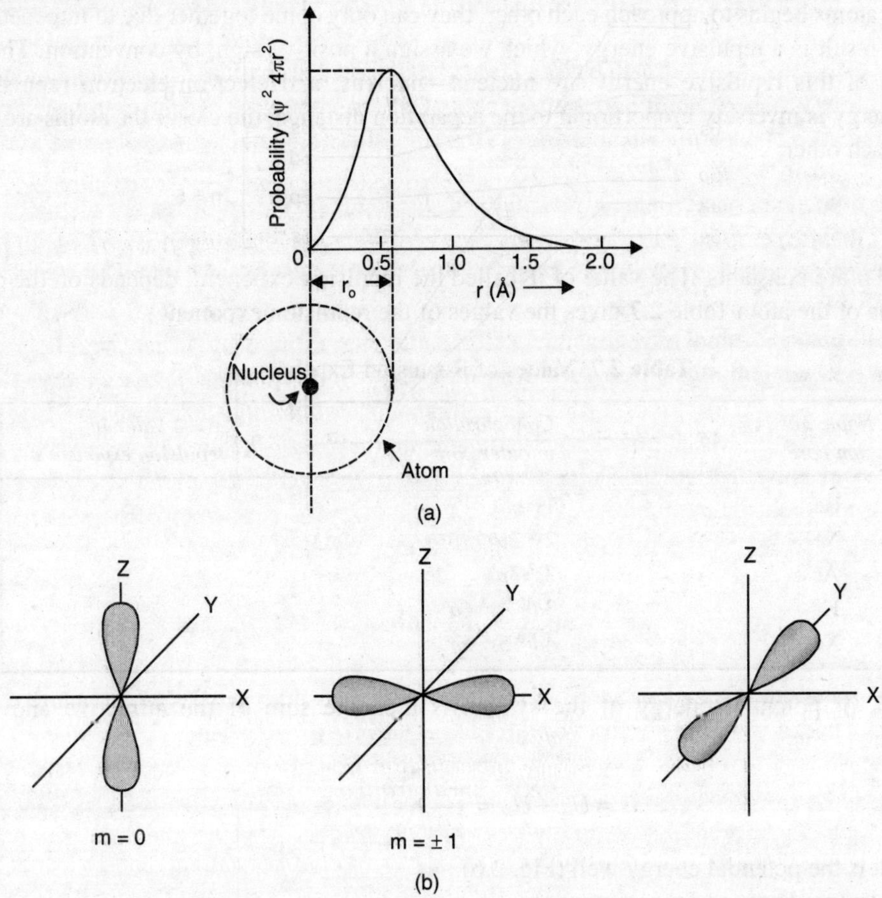

Fig. 2.5 (*a*) Probability distribution of electrons in 1*s* state. (*b*) Probability distribution for elements in 2*p* state.

2.4 ATOMIC BONDING IN SOLIDS

One can understand many of the physical properties of the materials from the knowledge of the *interatomic* forces that bind the atoms together. We know that atoms approach each other only to a certain distance, and then, if they form a compound, they will maintain some equilibrium separation distance known as the *bond length.* Hence, one expect that there is some *attractive energy* that brings them together, as well as some *repulsive energy* that keeps the atoms a certain distant apart.

Also known as *chemical affinity,* the attractive an energy between atoms is what causes them to approach each other. This attraction is due to electrostatic force between the nucleus and electron clouds of the separate atoms. The attractive energy (U_A) is inversely proportional to the separation distance, *r*; *i.e.,* the further the atoms are apart, the weaker the attraction:

$$U_A = -\frac{a}{r^m} \qquad ...(10)$$

where *a* is constant, and *m* is a constant with a value of 1 for ions and 6 for molecules. Negative sign is for attractive energy.

Once the atoms begins to approach each other, they can only come together due to impenetrability of matter. The result is a repulsive energy, which we assign a positive sign, by convention. The primary constituents of this repulsive energy are nucleus—nucleus, and electron-electron repulsions. The repulsive energy is inversely proportional to the separation distance; the closer the atoms are, the more they repel each other:

$$U_R = \frac{b}{r^n} \quad \text{...(11)}$$

where b and n are constants. The value of n, called the repulsion exponent, depends on the outer core configuration of the atom Table 2.7 gives the values of the repulsion exponent.

Table 2.7 Values of Repulsion Exponent (n)

Noble gas ion core	*Configuration in outer core*	*Value of repulsion exponent n*
He	$1s^2$	5
Ne	$2s^2\ 2p^6$	7
Ar	$3s^2\ 3p^6$	9
Kr	$3d^{10}\ 4s^2\ 4p^6$	10
Xe	$4d^{10}\ 5s^2\ 5p^6$	12

The total, or potential energy of the system is then the sum of the attractive and repulsive components:

$$U = U_A + U_R = \frac{-a}{r^m} + \frac{b}{r^n} \quad \text{...(12)}$$

The result is the potential energy well (Fig. 2.6).

The well-known Lennard-Jones potential

$$U = -\frac{a}{r^6} + \frac{b}{r^{12}} \quad \text{...(13)}$$

is a common potential energy function. It is a special case of Eqn. (12) with $n = 12$ (Xe configuration) and $m = 6$ (molecules).

Usually, it is useful to know the forces involved in bonding, as well as the energy. The force F and energy are related as

$$F = -\frac{dU}{dr} \quad \text{...(14)}$$

Using Eqn. (14), one can find the *equilibrium bond distance*, r_0 which occurs when forces are equal (the sum of attractive and repulsive forces is zero) or at *minimum potential energy*. Thus

$$F = -\frac{dU}{dr} = 0, \text{ at } r = r_0 \quad \text{...(15)}$$

Obviously, this is not the same thing as the *minimum attractive force,* which one gets by maximizing F:

$$F_{\max} = \frac{dF}{dr} = -\frac{d^2U}{dr^2} = 0 \quad \text{...(16)}$$

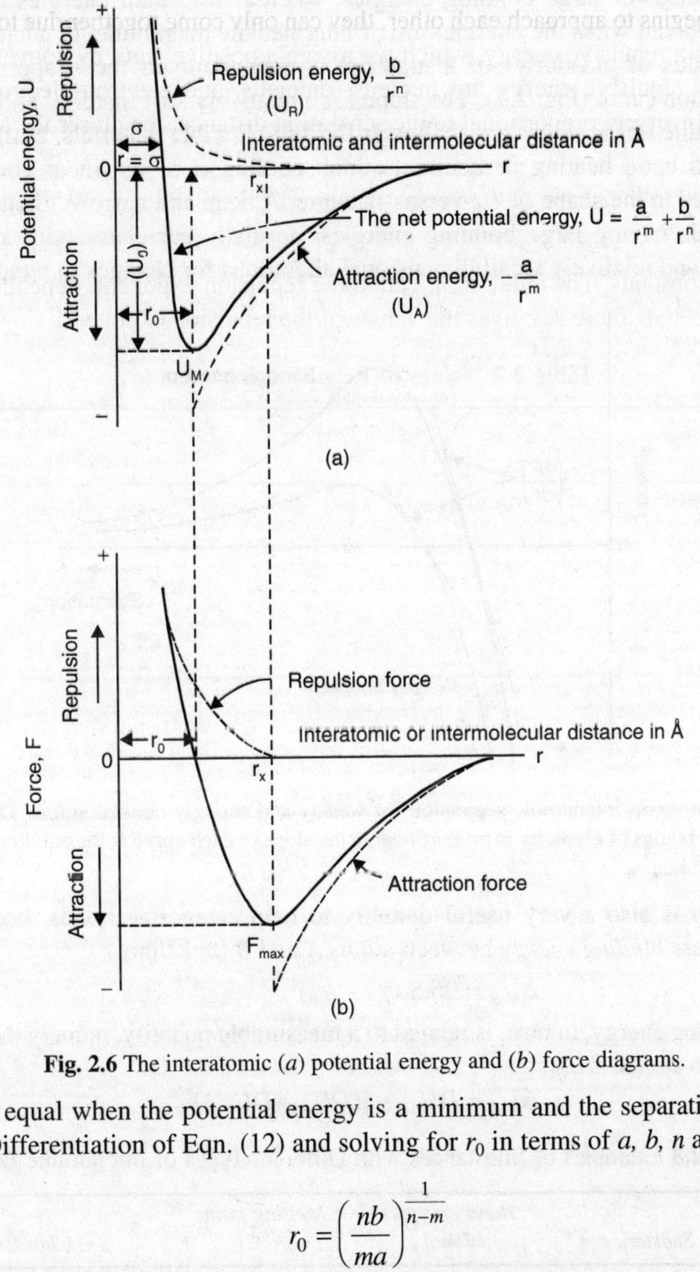

Fig. 2.6 The interatomic (*a*) potential energy and (*b*) force diagrams.

The forces are equal when the potential energy is a minimum and the separation distance is at the bond length, r_0. Differentiation of Eqn. (12) and solving for r_0 in terms of *a, b, n* and *m* gives

$$r_0 = \left(\frac{nb}{ma}\right)^{\frac{1}{n-m}} \qquad \text{...(17)}$$

So far we have dealt with an ideal situation involving only two atoms, a similar yet more complex condition exists for solid materials because force and energy interactions among many atoms must be considered. Nevertheless, a bonding energy, analogous to U_0 above, may be associated with each atom. The magnitude of this bonding energy and the shape of the energy-versus-interatomic separation curve vary from material to material, and they both depend on the type of the bonding. Furthermore, a number of material properties depend on U_0, the curve shape, and bonding type. For example, materials having large bonding energies typically also have high melting temperatures; at room temperature, solid substances are formed for large bonding energies, whereas for small energies the *gaseous* state is favoured; liquids prevail when the energies are of intermediate magnitude. In addition, the *mechanical stiffness* (or modulus of elasticity) of a material is dependent on the shape of its force-versus-interatomic separation curve (Fig. 2.7). The slope for a relatively stiff material at the $r = r_0$ position on the curve will be quite steep; slopes are shallower for more flexible materials. Furthermore, how much a material expands upon heating or contracts upon cooling (*i.e.*, its linear coefficient of thermal expansion) is related to the shape of U_0-versus-r_0 curve. A deep and narrow "though" which typically occurs for materials having large bonding energies, normally correlates with a low coefficient of thermal expansion and relatively small dimensional alterations for changes in temperature.

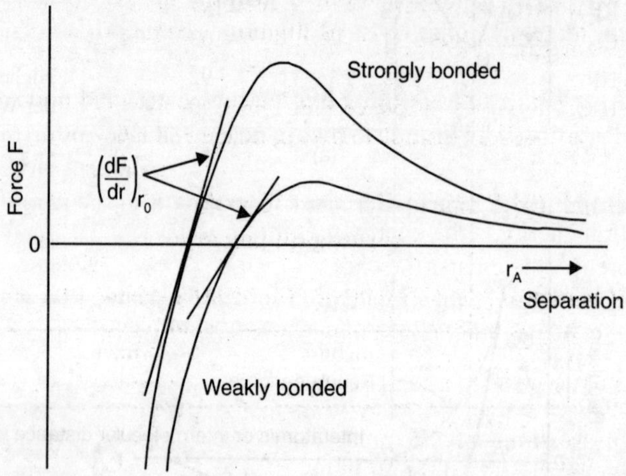

Fig. 2.7 Force versus interatomic separation for weakly and strongly bonded atoms. The magnitude of the modulus of elasticity is proportional to the slope of each curve at the equilibrium interatomic separation r_0.

Electronegativity is also a very useful quantity to help categorize bonds, because it provides a measure of the excess *binding energy* between atoms *A* and *B* (in kJ/mol):

$$\Delta_{A-B} = 96.5\,(\chi_A - \chi_B)^2 \qquad ...(18)$$

The excess binding energy, in turn, is related to a measurable quantity, namely the *bond dissociation energy* between two atoms, DU_{ij}:

$$\Delta_{A-B} = DU_{AB} - [(DU_{AA})(DU_{BB})]^{1/2} \qquad ...(19)$$

Table 2.8 Examples of Substances with Different Types of Interatomic Bonding

Type of bond	*Substance*	*Bond energy, kJ/mol*	*Melting point, (°C)*	*Characteristics*
Ionic	CaCl	651	646	Low electrical conductivity, transparent, brittle, high melting point
	NaCl	768	801	
	LiF	1008	870	
	CuF_2	2591	1360	
	Al_2O_3	15,192	3500	
Covalent	Ge	315	958	Low electrical conductivity very hard, very high melting point
	GaAs	~315	1238	
	Si	353	1420	

	SiC	1188	2600	
	Diamond	714	3550	
Metallic	Na	109	97.5	High electrical and thermal conductivity, easily deformable, opaque
	Al	311	660	
	Cu	340	1083	
	Fe	407	1535	
	W	844	3370	
van der Waals	Ne	2.5	−248.7	Weak binding, low melting and boiling points, very compressible
	Ar	7.6	−189.4	
	CH_4	10	−184	
	Kr	12	−157	
	Cl_2	31	−103	
Hydrogen bonding	HF	29	−92	Higher melting point than van der Waals bonding, tendency to form groups of many molecules
	H_2O	50	0	

The bond dissociation energy is the energy required to separate two bonded atoms. The greater the electronegativity difference, the greater the excess binding energy. These quantities give us a method of characterizing bond type as well as they relate to important physical properties, *e.g.* melting point (see Table 2.8). Now, we review the bond types and characteristics, then describe each in more detail.

2.4.1 Primary Bonds

These are also known as *strong bonds,* are created when there is a direct interaction of electrons between two or more atoms, either through transfer or as a result of sharing. The more electrons per atom that take place in this process, the higher the bond "order" (*e.g.* single, double, or triple bond) and the stronger the connection between atoms. There are four general categories of *primary bonds* are found: *ionic, covalent, polar covalent* and *metallic.* An ionic bond, also called a *heteropolar bond* results when electrons are transferred from the more electropositive atom to the more electronegative atom, as in sodium chloride, NaCl. Ionic bonds usually result when the electronegativity difference between two atoms in a diatomic molecule is greater than about 2.0. The classic example of ionic band is NaCl, but CaF_2 and MgO are also examples of molecules in which ionic bonding dominates.

A *covalent* bond, or *homopolar bond,* arises when electrons are shared between two atoms (*e.g.,* H–H). This means that a binding electron in a covalent diatomic molecule such as hydrogen (H_2) has equal likelihood of being found around either hydrogen atom. Covalent bonds are typically found in homonuclear diatomics such as O_2 and N_2, though the atoms need not be the same to have similar electronegativities. An example of a polar covalent bond can be found in HF. Though there is significant sharing of the electrons, some charge distribution exists that results in a polar or *partial ionic character* to the bond. The per cent ionic character of the bond can be related to the electronegativities of the individual atoms:

$$\% \text{ ionic character} = 100\,\{1 - \exp[-0.25(\chi_A - \chi_B)^2]\} \qquad ...(20)$$

The larger the electronegativity difference, the ionic character the bond has. Of course, if the electronegativity difference is greater than about 2.0, an ionic bond should result.

Metallic bond is found in an assembly of homonuclear atoms, such as copper or sodium. Here the bonding electrons become decentralized and are shared by the core of positive nuclei. Metallic bonds occur when elements of low electronegativity bond with each other to form a class of materials we call metals. Metals tend to have common characteristics such as ductility, luster, and high thermal and electrical conductivity.

2.4.2 Secondary Bonds or Weak Bonds

These occur due to indirect interaction of electrons in adjacent atoms or molecules. There are three main types of secondary bonding: *hydrogen bonding, dipole-dipole interactions, and van der Waal's forces.* These intermolecular forces are common, but very weak, and are found in inert gases where other types of bonding do not exist.

Hydrogen bonding is the attraction between hydrogen in a highly polar molecule and the electronegative atom in another polar molecule. In the water molecule, oxygen draws much of the electron density around it, creating positively charged centers at the two hydrogen atoms. These positively charged hydrogen atoms can interact with the negative center around the oxygen in adjacent water molecules. Although this type of bonding is of the same order of magnitude in strength as van der Waals bonding, it can have a profound influence on the properties of a material, such as boiling and melting points. Hydrogen bonding also plays an important role in many biological and environmental phenomena. It is responsible for causing ice to be less dense than water, an occurrence that allows fish to survive at the bottom of frozen lake.

There are some molecules which possess permanent charge separations, or dipoles, such as are found in water. In general, the interaction of any positive dipole with a negative dipole is called dipole-dipole interaction. Hydrogen bonding can be thought of as a special type of dipole-dipole interaction. A dipolar molecule like ammonia, NH_3, is able to dissolve other polar molecules, like water due to dipole-dipole interactions. In the case of NaCl in water, the dipole-dipole interactions are so strong as to break the intermolecular forces within the molecular solid.

Now, we study the formation of bonds in detail.

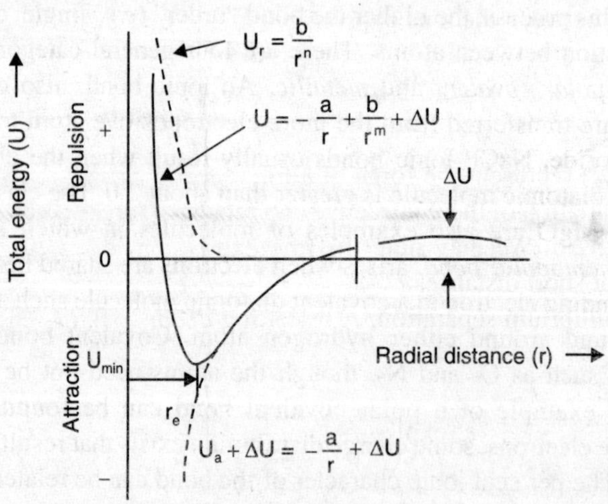

Fig. 2.8 Formation of NaCl molecule from the neutral atoms Na and Cl. ΔU denotes the energy required to form the ions. U_a and U_r are the coulombic interaction energies due to attraction and repulsion forces respectively. *a, b* and *m* are constants related to the interaction energies in this model.

(i) **Ionic bond:** This type of bond exists between two unlike atoms. If an electron is transferred from a metallic atom to a non-metallic atom, the two resulting ions are held together by electrostatic attraction, *e.g.* NaCl. Sodium atom give up an electron and becomes a positively charged ion,

Na^+, while chlorine atom takes the electron to fill its last orbit and become a negatively charged ion, Cl^-. These two ions attract each other and form a NaCl molecule. Figure 2.8 shows the energy changes associated with these processes.

One can write the process of formation of NaCl as:

$$Na + 5.14 \text{ eV} \rightarrow Na^+ + e^-$$

(ionization energy)

$$e^- + Cl \rightarrow Cl^- + 3.61 \text{ eV}$$

(electron affinity energy)

$$Na^+ + Cl^- \rightarrow NaCl + 7.9 \text{ eV}$$

(Coulombic energy per molecular unit of chloride due to interaction between the ions)

$$Na + Cl \rightarrow NaCl + 6.4 \text{ eV}$$

Clearly, the bond energy per molecular unit of sodium chloride relative to ions is 7.9 eV.

For a diatomic molecule, the attraction between the two ions is strictly due to opposite charges, so the attractive force is given by Coulomb's low = $(Z_1e \times Z_2e)/(4\pi\varepsilon_0 r^2)$, where ε_0 is a constant called the *electric permittivity* (= $8.854 \times 10^{-12} C^2/N$), Z is the respective number of charge of positive and negative ions. Thus, the interaction energy, U between the positive and negative ions, (say 1 and 2 respectively) may be expressed as

$$U = -\frac{Z_1 Z_2 e^2}{4\pi\varepsilon_0 r} + \lambda e^{-r/\rho} \qquad ...(21)$$

where λ and ρ are constants for a given pair of ions. The first term in Eqn. (21) represent the decrease in energy due to attraction between the ions and $\lambda \exp(-r/\rho)$ is the increase in energy due to repulsion between ions as a result of overlap of their ionic cores.

The force $F = -dU/dr$, between the ions is given by

$$F = -\frac{dU}{dr} = -\frac{Z_1 Z_2 e^2}{4\pi\varepsilon_0 r^2} + \left(\frac{\lambda}{\rho}\right)\exp\left(-\frac{r}{\rho}\right) \qquad ...(22)$$

where the first term in (22) represent the force of attraction and second term is the force of repulsion between the atoms. The variation of these attractive and repulsive forces between ions occur in the same manner as shown in Fig. 2.6. We may note that the expression for repulsive forces is such that it becomes important only at short distances.

One can obtain the equilibrium separation, r_e by letting $dU/dr = 0$, *i.e.* equating the net force zero. Thus obtains

$$r_e^2 \exp\left(-\frac{r_e}{\rho}\right) = \frac{\rho}{4\pi\varepsilon_0}\frac{Z_1 Z_2 e^2}{\lambda} \qquad ...(23)$$

or

$$r_e^2 = \frac{\rho}{4\pi\varepsilon_0}\frac{Z_1 Z_2 e^2}{\lambda}\exp\left(+\frac{r_0}{\rho}\right) \qquad ...(24)$$

one can obtain r_e by solving (24). r_e is considered as sum of ionic radii of the ions, *i.e.*

$$r_e = r_+ + r_- \qquad ...(25)$$

where r_e and r_- are the ionic radii of the positive and negative ions respectively. Ionic radii of few ions are given in Table 2.9.

Table 2.9 Ionic Radii of few Ions

Ion	*Radius (Å)*
Li^+	0.78
O^{2-}	1.32
Na^+	0.98
Mg^{2+}	0.78
Al^{3+}	0.57
Si^{4+}	0.39
P^{5+}	0.3 – 0.4
S^{2-}	1.74
Cl^-	1.81
K^+	1.33
Ca^{2+}	1.06

In an ionic solid, the number of nearest neighbours are such that they touch the central ion without overlapping each other so as to avoid repulsion. The number of nearest neighbours is called as the *coordination number.*

The energy of an ionic solid is due to the interaction between all ions. When we have an ordered assembly of atoms called a *lattice,* there is more than one bond per atom, and one must take into account interactions with adjacent atoms that result in an increased interionic spacing compared to an isolated atoms. We do this with the *Madelung constant,* $\alpha_M \left(= \Sigma \frac{\pm}{p_{ij}}\right.$, where $r_{ij} = p_{ij}\, r_e$, r_e is the nearest neighbour distance). This parameter depends on the structure of the ionic crystal, the charge on the ions, and the relative size of the ions. The lattice energy can be expressed as

$$U_L = \left(\alpha_M - \frac{1}{n}\right)\left(\frac{-e^2}{4\pi\varepsilon_0 r_0}\right) + \Delta E_{\text{ions}} \qquad ...(26)$$

where ΔE_{ions} is the energy required to form ions from two different atoms of a molecule. The values of α_M for different crystal structures are given in Table 2.10.

Table 2.10 Typical Values for Madelung Constant, α_M

Compound	*Crystal lattice*	α_M
NaCl	FCC	1.74756
CsCl	CsCl	1.76267
CaF_2	Cubic	2.51939
$CdCl_2$	Hexagonal	2.244
MgF_2	Tetragonal	2.381
ZnS (Wurtzite)	Hexagonal	1.64132
TiO_2	Tetragonal	2.408
β-SiO_2	Hexagonal	2.2197

Ionic crystals have poor electrical conductivity, high hardness and high melting point.

(ii) Covalent bond: This bond is formed by sharing of electrons by similar adjacent atoms. The simplest examples are of the hydrogen and chlorine molecules. The chlorine has 17 electrons ($1s^2$, $2s^2$,

$2p^6$, $3s^2$, $3p^6$). The chlorine atom can attain the stable configuration of argon atom $1s^2$, $2s^2$, $2p^6$, $3s^2$, $3p^6$ by accepting one electron. Suppose that two chlorine atoms come closer to each other such that one electron from the 3*p* state of each atom tend to become partly localised in a region between the two chlorine atoms, *i.e.* the two atoms share the two electrons. Such sharing of two electrons by two atoms stabilises the electronic shells and lowers the energy of the system. Consequently a bond is formed between the two atoms of Cl and the Cl_2 molecule is formed.

Molecular orbital theory helps to understand the formation of covalent bonding. Two atomic orbitals (1*s*) overlap to form two molecular orbitals (MOs) represented by σ: one *bonding orbital* (σ1*s*), and one *antibonding orbital* (σ*1*s*), where the asterisk superscript indicates antibonding. The antibonding orbitals are higher in energy than corresponding bonding orbitals.

Figure 2.9 shows the shapes of the electron cloud densities of various MOs. The overlap of two *s*-orbitals results in one σ-bonding orbital and σ-antibonding orbital. When two *p*-orbitals overlap in an end-to-end fashion (Fig. 2.9*b*), they are interacting in a manner similar to *s*-*s* overap, so one σ-bonding orbital and one σ-antibonding orbital once again are the result. We may note that all σ orbitals are symmetric about a plane between the two atoms. Side-to-side overlap of *p*-orbitals results in one π-bonding orbital and one π-antibonding orbital. There are a total of four π-orbitals: two for p_x and two for p_y. We may note that there is one more *node* (region of zero electron density) in an antibonding orbital than in the corresponding bonding orbital. This is what makes them higher in energy.

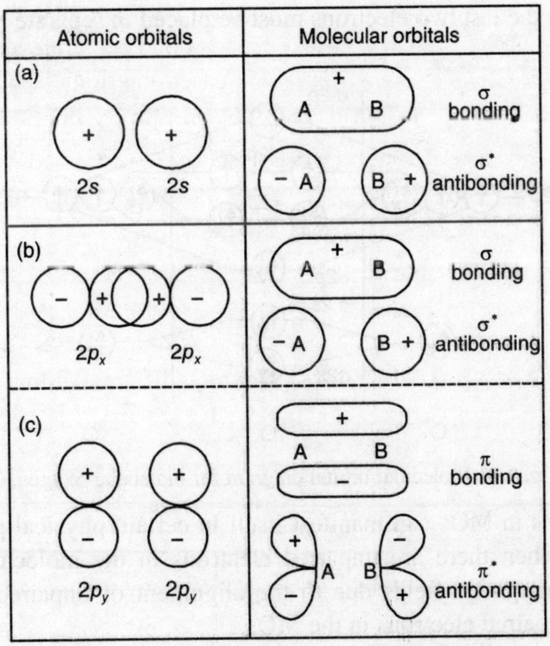

Fig. 2.9 The shape of selected molecular orbitals formed from the overlap of two atomic orbitals.

Similar to the case of ionic bonding, one can use a potential energy diagram to show how orbitals form as atoms approach each other (Fig. 2.10). The electrons from the isolated atoms are then placed in the MOs from bottom to top. *As long as the number of bonding electrons is greater than the number of antibonding electrons,* the molecule is stable. The diagrams for atoms with *p* and *d* orbitals, become more complex but the principles are the same. In all cases, there are the same number of molecular orbitals as atomic orbitals. However, there is some change in relative energies of the π and σ orbitals as one go down the periodic chart, particularly around O_2. As a result, we may see diagrams that have the $\pi 2p$ orbitals lower in energy than the $\sigma 2p$.

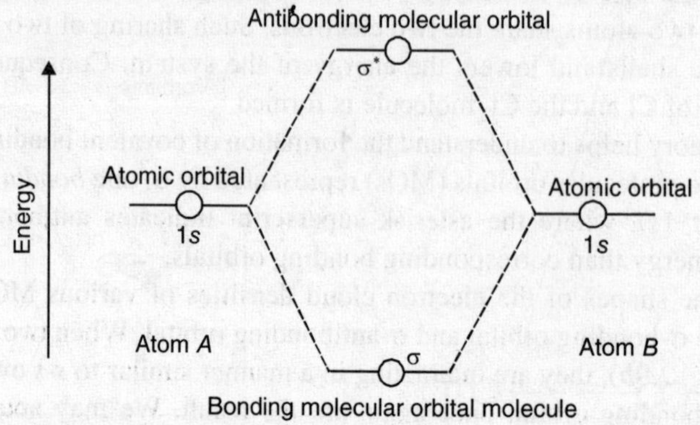

Fig. 2.10 Molecular orbital diagram for the hydrogen molecule.

Molecular orbital diagram is shown in Fig. 2.11. As shown in figure, each oxygen atom brings six outer-core electrons to the molecular orbitals. We may note that is orbitals are not involved in bonding and are thus not shown. 12 total electrons in the molecule are placed in the MOs from bottom to top; according to Hund's rule, the last two electrons must be placed in separate π^*2p orbitals prior they can be paired.

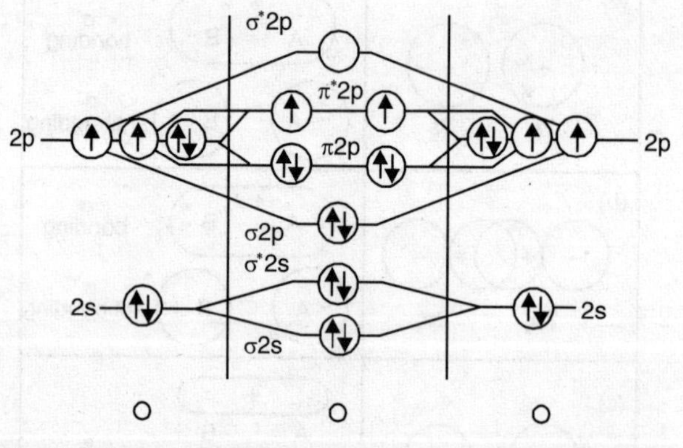

Fig. 2.11 Molecular orbital diagram for molecular oxygen, O_2.

The pairing of electrons in MOs can manifest itself in certain physical properties of the molecule. *Paramagnetism* results when there are unpaired electrons in the molecular orbitals. Paramagnetic molecules magnetize in magnetic fields due to the alignment of unpaired electrons. *Diamagnetism* occurs when there are all paired electrons in the MOs.

One can use molecular orbital theory to explain heteronuclear diatomic molecules as well. Figure 2.12 shows a relative molecular energy diagram for HF molecule. This is the case where the electronegativity of the atoms is useful. It qualitatively explains the relative energies of the atomic orbitals and the shape of the resulting MOs. Figure 2.13 shows the molecular energy level diagram for the general case of molecule *AB* where *B* is more electronegative than *A* and the corresponding molecular orbitals are shown in Fig. 2.14. In Fig. 2.13, we may note that how the B atomic orbitals are lower in energy than those of atom *A*. In Fig. 2.14, we may note how the number of nodes increases from bonding to antibonding orbitals, and also the electronegativity is greatest around the more electro-negative atom.

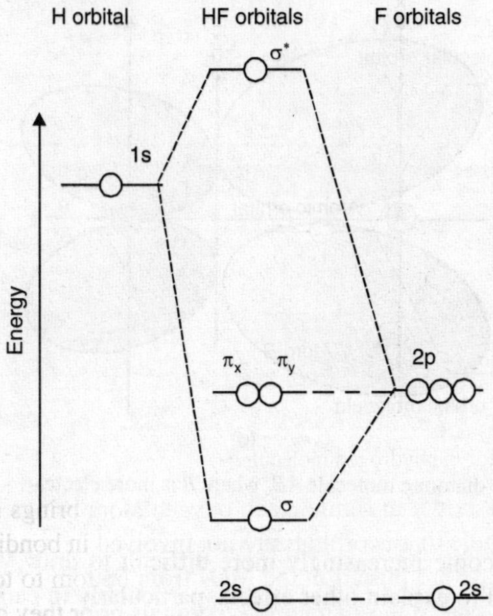

Fig. 2.12 Molecular orbital diagram for HF.

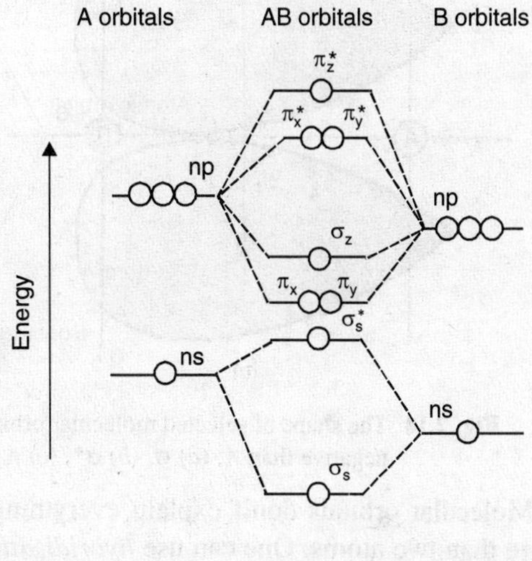

Fig. 2.13 Molecular orbital diagram for a general case of a diatomic molecule *AB*, where *B* is more electronegative than *A*.

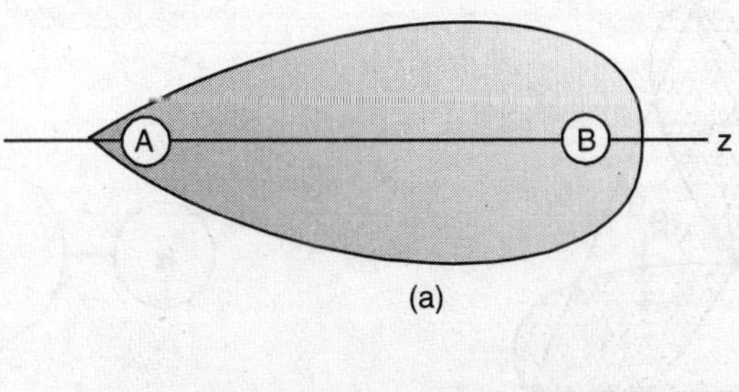

Fig. 2.14

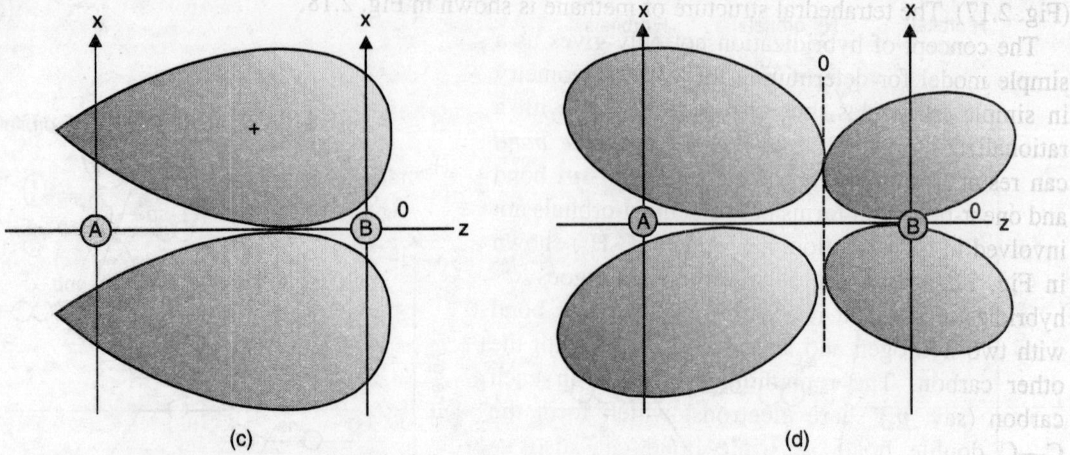

Fig. 2.14 The shape of selected molecular orbitals for the diatomic molecule *AB*, where *B* is more electronegative than *A*: *(a)* σ, *(b)* σ*, *(c)* π and *(d)* π*.

Molecular orbitals don't explain everything and become increasingly more difficult to draw with more than two atoms. One can use *hybridization* model to explain other effects, particularly in carbon compounds. Hybridization is a "mixing" of atomic orbitals to create new orbitals that have a geometry better suited. In pure carbon one has four equivalent sp^3 hybrid orbitals (Fig. 2.15). In sp^3 hybrids, when one-s and three *p*-orbitals are mixed, four sp^3 orbitals result, each having 1/4 *s* and 3/4 *p* characteristic. The tetrahedral orbitals creates a 109.5° bond angle, as found in methane (Fig. 2.18).

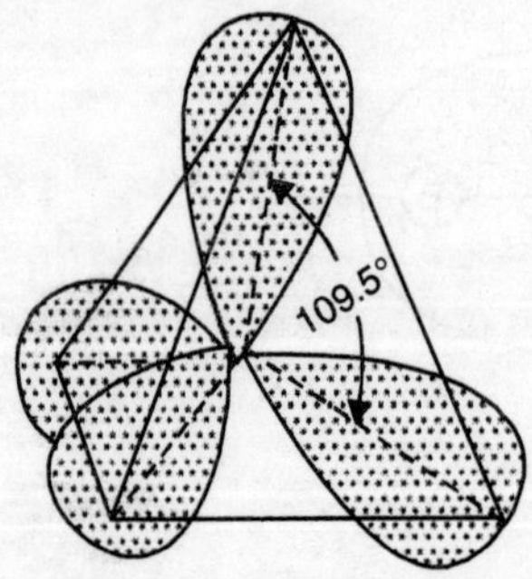

Fig. 2.15 sp^3 hybridization in carbon.

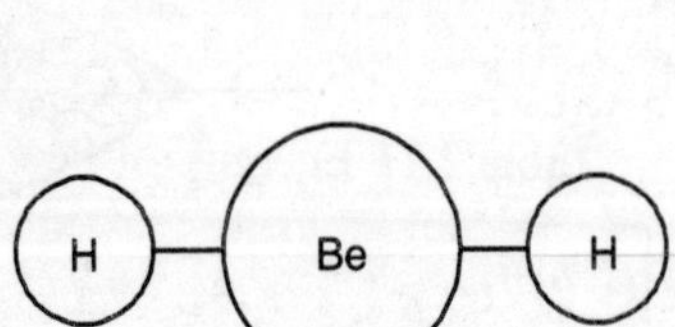

Fig. 2.16 The linear structure of BeH_2.

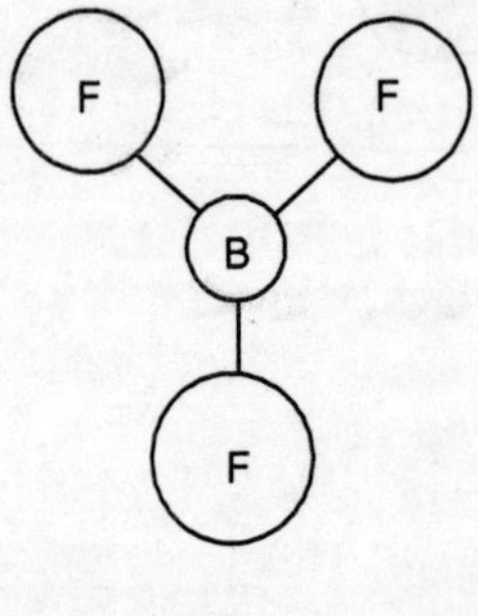

Fig. 2.17 The triagonal planar structure of BF_3.

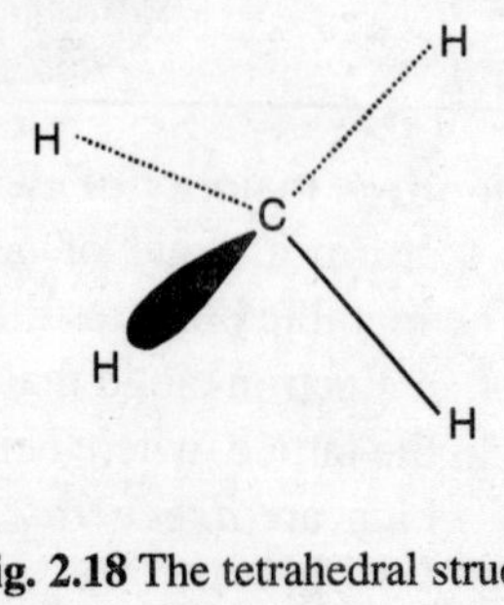

Fig. 2.18 The tetrahedral structure of CH_4.

Figure 2.16 shows the linear structure of BeH_2. An example of triangular planar molecule is BF_3 (Fig. 2.17). The tetrahedral structure of methane is shown in Fig. 2.18.

The concept of hybridization not only gives us a simple model for determining the correct geometry in simple molecules, but also provides us with a rationalization for multiple bonds. A *double bond* can result from sp^2 hybridization: one sp^2-sp^2 bond and one π-bond that forms between the *p*-orbitals not involved in hybridization, *e.g.* ethane (C_2H_4) shown in Fig. 2.19(*a*), where each carbon undergoes sp^2 hybridization so that it can form an $sp^2 - 1s$ bond with two hydrogen and an $sp^2 - sp^2$ bond with the other carbon. The remaining *p*-orbitals on each carbon (say, p_z) share electrons, which form the C—C double bond. A triple bond can also be explained in terms of *sp* hybridization. It is formed one *sp* – *sp* and two π-bonds which form between the two remaining *p*-orbitals aster hybridization, *e.g.* acetylene (C_2H_2) shown in Fig. 2.19(*b*).

Covalent bonds are directional and properties are not isotropic. Covalent crystals are characterized by poor electrical conductivity and high hardness. Covalent bonds are extremely strong bond. Diamond is an example, which is extremely hard and possesses high modulus of elasticity. Table 2.11 gives the energies of a number of covalent bonds and the bond lengths.

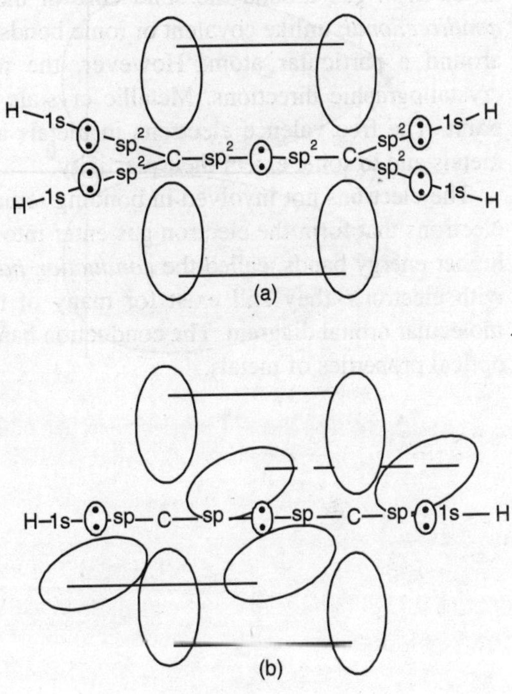

Fig. 2.19 Hybridisation resulting in (a) double bond and (b) triple bond.

Table 2.11 Energies of Few Selected Covalent Bonds and their Bond Lengths

Bond	*Bond energy (kJ/mole)*	*Bond length ($A = 10^{-10}m$)*
C—C	370	1.54
C—H	435	1.10
C—Cl	340	1.80
O—O	220	1.50
O—Si	375	1.60
H—H	435	0.74
C=C	680	1.33
C≡C	890	1.20

(iii) Metallic bond: A majority of elements are metals and in bulk metals each atom contributes its valence electrons to form a cloud of negative charge in which, the positive ions are arranged in a repetitive order. The metallic bond results when each atom of the metal contributes its valence electrons to the formation of an electron cloud that spreads throughout the solid metal [Fig. 2.20(*a*)]. In this case, there are N atoms in the lattice, where N is a large number. There are N atomic orbitals, so there must be N MOs, many of which are *degenerate,* or of the same energy. This leads to bands of electrons, as illustrated in Fig. 2.20(*b*) for Na. The characteristics of the metallic bond are that the valence electrons are not associated with any particular atom in the lattice, so they form what is figuratively referred to as an *electron gas* around the solid core of the metallic nuclei. As a result, the bonds in metals are *nondirectional,* unlike covalent or ionic bonds in which the electrons have a finite probability of being around a particular atom. However, the properties of metallic bond do vary somewhat with crystallographic directions. Metallic crystals are malleable and have variable hardness and melting point. The free valence electrons in metals account for high electrical and thermal conductivity of metals and to some extent their plasticity.

The electrons not involved in bonding remain in what is called the *core band,* whereas the valence electrons that form the electron gas enter into the *valence band.* The remaining unfilled orbitals form higher energy bands, called the *conduction band.* We may note that *d* and *f* orbitals may not be filled with electrons, they still exist for many of the heavier elements, so they must be included in the molecular orbital diagram. The conduction band plays very important role in the electrical, thermal and optical properties of metals.

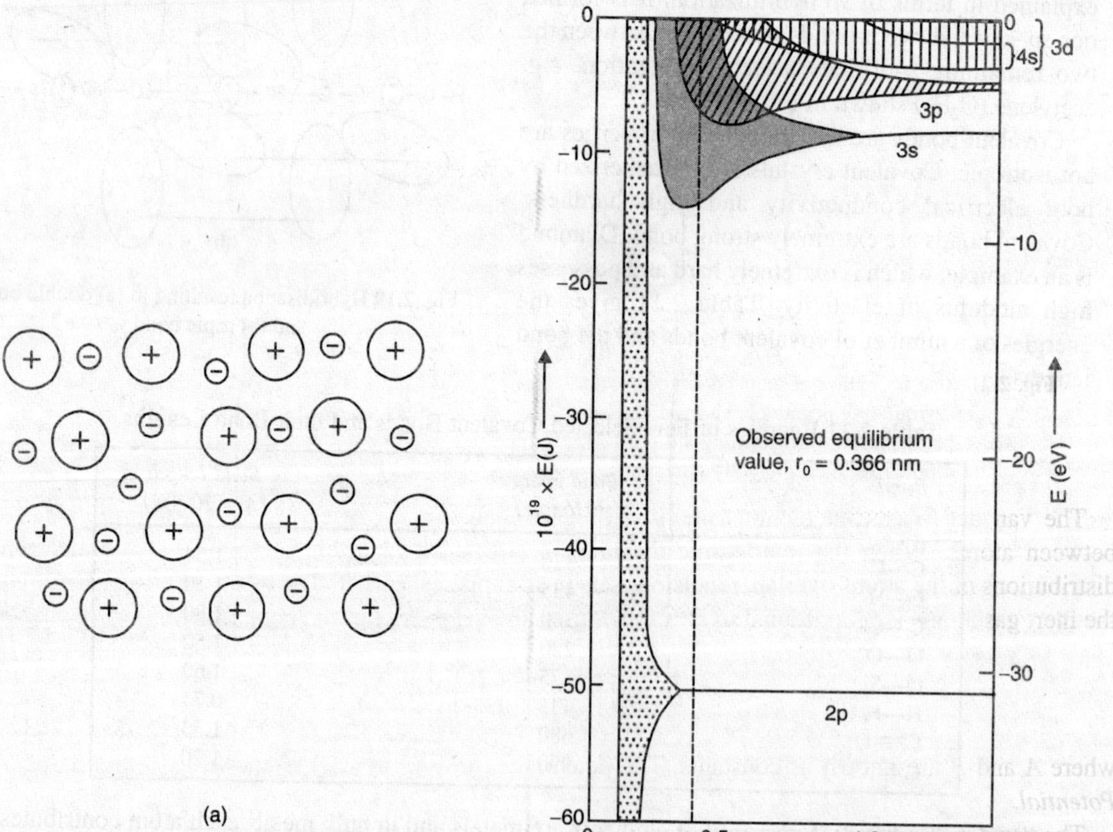

Fig. 2.20 (*a*) Schematic representation of metallic bonding. The valence electrons become disassociated with "their" atomic core and form an electron "sea" that acts as the binding medium between the positively charged ions. (*b*) Energy band diagram for a sodium lattice.

(iv) van der Waals bond: This is rather a weak bond and therefore called secondary bonds. This bond accounts for the weak attractive forces between gas molecules and involve the mutual attraction of dipoles. Inert gases and molecules like methane, which have no valence electrons available for crystalline binding, obtain a weak attractive force as a result of polarization of electric charges. We know that *polarization* is displacement of the centres of positive and negative charges in an electrically neutral atom or molecule when as it is brought close to its neighbouring atom. Its neighbours also get polarized. This can occur in all atoms and molecules due to statistical variations in the charge density. This resulting weak electrical attraction between neighbouring atoms or molecules is called the van der Waals force. Elements possessing van der Waals bond are soft, have poor electrical conductivity and low melting point.

When two atoms are brought closer, they induce in each other displacements of positive and negative charge centres relative to each other (Fig. 2.21). Whenever the centres of positive and negative charges are separated, an *electric dipole* exists. Clearly, electric dipoles are created. The induced dipoles cause an attraction between the atoms as positive charge centre of one atom attracts the negative charge centre of the other and vice versa. Such fluctuating dipoles bonds are termed as *van der Waals bonds.*

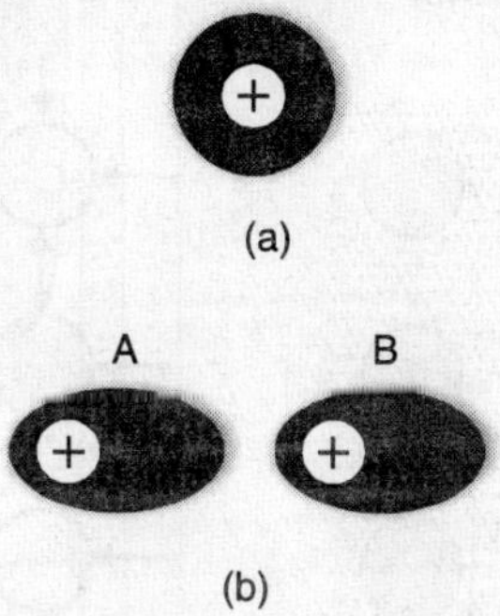

Fig. 2.21 The formation of induced dipole (*a*) An atom is represented by a positively charged core and a surrounding negatively charged electron cloud. (*b*) The electron cloud of atom 'A' is thought to be displaced, thus forming a dipole. This induces a similar dipole in a second atom, 'B'. Both dipoles are then mutually attracted, as proposed by van der Waals.

The van der Waals interaction energy is proportional to r^{-6}, where r is the separation distance between atoms. When the interatomic distances between the atoms is so small that the electron distributions of the atoms overlap, repulsions sets in as explained earlier. The repulsive energy between the inert gas atoms is proportional to r^{-12}. Obviously, the net energy of the bond is given by

$$U = A\left[\left(\frac{B}{r}\right)^{12} - \left(\frac{B}{r}\right)^{6}\right] \quad ...(27)$$

where A and B are known as constants. The potential expressed by (27) is known as *Lennard-Jones Potential.*

The van der Waals forces play important role in a large variety of solids. The planes of carbon atoms in graphite are bonded through van der Waals forces. Many polymeric chains which consist of covalently bonded atoms contain areas that are permanently polarized. The covalent bonding within the chains is quite strong. In contrast to this the individual chains are mutually attracted by weak van der Waals forces (Fig. 2.22). This results in smaller values of the melting points for the polymers. Moreover, many polymers can be deformed permanently since the chains slide effortlessly past each other when a

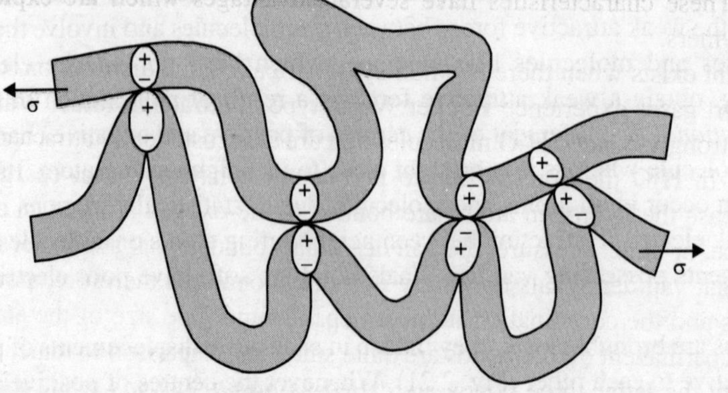

Fig. 2.22 Two polymer chains are mutually attracted by van der Waals forces. An applied external stress can easily slide the chains past each other.

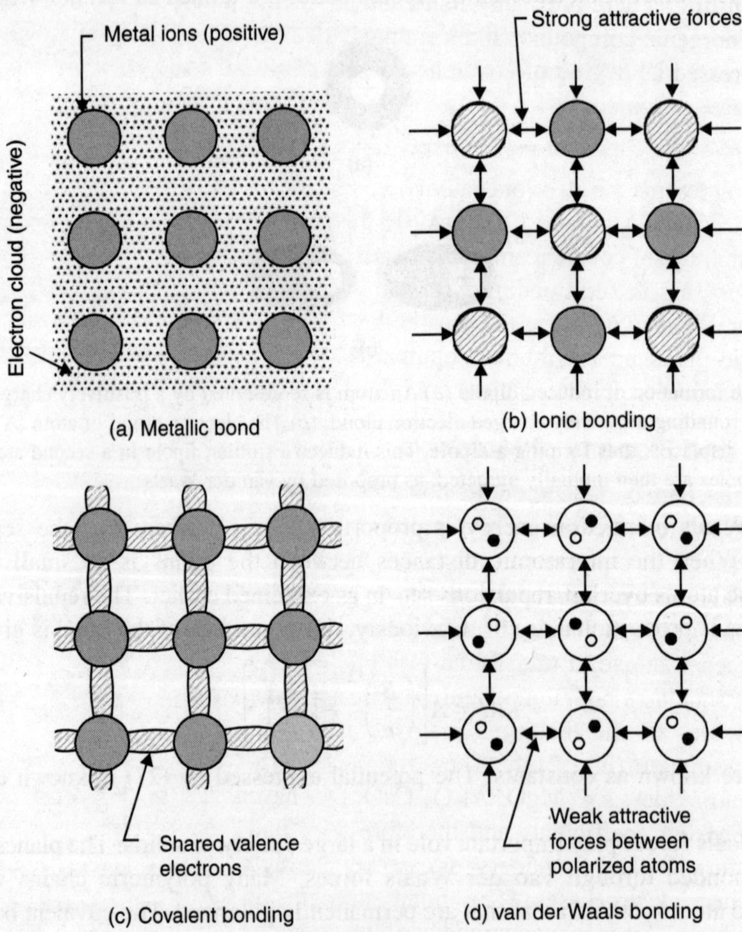

Fig. 2.23 Various types of bondings in solids: (*a*) Metallic bond, (*b*) Ionic bonding, (*c*) covalent bonding, and (*d*) van der Waals bonding.

force is applied. These characteristics have several advantages which are exploited in commercial application of polymers.

A dipole moment exists when there is a non-symmetrical charge distribution. On approaching each other, even the inert gases experience van der Waals type of attractive force. Water is an example ice crystal consist of strongly bonded H_2O molecules that are electrostatically attracted to each other by van der Waals forces. In H_2O the bond is due to permanent nonsymmetrical charge distribution in the molecule. In this case the hydrogen atoms are bonded to the oxygen atom at an angle of 105°. At the melting point of ice, or under pressure, the van der Waals bonds break and water is formed.

We may note that van der Waals bonds are non-directional and therefore, atoms attract each other from all directions and the coordination number depends upon the size of the atoms. The energies of van der Waals and permanent dipole bonds are quite small in comparison to ionic, covalent and metallic bonds. This is why the latter three bonds are called as *primary bonds* while the van der Waals and permanent dipole bonds are called as secondary bonds.

Various types of bonding in solids is schematically represented in Fig. 2.23.

(v) Mixed bonding: There are many materials possess atomic bonding involving more than one type, *e.g.* in the inorganic compounds like ceramics, the bond is neither pure covalent nor pure ionic. Here, this is expressed as degree of covalency or ionicity. Using the electronegativity system, Pauli expressed the degree of ionicity as:

$$\% \text{ ionic character} = [1 - \exp(-1/4)\,(X_A - X_B)] \times 100 \qquad ...(28)$$

where X_A and X_B represents the electronegativity of elements A and B respectively. The ionicity values in ceramics such as SiC and MgO are 12% and 70% respectively. Semiconductor compounds, *e.g.* GaAs are bonded by a mixture of covalent and ionic bonds. Intermetallic compounds, *e.g.* nickel and titanium aluminides also possess mixed bonding with some degree of ionicity and covalency mixed with the metallic character. We may note that the greater the difference in the electronegativity of the elements, the larger the ionic character. The titanium aluminides would be naturally having greater ionic character than the nickel aluminides. In SiO_2 (silica) the covalent and ionic bonds are in equal parts.

2.4.3 Bonding and Property of Materials

Materials properties largely depend on the bonding characteristics of materials.

The materials which are strongly bonded have high strength. With the rise in temperature, the bonds are broken down by thermal energy. When the thermal energy is sufficient the atoms or the molecules of solid become mobile and as a consequence the solid transforms to a liquid. This reveals that there should be high temperature stability and the melting temperatures of the materials may increase with increase in bond strength (Fig. 2.24). The strong and fairly high melting ionic oxides, *e.g.* MgO, Al_2O_3, CaO, etc. have low thermal expansion coefficients and therefore they find extensive use in furnaces and containers for carrying out high temperature industrial processes. The bond strength of diamond is quite high. This is why diamond has high strength, high melting point and quite high hardness. This is why diamond is widely used in cutting tools and drills employed for mineral and oil exploration.

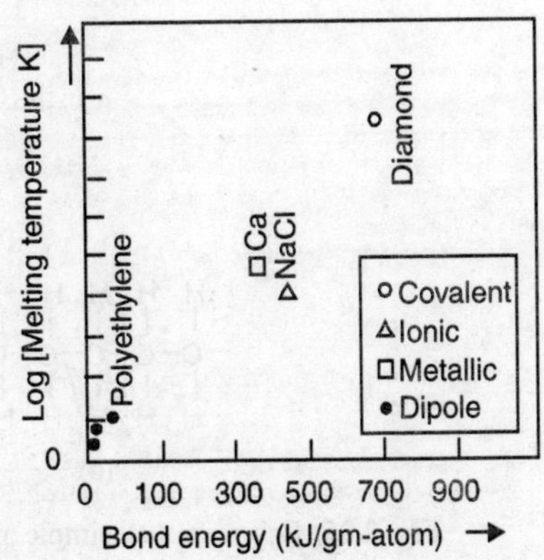

Fig. 2.24 For a number of materials, increase in melting temperature with increase in bond strength.

The shapes of many materials are changed by deformation for practical applications. During this process of shape change, atoms tend to break bonds with some of its neighbours and form bonds with other atoms. Materials with non-directional bonding tend to deform easily while those with directional bonding tend to the brittle. One can easily shaped the metals with their non-directional bonding. This is why that large high strength engineering components with complicated shapes are usually prepared from metals.

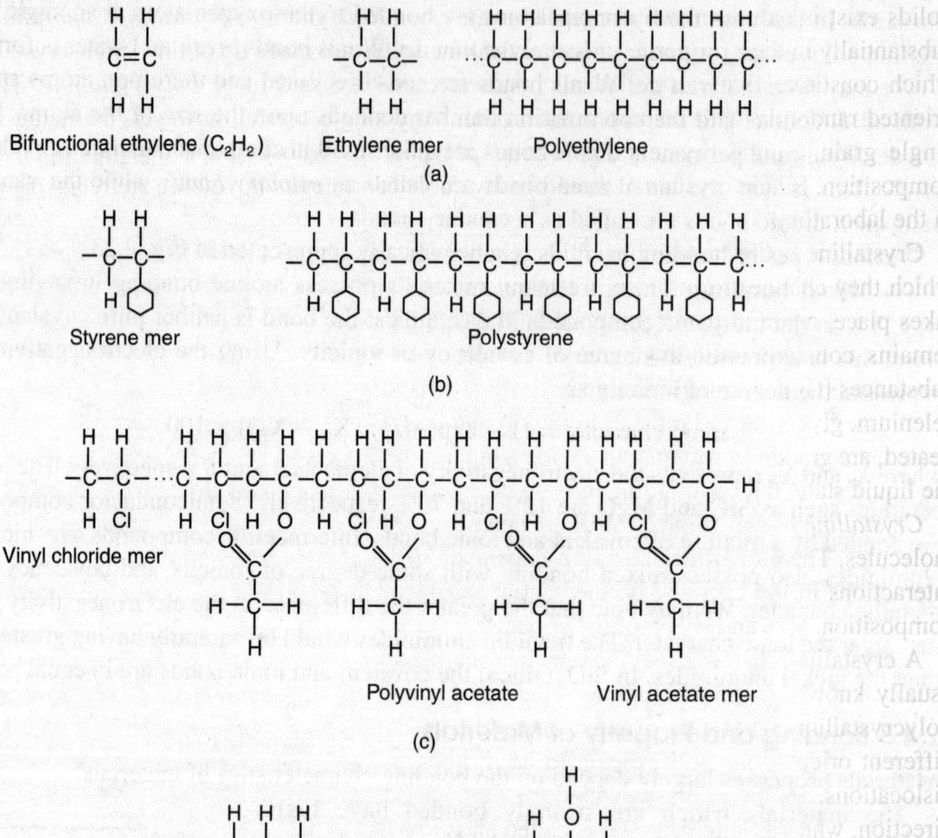

Fig. 2.25 Polymers (*a*) simple polymer, (*b*) homopolymers, (*c*) copolymers, (*d*) isomers, (*e*) isotactic stereoisomer, (*f*) syndiotactic stereoisomer, (*g*) atactic stereoisomer.

We may note that materials with metallic bonds have free electrons. This makes thermal and electrical conductivities of metals high, *e.g.* copper and aluminium. Cu and Al are widely used as conductors and heat exchangers in heavy industries. In comparison to metals, materials bonded with ionic, covalent and van der Waals bonds are good thermal and electrical insulators, *e.g.* mica, alumina and magnesia are widely used as insulators. Polymers (Fig. 2.25) are stable at low temperatures with insulating characteristics.

2.5 STRUCTURE OF CRYSTALLINE SOLIDS

Solids exist is nature in two principal forms: *crystalline* and *non-crystalline* (*amorphos*) which differ substantially in their properties. Most crystalline solids are made up of tiny single crystals called *grains* which constitutes what is called microstructure and are said to be *polycrystalline.* These grains are oriented randomly with respect to each other. Any single crystal, however, no matter how large, is a single grain. Single crystals are regular polyhedrons whose shape depends upon their chemical composition. Single crystals of metals many cubic centimeters in volume are relatively easy to prepare in the laboratory.

Crystalline bodies remain solid, *i.e.* retain their shape, upto a definite temperature (melting point) at which they change from the solid to liquid state. During cooling, the inverse process of solidification takes place, again at the definite solidifying temperature, or point. In both the cases, the temperature remains constant until the material is completely melted or respectively solidified. *Amorphous* substances [(Fig. 2.26(*a*)] have no crystalline structure in the condensed state. Ordinary glass, sulphur, selenium, glycerine and most of high polymers can exist in amorphous state. Amorphous bodies when heated, are gradually softened in a wide temperature range and become viscous and only then change in the liquid state. On cooling, the process takes place in the opposite direction.

Crystalline bodies [Fig. 2.26(*b*)] are characterized by an ordered arrangement of their ions, atoms, or molecules. The properties of crystals depend on the electronic structure of atoms and the nature of their interactions in the crystals, on the spatial arrangement of their ions, atoms or molecules, and on the composition, size and shape of the crystals.

A crystalline material may be either in the form of *single crystal* or an aggregate of many crystals usually known as polycrystalline separated by well-defined boundaries called as *grain boundaries.* Polycrystalline material is stronger than ordinary one because crystals in polycrystalline material have different orientations with respect to each other and the grain boundaries obstruct the movement of dislocations. Such crystals are called *isotropic*, because they exhibit same properties in every plane and direction, whereas single crystal is called an *isotropic.*

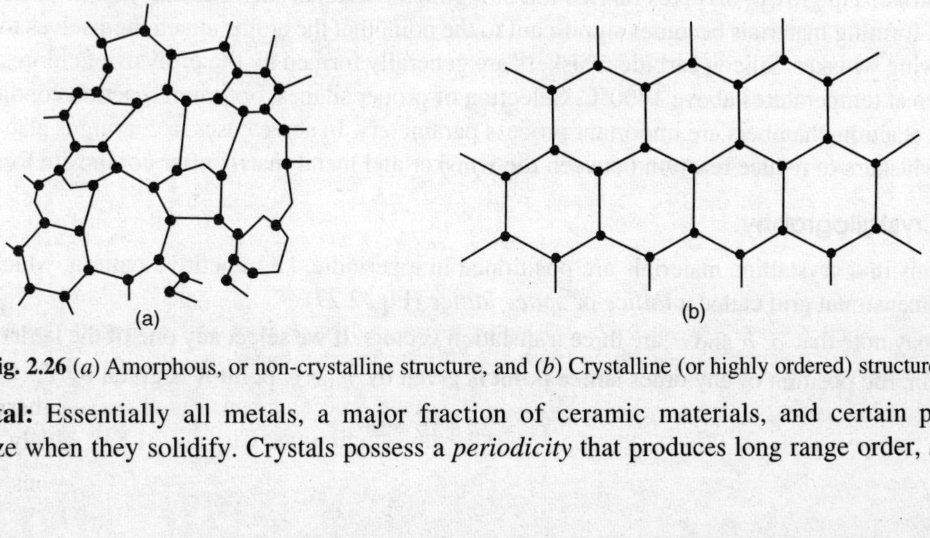

(a) (b)

Fig. 2.26 (*a*) Amorphous, or non-crystalline structure, and (*b*) Crystalline (or highly ordered) structure.

Crystal: Essentially all metals, a major fraction of ceramic materials, and certain polymers crystallize when they solidify. Crystals possess a *periodicity* that produces long range order, *i.e.* local atomic arrangement is repeated at regular intervals millions of times in the three dimensions of space. At temperatures below that of crystallization, the crystalline state is stable for all solids.

Single crystal: Single crystals represent a material in its ideal condition and are produced artificially from their vapour or liquid state. Single crystals help us in studying behaviour and defects of the material in ideal conditions.

Most of the materials exist in *polycrystalline form,* but there are some materials, which exist in the form of single crystals, *e.g.* sugar, sodium chloride (NaCl, *i.e.* common salt), diamond, etc.

Whiskers: These are very thin filaments, hair-like single crystals of about 13 mm length and perhaps 10^{-4} cm diameter. These are produced as dislocations of free crystals and are without any structural defect. This is why whiskers are far stronger than polycrystals of same material. These are used as reinforcements in materials to increase strength by embedding fibres of one material in a matrix of another.

Although the distinction between a *whisker* and a *fibre* is sometimes vague, whiskers are generally considered a subclass of fibres that have been grown in controlled conditions, often in the vapour phase, which leads to high purity crystals.

Whiskers can be formed from a variety of materials (Table 2.12), but there are only two general methods for forming them: *basal growth* and *tip growth.*

Table 2.12 Selected Material Properties of Some Common Whiskers

Material	*Density (10^3 kg m^{-3})*	*Melting point (°C)*	*Tensile Strength (GPa)*	*Young's modulus (GPa)*
Aluminum oxide	3.9	2082	14–28	550
Aluminum nitride	3.3	2198	14–21	335
Beryllium oxide	1.8	2549	14–21	700
Boron oxide	2.5	2449	7	450
Graphite	2.25	3593	21	980
Magnesium oxide	3.6	2799	7–14	310
Silicon carbide (α)	3.15	2316	7–35	485
Silicon carbide (β)	3.15	2316	7–35	620
Silicon nitride	3.2	1899	3–11	380

In basal growth, the atoms of growth migrate to the base of the whisker and extrude the whisker from the substrate. Tip growth involves fabrication at high temperatures such that the vapour pressure of the whisker-forming materials becomes significant to the point that the atoms attach themselves to the tip of the growing whisker. Silicon carbide whiskers are generally formed by the prolysis of chlorosilances in hydrogen at temperatures above 1400°C. Selection of proper silanes, optimum reaction conditions, and suitable reaction chambers are important process parameters. In some cases, a coating is also produced on the whiskers to reduce reaction between the whisker and metal matrix after composite formation.

2.5.1 Crystallography

The atoms in a crystalline materials are positioned in a periodic, *i.e.* repetitive pattern, which forms a three-dimensional grid called a *lattice* or *space lattice* (Fig. 2.27).

We may note that $\vec{a}$, $\vec{b}$ and $\vec{c}$ are three translation vectors. If we select any one of the lattice points as the origin, the position of any other lattice point is given by $\vec{t}$, *e.g.* point *A* is given by $\vec{t}_A$. Clearly,

$$\vec{t}_A = 2\vec{a} + 1\vec{b} + 1\vec{c} \qquad \text{...(29)}$$

In general, all the lattice points are given by

$$\vec{t} = u\vec{a} + v\vec{b} + w\vec{c} \quad ...(30)$$

where *u, v, w* are integers. Such a three-dimensional array of lattice points is known as *space lattice.*

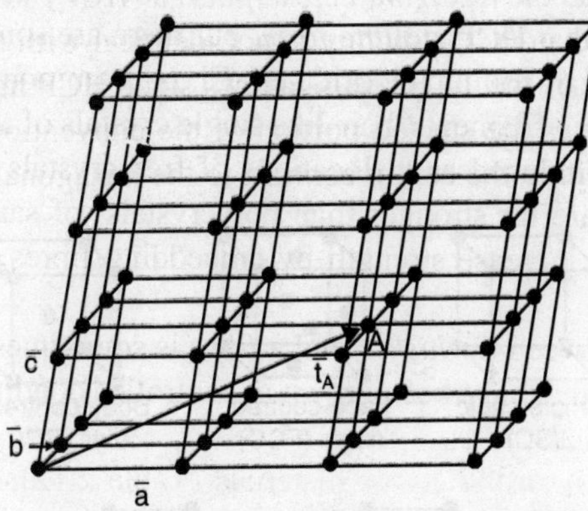

Fig. 2.27 A space lattice, $\vec{a}, \vec{b}$ and $\vec{c}$ are translational vectors. $\vec{t}_A$ is the position vector of lattice point A.

The intersections of lines are points of a space lattice. These points may be occupied by the atoms in crystals or they may be the points about which several atoms are clustered. However, the every point of a space lattice has identical surroundings.

Unit cell: The smallest unit of a lattice that still possesses the characteristic symmetry of the entire lattice, *i.e.* the grouping atoms whose repetition will produce the crystal is called a *unit cell.* Obviously, the entire lattice can be generated by translating the unit cell into three dimensional space.

Figure 2.28 shows the shape and size of the unit cell given by six lattice parameters *a, b, c,* α, β and γ.

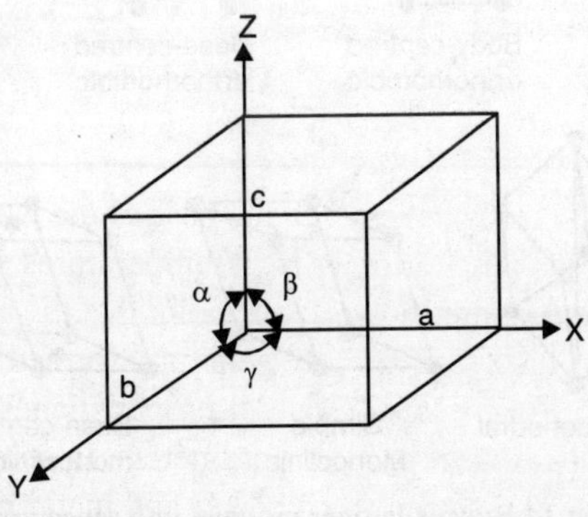

Fig. 2.28 Lattice parameters of a unit cell.

Bravais has identified 14 fundamental unit cells, often referred to as *Bravais lattices* or translational lattices, as shown in Fig. 2.29. They vary in length of their sides (called lattice constants *a, b* and *c*) and the angles between the axes (α, β and γ). As stated above the characteristic lengths and angles of a unit cell are termed *lattice parameters.* The arrangement of atoms into a regular, repeatable lattice is called a *crystal structure.*

Depending upon the relation between lattice parameters, the unit cells can be easily divided into seven groups, known as *crystal systems* (see bottom of the Fig. 2.29). The simplest possible relation which exists between the lattice parameters is when all the three axes are equal and also the angle between them are 90 degrees. This group is called as *cubic system.*

The most important crystal structures for metals *face centred cubic* (*FCC*) structure, which is typically found in the case of soft (ductile) materials, the *body centred cubic* (*BCC*) structure, which is common for strong materials, and the *hexagonal close-packed* (*HCP*) structure, which often is found in brittle materials. We may note that HCP structure is not identical with the simple hexagonal structure shown in Fig. 2.29 and not one of the 14 Bravais lattices since HCP has three extra atoms inside the hexagon. The unit cell for HCP is the shaded portion of the conventional cell shown in Fig. 2.30. It contains another "base" atom within the cell in contrast to the hexagonal cell shown in Fig. 2.29.

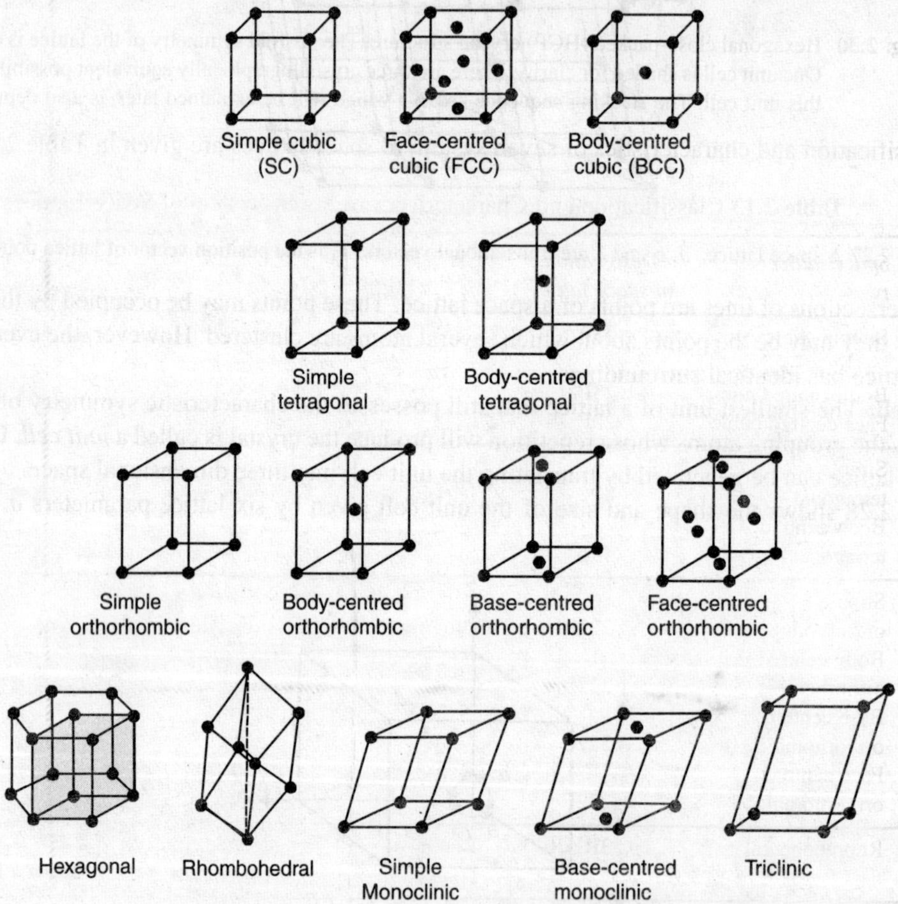

Fig. 2.29 The 14 Bravais lattices grouped into seven crystal systems:

First row: $a = b = c$, $\alpha = \beta = \gamma = 90°$ (cubic);
Second row: $a = b \neq c$, $\alpha = \beta = \gamma = 90°$ (tetragonal);
Third row: $a \neq b \neq c$, $\alpha = \beta = \gamma = 90°$ (orthorhombic);
Fourth row: at least one angle is $\neq 90°$. *Specifically;*
Hexagonal: $\alpha = \beta = 90°$, $\gamma = 120°$ $a = b \neq c$ (the unit cell is the shaded part of the structure);
Rhombohedral: $a = b = c$, $\alpha = \beta = \gamma \neq 90° \neq 60° \neq 109.5°$;
Monoclinic: $\alpha = \gamma = 90°$, $\beta \neq 90°$, $a \neq b \neq c$;
Triclinic: $\alpha \neq \beta \neq \gamma \neq 90°$, $a \neq b \neq c$.

We may note that FCC lattice with $\alpha = \beta = \gamma = 60°$ is rhombohedral.

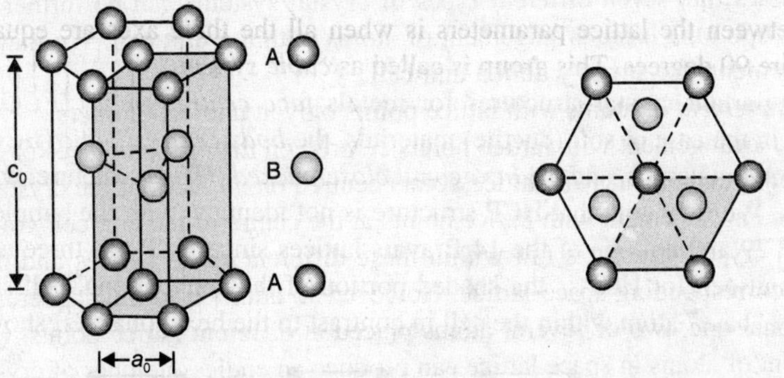

Fig. 2.30 Hexagonal close-packed (HCP) crystal structure. The six fold symmetry of the lattice is evident. One unit cell is shaded for clarity. There are three crystallographically equivalent possibilities for this unit cell. The stacking sequence (ABA), which will be explained later, is also depicted.

Classification and characteristics of seven system of space lattices are given in Table 2.13.

Table 2.13 Classification and Characteristics of Seven System of Space Lattices

S.No.	*Space lattice type*	*Designation of space lattice*	*System*	*Minimum symmetry*	*Axial lengths and angles*
1.	Simple cubic	23P	Cubic	4 three-fold rotation axes	$a = b = c$
2.	Body centred cubic	23I			$\alpha = \beta = \gamma = 90°$
3.	Face centred cubic	23F			
4.	Simple tetragonal	4P	Tetragonal	one four-fold rotation axis	$a \neq b \neq c$
5.	Body centred tetragonal	4I			$\alpha = \beta = \gamma = 90°$
6.	Simple orthorhombic	222P	Orthorhombic	3 two-fold mutually Perpendicular rotation axes	$a \neq b \neq c$
7.	Body centred orthorhombic	222I			$\alpha = \beta = \gamma = 90°$
8.	Base centred orthorhombic	222C			
9.	Face centred orthorhombic	222F			
10.	Rhombohedral	3R	Rhombohedral	one three-fold rotation axis	$a = b = c$ $\alpha = \beta = 90°$
11.	Hexagonal	3P	Hexagonal	One six-fold rotation axis	$a = b \neq c$ $\alpha = \beta = 90°$ $\gamma = 120°$
12.	Simple monoclinic	2P	Monoclinic	One two-fold rotation axis	$a \neq b \neq c$ $\alpha = \gamma = 90° \neq \beta$
13.	Base centred monoclinic	2C			
14.	Triclinic	1P	Triclinic	None	$a \neq b \neq c$ $\alpha \neq \beta \neq \gamma \neq 90°$

Table 2.13 shows that seven different types of crystal systems can be further subdivided into 14 types, depending upon the basic arrangement of atoms within a unit cell. We may note that there are four basic arrangements for atoms within a unit cell:

(*i*) *Simple arrangement* of atoms with lattice points only at unit cell corners,

(*ii*) *End centred* arrangement with lattice points centred on the opposite faces or ends of the crystal,

(*iii*) Face-centred arrangement with lattice points centred on all faces of the crystal.

(*iv*) Body-centred arrangement with lattice points at the centre of the unit cell volume.

All the known crystal structures fall within these different space lattices and possess the inherent geometry of the corresponding space lattice. However, in addition to this, different crystal structures may have additional one, two or several atoms placed at different lattice points. Obviously, different actual arrangement of atoms in space lattice can produce an endless number of crystal structures.

Table 2.14 provides some typical crystal structures of common metals.

Table 2.14 Typical Crystal Structures of Common Metals

S.No.	*Crystal structure*	*Examples (Typical metals)*
1.	Body centred cubic (BCC)	Cr, α-Fe, Mo, W
2.	Face centred cubic (FCC)	Al, Cu, Au, Ag, Pb
3.	Close-packed hexagonal (CPH)	Co, Cd, Zn, Mg, α-Ti
4.	Body centred tetragonal (BCT)	In, β-Sn
5.	Rhombohedral	Bi, Hg and Sb.

A table of crystal structures for the elements is shown in Table 2.15 (excluding the Lanthanide and Actinide series). Some elements can have multiple crystal structures, depending on temperature and pressure. This phenomenon is called *allotropy* and is very common in elemental metals (Table 2.16). Materials that undergo a transformation from one crystal structure to another are called *polymorphic* or, when referring to elements, *allotropic*. A well known example is the allotropic transformation of room temperature, ductile, diamond cubic, gray, α-modification. The transformation occurs slowly at 286.2 K and causes a 27% expansion and thus a disintegration into powder (tin plague). As a consequence of this transformation, old organ pipes made of tin (or tin alloys) may disintegrated in severe winter temperatures.

It is not unusual for closed packed crystals to transform from one stacking sequence to the other, simply through a shift in one of the layers of atoms. Other common allotropes include carbon (graphite at ambient conditions, diamond at high pressures and temperature), pure iron (BCC at room temperature, FCC at 912°C and back to BCC at 1394°C) and titanium (HCP to BCC at 882°C). The prevailing crystal structure also may depend on the external pressure, *e.g.* graphite is stable at ambient conditions whereas diamond is the high pressure modification of carbon.

We now define a few parameters that are linked to the mechanical properties of solids:

(i) c/a Ratio: The separation between the basal planes, Co, divided by the length of the lattice parameter, a_0 in HCP metals (Fig. 2.30) is theoretically $\sqrt{8/3} = 1.633$ assuming that the atoms are completely spherical in shape. c/a is called the axial ratio. In the extended structure, c/a is the ratio of the height of the cell to its base. Table 2.17 lists typical values of the axial ratio for some common HCP crystals.

We may note that deviations from ideal ratio, 1.633, result from mixed bondings and from non-spherical atom shapes. The c/a ratio influences the hardness and ductility of materials.

Table 2.15 Common Crystal Structures, Densities, and Lattice Parameters of the Elements

Struct. ρ, g/cc a, Å	1 **H** hcp n/a 4.70																	2 **He** fcc n/a 4.24
Struct. ρ, g/cc a, Å	3 **Li** bcc 0.533 3.51	4 **Be** hcp 1.85 2.29											5 **B** rhom 2.47 5.06	6 **C** diam 3.51 3.56	7 **N** hcp n/a 3.86	8 **O** mon n/a 5.40	9 **F** mon n/a 5.50	10 **Ne** fcc n/a 4.42
Struct. ρ, g/cc a, Å	11 **Na** bcc 0.966 4.29	12 **Mg** hcp 1.74 3.21											13 **Al** fcc 2.70 4.05	14 **Si** diam 2.33 5.43	15 **P** tricl 1.82 11.5	16 **S** orth 2.09 10.4	17 **Cl** orth n/a 6.22	18 **Ar** fcc n/a 5.26
Struct. ρ, g/cc a, Å	19 **K** bcc 0.862 5.33	20 **Ca** fcc 1.53 5.56	21 **Sc** hcp 2.99 3.31	22 **Ti** hcp 4.51 2.95	23 **V** bcc 6.09 3.03	24 **Cr** bcc 7.19 2.88	25 **Mn** cubic 7.47 8.91	26 **Fe** bcc 7.87 2.86	27 **Co** hcp 8.8 2.51	28 **Ni** fcc 8.91 3.52	29 **Cu** fcc 8.93 3.61	30 **Zn** hcp 7.13 2.66	31 **Ga** orth 5.91 4.52	32 **Ge** diam 5.32 5.65	33 **As** rhom 5.78 3.76	34 **Se** mon 4.81 9.05	35 **Br** orth n/a 6.72	36 **Kr** fcc n/a 5.71
Struct. ρ, g/cc a, Å	37 **Rb** bcc 1.53 5.59	38 **Sr** fcc 2.58 6.08	39 **Y** hcp 4.48 3.65	40 **Zr** hcp 6.51 3.61	41 **Nb** bcc 8.58 3.30	42 **Mo** bcc 10.22 3.14	43 **Tc** hcp 11.5 2.74	44 **Ru** hcp 12.36 2.71	45 **Rh** fcc 12.42 3.80	46 **Pd** fcc 12.0 3.88	47 **Ag** fcc 10.5 4.08	48 **Cd** hcp 8.65 2.98	49 **In** tetrag 7.29 3.25	50 **Sn** tetrag 7.29 5.83	51 **Sb** trig 6.69 4.31	52 **Te** trig 6.25 4.46	53 **I** orth 4.95 7.18	54 **Xe** fcc n/a 6.20
Struct. ρ, g/cc a, Å	55 **Cs** bcc 1.91 6.14	56 **Ba** bcc 3.59 5.01	57 **La** hcp 6.17 3.77	72 **Hf** hcp 13.3 3.20	73 **Ta** bcc 16.7 3.30	74 **W** bcc 19.3 3.17	75 **Re** hcp 21.0 2.76	76 **Os** hcp 22.58 2.73	77 **Ir** fcc 22.55 3.84	78 **Pt** fcc 21.4 3.92	79 **Au** fcc 19.28 4.07	80 **Hg** rhom n/a 3.01	81 **Tl** hcp 11.87 3.46	82 **Pb** fcc 11.34 4.95	83 **Bi** rhom 9.80 4.74	84 **Po** cub 9.2 3.36	85 **At** n/a n/a n/a	86 **Rn** n/a n/a n/a

Table 2.16 Some Metal Allotropes

Metal	*R.T. Crystal Structure*	*Structure at other temperatures*
Ca	FCC	BCC (> 447°C)
Co	HCP	FCC (> 427°C)
Hf	HCP	BCC (> 1742°C)
Fe	BCC	FCC (> 912°C)
		BCC (> 1394°C)
Li	BCC	BCC (< –193°C)
Na	BCC	BCC (< –233°C)
Sn	BCT	Cubic (< 13°C)
Tl	HCP	BCC (> 234°C)
Y	HCP	BCC (> 1481°C)
Zr	HCP	BCC (> 872°C)

Table 2.17 Axial Ratios (c/a) for Some HCP Metals

Metal	*c/a*
Be, Y	1.57
Hf, Os, Ru, Ti	1.58
Sc, Zr	1.59
Tc, Tl	1.60
La	1.61
Co, Re	1.62
Mg	1.63
Zn	1.85
Cd	1.89
Ideal (sphere packing)	1.633

(ii) Coordination number: This is the number of nearest neighbours to a given atom. The polyhedron which results from connecting the centres of atoms surrounding a central atom is called the *coordination polyhedron.* One can determine the coordination number by the relative sizes of the ions. A small ion of radius, *r*, surrounded by the large ion of radius, r^*, such that the ions touch each other without overlapping is shown in Fig. 2.31. The geometry of the configuration reveals.

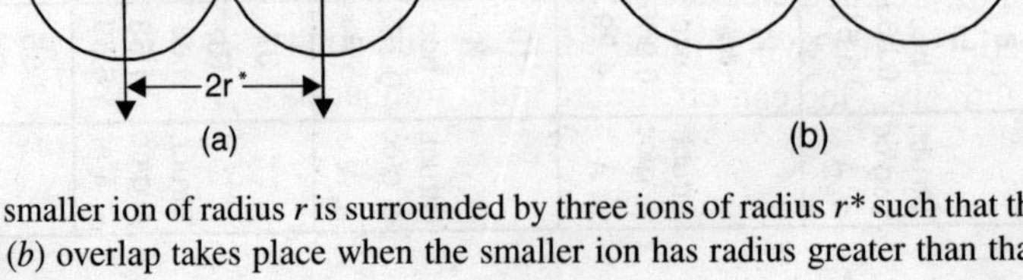

Fig. 2.31 (*a*) A smaller ion of radius *r* is surrounded by three ions of radius r^* such that the ions touch each other (*b*) overlap takes place when the smaller ion has radius greater than that shown in (*a*).

$$r^* = (r^* + r)\cos 30° \qquad ...(31)$$

and $r^*/r = 0.155$. When the number of neighbours increases beyond 3 or the ratio r^*/r exceeds 0.155, overlap takes place and excessive repulsion sets in (Fig. 2.31(*b*)).

More closely packed atoms in the lattice will have higher coordination number. BCC structure has (Fig. 2.28) eight nearest atoms. Its coordination number is therefore 8. Table 2.18 gives the number of expected nearest neighbours for various radius ratios (r/r^*).

Table 2.18 The Radius Ratios (r/r^*) of Ions in Ionic Solids and the Expected Coordination Number of the Smaller Ion

Ionic radius ratio, (r/r)*	*Coordination number of smaller ion*	*Geometrical arrangement*	
$0 < (r/r^*) < 0.155$	2	Linear	
$0.155 \le (r/r^*) < 0.225$	3	Triangular	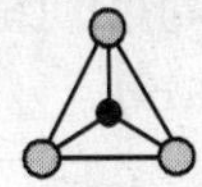
$0.225 \le (r/r^*) < 0.414$	4	Tetrahedral	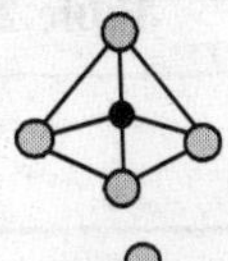
$0.414 \le (r/r^*) < 0.732$	6	Octahedral	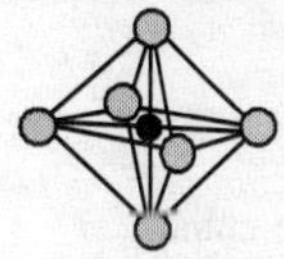
$0.732 \le (r/r^*) < 1$	8	Cubic (*e.g* Diamond)	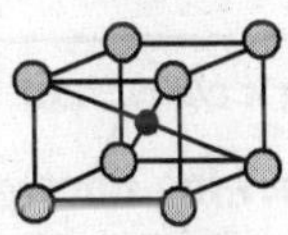
$(r/r^*) = 1$	12	(*a*) FCC	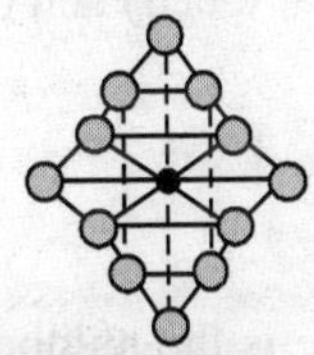
	12	(*b*) HCP	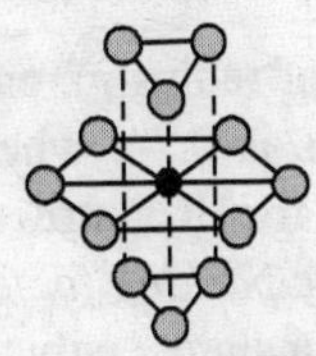

(iii) Atoms per unit cell: Usually, the corners of the unit cell are occupied by atoms. In additions to corners, a unit cell may also contain atoms located either at the faces, sides or interior of the unit cell. Obviously, an atom located at the corner of the unit cell is shared by eight unit cells and one at the face of the cube is shared between two unit cells. Clearly, the contributions of these atoms per unit cell are 1/8th and 1/2 respectively. One can obtain the total number of atoms per unit cell with the help of the following relation:

$$N = N_i + \frac{N_f}{2} + \frac{N_c}{8} \quad ...(32)$$

where N is the number of atoms per unit cell, N_i is the number of interior atoms, N_f is the number of atoms on faces and N_c is the number of atoms on corners.

We may note that unit cells containing only one atom per unit cell are called *primitive cells* and those containing more than one atom per unit cell are called *non-primitive*.

The *number of atoms per unit cell* is counted by taking into consideration that corner atoms in cubic crystals are shared by eight unit cells and face atoms are shared by two unit cells. They count therefore only 1/8 and 1/2 respectively. For example, the number of atoms associated with a BCC structure (assuming only one atom per lattice point) is $8 \times 1/8 = 1$ corner atom and one (not shared) center atom, yielding a total of two atoms per unit cell. In contrast to this, an FCC unit cell has 4 atoms $(8 \times 1/8 + 6 \times 1/2)$. The FCC unit cell is therefore more densely packed with atoms than the BCC unit cell (Table 2.19).

Table 2.19 Some Parameters and Properties of Different Crystal Structures

Crystal structure	*Coordination number*	*Atoms per unit cell*	*Packing factor*	*Mechanical properties*
HCP	12*	2	0.74*	Brittle
FCC	12	4	0.74	Ductile
BCC	8	2	0.68	Hard
Simple cubic (SC)	6	1	0.52	No representative materials

*Assuming a *c/a* ratio of 1.633.

(iv) Packing Factor (P): This is that portion of space within a unit cell which is filled with spherical atoms that touch each other, *i.e.*

$$P = \frac{A_{uc}^{*} V_A}{V_{uc}} \quad ...(33)$$

where A_{uc} is the number of atoms per unit cell, $V_A = \frac{4}{3}\pi r^3$ is the volume of an atom (assuming hard spheres of radius r), and V_{uc} is the volume of the unit cell. The packing factor (P) in FCC and HCP structures is 0.74, whereas a BCC crystal is less densely packed, having a packing factor of 0.68. $P = 0.74$ for HCP structures is only true if $c/a = 1.633$. Values of P are given in Table 2.19.

Atomic packing factor (P) can also be defined as the ratio of the volume of atoms to that of the unit cell. For a simple cubic

$$P = \frac{\text{Volume of atoms}}{\text{Volume of unit cell}}$$

$$= \frac{\text{Number of atoms per unit cell} \times \text{volume of one atom}}{\text{Volume of unit cell}} = \frac{1 \times 4/3\pi r^3}{(2r)^3} = 0.52$$

where r is the atomic radius.

(v) Linear and planar packing fractions: The linear packing fraction is the portion of a line through the centres of atoms in a specific *direction* that is filled by atoms. For example, the linear packing fraction in the direction of the face diagonal of an FCC unit cell in which the corner atoms and the face atom touch each other (Fig. 2.32) is 100%. Similarly, a *planar packing fraction* can be defined as that portion of a given *plane* that is filled by atoms.

(vi) Density (δ): The density of a material can be calculated by using

$$\delta = \frac{A_{uc^*} M_a}{V_{uc^*} N_o} = \frac{N_{a^*} M_a}{N_o} \quad ...(34)$$

where M_a is the atomic mass of the atom, $N_a{}^*$ is the number of atoms per unit volume, and N_o is Avogadro constant (6.022×10^{23} atoms/mol).

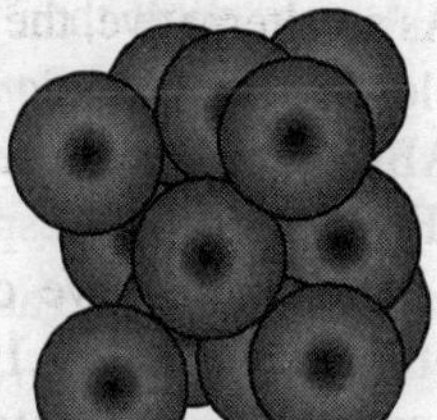

Fig. 2.32 FCC structure.

(vii) Stacking sequence: This describes the pattern in which closed-packed atomic planes are piled up in three dimensions. To make it clear, let us assume for simplicity that the atoms are represented by hard spheres and are arranged on a plane in such a way to take up the least amount of space [Fig. 2.33(*a*)]. The atoms are said to be situated in a close-packed plane we designate by plane A. We now attempt to construct a regular

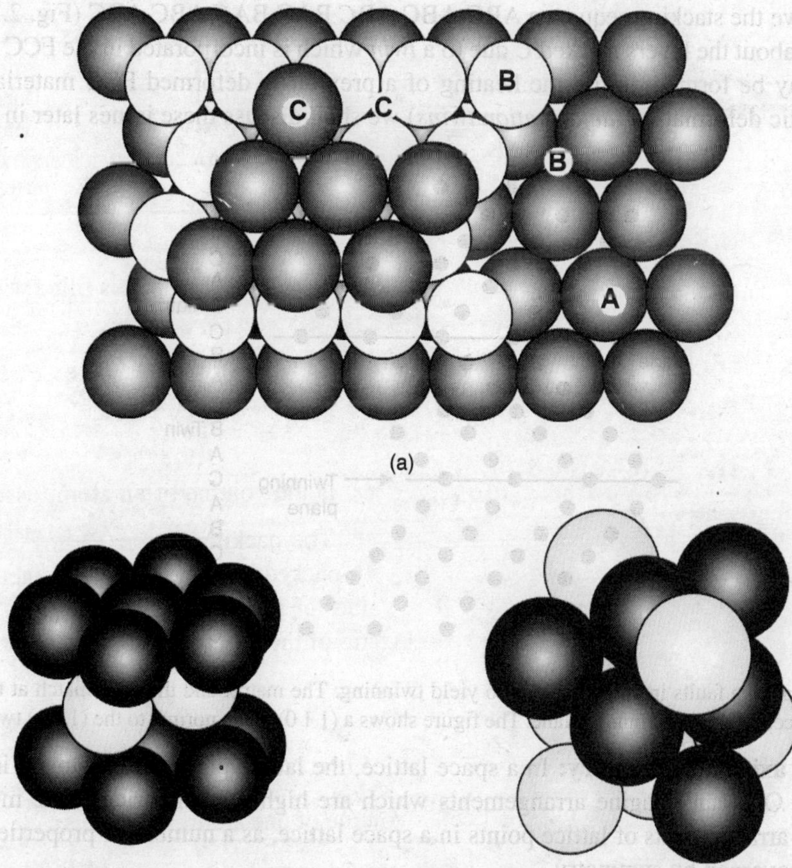

Fig. 2.33 (*a*) A two-dimensional representation of a three-dimensional stacking sequence for close-packed lattice planes. (*b*) Close-packed planes for an HCP crystal structure. (*c*) Close-packed planes in an FCC structure.

and repetitive array of atoms in three-dimensions. First, we fill the depressions marked B, which are formed by each set of three atoms in plane A. This constitutes the second plane of atoms. When filling the third level with atoms, one has two choices. One may arrange third level identical to the first one. The stacking sequence is then said to be ABAB, which represents the HCP crystal structure depicted in Figs. 2.30 and 2.33(*b*). In other words, atoms in the first and third levels are exactly on top of each other. As an alternative, the hitherto unoccupied spaces, C may be occupied on the third level. Only the fourth close-packed atomic plane is finally identical to the first one, which leads to a stacking sequence called ABCABC. We can see that the diagonal planes in an FCC crystal structure have this stacking sequence [Fig. 2.33(*c*)].

Concluding, we can say that HCP and FCC structures both have the highest possible packing efficiency (Table 2.19), but differ in their stacking sequence. Both types of crystal structures are quite common among metals. The preference of one type of structure over the other is rooted in the tendency of atomic systems to assume the lowest possible energy level.

(viii) Stacking faults and twinning: In some instances, the regular stacking sequence may be interrupted, *e.g.* a stacking sequence ABC AB ABC may occur instead of ABC ABC ABC. This configuration, where one layer is missing, is called an *intrinsic stacking fault.* On the other hand, the sequence ABC B ABC, where one B layer has been *added,* is termed as *extrinsic stacking fault.* Another defect may involve the stacking sequence ABC ABC AB*C* BAC BA*C* ABC ABC (Fig. 2.34). Here the order is inverted about the layers marked *C* due to a *twin* which is incorporated in the FCC crystal. Such configuration may be formed during the heating of a previously deformed FCC material (*annealing twins*) or by plastic deformation (*deformation twins*) we shall discuss these issues later in this chapter.

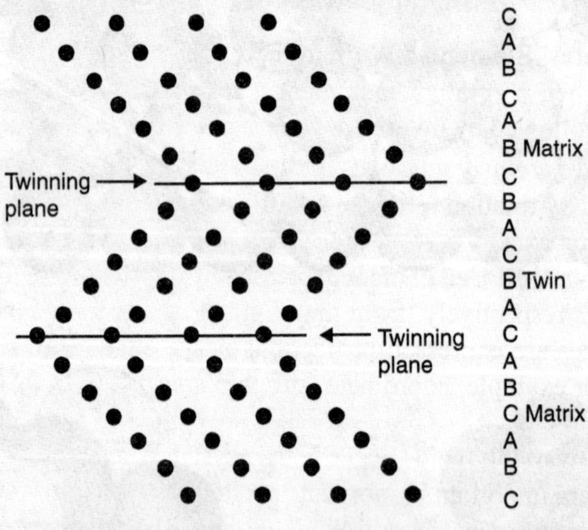

Fig. 2.34 Stacking faults in an FCC metal to yield twinning. The matrix and the twin match at the interface, called the twinning plane. The figure shows a (1 1 0) plane normal to the (1 1 1) twin plane.

(ix) Rotation axis and symmetry: In a space lattice, the lattice points are arranged in space in an ordered manner. One can imagine arrangements which are highly symmetrical. One must know the symmetry of the arrangements of lattice points in a space lattice, as a number of properties of practical importance are governed by symmetry.

Symmetry implies repetition of objects at regular intervals. A plane in a cubic crystal is said to have *four fold* symmetry because, when rotated about an axis normal to its faces, four rotations of 90° can be achieved at which the cell remains invariant. The cubic unit cell has three of those four-fold axes in addition to four three-fold axes that pass through opposite corners. Further, there are six two-fold axes that pass through two opposite edge corners. In general, an axis is termed *n*-fold when the angle of rotation is $2\pi/n$.

The operations which can describe the repetition of objects are called as *symmetry* operations. There are four symmetry operations: *translation, rotation, reflection* and *inversion.*

Translation: Translations can repeat points and objects at regular intervals. An object that is assymetric can occur in two varieties: the *left-handed variety* and *right-handed variety.* The translation repeats the variety present at the origin. We may note that the translation operation starting with a left-handed object cannot repeat a right-handed object and vice-versa.

Rotation: One can describe the operation of rotation in terms of an axis of rotation which is known as the *rotational symmetry element.* For example, a four-fold rotation axis implies that an object is repeated after angular rotation of $2\pi/4 = \pi/2$. The rotation axis repeats the same variety of object.

In a lattice, the operations of rotation and translation operate together in a lattice. One, two, three, four and six-fold rotation axes are the only possible symmetry elements.

Reflection: The operation of reflection involves a reflection plane, which is imagined to be a mirror and the repetition of an object is equivalent to the reflection of the object in the mirror. This is why a reflection mirror is also called as a mirror. In the operation of the reflection, the opposite variety of the object is repeated, *e.g.* a right-handed object when repeated becomes a left-handed object. The reflection plane is denoted by *m*.

Inversion: The operation of inversion involves a centre of inversion and each point located at $\overline{t}$ from the centre is repeated at $-\overline{t}$. When an object is subjected to the operation of inversion, its opposite variety appears.

Combination of Symmetry Elements: We may note that several symmetry operations may be satisfied simultaneously. When a rotation axis and an inversion centre occur simultaneously, the symmetry involves rotation followed by inversion. One can describe such symmetry operations by one, two, three, four and six-fold *rotoinversion* axes and denoted by $\overline{1}, \overline{2}, \overline{3}, \overline{4}$ and $\overline{6}$, respectively. In a similar way, objects repeated by rotation followed by reflection across a plane perpendicular to the axis are described by *rotoreflection* axes.

(x) Position of an atom in a Unit Cell: This is expressed by three coordinates, *i.e.* the three distances parallel to the *x, y* and *z* axes respectively, from the origin at one corner of the cell to the atom under consideration. These distances are expressed in fractions of the edge lengths *a, b* and *c* respectively, rather than in angstroms. For example, coordinates of atom are 0, 0, 0; 0, 0, 1/2; 1/2, 1/2, 0; 1/2, 1/2, 1/2, etc.

(xi) Voids in unit cells: Invariably voids occur in the crystal structure and other smaller atoms may fit in to these spaces. Voids are important in inorganic crystals, as well as in certain metals. Figure 2.35 (*a*), (*b*) shows tetrahedral voids and octahedral voids. Figure 2.35 (*c*) and (*d*) shows the voids in the cubic and hexagonal structures.

Any three dimensional array of spheres give rise to *void space* between the spheres. These voids are termed as *interstitial voids.* The two main types of interstitial voids in closed packed structures are: (*i*) tetrahedral voids, and octahedral voids as shown in Fig. 2.35 (*a*) and (*b*) respectively. There are eight tetrahedral voids in the unit cell. As the effective number of atoms in the unit is four and thus it gives a ratio 2 : 1 for the tetrahedral voids to the atoms.

An octahedral void is formed with three spheres on a close packed plane and three more spheres on an adjacent close packed plane [Fig. 2.35(*b*)]. We may note that there is one *octahedral void per sphere* in the three dimensional array. For example, in the FCC unit cell, the centres of the octahedral voids fall at the body centre and at the middle of the 12 cubic edges. The effective number of octahedral voids per unit cell is then $1+12\times 1/4=4$. Obviously, this is equal to the effective number of atoms in the unit cell. One can show that the largest sphere that can fit an octahedral void is 0.41 *r*, where *r* is the radius of the spheres of the closed packed array.

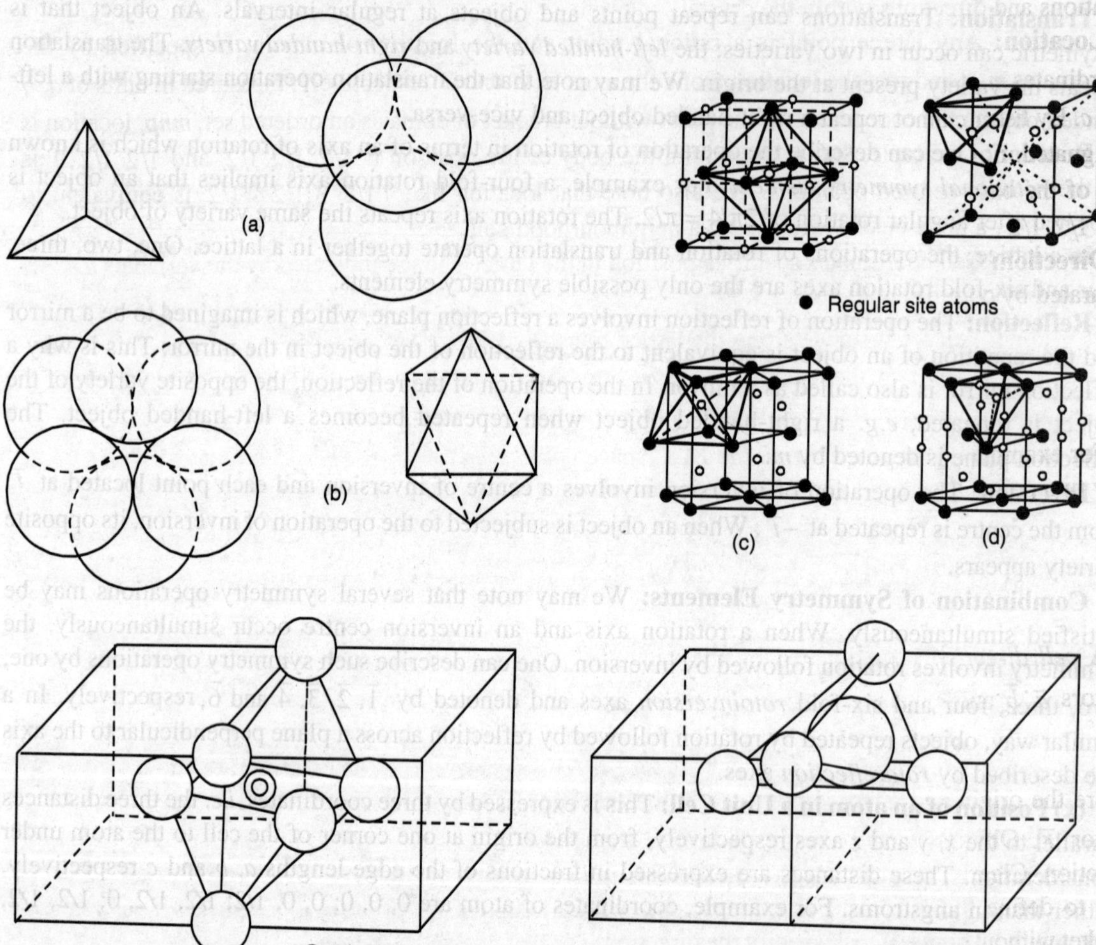

Fig. 2.35 Voids: (*a*) Tetrahedral voids, and (*b*) octahedral voids. Voids in closed packed unit cell structures: Top: FCC; bottom: HCP (*c*) octahedral void, (*d*) tetrahedral void. Voids in body centred cubic structure (*e*) octahedral and (*f*) tetrahedral.

In BCC structures the voids can be more clear by considering two unit cells. Octahedral voids occur at the centre of cube face and the centre of cube edges (Fig. 2.35(*e*) and (*f*). These are nevertheless irregularly shaped voids as two atoms are at a distance $1/2a$ and at $a/2\sqrt{2}=0.71a$. The tetrahedral voids occur in the space between the two cells. Obviously, there are thus four such voids in each cube face. We may note that tetrahedral voids are larger than the octahedral voids and can contain a sphere of radius 0.291 R. Moreover, the tetrahedral holes in BCC structures are slightly larger than the corresponding voids in the closed packed structures, but much smaller than octahedral voids in the latter. This means that the opportunity for elements with smaller atomic radius to occupy the voids in a BCC lattice is obviously reduced. This fact is of much importance in hardening of steels.

2.5.2 Lattice Planes and Directions

In order to calculate important quantities, *e.g.* cell volume and densities, we need to be able to specify locations and directions within the crystal.

Location: Any lattice point is a rational point and the location of a point is specified by the coordinates *m, n* and *p* of the location along the axes $\overline{a}$, $\overline{b}$ and $\overline{c}$ respectively, measured in units of *a, b* and *c*. By listing the coordinates of the point without bracket or coma as an ordered set, *mnp*, location is designated. For example, 000 is the origin, the body centre position is 1/2 1/2 1/2 and 1/2 1/2 0 is one of the base centred positions. We also note that location like 1 1/2 3 1/2 4 1/2 is equivalent to 1/2 1/2 1/2, a body centre location, in a neighbourhood cell.

Direction: Cell coordinates specify a position in the lattice and are indicated by the variables *u, v, w*, separated by commas with no brackets:

u distance along the lattice translation vector $\overline{a}$

v distance along the lattice translation vector $\overline{b}$

w distance along the lattice translation vector $\overline{c}$

For example, the centre atom in the BCC space lattice (Fig. 2.29) has cell coordinates 1/2, 1/2, 1/2. Any two points are equivalent if the fractional portions of their coordinates are equal:

$$\frac{1}{2}, \frac{1}{2}, \frac{1}{2} = -\frac{1}{2}, -\frac{1}{2}, -\frac{1}{2} \text{ (centre)}$$

$$0, 0, 0 = 1, 0, 1 \text{ (corner)}$$

A *cell direction* is designated by the vector $\overline{r}$, which is combination of the lattice translational vectors $\overline{a}$, $\overline{b}$ and $\overline{c}$:

$$\overline{r} = u\overline{a} + v\overline{b} + w\overline{c} \quad ...(35)$$

where the origin and a lattice point are at the two ends of the vector $\overline{r}$. Obviously, the vector, $\overline{r}$ defines the line and its direction. Clearly, one can use the coefficients *u, v, w* of Eqn. (35) to define the direction. The coefficients within a square bracket without coma, [*uvw*], specify the direction. The negative direction is indicated by putting a bar on the top of the coordinate. All directions are relative to the origin where the three lattice translation vectors originate (Fig. 2.28). We may note that [1 1 1], [2 2 2] and [3 3 3] define the same direction. Usually, the common factors from the coefficients is eliminated. Clearly, integers without any common factors are used to define the direction. Figure 2.36 shows some of the directions. As mentioned above, the lattice points are equivalent, any point can be taken as the origin. Obviously, all parallel directions are have the same sign.

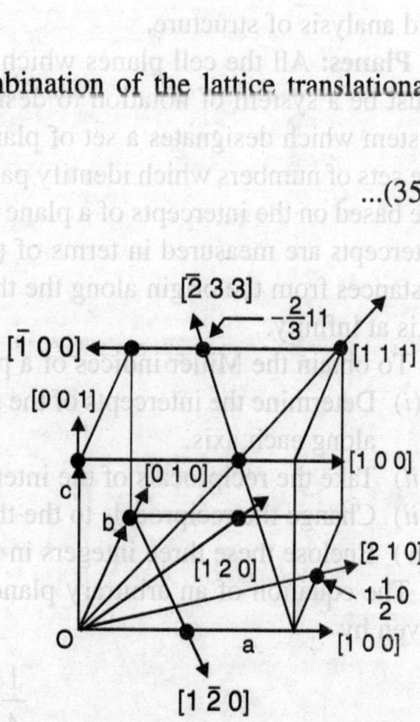

Fig 2.36 Figure showing some of indices of directions.

A set of directions may become equivalent depending upon the symmetry of a crystal, *e.g.* [1 0 0], $[0\bar{1}0]$, [0 1 0], [0 0 1], $[00\bar{1}]$ and $[\bar{1}00]$ are equivalent in a cubic system because of its four-fold axes. One can denote a set of equivalent directions as <*uvw*>, *e.g.* the cubic system just discussed above is denoted by <100>.

The cell volume (V): One can calculate V using the lattice translational vectors:

$$V = |\bar{a} \times \bar{b} \cdot \bar{c}| \qquad ...(36)$$

Mathematically, this is a triple scalar product and can be used to calculate the volume of any cell, with only a knowledge of the lattice translation vectors. If the lattice parameters and interaxial angles are known, one obtains the following expression for V.

$$V = abc(1 - \cos^2\alpha - \cos^2\beta - \cos^2\gamma + 2\cos\alpha\cos\beta\cos\gamma)^{1/2} \qquad ...(37)$$

For orthogonal systems: $\alpha = \beta = \gamma = 90°$. Eqn. (37) reduces,

$$V = abc \qquad ...(38)$$

Using cell volume, one can calculate density (ρ) of a material. For example, aluminium has an FCC space lattice. There are 4 atoms in the FCC unit cell. Atomic weight of aluminium is 27 g/mol. The cubic lattice parameter for Al is 4.05 Å or 0.405 nm (4.05×10^{-10}m). This gives us the volume, $V = 6.64 \times 10^{-23}$ cm^3.

Thus $$\rho = \frac{\left(4\dfrac{\text{atoms}}{\text{unit cell}}\right)(27\,\text{g/mol})}{(6.02\times10^{23}\,\text{atoms/mol})(6.64\times10^{-23}\,\text{cm}^3)}$$

= 2.70 g/cm^3 per unit cell.

In addition to cell coordinates and directions, *cell planes* are very important for the determination and analysis of structure.

Planes: All the cell planes which pass through lattice points, are rational planes. Obviously, there must be a system of notation to designate different planes and directions through the crystal. Such a system which designates a set of planes through the crystal is called *Miller indices, i.e.* Miller indices are sets of numbers which identify particular planes or sets of parallel planes in unit cells. Miller indices are based on the intercepts of a plane with the three crystal axes, *i.e.* the three edges of the unit cell. The intercepts are measured in terms of the edge lengths or dimensions of the unit cell which are the unit distances from the origin along the three axes. However, if plane is parallel to an axis, it intercepts the axis at infinity.

To obtain the Miller indices of a plane, following procedure is used:

(*i*) Determine the intercepts of the plane on the three axes in multiples or fractions of the axial lengths along each axis.

(*ii*) Take the reciprocals of the intercepts.

(*iii*) Change the reciprocals to the three smallest integers having the ratio as reciprocals.

(*iv*) Enclose these three integers in parenthesis as ( ).

The equation of an arbitrary plane with intercepts A, B and C, relative to the lattice parameters is given by

$$\frac{1}{A}\frac{x}{a} + \frac{1}{B}\frac{y}{b} + \frac{1}{C}\frac{z}{c} = 1 \qquad ...(39)$$

where *x, y, z* are cell's coordinate axes and *a, b, c* are lattice parameters (Fig. 2.37). We designate a plane by *Miller indices h, k* and *l* which are simply the reciprocals of the intercepts, *A, B* and *C*:

$$\left(\frac{1}{A}\ \frac{1}{B}\ \frac{1}{C}\right) = (hkl) \qquad \text{...(40)}$$

If a plane cuts any axis for example the y-axis, on the negative side of the origin, the corresponding index will be negative and is shown by placing a minus sign above the index $(h\bar{k}l)$.

To make it clear, let the intercepts of the plane along the axes of the crystal system are 3/4, 3/2, 1/2. The reciprocals of these intercepts are 4/3, 2/3, 2/1. Now, changing these into integers in the same ratio as the reciprocals and dividing by the largest common factor, one obtains 4/3, 2/3, 2/1 ≡ 2, 1, 3. Thus the indux of the plane is (2 1 3).

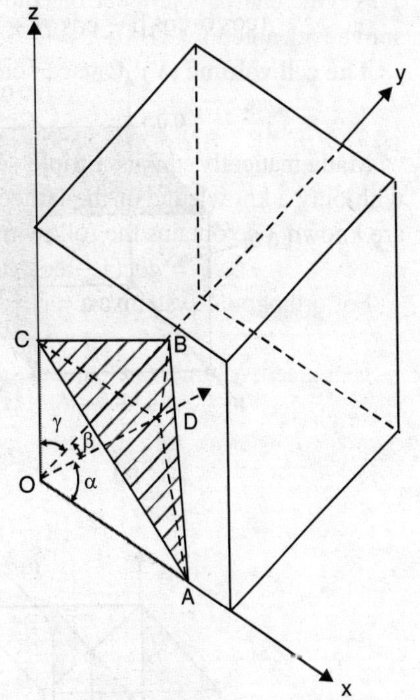

Fig. 2.37 Definition of Miller indices for an arbitrary plane (shaded area).

When a set of planes is equivalent due to symmetry, *e.g.* a cube has six equivalent faces having Miller induces of (1 0 0), (0 1 0), (0 0 1), $(\bar{1}\,0\,0)$, $(\bar{1}\,0\,0)$ and $(0\,0\,\bar{1})$. This set is written as {1 0 0}. Use of curly brackets avoid writing the indices of similar planes.

The face diagonal planes (1 1 0), (1 0 1), (0 1 1), $(1\,\bar{1}\,0)$, $(\bar{1}\,0\,1)$, $(0\,\bar{1}\,1)$ (and the equivalent planes having negative Miller indices) are summarized as {1 1 0}, where $\bar{1}$ signifies generally a shift of the coordinate axis and consequently an intercept at –1 [see Fig. 2.38(*c*)]. We may note that Miller indices and their negatives describe the same plane, *i.e.* $(1\,1\,0) \equiv (\overline{1\,1}\,0)\,\bar{1}$; however, the vectors normal to these planes point in opposite directions. Further multiples of Miller indices do not describe the same set of planes, even though the planes in question are parallel to each other [see Fig. 2.38(*b*)], where (1 1 1) describes a set of planes which has twice the interplanar spacing than the set designated by (2 2 2).

We may note that some planes may not necessarily be occupied by any atoms in certain crystal structures, *e.g.* in the cubic primitive structure, only every second (0 2 0) plane contain atoms. In contrast, in the FCC lattice every plane of the (0 2 0) set is occupied.

In the BCC lattice, the (111) planes pass only through the corner atoms but not through the centre of the body-centred ones. In order to accommodate the body-centred as well as the corner atoms, the (2 2 2) set of planes needs to be taken into account [Fig. 2.38(*b*)]. We may note that however, that the (2 2 2) plane sketched in Fig. 2.38 passes through the body centred atoms of adjacent cubes, for example, one immediately below the depicted one. Figure 2.38 shows selected examples.

One can also define *planar density, i.e.* density of atoms on a plane. If we know the lattice parameter or atomic radius, we can calculate the *area of the plane, A_p*, the area occupied by the atoms, A_c, and the corresponding *planar density*, P.D. = A_C/A_p. An important property of metals, ductility, is affected by planar density.

Interplanar spacings

The distance between the two (*hkl*) planes measured at right angles to them is called as *interplanar spacing* and designated as d_{hkl}. It is a function of (*i*) plane indices, and crystal system involved, *i.e.*

lattice parameters α, β, γ, a, b and c. The general expression for d_{hkl} in terms of lattice parameters and interaxial angles is somewhat complicated

$$d_{hkl} = [h^2b^2c^2 \sin^2\alpha + k^2a^2c^2 \sin^2\beta + l^2a^2b^2 \sin^2\gamma + 2hlab^2c(\cos\alpha\cos\gamma - \cos\beta) + 2hkabc^2 (\cos\alpha\cos\beta - \cos\gamma) + 2\,kla^2bc\,(\cos\beta\cos\gamma - \cos\alpha)]^{-1/2} \quad ...(41)$$

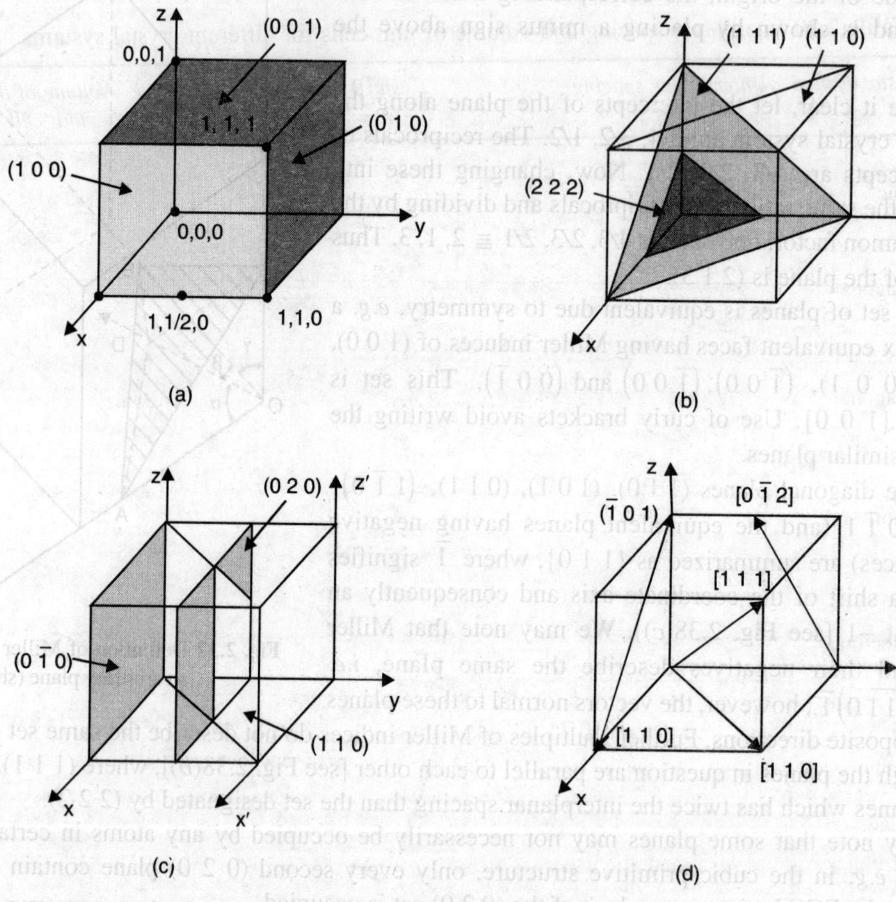

Fig. 2.38 (*a–c*) Examples for Miller indices for planes. [Note the primed coordinate system in (*c*).] (*d*) Indices for directions for a cubic unit cell.

where V is the volume of the cell given by Eqn. (38), we may note that the Miller indices of the plane under consideration are included here, to distinguish the distance between different planes in the same space lattice. For orthogonal systems (orthorhombic, tetragonal and cubic, $\alpha = \beta = \gamma = 90°$, and hence

$$\frac{1}{d_{hkl}} = \left(\frac{h^2}{a^2} + \frac{k^2}{b^2} + \frac{l^2}{c^2}\right)^{1/2}$$

or

$$d_{hkl} = \frac{1}{\sqrt{\dfrac{h^2}{a^2} + \dfrac{k^2}{b^2} + \dfrac{l^2}{c^2}}} \quad ...(42)$$

For a cubic system ($a = b = c$), (42) simplifies even further to

$$d_{hkl} = \frac{a}{\sqrt{h^2 + k^2 + l^2}} \quad \text{...(43)}$$

These results are summarized in Table 2.20.

Table 2.20 Interplanar spacing and volume of unit cells for different crystal systems

Sl. no.	*Crystal system*	*Interplanar spacing 'd_{hkl}'*	*$\sin^2\theta$ ($\lambda = 2d\sin\theta$)*	*Volume of the unit cell*
1.	Cubic	$\frac{a^*}{\sqrt{h^2+k^2+l^2}}$	$\frac{\lambda^2}{4}\left[\frac{h^2+k^2+l^2}{a^2}\right]$	a^3
2.	Tetragonal	$\frac{a}{\sqrt{h^2+k^2+l^2\frac{a^2}{c^2}}}$	$\frac{\lambda^2}{4}\left[\frac{h^2+k^2}{a^2}+\frac{l^2}{c^2}\right]$	a^2c
3.	Hexagonal	$\frac{a}{\sqrt{\frac{4}{3}(h^2+hk+k^2)+l^2\frac{a^2}{c^2}}}$	$\frac{\lambda^2}{4}\left[\frac{4}{3}\left(\frac{h^2+hk+k^2}{a^2}\right)+\frac{l^2}{c^2}\right]$	$\sqrt{\frac{3}{2}}a^2c$
4.	Orthorhombic	$\frac{1}{\sqrt{\frac{h^2}{a^2}+\frac{k^2}{b^2}+\frac{l^2}{c^2}}}$	$\frac{\lambda^2}{4}\left[\frac{h^2}{a^2}+\frac{k^2}{b^2}+\frac{l^2}{c^2}\right]$	abc
5.	Monoclinic	$\left[\frac{a^2\sin^2\beta}{h^2}+\frac{b^2}{k^2}+\frac{c^2\sin^2\beta}{l^2}-\frac{ac\sin^2\beta}{2hl\cos\beta}\right]^{\frac{1}{2}}$		
6.	Rhombohedral	$\left[\frac{a^2(1-3\cos^2\alpha+2\cos^3\alpha)}{(h^2+k^2+l^2)\sin^2\alpha+2(hk+kl+hl)(\cos^2\alpha-\cos\alpha)}\right]^{\frac{1}{2}}$		–

**hkl* are Miller indices of the planes and *a, b* and *c* are lattice parameters.

The designation of interplanar spacings through the crystal is shown in Fig. 2.39.

Identification of crystallographic planes

For identification of crystallographic planes, one may utilize the points where the planes intercept the three cartesian axes. One may follow the following rules:

(*i*) Locate the points where the crystal plane intersects the three axes. We may note that the plane cannot pass through the origin. However, if it does, then another plane must be constructed parallel to the original plane or the origin must be shifted. Usually, it is more convenient to shift the origin.

(*ii*) Take the reciprocal of intercepts. Now, a plane that parallels an axis may be considered to have an infinite intercept and therefore zero index when reciprocal of intercepts is taken.

(*iii*) Now, clear the fraction but take care that they do not reduce to lowest integers.

(*iv*) Enclose the numbers within the parentheses ( ). Negative numbers should be denoted with a *bar* over the number.

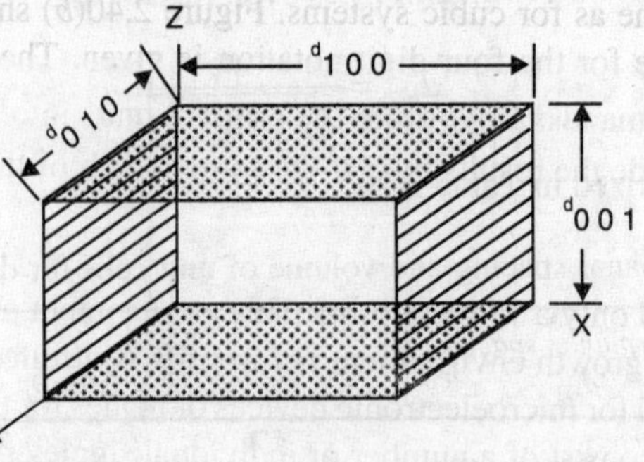

Fig. 2.39 Interplanar spacings through the crystal.

In a cubic lattice, the angle between two intersecting planes $(h_1\ k_1\ l_1)$ and $(h_2\ k_2\ l_2)$ can be obtained by the following expression:

$$\cos\theta = \frac{h_1 h_2 + k_1 k_2 + l_1 l_2}{\sqrt{\left(h_1^2 + k_1^2 + l_1^2\right)\left(h_2^2 + k_2^2 + l_2^2\right)}}$$

As stated earlier the family of planes is enclosed in braces { }.

2.5.3 Hexagonal Planes and Disections

The designation of planes in the *hexagonal* system is equivalent. However, because of the specific symmetry in hexagonal structures, one usually utilizes a coordinate system that contains four axes a_1, a_2, a_3 and c, and consequently four-digit Miller-Bravais indices (*hkil*), as depicted in Fig. 2.40(*a*). (The index *i* is defined by the relation $h + k + i = 0$). The Miller-Bravais indices for the top (or bottom) planes called *basal planes,* are therefore (0 0 0 1), whereas the indices for the six vertical, external surfaces (called prism planes) are of the type $\{1\ \bar{1}\ 0\ 0\}$.

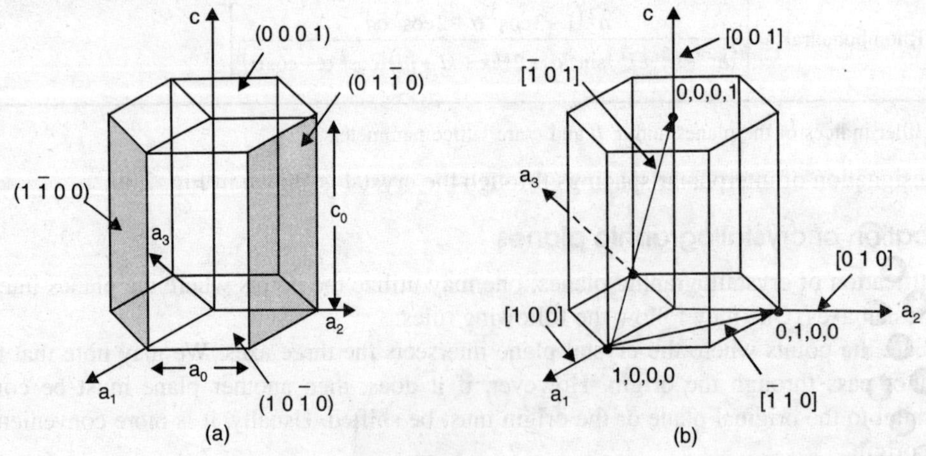

Fig. 2.40 (*a*) Four-digit Miller-Bravais indices for *planes* in a hexagonal crystal structure. (*b*) Selected examples for *directions* in the hexagonal crystal system using three-digit indices. The coordinate intercepts are given with commas separating them.

The *directions* in *hexagonal* systems are more conveniently expressed by three digit indices (whereas the a_3 axis is considered to be redundant and is therefore omitted). The procedure for arriving at these indices is the same as for cubic systems. Figure 2.40(*b*) shows some selected examples. (For completeness, an example for the four-digit notation is given. The [1 0 0] direction may be listed as 1/3 $\left[2\,\bar{1}\,\bar{1}\,0\right]$, which commands: move from the origin 2 units into the a_1 and −1 units each into the a_2 and a_3 directions and divide the results by 3 if the same length of the vector is desired).

2.5.4 Polycrystallinity

So far we have considered only a *single crystal.* This arrangement of atoms is, however, rarely found in practical cases unless the growth environment is carefully controlled, as, for example, when producing silicon single crystals used for microelectronic devices or blades for turbine engines made of superalloys. Instead, solids generally consist of a number of individual single crystallites, called grains. They range in size between millimeters and micrometers. (Recent discovery of nanostructured materials have made nanocrystals of interest). Their orientations are somewhat rotated with respect to the neighbouring grain (Fig. 2.41). These materials are then called *polycrystals.* The individual grains are separated by *grain boundaries* in which some atomic mismatch occurs. In essences the grain boundaries are less densely packed than the grains themselves, which result in important properties. One distinguishes between *large-angle grain boundaries* (Fig. 2.41(*a*)) and *small-angle grain boundaries.* For the later, the mismatch is relatively small, *i.e.* on the order of only a few degrees (θ in Fig. 2.41(*b*)). There are other types of grain boundaries also.

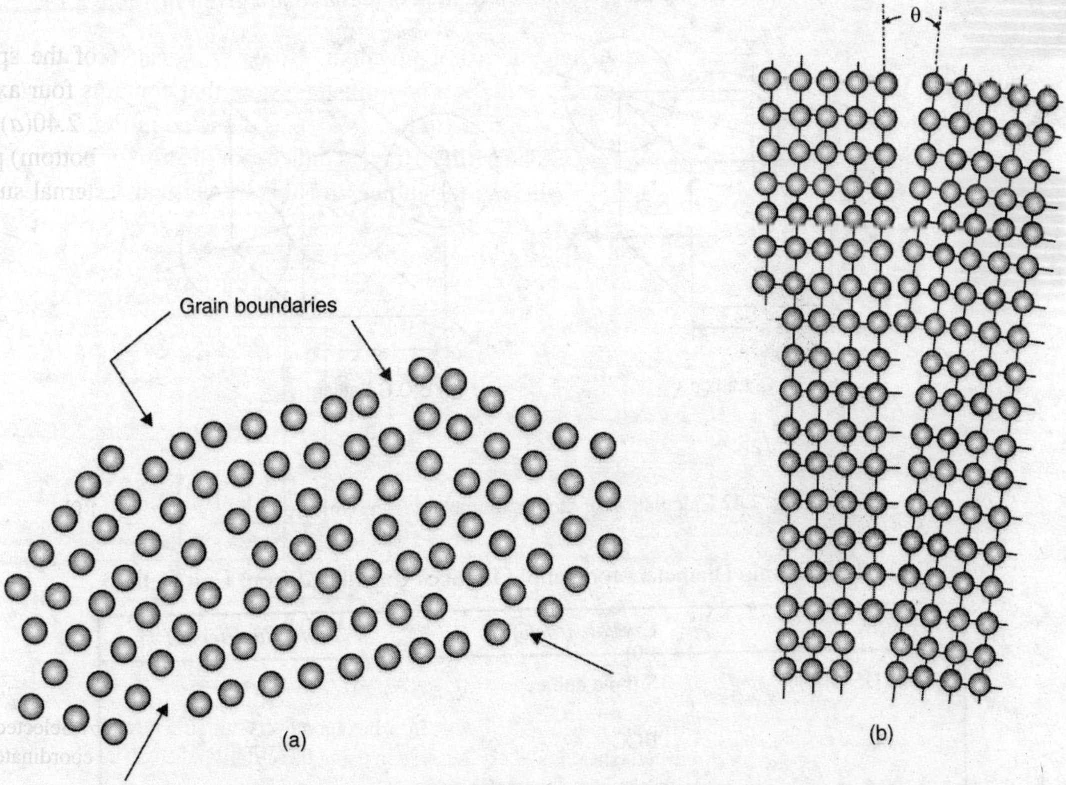

Fig. 2.41 Schematic representation of polycrystalline materials consisting of (*a*) three single crystals which meet on mutual large-angle grain boundaries. and (*b*) a small-angle grain boundary.

2.5.5 Anisotropy

Many physical properties of materials, such as the modulus of elasticity, or some optical, electrical, or magnetic constants may depend on the crystallographic direction in which they were measured. For example, the elastic modulus for copper varies between 66.7 × 10^9 P_a in the [1 1 0] direction and 191 × 10^9 P_a in the [1 1 1] direction, *i.e.,* by a factor of 3 (with an intermediate value of 130 × 10^9 P_a for the ([1 1 0] direction). This dependence of physical property on the crystallographic direction is called *anisotropy*. It is caused by the variation in distance between the atoms (or ions) in different directions. Materials for which the physical properties are identical in each direction are termed *isotropic*. For example, the electrical conductivity for materials having a cubic lattice structure is completely isotropic.

2.5.6 Atomic Diameter

When atoms come together and form a lattice, the forces holding the atoms together are determined by the force of repulsion between similar ions and electrons and the force of attraction between ions and free electrons. These two types of forces will have minimum energy of interaction at some distances. Obviously, this condition corresponds to the equilibrium condition of atoms in the lattice. If we visualize the atoms as spheres just touching one another, then the distance between the centres of the two spheres touching each other may be taken as the approximate *atomic diameter.* One can determine the distance of closest approach of the atom centres in a crystal from the lattice parameters of the unit cell (Fig. 2.42). For simple lattices, the atomic diameters thus calculated are given in Table 2.21.

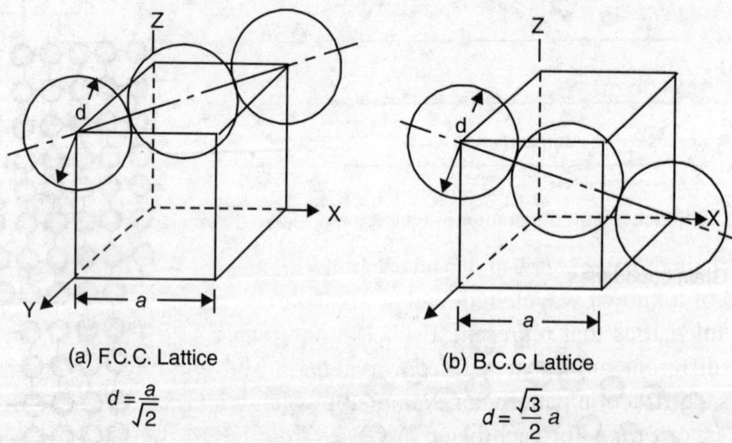

Fig. 2.42 Calculation of atomic diameters from Unit Cells.

Table 2.21 Atomic Diameters for Simple Lattices Calculated from Unit Cells

S. No.	*Crystal structure*	*Atom diameter (d)*
1.	Simple cubic	a*
2.	BCC	$\frac{\sqrt{3}}{2}a$
3.	FCC	$\frac{\sqrt{2}}{2}a$

* '*a*' is the lattice parameter.

2.6 X-RAY DIFFRACTION

The interplanar spacing between adjacent planes allows us to use a very powerful tool for structural determination called *X-ray diffraction* (*XRD*). If we bombard a crystal with X-rays of a certain wavelength, λ, at an incident angle, θ, the X-rays can either pass directly through the crystal or, depending on the angle of incidence, interact with certain atoms in the lattice. This interaction is best visualized as a "shooting gallery" in which the incident X-rays "bounce off" the hard-core spheres in the lattice (see Fig. 2.43). In reality, the X-ray photons are interacting with the electron density around the atoms, leading to diffraction. The path difference for rays reflected from adjacent planes is $2d(\sin\theta)$, and *constructive interference* occurs when the path difference is an integer, *n*, of the wavelength:

$$n\lambda = 2d\sin\theta \quad \text{...(44)}$$

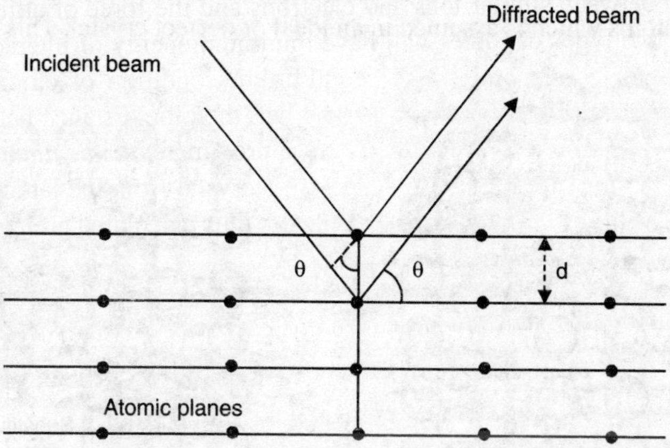

Fig. 2.43 Schematic illustration of incident radiation diffraction by a crystal lattice.

Equation (44) is called *Bragg's law* and is a very important result. It says that if we bombard a crystal lattice with X-rays of a known wavelength and at a known angle, we will be able to detect diffracted X-rays of various intensities that represent a specific interplanar spacing in the lattice. The result of scanning through different angles at a fixed wavelength and counting the number (intensity) of diffracted X-rays is a diffraction pattern, an example of which is shown in Fig. 2.44. Note how the sum of the Miller indices, or *order,* for each plane increases from left to right in the pattern, and also note

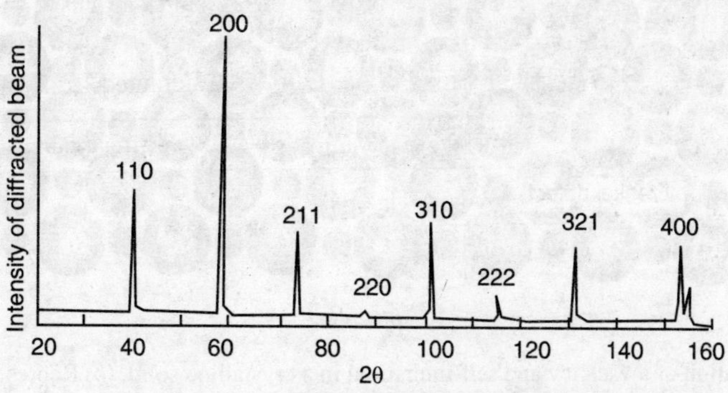

Fig. 2.44 X-Ray diffraction pattern for tungsten.

how different planes have different intensities. Also, not all planes result in diffracted X-rays. From the abscissa (which is by convention listed as 2θ for instrumentation reasons), the interplanar spacing, *d,* can be calculated using Bragg's law.

The study of crystals and their X-ray patterns is both fascinating and complex. From a known crystal structure, a theoretical diffraction pattern can be constructed using vector algebra, including such considerations as *structure factors* which account for disallowed interactions with the crystal lattice and determine the relative intensity of the diffracted beams. For our purposes, it is sufficient to know that a diffraction pattern can be obtained experimentally in a relatively routine fashion, and the resulting pattern is characteristic of a specific crystalline material.

2.7 DEFECTS IN CRYSTALS (CRYSTAL IMPERFECTION)

We have studied the most important aspects of perfect crystals. However, real crystals deviate from the perfect periodicity of atoms which is assumed in an ideal or perfect crystal. This is not necessarily a bad

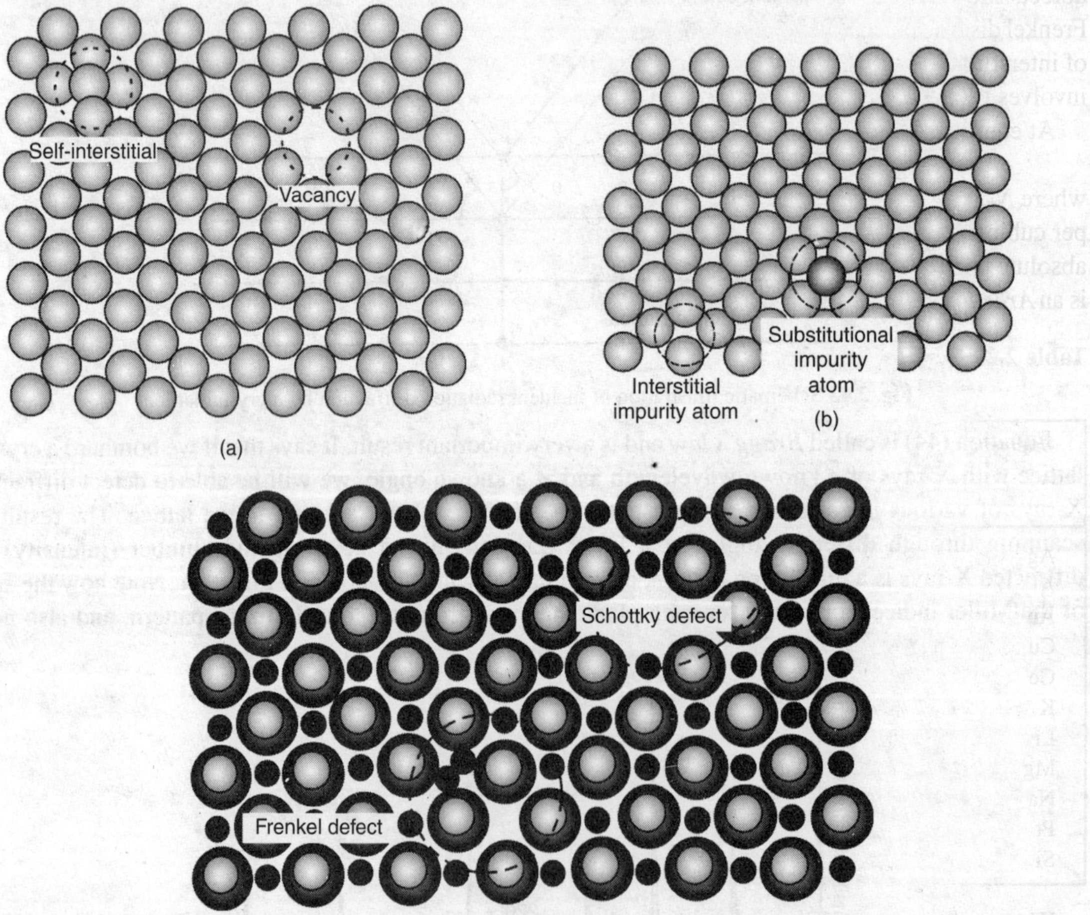

Fig. 2.45 (*a*) Representation of a vacancy and self-interstitial in a crystalline solid. (*b*) Representation of interstitial and substitutional impurity atoms in a crystalline solid. (*c*) Representation of Frenkel and Schottky defects in a crystalline solid.

thing. As we will see, many important materials phenomena that are based on defective structures can be exploited for very important uses. The defects, also known as *imperfections,* are grouped according to spatial extent. *Point defects* have zero dimension; *line defects,* also known as *dislocations,* are one dimensional; and *planar defects* such as *surface defects* and *grain boundary defects* and bulk or volume defects have two dimensions. These defects may occur individually or in combination.

(i) Point Defects: Figure 2.45 shows various types of point defects. The atom which formerly occupied the site of vacancy may be moved to the external surface of the crystal. Such type of crystal defect is termed as *Schottky defect.* The second type of point defect is called an *impurity.* Impurities can occur in two ways: as an *interstitial impurity,* in which an impurity atom occupies an interstitial site [Fig. 2.45(*b*)]; or when an impurity atom replaces an atom in the perfect lattice [Fig. 2.45(*b*)]. In the first instance, either the same atom as in the lattice, or an impurity atom, can occupy an interstitial site, causing considerable lattice strain as the atomic planes distort slightly to accommodate the misplaced atom. The amount of strain created depends on how large the atom is relative to lattice atoms. It is also possible for a lattice atom to move off a lattice site and occupy an interstitial site. In this case, both of the defects shown in Fig. 2.45(*a*) occur simultaneously, and a defect point is known as a *Frenkel defect* (or Frenkel disorder) occurs. We may note that in a pure Frenkel defect, there are always equal concentrations of interstitial impurities and vacancies. Frenkel defect differs from the Schottky defect in that the former involves two points whereas the later involves only one.

At equilibrium, the number of point defects at certain temperature is given by

$$N_d = N \exp[-E_d/k_B T] \qquad ...(45)$$

where N_d is the equilibrium number of point defects or vacancies, N is the total number of atomic sites per cubic meter (volume) or per mole, E_d is the activation energy to form the defect or vacancy, T is the absolute temperature and $k_B = 1.38 \times 10^{-23}$J/K $= 8.63 \times 10^{-5}$ eV/K is the Boltzmann constant. Eqn. (45) is an *Arhenius-type expression.* Values of N_d and E_d for some representative elements is given in Table 2.22.

Table 2.22 Formation Energy of Vacancies for Some Selected Elements and Equilibrium Concentration at Various Temperatures

Element	E_d *(kJ/mol)*	*Melting point* T_m*(°C)*	N_d *(vacancies/cm³)* 25°C	300°C	600°C	T_m
Ag	106.1	960	1.5×10^4	1.5×10^{13}	3.0×10^{16}	7.8×10^{17}
Al	73.3	660	1.0×10^{10}	1.2×10^{16}	2.4×10^{18}	5.0×10^{18}
Au	94.5	1063	1.5×10^6	1.5×10^{14}	1.5×10^{17}	1.2×10^{19}
Cu	96.5	1083	1.1×10^6	1.4×10^{14}	1.4×10^{17}	9.0×10^{18}
Ge	192.9	958	<1	1.3×10^5	1.3×10^{11}	8.2×10^{13}
K	38.6	63	2.1×10^{15}	–	–	1.3×10^{16}
Li	39.5	186	4.7×10^{15}	–	–	1.4×10^{18}
Mg	58.5	650	4.4×10^7	6.4×10^4	3.5×10^{17}	5.7×10^{17}
Na	38.6	98	4.0×10^{15}	–	–	1.0×10^{17}
Pt	125.4	1769	8.7	2.7×10^{11}	2.0×10^{15}	4.2×10^{19}
Si	221.8	1412	<1	3.1×10^2	2.5×10^2	8.0×10^{12}

We may note that the presence of a different type of atom either at a lattice position or an interstitial position, the point defect is said to be an impurity atom.

(ii) Line Defects and Dislocations: Linear defects are also known as dislocations The first type of one dimensional defect, or line defect, is also a dislocation. A dislocation is a linear disturbance of the atomic arrangement in a crystal caused by the displacement of one group of atoms from an adjacent group. The line defects obtain their name as they propagate as lines or as a two dimensional net in the crystal. There are three types of dislocations: *edge dislocations, screw dislocations,* and a combination of these two, termed *mixed dislocation.*

Edge dislocations: This occurs when a single atomic plane does not extend completely through the lattice. The termination of this half-plane of atoms creates a defect line (dislocation line) in the lattice (line DC in Fig. 2.46). The edge dislocation is designated by a ⊥ (perpendicular sign) if the plane is above the dislocation line or T if the plane is below the dislocation line. Edge dislocations can be

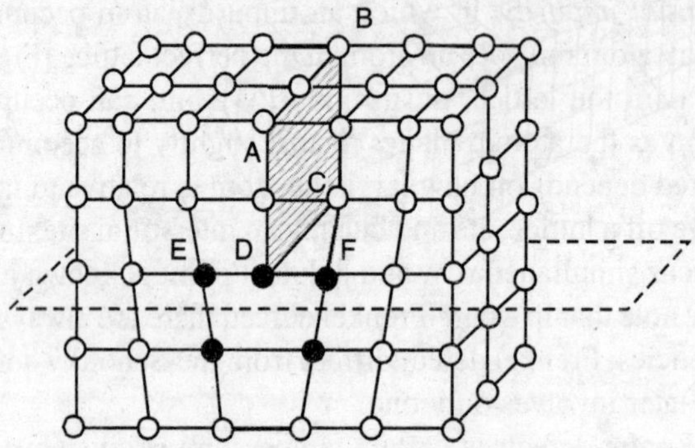

Fig. 2.46 Edge dislocation *DC* results from introducing an extra half-plane of atoms *ABCD,* where *DC* is considered a positive dislocation designated by ⊥.

quantified using a vector called *Burger's vector* $\vec{b}$, which represents the relative atomic displacement in the lattice due to dislocation (Fig. 2.47). One can determine the Burger's vector as follows:

- Define a positive direction along the dislocation line. This is usually done *into* the crystal.
- Construct a plane of atoms perpendicular to the dislocation line.
- Trace out a *clockwise* path around the dislocation line moving *n* lattice vectors in each of the four mutually perpendicular directions.
- The Burger's vector is drawn from the *finish* to the start of the path.

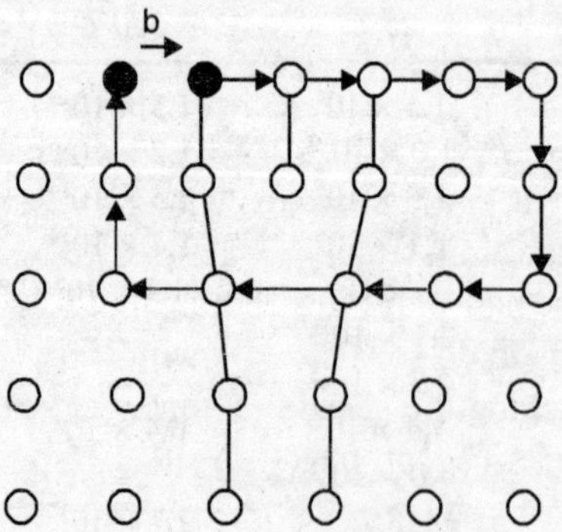

Fig. 2.47 Burger's circuit around an edge dislocation.

One can start the clockwise path, called the *Burger's circuit,* anywhere, as long as one entirely enclose the dislocation with the circuit.

Screw dislocation: This (second type of line defect) occurs when the Burger's vector is parallel to the dislocation line (OC in Fig. 2.48). This type of defect is called a screw's dislocation because the atomic structure that results is similar to a screw. The Burger's vector for a screw dislocation is constructed in the same way as with the edge dislocation.

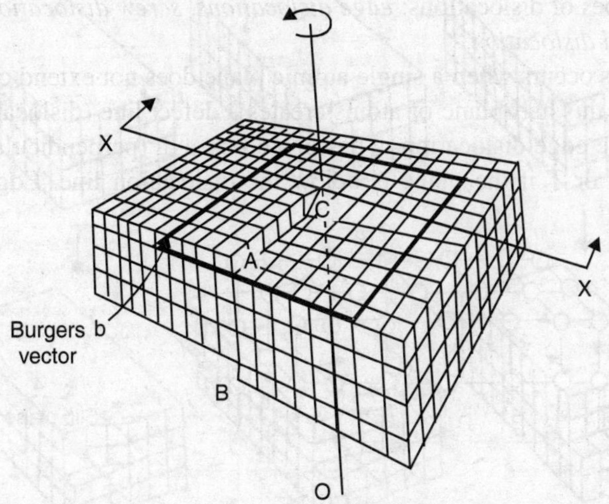

Fig. 2.48 Representation of defect line (*OC*). Burger's circuit and Burger's vector in a screw dislocation.

Mixed dislocations: When a line defect has both an edge and screw dislocation component, a mixed dislocation results. In this case, the Burger's vector is neither parallel nor perpendicular to the dislocation line, but can be resolved into edge and screw components.

The defect dislocation is mainly responsible for the observed low strength in purc metals. A dislocation moves easily on the application of a small amount of force and results in plastic deformation through the phenomenon of slip. A comparison of the three types of dislocations is given in Table 2.23.

Table 2.23 Comparison of Dislocation Characteristics

Dislocation type	*Burger's vector*	*Propagation direction*
Edge	⊥ to dislocation line	‖ to dislocation line, ‖ to Burger's vector
Screw	‖ to dislocation line	⊥ to dislocation line, ⊥ to Burger's vector
Mixed	Neither ‖ nor ⊥ to dislocation line	Neither ‖ nor ⊥ to dislocation line, Burger's vector

Dislocations, can move through a crystal lattice. This is most easily visualized in the edge dislocation, where the half-plane of the atom simply moves perpendicular to the defect line (line DC in Fig. 2.46). The lattice atoms below the half-plane incorporate the atoms above them into their plane, but must create a new half-plane of atoms to do so. This is how the half plane propagates. The dislocation movement can be very rapid-approaching the speed of sound. Screw and mixed dislocations move in more complex, but similar, fashions. The propagation of all three dislocation types is shown in Fig. 2.49.

A phenomenon of the movement in the direction of the extra half plane of atoms of an edge dislocation is called as *Dislocation climb.* We may note that dislocations can surmount obstacles in their path, by climbing into another slip plane and proceeding thence by normal slip. The addition or substraction of atoms from an edge dislocation need not be uniform across the half plane. In this case, one finds that the dislocation may operate on more than one slip plane when viewed three dimensionally, one or more steps, known as *jogs,* occurring between the various slip planes. A jog in an

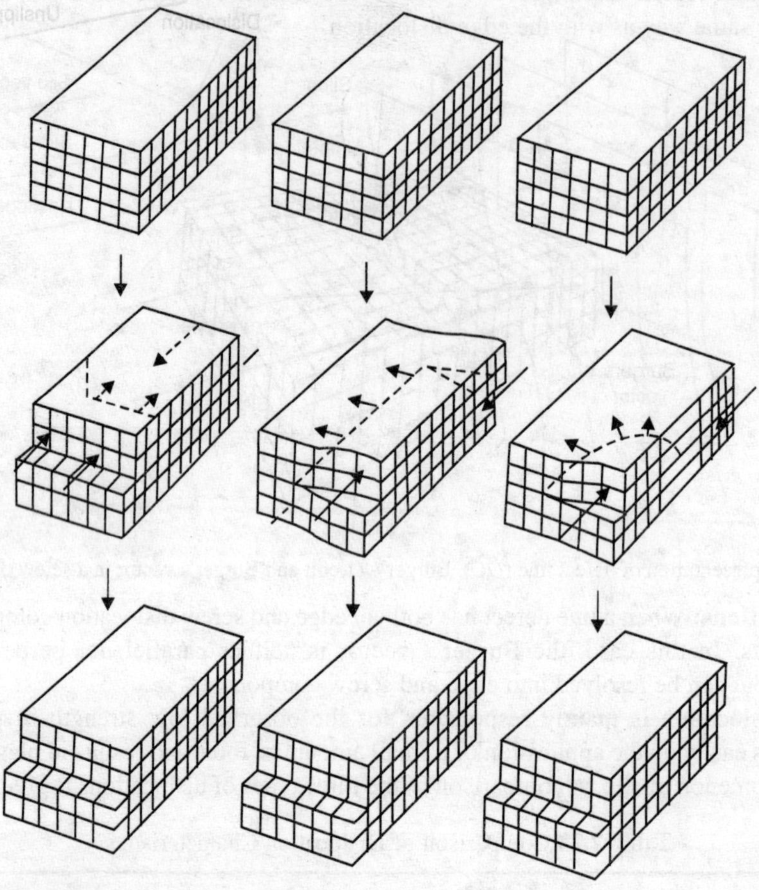

Fig. 2.49 Representation of propagation modes for (*a*) edge, (*b*) screw and (*c*) mixed dislocations.

edge dislocation is illustrated in Fig. 2.50. In these cases, the jog can move with the dislocation by the normal process of the slip. Screw dislocations may also have jogs, but since the slip direction is that of dislocation line, one finds that the jog resembles small piece of edge dislocation, and cannot move so easily. However, if jog climbs, by moving into another slip plane, point defects are created, but the screw dislocation containing the jog can continue to slip. Otherwise the jog must move sideways along the screw dislocation and attach it to an edge component of the dislocation line. However, these movements are more difficult and are expected to be involved in the work hardening or strain hardening of alloys and metals.

Similar to crystalline materials, semicrystalline polymers exhibit line defects. Screw dislocations with Burger vector parallel to the chains in semicrystalline polyethylene have been reported with the help of electron microscopy. These results clearly provide evidence that the dislocations were activated by stress to generate slip steps.

(iii) Planar or Surface Defects: These defects in crystals include external surfaces, grain boundaries, twin boundaries, stacking faults and phase boundaries.

Stacking faults arise due to imperfections in the stacking sequence of atomic planes. Recall that HCP structure is a result of ABCABC... stacking of closed-packed planes. If a plane of atoms is missing from

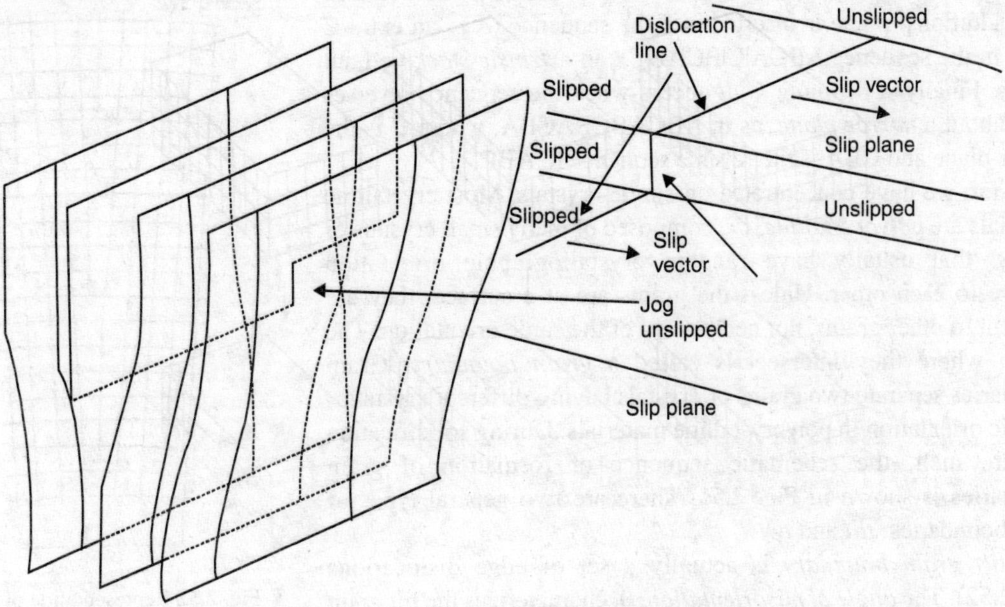

Fig. 2.50 Illustration of jog in an edge dislocation.

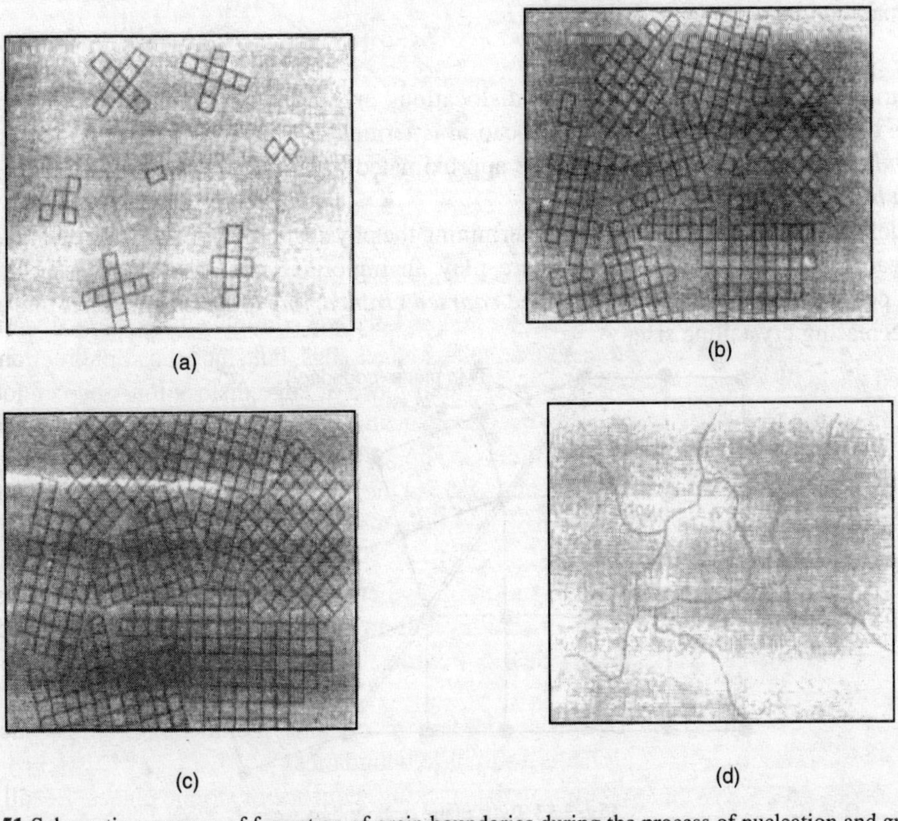

Fig. 2.51 Schematic sequence of formation of grain boundaries during the process of nucleation and growth.

this sequence, as in ABCAB|ABC, an *intrinsic stacking fault* results. If an additional plane is inserted into the sequence (*e.g.,* an extra C plane in the sequence ABCA|C|BCABC), an *extrinsic stacking* fault results. Finally, a *twinning fault* occurs when the sequence reverses itself about a *mirror plane,* as in ABCAB$\underline{C}$BACBA, where C is the mirror plane and CBA is the reverse sequence to ABC.

So far, we have concentrated on single crystals. Most crystalline materials are *polycrystalline, i.e.* composed of many small crystals or *grains,* that usually have random crystallographic orientation relative to each other. Unless the grains are at a surface, they are adjacent to other grains, not necessarily of the same orientation. The region where they intersect is called a *grain boundary.* Grain boundaries separate two grains or crystals having different crystallographic orientation in polycrystalline materials. During solidification of pure melt, the schematic sequence of formation of grain boundaries is shown in Fig. 2.51. There are two general types of grain boundaries: *tilt* and *twist.*

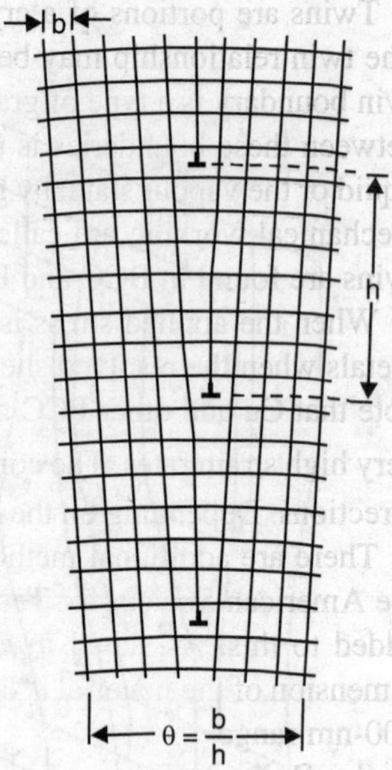

Fig. 2.52 Representation of a tilt grain boundary.

A *tilt grain boundary* is actually a set of edge dislocations (Fig. 2.52). The *angle of misorientation,* θ, characterizes the tilt grain boundary and is defined as the angle between the same directions in adjacent crystals. The angle of misorientation can be calculated from the Burger's vector and the vertical separation between edge dislocation, *h*:

$$\tan\theta = b/h \qquad ...(46)$$

Boundaries consisting entirely of edge dislocations are called as *tilt boundaries,* whereas the boundaries consisting entirely of screw dislocation is termed as a *twist boundary.* A height-angle *tilt grain boundary* results, and Eqn. (46) can be approximated to $\tan\theta \approx \theta = b/h$. In a similar manner, a *twist grain boundary* is a screw dislocations.

Planar defects play an important role in determining the physical properties of bulk crystalline solids. In this regard, not only will defect structure play an important role, but so will crystallize size. Generally, polycrystalline materials are termed *coarsed grained, fine grained;* and *ultra-fine-grained* in order of decreasing crystalline size.

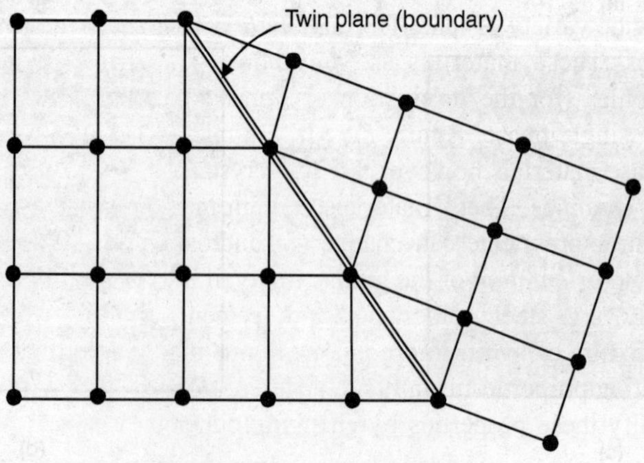

Fig. 2.53 Twin plane or boundary.

Twins are portions of a crystal which have certain specific orientations with respect to each other. The twin relationship may be such that the lattice of one part is the mirror image of the other and the twin boundary is a type of grain boundary (Fig. 2.53). As shown in the Fig. 2.53, the region of material between these boundaries is termed twin. Twins may occur frequently during crystallization from the liquid or the vapour state, by growth during annealing or during phase transformation. Twins formed by mechanical working are called *mechanical twins* or by annealing called *annealing twins*. Mechanical twins are found in BCC and HCP structures and annealing twins in FCC structures.

When the applied stress is high due to work hardening, or low temperature, or in the case of HCP metals when the resolved shear stress on the basal plane is low, *deformation twinning* occurs. We may note that Cu and other FCC metals can be made to deform by twinning at very low temperatures or at very high strain rates. The common twinning elements in HCP metals are the $\left(1\,1\,\bar{1}\,2\right)$ plane $\left[1\,0\,\bar{1}\,\bar{1}\right]$ directions. Depending on the c/a ratio, twinning results in a compression or elongation along the *c*-axis.

There are additional methods for classifying the grain size of materials, such as used for metals by the American Society for Testing materials (As TM). More recently the term *nanocrystalline* has been added to this. As stated in chapter 1, a nanostructural material is one in which at least one critical dimension of the material is less than 100 nm in size. The grain sizes of the nanomaterials fall below the 100-nm range.

(iv) Bulk or Volume Defects: Volume defects are of quite larger dimensions in comparison to the defects described so far. The volume defects include pores, cracks, inclusions and other phases. The volume defects may be present at microlevels or even at macrolevels. Volume defects usually originate due to poor chemistry of the material, *e.g.* steels or during processing.

2.8 MICROSTRUCTURE

A *microstructure* is the geometric arrangement of grains and phases in the materials. By studying microstructure, one may know the details of the grain shape, size and their distribution. Obviously, the variables include the *amount, size, shape* and *distribution* of these structural features. Looking to the dimensions, an *optical microscope* (upto × 2000)—or even an *electron microscope* (upto × 50,000)—is necessary for observation.

Microstructure of a material consists phase structure and grain structure. The phase structure is usually expressed in terms of various phases, their relative amount of distribution and alignment. Depending upon the number of phases present, microstructures are called either *single phase* or *multiphase* structures. A wide variety of materials are *single phase, e.g.* window glass and transparent polystyrene drinking cups. These materials are amorphous and therefore lack any conventional microstructure. Silicon boules (for the production of transistor circuits) and ruby gems are single crystals, and therefore are single phase. Again, they lack a microstructure in the conventional sense. The majority of single-phase materials however, are polycrystalline, and do possess a microstructure, *e.g.* brass products, MgO ceramics. These materials contain many *grains* of the same phase, with a variety of possible microstructures. One can change the microstructure of single-phase materials by change in the *size, shape* and *orientation* of the grains. Many, if not the majority, of technical materials possess *mixtures* of *phases, e.g.* steel, solder, portland cement, grinding wheels, points and glass-reinforced plastics. The mixture of two or more phases in one material permits interaction among the phases; therefore the resulting properties usually are different from the properties of individual phases. It is also possible to modify these properties by changing either the shape or the distribution of the phases.

The *grain structure* of a material reveals shape and size of the grains (crystals) which form the bulk material. The ASTM has prepared charts, all having different average grain sizes. To each is assigned a number ranging from 1 to 10, which is termed the *grain size number.* The larger this number, the smaller are the grains. Grain size is expressed as the grain size number of the chart that most nearly matches the grains in the micrograph.

Let n represent the grain size number, and N the average number of grains per square inch at a magnification of 100 X. These two parameters are related to each other through the expression

$$N = 2^{n-1} \quad ...(47)$$

The grains are actually three dimensional and therefore a two dimensional plane is not truly representative. Moreover, the varying shape of grains reveals that only a most probable grain size is used. If one measures many grains and the frequency of each grain diameter is plotted, one usually obtains a curve as shown in Fig. 2.54. The value chosen to represent the grain size is d_1, *i.e.* the value associated with the peak of the frequency plot.

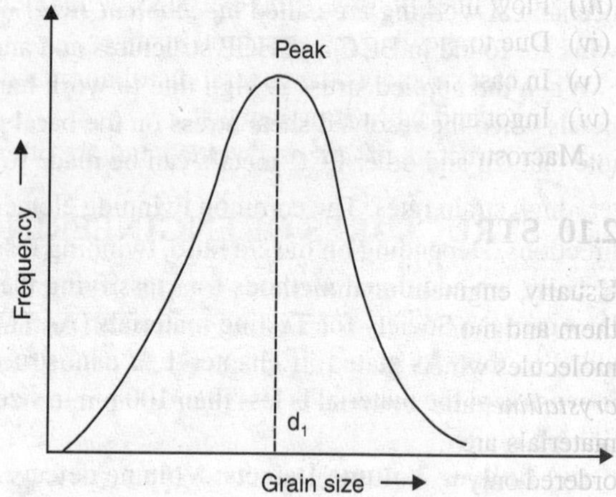

Fig. 2.54 Grain size distribution in a crystalline material.

2.9 MACROSTRUCTURE

The structure of a material examined with the naked eye or by using a small hand-magnifying glass is usually termed as *macrostructure.* This is useful in revealing the structural defects or inhomogenetics of

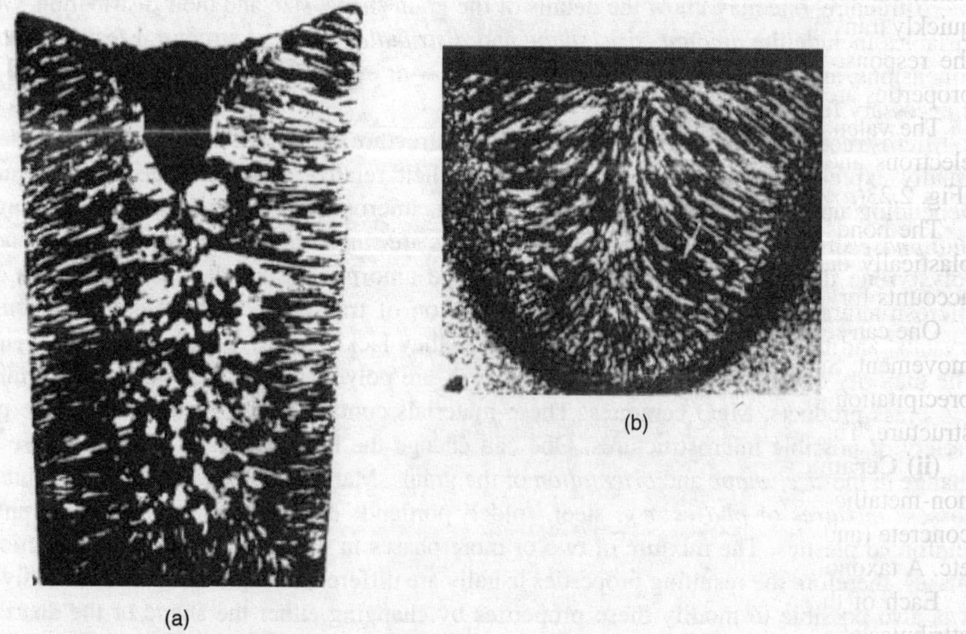

Fig. 2.55 Macrostructure of (*a*) etched polished section of aluminium alloy ingot, and (*b*) etched polished section of weld deposit in steel.

a large area. The method requires polishing and chemical etching of the surfaces to be examined. This helps to reveal the following features of the material:

(*i*) After solidification the grain structure,
(*ii*) Segregation of sulphur and phosphorus,
(*iii*) Flow lines in forged and extruded, components,
(*iv*) Due to welding the structural changes,
(*v*) In cast structures the general distribution and size of inclusions and porosity,
(*vi*) Ingot and fabricating defects.

Macrostructure of cast and welded part are often examined. Fig. 2.55 shows some examples.

2.10 STRUCTURES OF ENGINEERING MATERIALS

Usually, engineering materials are solid state substances. Engineers design product and systems, make them and monitor their usage. Every product is made with materials. Materials are made of atoms and molecules which are bonded together by definite forces. Solid materials form two type of structures: (*i*) *crystalline,* and (*ii*) *amorphous* depending upon the arrangement of atoms and molecules. Amorphous materials are characterized by atomic or molecular structures that are relatively complex and become ordered only with some difficulty. Furthermore, rapidly cooling through the freezing temperature favours the formation of a noncrystalline solid, since little time is allowed for the ordering process. Most metals and ceramics have crystalline structure, whereas others, the inorganic glasses, are amorphous. Polymers may be completely non-crystalline and semicrystalline consisting of varying degrees of crystallinity.

(i) Metals: These are characterized by their high thermal and electrical conductivities. Commonly, but not always, metals are relatively heavy and deformable.

Metals owe their metallic characteristics to the fact that the valence electrons are not bound, but rather can leave their "parent" atoms. Since some of the electrons are independent in metals, they can quickly transfer an electric charge and thermal energy. The opacity and reflectivity of a metal arise from the response of these unbound electrons to electromagnetic vibrations at light frequencies. These properties are another result of the partial independence of the electrons from their parent atom.

The valency of metallic materials are 1, 2 or 3. The atoms of these materials give up their valence electrons and held together by a *matrix* of electrons, resulting in the formation of a metallic bond [Fig. 2.23(*a*)].

The bonds between the atoms of these materials are not very strong. These metallic materials are plastically deformed under load, due to the movement of dislocation in the crystal structure. This accounts for their extensive use in structural applications.

One can strengthened the metals by doing things to their crystal structure that discourage dislocation movement. Such methods include, formation of alloys, *i.e.,* addition of other elements, cold working, precipitation of second phases through heat treatment, and by temperature-induced changes in crystal structure. The last method of quench hardening makes steels very strong and hard.

(ii) Ceramics and Glasses: Ceramics, in simplest terms, are compounds that contain metallic and non-metallic elements. There are many examples of ceramics materials ranging from the cement of concrete (and even the rocks themselves), to glass, to electrical insulators, and to permanent magnets, etc. A taxonomy of these several types is shown in Fig. 2.56(*a*).

Each of these materials is relatively hard and brittle. Indeed, hardness and brittleness are general attributes of ceramics, as is higher resistance than that of either metals or polymers to high temperatures and to severe environments. The basis for these characteristics is again the electronic behaviour of the constituent atoms. Consistent with their natural tendencies, the metallic elements release their outer-

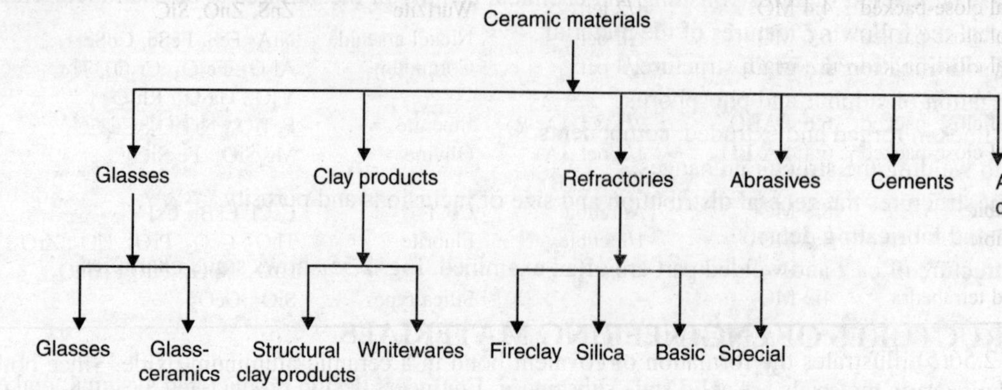

Fig. 2.56 (*a*) Classification of ceramic materials on the basis of application.

most electrons and give these electrons to the non-metallic atoms, which retain them. The result is that these electrons are immobilized, so the typical ceramic material is a good insulator, both electrically and thermally.

Equally important, the positive metallic ions (atoms that have lost electrons) and the negative non-metallic ions (atoms that have gained electrons) develop strong attractions for each other. Each *cation* (positive) surrounds itself with *anions* (negative). Ceramic materials tend to be hard (mechanically resistant), refractory (thermally resistant), and inert (chemically resistant).

Ceramic crystal structures: The structure of these materials is characterised by the presence of non-metallic element having a valence of 5, 6 or 7. The nature of the band between atoms really distinguishes these materials from other engineering materials. These materials have very rigid covalent or ionic bonds between adjacent atoms whereas metals have long-range-matrix bond. A summary of some simple ionic structure and their corresponding coordination numbers is given in Table 2.24 below:

Table 2.24 Simple Ionic Structures and Their Corresponding Coordination Numbers

Anion packing	*Coordination number of M and O*	*Sites by cations*	*Structure name*	*Examples*
Cubic close-packed	6.6 MO	All oct.	Rock salt	NaCl, KCl, LiF, KBr, MgO, CaO, SrO, BaO, CdO, VO, MnO, FeO, CoO, NiO
Cubic close-packed	4.4 MO	1/2 tet.	Zinc blende	ZnS, BeO. SiC
Cubic close-packed	4.8 M_2O	All tet.	Antifluorite	Li_2O, Na_2O, K_2O, Rb_2O, sulfides
Distorted cubic close-packed	6.3 MO_2	1/2 oct.	Rutile	TiO_2, GeO_2, SnO_2, PbO_2, VO_2, NbO_2, TeO_2, MnO_2, RuO_2, OsO_2, IrO_2
Cubic close-packed	12:6:6 ABO_3	1/4 oct. (B)	Perovskite	$CoTiO_3$, $SrTiO_3$, $SrSnO_3$, $SrZrO_3$, $SrHfO_3$, $BaTiO_3$,
Cubic close-packed	4:6:4 AB_2O_4	1/8 tet. (A) 1/2 oct. (B)	Spinel	$FeAl_2O_4$, $ZnAl_2O_4$, $MgAl_2O_4$
Cubic close-packed	4:6:4 $B(AB)O_4$	1/8 tet. (B) 1/2 oct. (A, B)	Spinel (inverse)	$FeMgFeO_4$, $MgTiMgO_4$

Hexagonal close-packed	4:4 MO	1/2 tet.	Wurtzite	ZnS, ZnO, SiC
Hexagonal close-packed	6:6 MO	All oct.	Nickel arsenide	NiAs,FeS, FeSe, CoSe
Hexagonal close-packed	6:4 M_2O_3	2/3 oct.	Corundum	Al_2O_3, Fe_2O_3, Cr_2O_3, Ti_2O_3 V_2O_3, Ga_2O_3, Rh_2O_3
Hexagonal close-packed	6:6:4 ABO_3	2/3 oct. (A, B)	Ilmenite	$FeTiO_3$, $NiTiO_3$, $CuTiO_3$
Hexagonal close-packed	6:4:4 A_2BO_4	1/2 oct. (A) 1/8 tet. (B)	Olivine	Mg_2SiO_4, Fe_2SiO_4
Simple cubic	8:8 MO	All cubic	CsCl	CsCl, CsBr, CsI
Simple cubic	8:4 MO_2	1/2 cubic	Fluorite	ThO_2, CeO_2, PrO_2, UO_2, ZrO_2, HfO_2, NpO_2, PuO_2, AmO_2
Connected tetrahedra	4:2 MO_2	–	Silica types	SiO_2, GeO_2

Figure 2.56(*b*) illustrates the formation of covalent bond in a ceramic-aluminum oxide. Since both ionic and covalent bonds involve very strong bonds between neighbouring atoms and therefore crystalline ceramics with this type of bond tend to be very brittle. Continued loading may eventually result in crystal cleavage. The strong bonds in ceramics results in high hardness, chemical inertness and electrical insulation. We may note that ceramics tend to be electrical insulators because the electrons are '*tied up*' in bonding and there are almost no free electrons to move the entire crystal.

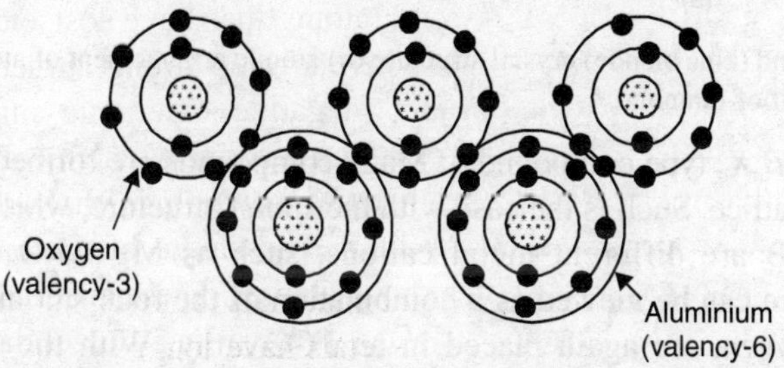

Fig. 2.56 (*b*) Formation of a covalent bond in Al_2O_3 a ceramic.

Many crystalline inorganic compounds are based on nearly close-packing of the anions (usually referred to as O or X, though oxygen is the most common anion) with metal atom cations (usually called M or A) placed interstitially within the anion lattice. We now describe the crystal structures of some of the most common ceramic and ionic compounds:

Rock salt structure: The compounds Nacl, MgO, MnS, LiF, FeO, and many other oxides such as NiO and CaO have the so-called *rock salt structure,* in which the anions, such as Cl or O, are arranged in an FCC array, with the cations placed in the octahedral interstitial sites, also creating an FCC array of cations (Fig. 2.57). The primitive cell of rock-salt structure, however, is simple cubic, with anions at four corners, and cations at alternating corners of the cube, as given by one of the quadrants in Fig. 2.57. Coordination number of both anions and cations is 6, so the cation-anion radius ratio is between 1.37 and 2.42.

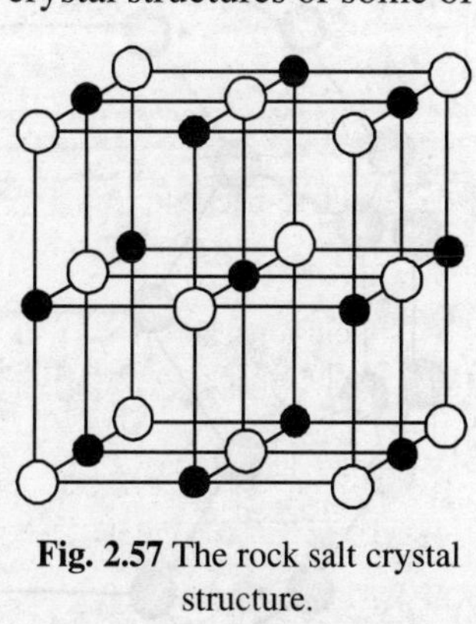

Fig. 2.57 The rock salt crystal structure.

Diamond structure: Another common ceramic structure arises when the tetrahedral sites in an FCC array of anions are occupied, *e.g.* an FCC array of sulphur anions with Zn ions in the tetrahedral positions gives the compound ZnS and results in the *zinc blende structure* (Fig. 2.58). The atoms can be all alike also, *e.g.* carbon (Fig. 2.59), in which case diamond results. In either case, the atoms in the tetrahedral sites have a coordination number of four. Many important compounds have the zinc blende or diamond structure, including SiC.

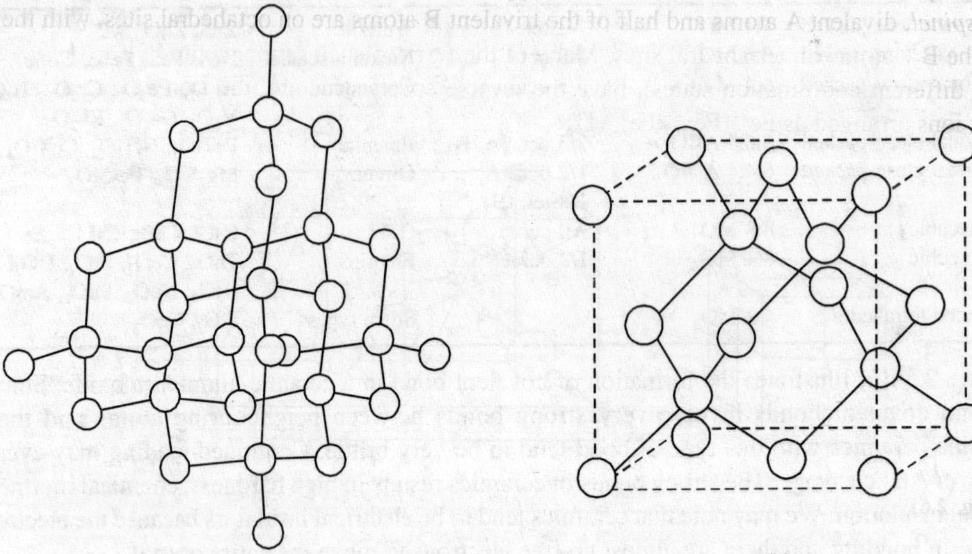

(a) (b)

Fig. 2.58 The diamond (zinc blende) crystal structure (*a*) atomic arrangement of atoms in diamond crystal (*b*) unit cell of diamond.

Spinel structure ($A_mB_nX_p$ type compounds): Many compounds are formed when there is more than one metal cation in the lattice. Such is the case with the *spinel* structure, which has the general formula AB_2O_4, where A and B are different metal cations, such as Mg Al_2O_4 (Magnesium aluminate) (Fig. 2.60). This structure can be viewed as a combination of the rock salt and zinc blende-structures. The anions, usually oxygen, are again placed in an FCC array. With the A and B ions occupying tetrahedral and octahedral sites. We may note that all interstitial sites are not filled, nor do the B and A

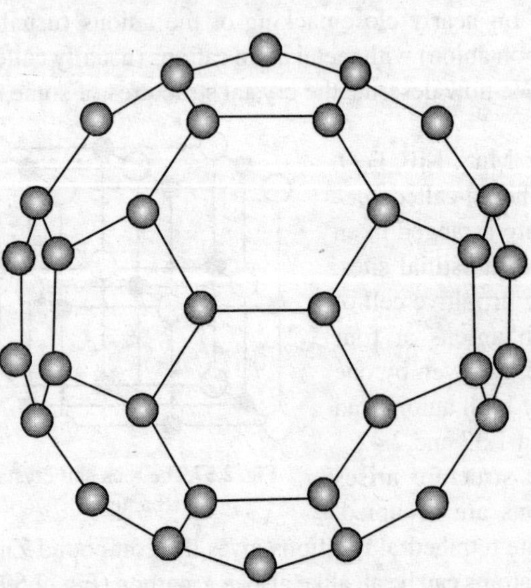

Fig. 2.59 A C_{60} molecule.

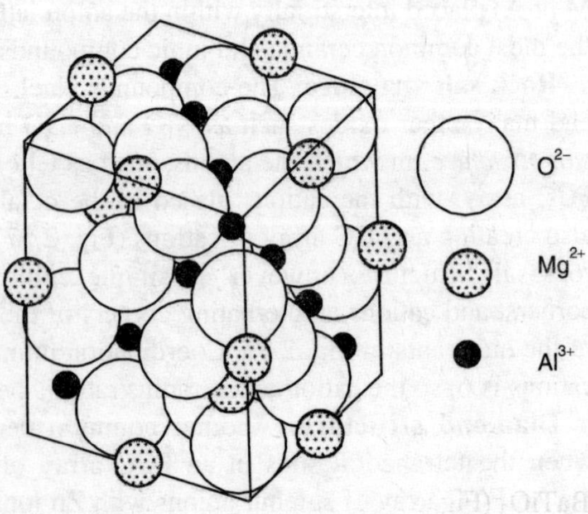

Fig. 2.60 The spinel structure of Mg Al_2O_4.

atoms always seek the same interstitial site. A typical unit cell of spinel is shown in Fig. 2.61. In an *inverse spinel*, divalent A atoms and half of the trivalent B atoms are on octahedral sites, with the other half of the B^{3+} atoms on tetrahedral sites. Many of the ferrites, such as magnetite Fe_3O_4 (in which iron has two different coordination states), have the inverse spinel structure, the iron having divalent and trivalent ions arranged as $Fe^{3+}(Fe^{2+} \cdot Fe^{3+})\ O_4$.

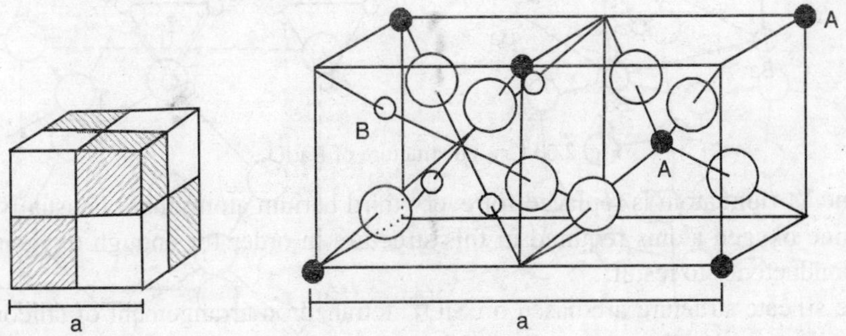

Fig. 2.61 Spinel structure. One can divide the unit cell into octants—tetrahedrally coordinated cations A, octahedrally coordinated cation B and oxygen atoms (large circles) shown in two octants only.

Other important ceramic structures: There are many additional crystal structures that arise due to compounds of both simple and complex stoichiometry. These structures generally have specific names, *e.g.* the *corundum* structure is common to Al_2O_3, the *rutile* structure comes from one of the forms of TiO_2, and a number of important ceramics, such as CaF_2, have the *fluorite* structure. One structure with current technological importance is *perovskite* structure (Fig. 2.62). Many important ceramics with unique electrical and dielectric properties have the perovskite structure, including barium titanate,

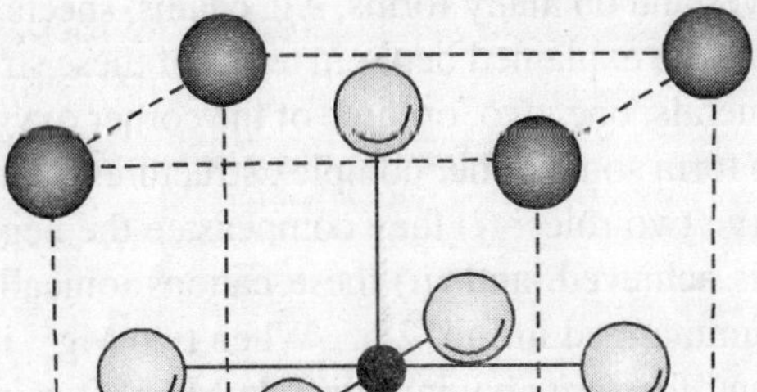

Fig. 2.62 The perovskite crystal structure of $CaTiO_3$.

$BaTiO_3$ (Fig. 2.63), and high temperature cuprate superconductors (HTSC). The perovskites has the general formula ABO_3, but in the case of most HTSCs, the A cation consists of more than one type of atom, such as in $Y_1\ Ba_2\ Cu_3\ O_{7-x}$, or the so-called 1-2-3 superconductors, in which the perovskite structure

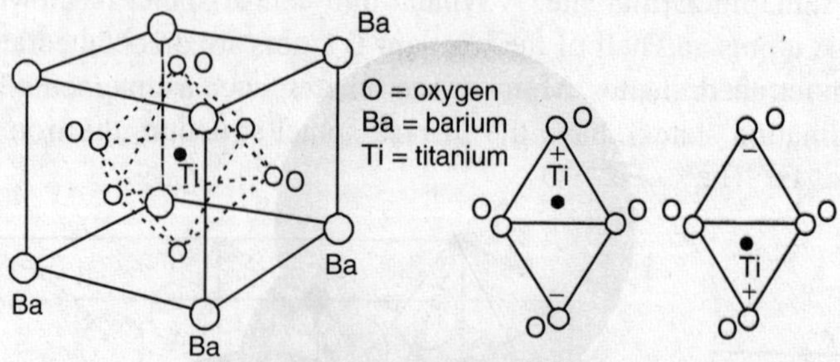

Fig. 2.63 Crystal structure of $BatiO_3$.

is tripled, and one Ytrrium atom is replaced for every third barium atom; there is usually less than the stoichiometric nine oxygen atoms required in this structure in order for enough oxygen vacancies to form and superconductivity to result.

Silicates: The silicate structure are based on SiO_4^{4-} tetrahedron arrangement of silicon and oxygen (Fig. 2.64). The silicates are an important class of ceramic compounds that can take on many structures. They are complex structures that can contain several additional atoms such as Mg, Na, K. What makes the silicates so important is that they can be either crystalline or amorphous (glassy) and provide an excellent opportunity to compare these two disparate types of structure.

The structural unit for the simplest silicate, SiO_2 also known as Silica, is the tetrahedron as shown in Fig. 2.62. Various structural units observed in crystalline silicates are given in Table 2.25. From table, the anion/cation ratio in SiO_2 is R_o/R_{si} = 1.32/0.39 = 3.3, which according to Fig. 2.62, dictates the tetrahedron as the base structural unit. We may note that the SiO_4 tetrahedron has a formal charge of –4, which must be neutralized with cations, such as other Si atoms, in real compounds. These tetrahedra can bond to metals and themselves and on many forms, *e.g.* chains, sheets, three dimensional networks, etc. Obviously, these materials can be explained better in terms of these structural forms rather than unit cells. For the various silicate minerals, one, two, or three of the corner oxygen atoms of SiO_4^{4-} tetrahedra are shared by other tetrahedra to form some rather complex structures. Note that positive charge cations such as Ca^{2+}, Mg^{2+} and Al^{+3} serve two roles: (*i*) they compensate the negative charges from the SiO^{4+} units so that charge neutrality is achieved, and (*ii*) these cations ionically bond the SiO_4^{4-} tetrahedra together. These structures are summarized in Fig. 2.65. When two Mg^{2+} ions are available to combine with one tetrahedron, a compound *forsterite* having formula Mg_2SiO_4 is produced. Similarly Fe^{2+} ions can combine silica tetrahedra to produce $Fe_2SiO_4 \cdot Mg_2SiO_4$ and Fe_2SiO_4 form a series of solid solutions and these class of materials are referred to as the *orthosilicates.* Two silicate tetrahedra can combine by sharing one corner to produce a double tetrahedra or a $Si_2O_7^{6-}$ ions. This ionic group can then combine

Table 2.25 Structural Units Observed in Crystalline-Silicates

O/Si Ratio	*Silicon-oxygen groups*	*Structural units*	*Examples*
2	SiO_2	Three-dimensional network	Quartz
2.5	Si_4O_{10}	Sheets	Talc
2.75	Si_4O_{11}	Chains	Amphiboles
3.0	SiO_3	Chains, rings	Pyroxenes, beryl
3.5	Si_2O_7	Tetrahedra sharing one oxygen ion	Pyrosilicates
4.0	SiO_4	Isolated orthosilicate tetrahedra	Orthosilicates

Fig. 2.64 The $(SiO_4)^{4-}$ tetrahedron. The silicon atom is the solid circle at the centre of the tetrahedron; large open circles are oxygens.

with other ions to produce a class of compounds called the *pyrosilicates.* Sharing two out of the four corners of SiO_4 tetrahedra results in *chains.* The angle formed between adjacent tetrahedra can vary widely, resulting in unique structures such as rings (Fig. 2.64). In all cases, when only two corners are shared, the repeat unit is $(SiO_3)^{2-}$ having general formula $(SiO_3)_n^{2n-}$, where n gives the number of $(SiO_3)^{2-}$ groups in the ring or chain. A large number of ceramic materials have this *metasilicate* structure. Wallastonite having formula $CaSiO_3$ is built from Si_3O_9 rings; beryl having formula $Be_3Al_2Si_6O_{18}$ contains larger Si_6O_{18} rings, and enstatite with formula $MgSiO_3$ has a chain structure.

When all four corners of the SiO_4 tetrahedra are shared, a highly ordered array of networked tetrahedra results (Fig. 2.65). This is the structure of *quartz,* (Fig. 2.66) one of the crystalline forms of

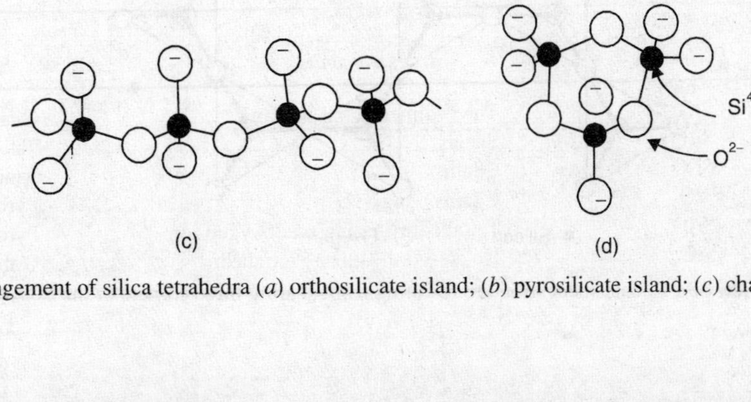

Fig. 2.65 Arrangement of silica tetrahedra (*a*) orthosilicate island; (*b*) pyrosilicate island; (*c*) chains; and (*d*) ring.

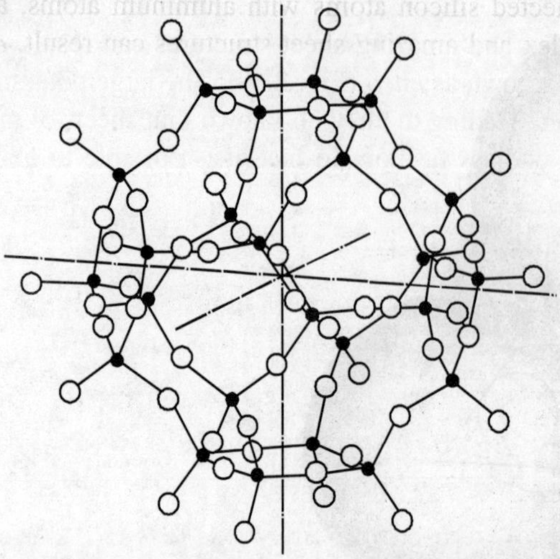

Fig. 2.66 The structure of quartz, showing the three-dimensional network of SiO_4 tetrahedra.

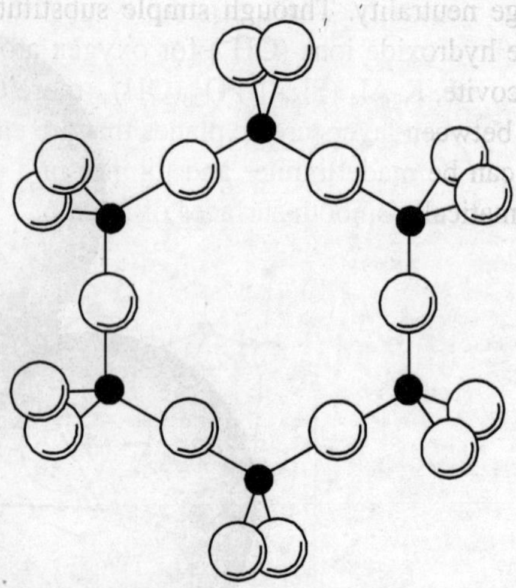

Fig. 2.67 A silicate ring, beryl, with two corners of the SiO_4 tetrahedra shared.

SiO_2, (Fig. 2.67) having density 2.655 g/cm^3, is stable upto 870°, at which point it transforms into *tridymite,* with a density of 2.27 g/cm^3. At 1470°C, tridymite transforms to *cristobalite* (density = 2.30 g/cm^3), which melts at around 1710°C. There are "high" and "low" forms of each of these structures, which result from slight, albeit rapid, rotation of the silicon tetrahedra relative to one another.

If three of the four corners of the $(SiO_4)^{4-}$ tetrahedron are shared, repeat units of $(Si_2O_5)^{2-}$ or $(Si_4O_{10})^{4-}$ result, with O/Si of 2.5. Figure 2.68 shows that *sheet structures* are the result of sharing three corners. In these structures, additional cations or network modifiers, such as Al^{3+}, K^+, and Na^+, preserve

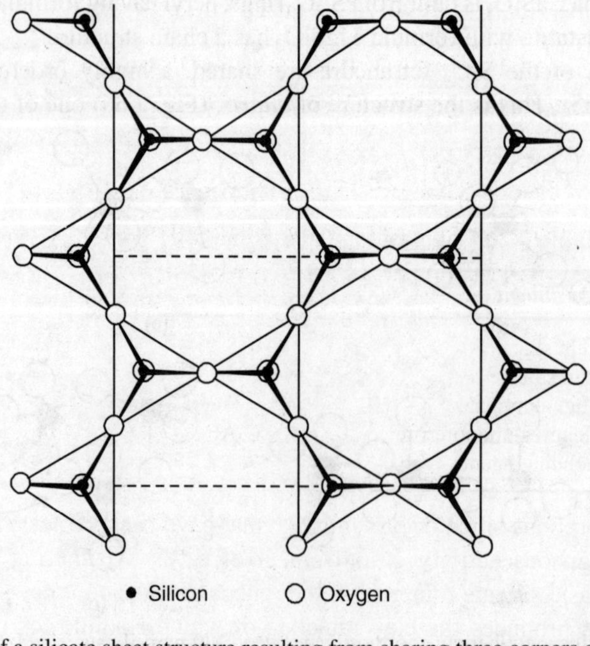

Fig. 2.68 Top view of a silicate sheet structure resulting from sharing three corners of the SiO_4 tetrahedra.

charge neutrality. Through simple substitution of selected silicon atoms with aluminum atoms, and some hydroxide ions (OH^-) for oxygen atoms, complex and amazing sheet structures can result, *e.g.* muscovite, $K_2Al_4(Si_6Al_2)O_{20}(OH)_4$, more commonly known as *mica* (Fig. 2.69). The large potassium ions between layers create planes that are easily cleaved, leading to the well-known thin sheets of mica that can be made thinner and thinner in a seemingly endless fashion. In fact, it is possible to obtain automatically smooth surfaces of mica.

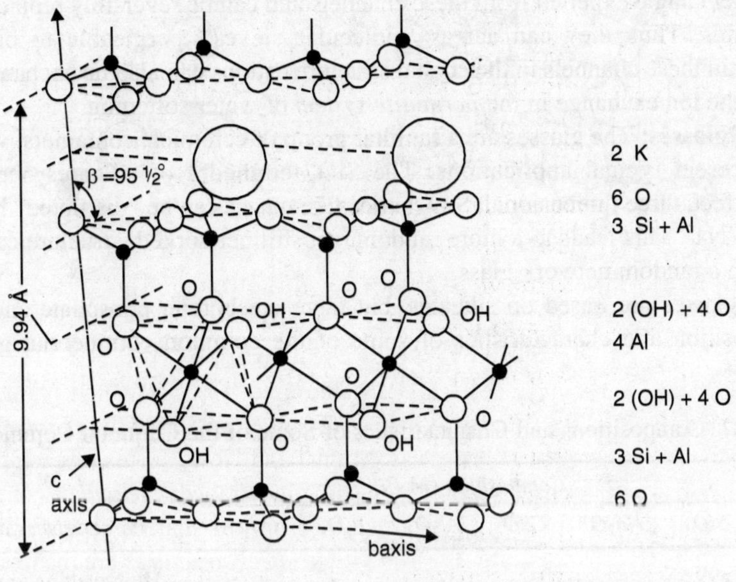

Fig. 2.69 The structure of muscovite (mica), a sheet silicate.

Cement and *Concrete* are widely used common building construction materials in huge quantities. Portland cement consists of a number of silicate minerals. The approximate composition of Portland cement is given in Table 2.26. Cement mixed with water sets as a function of time due to several hydration reactions. The function of water is to bind the sheet-like silicate molecules together thereby hardening the cement. Concrete is a sized aggregate of rocks embedded in a cement matrix, which binds the rock pieces together.

Table 2.26 Composition (approximate) of Portland Cement

Constituent	*Symbol*	%
Tricalcium silicate	C_3S	45
Dicalcium silicate	C_2S	30
Tricalcium aluminate	C_3A	10
Tetracalcium aluminoferrite	C_4AF	8
Other bonding agents	–	7

Natural form of amorphous silica occurs in rocks that have been subjected to high temperature and rapid cooling due to igneous activity or movement of earth. Weathering of rocks produces some decomposition of the basic silicate minerals of the igneous rocks and a new set of silicates result, *e.g.* weathering of feldspars produces the clay minerals of which Kaoline (china clay) is pure hydrated aluminium silicate. The clay minerals owe their plasticity to the water of hydration and the hydrogen-hydroxyl bonding that results. Once the clay mineral looses its water, as it does when fired at high temperature, the bonding is changed, and becomes amorphous (vitrified). Since the water cannot be replaced easily and hence fired pottery remains reasonably strong but porous.

Zeolites is another example of silicates which also occur as natural minerals, and referred as 'Frame-work silicates'. Zeolites are hydrated silicates, such as $NaCa_2\ (Al_5Si_5)\ O_{20}.\ 6H_2O$ with an open framework intersected by wide channels in which the cations and water molecules are located. By heating, the water can be expelled from these channels and can be reversibly replaced or substituted by neutral molecules. Thus they can act as 'molecular sieves', permeable to only certain smaller molecules. Within these channels in the crystal structure cations can also be exchanged reversibly. This is the basis for the ion exchange in the *permutite system* of water softening.

Structure of glasses: The glasses are a familiar group of ceramics; containers, windows, lenses and fibre-glass represent typical applications. The SiO_4 tetrahedra of silicates need not be arranged regularly. A perfect, three dimensional, SiO_2 lattice like quartz can be "disrupted" by introducing other cations, such as Na^+. This leads to a more random, yet still networked, structure called a *glass.* Figure 2.70 shows such a random network glass.

Majority of glasses are based on silicates, but there are borate, phosphate and aluminate glasses as well. Composition and characteristics of some of the common commercial glasses are given in Table 2.27.

Table 2.27 Compositions and Characteristics of Some of the Common Commercial Glasses

	Composition (wt%)						
Glass type	*SiO_2*	*Na_2O*	*CaO*	*Al_2O_3*	*B_2O_3*	*Other*	*Characteristics and applications*
Fused silica	>99.5						High melting temperature, very low coefficient of expansion (thermally shock resistant)
96% silica (Vycor)	96				4		Thermally shock and chemically resistant—laboratory ware
Borosilicate (Pyrex)	81	3.5		2.5	13		Thermally shock and chemically resistant—ovenware
Container (soda-lime)	74	16	5	1		4MgO	Low melting temperature, easily worked, also durable
Fibre-glass	55		16	15	10	4MgO	Easily drawn into fibres-glass-resin composites
Optical flint	54	1				37PbO, $8K_2O$	High density and high index of refraction—optical lenses
Glass-ceramic (Pyroceram)	43.5	14		30	5.5	$6.5TiO_2$, $0.5As_2O_3$	Easily fabricated; strong, resists thermal shock—ovenware

The term glass applies to those materials that have the expansion characteristics. Glasses may be either inorganic or organic and are characterized by a *short-range order* (and an absence of long-range order). Some metallic oxides can be produced at least partially in the glassy form by rapid cooling from the melt. The replacement of silicon ion in the silica tetrahedra by Na, K, Ca or Pb can readily lead to production of glasses. Table 2.28 classifies cations into three categories: *glass formers, intermediates* and *modifiers.* Glass formers are cations with a valence greater than or equal to three and a coordination number less than 4. We may note that there are many more glass-forming cations than just silicon.

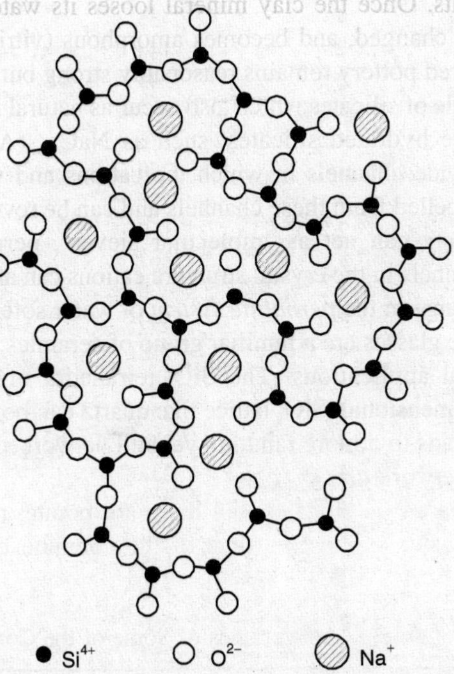

Fig. 2.70 Schematic representation of a random network sodium silicate glass.

Intermediates are cations of lower valence and higher coordination number that can sometimes act as a glass formers, *e.g.* aluminum, but can also act network modifiers. A network modifier is a cation that serves to interrupt the random, glass network, partly by being of high enough valence to provide additional oxygen to the network. We may note that some cations can be in several categories, such as Pb, which can have multiple oxidation states. Leaded glass exist for important applications as television screens, but the role of lead can be that of either an intermediate or a network modifier. Na is well-known modifier and is added to sand (quartz) in the form of Na_2O to form sodium silicates, which constitute a large class of glasses and, in the aqueous solution form, a large class of adhesives.

Table 2.28 Coordination Number and Bond Strength of Oxides

	M in MO_x	*Valence*	*Dissociation energy per MO_x (kcal/g-atom)*	*Coordination number*	*Single-bond strength (kcal/g-atom)*
Glass formers	B	3	356	3	119
	Si	4	424	4	106
	Ge	4	431	4	108
	Al	3	402–317	4	101–79
	B	3	356	4	89
	P	5	442	4	111–88
	V	5	449	4	112–90
	As	5	349	4	87–70
	Sb	5	339	4	85-68
	Zr	4	485	6	

Intermediates	Ti	4	435	6	73
	Zn	2	144	2	72
	Pb	2	145	2	73
	Al	3	317–402	6	53–67
	Th	4	516	8	64
	Be	2	250	4	63
	Zr	4	485	8	61
	Cd	2	119	2	60
Modifiers	Sc	3	362	6	60
	La	3	406	7	58
	Y	3	399	8	50
	Sn	4	278	6	46
	Ga	3	267	6	45
	In	3	259	6	43
	Th	4	516	12	43
	Pb	4	232	6	39
	Mg	2	222	6	37
	Li	1	144	4	36
	Pb	2	145	4	36
	Zn	2	144	4	36
	Ba	2	260	8	33
	Ca	2	257	8	32
	Sr	2	256	8	32
	Cd	2	119	4	30
	Na	1	120	6	20
	Cd	2	119	6	20
	K	1	115	9	13
	Rb	1	115	10	12
	Hg	2	68	6	11
	Cs	1	114	12	10

Glass Ceramics: There are materials that are hybrid between glasses and ceramics. *Glass ceramics* are a family of fine-grained crystalline materials obtained through controlled crystallization of glasses.

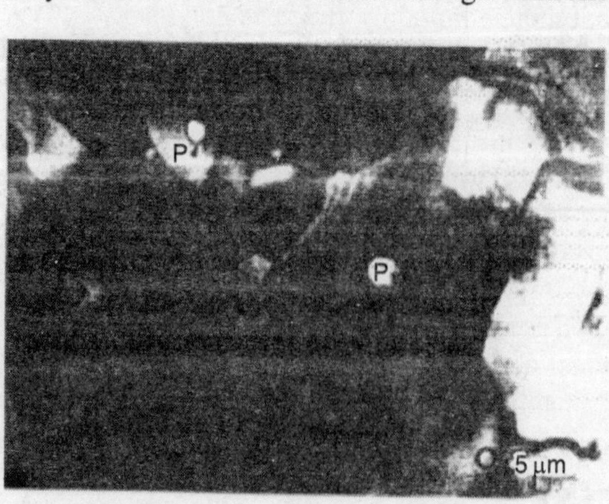

Fig. 2.71 Microstructure of a typical glass ceramic with a fine grain size, minimal residual glass, and little porosity (P).

They are nonporous and are either *opaque* or *transparent.* Their optical transparency arises due to the fact that the crystals are smaller than the wavelength of visible light such that the light waves are *transmitted* without significant scattering (Fig. 2.71). Like glasses, glass ceramics have excellent corrosion and oxidation resistance, but have improved impact strength and dimensional stability over both glasses and ceramics. Glass ceramics are commonly made of compositions in $MgO - Al_2O_3$-SiO_2, and $Li_2O - Al_2O_3 - SiO_2$ systems, with the main crystal phases being *cordierite* ($Mg_2Al_4Si_5G_{18}$) and β-quartz and β-*spudomene* ($Li_2O \cdot Al_2O_3 \cdot 4\ SiO_2$) solid solutions. The residual glass phase is usually 1-10% by volume.

One can achieve the desired crystalline microstructure through the addition of *nucleating agents,* usually titanium oxide, to the glass prior to heat treating. The glass is heated to an appropriate nucleating temperature that allows the maximum number of crystal nuclei to form, followed by heat treatment at higher temperature to grow the crystals. The result is a highly crystallite material with tiny crystallites, on the order of a few nanometers. Glass ceramics are used as ovenware and tableware, electrical insulators, substrates for circuit boards, architectural cladding, and for heat exchangers and regenerators.

We may note that like metals, ceramic crystals are also not perfect. Ceramic crystals also contain all of the same types of defects. What is unique about ceramic crystals, particularly oxide ceramics, is that the concentration of point defects, such as vacancies and interstitials, is not only determine not only by temperature, pressure, and composition, but can be influenced greatly by the concentration of gaseous species in which they come in contact (*e.g.*, gaseous oxygen). The concentration of gaseous species affects the crystal structure, which in turn can affect physical properties such as conductivity and thermal expansion. This opens up a wide scope of applications for ceramic materials, ranging from *sensors* and *actuators to nanoscale reactors.*

(iii) Structure of Polymers: Polymers are substances composed of long chain of repeating molecules. In most cases the element carbon forms the back-bone of the chain. Polymers may be in semicrystalline, amorphous or in liquid states. The term *polymer* comes from 'poly', meaning many, and "mer" meaning units. Hence, polymers are composed of many units—in this case, structural units called *monomers.* A monomer is any unit which can be converted into a polymer. Similarly, a *dimer* is a combination of two monomers, a *trimer* is a combination of three monomers, and so on. Polymer chains are formed from the reaction of many monomers to form long chain hydrocarbons, sometimes called *macromolecules,* but more commonly called as polymers.

Review of organic chemistry: *Alkanes*, also called *paraffins,* are composed of all C—C, saturated bonds and have the general formula CnH_{2n+2}. The alkanes have one C═C double bond, and alkynes have a carbon-carbon triple bond.

Many chemical reactions involving organic molecules, particularly polymerization reactions. involve *functional* groups. A functional group is an atom or group of atoms that show a relative constancy of properties when attached to different carbon chains, *e.g.,* the attachment of a hydroxyl group—OH, to hydrocarbons, leads to a group of compounds called *aldohols,* which taken as a class, have similar properties, such as a more polar nature than the parent hydrocarbon. Some of the important functional groups in polymer chemistry are given in Table 2.29.

Isomerism is another important concept in organic chemistry which has an impact on polymer structure. *Structural isomers* are molecules that have the same chemical formula, but different molecular architectures, *e.g.* there are two different types of propyl alcohols, both with the same formula, depending on where the –OH functional group is placed on the carbon backbone (Fig. 2.72). Another types of isomerism results in *stereoisomers,* in which the functional groups are in the same

Table 2.29 Some of the More Important Functional Groups in Polymer Chemistry

Group	*General formula*	*Compound class*	*Example*
—Cl	R—Cl	Chlorides	Methyl chloride, CH_3Cl
—OH	R—OH	Alcohols	Ethyl Alcohol, CH_3CH_2OH
$-C(=O)OH$	$R-C(=O)OH$	Acids (or carboxylic acids)	Acetic acid, CH_3COOH
$>C=O$	$R_2C=O$	Ketones	Acetone, CH_3COCH_3
$H(>C=O)$	$H(R)C=O$	Aldehydes	Propionaldehyde, CH_3CH_2CHO
— O —	R — O — R	Ethers	Ethyl ether, $CH_3CH_2OCH_2CH_3$
$-NH_2-$	$R-NH_2$	Amines	Propylamine, $CH_3(CH_2)_2NH_2$
$-C(=O)NH_2$	$R-C(=O)NH_2$	Amides	Acetamide, CH_3CONH_2
$-C(=O)O-R'$	$R-C(=O)O-R'$	Esters	Ethyl acetate, $CH_3COOC_2H_6$

position on the chain but occupy different geometric positions, *e.g.*, the *cis* and *trans* forms of butene are *stereoisomers* (Fig. 2.73). An asymmetric carbon is required for stereoisomers to form. Asymmetric Carbon play an important role in determining the tacticity of polymers. Finally, two terms related to the structure and chemistry of the repeat units are *confirmation* and configuration.

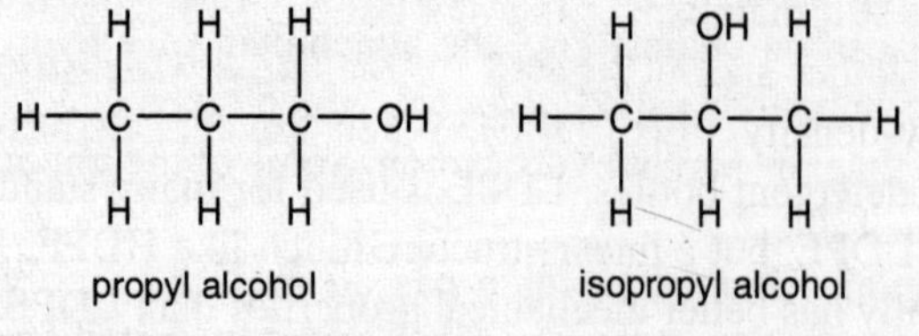

Fig. 2.72 An example of polymer structural isomers.

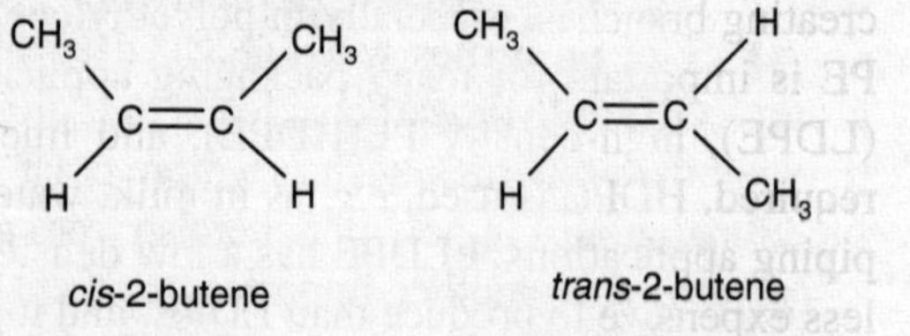

Fig. 2.73 An example of polymer stereoisomers.

Polymer Classification: In order to make generalizations regarding physical properties, formability and reactivity, it is useful to classify polymers. However, the appropriate classification scheme can change, as there are different ways in which to classify polymers. The first scheme groups polymers according to their chain chemistry. *Carbon-chain polymers* have a backbone composed entirely of carbon atoms. In contrast, *heterochain polymers* have other elements in the backbone, such as oxygen in a polyther, —C—O—C—. One can also classify polymers according to their macroscopic structure, *i.e.,* independent of the chemistry of the chain or functional groups. According to this scheme, there are three categories of polymers: *linear, branched* and *networked* (cross linked polymers) (Fig. 2.74). Finally, polymers can be classified according to their formability. Polymers that can be repeatedly shaped and reshaped are called *thermoplastics,* whereas these polymers that cannot be reshaped at any temperature once they are set are termed *thermosets.* Often times network polymers are thermosets, and linear and branched polymers are thermoplastics.

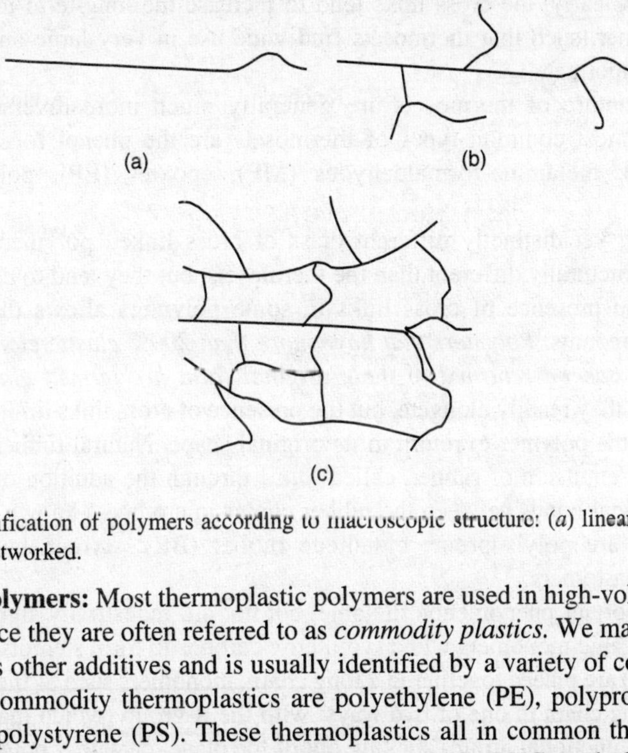

Fig. 2.74 Classification of polymers according to macroscopic structure: (*a*) linear, (*b*) branched and (*c*) networked.

Thermoplastic polymers: Most thermoplastic polymers are used in high-volume widely recognised applications, and hence they are often referred to as *commodity plastics.* We may note that a plastic is a polymer that contains other additives and is usually identified by a variety of commercial trade names. The most common commodity thermoplastics are polyethylene (PE), polypropylene (PP), polyvinyl chloride (PVC) and polystyrene (PS). These thermoplastics all in common the general repeat unit—$(CHX-CH_2)$—, where X is —H for PE, $-CH_3$ for PP and —Cl for PVC, and a benzene ring for PS. We may note that all these thermoplastics can be produced by the same type of reaction.

In their simplest forms, the thermoplastics are linear, carbon chain polymers. There are methods for creating branches, especially in polyethylene, while still maintaining thermoplasticity. Thus, branched PE is important for many packaging applications. Other special type of PE include low-density PE (LDPE), high-density PE(HDPE), and linear low-density PE(LLDPE). When greater stiffness is required, HDPE is used, *e.g.* as in milk, water and detergent bottles. LDPE is used for many standard piping applications. LLDPE has a low density like LDPE, but a linear structure much like HDPE. It is less expensive to produce than LDPE, and it generally has better mechanical properties than LDPE.

PP, PVC and PS have in common an asymmetric carbon atom in their backbone, a fact that leads to interesting structural properties. These are linear thermoplastics, whose properties can be tailored through blending, branching, and additives. We may note that the ability of thermoplastics to soften when heated and harden upon cooling leads to the principle of *recycling.*

Thermoset polymers: A subset of these are sometimes referred as *resins.* The thermoplastic PE can be treated by electron radiation or chemical means to form chemical bonds between adjacent called *cross links*. In this process, some of the carbon-hydrogen or carbon-carbon bonds in the linear chain are broken, creating free radicals, which react with free radicals on other chains to form bridges between the chains. When the cross linking is brought about by chemical means, the term *curing* is often used. The result is that unlike linear PE, cross linked PE has no distinct melting point due to limited chain mobility and will eventually degrade upon heating. This is a common characteristic of thermoset plastics. Due to their three-dimensional structure, they cannot be reshaped upon heating. Instead, they tend to remain rigid, or soften only slightly, until the backbone begins to break, leading to *polymer degradation.* This seems not a desirable quantity. It means that thermoset polymers are generally stiffer than thermoplastics. Clearly, the cross links tend to increase the long-term thermal and dimensional stability of the polymer, such that thermosets find wide use in very large and complex parts where thermal stability is important.

The chemical structure of thermosets are generally much more diverse than the commodity thermoplastics. The most common types of thermosets are the phenol formaldehydes (PF), Urea-formaldehydes (UF), melamine-formaldehydes (MF), epoxies (EP), polyurethanes (PU), and polyimides (PI).

There is a related, yet distinctly different class of cross linked polymers called as *elastomers.* Although, they are structurally different than the thermosets, but they tend to decompose when heated, rather than flow. The presence of cross links in some polymers allows them to be stretched, or elongated, by large amounts. *Polymers that have more than 200% elastic elongation (three times the original length) and can be returned to their original form are termed elastomers.* They are like thermoplastics in that they readily elongate, but the presence of cross links limits the elongation prior to breakage and allows the polymer to return to its original shape. Natural rubber is an elastomer. Cross links are added to an emulsion of rubber, called *latex,* through the addition of heat and sulphur. The sulphur creates chemical bonds between the rubber chains in a process known as *vulcanization.* Other common elastomers are polyisoprene, butadiene rubber (BR), styrene butadiene rubber (SBR), silicones and fluoroelastomers.

Tacticity: An important phenomenon in some, but not all, industrially significant polymers arises from the ability of organic monomers with asymmetric carbons to form stereoisomers. When the repeat units (monomer units) are placed together in a long chain, monomers such as the one shown in Fig. 2.75 can add to the growing chain in one of two ways: with the R group (which may be a chlorine, methyl group, or some other functional group) sticking out of the plane, or with it pointing back into the plane

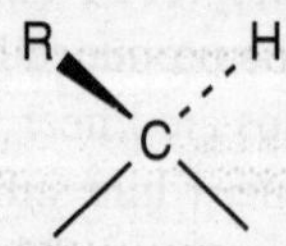

Fig. 2.75 An asymmetric carbon atom.

of the plane of the paper. The variation in how the functional groups are arranged leads to differences in *tacticity.* Tacticity is variations in the configuration (not confirmation!) of polymer chains as a result of the sequence of asymmetric carbon centres in the repeat units. There are three types of tacticity in polymers: *isotactic, syndiotactic* and *atactic.* In *isotactic* polymers, the functional groups are all arranged on the same side of the asymmetric carbon (Fig. 2.76*a*). Figure 2.76(*b*) shows the *syndiotactic* arrangement, in which the R groups perfectly alternate between sticking in and out of the plane of the carbon backbone. Finally, it is possible (and common) to have a perfectly random arrangement of functional groups (Fig. 2.76*c*), in which case an *atactic* polymer results. For example, polymers that exhibits tacticity are polyvinyl chloride (PVC, R = Cl), polystyrene (PS, R = Ph, phenyl group), and polyproplene (PP, R = CH_3).

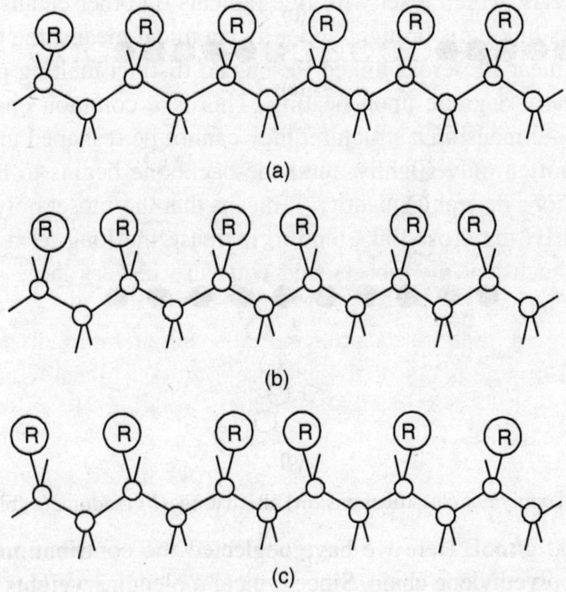

Fig. 2.76 Different types of tacticity, where R represents an atom other than hydrogen: (*a*) isotactic; (*b*) syndiotactic and (*c*) atactic.

Copolymers: Polymers can be made that contain more than one type of repeat unit, *e.g.* the R group on the asymmetric carbon in Fig. 2.76 could be chlorine in some of the monomer units and fluorine in the rest. Such polymers are termed as *copolymers.* The ratio of the two types of monomers can vary from 0 to 1, and there can be more than two types of monomers in a copolymer. The presence of more than one type of repeat unit opens up many possibilities for variation in the structure of the polymer, or *chain architecture* as it is sometimes referred. One may categorize copolymers in some broad, structural units.

If the two moments of bi-component copolymers are perfectly alternating, an *alternating copolymers* results (Fig. 2.77*a*). If the monomer units alternate in a random fashion, a *random copolymer* results (Fig. 2.77*b*). Figure 2.77(*c*) shows a *block copolymer* in which the different monomers prefer to add to each other, resulting in large segments of one type of monomer followed by large segments of the other type of monomer. Finally, a second type of monomer can be added as a branch to a polymer backbone (Fig. 2.77*d*), resulting in a *graft copolymers.* These different types of copolymers can have markedly different physical properties and are the basis for many important materials in the healthcare and consumer product industries.

Degree of Polymerization: The number of repeat units in an isolated polymer chain is called the degree of *polymerization,* x_n. If the monomer from which our polymer is formed has an initial molecular weight of M_o, then the molecular weight of M_o, then the molecular weight of an isolated polymer chain is X_nM_o, *e.g.* polyethylene, —(H_2C—CH_2)—, is made of ethylene repeat units, each of which has a molecular weight M_o = 2(12) + 4(1) = 28 g/mol. Let us assume that an isolated polyethylene chain has 1000 repeat units, so X_n = 1000. The molecular weight of this isolated polyethylene chain would be

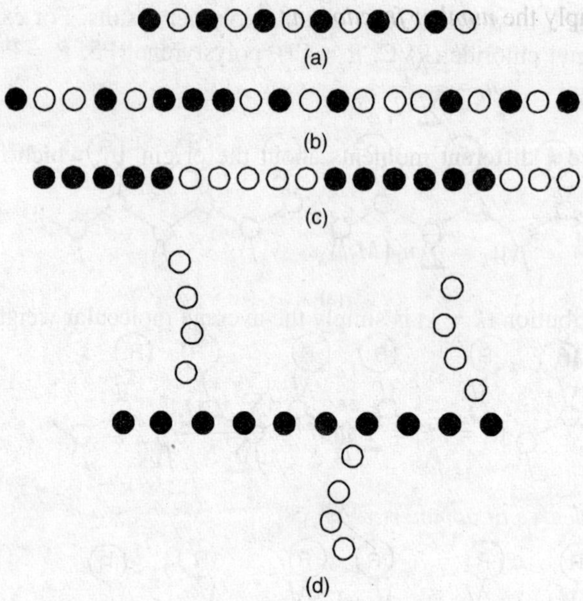

Fig. 2.77 Summary of copolymer classifications: (*a*) alternating, (*b*) random, (*c*) block and (*d*) graft.

1000 × (28 g/mol) = 28,000 g/mol. Here we have neglected the contribution of terminal hydrogens located at both ends of the polyethylene chain. Since typical molecular weights are of the order 10^6 and larger, terminal groups are not a significant contribution to the molecular weight, and neglecting them does not introduce substantial error in calculations. We will see that the molecular weight of the repeat unit in the polymer chain is not necessarily exactly the same as the molecular weight of the monomer. Sometimes, small molecules may be produced during polymerization reactions, *e.g.* condensation reactions in which water is formed as a by-product.

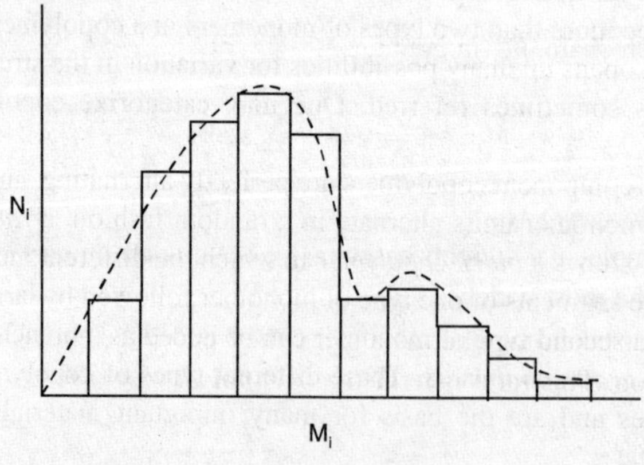

Fig. 2.78 Histogram exhibiting the number of molecules, N_i, having the molecular weight.

Average Molecular Weight: In a typical polymer, not all chains have the same length. Thus, there is a distribution of X_n values, and the molecular weight M_i, varies from chain to chain. One could construct a histogram of the distribution of number of chains, N_i, with molecular weight M_i (Fig. 2.78). Molecular weights have discrete values of course, and it is possible to find some mathematical function, $f(M_i)$ that appropriately describes the curve represented by the dotted line shown in Fig. 2.78. For molecular weights, this function is simply the *number fraction, n_i*:

$$n_i = \frac{N_i}{\sum N_i} \qquad ...(48)$$

The distribution can have k different moments about the origin, μ_k, which for a discrete random variable, M_i, are given by

$$\mu_k = \sum_i n_i (M_i)^k \qquad ...(49)$$

The first moment of distribution (k = 1) is simply the average molecular weight, called the number-average molecular weight, $\bar{M}_n$,

$$\mu_1' = \bar{M}_n = \sum n_i M_i = \frac{\sum N_i M_i}{\sum N_i} \qquad ...(50)$$

or, in terms of the *average degree of polymerization, $\bar{X}_n$*

$$\bar{M}_n = \bar{X}_n M_o \qquad ...(51)$$

where M_o is the molecular weight of a repeat unit. The second moment of the molecular weight distribution is then

$$\mu_2' = \sum n_i M_i^2 \qquad ...(52)$$

A ratio of the second moment to the first moment (number average molecular weight) is called the *weight-average molecular weight. $\bar{M}_w$*, which is the average based on the *weight* of each polymer chain:

$$\bar{M}_w = \frac{\sum N_i M_i^2}{\sum N_i M_i} = \frac{\sum n_i M_i^2}{\sum n_i M_i} \qquad ...(53)$$

or, in terms of weight of each chain, m_i (in grams)

$$\bar{M}_w = \frac{\sum m_i M_i}{\sum m_i} = \sum w_i M_i \qquad ...(54)$$

where w_i is the weight fraction, $m_i/\sum M_i$,

There are other types of molecular weights based on higher-order moments to the distribution (Table 2.30), but $\bar{M}_n$ and $\bar{M}_w$ are the most widely used quantities.

In general, $\bar{M}_n \neq \bar{M}_w$, but taken together as a ratio, they provide a measure of the breadth of the molecular weight distribution. This ratio is termed as the *polydispersity index*, and it takes on values greater than or equal to 1.0:

$$\frac{\bar{M}_w}{\bar{M}_n} \geq 1.0 \qquad ...(55)$$

If all chains have exactly the same weight and number of repeat units, then the system is termed *monodispersed*, and the polydispersity index is exactly 1.0. Most real polymers have rather large polydispersity indexes, but some standards used for chromatography calibration, such as polystyrene, can have values approaching unity.

Table 2.30 Most Common Molecular Weight-Averages

Average	*Definition*	*Methods of determination*
$\bar{M}_n$	$\frac{\sum N_i M_i}{\sum N_i}$	Osmotic pressure, colligative properties, end group analysis.
$\bar{M}_w$	$\frac{\sum N_i M_i^2}{\sum N_i M_i}$	Light scattering sedimentation velocity.
$\bar{M}_z$	$\frac{\sum N_i M_i^3}{\sum N_i M_i^2}$	Sedimentation equilibrium.
$\bar{M}_v$	$\left(\frac{\sum N_i M_i^{1+a}}{\sum N_i M_i}\right)^{1/a}$	Intrinsic viscosity (a is a characteristic of a system and lies between 0.5 and 1.0).

Polymer Crystallinity: An interesting and important structural characteristic of many polymers is that they are easily transformed from the amorphous to the crystalline state. Unlike inorganic glasses which generally need a great deal of energy to devitrify (crystallize), or metals that require non-equilibrium conditions to form amorphous structures, many polymers have amorphous to crystalline transition temperatures that are near room temperature. However, not all polymers crystallize readily. Intuitively, one expects that a jumbled mass of spaghetti-like strands that make up a polymer solution or melt would tend to be amorphous, and this is indeed we find. However, there are a number of structural factors that contribute to the ability of amorphous polymer chains to rearrange themselves into an ordered structure. These factors include *chain architecture* (the chemical constituents and bond angles of the backbone and side groups), order and regularity (*e.g.* tacticity), intermolecular forces (both within an individual chain and between adjacent chains, such as hydrogen bonding, and steric effects (the size of groups and branches).

Types of Bulk Polymer Crystallinity: Normally, polymer crystallinity in bulk polymers is classified under two general categories: *extended chain* and *folded chain.* The first type of crystallity, *i.e.* extended chain arises in many polymers with highly regular structures, such as polyethylene, poly (vinyl alcohol), syndiotactic polymers of poly (vinyl chloride) and poly (1, 2-butadiene), most polyamides, and cellulose. In these molecules, the so-called "planar zigzag" structure shown in Fig. 2.79(*a*) possesses the minimum energy for an isolated section of chain and is therefore the thermodynamically favoured confirmation. Side groups, if they are small enough and arranged in a regular fashion, as in the syndiotactic structure, need not prevent crystallinity (Fig. 2.79*b*), but as the bulkiness and irregularity of the side groups grow, crystallization becomes more and more difficult. As a result, highly branched molecules, such as branched polyethlene, do not crystallize, even though polyethylene itself is easily crystallized. Similarly, networked polymers do not have the freedom to move in a way such that extended chain crystallinity can occur.

Like other classes of materials, polymers can be either *single crystals* or *polycrystalline.* Polycrystalline polymers are more appropriately turned *semi-crystalline polymers,* since the region between crystalline domains in polymers can be quite large and result in a significant amorphous component to the polymers. The crystalline regions in semi-crystalline polymers are called *crystallites,* which have dimensions of several hundred angstroms, but the length of polymer chains is generally much larger than this, *e.g.* a polyethylene chain with the extended chain structure (Fig. 2.79*a*) with

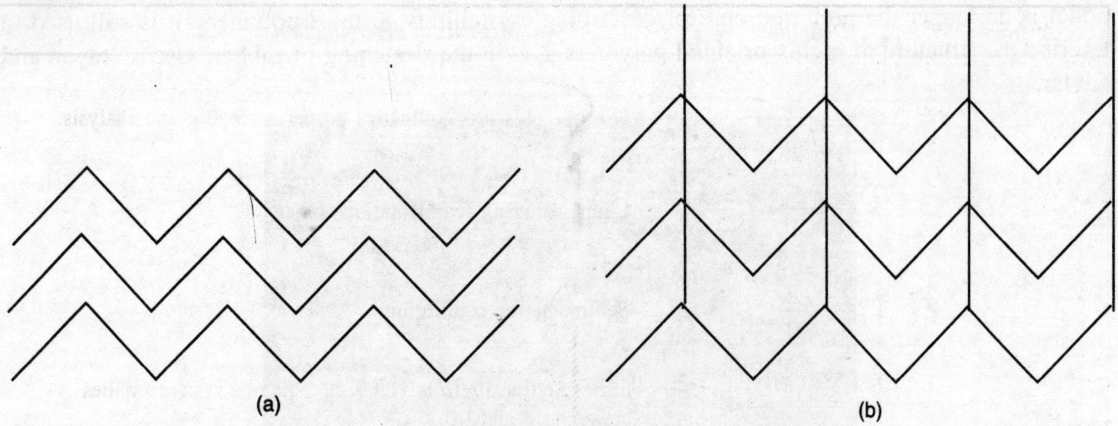

Fig. 2.79 Schematic illustration of extended chain crystallinity in polymers (*a*) polyethylene and (*b*) polypropylene.

molecular weight 50,000 have an end-to-end length of about 4500 Å. How can this be? The second type of crystalline structure in polymers chain folding provides an explanation. Polymer chains can-fold in a regular fashion to form plate-like crystallites called lamellae (shown in the insert of Fig. 2.80). We may note that polymer chains not only fold, but can extend from one lamella to another to form amorphous regions. In polymers crystallized from the melt, these lamellae often radiate from a central nucleation site, forming, three-dimensional special structures called *spherulities* (Figs. 2.80 and 2.81). In cross-polarized light, the spherulites form a characteristic *Maltese cross* pattern due to birefringent effects associated with the lamellar structures.

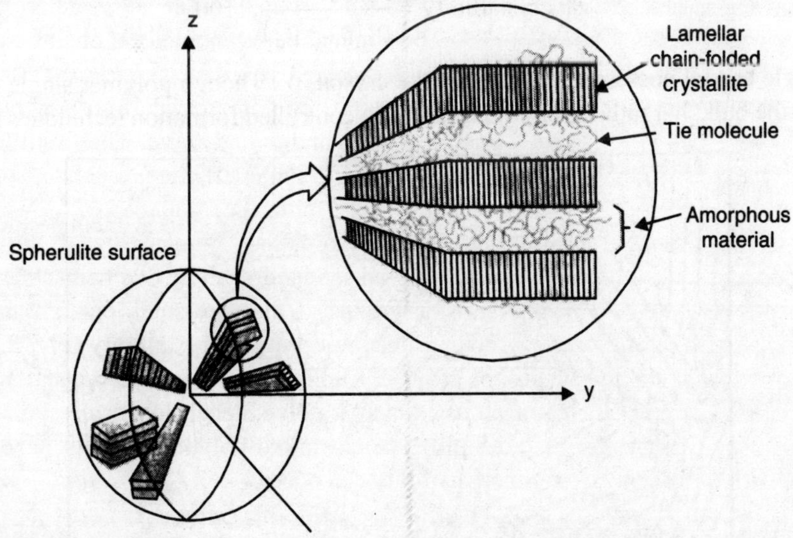

Fig. 2.80 Schematic illustration of chain-folding leading to lamellar crystallites (*inset*) and lamellar stacking to form spherulites.

A polymer crystal structure related to chain-folding is called the *fringed micelle model,* in which the polymer chains do not fold in a regular fashion but extend from one crystalline region to another, again forming amorphous regions between the crystallites (Fig. 2.82). We may note that the fringed micelle model is no longer the preferred one for describing crystallinity in most polymers, it is still used to describe the structure of highly oriented polymers, *e.g.* in the stretching of rubber, viscose rayon and Kevlar.

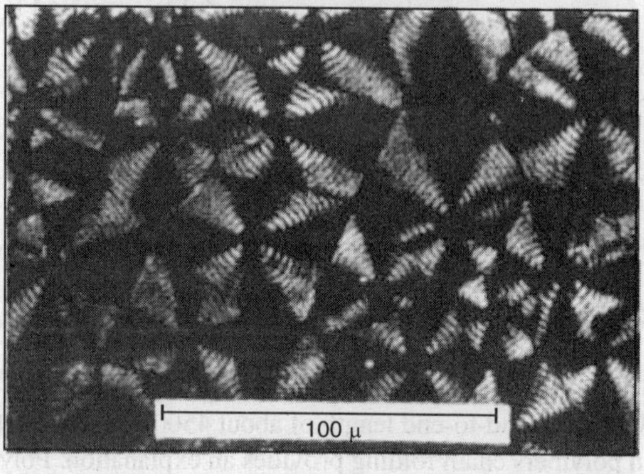

Fig. 2.81 Cross-polarized micrograph of polyethylene showing spherulitic structure.

The amorphous and crystalline regions each have different densities, with the crystalline density ρ_c being higher than the amorphous density ρ_a due to more compact structure. The per cent crystallinity in a semi-crystalline polymer with bulk density ρ_s can be calculated from the respective crystalline and amorphous densities:

$$\% \text{ crystallinity} = \frac{\rho_c(\rho_s - \rho_a)}{\rho_s(\rho_c - \rho_a)} \times 100 \qquad ...(56)$$

Polymer single crystal possess the density of the crystal, ρ_c. Though polymer single crystals do not usually form in the bulk, but rather from more carefully controlled formation techniques such as *vapour*

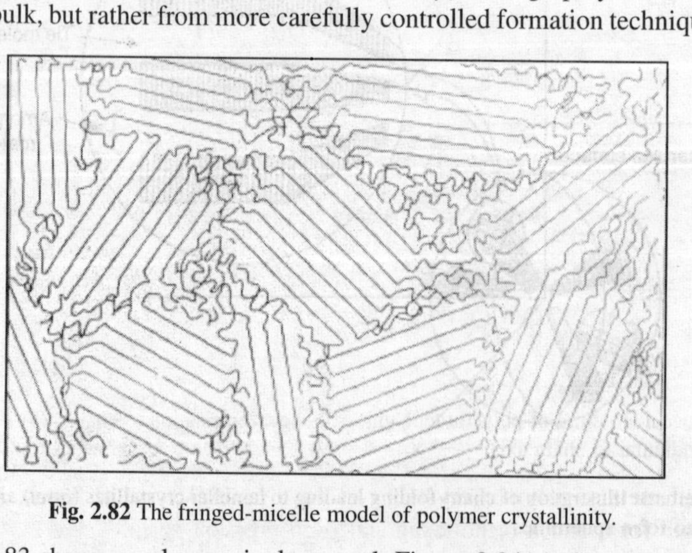

Fig. 2.82 The fringed-micelle model of polymer crystallinity.

deposition. Figure 2.83 shows a polymer single crystal. Figure 2.84(*a*) shows the crystal structure of polyethylene unit cell in relation to chains and 2.84(*b*) shows the view of unit cell perpendicular to chain axis. It is found that *small-angle X-ray scattering* (SAXS) is often used for polymer structural characterisation instead of traditional *wide-angle* X-ray scattering (WAXS).

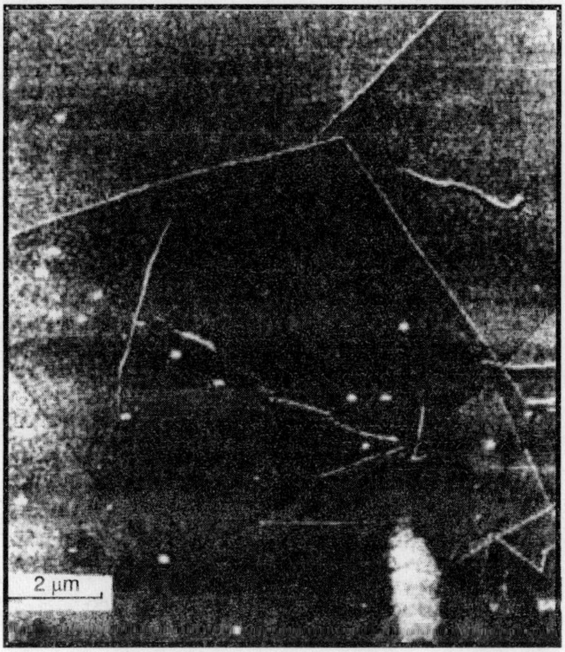

Fig. 2.83 Polyethylene single crystals.

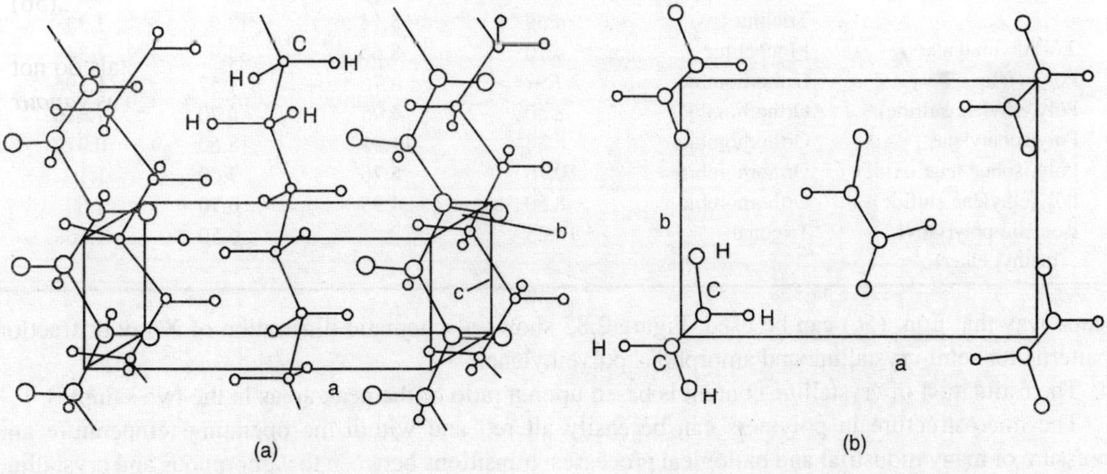

Fig. 2.84 (*a*) Crystal structure of polyethylene unit cell shown in relation to chains. (*b*) View of unit cell perpendicular to chain axis.

Like ceramics and metals, polymer crystals can have multiple crystal forms. Polyethylene has a metastable monoclinic form and a orthohexagonal high pressure form. Table 2.31 provides a list of some of the more common polymers and their corresponding crystal structure. One can use X-ray diffraction to determine the amorphous to crystalline ratio in semi-crystalline polymers in much the

Table 2.31 The Crystal Structures of Selected More Common Polymers

Polymer	*Crystal system*	*Lattice constants, Å* a	b	c	*Crystal density* (g/cm^3)
Polyethylene	Orthorhombic	7.417	4.945	2.547	1.00
Polytetrafluoroethylene	Trigonal (>19°C)	5.66	–	19.50	2.30
Isotactic	α-Monoclinic	6.65	20.96	6.50	0.936
polypropylene	β-Hexagonal	19.08	—	6.49	0.922
Syndiotactic polypropylene	Orthorhombic	14.50	5.60	7.40	0.93
Isotactic polystyrene	Trigonal	21.90	–	6.65	1.13
Poly(vinyl chloride)	Orthorhombic	10.6	5.4	5.1	1.42
Poly(vinyl alcohol)	Monoclinic	7.81	2.25	5.51	1.35
Poly(vinyl fluoride)	Orthorhombic	8.57	4.95	2.52	1.430
Poly(vinylidine	α-Monoclinic	4.96	9.64	4.62	1.925
fluoride)	β-Orthorhombic	8.58	4.91	2.56	1.973
Isotactic poly(methyl methacrylate)	Orthorhombic	20.98	12.06	10.40	1.26
trans-1, 4-Polybutadiene	Monoclinic	8.63	9.11	4.83	1.04
cis-1, 4-Polybutadiene	Monoclinic	4.60	9.50	8.60	1.01
Poly(ethylene oxide)	Monoclinic	8.05	13.04	19.48	1.228
	triclinic	4.71	4.44	7.12	1.197
Isotactic poly(propylene oxide)	Orthorhombic	10.46	4.66	7.03	1.126
Nylon 66	α-Triclinic	4.9	5.4	17.2	1.24
	β-Triclinic	4.9	8.0	17.2	1.248
2,6-Polyurethane	Triclinic	4.93	4.58	16.8	1.27
	Triclinic	4.59	5.14	13.9	1.33
3,6-Polyurethane	Monoclinic	4.70	8.66	33.9	1.34
Polyketone	Orthorhombic	7.97	4.76	7.57	1.296
Poly(ethylele sulfide)	Orthorhombic	8.50	4.95	6.70	1.416
Polyisobutylene	Orthorhombic	6.88	11.91	18.60	0.972
Poly(isobutylene oxide)	Orthorhombic	10.76	5.76	7.00	1.10
Poly(ethylene sulfide)	Orthorhombic	8.50	4.95	6.70	1.60
Isotactic poly(vinyl methyl ether)	Trigonal	16.25	–	6.50	1.168

same way that Eqn. (56) can be used. Figure 2.85 shows a schematic illustration of X-ray diffraction patterns for semi-crystalline and amorphous polyethylene.

The estimation of crystalline content is based upon a ratio of the peak areas in the two samples.

The microstructure in polymers can be easily altered and within the operating temperature and pressure of many industrial and biological processes, transitions between the amorphous and crystalline state, and the ratio between the amorphous and crystalline components, can easily take place. The amorphous component of a polymer is less dense than the crystalline component. Conversely, the *specific volume,* or volume per mole in CC, is lower for polymer crystals than it is for amorphous polymers. One can best understood this distinction by observing the volume change of a polymer melt as in cools. Let us consider a molten polymer at point A (Fig. 2.86). If we cool the polymer melt slowly, as in path ABG, the polymer chains have sufficient time to rearrange, fold, and form lamellar structures, resulting in a crystalline polymer, provided, of course, that they have the propensity to fold in the first

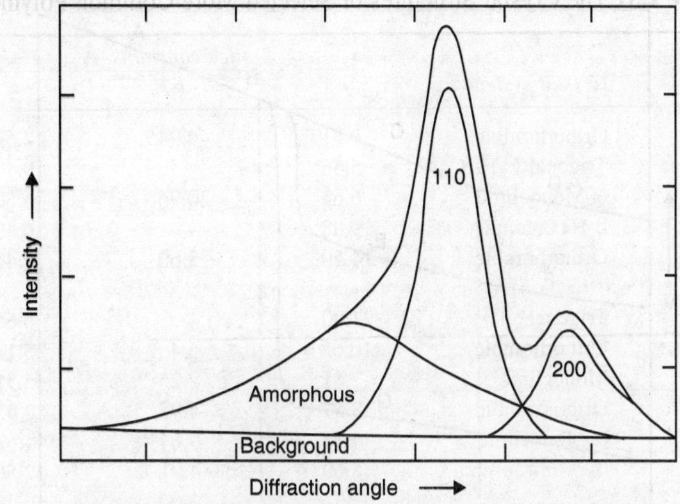

Fig. 2.85 Resolution of the X-ray scattering curve of a semi-crystalline polyethylene sample into contributions from crystalline (110 and 200 planes) and amorphous components.

place, and are not prevented from doing so by such factors as steric effects. The point at which the melt solidifies in the form of crystals is called the *crystalline melting point,* T_m, and is characterised by a sharp decrease in the specific volume and an increase in the density. If the same polymer melt is cooled rapidly, as in the path ABCD, a supercooled liquid is first obtained at point C, followed by an amorphous solid, or *glassy polymer*, at point D, due to insufficient time for the large molecules to arrange themselves in an ordered structure. The point at which the slope of the specific volume with temperature curve decreases (between C and D), representing solidification, is called the *glass transition temperature,* T_g. At some points, there is sufficient mobility in the polymer chains for them to begin to align themselves in a regular array and form crystallites. The polymer is not yet molten—there is simply short range chain movement that results in an amorphous to crystalline transformation in the solid state. This point is shown by T_g. On further heating the sample, the non-crystalline polymer eventually reaches T_m and melts. Most polymers are a combination of amorphous and crystalline components and are the result of intermediate cooling paths such as ABEF (Fig. 2.86). There are some polymers which cannot be crystallize, no matter how slowly we cool them, and they follow the *glassy path* ABCD all the time.

Liquid Crystalline Polymers (LCPS): These are non-isotropic materials that are composed of long molecules parallel to each other in large clusters and that have properties intermediate between those of crystalline solids and liquids. Because they are neither completely liquids nor solids. LCPs are called *mesophase* (intermediate phase) materials. Mesophase materials have liquid-like properties, so that they can flow; but under certain conditions, they also have long-range order and crystal structures. Because they are liquid-like, LCPs have a traditional degree of freedom that most solid crystals do not have. That is, crystals have three-dimensional order, whereas LCPs have only one or two-dimensional order. Nevertheless, they are called crystals.

In many cases, these polymer chains take on a rod-like (calamite LCPs) or disc-like (discotic LCPs) confirmation, but this does affect the overall structural classification scheme. We may note that there are many organic compounds, though not polymeric in nature, that exhibit liquid crystallinity and play important roles in biological processes *e.g.* arteriosclerosis is possibly caused by the formation of a cholesterol containing liquid crystal in the arteries of the heart. Similarly, cell wall membranes are generally considered to have liquid crystalline properties.

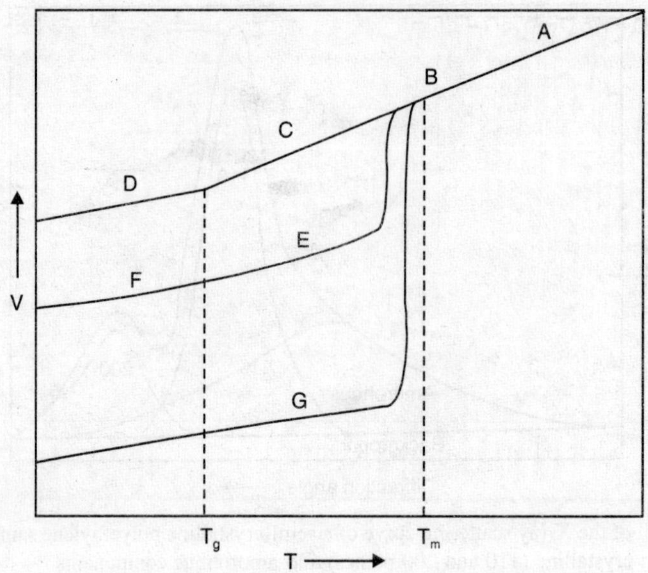

Fig. 2.86 Specific volume as a function of temperature on cooling from the melt for a polymer that tends to crystallize. Region *A* is liquid. *B* liquid with elastic response. *C* supercooled liquid, *D* glass, *E* crystallites in a supercooled liquid matrix, *F* crystallites in a glassy matrix, and *G* completely crystalline. Paths *ABCD, ABEF* and *ABG* represent fast, intermediate, and very slow cooling rates, respectively.

There are three categories of LCPs, grouped in accordance with the molecules: smectic, nematic, and cholesteric. Nematic (meaning "thread like") LCPs have their molecules aligned along the chain axis (Fig. 2.87*a*). Nematic liquids have low viscosity, and tend to be turbid, or "cloudy". *Smectic* (meaning "soap") LCPs have an additional level of structure, in that the polymer chains are also aligned along the chain axis, but they also segregate into layers. Smectic liquids are also turbid, but tend to be highly viscous. Finally, cholesteric (meaning "bile") LCPs have layered structures, but the aligned polymer chains in one layer rotated from the aligned polymer chains in adjacent layers. Cholesteric LCPs are also highly viscous, but often possess novel photochromic, optical, thermochromic, and electro-optical properties.

Obviously, not all polymeric molecules possess the ability to form LCPs. Generally, LCPs have a molecular structure in which there are two regions with dissimilar chemical properties, *e.g.* chains that consist of aliphatic-aromatic, dipolar—non-polar, hydrophobic-hydrophilic, flexible-rigid, or hydrocarbon-fluorocarbon combinations of substantial size have a propensity to form LCPs. The two different portions of the chains can interact locally with similar regions of adjacent chains, leading to ordering. Liquid crystalline polymers that rely on structural units in the backbone or main chain, to import their crystallinity are called *main-chain LCPs.* Figure 2.88 shows the building blocks of typical main-chain LCPs. Side-chain LCPs Crystallize due to interaction of the side-chains, or branches (Fig. 2.88). Side-chains play an important role in determining not only the liquid crystalline activity of polymers, but the type of structure that will form as well, *e.g.* in Table 2.32. We note that the number of hydrocarbon units in the side-chain (not the backbone repeat unit) affects whether the resulting LCP is *nematic* or *smectic.*

Obviously, we have used the general term 'liquid' to describe LCPs, *i.e.* this special class of polymers, but we know that a liquid can be either a melt or a solution. However, in the case of LCPs,

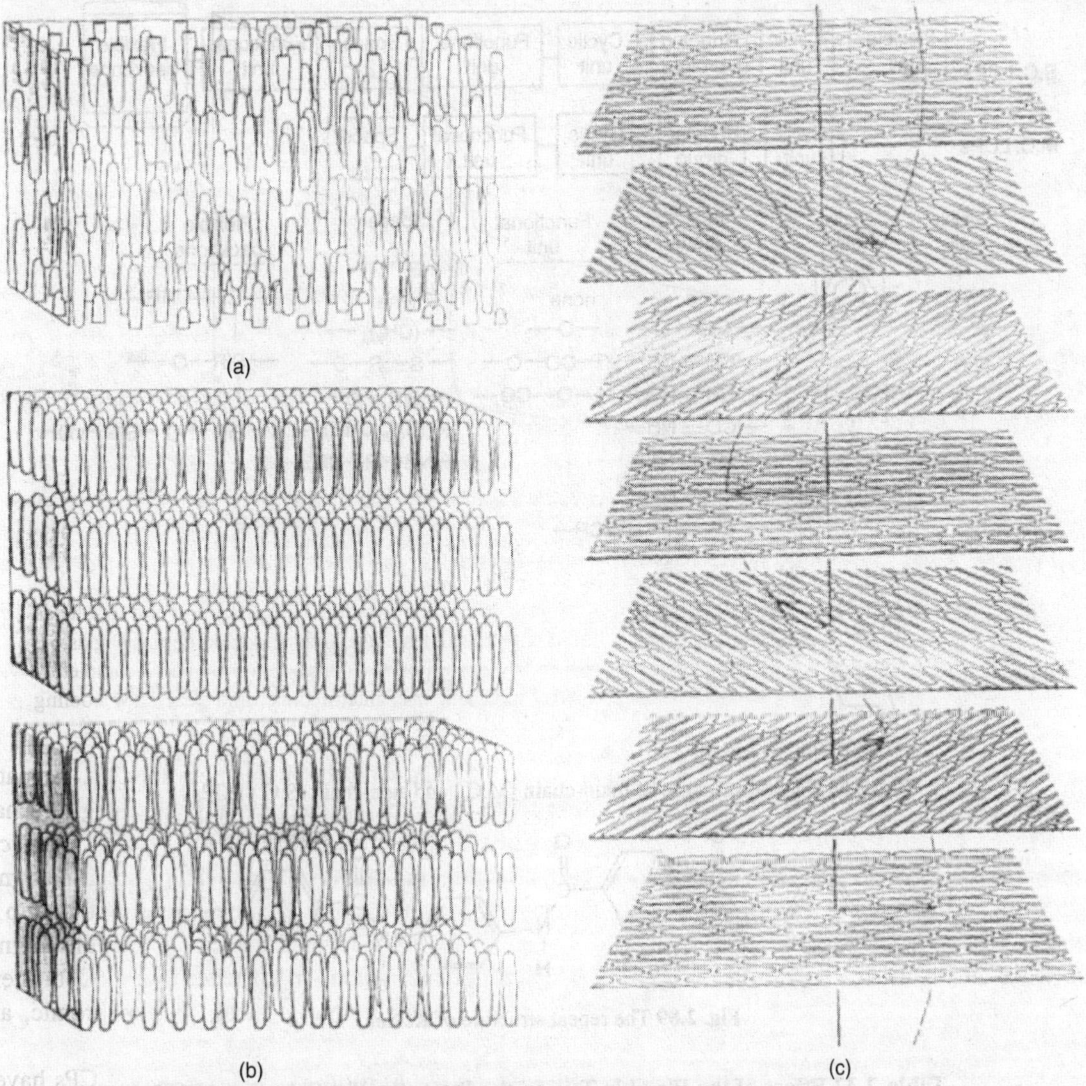

Fig. 2.87 The structure of liquid crystalline polymers (*a*) nematic, (*b*) smectic and (*c*) cholesteric.

both types of liquids can occur. A polymer that exhibits crystallinity in the melt and that undergoes an ordered-disordered transformation as a result of thermal effects is called a *thermotropic* LCP. A polymer requiring a small molecule solvent in order to exhibit crystallinity is termed a *lyotropic.* All the three types of LCP structures can occur in either thermotropic or lyotropic polymers, and industrially, both are relevant materials. Perhaps the mass widely utilized and studied lyotropic LCP is poly *p*-phenylene terephthalamide (PPTA), more commonly known as *Kevlar* (Fig. 2.89). Kevlar belongs to the class of aramids that are well known for their LCP properties. Since these polymers are crystalline in solution, they are often spun into filaments, from which the solvent is subsequently removed in order to retain the aligned polymer structure. The result is a highly oriented, strong filament that can be used for a wide variety of structural applications. We may note that most thermotropic LCPs are polyesters or copolymers that can be melted and molded into strong, durable objects.

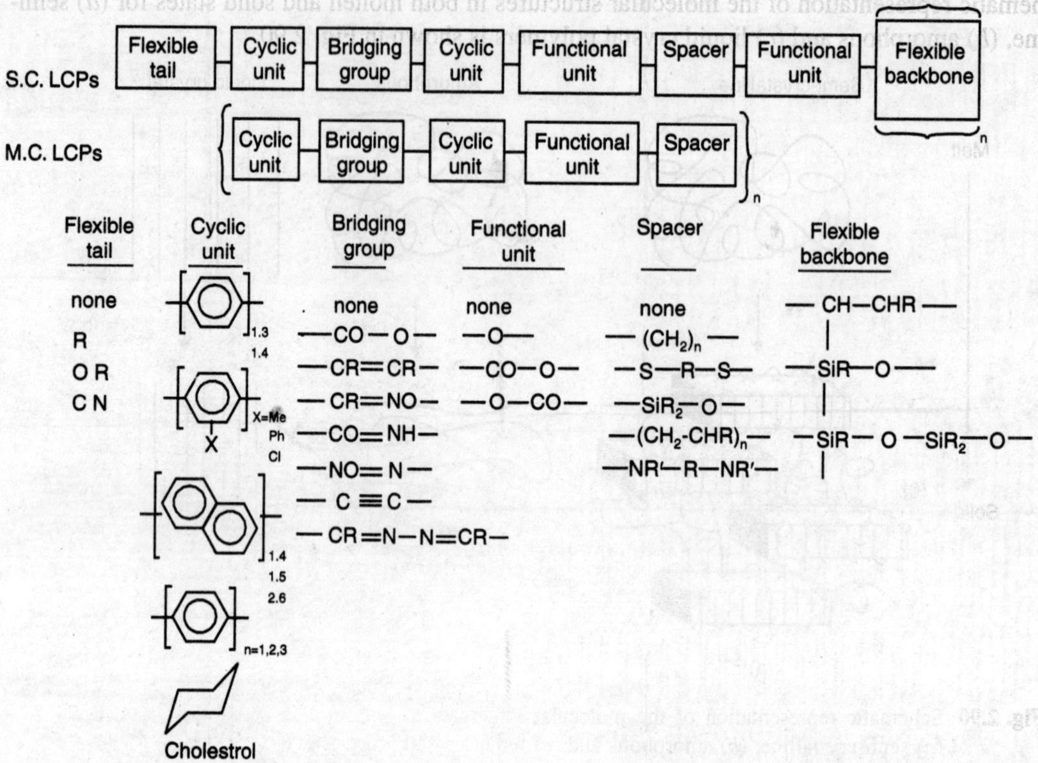

Fig. 2.88 General structure of main-chain (M.C.) and side-chain (S.C.) LCPs.

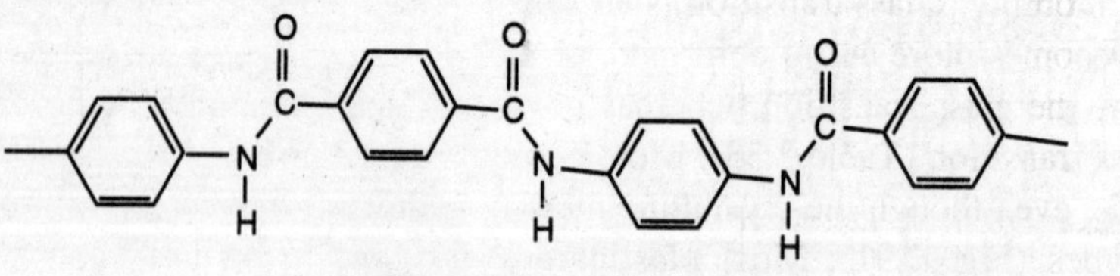

Fig. 2.89 The repeat structure of Kevlar.

Table 2.32 Effect of the Flexible Tail on the Structure of a Side-chain LCP

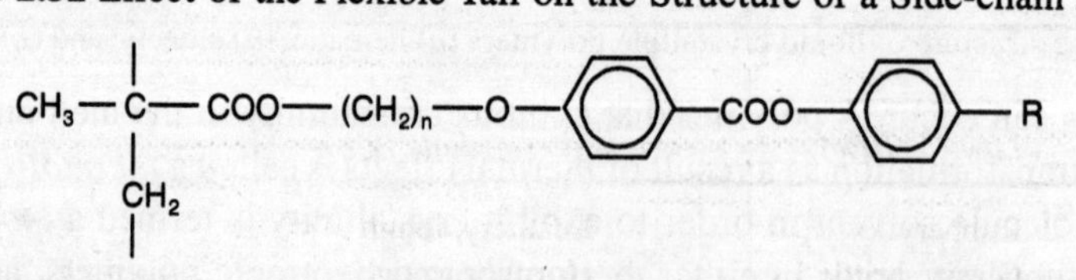

Number	*n*	*R*	*Structure*
1	2	OCH_3	Nematic
2	2	OC_3H_7	Smectic
3	6	OCH_3	Nematic
4	6	OC_6H_{13}	Smectic

A schematic representation of the molecular structures in both molten and solid states for (*a*) semi-crystalline, (*b*) amorphous and (*c*) liquid crystal polymers is shown in Fig. 2.90.

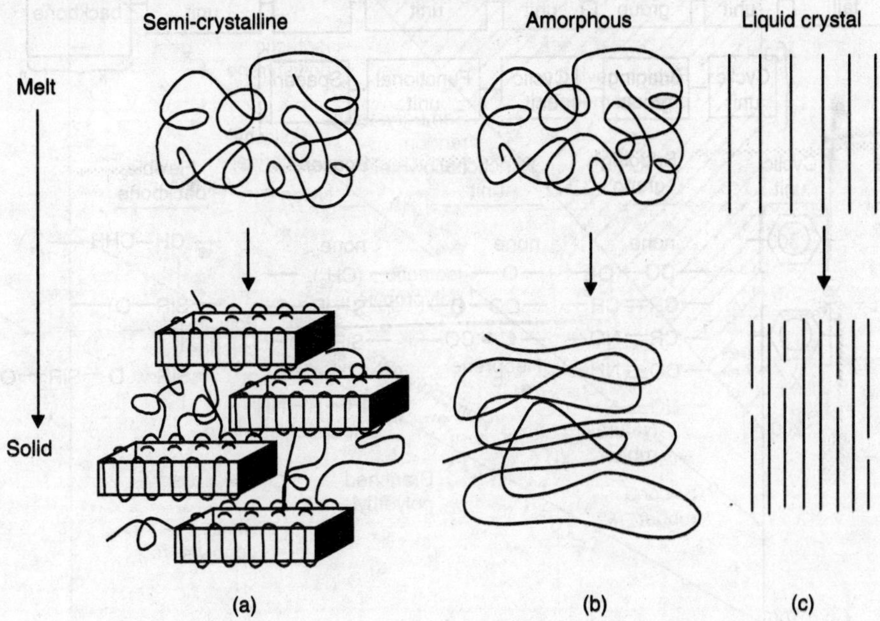

Fig. 2.90 Schematic representation of the molecular structures in both molten and solid states for (*a*) semi-crystalline, (*b*) amorphous and (*c*) liquid crystal polymers.

Glass Transition: This is an important parameter, since many physical properties change when polymer chains gain mobility. Glass transition is an important temperature because it marks the point where the polymer becomes more easily deformed, or more ductile. Totally non-crystalline polymers are very brittle below the glass transition, whereas polymers with intermediate crystallinity are very tough above the glass transition (Table 2.33). Most homopolymers have glass transition temperatures that are 0.5 – 0.75 T_m, even though the crystalline melting point can vary by several hundred degrees between polymer types (Fig. 2.91). Most plastics are designed to be used between T_m and T_g (Fig. 2.91A).

Table 2.33 Predominant Properties of Crystalline Polymers

	Degree of crystallinity		
Temperature range	*Low (5–10%)*	*Intermediate (20–60%)*	*High (70–90%)*
Above T_g	Rubbery	Leathery, tough	Stiff, hard, brittle
Below T_g	Glassy, brittle	Hornlike, tough	Stiff, hard, brittle

Structure of Polymer Chain: In a polymer chain the atoms of backbone chain are connected by covalent bonds forming definite angles that may vary depending on the type of substituent atoms or group of atoms along the backbone chain as shown in Fig. 2.91A. Such an arrangement is referred as the configuration of the polymer chain, which can be altered only by breaking or reforming covalent bonds. However, there is a continuous rotation or vibration of atoms or groups of atoms about the main-chain. This causes a number of conformations of the polymer chain constantly changing with time.

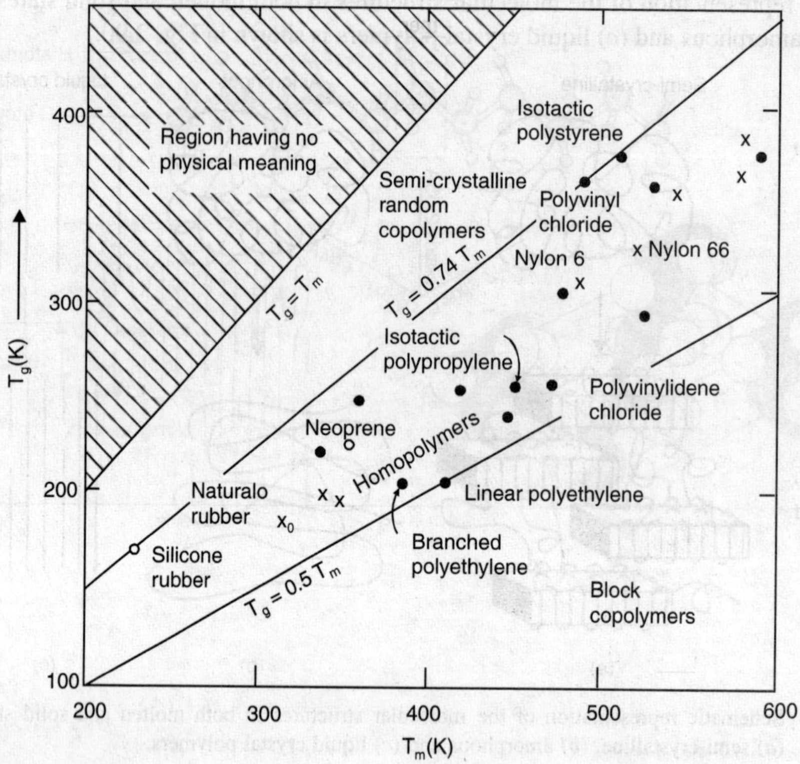

Fig. 2.91 The glass transition temperature, T_g, as a function of crystalline melting point, T_m, for homopolymers. Filled circles are addition homopolymers, open circles are elastomers, and crosses are condensation homopolymers.

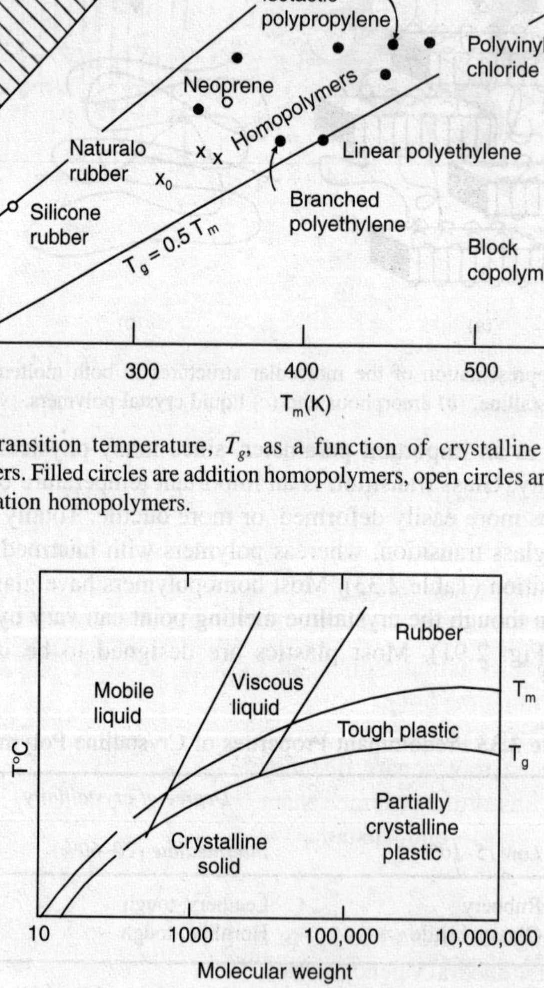

Fig. 2.91A Approximate relations among molecular weight, T_g, T_m and polymer properties.

If a polymer chain is fully extended the carbon atoms of the backbone chain assumes a zigzag planar arrangement as shown in Fig. 2.92(*a*). One can determine the length of such an extended chain L as

$$L = Nd \sin \theta/2 \quad ...(57)$$

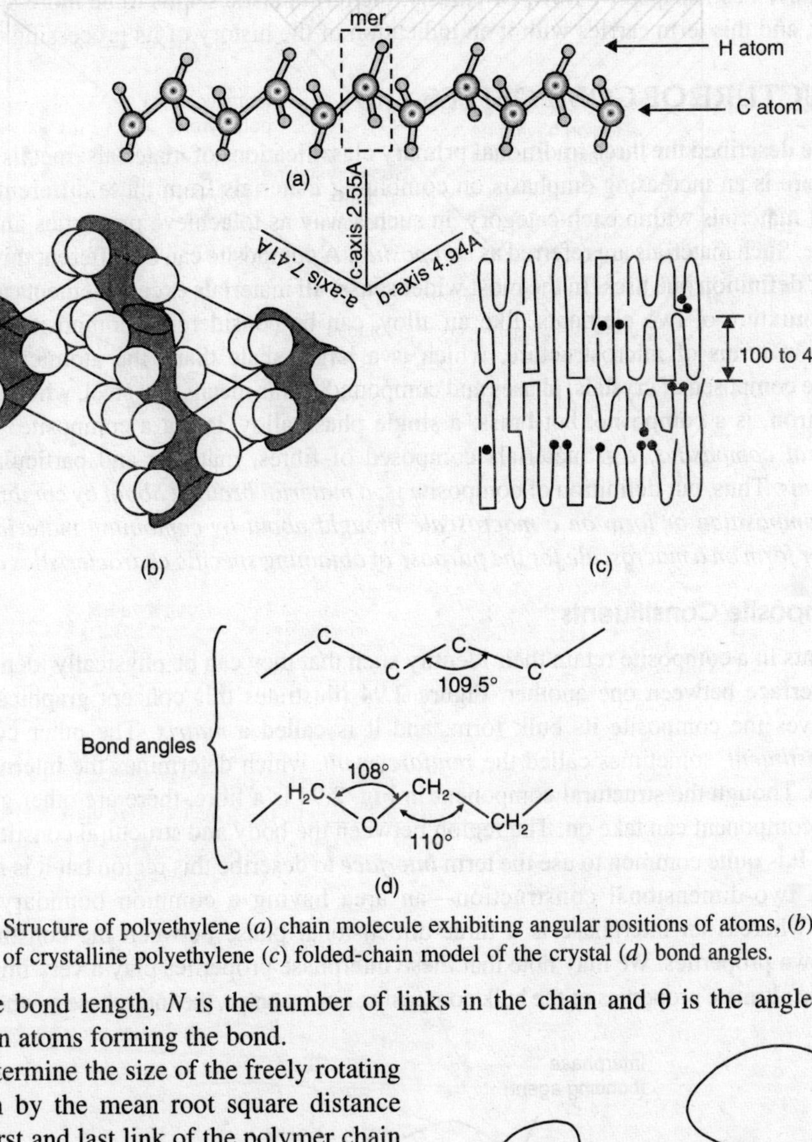

Fig. 2.92 Structure of polyethylene (*a*) chain molecule exhibiting angular positions of atoms, (*b*) unit cell of crystalline polyethylene (*c*) folded-chain model of the crystal (*d*) bond angles.

where d is the bond length, N is the number of links in the chain and θ is the angle between the backbone chain atoms forming the bond.

One can determine the size of the freely rotating polymer chain by the mean root square distance between the first and last link of the polymer chain (Fig. 2.93). The mean end-to-end distance is defined from statistical analysis as

$$L^2 = cNd^2 \qquad ...(58)$$

where C is constant and may vary from 1 to 2. For a freely rotating chain with tetrahedral valency angles, ideally $C = 2$.

We must note that we have used the term *polymer* almost exclusively to refer to organic macromolecules. The term *plastic* refers not only to organic substances of high molecular weight, but also to such substances that at some point in their

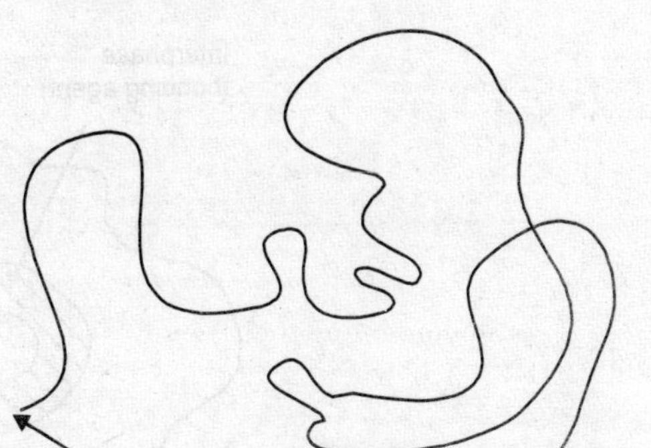

Fig. 2.93 In a freely rotating chain, the end-to end distortion.

manufacture have been shaped by flow. Obviously, the term plastic seems to be more specific than the term polymer, and this term carries with it an indication of the history of its processing.

2.11 STRUCTURE OF COMPOSITES

So far we have described the three traditional primary classifications of materials: metals, ceramics, and polymers. There is an increasing emphasis on combining materials from these different categories, or even different materials within each category, in such a way as to achieve properties and performance that are unique. Such materials are referred as *composites.* A composite can be different things, depending on the level of definition one uses. In the most wider sense, all materials except elements are composites, *e.g.* a binary mixture of two elements, like an alloy, can be considered a composite structure on an atomic scale. In terms of microstructure, which is a larger scale than the atomic level definition, composites are composed of crystals, phases and compounds. This means that steel, which is a suspension of carbon in iron, is a composite, but brass, a single phase alloy, is not a composite. There are also *macrostructural composites, e.g.* materials composed of fibres, matrices and particulates—they are *material systems.* Thus, our definition of composite is: *a material brought about by combining materials differing in composition or form on a macroscale brought about by combining materials differing in composition or form on a macroscale for the purpose of obtaining specific characteristics and properties.*

2.11.1 Composite Constituents

The constituents in a composite retain their identity such that they can be physically identified and they exhibit an interface between one another. Figure 2.94 illustrates this concept graphically. The body constituent gives the composite its bulk form, and it is called a *matrix.* The other component is a *structural constituent,* sometimes called the *reinforcement,* which determines the internal structure of the composite. Though the structural component in Fig. 2.94 is a fibre, there are other geometries that the structural component can take on. The region between the body and structural constituent is termed as *interphase.* It is quite common to use the term *interface* to describe this region but it is not correct. An interface is a two-dimensional construction—an area having a common boundary between the constituents—whereas an interphase is a three-dimensional phase between the constituents and, as such, has its own properties. We may note that these interphase properties play a very important role in determining the ultimate properties of the bulk composite. For instance, the interphase is where mechanical

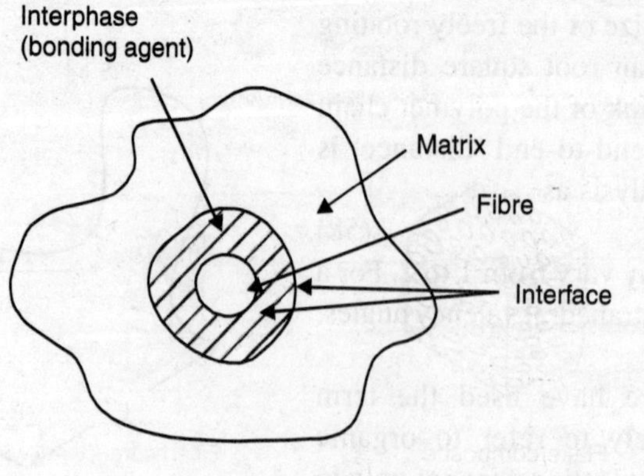

Fig. 2.94 Schematic illustration of composite constituents.

stresses are transferred between the matrix and the reinforcement. The interphase is also critical to the long-term stability of a composite. We assume that there is always an interphase present in a composite, even though it may have a thickness of only an atomic dimension.

We may note that the chemical composition of the composite constituents and the interphase is not limited to any particular material class. There are metal-matrix, ceramic matrix and polymer matrix composites, all of which find industrially relevant applications. Similarly, reinforcements in important commercial composites are made of such materials as steel, E-glass, and Kevlar. Many times a binding agent is added to the fibres prior to compounding to create an interphase of a specified chemistry.

2.11.2 Composite Classification

One can classify composites in many ways, including schemes based upon: (*i*) material combinations, such as metal-matrix, or glass-fibre-reinforced composites: (*ii*) bulk form characteristics, such as laminar composites or matrix composites; (*iii*) distribution of constituents, such as continuous or discontinuous; or (*iv*) function, like structural or electrical composites. Scheme (*ii*) is most general and we will make use of it. Other classification schemes are also useful and we will make use of them in later sections.

There are five general types of composites when categorized by bulk form (Fig. 2.95). *Fibre composites* consists of fibres; with or without a matrix. A fibre is a particle longer than 100 μm with a length-to-diameter ratio (aspect ratio) greater than 10:1. *Flake composites* consists of fibres; with or without a matrix. A flake is a flat, plate like material. *Particulate composites* can also have either a matrix or no matrix alongwith the particulate reinforcement. Particulates are roughly special in comparison to fibres or flakes. In a filled composite, the reinforcement, which may be a three-dimensional fibrous or porous structure, is continuous and often considered the primary case, with a second material added through such processes as chemical vapour infiltration (CVI). Lastly, *laminar composites* are composed of distinct layers. The layers may be variations of these classifications, of course, and one may see that the components in fibre, flake and particulate composites need not be disturbed uniformly and may even be arranged with specific orientations.

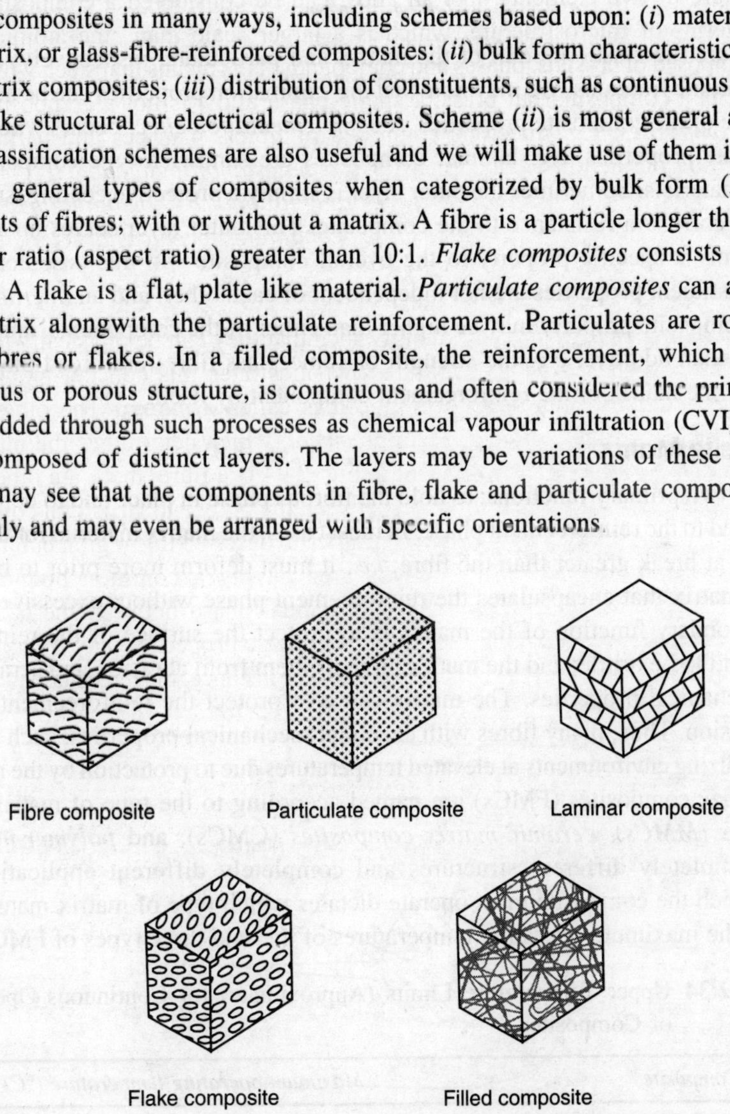

Fig. 2.95 Classes of composites.

2.11.3 Combination Effects in Composites

There are three ways that a composite can offer improved properties over the individual components, collectively called *combination effects.* A *summation effect* arises when the combination of each constituent is independent of the others, *e.g.* the density of a composite is independent of the other components. Elastic modulus is also a summation effect, with the upper limit $E_c(u)$ given by

$$E_c(u) = V_p E_p + V_m E_m \quad ...(59)$$

and the lower limit $E_c(l)$ given by

$$E_c(l) = \frac{E_m E_p}{V_m E_p + V_p E_m} \quad ...(60)$$

where E_m and E_p are the elastic moduli of the matrix and particulate, respectively; and V_m and V_p are their respective volume fractions. We may note that the summation properties can be added appropriately to give an estimate of the composite properties. A *complementation effect* occurs when each constituent contributes separate properties, *e.g.* laminar composites are sandwich-type composites composed of several layers of materials. Sometimes the outer layer is simply a protective coating, such as a polymeric film, that imparts corrosion resistance to the composite. This outer layer serves no structural purpose, and contributes only a specific property to the overall composite—in this case corrosion resistance. Further, some constituent properties are not independent of each other, and an *interaction effect* occurs. In this case, the composite property may be higher than either of the components, and the effect may be synergistic rather than additive, *e.g.* the strength of some glass-fibre-reinforced plastic composites is greater than either the matrix or the reinforcement component by itself.

2.11.4 Composite Matrix

The matrix serves two primary functions: to hold the fibrous phase in place and to deform and distribute the stress under load to the reinforcement phase. In most cases, the matrix material for the fibre composite has an elongation at break greater than the fibre; *i.e.*, it must deform more prior to breaking. It is also useful to have a matrix that encapsulates the reinforcement phase without excessive shrinkage during processing. A secondary function of the matrix is to protect the surface of the reinforcement. Many reinforcements tend to be brittle, and the matrix protects them from abrasion and scratching, which can degrade their mechanical properties. The matrix can also protect the reinforcement component from oxidation or corrosion. Thus, many fibres with excellent mechanical properties, such as graphite fibres, can be used in oxidizing environments at elevated temperatures due to protection by the matrix constituent.

Most *fibre-matrix* composites (FMCs) are named according to the type of matrix involved. *Metal matrix composites (MMCs), ceramic-matrix composites* (CMCs), and *polymer-matrix composites* (PMCs) have completely different structures and completely different applications. Usually the temperature at which the composite must operate dictates which type of matrix material is to be used. Table 2.34 gives the maximum operating temperatures of the said three types of FMCs.

Table 2.34 Upper Temperature Limits (Approximate) for Continuous Operation of Composites

Composite	*Maximum operating temperature (°C)*
Polymer matrix	400
Metal matrix	1000
Ceramic matrix	1300

Most structural PMCs consist of a relatively soft matrix, such as a thermosetting plastic of polyester, phenolic, or epoxy, sometimes referred to as resin-matrix composites. Some typical polymers used as matrices in PMCs are given in Table 2.35. We may note that the list of metals used in MMCs is much shorter. Aluminium, magnesium, titanium, and iron-and nickel-based alloys are the most common (Table 2.36). These metals are typically utilized due to their combination of low density and good

Table 2.35 Few Typical Polymers used as Matrices in PMCs

Matrix material	*Process temperature (°C)*	*Upper use temperature (°C)*
Thermosetting		
Polyester (pthalic/maleic type)	RT	70
Vinyl ester	RT	125
Epoxy	150	125
Epoxy	200	175
Phenolic	250	200
Cyanates (triazines)	250	200
Bis maleimides	250	225
Nadic end-capped polyimides (*e.g.* PMR - 15)	316	316
Thermoplastic		
Polysulfone	325	180
Polyamide	250	100
Polycarbonate	280	100
Polyphenylene oxide (PPO)	280	100
Polysulfides (PPS)	300	150
Polyether ether Ketone (PEEK)	370	175
Polyether sulfone (PES)	350	175
Polyamide-imides	325	200
Polyetherimide	400	275
Polimide	370	316
Polyarylate	400	300
Polyester (liquid crystalline)	300	150

Table 2.36 Common Metals for Metal-Matrix Composites

Matrix material	*Fabrication method*	*Typical composite density (g/cm³)*	*Use temperature (°C)*
Aluminium	Diffusion bonding		350
	Hot molding	2.62–345	
	Powder metallurgy		
	Liquid processing		
Magnesium	Liquid processing	1.82–2.80	300
	Diffusion bonding		
Titanium	Diffusion bonding	3.76–4.00	650
Iron, nickel-based alloys	Diffusion bonding	5.41–11.7	800–1150

mechanical properties. Matrix materials for CMCs usually fall into four categories: glass ceramics like lithium aluminosilicate; oxide ceramics like aluminum oxide (alumina) and mullite; nitride ceramics such as silicon nitride; and carbide ceramics such as silicon carbide.

The processing techniques used for CMCs can be quite exotic (and expensive), such as chemical vapour infiltration (CVI), or through prolysis of polymeric precusors. Their maximum use temperatures are theoretically much higher than most MMCs or PMCs, exceeding 1800°C, although the practical use temperature is often lower due to creep deformation (Table 2.37). The attraction of CMCs, then, is for use in high-temperature applications, *e.g.* combustion engines.

Table 2.37 Common Ceramics for Ceramic-Matrix Composites

Matrix material	*Density (g/cm^3)*	*Use temperature (°C)*
Alumina, Al_2O_3	4.0	~1000
Glass ceramics	2.7	900
Si_3N_4	3.1	~1300
SiC	3.2	~1300

2.11.5 Composite Reinforcement

We have discussed several classification schemes for composites, including one that is based upon the distribution of the constituents. For reinforced composites this scheme is quite useful (Fig. 2.96). In reinforced composites, the reinforcement is the structural constituent and determines the internal structure of the composite. The reinforcement may take on the form of particulates flakes, lamina, or fibres, or may be generally referred to as "filler". Fibres are the most common type of reinforcement, resulting in fibre-matrix composites (FMCs). We now examine some of these reinforcement constituents in detail:

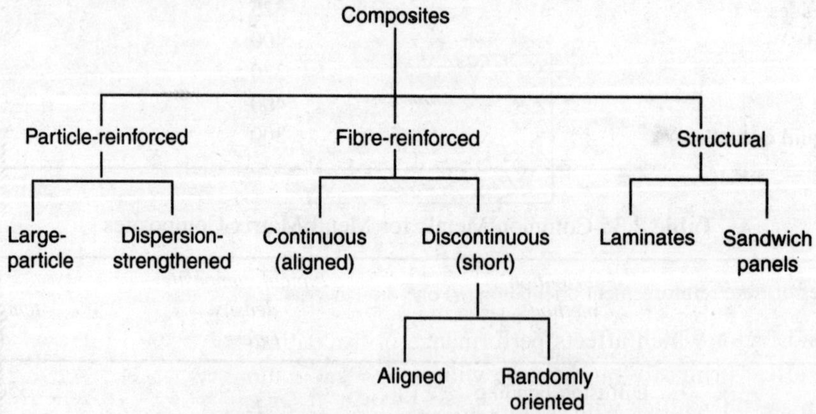

Fig. 2.96 Types of reinforced composites.

(i) Fibre-matrix composites: There are two main classes of FMCs (Fig. 2.96): those with continuous fibre reinforcement and those with discontinuous fibre reinforcement. The first type of composites are made from fibre *rovings* (bundles of twisted filaments) that have been woven into two dimensional sheets resembling a cloth fabric. These sheets can be cut and formed to a desired shape, or *perform, i.e.* then incorporated into a composite matrix, typically a thermosetting resin such as epoxy.

Metallic, ceramic and polymeric fibres of specific components can all be produced in continuous fashions, and the properties of the resulting composite are highly dependent not only on the type of fibre and matrix used, but also on the processing techniques with which they are formed.

(ii) Discontinuous-fibre-reinforced composites: These are much more widely used, and there are some general underlying principles that affect their overall properties. Four main factors contribute to the performance level of fibre in discontinuously reinforced FMCs. The first factor is fibre orientation. There are several ways that short fibres can be oriented within the matrix (Fig. 2.97). One-dimensional reinforcement occurs when the fibres are oriented along their primary axis. This offers maximum mechanical strength along the orientation axis, but results in *anisotropic composites; i.e.,* the mechanical and physical properties are not the same in all directions. *Planar reinforcement* occurs with two-dimensional orienting of the fibres, as often occurs with woven fabrics. The fabric, as is common in woven glass fibres, is produced in sheets, and it is laid down (much like a laminate) to produce a *two-dimensional* reinforcement structure. *Three-dimensional reinforcement* results from the random orientation of the fibres. This creates an isotropic composite, in which the properties are same in all directions, but the reinforcing value is often decreased compared to the aligned fibres.

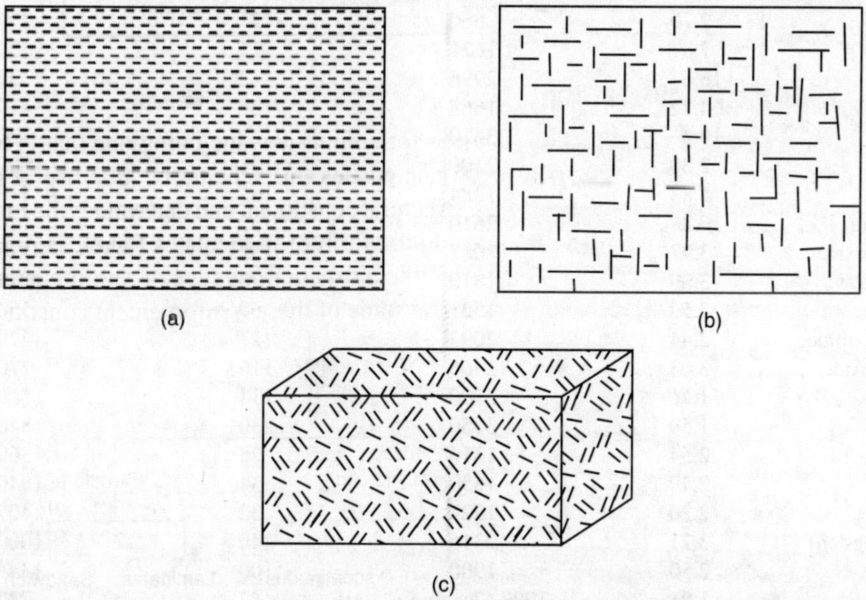

Fig. 2.97 Types of fibre reinforcement orientation (*a*) one-dimensional, (*b*) two-dimensional, and (*c*) three-dimensional.

The second factor which affects performance in discontinuously reinforced FMCs is fibre length. This has an affect primarily on the ease with which the composite can be manufactured. Very long fibres can create difficulties with methods used to create discontinuously reinforced FMCs and can result in non-uniform mechanical properties. The third factor, namely, the *fibre shape* is also related to fibre geometry. We know that *fibre is a particulate with length greater than 100 μm and an aspect ratio greater than 10:1.* This definition of a fibre allows for a great deal of flexibility in the geometry of the fibre, *e.g.* aspect ratio can vary widely; many reinforcement filaments have aspect ratios much larger than 10:1. It is also not necessary that the fibre cross-section be exactly circular. Hexagonal, ellipsoidal, and even annular (hollow fibre) cross-sections are quite common. Finally, it is also not necessary that the fibre, even those with circular cross-sections, be exactly cylindrical. "Dumbbell" shaped fibres are very common, and even preferable since mechanical stresses tend to concentrate at the fibre ends. The dumbell shape helps distribute these stresses.

The final factor affecting the reinforcement performance is its '*composition chemistry*' affects properties, and strength is usually the most important property of a reinforcing fibre. Table 2.38 summarizes selected properties of some common reinforcing fibres. We may note offentimes the important design consideration for reinforcement materials (or the matrix, for that matter) is not the absolute value of a particular design criterion, such as tensile strength or modulus, but the value per unit weight. such as *specific strength* or *specific modulus.* To make this point clear, we have listed these values in Table 2.38 rather than absolute values. This fact is quite important for several applications, *e.g.* automotive and aerospace composites, for which weight savings is paramount. We may note that reinforcing fibres come all material classes. Let us examine the more common fibres in detail.

Table 2.38 Selected Properties of Some Common Reinforcing Fibres

Fibre	*Density* *(g/cm^3)*	*Melting point* *(°C)*	*Specific modulus* *($GPa \cdot cm^3/g$)*	*Specific strength* *($MPa \cdot cm^3/g$)*
Aluminum	2.70	660	27	230
Steel	7.87	1621	25	530
Tantalum	16.6	2996	12	37
Titanium	4.72	1668	24	410
Tungsten	19.3	3410	21	220
Boron	2.30	2100	192	1500
Beryllium	1.84	1284	165	710
Molybdenum	10.2	2610	35	140
Aluminum oxide	3.97	2082	132	170
Aluminosilicate	3.90	1816	26	1060
Asbestos	2.50	1521	60	550
Beryllium carbide	2.44	2093	127	420
Beryllium oxide	3.03	2566	116	170
Carbon	1.76	3700	114	1570
Graphite	1.50	3650	230	1840
E-glass	2.54	1316	28	1360
S-glass	2.49	1650	34	1940
Quartz	2.20	1927	32	407
SiC (on tungsten)	3.21	2316	140	1000
Si_3N_4	2.50	1900	100	1344
BN	1.91	3000 (decomp.)	47	722
ZrO_2	4.84	2760	71	427
Wood	0.4–0.8	–	17	–
Polyamide (Kevlar)	1.14	249	2.5	730
Polyester	1.40	249	2.9	490
Polypropylene	0.9	154	1.8	77

Organic fibres usually have very low specific gravities and therefore they are attractive for applications where strength/weight ratio is important. They are also very common for *textile* applications, since some organic fibres are easily woven. Carbon fibres offer excellent thermal shock resistance and very high strength-to-weight ratio. Carbon fibres are of two general types: *PAN-based carbon fibres* and *pitch based carbon fibres.* The first type of fibres are manufactured by the prolysis of *polyacrylonitrile* (PAN) (Fig. 2.98). The PAN is polymerized and fibres spun from the pre-polymer.

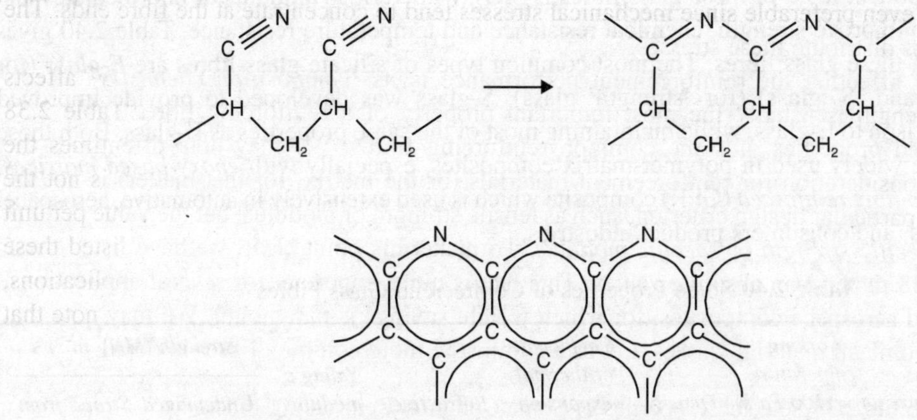

Fig. 2.98 Phrolysis of polyacylonitrile (PAN) to from carbon fibres

The PAN fibre are then pyrolyzed to remove the hydrogens and from benzene-ring structures. *Pitch-based fibres* are produced in a similar pyrolysis process of a precursor fibre, except that the precursor in this case is *pitch.* Pitch is actually a liquid crystalline material, often termed as *mesophase pitch,* and is composed of a complex mixture of thousands of different species of hydrocarbon and heterocyclic molecules. It is the residual product of petroleum refining operations.

Wood fibres are technically organic and it is an important structural material. Wood fibres, often in the form of wood flour, possess a variety of properties depending upon the type of tree from which they are derived, but are used extensively in low-cost composites. Wood fibre has a good strength/weight ratio and provides a use for recycled paper products. Synthetic organic fibres, *e.g.* polymide fibres, such as Kevlar, offer excellent specific mechanical properties. Kevlar is used in several applications where high toughness is desired, *e.g.* ropes and ballistic clothes. In addition to polyamides, polyesters, and polyproplene fibres listed in Table 2.38, nylon, and polyethylene are other common polymeric fibres used for composites In all cases, an added attraction of synthetic fibres is their chemical inertness in most matrix materials.

Glass fibres are the most common reinforcing fibre due to their excellent combination of mechanical properties, dielectric properties, thermal stability and relatively low cost. As a result, there are several different types of silicate glass fibres, all with varying properties designated for various applications (Table 2.39).

Table 2.39 Composition of Commercial Glass Fibres

	Composition (wt%)											
	SiO_2	Al_2O_3	Fe_2O_3	B_2O_3	ZrO_2	MgO	CaO	Na_2O	K_2O	Li_2O	TiO_2	F_2
A-glass (typical)	73	1	0.1			4	8	13	0.5			
E-glass (range)	52–56	12–16	0–05	8–13		0–6	16–25		←<1 total→		0	0–1.5
AR glass (range)	60–70	0–5			15–20		0–10	10–15			0–5	
C-glass (range)	59–64	3.5–5.5	0.1–0.3	6.5–7		2.5–3.5	13.5–14.5	8.5–10.5	0.4–0.7			
S and R glasses (range)	50–85	10–35					← 4–25 total →	0				

The majority component in all of these glass fibres is SiO_2, with various amounts of intermediates and modifiers added to improve strength, chemical resistance and temperature resistance. Table 2.40 gives some properties of these glass fibres. The most common types of silicate glass-fibres are *E-glass* (for 'electronic' glass and 'S' glass) (for 'strength' glass). S-glass was developed to provide improved strength in comparison to E-glass, while maintaining most of the same properties as E-glass. Both these glass filaments are widely used in polymer-matrix composites, especially with *epoxy-based matrices.* The result is a *glass-fibre reinforced* (GFR) composite which is used extensively in automative, aerospace, marine, electronics, and consumers product industries.

Table 2.40 Some Properties of Commercial Glass Fibres

	Liquidus temperature (°C)	*Working temperature (η = 100 Pa s) (°C)*	*Density (g cm^{-3})*	*Coefficient of thermal expansion (°C^{-1})*	*Refractive index*	*Young's modulus (GN · m^{-2})*	*Strength (MN · m^{-2})*	
							Undamaged filament	*Strand from roving*
A-glass	1140	1220	2.46	7.8×10^{-6}	1.52	72	3500	
E-glass	1400	1210	2.54	4.9×10^{-4}	1.55	72	3600	1700–2700
AR glass	1180–1200	1280–1320	2.7	7.5×10^{-6}	1.56	70–75	3600	1500–1900
C-glass	–	–	~2.5					
S and R glasses			~2.5			~85	~4500	2000–3000

As a class, ceramic fibres offer better thermal resistance than glass fibres and they are the preferred reinforcement in high-temperature structural composites. There are a number of commercial oxide-based fibre available, such as saffil (SiO_2/Al_2O_3), Nexel ($SiO_2/B_2O_3/Al_2O_3$). Fibre frex (SiO_2/Al_2O_3), and Kaowool (SiO_2/Al_2O_3), as well as nearly pure single-component metal oxide fibres of Al_2O_3 and ZrO_2. There are also a number of slag-based fibres of varying composition, based upon SiO_2, Al_2O_3, MgO and CaO, which are recovered from smelting operations. The more refractory fibres consists of nitrides and carbides, *e.g.* Si_3N_4 and SiC. These fibres are usually produced by more exotic techniques. For example, Si_3N_4 fibres are produced from polymeric precursors such as polysilazanes (Si–N), polycarbosilanes (Si–CH_2), or polysilanes (Si–Si). Silicon carbide fibres are currently substrate-based and are formed by the deposition of SiC on a metallic filament such as tungsten or carbon. Silicon carbide whiskers can also be produced by the pyrolysis of polycarbosilane precursors.

Finally, metallic fibres find some limited applications as reinforcement in composites. They are generally not desirable due to their inherently high densities and because they present difficulties in coupling to the matrix. Nonetheless, tungsten fibres are used in metal-matrix composites, as are steel fibres in cement composites. There is increasing interest in *shape memory alloy* filaments such as Ti–Ni (Nitanol) for use in piezoelectric composites.

2.11.6 Particulate Composites

These encompass a wide range of materials, from cement reinforced with rock aggregates (concrete) to mixtures of ceramic particles in metals, called *cermets.* In all cases, however, the particulate composite consists of a reinforcement that has similar dimensions in all directions (roughly spherical), and all phases in the composite bear a proportion of an applied load. The percentage of particulates in this class of composites range from a few per cent to 70%.

One particular and important class of particulate composites in *dispersion-hardened alloys.* These composites consist of a hard particle constituent in a softer metal matrix. The particle constituent seldom exceeds 3% by volume, and the particles are very small, below micrometer sizes. The characteristics of the particles largely control the property of the alloy, and a spacing of 0.2 – 0.3 μm between particles usually helps optimize properties. As particle size increases, less material is required to achieve the desired interparticle spacing. Refractory oxide particles are often used, although intermetallics such as $AlFe_3$ also find use. Dispersion hardened composites are formed in several ways, including surface oxidation of ultrafine metal powders, resulting in trapped metal oxide particles within the metal matrix. Aluminium, nickel, and tungsten are the metals of commercial interest for dispersion-hardened alloys.

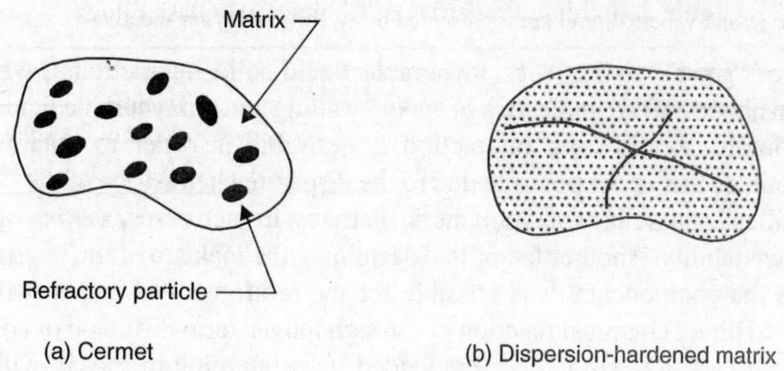

Fig. 2.99 Comparison of (*a*) cermet and (*b*) dispersion-hardened alloy.

A cermet is a particulate composite similar to a dispersion-hardened alloy, but consists of larger ceramic grains (cer-) held in a metal matrix (Fig. 2.99). The refractory particulates can be from the oxide category, *e.g.* alumina (Al_2O_3), magnesia (MgO), or thoria (ThO_2), or from the carbide category, like tungsten carbide (WC), chromium carbide (Cr_3C_2) or titanium carbide (TiC). Cremets are formed by traditional powder-metallurgical techniques, although bonding agents must sometimes be added to improve bonding between the ceramic particulate and the metal matrix. Cermets are used in a wide number of applications, including (*i*) fuel elements and control rods in the nuclear industry, (*b*) pulse magnetrons, and (*c*) cutting tools.

Finally, metal-and resin bounded composites are also classified as particulate composites. *Metal-bonded composites* included structural parts, electrical contact materials, metal-cutting tools, and magnet materials are formed by incorporating metallic or ceramic particulates such as WC, TiC, W or Mo in metal mixtures through traditional powder metullargical or casting techniques. *Resin-bonded composites* are composed of particulate fillers such as silica flour, wood flour, mica, or glass spheres in phenol formaldehyde (Bakelite), epoxy, polyester, or thermoplastic matrixes.

2.11.7 Composite Interphase

The interphase, or the region between the matrix and the reinforcement, is often the primary determinant of mechanical properties of the composites. It can have an effect on other properties as well, but since it serves to transfer loads from the matrix to the reinforcement, its primary impact is on the strength of the composite.

There are many factors that affect the composition and spatial extent of the interphase: (*i*) *Wettability,* or the ease with which the liquid matrix wets the reinforcing constituent prior to solidification. (There are other ways to form composites, of course, but incorporation of a solid particle into a liquid matrix is common and is germane to the description of wetting). The ability of a liquid to wet a solid is measured by an interfacial *contact angle* (Fig. 2.100). We may note that low contact angles

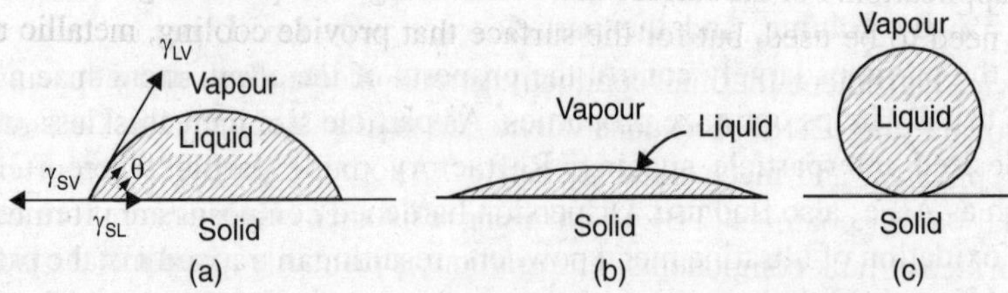

Fig. 2.100 Interaction of a liquid droplet with a solid substrate: (*a*) Partially wetting (0° < θ < 180°), (*b*) completely wetting (θ ≈ 0°), and (*c*) completely non-wetting (θ = 180°). Interfacial energies between the solid-liquid, solid-vapour, and vapour-liquid are represented by γ_{SL}, γ_{SV} and γ_{LV} respectively.

are representative of "good" wetting (*i.e.,* favourable liquid-solid interactions), whereas high contact angles (greater than about 90°) are indicative of "poor" wetting and unfavourable liquid-solid interactions. Obviously, a favourable liquid-solid interaction is desirable in order to obtain good matrix-fibre interactions. Sometimes this is not possible due to the disparate chemistries of the constituents, such as is found between silicate glass fibres and polymeric matrixes. In such cases, *wetting agents* are sometimes added to improve wettability. Another factor in determining the makeup of the interphase is the inherent reactivity between the components. It is possible for the reinforcement and matrix to react with one another, either due to direct chemical reaction or through longer-term diffusion of components from one phase to another, *e.g.* a CaO–Al_2O_3 fibre embedded in an aluminium matrix with traces of silicon. Elemental analysis of the interphase reveals that it contains silicon, which has diffused into the fibre surface during melt processing of the aluminium matrix to form an aluminium silicon oxide compound.

Despite the emphasis on favourable interactions between the matrix and reinforcement and compound formation between them, it may be beneficial in certain circumstances for the interaction between the two primary constituents to be relatively, weak. This is especially true in ceramic-ceramic composites, where both constituents are brittle, and the only way to impart some ductility on the composite is for the interphase to fail 'gracefully', *i.e.* for the fibres to actually "pull out" of the matrix in a controlled manner. Optimization of the interphase properties in advanced composites is currently the focus of research.

2.11.8 Functionally Graded Materials (FGMs)

These feature gradual transitions in microstructure and composition that impart functional performance to the component. This term came into common use in 1980s as new materials for *space plane* were being conceived and developed. It was recognized early on that the material requirements would be

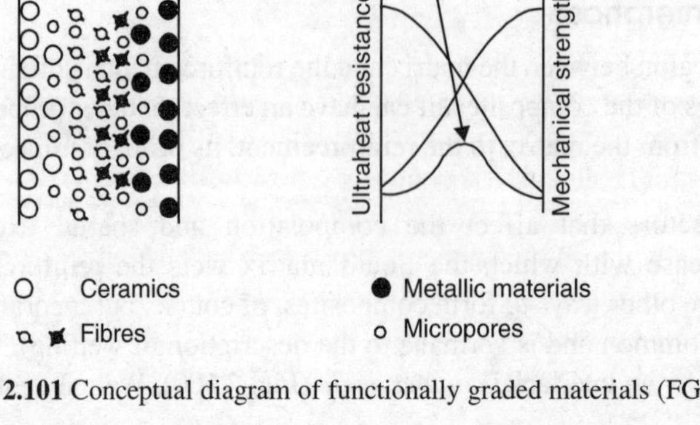

Fig. 2.101 Conceptual diagram of functionally graded materials (FGMs).

extreme for this application. For the surface that contacts high-temperature gases at thousands of degrees, ceramics would need to be used, but for the surface that provide cooling, metallic materials would be required to provide sufficient thermal conductivity and mechanical strength, *e.g.* an FGM for this application is shown in Fig. 2.101. The constituents are graded along the thickness of this composite, as are the resulting physical and mechanical properties. Another example, where FGMs are common is coatings. Thermal barrier coatings (TBCs) and wear resistance coatings are often formed in a gradient fashion on the surface of bulk components. The gradient is primarily a result of the processing technique. *Plasma vapour deposition* (PVD), Chemical vapour deposition (CVD), thermal spray processings are few of the techniques used for the FGMs.

We may note that the term '*functionally graded materials*' is relatively new, but the concept is not new. Nature has been producing FGMs for a long time. A cross-section of corn plant is shown in Fig. 2.102. The bundle sheaths on the outside of the stalk provide structural integrity to the plant. The fibre strength in the bundles is about 10 times that of the matrix. However, the fibre distribution is dense in the outer region and sparse in the inner region. Materials developed as a results of adaptation of structural elements found in nature are termed *biomimetic structures.* This area is receiving considerable attention by the researchers today.

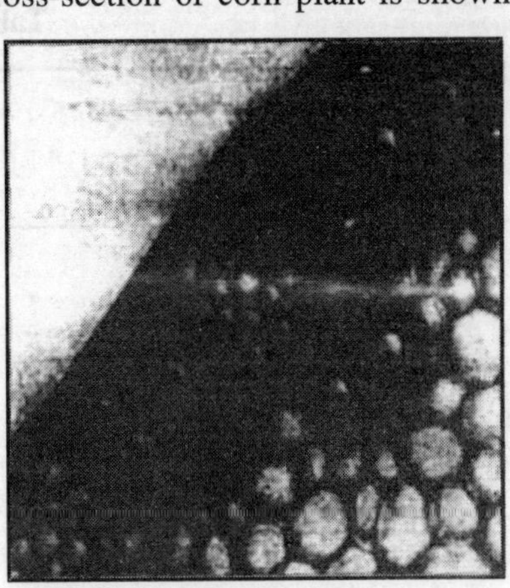

Fig. 2.102 Cross-section of corn stalk exhibiting functionally graded structure.

2.12 STRUCTURE OF BIOLOGICS

Nowadays, the boundary between the disciplines of biology and materials engineering is becoming almost blurred. Molecules that were once thought of as being biological 'macromolecules' such as *DNA* are now being used as templates to produce molecules with desired *gene* sequences, and live *tissue cultures* are being used to reconstruct human body parts similar to an auto body is repaired with fibre-glass and epoxy. Now, the scientific and engineering principles behind the application and development of biological materials are becoming more well understood.

2.12.1 Biological Molecules

(i) Amino Acids and Proteins: Proteins are the molecules that perform the functions of life. They can be enzymes that catalyze biological reactions, or they can be the receptor site on a membrane that binds a specific substance. Proteins are important part of both bones—the so-called *hard biologics* and of the *soft biologics* such as muscle and skin.

Proteins are composed of *amino acids.* Figure 2.103 shows the general structure of an amino acid. An amino acid has a carboxyl group, an amino group (Table 2.41), a central atom identified as the alpha (α) carbon, and the variable part known as the side-chain (R). Normally the amino acid is in its dissociated state, such that the terminal hydrogen on the carboxyl group moves to the amino group thereby creating a carboxylate group (COO^-) and an ammonia group (NH_3^+) known as *zwitterion.* The molecule

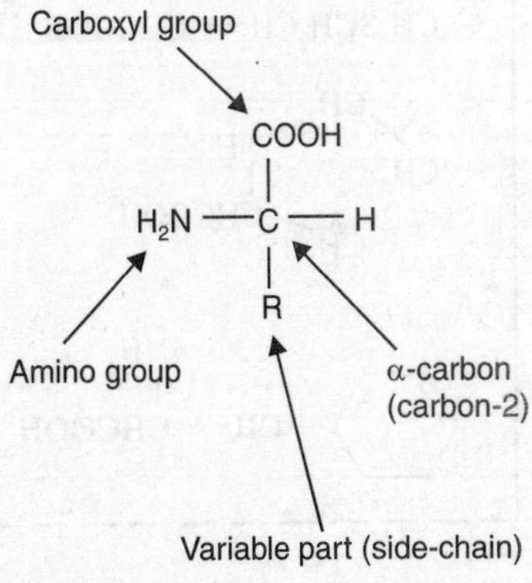

Fig. 2.103 General structure of an amino acid.

remains neutral overall and there may be several stable zwitterion forms, depending on the amino acid. We may note that the α-carbon is an asymmetric carbon, resulting in a chiral molecule for all but one of the naturally occurring amino acids. All the amino acids are found in only one of the stereoisomer configurations (the L-configuration).

There are only 20 amino acids which make up proteins (technically, 19 amino acids and one imino acid), differing only in the type of R group they contain (Table 2.41). Each amino acid has a specific name and three-letter designation. Most amino acids decompose instead of melting time to the strong intermolecular electrostatic attractions, with decomposition temperatures ranging from 185–315°C, and they are only sparingly soluble in water, with the exceptions of glycine, alanine, proline, lysine and arginine.

Table 2.41 Common Amino Acids

$R-\underset{NH_2}{CH}COOH$	*Name*	*Abbreviation*
$H-\underset{NH_2}{CH}COOH$	Glycine	Gly
$CH_3-\underset{NH_2}{CH}COOH$	Alanine	Ala
$CH_3\underset{CH_3}{CH}-\underset{NH_2}{CH}COOH$	Valine	Val
$CH_3\underset{CH_3}{CH}CH_2-\underset{NH_2}{CH}COOH$	Leucine	Leu
$CH_3CH_2\underset{CH_3}{CH}-\underset{NH_2}{CH}COOH$	Isoleucine	Ile
$CH_3SCH_2CH_2-\underset{NH}{CH}COOH$	Methionine	Met
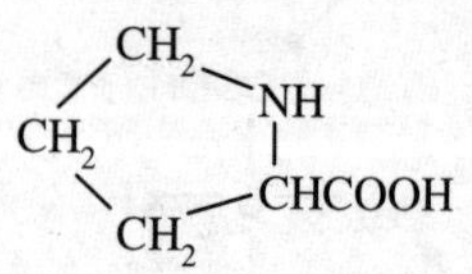	Proline	Pro
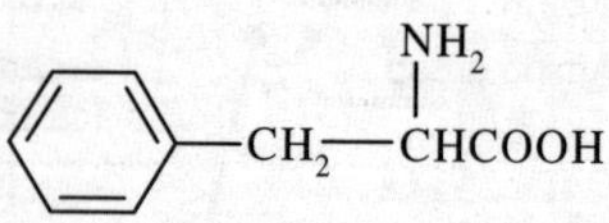	Phenylalanine	Phe

(*Contd.*)

Structure	Name	Abbreviation
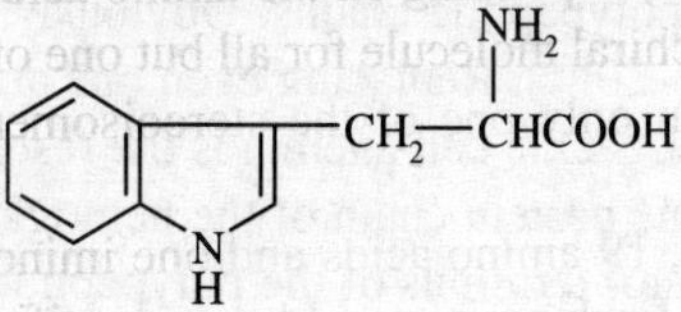	Tryptophan	Trp
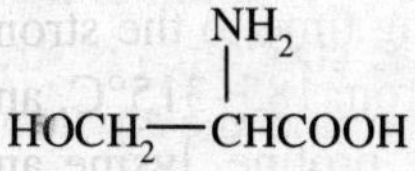	Serine	Ser
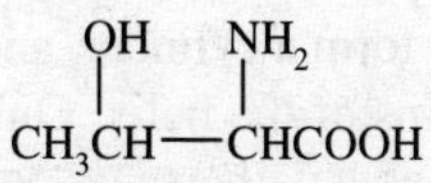	Threonine	Thr
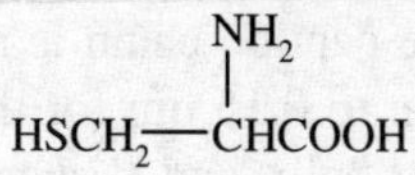	Cysteine	Cys
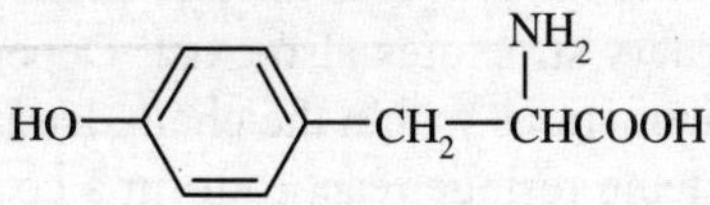	Tyrosine	Tyr
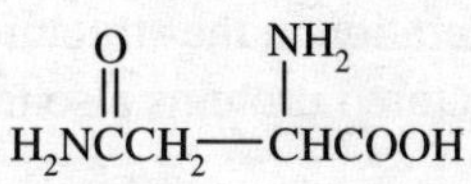	Asparagine	Asn
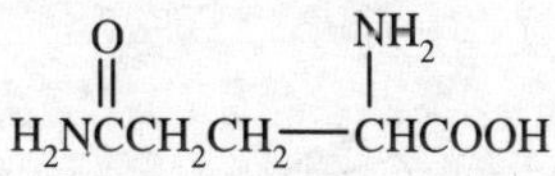	Glutamine	Gln
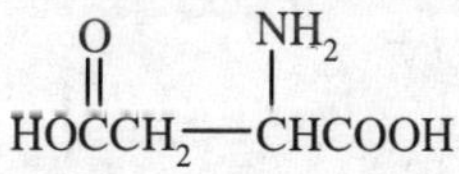	Aspartic acid	Asp
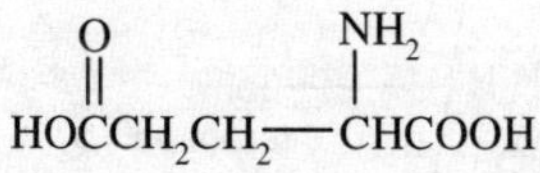	Glutamic acid	Glu
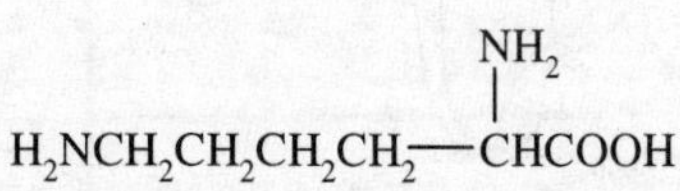	Lysine	Lys
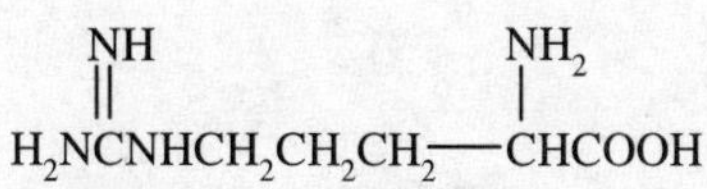	Arginine	Arg
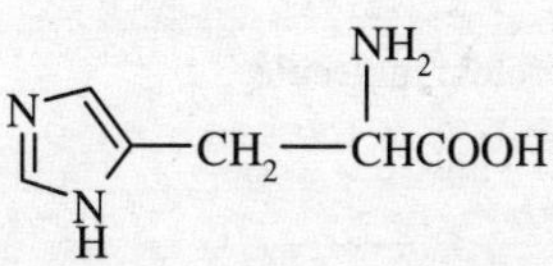	Histidine	His

To form proteins, the carboxylic acid group on one amino acid reacts with the amine group on another molecule in a condensation reaction that forms one water molecule and a—CO—NH—CHR— linkage known as a *peptide bond.* A molecule containing more than about 100 amino acid sequences is called a *polypeptide,* and a protein is composed of one (or more) polypeptide chains. Obviously, the number of possible proteins from the 20 amino acids is enormous (~20^{100}). Replacing even one amino acid in the sequence of a protein can change its function completely. Sickle cell anemia is the result of replacing only one valine amino acid with a glutamic acid unit in one protein chain of the hemoglobin molecule. The peptide sequence is named by starting at the N (-amino) terminus of the polypeptide.

With the polymers, peptides and proteins can possess complex confirmations by rotation of bonds in the backbone and the interaction between the side chains. The stability of these structures is strongly dependent upon the R groups and, hence, the specific amino acid sequence, as well as the environment in which the protein finds itself. A few of the common confirmations found in proteins are illustrated in Fig. 2.104. The *α-helix* occurs when the chain coils like a right-hand screw to form a cylinder, and it is the result of hydrogen bonding between the C═O and N—H in adjacent turns of the helix. Only the right-handed α-helix occurs in nature, and its presence results in an electric dipole with excess positive charge at one end and excess negative charge at the other. In the β-sheet, the peptide chain is much more extended, with 0.35 nm between adjacent peptide groups, in comparison to 0.15 nm for the α-helix. These sheet structures are also the result of hydrogen bonding. The α-helix and β-sheet are examples of *secondary structure* in proteins. Tertiary and quaternary structures also exist. Depending on the nature of the side-chains, there can also be hydrophobic interactions within the chain, leading to chain extension. Finally, disulfide bond can occur when two cysteine residue react to form a covalent S–S–bond. The breaking of disulfide bonds, or any action that leads to an alteration in the structure of a protein as to render it anactive, leads to *denaturation.* We may note that denatured proteins also tend to have decreased solubility.

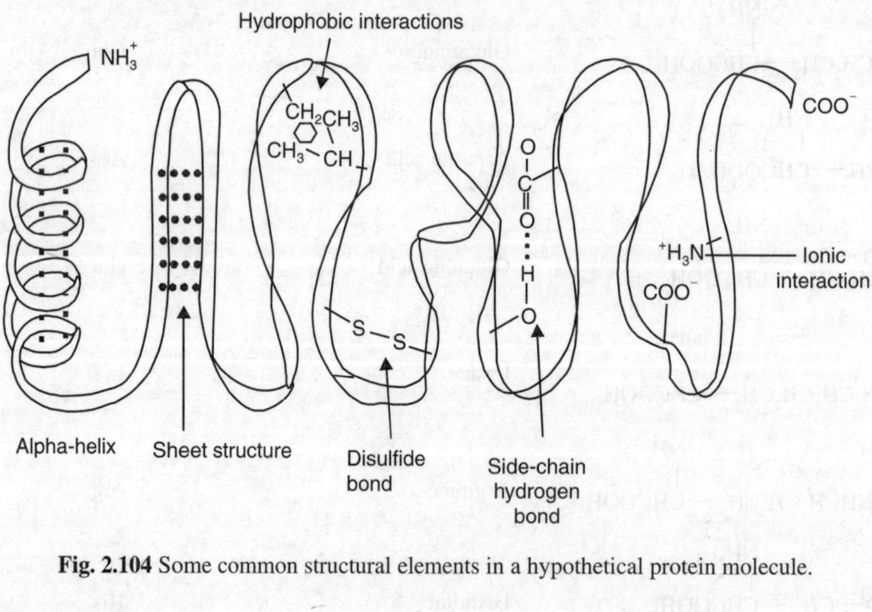

Fig. 2.104 Some common structural elements in a hypothetical protein molecule.

2.12.2 DNA and RNA

Like proteins, *deoxyribonucleic* acid (DNA) and ribonucleic acid (RNA) are polymers, but instead of amino acids as repeat units, they are composed of a chain of nucleotides. Each nucleotide is composed of three basic structural units: a *base,* a *sugar* and a *phosphate* group (Fig. 2.105) one base with its sugar (and without the phosphate group) is called a nucleoside. The sugar in RNA is called *ribose,* which is

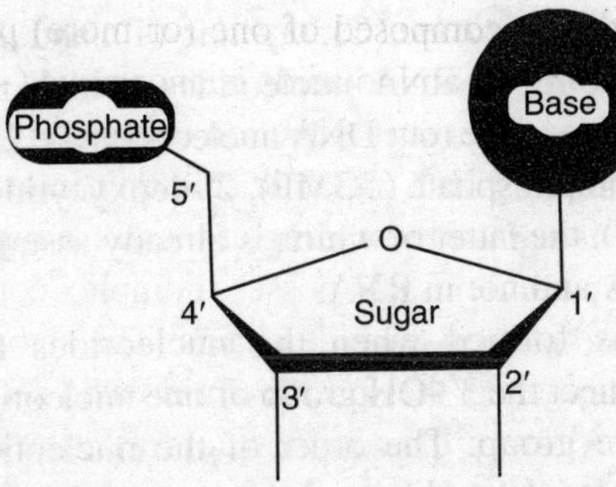

Fig. 2.105 The chemical structure of DNA and RNA nucleotides.

reduced in DNA by a loss of oxygen at the 2′ carbon to form deoxyribose (Fig. 2.106). There are only five primary bases found in polynucleotides: two purines represented by *adenine* (A) and *guanine* (G);

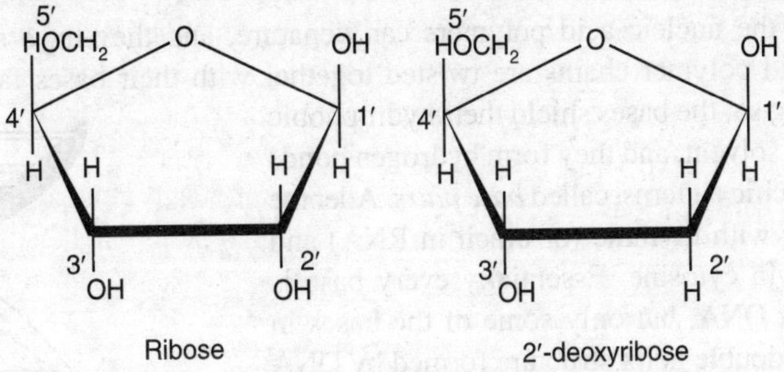

Fig. 2.106 The chemical structure of the sugars in RNA (ribose) and DNA (deoxyribose).

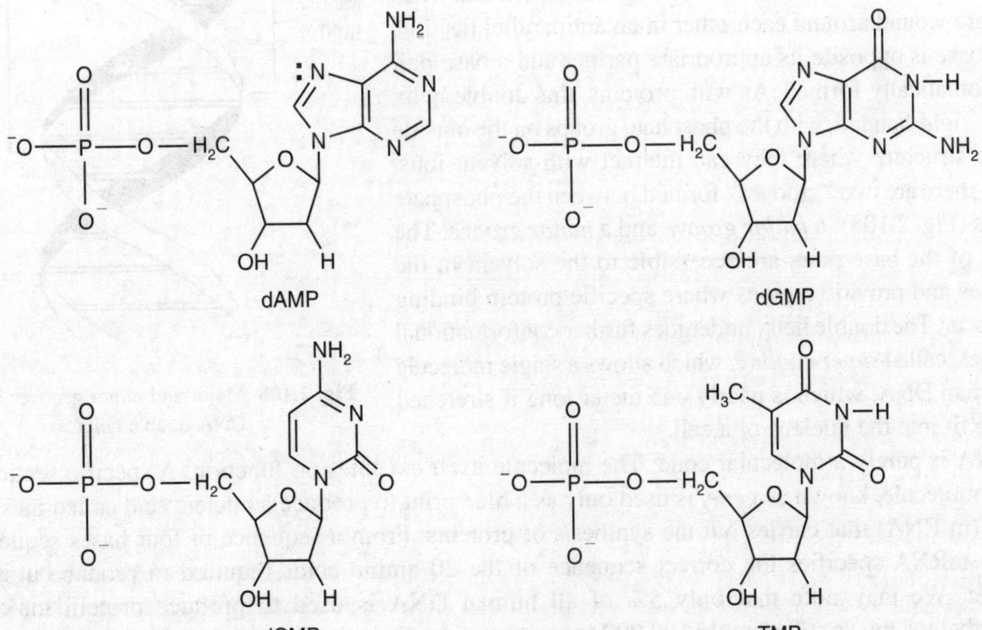

Fig. 2.107 The four DNA nucleotides.

and three pyrimidines represented by *cytosine* (C), *thymine* (T), and *uracil* (U). Thymine is found only in DNA nucleotides, and uracil only in RNA nucleotides, which results in four DNA nucleotides (Fig. 2.107) and four RNA nucleotides. The four DNA nucleotides are 2′-deoxyadenosine monophosphate (dAMP), 2′-deoxy guanoside monophosphate (dGMP), 2′-deoxycytidine monophosphate (dCMP), and thymidine monophosphate (TMP), the latter of which is already assumed to have 2′-deoxyribose as the sugar since it occurs only in DNA and not in RNA.

The nucleic acid polymer is formed when the nucleotides attach to one another through phosphodiester bonds, which connect the 3′–OH group of one nucleotide to the 5′–OH group of another nucleotide through the phosphate group: The order of the nucleotides in the chain is the primary structure of the DNA or RNA molecule, and it can be represented in short hand notation with only the base pair designation (A, G, C, T or U). Some sequences are highly repetitive, such as Alu sequence, which is a sequence of 1 2 3 base pairs that occurs millions of times in the chain. Some sequences are specific binding sites for certain proteins, as in the TATAA sequence found near the start of many DNA molecules.

As with proteins, the nucleic acid polymers can denature, and they have secondary structure. In DNA, two nuclei acid polymer chains are twisted together with their bases facing inward to form a *double helix.* In doing so, the bases shield their hydrophobic components from the solvent, and they form hydrogen bonds in one of only two specific patterns, called *base pairs.* Adenine hydrogen bonds only with thymine (or uracil in RNA) and guanine pairs only win cytosine. Essentially every base is part of a base pair in DNA, but only some of the bases in RNA are paired. The double helix structure formed by DNA is comprised not only of two nucleic acid polymer chains, but the chains have complementary sequences such that when they are wound around each other in an antiparallel fashion, each base is opposite its appropriate partner and a base pair is automatically formed. As with proteins, this double helix is also right-handed, with the phosphate groups on the outside of the structure where they can interact with solvent ions. Thus, there are two "grooves" formed between the phosphate chains (Fig. 2.108): a *major groove* and a *minor groove.* The edges of the base pairs are accessible to the solvent in the grooves and provide regions where specific protein binding can occur. The double helix undergoes further conformational changes, called *super cooling,* which allows a single molecule of human DNA, which is nearly one meter long if stretched out, to fit into the nucleus of a cell.

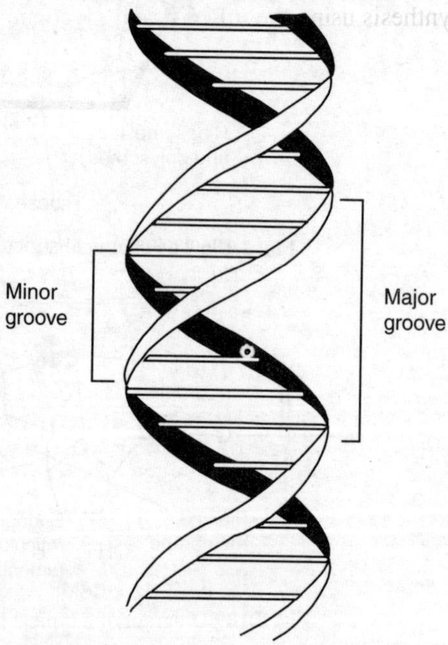

Fig. 2.108 Major and minor grooves in the DNA double helix.

DNA is purely a molecular code: The molecule itself executes no function. A specific section of a DNA molecule, known as *gene,* is used only as a blue print to produce a nucleic acid called messenger RNA (m-RNA) that carries out the synthesis of proteins. From a sequence of four bases sequence in DNA, mRNA specifies the correct sequence of the 20 amino acids required to produce a specific protein. We may note that only 5% of all human DNA is used to produce protein molecules. Nevertheless, this is still roughly 100,000 gene sequences. There is currently an effort underway to map all of human genes, called the Human Genome Project.

2.12.3 Cells

Cells and assemblies of molecules enclosed within a *plasma membrane* that carry out specific functions. The human body contains over 10^{14} cells, all of which take in nutrients, oxidize fuels, and excrete waste products. Despite their varied functions, all cells have a similar internal organization.

Surrounding the outside of all cells is the *plasma membrane* (Fig. 2.109). It is composed primarily of *lipids* and is selectively permeable, limiting the exchange of molecules between the inside and outside of the cell. The outside of the plasma membrane contains all the carbohydrates and receptor sites. The *cytoplasm* includes everything inside the plasma membrane except for the nucleus. Energy is generated in the cell by the *mitochondria,* where cellular fuel is oxidized. There may be several mitochondria per cell. The *Golgi complex* is composed of parallel membrane sacs and is used to secrete proteins. The *endoplasmic reticulum* (ER) is a system of membranes that store, segregate, and transport substances within the cell. The ER is continuous with the Golgi complex and the nuclear membrane. The nuclear membrane, or envelope, surrounds the nucleolus, which contains RNA, as well as DNA and other proteins. Normally, the DNA is a diffuse tangle of fine threads called *chromatin.* The rest of the cytoplasm is composed of lysosomes which degrade nuclei acids, proteins, and complex carbohydrates; peroxisomes, which contain a variety of oxidation enzymes; *polysomes,* which are engaged in protein synthesis using the mRNA; and glycogen, which is a polymer of glucose, used for energy.

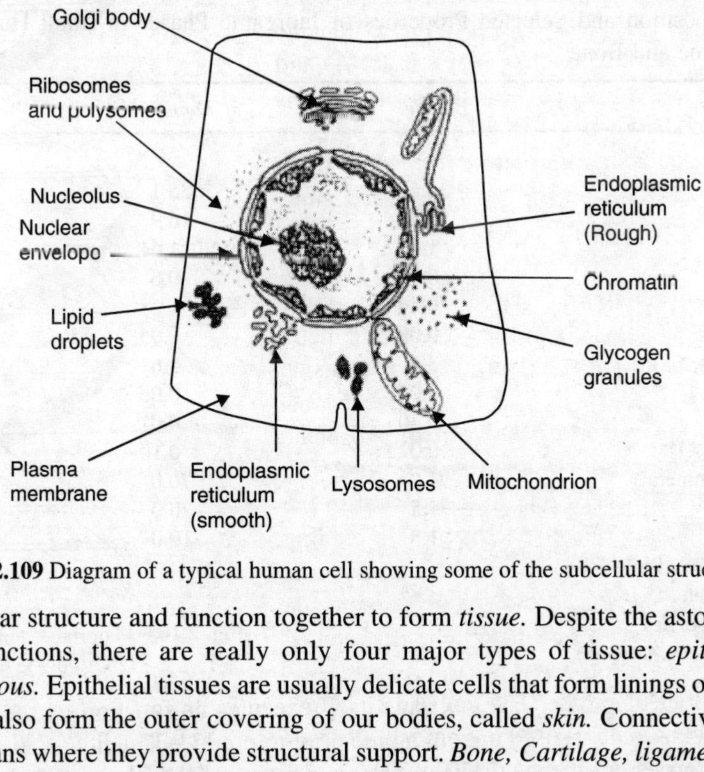

Fig. 2.109 Diagram of a typical human cell showing some of the subcellular structures.

Cells with similar structure and function together to form *tissue.* Despite the astounding diversity of cell types and functions, there are really only four major types of tissue: *epithelial, connective, muscular* and *nervous.* Epithelial tissues are usually delicate cells that form linings of internal structures and organs. They also form the outer covering of our bodies, called *skin.* Connective tissues are found in the walls of organs where they provide structural support. *Bone, Cartilage, ligaments* and tendons are all types of connective tissue. Muscle tissue unables the body to move, and it is characterized by its ability to contract. Nervous tissue is composed of highly specialized cells called *neurons,* and it is characterized by its ability to translate stimulation into electrochemical nervous impulses. The region between cells in a tissue is equally important. It is usually termed the *extracellular matrix,* and it contains proteins and ions that perform vital functions. The extracellular matrix components have a profound effect on how well foreign materials are accepted (or rejected) by the body. Biological materials can also be generalized as either "hard" or "soft" materials.

2.12.4 Hard Biologics

Though most (by volume) of the human body is composed of soft materials such as skin and muscle, one would have limited abilities to eat, move and protect ourselves were it not for hard biologics such as bone and teeth.

The delineation between hard and soft biologics is not always clear, such as in the case of connecting tissue between bone and muscle called tendons, but for our purposes, hard biologics contain some significant fraction of inorganic components. For both bone and teeth, the primary inorganic constituent belongs to the *calcium phosphate* family, $CaO–P_2O_5$. Actually, most calcium phosphates found in biological environments have water associated with them, so that the calcium phosphates belong to the ternary system $CaO—P_2O_5—H_2O$. The specific calcium phosphate phase depends on such factors as temperature and pH of the environment, as well as the chemical reactivity of that phase with its surroundings. The most prevalent form of calcium phosphate in human body is *hydroxyapatite* (sometimes called *hydroxyl* apatite and abbreviate HA and HAp), $Ca_{10}(PO_4)_6(OH)_2$. Hydroxyapatite can occur both naturally and synthetically and has a hexagonal crystal structure (Fig. 2.110), with

Table 2.42 Composition and Selected Properties of Inorganic Phases in Adult Human Enamel, Dentine and Bone

	Enamel	*Dentine*	*Bone*
Composition[a]			
Calcium, Ca^{2+}[b]	36.5	35.1	34.8
Phosphorus, as P	17.7	16.9	15.2
(Ca/P) molar[b]	1.63	1.61	1.71
Sodium, Na^{+}[b]	0.5	0.6	0.9
Magnesium, Mg^{2+}[b]	0.44	1.23	0.72
Potassium, K^{+}[b]	0.08	0.05	0.03
Carbonate, as CO_3^{2-}[c]	3.5	5.6	7.4
Fluoride, F^{-}[b]	0.01	0.06	0.03
Chloride, Cl^{-}[b]	0.30	0.01	0.13
Pyrophosphate, $P_2O_7^{4-}$	0.022	0.10	0.07
Total inorganic (mineral)	97.0	70.0	65.0
Total organic[d]	1.5	20.0	25.0
Absorbed H_2O	1.5	10.0	10.0
Trace elements: Sr^{2+}, Pb^{2+}, Zn^{2+}, Cu^{2+}, Fe^{3+}, etc.			
Crystallographic properties			
Lattice Parameters (± 0.003 Å)			
a axis	9.441	9.42	9.41
c axis	6.880	6.88	6.89
Crystallinity index[e]	70–75	33–37	33–37
Crystallite size (average), Å	1300 × 300	200 × 40	250 × 30
Ignition products (800°C)	β–TCP + HAp	β–TCP+ HAp	HAp + CaO

[a] Weight % [b] Ashed sample [c] Unashed sample, IR method

[d] Principal organic component: enamel, noncollagenous; dentine and bone, collagenous.

[e] Calculated from ratio of coherent/incoherent scattering, mineral, HAp = 100.

Ca^{2+} and $(PO_4)^{3-}$ ions arranged about columns of $(OH)^-$ ions. Other biologically calcium phosphates include dicalcium phosphate, $CaHPO_4$, and tricalcium phosphate (TCP), $Ca_3(PO_4)_2$. Table 2.42 gives the remaining chemical constituents in teeth and bone.

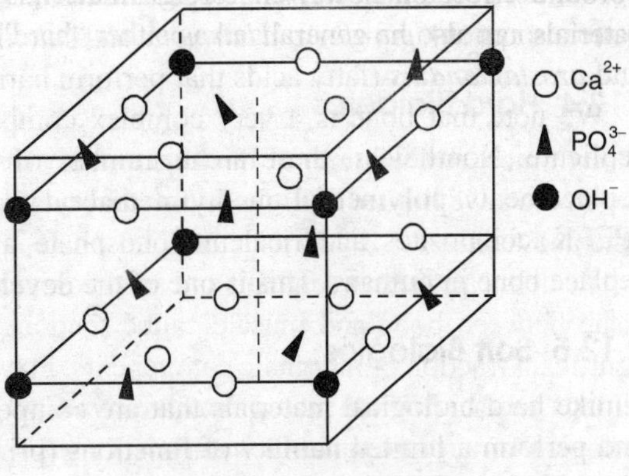

Fig. 2.110 The unit cell perspective of hydroxyapatite.

Figure 2.111 shows the primary components of a human tooth. The outer surface of a tooth is composed entirely of *enamel.* Enamel is a smooth, white, semi-transparent material that varies in thickness from 0.01 mm to about 2.5 mm over the grinding surfaces of molars. It is extremely hard and brittle and not very strong. The main constituent inorganic phase (95 wt%) is HA, with an organic phase (1 wt%) made up of protein, carbohydrates, lipids, and other matter. The remaining phase (about 4 wt%) is water, mostly present as a shell surrounding the HA crystallites. *Dentin,* the more important part of the tooth structure is covered by the enamel in the crown and the *cementum* in the root. Dentin is a calcified mesh of collagen fibrils surrounding, the cellular pathways of the dentin-forming cells. The resultant tubular structure contains about 50% HA (but the crystal size is smaller and the orientation different from that in enamel), 32% collagen, 8% mucopolysaccharides and 10% water.

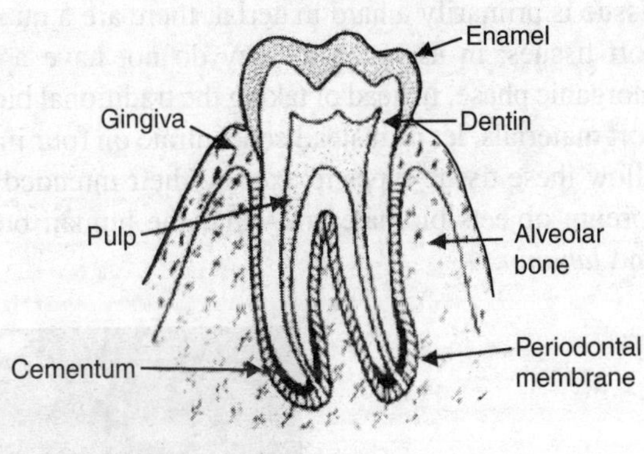

Fig. 2.111 Schematic diagram of a human tooth.

Bone is also a composite material composed of about 70% HA, as well as collagen, but in contrast to enamel, bones also contain blood vessels that help the bone regenerate itself (Fig. 2.112). The collagen fibres are a bundled array of cross linked helical peptides. Bone tissues replaces itself through the action of cells called osteoclasts. Osteoclasts produce acids that reabsorb HA and breakdown collagen. The resulting calcium and proteins cause other cells called *osteoblasts* to lay down new matrix that mineralizes and forms HA and collagen. Bone cells produce growth factors, such as bone *morphogenetic proteins,* to increase or decrease bone regrowth. The blood vessels navigate through the

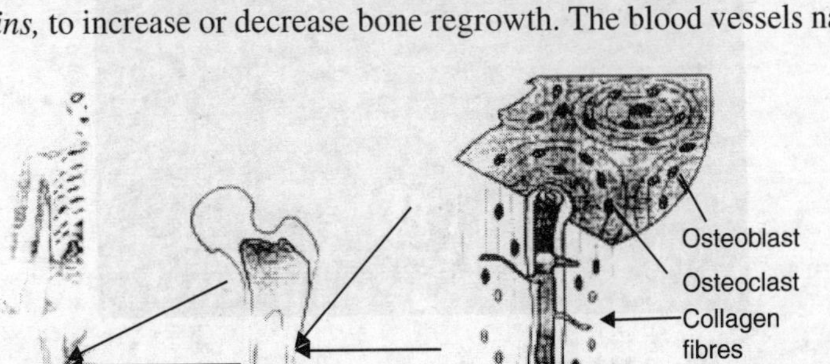

Fig. 2.112 The structure of human bone.

porous bone structure, carrying important compounds such as *calcitonin* (a thyroid harmone that prevents bone resorption), *parathyroid harmone* and *calcitrol* (which regulate calcium and phosphate metabolism), and *prostaglandins* (fatty acids that perform harmone-like functions).

We note that bone is a very complex composite material and, as such, is difficult to artificially replicate. Nontheless, there are a number of materials of all types that are being used as bone replacements polymethyl methylmethacrylate (PMMA), titanium, graphite/polyethyletherketone (PEEK) composites, and tricalcium phosphate, among numerous others, have all been used to repair or replace bone in humans. This is one of the developing area of engineering materials.

2.12.5 Soft Biologics

Unlike hard biological materials that are composed primarily of one compound (calcium phosphate) and perform a limited number of functions (primarily structural), soft biological materials are highly diversified, performing a myriad of highly specialized and/or combinatorial functions. We have classified tissue into four general categories (muscle, nervous, epithelial and connective). Although connective tissue is primarily a hard material, there are a number of connective tissues that are more like the other soft tissues, in as much as they do not have a significant fraction of the calcium phosphate-based inorganic phase. Instead of taking the traditional biology-based approach to classifying and characterizing soft materials, let us instead concentrate on four important proteins found in the extracellular matrix that allow these tissue types to execute their intended functions, as well as control our ability to introduce foreign objects-biomaterials—into the human body. These proteins are *collagen, elastin, fibronectin* and *laminin.*

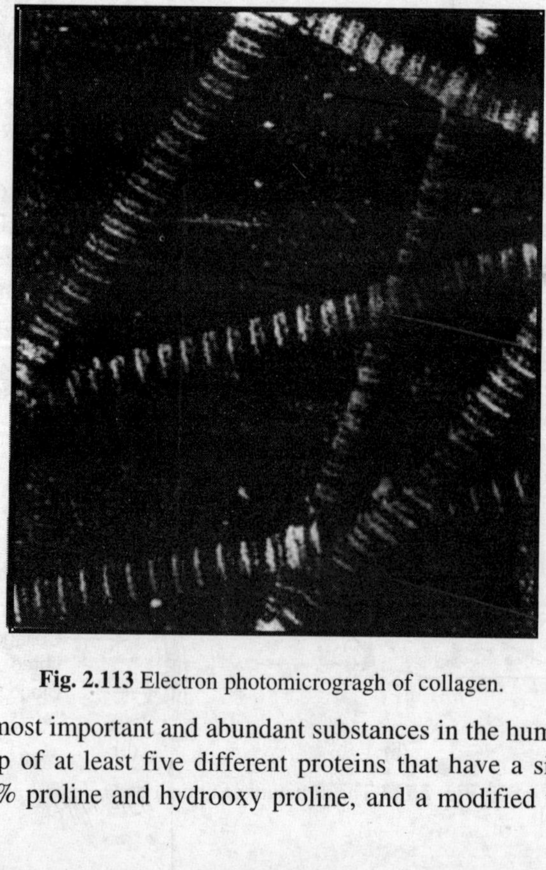

Fig. 2.113 Electron photomicrogragh of collagen.

Collagen is one of the most important and abundant substances in the human body. It is not a single protein, but rather a group of at least five different proteins that have a similar structure. Collagen contains 30% glycine, 20% proline and hydrooxy proline, and a modified version of hydroxylysine.

The secondary structure of collagen in a triple helix (Fig. 2.113), but not an α-helix, because the high proline content prevents the formation of the α-helix. The three chains in the helix may be identical or different. There are at least 10 different types of collagen found in connective tissue (Table 2.43), with types I–III having the ability to form fibres called *fibrils*. Type I collagen is the principal structural component of most tissue. Types II and III collagen are found in cartilage (II) and cardiovascular tissue (III), among other places. Type IV collagen is found in basement membranes, which are sheet-like structures found beneath epithelial cells or blood vessel linings (endothelial cells).

Table 2.43 Types of Collagen Found in Tissue[a]

Collagen type	*Tissue or Organ*	*Location*
I	Tendon, skin, bone and fascia	Thick extracellular fibrils and fibres
II	Cartilage	Thin fibrils around cartilage cells
III	Cardiovascular tissue	Intermediate-size extracellular fibrils
IV	Basement membranes	Network-forming component
V	Tendon, skin and cardiovascular tissue	Pericellular matrix around cells
VI	Cardiovascular tissue, placenta, uterus, liver, kidney, skin, ligament and cornea	Extracellular matrix
VII	Skin	Anchoring fibrils
VIII	Cardiovascular tissue	Around endothelial cells
IX	Cartilage	Extracellular matrix
X	Cartilage	Extracellular matrix
1α, 2α, 3α	Cartilage	Extracellular matrix

Elastin is a protein also found in connective tissue that imparts an ability for these tissues to undergo large shape and size changes without permanent damage to the tissue. Elastin is the amorphous component (up to 90%) of elastic fibres that are found in the extracellular matrix of most tissue, *e.g.* the arota has an elastin content as high as 30–60% of the dry weight-higher than that of any other tissue. Skin, in contrast, contains less than 5% elastin by weight. Elastin is similar to collagen in that it is composed of about 33% glycine and 13% proline, but it contains no hydroxy proline, and contains 10–14% valine and 21–24% alanine. The later two amino acids are non-polar and do not form hydrogen bonds with water molecules. There are a number of cross links between the elastin chains, as one would expect for an elastic substance.

Fibronectin and Laminin: These are adhesion proteins responsible for linking the outer surface of cells to collagens and other components in the extracellular matrix. Both fibronectin and laminin are glycoproteins-proteins that contain polysaccharide residues. Fibronectin binds to cells through a *tripeptide sequence* (—Arg—Gly—Glu) called RLD, which binds to *fibronectin receptor,* one of the family of cellular transmembrane proteins called intergrins (Fig. 2.114). Intergrins link the extracellular matrix to the cytoskeleton.

Smart Biomaterials: With the development of synthetic polymer systems in the past few decades that biomaterials have begun to find broad applications in modern medicine. A new wave of advances in cell biology, chemistry and material science is enabling the production of a new generation of smart biomaterials.

Minute nanofibres of various structures and chemistries are formed through simple self-association and organisation of peptides and proteins. Several bioactive extracellular protein domains have been identified that can be incorporated as small peptides into nanofibres through simple modification of the peptide amino acid sequence. Nanofibres can be designed to present these peptide sequences at high density. Three dimensional (3D) macroscopic gel-like solids can also present high densities of such bioactive peptides.

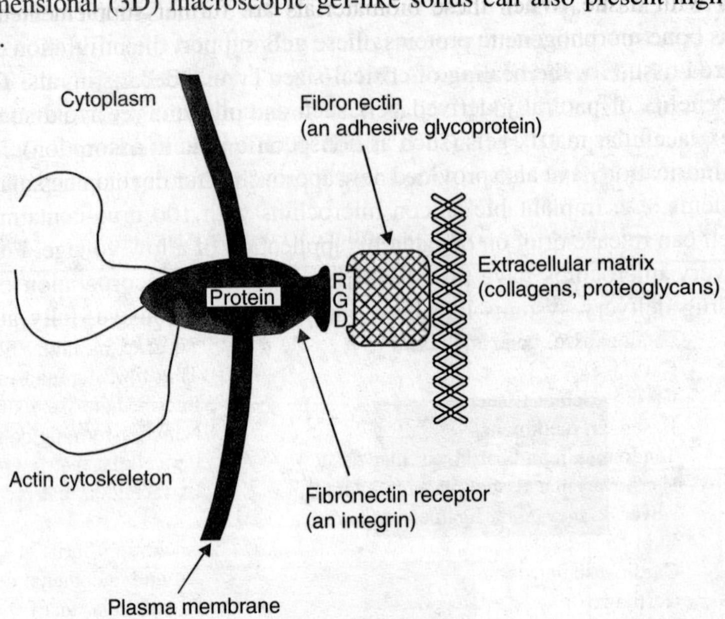

Fig. 2.114 The function of integrin.

Applying molecular self-assembly, Silva *et al.* (Science 303, 1352, (2004)) report a new 3D material capable of directing the differentiation of neutral progenitor cells to a specific lineage without the help of growth factors (see Fig. 2.115). Stem and progenitor cells have the ability to differentiate into derivative tissues and have great potential for tissue repair or replacement. Typically, differentiation is controlled by soluble compounds such as growth factors. Silva *et. al.* synthesized self-assembling peptide amphiphiles that present the pentapeptide eqitope isoleucine-lysine-valine-alanine-valine (IKVAV). IKVAV is an amino acid sequence found in laminin, which promotes neurite adhesion, sprouting, and growth. These peptide amphiphiles self-assemble in aqueous media to form nanofibres with diameters of 5 to 8 nm and lengths several orders of magnitude higher. Macroscopically, these intermeshed fibres form highly hydrated 3 D gels that are able to direct the rapid differentiation of encapsulated neural progenitors into neutrons while discouraging the production of astrocytes. The inhibition of astrocyte proliferation may prevent glial scar formation, which inhibits the regeneration and elongation of axons after central nervous system trauma.

Although the discovery of specific bioactive peptides has enabled the rational design of materials with the ability to control cell behaviour, it is often unclear which chemical properties are necessary to provide this control. A high-through put synthesis and screening platform for the testing of polymer-cell interactions can accelerate the discovery of such materials (Fig. 2.112b). Researchers have screened a library of polymers synthesized in nonliter volumes for their effects on human embryonic stem (hES) cell growth and differentiation. There were numerous unexpected interactions: some materials supported high levels of hES cell differentiation into epithelial-like cells and others supported hES cell growth only in the absence of certain growth factors.

In addition to biometers that direct specific cellular behaviours. researchers are also developing smart biomaterials that respond to specific cellular signals. Hydrogels containing both *matrix metallo proteinase* (MMP) degradeble sites and tethered adhesive ligands have been manfactured. The presence of these MMP sites allows native cells to control gel remodeling such that these cells replace the synthetic gel material with tissue. When these biomaterials are further supplemented with specific growth factors, such as bone morphogenetic proteins, these gels support the infiltration of cells and the formation of mineralized tissue for the healing of critical-sized cranial defects in rats. These materials exhibit many of the benefits of naturally derived gels such as collagen; yet avoid some undesirable properties of natural extracellular matrix gels (such as nonspecific protein adsorption).

Advances in micro fabrication have also provided new approaches for developing smart biomaterials and drug delivery systems, *e.g.* implantable silicon microchips with 100 drug-containing wells have been created. Each well can release drug on demand by application of a low voltage. Fully degradable versions of drug delivery microchips have also been developed. The incorporation of sensors into computer-controlled drug delivery systems like these may lead to responsive, fully automated drug therapies.

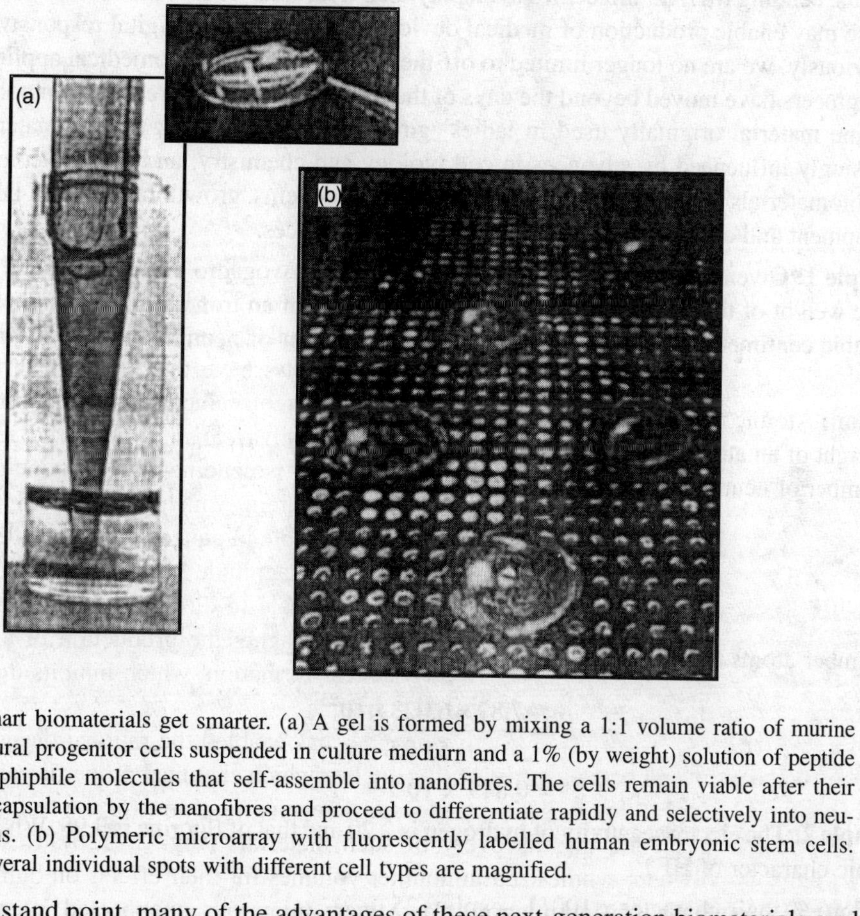

Fig. 2.115 Smart biomaterials get smarter. (a) A gel is formed by mixing a 1:1 volume ratio of murine neural progenitor cells suspended in culture mediurn and a 1% (by weight) solution of peptide amphiphile molecules that self-assemble into nanofibres. The cells remain viable after their encapsulation by the nanofibres and proceed to differentiate rapidly and selectively into neurons. (b) Polymeric microarray with fluorescently labelled human embryonic stem cells. Several individual spots with different cell types are magnified.

From a clinical stand point, many of the advantages of these next-generation biomaterials are lost if the material cannot be implemented correctly, *e.g.* proper contact between bone and implant is extremely important for the apposition and integration of bone tissue. To this end, biomaterials are being developed to allow easy application in one form (such as a liquid) that is rapidly converted into another (a solid or gel) at the appropriate destination. Few examples of this are biomaterials that solidify in vivo with exposure to light, temperature, or pH changes. Biodegradable shape-memory polymers are also underdevelopment. These materials are composed of at least two separated phases, each with a different thermal transition temperature. These distinct phase transition temperatures allow the material to 'memorize' a permanent shape at body temperature that can be substantially different from a temporary shape at room temperature. Materials such as these may be greatly simplified the use of biomaterials in surgical procedures, as they can be implanted at room temperature in a minimally invasive form (*e.g.* laparoscopically) and then expand to a final shape after reaching body temperature.

In addition to materials with responsive bulk properties, researchers are also developing *smart surfaces, e.g.,* a surface that can reversibly switch between hydrophilic and hydrophobic in response to an electric potential. A key design feature of this surface is the carefully controlled intermolecular spacing of a molecule with a negative terminus (16-mercapto) hexadecanoic acid (2-chlorophenyl) diphenylmethyl ester (MAHE) —on a gold surface. With the proper spacing, the surface responds to negative electric potential by extending itself to display the hydrophilic, negatively charged terminus. Upon application of a positive of the positive charge, the gold surface attracts the negatively charged terminus, bending MHAE molecules to display their hydrophobic moieties. In the future, coatings such as these may enable production of medical devices and sensors with digital responsive surfaces.

Obviously, we are no longer limited to off-the shelf materials for biomedical applications. Scientists and engineers have moved beyond the days of the first artificial heart created from poly-etherurethanes, the same material originally used in ladies' girdles. One may expect that biomaterials will become increasingly influenced by advances in cell biology and chemistry, and that the combination of these small biomaterials with biosensers, new drug delivery systems, growth factors, and DNA will boost the development and clinical application of new medical devices.

Example 1: Given that the density of iron is 7.87 g/cm^3, Avogadro's number (N_A) = 6.023×10^{23} and atomic weight of the iron in 55.85. Calculate the weight of an iron atom, the number of iron atoms in one cubic centimetre of solid iron at 25°C and the number of neutrons present in an atom of iron. [B.E.]

Solution: Atomic number of iron = 26

Weight of an atom of iron = $55.85/6.023 \times 10^{23} = 0.927 \times 10^{-22}$g

Number of neutrons in an atom of iron

$$= \frac{0.927 \times 10^{-22}}{1.673 \times 10^{-24}} - 26$$

$$= 29.4$$

∴ Number atoms of iron in one cm^3

$$= \frac{7.87 \times 6.023 \times 10^{23}}{55.85}$$

$$= 0.849 \times 10^{23}$$

Example 2: The electronegativity of hydrogen is 2.20 and that of fluorine is 3.98. What is the percentage of ionic character of HF?

Solution: % ionic character = $100\{1 - \exp[-0.25\,(\chi_A - \chi_B)^2$

∴ % ionic character of HF = $100[1 - \exp\{-0.25(2.20 - 3.98)^2\}] = 55\%$

Example 3: Figure 2.116 shows the force separation curves for two crystalline materials. Out of these two materials, one is weakly bonded, while the other is strongly bonded. Which one of these two materials will have the higher values of Young's modulus? [B.E.]

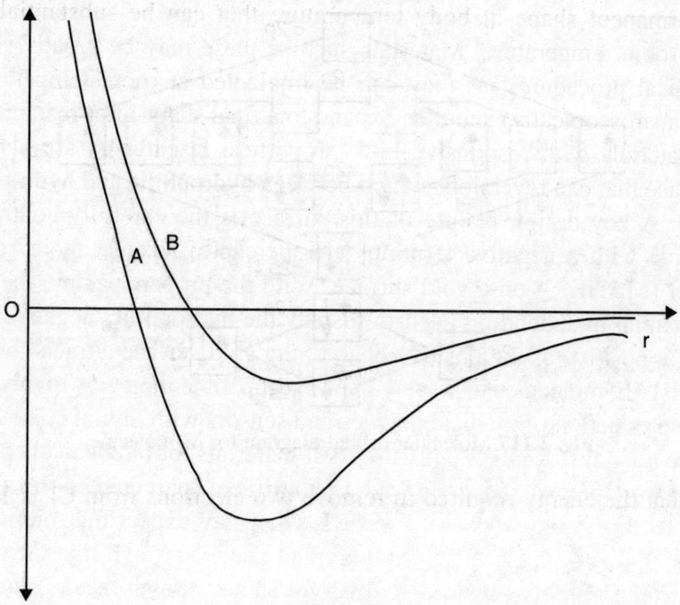

Fig. 2.116

Solution: From figure it is clear that the force-atomic separation curve is steeper at the equilibrium separation for A and hence it will have the higher value of Young's modulus.

One can approximately represent the body force between atoms by the following general equation:

$$F = \frac{A}{r^M} - \frac{B}{r^N} \quad \text{...(1)}$$

where r is the centre to centre spacing between atoms and M, N, A and B are constants. The first term represents the attractive force and the second term the repulsive force. At $F(r) = 0$, these are equal, *i.e.* atoms are at their equilibrium space, r_o. One can express the equilibrium spacing r_o in terms of the said constants as

at $$r = r_o, F = 0$$

i.e. $$\frac{A}{r_o^M} = \frac{B}{r_o^N}$$

or $$r_o^{N-M} = \frac{B}{A} \text{ or } r_o = \left[\frac{B}{A}\right]^{\frac{1}{N-M}}$$

Example 4: Construct the molecular orbital diagram for diatomic nitrogen (N_2). Is this molecule of N_2 paramagnetic or diamagnetic?

Solution: Figure 2.117 shows the molecular orbital diagram for N_2 molecule. Obviously, diamagnetic nitrogen has all paired electrons and hence it is *diamagnetic.*

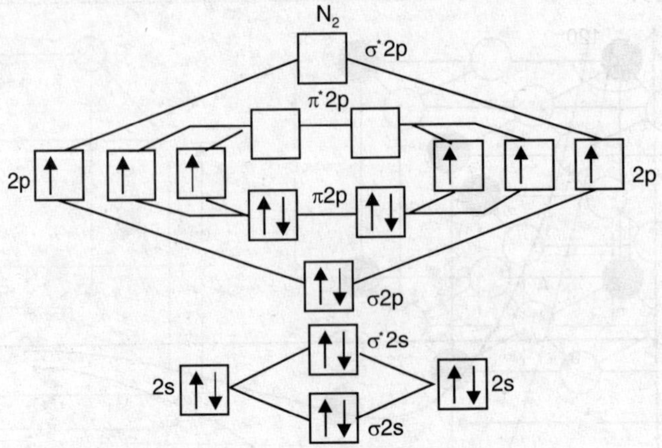

Fig. 2.117 Molecular orbital diagram for N_2 molecule.

Example 5: Show that the energy required to remove two electrons from Cl^- is 1255 kJ/molecule: [B.E.]

Solution: We have

$$Cl^- + 3.61 \text{ eV} \rightarrow Cl + e^-$$
(electron affinity energy)

Now energy required = 3.61 × 96.49 = 348 kJ/mole

Further $Cl + 13.01 \text{ eV} \rightarrow Cl^+ + e^-$
(ionisation energy)

∴ Energy required = 13.01 × 96.49 = 1255 kJ mole

Obviously, a lot of energy is needed to produce Cl^+. We may note that to produce Cl^- energy is given off.

Example 6: The structure of naphthalene is a double benzene ring with two carbons of each ring in common. (*a*) How many delocalized electrons are there? (*b*) How many wave patterns (electron states) are available to the delocalized electrons? [B.E.]

Solution: (*a*) Formula of Naphthalene is $C_{10}H_8$ (Fig. 2.118) valence electron balance

Total valence electrons:

$$(4/C)(10C) + (1/H)(8H) = 48$$

Electrons in bonds: C–H = 16

C–C(σ) = 22

∴ Remainder (delocalized) = 10
(48 – 38)

(*b*) Electron states = Number of carbon atoms = 10

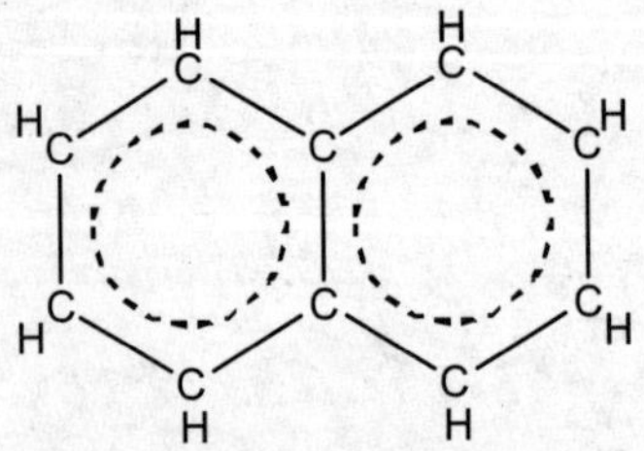

Fig. 2.118 Naphthalene. A carbon is located at each hexagon corner.

Example 7: Indicate the positions of the effective atoms of Al and Ti in unit cells of (*a*) Ti_3Al and (*b*) TiAl, Ti_3Al in the ordered form possesses a hexagonal structure with a_0 = 5.782 Å and c = 4.6885 Å at stoichiometry and TiAl (γ) phase stable from 49 to 66 at % Al, is considered to be ordered upto its melting point having FCT structure with a = 4.005 Å and c = 4.07 Å at stoichiometry.

Solution: Fig. 2.119.

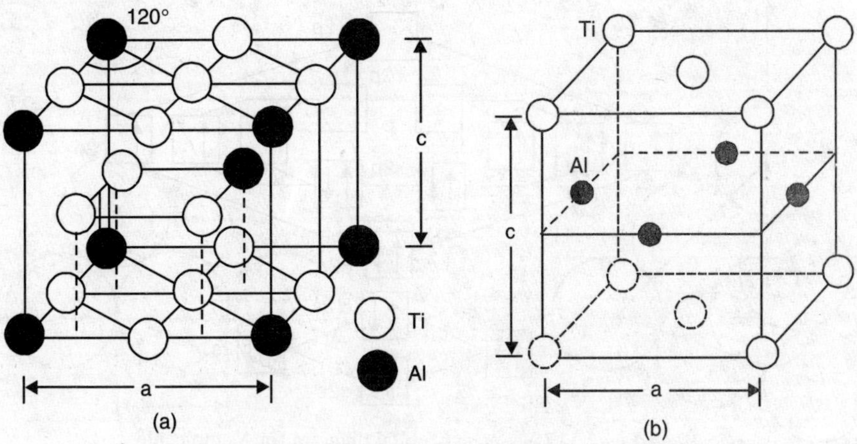

Fig. 2.119 Unit cells in (*a*) Ti_3Al and (*b*) TiAl.

Example 8: Molybdenum has a BCC structure with an atomic radius of 1.36 Å. Calculate the lattice parameter for BCC Mo.

Solution: We know that the molybdenum atoms touch along the body diagonal in BCC, as shown in the projection at right. The length of the body diagonal, then, is 4*r*, and is related to the lattice parameter, *a* (which is the length of the cube edge, not the length of the face diagonal, which is $a\sqrt{2}$) by application of the Pythagorean theorem.

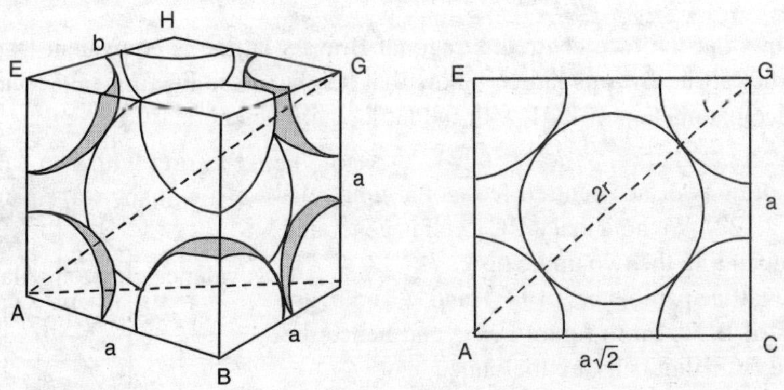

Fig. 2.120

$$(4r)^2 = a^2 + \left(a\sqrt{2}\right)^2 = 3a^2$$

$$a = 4r/\sqrt{3} = 4(1.36)/\sqrt{3} = 3.14$$

The lattice parameter for BCC Mo is 3.14 Å.

Example 9: Show that the planar density of atoms in $(\dot{1}11)$ plane for copper is 1.77×10^{-19} m^{-2}. Given the atomic radius of copper is 1.28 Å. [BE]

Solution: Number of atoms in the triangular area = 3/6 = 1/2 and $h = \sqrt{3}r$

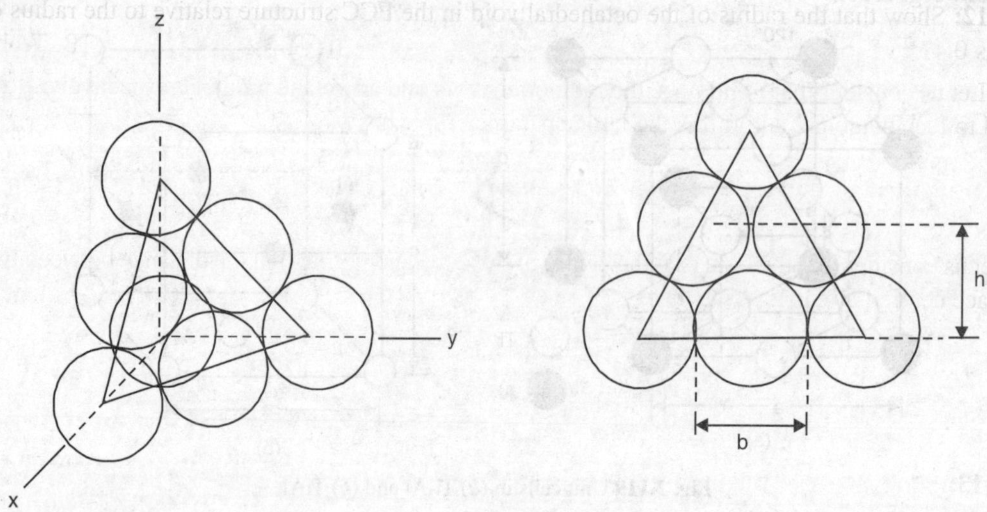

Fig. 2.121

$$\therefore \text{ Number of atoms per unit area} = \frac{1/2}{1/2\,\text{bh}}$$

or $$\text{number of atoms/Area} = \frac{1/2}{1/2(2\times 1.28)\times\sqrt{3}(1.28)}$$

$$= 1.77\times 10^{19}\ \text{m}^{-2}.$$

Example 10: Show that the face centred tetragonal Bravais lattice is equivalent to the body centred tetragonal and is not a true Bravais lattice. Show that height of the FCT and BCT cells are the same. Compare the lattice parameters of FCT with the lattice parameters of BCT. [B.E.]

Solution: Let us draw two face centred tetragonal unit cells side by side (Fig. 2.122). We note that a BCT cell is generated along the plane separating the two unit cells.

Let a and c are lattice parameters of FCT and a' and c' are lattice parameters of BCT. From figure $c = c'$ and hence the height of FCT and BCT unit cells are the same.

Now $a' = \dfrac{a\sqrt{2}}{2}$, *i.e.* lattice parameter a' is half of the face diagonal of the FCT cell.

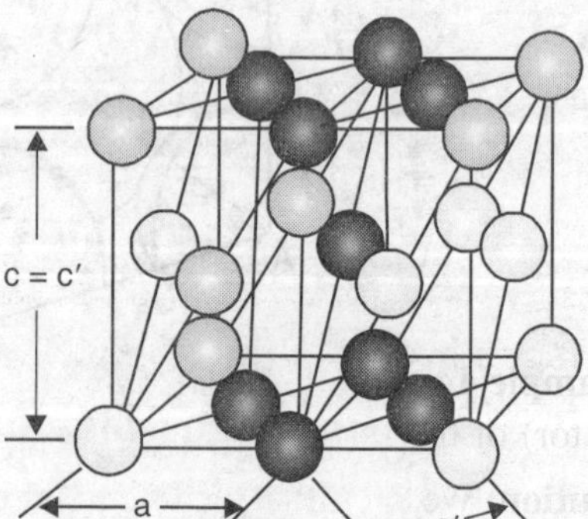

Fig. 2.122

Example 11: Consider the molecule NF_3. How can one explain the observation that the F-N-F bond angles in NF_3 are 107.3° and not 120°, as one might predict? [B.E.]

Solution: Nitrogen goes sp^3 hybridization, not sp^2, so it is tetrahedral. The additional sp^3 orbital is occupied by a lone pair of electrons from the nitrogen. This lone pair results in electron-electron repulsion that causes the other sp^3 orbitals bonded to fluorines to be closer together than the normal 109° tetrahedral bond angle, hence the F-N-F bond angle is 107.3°.

Example 12: Show that the radius of the octahedral void in the FCC structure relative to the radius of the atom is 0.41. [B. Tech.]

Solution: Let us consider that r and r_h be the radii of an atom and an octahedral hole respectively. Now, $4r$ is equal to face diagonal, and if a is the lattice parameter, we have

$$r = \frac{\sqrt{a^2 + a^2}}{4} = \frac{\sqrt{2}\,a}{4}$$

Now, let us consider that the hole is at the centre of the cube and noting that the distance between two opposite face centred atoms is a, we have

$$2r_h + 2r = a$$

or
$$\left(\frac{r_h}{r}\right) = \left(\frac{a}{2\sqrt{2}\,a/4}\right) - 1 = 0.41$$

Example 13: Show that the packing factor of any crystal having the BCC structure is equal to 0.68. [AMIE, B.E.]

Solution: We have the effective number of atoms per unit BCC lattice = 2

($\because$ BCC lattice = central atom + $\frac{1}{8}$ (8 corner atoms = 2)

From geometric consideration, we have

$$a\sqrt{3} = 4R$$

where R is the atomic radius and a is the lattice parameter.

$\therefore$ Volume of 2 atoms $= 2\left(\frac{4\pi}{3}R^3\right)$

and Volume of unit cell $= a^3 = (4\pi R/3)^3$

$\therefore$ Packing factor $= \frac{\text{Volume of two atoms}}{\text{Volume of a unit cell}}$

$$= 2\left(\frac{4\pi}{3}R^3\right)\Big/\left(4R/\sqrt{3}\right)^3$$

$$= 0.68.$$

Example 14: Figure 2.123(b) shows the packing in BCC structure. Show that the packing efficiency (factor) of the BCC structure Fig. 2.123(a) is 0.68. [AMIE, B.E.]

Solution: We know that the minimum interatomic distance in the BCC structure is along the body diagonal. Therefore, we consider the atoms as touching each other along this direction. Now, if a is the lattice parameter and r is the radius of the atom, then $4r = \sqrt{3}\,a$. Thus

$$\frac{\text{Volume}}{\text{atom}} = \frac{4}{3}\pi r^3 = \frac{4}{3}\pi\left(\frac{\sqrt{3}a}{4}\right)^3 = \frac{\sqrt{3}}{16}\pi a^3$$

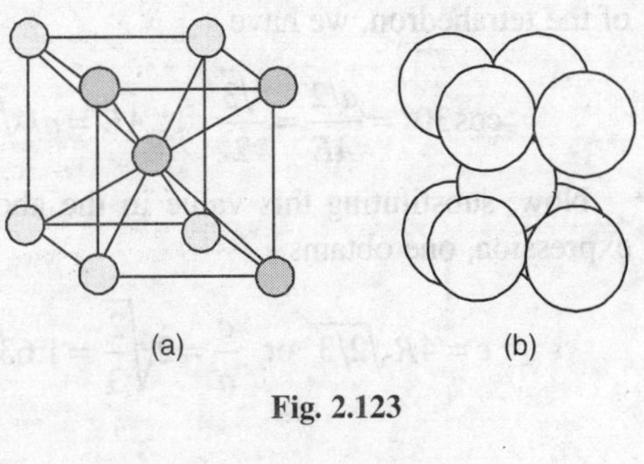

Fig. 2.123

$\because$ Number of atoms per unit cell = 2

$$\therefore \text{ Packing efficiency (factor)} = \frac{\text{Volume of two atoms}}{\text{Volume of the unit cell}}$$

$$= \frac{\sqrt{3}\pi}{a^3} = \frac{\sqrt{3}\,\pi}{8} = 0.68.$$

Example 15: The crystal structure of hydroxyapatite (Fig. 2.110) is a hexagonal unit cell with a = 9.42 Å and c = 6.88 Å. The relationship between interplanar diffraction spacing, d, and the lattice parameter for the HCP structure is

$$d = \frac{1}{\sqrt{\frac{4}{3}\left(\frac{h^2 + hk + k^2}{a^2} + \frac{l^2}{c^2}\right)}}$$

The diffraction pattern for various forms of hydroxyapatite is shown in Fig. 2.123. Calculate:

(*i*) d-spacing (in nm) for the first diffraction peak in hydroxyapatite. Assume a first-order diffraction and an X-ray source of $\lambda = 0.1537$ nm.

(*ii*) Derive a relationship in simplest terms for the d-spacing of hydroxyapatite in terms of the Miller indices only (h, k, l). Use the cell parameters in nm.

Combine your information to determine the Miller indices of the first diffraction peak for hydroxyapatite.

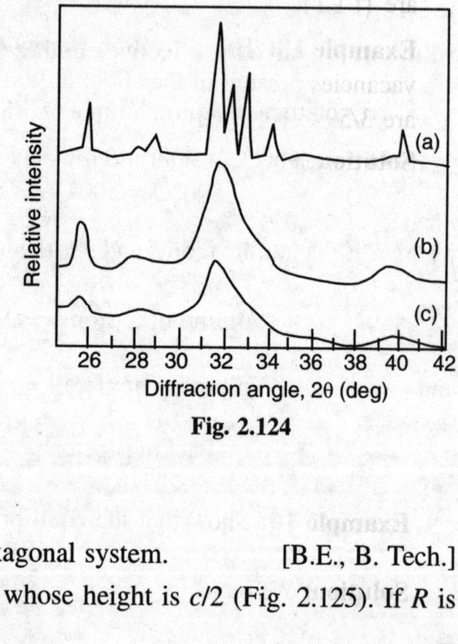

Fig. 2.124

Solution: $2d \sin\theta = n\lambda$

$n = 1, \lambda = 0.1537$ nm

$\therefore \quad d = 0.355$ nm

$$\frac{1}{d^2} = 1.51(h^2 + hk + k^2) + 2.82l^2$$

$\therefore \quad (hkl) = (111) \quad$ for $2\theta = 25°$

Example 16: Determine the c/a ratio in a closed packed hexagonal system. [B.E., B. Tech.]

Solution: Construct a tetrahedron along the C-axis of HCP, whose height is $c/2$ (Fig. 2.125). If R is atomic radius, then by simple geometry, we have

$$(c/2)^2 = (2R)^2 - (AE)^2$$

The line AE bisects the angle at the corner of the tetrahedron, we have

$$\cos 30° = \frac{a/2}{AE} = \frac{\sqrt{3}}{2} \quad \therefore AE = a/\sqrt{3}$$

Now, substituting this value in the above expression, one obtains

$$c = 4R\sqrt{2/3} \text{ or } \frac{c}{a} = 2\sqrt{\frac{2}{3}} = 1.633$$

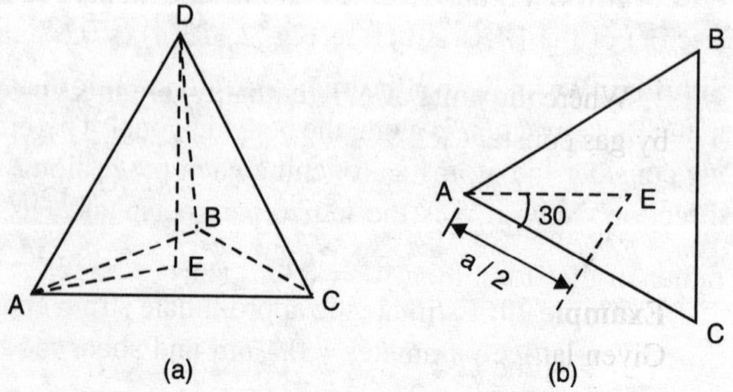

Fig. 2.125

Example 17: Show that the indices for location 1 which is one-quarter of the length along the body diagonal (Fig. 2.126) is [1/4, 1/4, 3/4] and for the directions 2 and 3 and for the plane 4 are [011], [011] and $\left(1\,1\,\overline{1}\right)$ respectively. [B.E., AMIE]

Solution: We see that the coordinates of location 1 along $\overline{a}$, $\overline{b}$ and $\overline{c}$ are 1/4, 1/4 and 3/4 respectively. Obviously, the notation for the location is 1/4 1/4 3/4.

We have for the direction 2, the vector

$$\overline{t} = 0 \cdot \overline{a} + 1 \cdot \overline{b} + 1 \cdot \overline{c}$$

Clearly, the direction is [0 1 1]

Now, the direction 2 and 3 are parallel and hence the index of direction 3 is also [0 1 1].

Now, the plane 4 cuts $\overline{a}$ and $\overline{b}$ axes at 1 and the $\overline{c}$ axis extended in the negative direction at –1. Thus the reciprocals of the intercepts are $\left(1\,1\,\overline{1}\right)$.

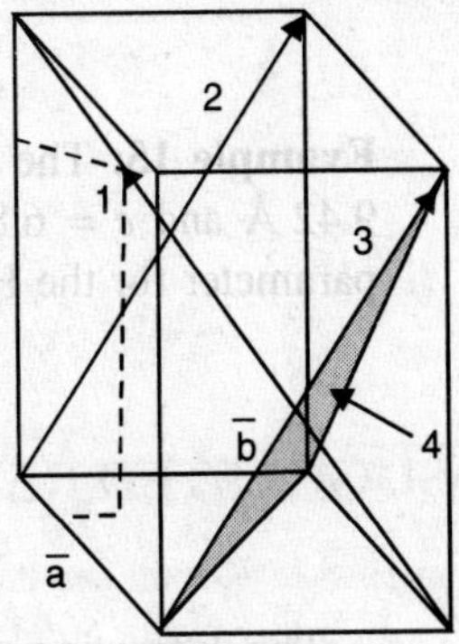

Fig. 2.126

Example 18: The activation energy for vacancy formation in silver is 1.1 eV/atom. If the number of vacancies present in the silver at 900°C is 10^{24} m^3, then shown that the number of vacancies at 500°C are 3/59 × 10^{21}/m^3. [B.Tech.]

Solution: 900°C = 900° + 273° = 1173K

500°C = 500° + 273° = 773K

$$N = 10^{24} \exp\left[\frac{1.1\text{eV}}{8.62 \times 10^{-5}\text{eV/K} \times 1173}\right]$$

$$= 5.3 \times 10^{28}/\text{m}^3$$

$$\therefore \quad \left(n_V\right)_{773\text{ K}} = 5.3 \times 10^{28}\left[\frac{1.1}{8.62 \times 10^{-5} \times 773}\right]$$

$$= 3.59 \times 10^{21}/\text{m}^3$$

Example 19: Show that the equilibrium concentration of vacancies in Cu at 1000°C is 1.2 × 10^{-3}. [B.E.]

Solution: We have

$$C = \frac{n}{N} = e^{-\frac{\Delta H_f}{RT}}$$

Where the units of ΔH_f (enthalpy) are in kJ/mol. This is why Boltzmann constant k has been replaced by gas constant R.

$$\therefore \quad C = e - \frac{120 \times 10^3}{8.314 \times 1273} = 1.2 \times 10^{-3}$$

Example 20: Estimate the approximate strain energy per atom length of a dislocation in an FCC crystal. Given lattice parameter = 0.4 nm and shear modulus = 30 GPa.

Solution: For an FCC crystal, $b = a/2$<110> and magnitude $b = \frac{a}{V^2} = 0.2828$

$$\therefore \qquad E(\text{per atom length}) = \frac{Gb^3}{2} = \frac{30\times10^9\times\left(0.2828\times10^{-9}\right)^3}{2}$$

$$= 339.2\times10^{-21}\,\text{J}$$

$$= \frac{339.2\times10^{-21}}{1.6\times10^{-19}}\,\text{eV}$$

$$= 2.12\ \text{eV/atom length.}$$

SUGGESTED READINGS

1. S.L. Kakani and Amit Kakani, *Material Science*, New Age International, New Delhi (2nd edn. 2010).
2. S.L. Kakani, *Solid State Physics*, Sultan Chand, New Delhi (4th edn. 2005).
3. Z. Jastrzebski, *The Nature and Properties of Engineering Materials*, John Wiley (2nd edn. 1976).
4. D. William, Callister, *Material Science and Engineering, An Introduction* (5th edn. 2000), John Wiley.
5. *Treatise on Material Science and Technology*, Vol. 1-19, Academic Press.
6. G.S. Brady and H.R. Clauser (eds), *Materials*, Handbook, McGraw Hill, (13th edn. 1991).
7. *Handbook of Industrial Materials*, Elsevier, (2nd edn. 1992).
8. K.C. Dee and R. Bizios, *Proactive Biomaterials and Bone Tissue Engineering*, Biotech Bioeng. 50, 438 (1996).
9. C.M. Mansur et al., *Ceramic Int. 24*, 77 (1998).
10. R. Langer and D.A. Tirrell, *Nature, 428*, 487 (2004).
11. D.G. Anderson et al., *Science*, 305, 1923 (2004).

REVIEW QUESTIONS

1. What is a crystalline substance? How does it differ from an amorphous material?
2. Write the three most common space lattices observed in metals.
3. Why ceramic crystals have very little plasticity?
4. What type of lattice structure is usually indicative of ductility in metals?
5. Explain the difference between a crystal structure and a crystal system.
6. One can form cubic BN structure by replacing Zn^{2+} and S^{2-} in the cell of ZnS by boron and nitrogen atom. Explain, why the properties of BN is not similar to that of ZnS, but akin to diamond? [B.E.]
7. Explain why the properties of polycrystalline materials are most often isotropic?
8. Would you expect a material in which the atomic bonding is predominantly ionic in nature to be more or less likely to form a noncrystalline solid upon solidification than a covalent material? [B. Tech.]
9. Define a space lattice and mention its important characteristics.
10. Boron nitride can be easily machined, silicon nitride cannot: What structural differences account for this? [B.E., AMIE]
11. What is the difference between edge and screw dislocations?
12. Explain how surface, grain boundaries and dislocations (edge and screw) can act as source or sinks for vacancies. [B.E.]
13. Explain how be mere examination of microstructure of an engineering component one can assess its quality?

14. What is the effect of two edge dislocation of same and opposite signs on the same glide plane in a crystal? [B.E.]
15. Is there a unique way of classifying solids?
[**Ans.** No. The method chosen depends in a great measure on the purpose in hand. Normally, solids are classified according to the type of bond which holds the constituent particles of the solid together.]
16. Many ionic crystals can be described as a close-packing of anions with cations in the interstitial voids of the structure. Give three examples. [**Ans.** NaCl, CsCl, ZnS]
17. Draw a hcp unit cell and fcc unit cell. Identify the atom sites and lattice points in these cells.
18. What do you understand by Miller Bravais system? How is this system used to find the index of different planes?
19. What are point defects? How these defects affect the properties of a material?
20. Explain Schottky and stacking faults.
21. In case of hexagonal crystals, four axes are used to specify its planes and disections. What is its name? [**Ans.** Miller Bravais system]
22. How many types of fundamental symmetry elements are?
[**Ans.** Three: plane of symmetry, axis of symmetry and centre of symmetry]
23. The crystal structure can be described in terms of a three-dimensional lattice by repeated translation of any of its elements. What are these elements?
[**Ans.** Lattice points, unit cells, crystallographic directions and crystallographic planes]
24. What information you get from the coordination number of a given crystal structure?
[**Ans.** Closeness of the packing of atoms]
25. What are biomaterials? Explain their importance.
26. Explain smart biomaterials.
27. Draw structural formulas comparing starch with cellulose.

PROBLEMS

1. If the atomic radius of aluminium is 0.143 nm, calculate the volume of its unit cell in cubic meters. [**Ans.** $6.62 \times 10^{-29} m^3$]
2. Calculate the radius of an iridium atom, given that Ir has an FCC crystal structure, a density of 22.4 g/cm^3, and an atomic weight of 192.2 g/mol. [**Ans.** $R = 0.136$ nm]
3. Show for the body-centred cubic crystal structure that the unit cell edge length a and the atomic radius R are related through $a = 4R/\sqrt{3}$.
4. An Al–Zn alloy constaints 10.17 weight per cent Zinc. What is the atomic percentage of Zinc? Atomic weights of Zn and Al are 65.38 and 26.98 respectively. [**Ans.** 4.445%]
5. Estimate the % ionic character of the interatomic bonds in the following compounds: TiO_2, Zn Te, Cs Cl, In Sb, and Mg Cl_2. [**Ans.** 63.2% for TiO_2 1% for In Sb]
6. Cite the indices of the direction that results from the intersection of each of the following pair of planes within a cubic crystal: (*a*) (110) and (111) planes; (*b*) (110) and $(1\bar{1}0)$ planes; and (*c*) $(10\bar{1})$ and (001) planes [(*c*) [010]].
7. Titanium has an HCP crystal structure and a density of 4.51 g/cm^3. (*a*) What is the volume of its unit cell in cubic meters? (*b*) If the c/a ratio is 1.58, compute the values of c and a.
[**Ans.** (*a*) $V_c = 1.06 \times 10^{-28}$ m^3 (*b*) $a = 0.296$ nm, $c = 0.468$ nm]

8. (*a*) Can fully cured bakelite be ground up and reused? Explain, (*b*) can polyethylene be ground up and reused? Explain. **[Ans.** (*a*) No, it is a thermostat (*b*) yes, it is a thermoplastic]
9. Cobalt has an HCP crystal structure, an atomic radius of 0.1253 nm, and a c/a ratio of 1.623. Find the volume of the unit cell for CO. **[Ans.** $V_c = 6.64 \times 10^{-2} nm^3$]
10. α-uranium has the base-centred orthorhombic lattice. The atoms are located at $0\,y\,\frac{1}{4}$, and $0\,\bar{y}\,\frac{1}{4}$ and $Y = 0.105$. (*i*) Draw the unit cell of the structure. (*ii*) How many atoms are associated with each lattice point? Also find the number of atoms per unit cell. [BE] **[Ans.** (*ii*) 2, (*iii*) 4]
11. Estimate the coordination number for the cation in each of these ceramic oxides: Al_2O_3, B_2O_3, CaO, MgO, SiO_2 and TiO_2.
[Ans. 6, 2 (actually 3), 8 (actually 6), 6, 4, 6. Discrepancies are due to a combination of uncertainty in the estimation of ionic radii and bond directionality due to partially covalent character].
12. Which of the following substitutions in an α-helical part of a protein is most likely to affect the function of the protein: Gln → Asp; Lys → Arg; Val → Phe; Ser → Cys; or Gln → Pro? **[Ans.** Gln → Pro, the others are structurally similar.]
13. (*a*) Derive linear density expressions for FCC [100] and [111] directions in terms of the atomic radius *R*.
(*b*) Compute and Compare linear density values for these same two planes for copper.
[Ans. (*a*) $LD_{100} = \frac{1}{2R\sqrt{2}}$]
14. Indium has a tetragonal unit cell for which the *a* and *c* lattice parameters are 0.459 and 0.495 nm respectively.
If the atomic packing factor and atomic radius are 0.693 and 0.1625 nm, respectively, determine the number of atoms in each unit cell. [BE] **[Ans.** $n = 4$, $\rho = 7.31$ g/cm^3]
15. Derive planar density expression for FCC(100) and (111) planes in terms of the radius R. [BE] **[Ans.** $PD_{100} = 1/4R^2$]
16. Determine the expected diffraction angle for the first order reflection from the (310) set of planes for BCC chromium when monochromatic radiation of wavelength 0.0711 nm is used. **[Ans.** $2\theta = 45.88°$]
17. The metal niobium has a BCC crystal structure. If the angle of diffraction for the (211) set of planes occurs at 75.99° (first-order reflection) when monochromatic X-radiation having a wavelength of 0.1659 nm is used, calculate (*a*) the interplanar spacing for this set of planes, and (*b*) the atomic radius for the niobium atom. [BE] **[Ans.** (*a*) $d_{211} = 0.1348$ nm, $R = 0.1429$ nm]
18. Below are listed the atomic weight, density, and atomic radius for three hypothetical alloys. For each determine whether its crystal structure is FCC, BCC, or simple cubic and then justify your determination. A simple cubic unit cell is shown in Fig. 2.127.

Alloy	Atomic weight (g/mol)	Density (g/cm^3)	Atmic radius (nm)
A	43.1	6.40	0.122
B	184.4	12.30	0.146
C	91.6	9.60	0.137

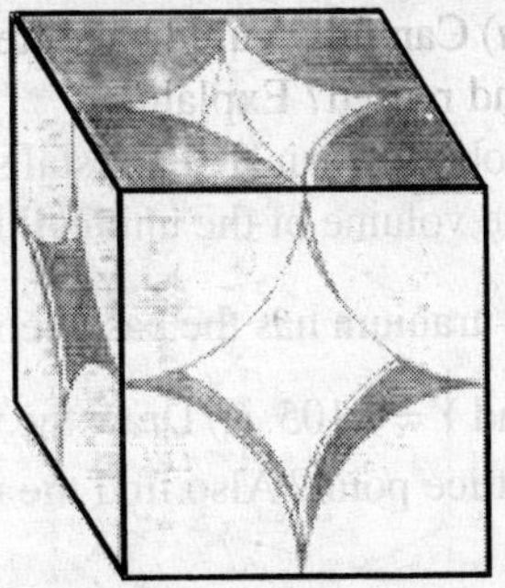

Fig. 2.127 Hard-sphere unit cell of the simple cubic crystal structure.

[Ans. Metal B: Simple Cubic]

19. Here are unit cells for two hypothetical metals:
    (*a*) What are the indices for the directions indicated by the two vectors in sketch (*a*)?
    (*b*) What are the indices for the two planes drawn in sketch (*b*)?

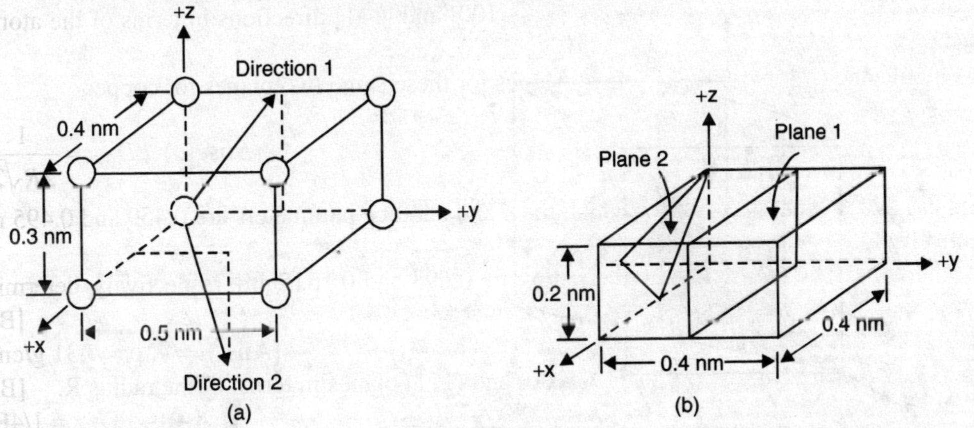

Fig. 2.128

[Ans. (*a*) Direction 1:[012]; (*b*) Plane 1: (020)]

20. Determine the indices for the directions shown in the following cubic unit cell:

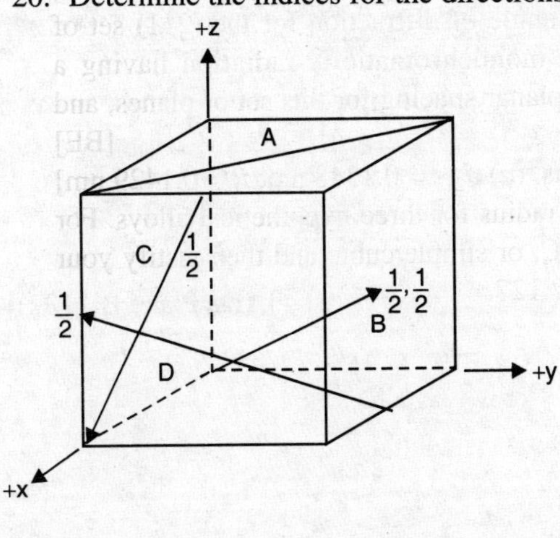

[Ans. Direction A: $[\bar{1}\,1\,0]$; Direction C: $[0\,\bar{1}\,\bar{2}]$]

Fig. 2.129

21. Determine the indices for the directions shown in the following cubic unit cell:

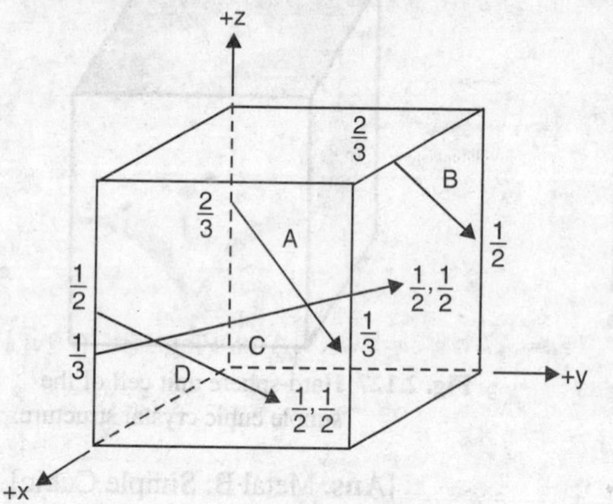

Fig. 2.130

[**Ans.** Direction B: $[\bar{4}\,0\,\bar{3}]$; Direction D: $[\bar{1}\,\bar{1}\,\bar{1}]$]

22. Determine the Miller indices for the planes shown in the following unit cell:

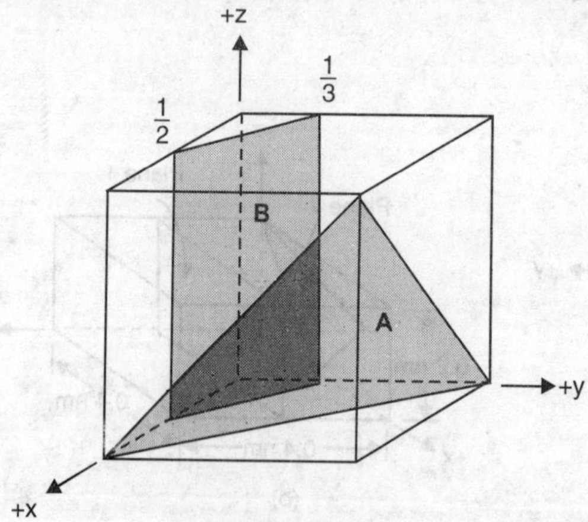

Fig. 2.131

[**Ans.** Plane A: $(1\,1\,\bar{1})$ or $(\bar{1}\,\bar{1}\,1)$]

23. Determine the Miller indices for the planes shown in the following unit cell:

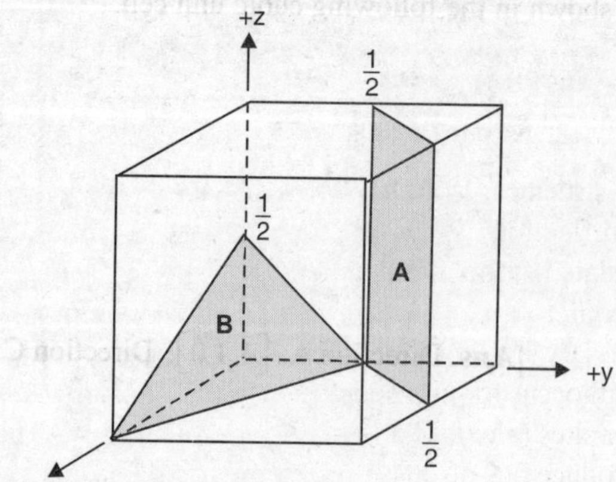

Fig. 2.132

[**Ans**. Plane B: (122)]

24. Determine the Miller indices for the planes shown in the following unit cell:

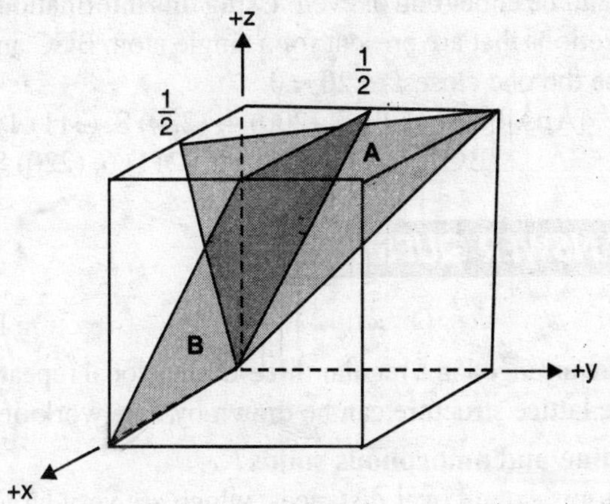

Fig. 2.133

[**Ans.** Plane B: $(0\ 2\ \bar{1})$]

25. Determine the indices for the planes shown in the hexagonal unit cells shown below:

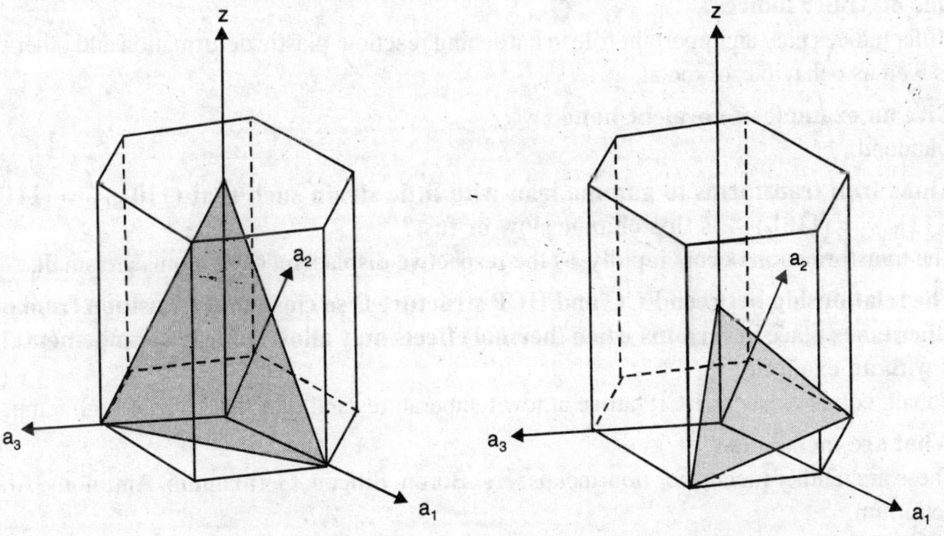

Fig. 2.134

[**Ans.** (*a*) $(1\ \bar{2}\ 1\ 1)$]

26. The metal niobium has a BCC crystal structure. If the angle of diffraction for the (211) set of planes occurs at 75.99° (first-order reflection) when monochromatic X-radiation having a wavelength of 0.1659 nm is used, calculate (*a*) the interplanar spacing for this set of planes, and (*b*) the atomic radius for the niobium atom. [**Ans.** (*a*) $d_{211} = 0.1348$ nm, $R = 0.1429$ nm]

27. Bragg's law ($2d \sin \theta = n\lambda$) is a necessary but not sufficient condition for diffraction by real crystals. It specifies when diffraction will occur for unit cells having atoms positioned only at cell corners. However, atom situated at other sites (*e.g.* face and interior positions in FCC or BCC) act as extra scattering centres, which can produce out-of-phase scattering at certain Bragg angles. The net result is the absence of some diffracted beams that, according to equation, $n\lambda = 2d \sin\theta$, should

be present. For example, for the BCC crystal structure, $h + k + l$ must be even if diffraction is to occur, whereas for FCC, h, k and l must all be either odd or even. Using this information, determine the Miller indices for the first five reflections that are present for a single atom BCC and FCC unit cell. The first reflection is defined to be the one closest to $2\theta = 0$.

[**Ans.** FCC: (111) 3, (200) 4, (220) 8, (311) 11, (222) 12;
BCC (110) 2, (200) 4, (211) 6, (220) 8, (310) 10]

SHORT ANSWER QUESTIONS

1. What is a lattice structure?

Ans. The atoms in the crystalline materials are arranged in a regular three-dimensional repeating pattern, which is called as lattice structure. The lattice structure can be drawn by a network of lines.

2. What is differences between crystalline and amorphous solids?

Ans. In crystals, a regular arrangement of atoms extend over distances, which are very large compared with the diameter of atoms. In amorphous solids the atoms occupy random positions.

3. Crystal planes and crystal direction are specified by Miller indices. What is the important role of Miller indices?

Ans. Miller indices play an important role in hardening reaction, plastic deformation and other properties as well as behaviour of metal.

4. Give an example of covalent bond.

Ans. Diamond.

5. Alpha iron transforms to gamma iron with little strain such that $(110)_{BCC} = (111)_{FCC}$, and $[111]_{BCC} = [110]_{FCC}$. Is this change slow or fast?

Ans. The transformation occurs rapidly, as the respective displacement of atoms are small.

6. The relationship between FCC and HCP structure is so close that transition from one to the other takes place in certains when thermal effects may allow some rearrangement. Illustrate it with an example.

Ans. Cobalt, cobalt exists as HCP lattice at low temperatures and as a FCC one at high temperatures.

7. What are metalloids?

Ans. These are neither metals nor non-metals; *e.g.* Boron, Silicon, Germanium, Antimony, Arsenic and Tellurium.

8. How metals are classified?

Ans. There are mainly two major groups, *e.g.* ferrous and non-ferrous metals. Usually iron is considered to be ferrous and other metals as non-ferrous metals. Based on this scheme, iron base alloys are known as ferrous materials and alloys of other metals as non-ferrous materials. Metal can also be grouped based on their salient feathers as: (*i*) light metals, (*ii*) alkali-earth metals, (*iii*) rare-earth metals, (*iv*) refractory metals, (*v*) nobel metals, (*vi*) nuclear metals and (*vii*) low melting metals.

9. In general, which class of metals has lowest density.

Ans. Alkali-earth metals.

10. State some of the important characteristics of ceramics.

Ans. Ceramics are strong and hard but brittle in nature and possess good thermal and electrical insulating properties. They exhibit excellent resistance to high temperatures and certain corrosive media. They also possess unusual optical and electrical properties.

11. Mention some important characteristics of polymers.

Ans. (*i*) Poor thermal conductivity, (*ii*) poor electrical conductivity, (*iii*) low density, (*iv*) low melting temperatures, (*v*) high degree of plasticity, and (*vi*) low strength levels.

12. What do you understand by *cleavage*?

Ans. Brittle fracture takes place without any appreciable deformation, and by rapid crack propagation. The direction of crack motion is very nearly perpendicular to the direction of the tensile stress and yields a relatively flat fracture surface.

Fracture surfaces of materials that failed in a brittle manner will have their own distinctive patterns; any signs of gross plastic deformation will be absent, *e.g.* in some steel pieces, a series of V-shaped "chevron" markings may form near the centre of the fracture cross-section that point back toward the crack initiation site. Other brittle fracture surfaces contain lines or ridges that radiate from the origin of the crack in a fanlike pattern. Often, both of the marking patterns will be sufficiently coarse to be discerned with the naked eye. For very hard and fine-grained metals, there will be no discernible fracture pattern. Brittle fracture in amorphous materials, such as ceramic glasses, yields a relatively shiny and smooth surfaces.

For most brittle crystalline materials, crack propagation corresponds to the successive and repeated breaking of atomic bonds along specific crystallographic planes; such a process is termed *cleavage.* This type of fracture is said to be *trans granular* (or *transcrystalline,* because the fracture cracks pass through the grains.

CHAPTER 3

Electronic Materials

3.1 INTRODUCTION

The electronics industry employs several types of materials in the circuitry and devices that operate on electronic impulses. Research and development efforts of both materials scientists and electronic engineers aim for reduction of the size of electronic components, elimination of moving parts, reduction of energy to operate electronic components, and an increase in speed of their operation.

One of the principal characteristic of materials is their ability (or lack of ability) to conduct electrical current. Indeed, materials are classified by this property, *i.e.*, they are divided into conductors, semiconductors and nonconductors (insulators or dielectrics). The conductivity, σ, of different materials at room-temperature spans more than 25 orders of magnitude (Fig. 3.1). Moreover, when we take into consideration the conductivity of superconductors, measured at low temperatures, into consideration, this span extends to 40 orders of magnitude (taking an estimated conductivity for superconductors of about 10^{20} 1/Ω cm). We may note that this is the largest known variation in a physical property and is comparable to the ratio between the diameter of the universe (~ 10^{26} m) and the radius of an electron (10^{-14} m).

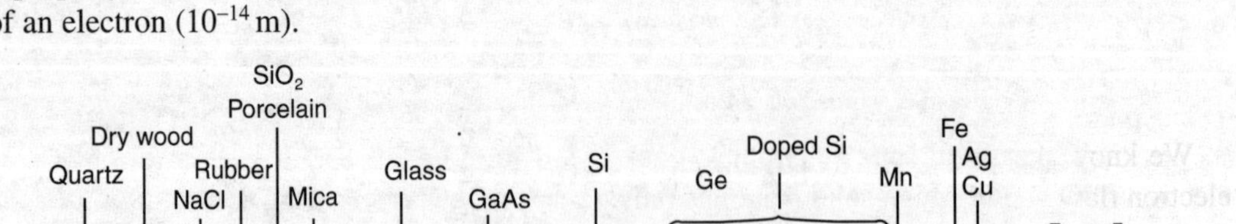

Fig. 3.1 Room-temperature conductivity of various materials. (Superconductors, having conductivities of many orders of magnitude larger than copper, near 0 K, are not shown. The conductivity of semiconductors varies substantially with temperature and purity). The reciprocal of the ohm (Ω) is defined to be 1 siemens (S).

Electronic materials include conductors, semiconductors, insulators, and superconductors. This family of materials has revolutionized the entire world. From spark plugs made from alumina, and copper wires for transmission of electric power to components for wireless communications, high-power magnets used in magnetic resonance imaging (MRI), capacitors, insulators, solar cells, active matrix displays, silicon, and gallium arsenide based computer chips, electronic materials are found in numerous applications. Recent advances in materials have led to several breakthroughs in the development of new electronic materials. We now have ceramics that are not just excellent insulators, but also semiconductors and superconductors. Similarly, we have polymers which exhibit semiconducting and also superconducting properties. Figure 3.2 presents a simple and broad classification of some of the electronic materials.

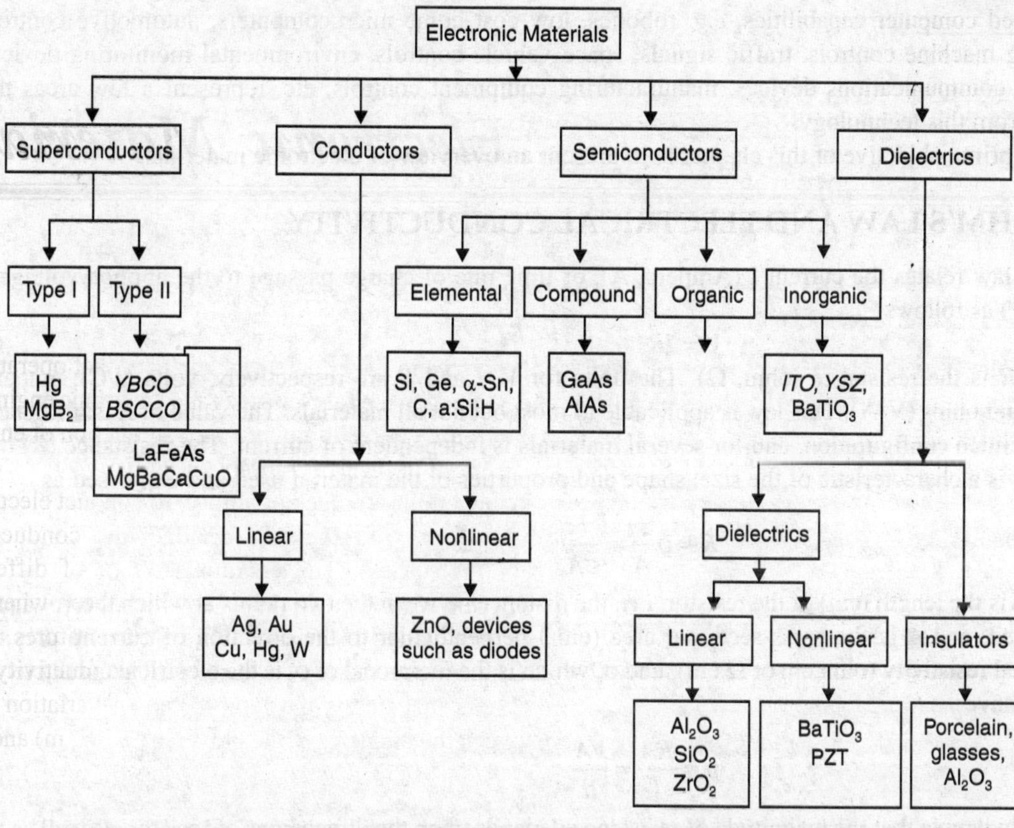

Fig. 3.2 A broad classification of technologically useful electronic materials.

We know that solid state electronics, got its start in the early 1950s, deals with devices in which electron flow goes through solid materials rather than the vacuum used in vacuum tubes. Vacuum tubes dominated electronics prior to solid-state devices and created problems of size as well as heat but are almost eliminated in modern electronic devices. Vacuum tube computers filled entire buildings, while today's computers have replaced the former with memory *chip* less than a square centimeters in size. *Printed circuit* boards by making use of transistors, resistors, diodes, and capacitors mounted on reinforced plastic boards with etched copper conductors connect the components.

Integrated circuits (IC) developed from solid-state electronics bring together several electronic devices, *e.g.* transistors, conductors, diodes, and resistors, into a single chip. Semiconducting materials form the backbone for IC devices. ICs allow the reduction of complex circuit boards into small microprocessor chips, which perform the functions of a circuit board that can be hundreds of times larger.

The IC chip may contain several different types of electronic circuits. During the initial stages of development, IC memory chips stored about 1000 bits of information. The information is stored in the form of simple electronic charges. Through progress in the field of IC technology, LSI (large scale integration) evolved so that memory chips only one centimeter square have a memory capacity of more than 10,000 bits of information. The development of dynamic random access memory (DRAM) chip is a major breakthrough in very large scale integration (VLSI) techniques. Silicon and aluminium metal oxide semiconductor (SAMOS) technology can produce these one megabit chips.

Microprocessors hold promise for several technological developments in all phases of society. Improved computer capabilities, *e.g.* robotics, low cost home microcomputers, automotive controls, vending machine controls, traffic signals, space vehicle controls, environmental monitoring devices, several communications devices, manufacturing equipment controls, etc. represent a few areas that profit from this technology.

The prime objective of this chapter is to present an overview of electronic materials.

3.2 OHM'S LAW AND ELECTRICAL CONDUCTIVITY

Ohm's law relates the current I (Ampere, A), or time rate of charge passage to the applied voltage V (Volt, *V*) as follows:

$$V = IR \qquad \ldots(1)$$

where R is the resistance (ohm, Ω). The units for V, I and R are respectively, volts (J/C), amperes (C/s), and ohms (V/A). This law is applicable to most but not all materials. The value of R is influenced by specimen configuration, and for several materials is independent of current. The resistance (R) of a resistor is a characteristic of the size, shape and properties of the material used and expressed as

$$R = \rho\frac{l}{A} = \frac{l}{\sigma A} \qquad \ldots(2)$$

where l is the length (cm) of the resistor, *i.e.*, the distance between the two points at which the voltage is measured, and A is the cross-sectional area (cm^2) perpendicular to the direction of current. ρ is the electrical resistivity (ohm.cm or Ω.cm), and σ, which is the reciprocal of ρ, is the electrical conductivity.

We have

$$\rho = \frac{RA}{l} = \frac{VA}{Il} \qquad \ldots(3)$$

We may note that the magnitude of resistance depends upon the dimensions of resistor, as well as the microstructure and composition of the material. The electrical conductivity (σ) is used to specify the electrical character of a material. It is reciprocal of the resistivity, *i.e.*

$$\sigma = \frac{1}{\rho} \qquad \ldots(4)$$

Thus, resistivity (ρ) or conductivity (σ) allows one to compare different materials. Resistivity is a microstructure sensitive property.

In addition to Eqn. (1) ohm's law may be expressed as

$$J = \sigma E \qquad \ldots(5)$$

where J (A/cm^2) is the current density (I/A), the current per unit of specimen area, and E is the electric field intensity or the voltage difference between two points divided by the distance (l) separating them, *i.e.*

$$E = \frac{V}{l} \qquad \ldots(6)$$

One can also determine the current density J is

$$J = n.q.\overline{\text{v}} \qquad \ldots(7)$$

where n is number of charge carriers (carriers/cm^3), q is the charge on each carrier (1.6×10^{-19} C), and $\overline{\text{v}}$ is the average drift velocity (cm/s) at which the charge carriers move. Thus, we have

$$\sigma E = nq\overline{\text{v}} \text{ or } \sigma = nq\frac{\overline{\text{v}}}{E}$$

We may note that *diffusion* occurs as a result of temperature and concentration gradients, and drifts occurs as a result of an electric or magnetic field.

The term $\frac{\overline{\text{v}}}{E}$ is the *mobility* of carriers and denoted by $\mu\left(\frac{\text{cm}^2}{V \cdot s}\right)$. In the case of metals μ is the mobility of electrons.

We have

$$\mu = \frac{\overline{\text{v}}}{E}$$

Thus, we have

$$\sigma = nq\mu \qquad \ldots(8)$$

Here q is the charge. Equation (8) reveals that one can control the electrical conductivity of materials by (*i*) controlling the number of charge carriers in the material, (*ii*) controlling the mobility or ease of movement of the charge carriers. We may note that the mobility is particularly important in metals, whereas the number of carriers is more important in semiconductors and insulators.

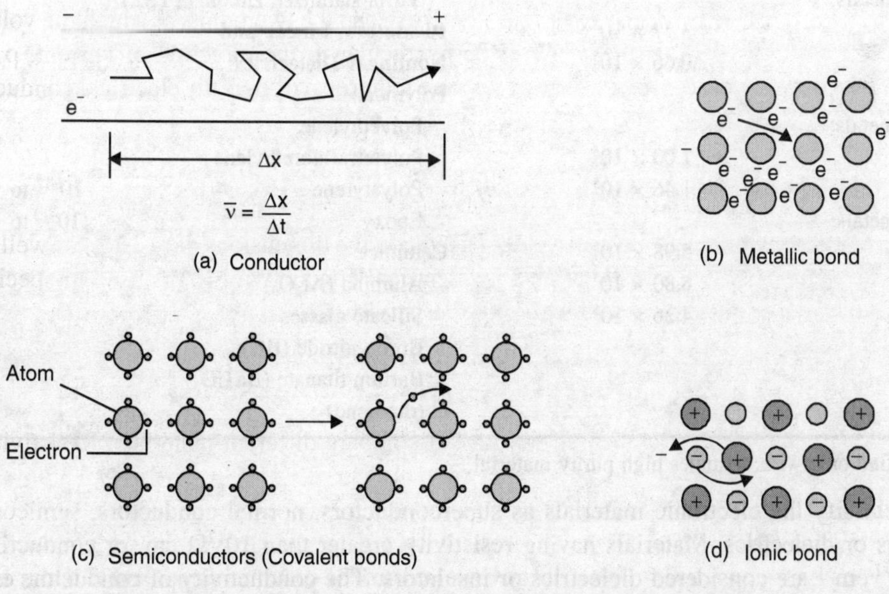

Fig. 3.3 (*a*) Charge carriers, such as electrons, are deflected by atoms or defects and take an irregular (zig-zag) path through a conductor. The average rate at which the carriers move is called the drift velocity $\overline{v}$. (*b*) Valence electrons in the metallic bond move easily. (*c*) Covalent bonds must be broken in semiconductors and insulators for an electron to be able to move. (*d*) Entire ions must diffuse to carry charge in many ionically bonded materials.

Charge carriers in metals are electrons, whereas charge carriers in semiconductors are both electrons and holes. Equation (8) gets modified for expressing conductivity of semiconductors:

$$\sigma = nq\mu_n + pq\mu_p \quad \ldots(9)$$

where μ_n and μ_p are the mobilities of electrons and holes, respectively. The terms n and p in (9) represents the concentrations of free electrons and holes in a semiconductor.

Electrical conductivity of materials varies tremendously (Table 3.1). We may note that in ceramics, when conduction does occur, it can be the result of electrons that 'hop' from one defect to another or movement of ions [Fig. 3.3(*b*)–(*d*)]. We may note that the mobility depends on atomic bonding, microstructure, imperfections, and in ionic compounds, diffusion rates.

Table 3.1 Electrical Conductivity of Few Selected Materials (at T = 300 K*)

Material	*Conductivity ($ohm^{-1}. cm^{-1}$)*	*Material*	*Conductivity ($ohm^{-1}. cm^{-1}$)*
Superconductors		**Semiconductors**	
Hg, Nb_3Sn,		Group 4B elements:	
$YBa_2Cu_3O_{7-x}$,	Infinite (under certain conditions	Si	5×10^{-6}
MgB_2	such as low temperatures)	Ge	0.02
Metals		α-Sn	0.9×10^5
Alkali metals:		Compound semiconductors	
Na	2.13×10^5	GaAs	2.5×10^{-9}
K	1.64×10^5	AlAs	0.1
Alkali earth metals:		SiC	10^{-10}
Mg	2.25×10^5	**Ionic Conductors**	
Ca	3.16×10^5	Indium tin oxide (ITO)	
Group 3B metals:		Yttria-stabilized zirconia (YSZ)	
Al	3.77×10^5	**Insulators, Linear and**	
Ga	0.66×10^5	**Nonlinear Dielectrics**	
		Polymers:	
Transition metals:		Polyethylene	10^{-15}
Fe	1.00×10^5	Polytetrafluorethylene	10^{-18}
Ni	1.46×10^5	Polystyrene	10^{-17} to 10^{-19}
Group 1B metals:		Epoxy	10^{-12} to 10^{-17}
Cu	5.98×10^5	Ceramics:	
Ag	6.80×10^5	Alumina (Al_2O_3)	10^{-14}
Au	4.26×10^5	Silicate glasses	10^{-17}
		Boron nitride (BN)	10^{-13}
		Barium titanate ($BaTiO_3$)	10^{-14}
		C (diamond)	$<10^{-18}$

* Unless specified otherwise, assumes high purity material.

One can classify the electronic materials as superconductors, normal conductors, semiconductors and insulators or dielectrics. Materials having resistivity greater than $10^{12}\Omega$ cm, or conductivity less than $10^{-12}\ \Omega^{-1}\ cm^{-1}$ are considered dielectrics or insulators. The conductivity of conductors is greater than $10^3\ \Omega^{-1}\ cm^{-1}$, or resistivity less than $10^{-3}\ \Omega$ cm. Semiconductors are materials having conductivity less than $10^3\ \Omega^{-1}\ cm^{-1}$ but greater than $10^{-12}\ \Omega^{-1}\ cm^{-1}$.

Dielectric materials are used in applications where the dielectric constant (K) is important.

3.3 ENERGY BAND STRUCTURE IN SOLIDS

In all conductors, semiconductors and several insulating materials, only electronic conduction exists and the magnitude of the electrical conductivity is found to be strongly dependent on the number of electrons available to participate in the conduction process. However, not all electrons in every atom participate in conduction. The number of electrons available for conduction in a particular material is related to the arrangement of electron states or energy levels with respect to energy and then the manner in which these states are occupied by electrons. Electrons in a single atom occupy discrete energy levels. According to *Pauli exclusion principle*, each energy level can contain only two electrons, *e.g.* the $2s$ level of a single atom contains one energy level and two electrons. The $2p$ level contains three energy levels and a total of six electrons.

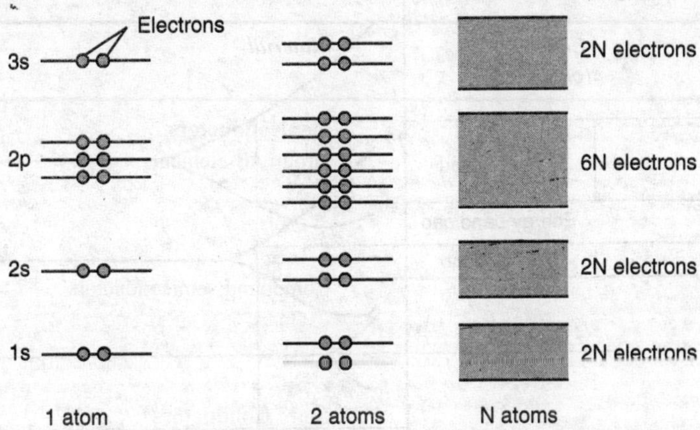

Fig. 3.4 Formation of energy bands as the number of electrons grouped together increases.

When a solid is formed by atoms (say N) coming together, the Pauli principle still demands that only two electrons in the entire solid have the same energy. When a solid is formed, the different split energy levels of electrons come together to form the continuous bands of energies as shown in Fig. 3.4. We can see that as a consequence $2s$ band in a solid contains N discrete energy levels and $2N$ electrons, *i.e.* two electrons in each energy level. Further, each of the $2p$ levels contain N energy levels and $2N$ electrons. We may note that three $2p$ bands overlap, one could alternatively describe a single, broad $2p$ band containing $3N$ energy levels and $6N$ electrons.

We consider an example of sodium. Sodium has 11 electrons and has an electronic structure of $1s^2\,2s^2\,2p^6\,3s^1$. Figure 3.5 shows an idealized picture of band structure of $^{23}_{11}$Na. We see that the energies within the bands depend on the spacing between the atoms; the vertical line represents the

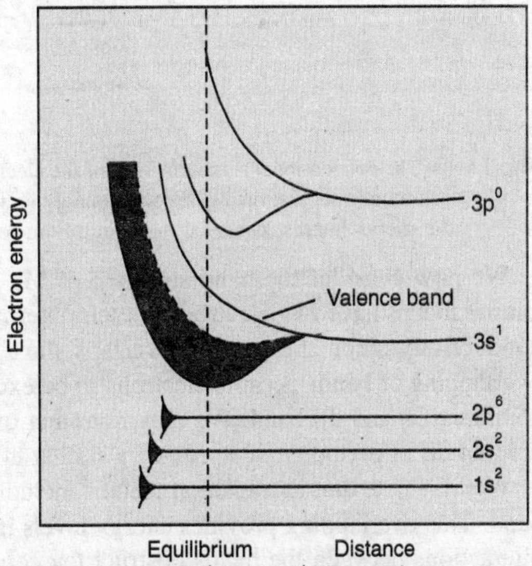

Fig. 3.5 An idealized band structure for sodium ($1s^2\,2s^2\,2p^6\,3s^1$). The energy levels broaden into bands. The $3s$ band, which is only half filled with electrons, is responsible for conduction in sodium.

equilibrium interatomic spacing of atom in solid sodium. We can see that 3*s* energy levels are the *valence band* and the empty 3*p* energy levels, which are separated from the 3*s* band by an energy gap, form the *conduction band*.

We may note that the extent of splitting depends on interatomic separation and begins with the outermost electron shells, since they are the first to be perturbed as the atoms coalesce. Within each band, the energy states are discrete, yet the difference between adjacent states in exceedingly small.

At the equilibrium spacing, band formation may not occur for the electron subshells nearest the nucleus [Fig. 3.6(*b*)]. Moreover, gaps may exists between adjacent bands, as shown in figure; normally energies lying within these band gaps are not available for electron occupancy. Figure 3.6(*a*) shows the conventional way of representing energy band structure in solids.

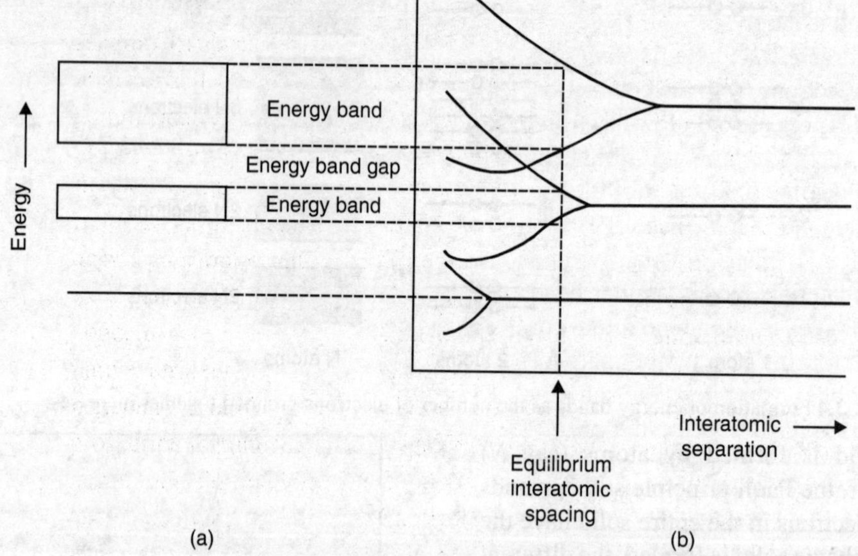

Fig. 3.6 (*a*) The conventional representation of the electron energy band structure for a solid material at the equilibrium interatomic separation. (*b*) Electron energy *vs.* interatomic separation for an aggregate of atoms, exhibiting how the energy band structure at the equilibrium separation in (*a*) is generated.

We now consider the band structure of Mg and other metals in the periodic table. Mg and other similar metals have two electrons in their outermost *s* band. We may note that these metals have a high conductivity since the *p* band overlaps the *s* band at the equilibrium interatomic spacing. This overlapping of bands permits electrons to be excited into the large number of unoccupied levels in the combined 3*s* and 3*p* band. We may note that overlapping 3*s* and 3*p* bands in Al and other metals in column 3B of periodic table provide a similar effect.

We may note that in transition metals, including Na through Ni, an unfilled 3*d* band overlaps the 4*s* band. This overlapping provides energy levels into which electrons can be excited; however, complex interactions between the bands obstruct the conductivity from being as high as in some of the better conductors. In case of Cu, the inner 3*d* band is full, and the atom core tightly holds these electrons. This is why, there is little interaction between the electrons in the 4*s* and 3*d* bands, and Cu has a high conductivity. Ag and Au also have the similar situation.

3.4 ENERGY BAND STRUCTURE OF SEMICONDUCTORS AND INSULATORS

In group 4 of periodic table–carbon (diamond), Si, Ge and Sn contain two electrons in their *p*-shell and have a valence of four.

These elements are covalently bonded, *i.e.* the electrons in the outer *s* and *p* bands are rigidly bound to the atoms. We may note that the covalent bonding produces a complex change in the band structure, or **hybridization**. We know that 2*s* and 2*p* energy levels of the C-atoms in diamond can contain upto eight electrons, but there are only four valence electrons available. When carbon atoms form solid diamond, the 2*s* and 2*p* levels interact and produce two bands as shown in Fig. 3.7. Each hybrid band can contain 4N electrons. Now, since there are only 4N electrons available, the lower, *i.e.* valence band is completely filled and the upper, *i.e.* conduction band is empty.

In diamond, a large *energy gap* or *band gap* (E_g) separates the electrons from the conduction band ($E_g \sim 5.5$ eV). There are only few electrons possess sufficient energy to jump the forbidden zone to the conduction. As a result, the conductivity of diamond is less than $10^{-18}\ \Omega^{-1}$ cm^{-1}. Like diamond, other covalently and ionically bonded materials have a similar band structure and behave as nonconductors of electricity. On increasing the temperature or applying on voltages supplies the energy required for electrons to overcome the energy gap, *e.g.* the electrical conduction of boron nitride increases from about 10^{-13} at room-temperature to $10^{-4}\ \Omega^{-1}$-cm^{-1} at 800 °C.

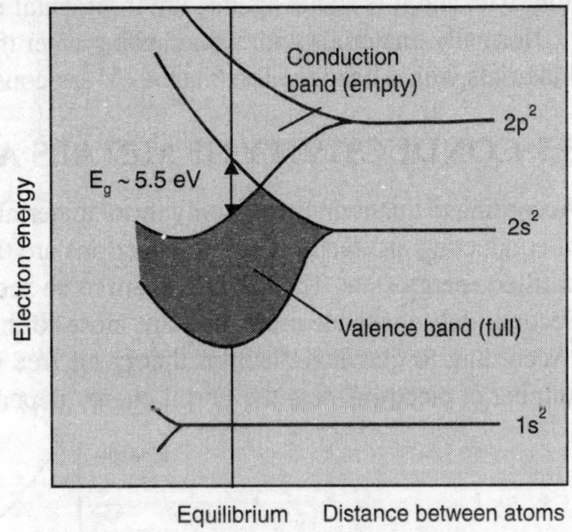

Fig. 3.7 The energy band structure of carbon in the diamond form. The 2*s* and 2*p* levels combine to form two hybrid bands separated by an energy gap, E_g.

Figure 3.8(A) shows a schematic of band structure at 0 K (*a*) and (*b*) for conductors, (*c*) for insulators and (*d*) for semiconductors. We see that important distinction between metals and semiconductors is that the conductivity of semiconductors increases with size in temperature, as more and more electrons can

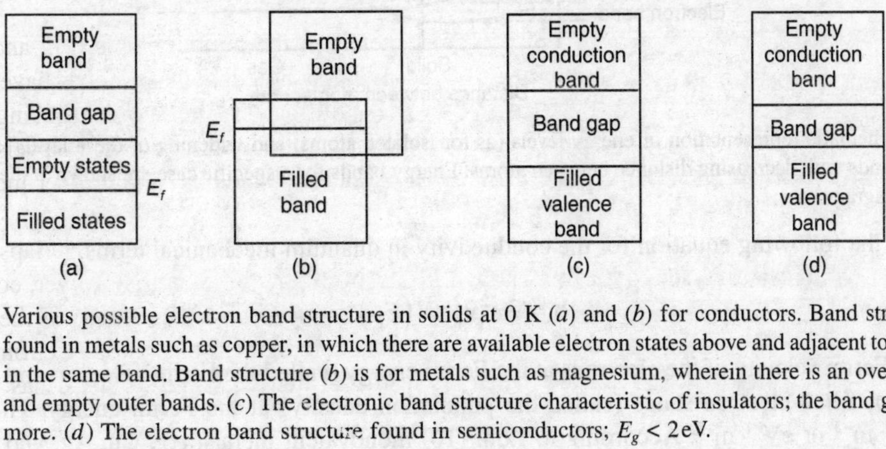

Fig. 3.8(A) Various possible electron band structure in solids at 0 K (*a*) and (*b*) for conductors. Band structure (*a*) is found in metals such as copper, in which there are available electron states above and adjacent to filled states, in the same band. Band structure (*b*) is for metals such as magnesium, wherein there is an overlap of filled and empty outer bands. (*c*) The electronic band structure characteristic of insulators; the band gap > 2 eV or more. (*d*) The electron band structure found in semiconductors; $E_g < 2$ eV.

make it to the conduction band from the valence band. Obviously, an increasing of electrons from the covalent bonds in a semiconductor is freed and gets available for conduction. On the other hand, the conductivity of most metals decreases with rise in temperature. This means, the number of electrons that are already available begin to scatter more, *i.e.* increasing temperature of a metal reduces mobility.

The crystal structure of Ge, Si and α-Sn is same and band structure as diamond, the energy is smaller. Interestingly, the energy gap (E_g) in α-Sn is so small (E_g = 0.1 eV) that α-Sn behaves as metal. The energy gap in Si (E_g = 1.1 eV) and Ge (E_g = 0.67 eV) is somewhat larger. Semiconductors have proved in recent years to be the lifeblood for electronic industries. Si the major species of semiconducting materials is abundant (28% of earth's crust consists of it); the raw material (SiO_2 or sand) is inexpensive. Si forms a natural, insulating oxide; its heat conduction is reasonable; it is nontoxic; and it is stable against environmental influences.

Normally, materials with a band gap greater than 4.0 eV are insulators, dielectrics or nonconductors. Materials with a bandgap less than 4 eV are considered semiconductors.

3.5 CONDUCTIVITY OF METALS AND ALLOYS

According to quantum theory, only those materials that possess partially field electron bands are capable of conducting an electric current. Electrons are then lifted above the Fermi energy into an allowed and unfilled energy state. This permits them to be accelerated by an electric field, thus producing a current. Second, only those electrons that are close to the Fermi energy participate in the electric conduction. (According to classical electron theory all free electrons would contribute to the current). Third, the number of electrons near the Fermi energy depends on the available electron states Fig. 3.8(B).

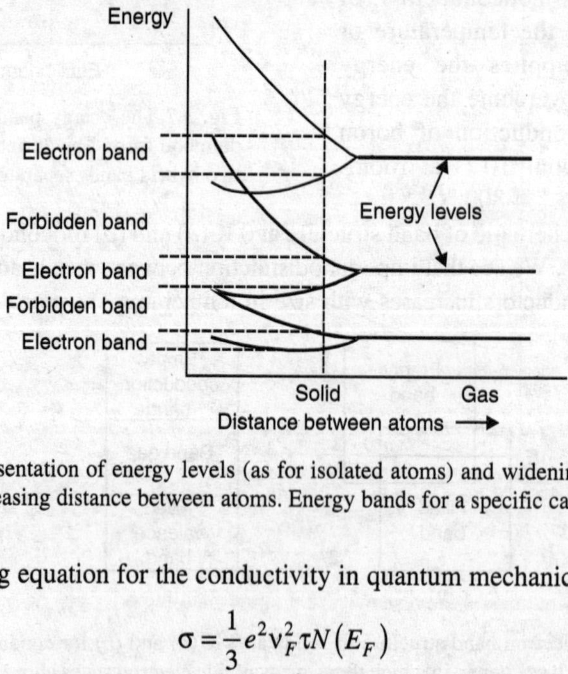

Fig. 3.8(B) Schematic representation of energy levels (as for isolated atoms) and widening of these levels into energy bands with decreasing distance between atoms. Energy bands for a specific case are shown at the left of the diagram.

One obtains the following equation for the conductivity in quantum mechanical terms.

$$\sigma = \frac{1}{3} e^2 v_F^2 \tau N(E_F) \quad ...(10)$$

where V_F is Fermi velocity of the electrons at the Fermi energy (called the Fermi velocity) and $N(E_F)$ is the density of filled electron states (called the population density) at the Fermi energy. The unit of $N(E_F)$ is $J^{-1}\ m^{-3}$ or $eV^{-1}\ m^{-3}$. According to Eqn. (10) monovalent metals, *e.g.* Cu, Ag, and Au have partially filled bands. Their electron population density near the Fermi energy is high, which results in a large conductivity. Bivalent metals, on the other hand, are distinguished by an overlapping of the upper bands and by a small electron concentration near the bottom of the valence band. As a consequence, the electron population near the Fermi energy is small, which leads to a comparatively low conductivity.

One can change the conductivity by influencing the mobility, μ, of the carriers. The mobility (μ) is proportional to the average drift velocity ($\bar{v}$), which is low if the electrons collide with imperfections in the lattice. We have the mean free path of the electron.

$$\lambda_e = \tau\bar{v} \quad \text{...(11)}$$

where τ is the average time between collisions. The mean free path is the average distance between collisions. A longer mean free path permits high mobility and high conductivities. The conductivity σ of most materials may be expressed as

$$\sigma = n \mid e \mid \mu_e \quad \text{...(12)}$$

where n is the number of free or conducting electrons per unit volume (per cm^3), e is the absolute magnitude of electronic charge and μ_e is electron mobility.

Influence of Temperature

For the pure metal, the resistivity rises linearly with temperature (Fig. 3.9).

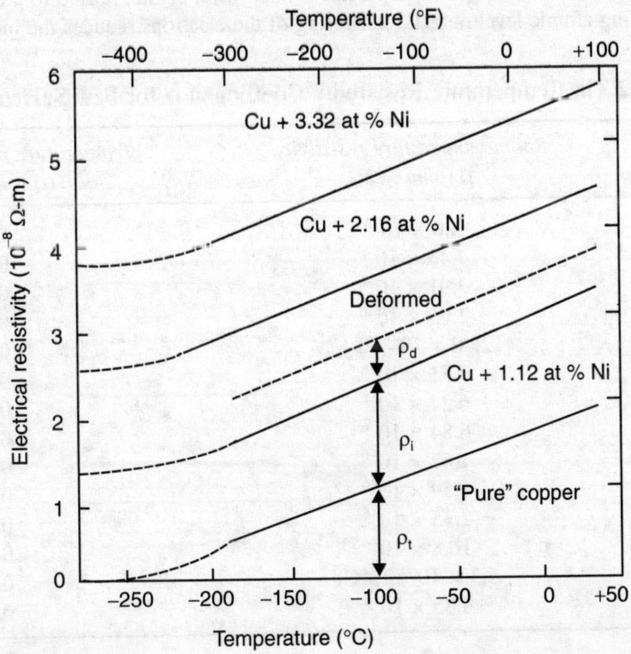

Fig. 3.9 The electrical resistivity *vs.* temperature curves for Cu and three copper-nickel alloys, one of which has been deformed. The contributions to the resistivity due to thermal, impurity and deformation are shown at −100°C.

Thus,

$$\rho_t = \rho_o + \alpha T \quad \text{...(13)}$$

where ρ_o and α are constant for each particular metal. α is the temperature resistivity coefficient. This dependence of thermal resistivity component on temperature is due to the increase with temperature the thermal vibrations and other irregularities (Fig. 3.10). At any instant, the atom may not be in its equilibrium position, and it therefore interacts with and scatters electrons. The mean free path decreases, the mobility of electrons is reduced and the resistivity increases. For few selected metals, the temperature resistivity coefficient (α) is given in Table 3.2.

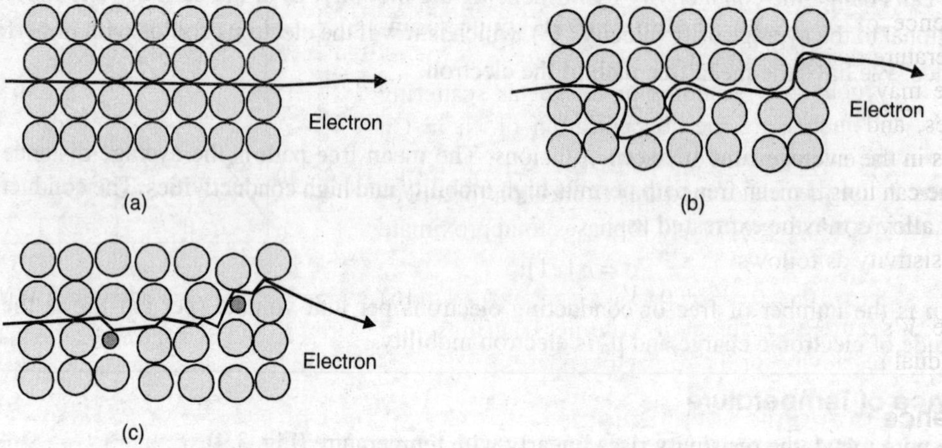

Fig. 3.10 Movement of an electron through (*a*) a perfect crystal, (*b*) a crystal heated to a high temperature, and (*c*) a crystal containing atomic level defects. Scattering of the electrons reduces the mobility and conductivity.

Table 3.2 The Temperature Resistivity Coefficient α for Few Selected Metals

Metal	*Room temperature resistivity* ρ *(ohm . cm)*	*Temperature resistivity coefficient* (α) *(ohm/ohm. °C)*
Be	4.0×10^{-6}	0.0250
Mg	4.45×10^{-6}	0.0037
Ca	3.91×10^{-6}	0.0042
Al	2.65×10^{-6}	0.0043
Cr	12.90×10^{-6} (0°C)	0.0030
Fe	9.71×10^{-6}	0.0065
Co	6.24×10^{-6}	0.0053
Ni	6.84×10^{-6}	0.0069
Cu	1.67×10^{-6}	0.0043
Ag	1.59×10^{-6}	0.0041
Au	2.35×10^{-6}	0.0035
Pd	10.8×10^{-6}	0.0037
W	5.3×10^{-6}(27°C)	0.0045
Pt	9.85×10^{-6}	0.0039

Influence of Impurities

Another factor which reduces the mean free path of the electrons is the impurity or solute atoms. The solute atom provide the breakage in the regular crystalline structure, *i.e.*, presenting an obstacle in the movement of electron waves. A solid solution alloy will always have lower conductivity in comparison to its pure components though both individual components have higher conductivity than the alloy. For additions of a single impurity that forms a solid solution, the impurity resistivity ρ_i is related to the impurity concentration C_i in terms of the atom fraction (at % 100) through the relation

$$\rho_i = AC_i\,(1 - C_i) \quad ...(14)$$

where A is a composition constant that is a function of both the impurity and host metals. Figure 3.11 shows the influence of Ni impurity additions at about room-temperature resistivity of Cu.

We may note that Ni atoms in Cu act as scattering centres, and increasing the concentration of Ni in Cu results in the enhancement of resistivity.

One can use the rule of mixtures expression for a two-phase alloy consisting of α and β phases to approximate the resistivity as follows:

$$\rho_i = \rho_\alpha V_\alpha + \rho_\beta V_\beta \quad ...(15)$$

where V's and ρ's represent volume fractions and individual resistivity for the respective phases.

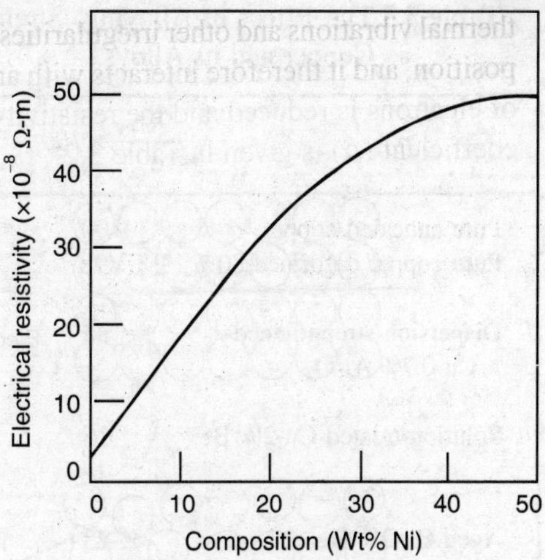

Fig. 3.11 Room-temperature electrical resistivity *vs.* composition for Cu-Ni alloys.

Influence of Atomic Level Defects

Imperfections in crystal structures scatter electrons, reducing the mobility and conductivity of metals (Fig. 3.10(*c*)). One can express the increase in the resistivity due to solid solution atoms as

$$\rho_d = b(1 - x)x \quad ...(16)$$

where ρ_d is the increase in resistivity due to the defects, x is the atomic fraction of the impurity or solid solution atoms present, and b is the coefficient of the defect resistivity. Similarly, vacancies, dislocations, and grain boundaries reduce the conductivity of the metal. Each defect contribute to an increase in the resistivity of the metal. One can express the overall resistivity as

$$\rho = \rho_T + \rho_d \quad ...(17)$$

where ρ_d represents the contribution from all defects. Equation (17) is called **Matthiessen's rule**. We may note that the effect of the defects is independent of the temperature (Fig. 3.12).

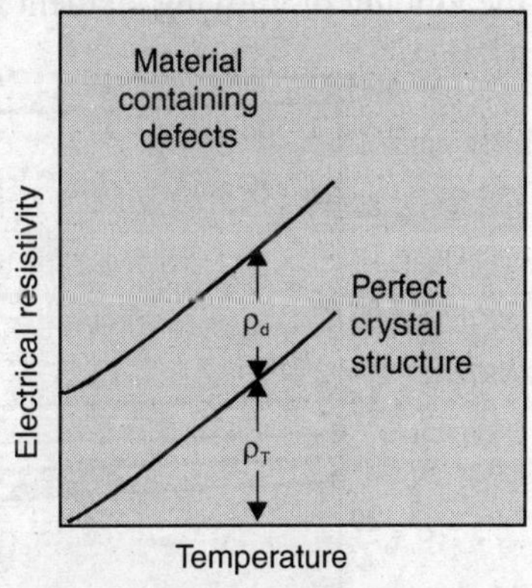

Fig. 3.12 The electrical resistivity of a metal is sum of a constant defect contribution ρ_d and a variable temperature contribution ρ_T.

Influence of plastic deformation: Plastic deformation also raises the electrical resistivity as a result of increased number of electron-scattering dislocations. Figure 3.9 also shows the effect of deformation.

Effect of pressure: At room-temperature the general behaviour of ρ of metal is to decrease initially with increasing pressure and it may pass through a minimum. The initial decrease is due to the effect of pressure in reducing the amplitude of lattice vibrations. The subsequent increase is probably due to modification of the electron band structure which leads to increased phonon scattering.

Effect of processing and strengthening: It is observed that strengthening mechanisms and metal processing techniques affect the electrical properties in a different ways. Table 3.3 shows the effect of alloying, strengthening and processing on the electrical conductivity of copper. It is reported that solid-solution strengthening is not a good way to get high strength in materials intended to have high electrical conductivities. Due to interstitial or substitutional atoms, the mean free paths are very short. The effect

Table 3.3 The Effect of Alloying, Strengthening, and Processing on the Electrical Conductivity of Copper and its Alloys.

Alloy	$\frac{\sigma_{alloy}}{\sigma_{Cu}} \times 100$	*Remarks*
Pure annealed copper	100	Few defects to scatter electrons; the mean free path is long.
Pure copper deformed 80%	98	Many dislocations, but because of the tangled nature of the dislocation networks, the mean free path is still long.
Dispersion-strengthened Cu-0.7% Al_2O_3	85	The dispersed phase is not as closely spaced as solid-solution. atoms, nor is it coherent, as in age hardening. Thus, the effect on conductivity is small.
Solution-treated Cu-2% Be	18	The alloy is single phase; however, the small amount of solid-solution strengthening from the supersaturated beryllium greatly decreases conductivity.
Aged Cu-2% Be	23	During aging, the beryllium leaves the copper lattice to produce a coherent precipitate. The precipitate does not interfere with conductivity as much as the solid-solution atoms.
Cu-35% Zn	28	This alloy is solid-solution strengthened by zinc, which has an atomic radius near that of copper. The conductivity is low, but not as low as when beryllium is present.

of zinc and other alloying elements on the conductivity of copper is shown in Fig. 3.13. We note that as the amount of alloying element increases the conductivity decreases substantially.

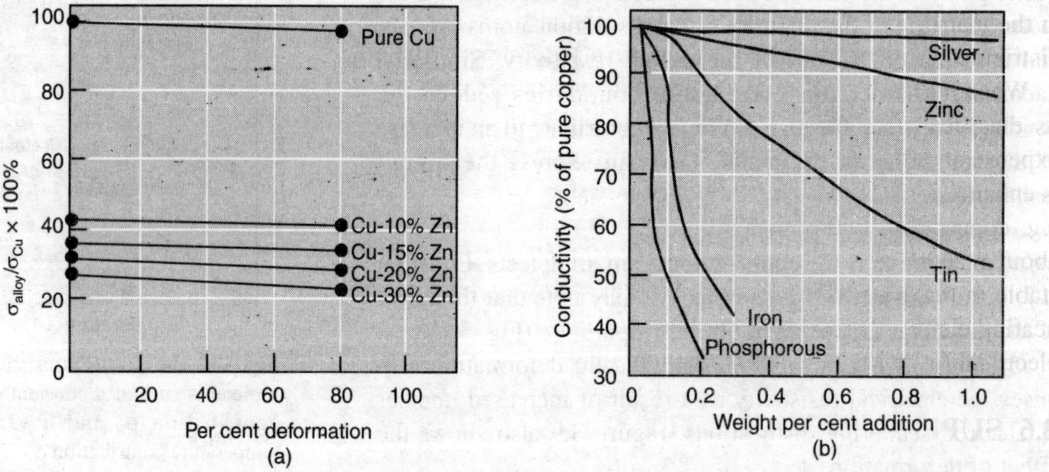

Fig. 3.13 (*a*) The effect of solid-solution strengthening and cold working on the electrical conductivity of copper, and (*b*) the effect of addition of few selected elements (Ag, Zn, Sn) on the electrical conductivity of copper.

We may note that age hardening and dispersion strengthening reduce the conductivity to an extent that it is less than solid-solution strengthening, since there is a longer mean free path between precipitants in comparison to the path between point defects. Interestingly strain hardening and grain size control have less effect on conductivity. Since dislocations and grain boundaries are further apart than solid-solution atoms, there are large volumes of metal that have a long mean free path. This means, cold working is an effective way to increase the strength of a metallic conductor without seriously affecting the electrical properties. Moreover, one can eliminate the effects of cold working on conductivity by the low temperature recovery heat treatment, in which good conductivity is restored while the strength is retained.

3.5.1 Conductivity of Alloys

The residual resistivity of alloys increases with increasing amount of solute content (Fig. 3.9 and Fig. 3.14). The slopes of individual ρ *vs.* T lines remains, however, essentially constant (Fig. 3.9). Small additions of solute cause a linear shift of ρ *vs.* T curves to higher resistivity values in accordance with the Matthiessen rule. Various solute elements might alter the resistivity of the host material to different degrees. This is depicted in Fig. 3.14 for Ag, which shows that the residual resistivity increases with increasing atomic number of the solute. This may be due to increase in interactions between electrons and impurity atoms when the solute has a larger atomic size, as is encountered by proceeding from Cd to antimony. However, ordering of atoms in alloys by heat treatment can decrease their resistivity.

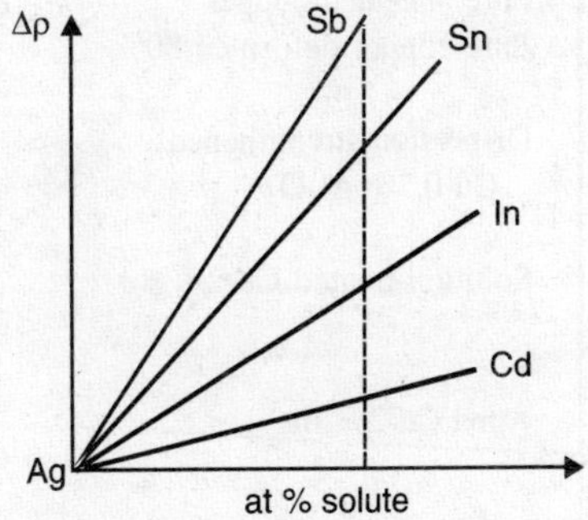

Fig. 3.14 Resistivity change of various dilute silver alloys (schematic). Solvent and solute are all from the fifth period.

The resistivity of *two phase* alloys is, in several instances increases, the sum of the resistivity of each components, taking the volume fractions of each phase into consideration. However, additional factors, such as the crystal structure and the kind of the phases in each other, also have to be considered. Some alloys, when in the ordered state, *i.e.* when the solute atoms are periodically arranged in the matrix, have a distinctly smaller resistivity compared to the case when the atoms are randomly distributed. Slowly cooled Cu_3Au or CuAu are common examples of ordered structure.

When it is desired to improve the mechanical strength of a metal alloy without imparing significantly its electrical conductivity. Both solid solution alloying and cold working improve strength at the expense of conductivity, and thus, one must made trade off for these two properties. Most often strength is enhanced by introducing a second phase that does not have so adverse an effect on the conductivity, *e.g.* copper-beryllium alloys are precipitation hardened, but even so, the conductivity is reduced by about a factor of 5 over high purity copper. Compared to pure metals, the resistivity of alloys tends to be stable in regards to temperature variation. Alloys, *e.g.* nichrome (~ 80% Ni-20% Cr) can be used for heating elements or for making thermocouple wires. Certain alloys of Bi-Sn-Pb-Cd are used to make electrical fuses due to their low melting ranges.

3.6 SUPERCONDUCTIVITY

Superconductivity is the ability of certain metals, alloys and ceramic materials to let electric current flow with no electrical resistance and energy dissipation. Superconductivity appears below a critical temperature, T_c as shown in Fig. 3.15. About 27 elements, numerous alloys, ceramic metals, MgB_2, uranium compounds, iron arsenides (FeAs), etc. have been found to possess this property (Table 3.4 and Fig. 3.16). It is established that the conductivity of superconductors below T_c is about 10^{20} 1/Ω-cm). The transition temperatures where superconductivity starts range from 0.01 K (for tungsten) up to about 134 K for mercury ceramic superconductors. Of particular interest are superconducting materials whose T_c is above 77 K, *i.e.* the boiling point of liquid nitrogen, which is more readily available than other coolants. Among the so-called *high-T_c cuprate superconductors* (HTSC) are the 1-2-3 compounds such as $YBa_2Cu_3O_{7-x}$ whose molar ratios of rare earth to alkaline earth to copper relate are 1:2:3. The transition temperature of HTSC cuprates range from 40 to 134 K. Ceramic superconductors have an orthorhombic, layered, perovskite crystal structure (similar to $BaTiO_3$).

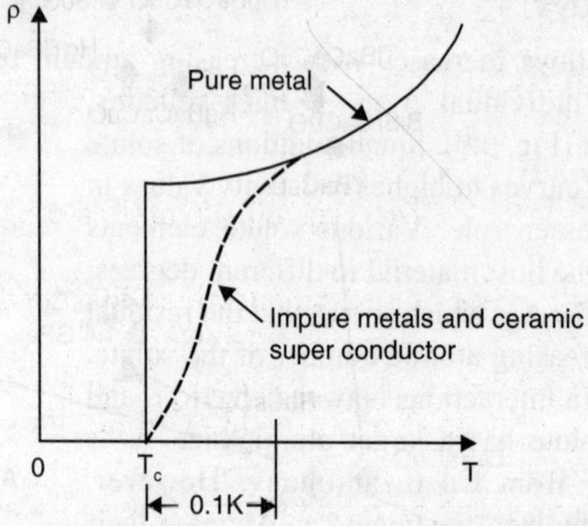

Fig. 3.15 Schematic representation of the resistivity of pure and compound superconducting materials. T_c is the critical or transition temperature, below which superconductivity commences.

Table 3.4 Critical temperatures of some superconducting materials

Materials	T_c *[K]*	*Remarks*
Tungsten	0.01	–
Mercury	4.15	H.K. Onnes (1911)
Sulfur-based organic superconductor	8	S.S.P. Parkin et al. (1983)
Nb_3Sn and Nb–Ti	9	Bell Labs (1961), Type II
V_3Si	17.1	J.K. Hulm (1953)
Nb_3Ge	23.2	(1973)
La–Ba–Cu–O	40	Bednorz and Müller (1986)
$YBa_2Cu_3O_{7-x}$	92	Wu, Chu and others (1987)
$RBa_2Cu_3O_{7-x}$	~92	R = Gd, Dy, Ho, Er, Tm, Yb, Lu
$Bi_2Sr_2Ca_2Cu_3O_{10-\delta}$	113	Maeda et al. (1988)
$Tl_2CaBa_2Cu_2O_{10-\delta}$	125	Hermann et al. (1988)
$HgBa_2Ca_2Cu_3O_{8-\delta}$	134	R. Ott et al. (1995)

A high magnetic field or a high current density (J) may eliminate superconductivity. Thus, for superconductivity to exist, the temperature must be below T_c, magnetic field must be below a certain value (H_c: called critical field). The current density above which superconducting state cannot be maintained is called critical current density (J_c). These three material parameters T_c, H_c and J_c are of extreme importance in the practical applications superconductivity. Figure 3.17 shows schematically the boundary in temperature, magnetic field, current density space separating normal and superconducting states. The position of this boundary will of course, depend on the material. For temperature, magnetic field and current density values lying between the origin and this boundary, the material will be superconductive, outside the boundary the conduction is normal.

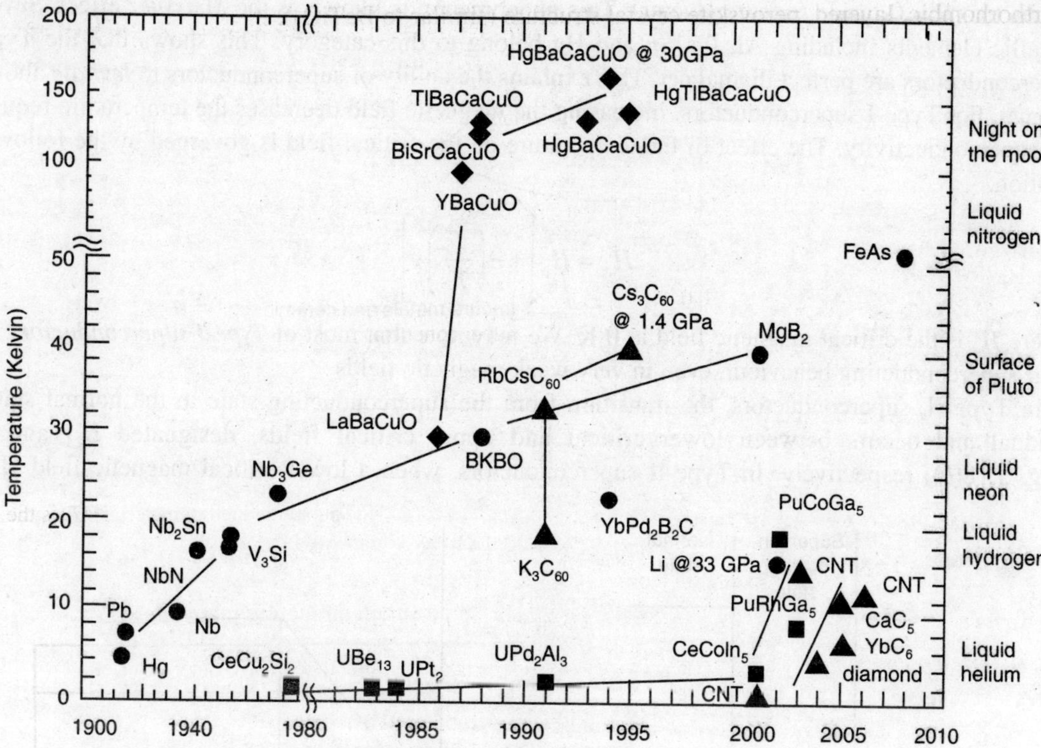

Fig. 3.16 Superconducting materials
Source: CCAS

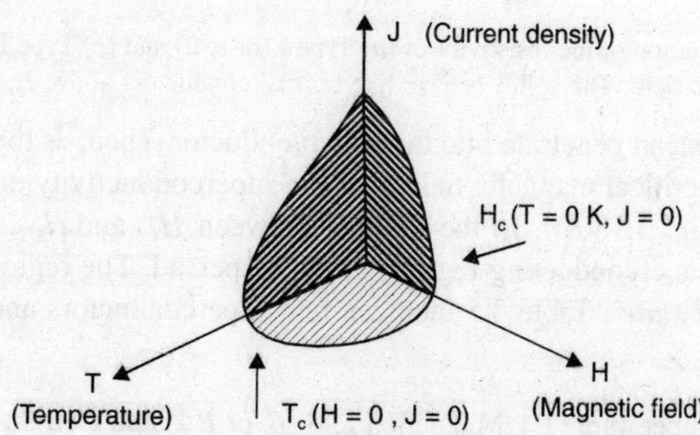

Fig. 3.17 Critical temperature (T_c), current density (J_c) and magnetic field boundary separating superconducting and normal conducting states.

The applied magnetic field (H) induces a magnetic flux density (B). The unit of B is Tesla. One can also represent the critical field in the form of a critical-flux density (B).

On the basis of magnetic response, superconducting materials may be classified as Type-I and Type-II materials. In Type-I superconductors, superconductivity is stable upto a magnetic field, H_c. Obviously, for superconductivity to exist, the temperature must be below T_c and magnetic field must be below a certain value (H_c). In Type-I superconductors, the annihilation of the superconducting state by a magnetic field, *i.e.* the transition between superconducting and normal states, occurs sharply (Fig. 3.18(a)). In order for a material to be a superconductor, the magnetic must be excluded from the conductor ($H = 0$ inside superconductor). This phenomenon is known as the *Meissner* effect. Several metallic elements including Al, Pb, Sn and Hg belong to this category. This shows that the Type-I superconductors are perfect diamagnet. This explains the ability of superconductors to *levitate* above a magnet. For Type-I superconductors, increasing the magnetic field decreases the temperature required for superconductivity. The effect of the temperature on the critical field is governed by the following relation:

$$H_c = H_o\left[1-\left(\frac{T}{T_c}\right)^2\right] \quad ...(18)$$

where H_o is the critical magnetic field at 0 K. We may note that most of *Type-I superconductors* lose their superconducting behaviour, even in very weak magnetic fields.

In Type II, superconductors, the transition from the superconducting state to the normal state is gradual and occurs between lower critical and upper critical fields, designated H_{c1} and H_{c2} (Fig. 3.18(*b*)) respectively. In Type-II superconductors, when a lower-critical magnetic field H_{c1} is

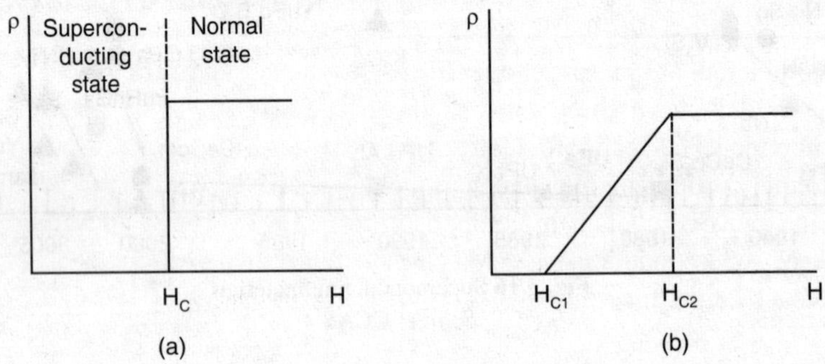

Fig. 3.18 Schematic representation of the resistivity of (*a*) Type I (or soft) and (*b*) Type II (or hard) superconductors in an external magnetic field. The solids behave like normal conductors above H_c and H_{c2} respectively.

applied, the magnetic flux can penetrate into the superconductor. Then, as the applied field increases to a higher value, *i.e.* upper critical magnetic field H_{C2}, the superconductivity disappears. H_{C2} is often 100 times larger than H_{C1} (Fig. 3.18(*b*)). In the interval between H_{C1} and H_{C2}, normal conducting areas, called *vortices*, and the superconducting regions are interspersed. The region between H_{C1} and H_{C2} is usually termed as '*mixed state*'. Table 3.5 includes few superconductors and their T_C and B_C (or B_{C2}) values.

Table 3.5 The Critical Temperature (T_c), Magnetic Field (B_c or B_{c2}) and Current Density (J_c) for Superconductivity of Few Selected Materials

Material	*T_c (K)*	*B_c or B_{c2} (Tesla)*	*J_c (A/cm²) at 0 K*
Type-I superconductors:			
W	0.015	–	
Al	1.180	0.0099	
Sn	3.720	0.0300	
Hg	4.150	0.0410	
Pb	7.190	0.0800	

Type-II superconductors:			
Nb	9.250	0.1990	~10^5
Nb_3Sn	18.050	24.5 at 0 K	10^7
Nb_3Ge	23.2	38	
MgB_2 (Ref.[5])	39	~10 at 4.2 K	
Ceramic superconductors:			
$(La, Sr)_2CuO_4$	40.000		
$YBa_2Cu_3O_{7-x}$ (*YBCO*)	93.000	300	10^4–10^7
$TiBa_2Ca_3Cu_4O_{11}$	122.000		
$Bi_2Sr_2Ca_2Cu_3O_{10}$ (*BSCCO2223*)	110		
$Bi_2Sr_2CaCu_2O_8$ (*BSCCO2212*)	92		
$HgBa_2Ca_2Cu_3O_8$	133–135		
$Hg_{0.8}Tl_{0.2}Ba_2Ca_2Cu_3O_{8.33}$	138 (Highest recorded at ambient pressure)		

Type-II superconductors are preferred over Type-I for most practical applications by virtue of their high T_c and H_{c2}. Widely used three most superconductors are Nb-Zr, Nb-Ti alloys and Nb_3Sn intermetallic compound. The technological potential of recently discovered HTSC Cuprate superconductors is extremely promising in as much as their T_c is above 77 K, which permits the use of liquid nitrogen, a very inexpensive coolant in comparison to liquid hydrogen and liquid helium. However, new HTSC Cuprate superconductors are not without the drawbacks, main of which is their brittle nature. This drawback limits the ability of these materials to be fabricated into useful forms such as wires.

We may note that the higher the value of magnetic field present, the lower the critical current density (J_c) or maximum amount of current per unit area. Even at zero magnetic fields, the superconductor cannot carry an infinite amount of current because, in polycrystalline superconductors, the current cannot flow without resistance from one *grain* to another. Obviously, grain boundaries and grain orientation or *texture* are quite important in both the bulk and thin-flim forms of superconductors. However, the motion of magnetic flux-lines limits the current transport within the grains.

There may be certain atomic level defects in superconductors can pin the magnetic vortices and these, in turn, help increase the maximum current carrying capacity of the superconductor. Moreover, the transport of grains across grain boundaries (termed as '*weak link*' effect), the transport of current within the grains, or the transport of current, may also be limited by percolation. However, the problem arises if there are pores of grains of second phases that can block the super current. This reveals that current density in a superconductors is indeed a micro-structure-sensitive property.

The quantum-mechanical *BCS theory* explains the origin of superconductivity. The theory is based on electron-phonon coupling and the resultant pairing of conduction electrons. According to BCS theory, superconductivity occurs as a result of the formation of Cooper pairs of electrons having opposite momenta and opposite spin ($K\uparrow$–$K\downarrow$). BCS theory explains the properties of superconductivity in metallic superconductors. However, a proper and acceptable theory of HTSC curprates and other anisotropic superconductors is still lacking.

Flux quantization: This refers to the fact that the magnetic flux threading a superconducting loop cannot have an arbitrary value; it has to be a multiple of $\phi_0 = h/2e$. The flux quantum is defined as

$$\phi = \frac{h}{2e} = 2.0678 \times 10^{-15} \text{ Weber} \quad \ldots(19)$$

Superconducting devices can measure this tiny variation of magnetic flux which is exceedingly important in metrology and advanced instrumentation.

3.7 JOSEPHSON EFFECTS

In 1962 B.D. Josephson predicted theoretically that Cooper pair tunnelling through a very thin insulating layer (~2 nanometer) is possible. His theory predicted in addition to the Giaever current, the existence of supercurrent, arising from tunnelling of the bound electron pairs. Such a junction is called *weak link*. The effects of pair tunnelling include:

(*i*) *D.C. Josephson effect*: A d.c. current flows across the junction resulting from cooper pair tunnelling in the absence of any electric or magnetic field, *i.e.* at zero voltage across the insulating layer. One obtains the following expression for current through the contact for two identical superconductors as

$$I = I_c \sin \delta$$

where I_c is the critical current for the Josephson junction and δ is the phase difference of the two cooper wave functions. Theoretical results for voltage drop across the junction yield.

$$V_t = \frac{\pi\Delta}{2e} \quad \ldots(20)$$

For metals most commonly used as electrodes, lead and niobium. V_t is of the order of ~2.5 mV.

(*ii*) *A.C. Josephson effect*: Josephson also predicted that if a constant non zero voltage V were maintained across the tunnel barrier, an alternating supercurrent would flow through the barrier in addition to the *dc* current produced by the tunnelling of unpaired electrons. This effect has been utilized in a precision determination of $\hbar/e$. Further, an r_f voltage applied with the dc voltage can then cause a dc current across the junction. The frequency of the ac Josephson current given by the Josephson relation

$$\omega = \frac{2\,eV}{\hbar} \quad \ldots(21)$$

where V is the dc voltage applied across the junction.

Josephson Junctions

These are sophisticated sandwich structures of superconducting films (usually of Nb) separated by extremely thin (~10 Å) insulating oxide layers. In a suitable submicroscopic circuit they act as the fastest switching elements available today.

Josephson junction have also the lowest power consumption at their operating temperature, so they might provide the basis for the architecture of some of the fastest computers of the future. They are already used in the fastest commercially available oscilloscopes that operate at 10 GHz.

Applications of Superconductors

Superconductors offers several advantages over conventional electrical conductors. They enable the manufacturing of components (*e.g.* high field magnets) that could not be feasible using conventional conductors. Under suitable conditions superconductors offer zero resistance, which means one can consider superconductors for electrical applications (motors, cables for transmission of power, etc.). The energy saving due to the absence of electrical resistance more than compensate for the energy required to maintain superconductors' operation temperature. Superconducting devices are typically 50% smaller and lighter than equivalent conventional components and their manufacturing process generates no incremental emissions of greenhouse gases. Their cooling is ensured by non-flammable liquid nitrogen or helium, as opposed to flammable and/or toxic oil Coolant used in high performance conventional components. Apart from the cooling system, the typical cost of low temperature superconductors (LTSC) per unit of carried electrical current is at least than ten times lower than the cost of an equivalent conventional conductors. All these advantages translate into technical and economical benefits. However, considering the cost of superconductors' cooling system, superconductivity is not economically competitive with conventional conductors in most applications.

The special properties of superconductors:

(*i*) Zero resistance to direct current.
(*ii*) High current density (J_c).
(*iii*) Low resistance at high frequency.
(*iv*) Low signal dispersion.
(*v*) High sensitivity to magnetic field.
(*vi*) No penetration of externally applied magnetic field.
(*vii*) Rapid single flux quantum transfer.
(*viii*) Close to speed of light signal transmission.

These properties enable important applications: (*i*) Low resistance at high frequency and low signal dispersion play a fundamental role for microwave components and in communication technologies. (*ii*) High sensitivity to magnetic fields makes it possible to produce superconductive sensors with a sensing capability 1000 times higher than conventional devices. (*iii*) The capability of driving off external magnetic fields holds the potential for applications in magnetic levitation systems for transportation. (*iv*) Important applications in digital electronics and high speed computing derive from other peculiar superconducting properties such as the Josephson effect and the sharp transition to superconducting status. A comprehensive list of superconductivity applications is provided in Tables 3.6 and 3.7.

Table 3.6 Large-scale Applications of Superconductivity

Application	*Major technical features*
Power cables	Higher current densities, smaller cable diameters, lower transmission losses.
Current limiters	Highly non-linear super-normal conductor transition, self-controlled current limitation.
Transformers	Higher current densities, smaller size, lower weight, lower losses.
Motors/generators	Higher current densities, higher magnetic fields, smaller size, low weight and losses.
Magnets for RTD, magnetic energy, storage, magnetic, separation, NMR, spectroscopy, MRI, magnetic levitation	Higher current densities, higher and ultra-higher magnetic fields, higher magnetic field gradients, smaller size, lower weight lower losses, persistent currents, ultra-high temporal field stabilities, stronger levitation forces, larger air gaps.
Cavities for Accelerators	Lower surface resistance, higher quality factors, higher microwave-power handling.
Magnetic bearings (based on HTS bulk materials)	Higher current densities, lower losses, stronger levitation forces, self-controlled autostable levitation.

Table 3.7 Electronics applications of superconductivity

Application	*Major technical features*
High frequency sensor coils for NMR	Lower resistive losses, higher quality factors, smaller size.
Microwave filters for wireless communication	Lower surface resistance, smaller size, lower transmission losses, higher quality factors.
Resonator for oscillators and other passive microwave devices	Lower resistive losses, higher quality factors, lower transmission losses, smaller size.
Far-infrared bolometers	Highly non-linear super-normal conductor transition, higher irradiation-mediated temperature sensitivities.
Microwave detectors	Highly non-linear junction characteristic, higher conversion efficiency for frequency-mixing.
X-Ray detectors	Lower particle excitation energies, higher photon energy resolution.
SQUID sensors for RTD, medical diagnosis and non-destructive testing SQUID amplifiers	Persistent currents, quantum interference effects, ultra-high magnetic field sensitivities, low noise low-signal amplification.
Voltage standards for metrology and industry	Voltage steps in microwave irradiated junction arrays, quantum precision output voltages.
Digital circuit and micro-processors	Persistent current, single flux quantum signal levels, ultra-fast ultra-low power data transfer and processing.

Boosted by the growing demand for sustainable technologies and most efficient use of resources, the integration of superconductors in end-use technologies is one of the technology challenges of this century. Superconductor technology challenges can in fact provide cost-efficient and environmentally favourable solutions to many current and future needs (Tables 3.6 and 3.7).

At present, high field magnets, with high current density and zero DC resistance are mostly used in research and technology development (RTD) and in biomedical applications such as the *magnetic resonance imaging* (MRI). Over the past decades, the extraordinary development in cooling systems made it possible the introduction of superconducting magnets in MRI and enabled impressive improvements in the image resolution for medical diagnosis. LTSC magnets are also used for research devices and for large-scale, high-field magnetic systems used in nuclear fusion and particle accelerators and detectors. Current application fields of superconductors also include electronics and industrial processes. An example of industrial application is the magnetic separation of Kaolin clay where the use of superconductors enables a 95% reduction of energy use. Examples of electronic applications are outstanding *sharp and low-noise microwave filters* used in radio communication systems, and *superconducting quantum interference devices* (SQUIDs) based on weakly coupled functions.

SQUIDs are able to monitor magnetic fields billion times weaker than the earth magnetic field. They are used for monitoring and recording heart and brain functions. Notably, multi-channel SQUIDS for magneto-cardiogram (MCG) and magneto-encephalogram (MEG) devices have been developed by improving the sensitivity of LTSC SQUIDs above a few *fT*/(Hz) 1/2.

Superconductors with ultralow losses in alternate current (AC) are underdevelopment for power applications such as electricity transmission, large transformers and motors. HTSC multi-channel

SQUIDs are underdevelopment for next generation electronic devices to save space and provide mobility. A large R and D effort is devoted to *Quantum computers* based on SQUIDs and Josephson junctions. Much faster than existing computers, quantum computers could solve problems with exponential complexity in polynomial times, with important consequences in cryptan analysis and both civilian and military applications.

Superconductors could replace copper and permanent magnets in new design stators, rotors and transformers with improved performance and reliability and reduced energy losses and weight. It is estimated that a 10 MW wind turbine using HTSC technology could weigh one-third for a conventional wind generator with equivalent power. Further prospective applications include information and communication technologies, industrial processes, transportation and medical applications.

New discoveries, such as the superconductivity in FeAs, MgB_2 and carbon nanotubes, may also bring new applications so far not considered possible.

HTSC cost-to-performance ratio as well as cost and technical development of commercial cooling systems for both LTSC and HTSC are currently the main barriers to large superconductivity deployment.

3.8 ELECTRICAL CONDUCTIVITY IN OTHER MATERIALS

It is found that electrical conductivity in most ceramics and polymers is normally low. However, there are special materials which provide limited or even good conduction. Using dopants, several ceramics, *e.g.* $BaTiO_3$, TiO_2, ZrO_2, that are normally insulating, can be converted into conductive oxides. The conduction in these ceramic oxides occur as a result of movement of ions or electrons and holes.

3.8.1 Electrical Conduction in Ionic Materials

In ionic materials, conduction often occurs by movement of entire ions, since the energy gap is too large for electrons to enter the conduction band. This is why, most ionic materials behave as insulators.

The mobility of the charge carriers or ions in ionic materials is given by

$$\mu = \frac{ZqD}{k_B T} \qquad \ldots(22)$$

where D is coefficient of diffusion, k_B is Boltzmann constant and Z is the valence of the ion. We may note that the mobility is many orders of magnitude lower than the mobility of electrons. This means the conductivity of ionic materials is very small:

$$\sigma = nqz\mu \qquad \ldots(23)$$

It is observed that impurities and vacancies increase conductivity. Vacancies are necessary for diffusion in substitutional types of crystal structures and impurities can diffuse and help carry the current. With the rise in temperature conductivity increases due to increase in rate of diffusion.

Indium tin oxide (ITO) is most widely used conductive and transparent oxide. This is used as a transparent conductive coating on plate glass. Other applications of ITO include touch screen displays for computers and devices, *e.g.* automated teller machines. Other conductive oxides include YSZ (Ytrria-stabilized Zirconia). This oxide is used as a solid electrolyte in solid oxide fuel cells. Another oxide, lithium cobalt oxide is used as a solid electrolyte in lithium ion batteries. We may note that, although most ceramic materials behave as electrical insulators, but one can convert many of them into semiconductors by properly engineering the point defects.

Polymers have low electrical conductivity due to involvement of their valence electrons in covalent bonding and also having a band structure with a large energy gap. Polymers find frequent use in applications that require electrical insulation to prevent short circuits, arching and safety hazards. Conductivity of four common polymers are given in Table 3.1. However, in some cases, the low conductivity of polymers is a hindrance, *e.g.* static electricity can accumulate on housings for electronic equipment, making the polymer transparent to electromagnetic radiation that damages the internal solid-state devices. In case of lightening strikes the polymer-matrix composite wing of an airplane, severe damage can occur. One can solve these problems by two approaches: (*i*) by introducing an additive to the polymer to improve the conductivity, and (*ii*) by creating polymers that inherently have good conductivity. There are some applications, such as liquid crystal display (LCD), one require polymers that respond to application of an electric field.

It is observed that adding ionic compounds to the *polymer* can reduce resistivity because ions migrate to the polymer surface and attract moisture, which in turn dissipates static charges. One can also dissipate static charge by introducing conductive filler materials. Polymer matrix composites containing carbon or nickel-plated carbon fibers combine high stiffness with improved conductivity; hybrid composites containing metal fibers, along with normal carbon, glass, or aramid fibers, also produce lightening safe aircraft skins. Effect of carbon fibers on the electrical resistivity of nylon is shown in Fig. 3.19.

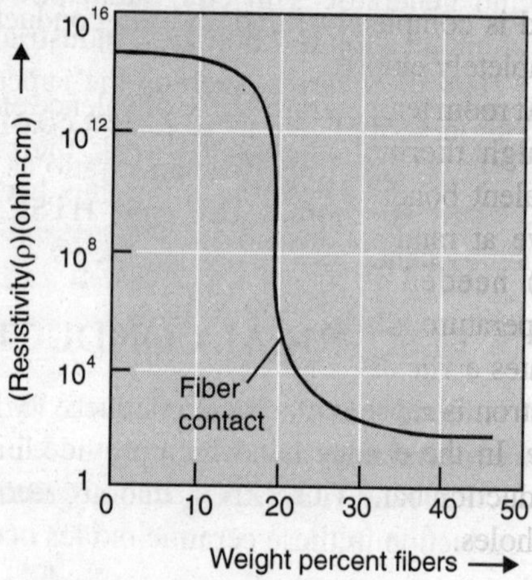

Fig. 3.19 Effect of carbon fibers on the electrical resistivity of nylon.

From Fig. 3.19, we note that when enough carbon fibers are introduced to nylon in order to assure fiber to fiber contact, the resistivity of nylon reduces by about 13 orders of magnitude. One also uses conductive fillers and fibers to produce polymers that shield against electromagnetic radiation.

As a result of doping or processing techniques, some polymers inherently have good conductivity. When acetal polymers are doped with agents, *e.g.* arsenic pentafluoride, electrons or holes are able to jump freely from one atom to another along the backbone of the chain. This helps to increase the conductivity to near that of metals. There are some polymers, *e.g.* polyphthalocyanine, can be cross-linked by special cutting processes to enhance the conductivity to as high as $10^2\ \Omega^{-1} \cdot cm^{-1}$, a process that allows the polymer to behave as a semiconductor. Cross-linking helps electrons to move more easily from one chain to another.

Carbon nanotubes is another example of polymeric materials used in electronics. Carbon nanotubes can be made either conducting or insulating carbon nanotubes are being developed for electronics, hydrogen and several other applications.

3.9 SEMICONDUCTORS

Semiconductors such as Si or Ge are neither good conductors nor good insulators. This may seem to make semiconductors to be of little interest. Their usefulness results, however, from a completely different property, namely that extremely small amounts of certain impurity elements, called *dopants*, remarkably change the electrical behaviour of semiconductors. Pure elemental Si and Ge are *intrinsic semiconductors*. These elements which are in group IV A in the periodic table, have the diamond cubic structure with highly directional covalent bonds. Tetrahedral sp^3 hybrid bonding orbitals consisting of electron pairs bond the atoms together in the crystal lattice. In this structure each Si and Ge atom contributes four valence electrons. A simplified two dimensional representation of a Ge crystal at absolute zero temperature is shown in Fig. 3.20. Since the valence electrons bind one atom to another, they are not available for electrical conduction in the absence of any thermal disruption of chemical bonds at 0 K. Obviously at 0 K an intrinsic (pure) semiconductor behaves like an insulator. In the band picture, we can describe the situation by saying that at 0 K the valence bond is completely filled and the conduction band is completely empty.

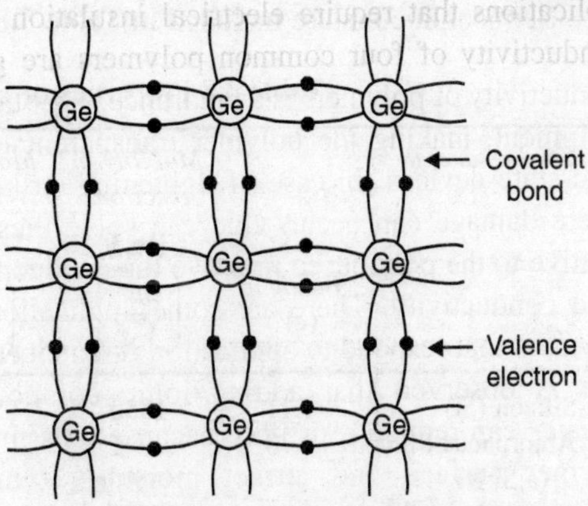

Fig. 3.20 A simplified representation of the crystalline structure of a semiconductor germanium (Ge) crystal at absolute zero temperature.

At room temperature, some of valence electrons have enough thermal energy to break their valence or covalent bonds. These electrons thus become *free* to move at random throughout the crystal. The energy (E_g) needed to break a covalent bond at room temperature is about 1.1 eV in silicon and 0.72 eV in Ge. Each electron that leaves the covalent bond creates a *vacant site* or *hole* at its original position. A vacancy in the covalent bond caused by a free electron is represented by a tiny circle (Fig. 3.21(*a*)). This vacancy or deficiency of electron is called a *hole*. In the energy band picture, the electrons are thermally excited from the valence band into the conduction band where they become free (Fig. 3.21(*b*)). The vacancies thus formed in the valence band are holes.

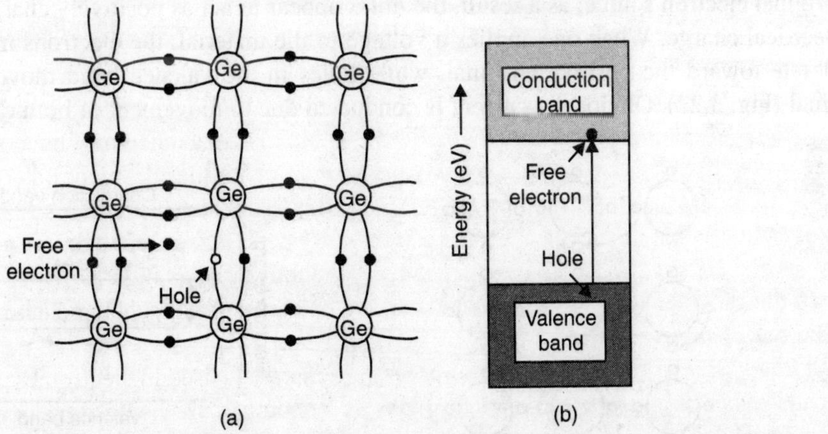

Fig. 3.21 Creation of an electron-hole pair by the rupture of a covalent bond at room temperature (*a*) two-dimensional lattice of Ge crystal at room temperature and (*b*) energy band picture of creation of electron-hole pair in Ge crystal at room temperature.

Ge and Si are elemental semiconductors. There are also compound semiconductors formed from elements in groups 2*B* and 6*B* of the periodic table (*e.g.* CdS, CdSe, CdTe, HgCdTe, etc.) and are known as II-VI (two-six) semiconductors. Compound semiconductors can also be formed combining elements from groups 3*B* and 5*B* of periodic table (*e.g.* GaN, GaAs, AlAs, AlP, InP, etc.). These compound semiconductors are known as III-V (three-five) semiconductors. These semiconducting materials including Si and Ge, provide the building blocks for many electronic devices. The properties of some of the commonly encountered.

Semiconductors are included in Table 3.8.

Table 3.8 Properties of Few Commonly Known Semiconductors

Semiconductor	*Bandgap (eV)*	*Mobility of electrons* (μ_n) $\left(\frac{cm^2}{V\text{-}s}\right)$	*Mobility of holes* (μ_p) $\left(\frac{cm^2}{V\text{-}s}\right)$	*Dielectric constant (k)*	*Resistivity* $(\Omega \cdot cm)$	*Density* $\left(\frac{gm}{cm^3}\right)$	*Melting temperature (°C)*
Silicon (Si)	1.11	1350	480	11.8	2.5×10^5	2.33	1415
Amorphous Silicon (a:Si:H)	1.70	1	10^{-2}	~11.8	10^{10}	~2.30	–
Germanium (Ge)	0.67	3900	1900	16.0	43	5.32	936
SiC(α)	2.86	500	–	10.2	10^{10}	3.21	2830
Gallium Arsenide (GaAs)	1.43	8500	400	13.2	4×10^8	5.31	1238
Diamond	~5.50	1800	1500	5.7	$>10^{18}$	3.52	~4200
α-Sn	0.10	2000	1000	–	10^{-4}	5.80	232

The energy gap (E_g) between the valence band and conduction band in semiconductors is relatively small (Fig. 3.8A). Consequently, some electrons possess enough thermal energy to exceed the gap and enter the conduction band. As mentioned earlier, the excited electrons leave behind unoccupied energy levels, or holes, in the valence band. When an electron moves to fill a hole, another hole is created from the original electron source; as a result, the *holes* appear to act as positively charge electrons and carry an electrical charge. When one applies a voltage to the material, the electrons in the conduction band accelerate toward the positive terminal, while holes in the valence band move toward the negative terminal (Fig. 3.22). Obviously, current is conducted due to movement of both electrons and holes.

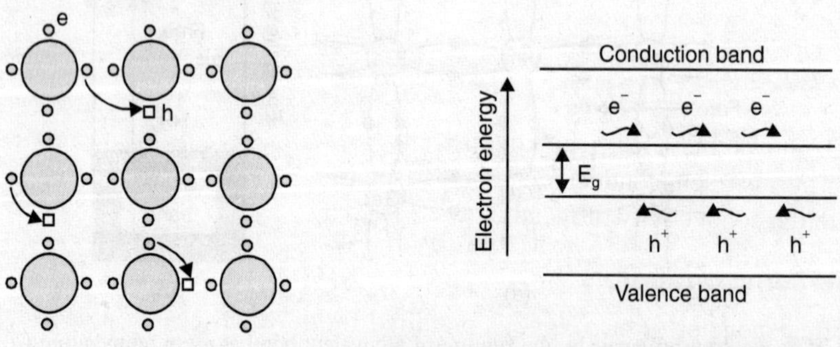

Fig. 3.22 When a voltage is applied to a semiconductor, the electrons move through the conduction band, while the electron holes (h^+) move through the valence band in the opposite direction. E_g is the gap energy between the valence and conduction bands.

One can determine the conductivity by the number of electron-hole pairs,

$$\sigma = nq\mu_n + pq\mu_p \qquad \text{...(24)}$$

Here n is the concentration of electrons in the conduction band, p is the concentration of holes in the valence band, and μ_n and μ_p are the mobilities of the electrons and holes, respectively (Table 3.8).

In intrinsic superconductors, we can see that for every conduction electron promoted to the conduction band, there is a hole in the valence band. Thus

$$n_i = p_i$$

This means, the conductivity of intrinsic semiconductor becomes

$$\sigma = qn_i(\mu_n + \mu_p) \quad \text{...(25)}$$

We may note that in intrinsic semiconductors, we control the number of charge carriers and hence, the electrical conductivity by controlling the temperature. With the rise in temperature, there is a greater probability that an energy level in the conduction band is occupied (and an equal probability that a level in the valence band is unoccupied, *i.e.*, a hole is present). The number of electrons in the conduction band, which is equal to the number of holes in the valence band is expressed as

$$n = n_i = p_i = n_o \exp\left(\frac{-E_g}{2k_BT}\right) \quad \text{...(26)}$$

where n_o is related to the density of states and the effective masses of electrons (m_n^*) and holes (m_p^*) as follows:

$$n_o = 2\left(\frac{2\pi k_BT}{h^2}\right)^{3/2}\left(m_n^* m_p^*\right)^{3/4} \quad \text{...(27)}$$

where k_B and h are Boltzmann and Planck's constants. For Ge, Si and GaAs the values of n_i at room temperature are: 2.5×10^{13}, 1.5×10^{10} and 2×10^6 electrons per cm^3, respectively. We may note that the product n_ip_i remains constant at any given temperature and a given semiconductor. This permits one to calculate n_i or p_i values at different temperatures.

At higher temperatures, more electrons may cross the forbidden zone and, hence, the conductivity of a semiconductor increases:

$$\sigma = n_o q(\mu_e + \mu_n)\exp\left(\frac{-E_g}{2k_BT}\right) \quad \text{...(28)}$$

Figure 3.23 shows the electrical conductivity *vs.* temperature for intrinsic semiconductor. A comparison is also shown with the conductivity of metals.

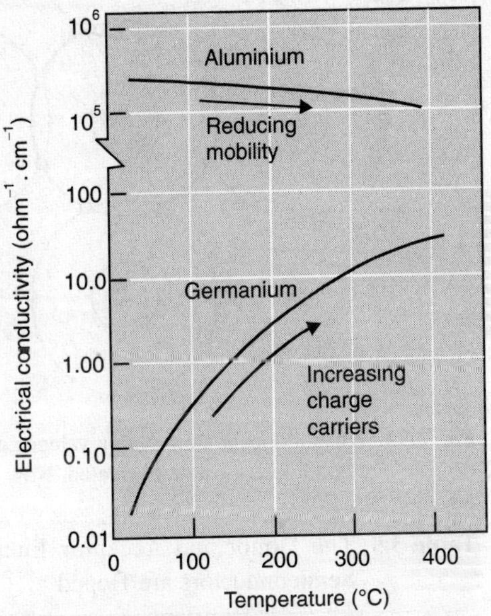

Fig. 3.23 The electrical conductivity *versus* temperature for intrinsic semiconductors compared with metals. We may note the break in *y*-axis scale.

3.9.1 Extrinsic Semiconductors

An extrinsic semiconductor (*n* or *p*) type is preferred for devices since its properties are temperature stable and can be controlled using ion implantation or diffusion of dopants. Extrinsic superconductors are very dilute substitutional solid solutions in which the solute impurity atoms have different valence characteristics from the solvent atomic lattice. The concentrations of the added impurity atoms in these semiconductors are usually in the range of 100 to 1000 parts per million (ppm).

n-Type (Negative-Type) Extrinsic Semiconductors

When group *VA* impurity atoms such as P, As and Sb are added to Si or Ge, they provide easily ionized electrons for electrical conduction. Figure 3.24 shows that four of the electrons from the Sb atom participate in the covalent-bonding process, while the extra electron enters an energy level in a donor state just below the conduction band. *Since the extra* electron is not rigidly bound to the atoms, only a small increase in energy E_d, is required for the electron to enter the conduction band. We usually define, E_d as the energy difference between the top of the valence band and donor band. In this case, the energy difference required $=(E_g - E_d)$. Obviously, the energy gap controlling is now E_d rather than E_g (Table 3.9). We may note that no corresponding holes are created when the donor electrons enter the conduction band.

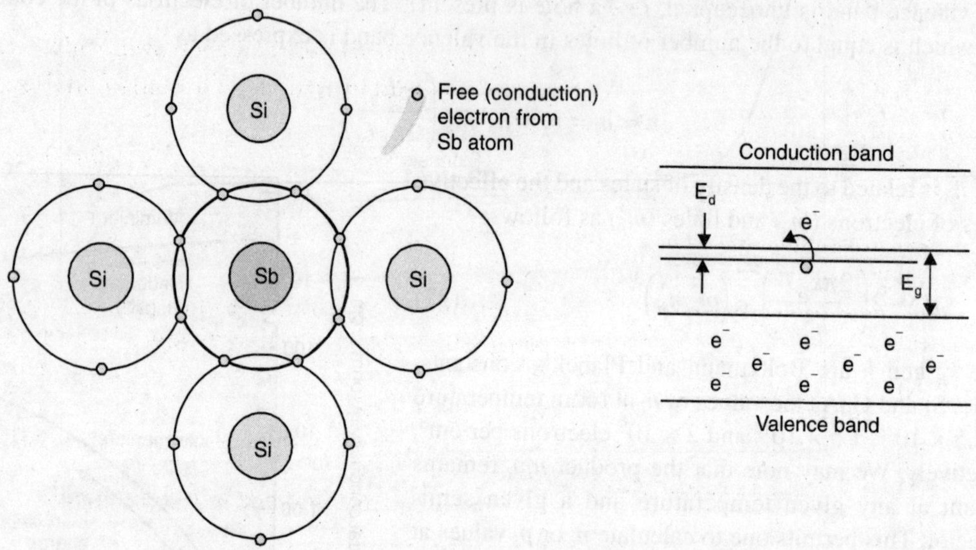

Fig. 3.24 When a dopant atom with a valence greater than four is added to silicon, an extra electron is introduced and a donor energy state is created. Now electrons are more easily excited into the conduction band.

Table 3.9 The Donor and Acceptor Energy Gaps (in Electron Volts) when Silicon and Germanium Semiconductors are Doped

	Silicon		*Germanium*	
Dopant	E_d	E_a	E_d	E_a
P	0.045		0.0120	
As	0.049		0.0127	
Sb	0.039		0.0096	
B		0.045		0.0104
Al		0.057		0.0102
Ga		0.065		0.0108
In		0.160		0.0112

Some intrinsic semiconductors still occurs, with a few electrons gaining enough energy to jump the large E_g gap. However, this becomes important only at higher temperatures. We may remember that in an extrinsic semiconductor, there has to be overall electrical neutrality. This means, the number of donor atoms (N_d) and holes per unit volume (p_o) (both are positively charged) is equal to the number of acceptor atoms (N_a) and electrons per unit volume (both are negatively charged). Thus,

$$p_o + N_d = n_o + N_a \quad \ldots(29)$$

where n_o and p_o are the concentrations of electrons and holes in the extrinsic semiconductor.

When the extrinsic semiconductor is heavily doped (*i.e.* $N_d >> n_i$), then $n_o \sim N_d$. Similarly, if it is a heavily acceptor doped (*p*-type) semiconductor, then $N_a >> p_i$ and hence $p_o \sim N_a$. This is very important, since this implies that by adding a considerable amount of dopant, one can, dominate the conductivity of a superconductor by controlling the concentration. This clearly reveals that why extrinsic semiconductors are most useful for making controllable semiconductor devices, *e.g.* transistors.

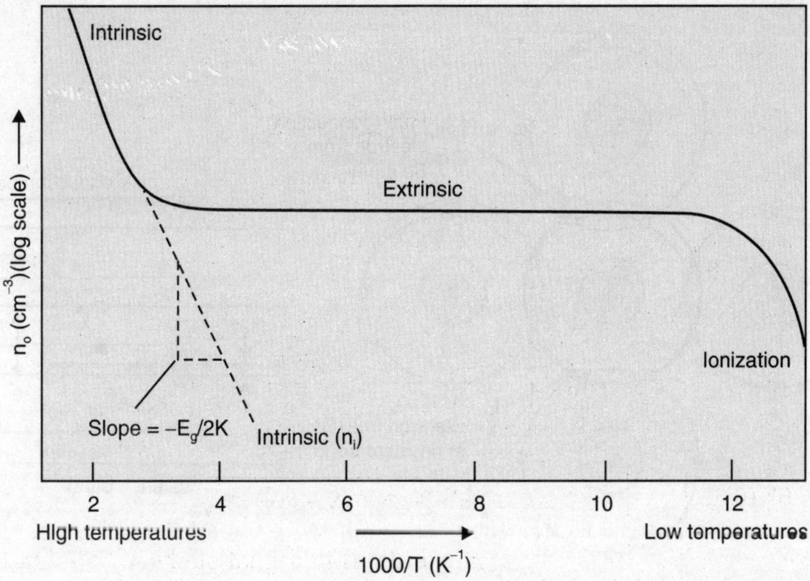

Fig. 3.25 The effect of temperature on the carrier concentration of an *n*-type semiconductor.

Figure 3.25 shows the changes in concentration with temperature. From this curve, one can easily follow the approximate conductivity changes in an extrinsic semiconductor. At too low temperature, the donor or acceptor ions are not ionized and hence the conductivity is very small. With the increase in temperature, ionization process is complete and electrons (holes) contributed by donors (or acceptors) become available for conduction. In the region labelled extrinsic, the conductivity remains nearly independent of temperature. The value of conductivity at which plateau occurs would depend on the level of doping. When temperature become too high, the behaviour now approaches that of an intrinsic semiconductor as the effect of dopant is almost lost. In this analysis, we have not accounted for the effect of dopant concentrations on the mobility of electrons and holes and the temperature dependence of bandgap. At very light temperatures (not shown in Figure 3.25) the conductivity *decreases* again as scattering of carriers dominates.

p-Type (Positive-Type) Extrinsic Semiconductors

When a trivalent group IIIA element, such as (B^{3+}) is substitutionally introduced in the Si or Ge tetrahedrally bonded lattice, there are not enough electrons to complete the covalent bonding process, *i.e.* one of the bonding orbitals is missing, and a hole exists in the bonding structure of Si (Fig. 3.26(*a*)). The holes act as acceptor of electrons. These hole sites have a somewhat higher energy than normal energy and create an acceptor level of possible electron energies just above the valence bond (Table 3.9). In order to create a hole in the valence band, an electron must gain an energy of only E_g. When the hole associated with the B atom is filled by an electron from a neighbouring Si atom, the B atom becomes ionized and acquires a negative charge of –1. The binding energy associated with the removal of an electron from a Si atom, thereby creating a hole, and subsequent transfer of the electron to the B atom is only 0.045 eV. We may note that this amount of energy is small compared to the 1.1 eV required to transfer an electron from the valence band to the conduction band. In the presence of an *applied electric field,* the hole created by the ionization of a boron atom behaves as a positive charge carrier and migrates in the Si lattice toward the negative terminal.

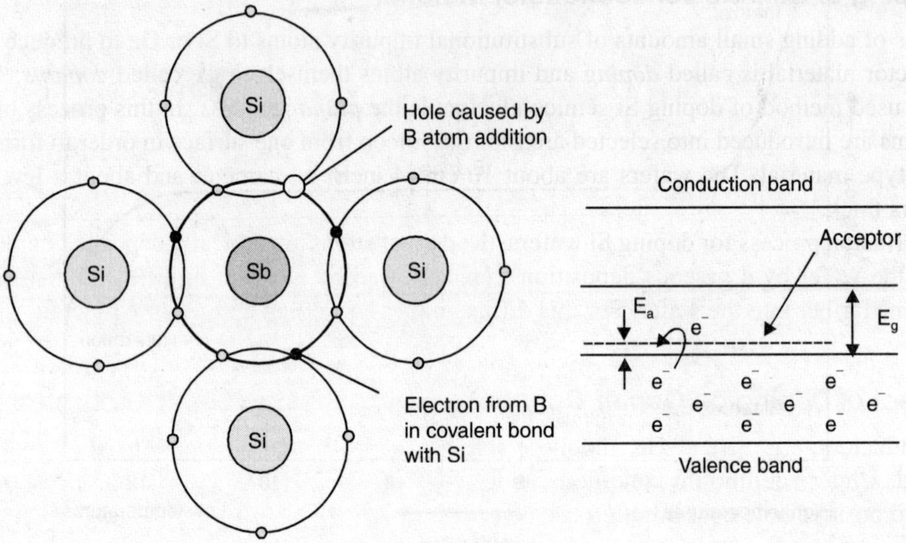

(a)

Fig. 3.26 (*a*) When a dopant atom with a valence of less than four is substituted into the silicon structure, a hole is created in the structure and an acceptor energy level is created just above the valence band.

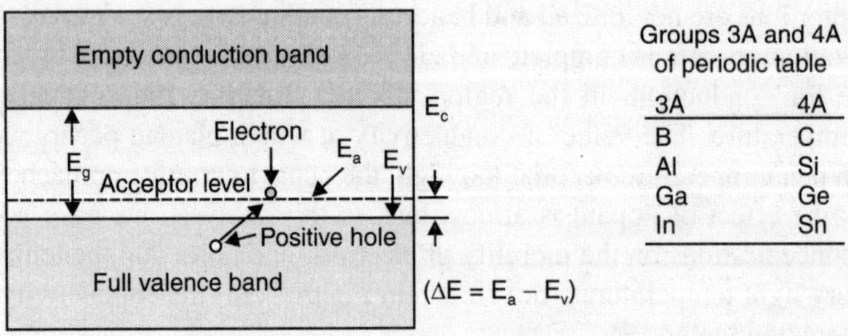

(b)

Fig. 3.26 (*b*) Energy-band diagram for a *p*-type extrinsic semiconductor showing the position of the acceptor level created by the addition of an atom of a group III A element such as Al, B or Ga to replace a silicon atom in the silicon lattice. Only a small amount of energy ($\Delta E = E_a - E_v$) is necessary to excite an electron from the valence band to the acceptor level, thereby creating an electron hole (charge carrier) in the valence band.

In terms of the energy band diagram, the B atom provides an energy level called an *acceptor level* that is slightly higher (≈ 0.045 eV) than the upper most level of the full valence band of silicon. When a valence electron of a Si atom near a boron atom fills a missing electron hole in a B–Si valence band, this electron is elevated to the acceptor level and creates a negative boron ion. In this process an electron-hole is created in the silicon lattice that acts as a positive charge carrier. Atoms of group IIIA elements such as B, Al and Ga provide acceptor levels in Si semiconductors and are called acceptor atoms. Since the majority carriers in these extrinsic semiconductors are holes in the valence band structure, they are designated as *p*-type semiconductors or *p-type extrinsic semiconductors* (Fig. 3.26(*b*)).

3.9.2 Doping of Extrinsic Semiconductor Material

The process of adding small amounts of substitutional impurity atoms to Si or Ge to produce extrinsic semiconductor material is called *doping* and impurity atoms themselves are called *dopants*. The most commonly used method of doping Si semiconductors is the *planar process*. In this process of doping, dopant atoms are introduced into selected areas of the silicon from one surface in order to form regions of *p*- or *n*-type material. The wafers are about 10 cm (4 inch) in diameter and about a few hundred micrometers thick.

In the diffusion process for doping Si wafers, the dopant atoms are typically deposited on or near the surface of the wafer by a gaseous deposition step, followed by a driven in diffusion that moves the dopant atoms farther into the wafer. For this diffusion process, a high temperature of about 1100 °C is required.

3.9.3 Effect of Doping on Carrier Concentrations in Extrinsic Semiconductors

In semiconductors, *e.g.* Si and Ge, mobile electrons and holes are constantly being generated and recombined. Under equilibrium conditions under constant temperature, the product of negative free electron and positive hole concentration is a constant. We have

$$np = n_i^2 \qquad \ldots(30)$$

where n_i is the intrinsic concentration of carriers in a semiconductor and is a constant at a given temperature. We may note that the relation (30) is valid for both intrinsic and extrinsic semiconductors. In an extrinsic semiconductor the increase in one type of carrier (*n* or *p*) reduces the concentration of the other through recombination so that the product of the two, *i.e.* *n* and *p* is a constant at any given temperature.

The carriers whose concentration in extrinsic semiconductor is the larger are designated the *majority carriers* and those carriers whose concentration is the smaller and termed the *minority* carriers (Table 3.10).

Table 3.10 Carrier Concentrations in Extrinsic Semiconductors

Semiconductor	*Majority-carrier concentrations*	*Minority-carrier concentrations*
n-type	n_n (concentration of electrons in *n*-type material)	p_n (concentration of holes in *n*-type material)
p-type	p_p (concentration of holes in *p*-type material)	n_p (concentration of electrons in *p*-type material)

n_n and p_n denotes the concentration of electrons and holes in an *n*-type semiconductor respectively. Similarly in a *p*-type semiconductor p_p and n_p denotes the concentrations of holes and electrons respectively.

Charge densities in extrinsic semiconductors

The total crystal of an extrinsic semiconductor must be electrically neutral. This means that the charge density in each volume element must be zero. There are two types charged particles in extrinsic semiconductors, *e.g.* Si and Ge: immobile ions and mobile charge carriers. The immobile ions originate due to ionization of donor or acceptor impurity atoms in Si or Ge. The concentration of the positive donor ions is denoted by N_d and that of the negative acceptor ions by N_a. The mobile charge carriers originate mainly from the ionization of the impurity atoms in the Si or Ge, and their concentrations are denoted by n for negatively charged electrons and p for positively charged holes.

The electrical neutrality of a semiconductor demands that the magnitude of the total negative charge density must equal the total positive charge density. This means the total negative charge density is equal to the sum of the negative acceptor ions N_a and the electrons, or $N_a + n$. The positive charge density is equal to the sum of the positive donor ions N_d and the holes, or $N_d + p$. Thus, we have

$$N_a + n = N_d + p \quad \text{...(31)}$$

In n-type semiconductor created by adding donor impurity atoms to intrinsic Si, $N_a = 0$. In n-type semiconductor, the number of electrons is much greater than the number of holes, *i.e.* $n >> p$, then eqn. (31) reduces to

$$n_n \approx N_d \quad \text{...(32)}$$

Obviously, in an n-type semiconductor the free-electron concentration is approximately equal to the concentration of donor atoms. One can obtain the concentration of holes in an n-type semiconductor from eqn. (30). Thus, we have

$$p_n = \frac{n_i^2}{n_n} \approx \frac{n_i^2}{N_d} \quad \text{...(33)}$$

The corresponding equations for p-type semiconductors of Si and Ge are

$$p_p \approx N_a \quad \text{...(34)}$$

and

$$n_p = \frac{n_i^2}{p_p} \approx \frac{n_i^2}{N_a} \quad \text{...(35)}$$

We may note that for Si at 300K the intrinsic carrier concentration $n_i = 1.5 \times 10^{16}$ carriers/m^3. For extrinsic Si doped with arsenic at a typical concentration of 10^{21} impurity atoms/m^3, we have concentration of major carriers $n_n = 10^{21}$ electrons/m^3 concentration of minority carriers $p_n = 2.25 \times 10^{11}$ holes/m^3. Obviously, for extrinsic semiconductors the concentration of the majority carriers is normally much larger than that of the minority carriers.

The mobilities of electrons and holes in Si at room temperature are at a maximum at low impurity concentrations and then decrease with impurity concentration, reaching a maximum value at high concentrations (Fig. 3.27).

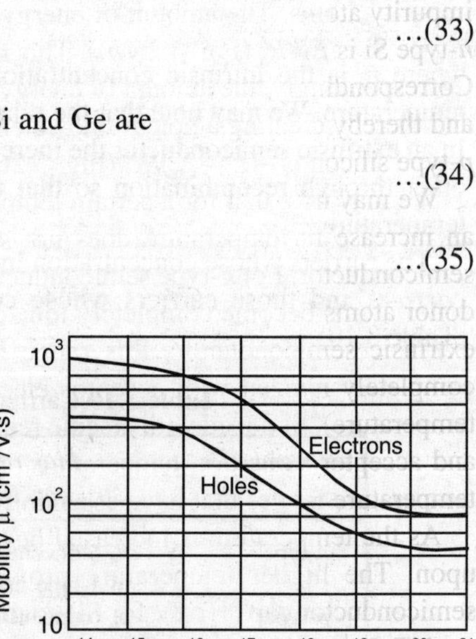

Fig. 3.27 The effect of total ionized impurity concentration on the mobility of charge carriers in silicon at room temperature (300 K).

Effect of temperature on the electrical conductivity of extrinsic semiconductors

The electrical conductivity of doped Si is affected by temperature (Fig. 3.28). At lower temperatures the number of impurity atoms per unit volume activated (ionized) determines the electrical conductivity of the Si. With the rise in temperature, more and more impurity atoms are ionized, and thus the electrical conductivity of extrinsic Si increases with increasing temperature in the extrinsic range.

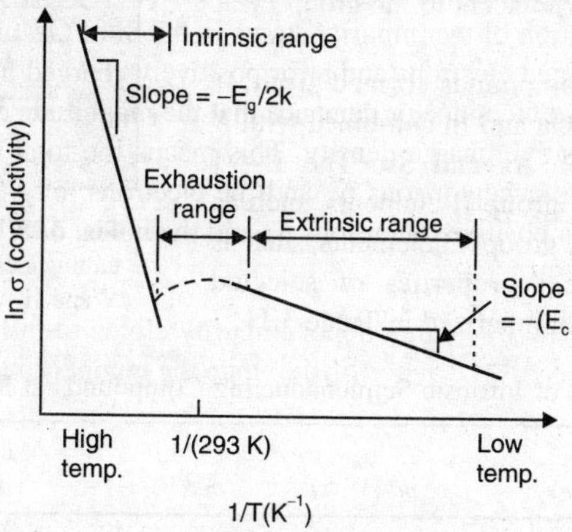

Fig. 3.28 Schematic plot of in σ (conductivity) *vs.* 1/*T*, *K*$^{-1}$ for an *n*-type extrinsic semiconductor.

In this extrinsic range only a relatively small amount of energy (≈ 0.04 eV) is required to ionize the impurity atoms. The amount of energy required to excite a donor electron into the conduction band in *n*-type Si is E_c–E_d (Fig. 3.26(*b*)). This means the slope of *ln* σ *vs.* 1/*T*, K^{-1} for *n*-type Si is $-(E_c - E_d)/K$. Correspondingly, the amount of energy required to excite an electron in *p*-type Si into an acceptor level and thereby creating a hole in the valence band is $E_a - E_v$. Obviously, the slope of *ln* σ vs. 1/T, K^{-1} for *p*-type silicon is $-(E_a - E_v)/K$. (Fig. 3.26(*b*)).

We may note that for a certain temperature range above that where complete ionization takes place, an increase in temperature does not substantially change the electrical conductivity of an extrinsic semiconductor. For *n*-type semiconductor this temperature range is called as the *exhaustion range* since donor atoms become completely ionized after the loss of their donor electrons (Fig. 3.28). For *p*-type extrinsic semiconductors this range is called as *saturation range*. Since acceptor atoms become completely ionized with acceptor electrons. To provide an exhaustion range at about 300 K (room temperature), Si doped with arsenic requires about 10^{21} carriers/m^3. We may note that donor exhaustion and acceptor saturation temperature ranges are important for semiconductor devices as they provide temperature ranges that have essentially constant electrical conductivities for operation.

As the temperature is increased beyond that of the exhaustion range, the intrinsic range is entered upon. The higher temperature provide sufficient activation energies for electrons to jump the semiconductor gap (1.1 eV for Si) so that intrinsic conduction becomes dominant. The slope of the *ln* σ *vs.* 1/*T*, *K*$^{-1}$ plot becomes much steeper and is $-E_g/2k$. For Si-based semiconductors with an energy gap of 1.1 eV, extrinsic semiconductors can be used up to about 200 °C. We may note that the upper limit for the use of extrinsic conduction is determined by the temperature at which intrinsic conductivity becomes important.

3.9.4 Compound Semiconductors

There are many compounds of different elements that are normally insulating (because the band gap is too large) are semiconductors. Among the major type of semiconducting compounds are the *MX* ones, where *M* is a more electropositive element and *X* a more electronegative element. Of the *MX* semiconducting compounds, two important groups are the III–V (3–5) and II–VI (2–6) compounds formed by elements adjacent to the group IV A of the periodic table (Fig. 3.29).

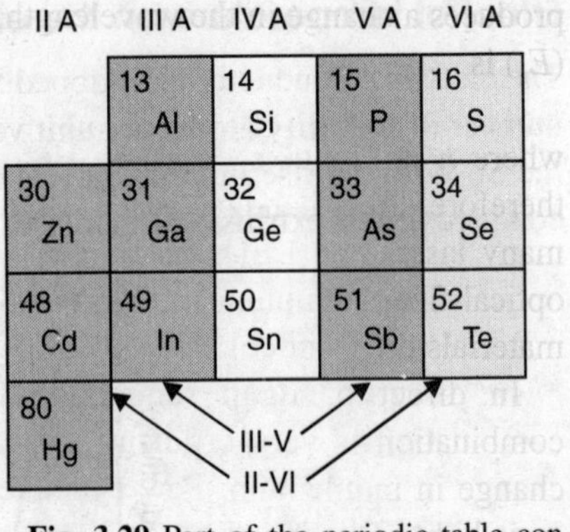

Fig. 3.29 Part of the periodic table containing elements which form *MX*-type III–V and II–VI semiconductor compounds.

The III–V semiconductor compounds formed from *M* group III elements, such as Al, Ga and In combined with *X* group V elements, such as P, As and Sb. The II–VI compounds are formed from M group II elements, such as Zn, Cd and Hg combined with X group VI elements, such as S, Se and Te. Some electrical properties of selected compound semiconductors are summarized in Table 3.11.

Table 3.11 Electrical Properties of Intrinsic Semiconducting Compounds at Room-temperature (300 K)

Group	*Material*	E_g *eV*	μ_n $m^2/(V \cdot s)$	μ_p $m^2/(V \cdot s)$	*Lattice constant*	n_i *carriers/m*^{3}
IV A	Si	1.10	0.135	0.048	5.4307	1.50×10^{16}
	Ge	0.67	0.390	0.190	5.257	2.4×10^{19}
III A – V A	GaP	2.25	0.030	0.015	5.450	–
	GaAs	1.47	0.720	0.020	5.653	1.4×10^{12}
	GaSb	0.68	0.500	0.100	6.096	–
	InP	1.27	0.460	0.010	5.869	–
	InAs	0.36	3.300	0.045	6.058	–
	InSb	0.17	8.000	0.045	6.479	1.35×10^{22}
II A – VI A	ZnSe	2.67	0.053	0.002	5.669	–
	ZnTe	2.26	0.053	0.090	6.104	–
	CdSe	2.59	0.034	0.002	5.820	–
	CdTe	1.50	0.070	0.007	6.481	–

GaAs is the most important of all the semiconductor compounds and is used in several electronic devices. There are other materials that are normally insulating (due to too large bandgap) can also be turned into semiconductors by doping. For example, $BaTiO_3$, ZnO, TiO_2 and many other oxides. Obviously, the concept of *n*– and *p*– type dopants is not limited to Si, Ge, GaAs, etc. One can dope $BaTiO_3$ for example, and make *n*– or *p*– type $BaTiO_3$. Such semiconductor compounds are useful in several sensor applications such as *thermistors*.

3.9.5 Direct and Indirect Bandgap Semiconductors

One can promote an electron from the conduction band to the valence band without changing the momentum of the electron. Such semiconductors are referred as *direct band semiconductors, e.g.* GaAs When the excited falls back into the valence band, light is produced due to combination of electrons and holes, *i.e.*

$$\text{Electron + hole} \rightarrow h\nu \qquad \ldots(36)$$

This is known as *radiative recombination*. The direct bandgap materials, *e.g.* GaAs and solid solutions of these, *e.g.* GaAs-AlAs, etc. are used to make light emitting diodes (LEDs) of different colours. One can tune the bandgap of semiconductors using solid solutions. The change in bandgap produces a change in the wavelength, *i.e.* the relation between the frequency of colour (ν) and bandgap (E_g) is

$$E_g = h\nu \qquad \ldots(37)$$

where h is Planck's constant. Here, an optical effect is obtained using an electronic material and therefore direct bandgap materials are also known as *optoelectronic materials*. Using these materials, many lasers and LEDs have been developed LEDs that emit light in the infrared range are used in optical-fiber communication systems to convert light waves into electrical pulses. Using direct bandgap materials different coloured lasers, such as the newest blue laser using GaN, have been developed.

In direct bandgap semiconductor materials are such materials in which the electron-hole combination is very efficient and the electrons cannot be promoted to the valence band without a change in momentum. Few examples of these materials are Si, Ge and GaP. These materials cannot emit light. They produce heat due to combination of electrons and holes and the heat is dissipated within the material. This type of combination is called *nonradiative recombination*:

$$\text{electron} + \text{hole} \rightarrow \text{heat} \qquad \ldots(38)$$

We may note that both direct and indirect bandgap materials can be doped to form *n*– or *p*– type of semiconductors.

3.9.6 Applications of Semiconductors

Semiconductor devices

Semiconductors are at the heart of electronics industry and all computer chips. The ability of semiconductor manufacturers to put extremely complex electrical circuits on a single chip of Si of about 1 cm square or less and having thickness 200 μm has revolutionized the design and manufacture of countless products, *e.g.* an advanced microprocessor or "Computer on a chip". We may note that the microprocessor forms the basis for many of the latest products that use the progressive miniaturization of Si-based semiconductor technology.

Semiconductors are used for making diodes, transistors, lasers and LEDs, etc. Silicon is the workhouse of very large scale integrated circuits (ICs). ICs consist of millions of transistors integrated into an area as small as 1 cm^2.

pn Junction

Most common semiconductor devices depend on the properties of the boundary between *p*-type and *n*-type materials. A *p-n* junction diode can be produced by growing a single crystal of intrinsic Si and doping it first with *n* type material and then with *p*-type material (Fig. 3.30(*a*)). Most commonly, however, the *pn* junction is produced by solid state diffusion of one type impurity (for *e.g.*; *p*-type) into existing *n*-type material (Fig. 3.30(*b*)).

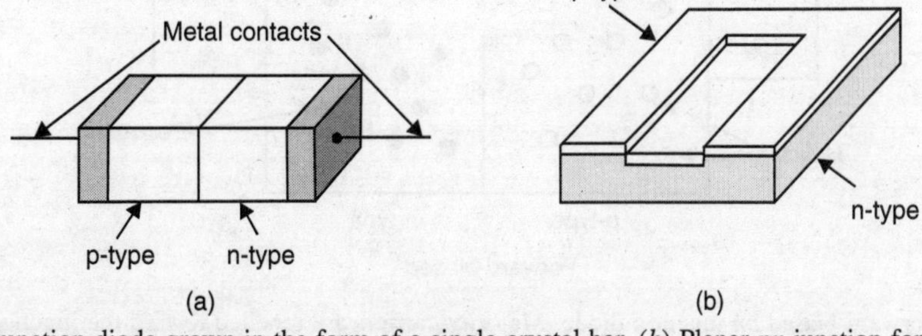

Fig. 3.30 (*a*) *pn* junction diode grown in the form of a single crystal bar. (*b*) Planar *pn* junction formed by selectively diffusing a *p*-type impurity into an *n*-type semiconductor crystal.

In *p-n* junction the majority carriers near or at the junction diffuse across the junction and recombine (Fig. 3.31(*a*)). Since the remaining ions near or at the junction are physically larger and heavier than the electrons and holes, they remain in their positions in the Si lattice (Fig. 3.31(*b*)). After a few recombinations of majority carriers at the *pn* junction, the process stops due to the electrons crossing over *pn* junction into the *p*-type materials are repelled by the large negative ions. Similarly, holes crossing over the *pn* junction are repelled by the large positive ions in the *n*-type material. The immobile ions at the *pn* junction create a zone depleted of majority carriers called a *depletion region.* Under equilibrium conditions, *i.e.* open-circuit conditions) there exists a potential difference or barrier to majority-carrier flow. Clearly, no net current flow under open-circuit conditions.

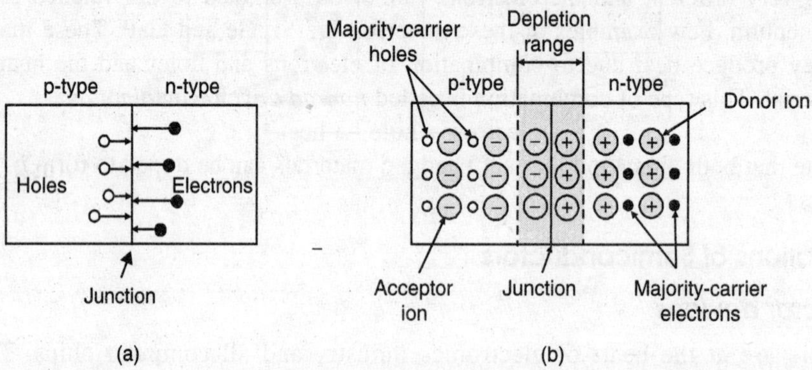

Fig. 3.31 (*a*) *pn* junction diode showing majority-carriers (holes in *p*-type material and electrons in *n*-type material) diffusing toward junction. (*b*) Formation of depletion region at and near *pn* junction due to loss of majority carriers in this region by recombination. Only ions remain in this region in their positions in the crystal structure.

The *p-n* junction is used to make diodes and many other devices such as transistors. Electrically, the *pn* junction is conducting when the *p*-side is connected to a positive voltage (known as *forward bias* as shown in Fig. 3.32).

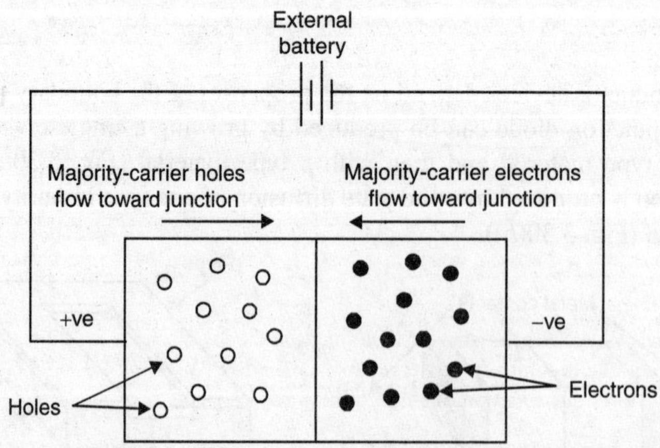

Fig. 3.32 Forward-biased *pn* junction diode. Majority-carriers are repelled toward the junction and cross over it to recombine so that a large current flows.

When a negative bias is applied to the *p*-side of a *pn* junction (*reverse bias* as shown in Fig. 3.33) the *p-n* junction does not carry much current.

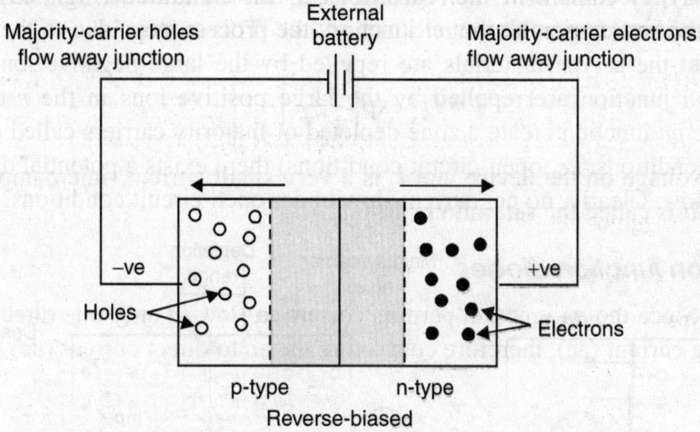

Fig. 3.33 Reverse-biased *pn* junction diode. Majority-carriers are attracted away from the junction, creating a wider depletion region than when the junction is at equilibrium. Current flow due to majority carriers is reduced to near zero. However, minority carriers are biased forward, creating a small leakage current.

The forward current can be upto a few milliamperes, while the reverse-bias current is few nano-amperes. We may note that in a *pn* junction under no bias, the diffusion and drift currents due to electron and hole motions cancels out.

The *V–I characteristics* of a *pn* junction diode are shown in Fig. 3.34.

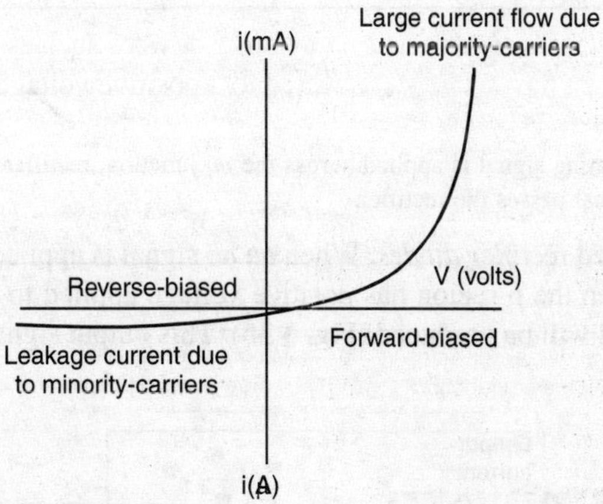

Fig. 3.34 Current-voltage (V–I) characteristics of a *pn* junction diode. When the *pn* junction diode is reverse-biased, a leakage current due to minority carriers combining exists. When the *pn* junction diode is forward-biased, a large current flows due to recombination of majority-carriers.

Under forward-bias — *i.e.*, forward bias w.r.t. majority-carriers — the energy barrier at the junction is reduced so that some electrons and holes can cross the junction and subsequently recombine. During the forward biasing of a *pn* junction, electrons from the battery enter the negative material of the diode (Fig. 3.32). For every electron that crosses the junction and recombines with a hole, another electron enters from the battery. Moreover, for every hole that recombines with an electron leaves the *p*-type material and flows toward the positive terminal of the battery. Since the energy barrier to electron flow is reduced when the *pn* junction is forward biased, considerable current can flow through the junction (Fig. 3.34). Obviously, electron flow, *i.e.* current flow can continue as long as the *pn* junction is forward-biased and the battery provides an electron source. This current is

$$I = I_S \left[\exp\left(\frac{eV}{k_B T} \right) - 1 \right] \quad \ldots(39)$$

where V is applied voltage on the device and I_s is a very small current (microamp) which still flows during reverse-bias. It is called the saturation current.

Applications for pn junction diodes

(*i*) *Rectifier diodes:* Since the *pn* junction permits current to flow in only one direction, it passes only half of an alternating current (ac), therefore converting the ac to direct current (dc) (Fig. 3.35). Diodes

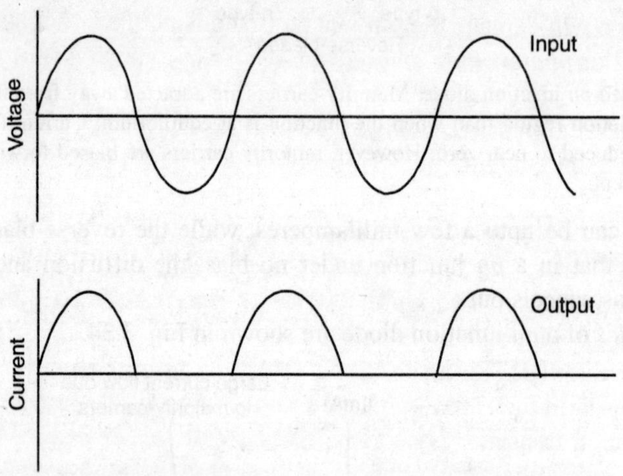

Fig. 3.35 When an alternating signal is applied across the *pn* junction, rectification occurs and only half of the input signal passes the rectifier.

used for this process are called *rectifier diodes*. When an *ac* signal is applied to a *pn* junction diode, the diode will conduct only when the *p* region has positive voltage applied to it relative to *n* region. As a result, half wave rectification will be produced (Fig. 3.36). This output signal can be smoothed out with

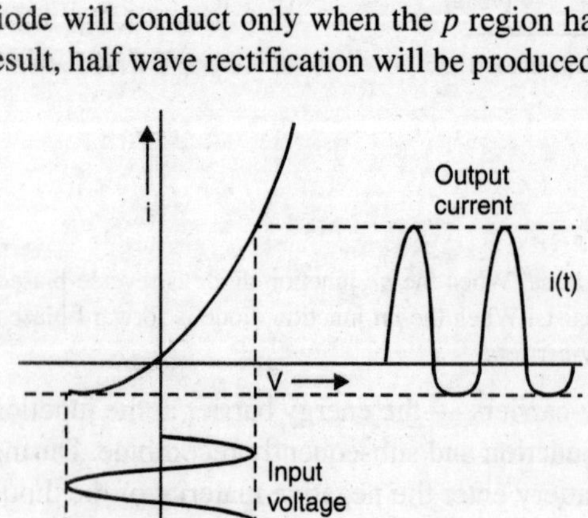

Fig. 3.36 Voltage-current diagram illustrating the rectifying action of a *pn* junction diode to convert alternating current (ac) to direct current (dc). The output current is not completely direct current but is mostly positive. This dc signal can be smoothed out by using other electronic devices.

other electronic devices and circuits so that a steady dc signal can be produced. In a wide range of current capacities solid-state silicon rectifiers are used that can be from tenths of an ampere to several hundred amperes or more. Voltages, too can be as high as 1000 V or more.

Breakdown diodes or Zener diodes

These are silicon rectifiers in which the reverse current (leakage current) is small, and then with only slightly more reverse-bias voltage, a breakdown voltage is reached upon which the reverse current increases very rapidly (Fig. 3.34). The breakdown under reverse bias can be due to the tunnelling of electrons. This is called as *Zener breakdown*, which occurs in heavily doped *p-n* junctions with smaller depletion layer widths. Breakdown diodes can be made with breakdown voltages from a few to several hundred volts and are used for voltage-limiting applications and for voltage stabilizing under conditions of widely varying current. We may note that Zener diodes are always used in *reverse-bias*. For lightly doped *p-n* junctions, breakdown can occur through the *avalanche breakdown* mechanism. We may note that, when the voltage across *p-n* junctions becomes too high, electron hole pairs are created by *impact ionization*. This breakdown will damage the *p-n* junctions and is normally not useful.

Solar cell: A solar cell or *photovoltaic device* consists of a *p-n* junction. If light of sufficient large energy ($E = h\nu$) falls on this device, electrons are lifted from the valence band into the conduction band. The electrons in the barrier region drift quickly to the *n*-side, whereas the concomitantly created holes in the valence band migrate into the *p*-region. This provides a potential difference between the *n* and *p* terminals.

Bipolar Junction Transistor

The term transistor is derived from two words "transfer" and "resistance". There are two types of transistors fabricated using *p-n* junction: *pnp* and *npn*.

A bipolar junction transistor (BJT) is an electronic device that can serve as a '*switch*' or as a '*current amplifier*'. BJT is often used in the central processing units of computers due to their rapid switching response. A bipolar junction transistor is a sandwich of either *npn* or *pnp* semiconductor materials. Figure 3.37 shows schematically an *npn* type bipolar junction transistor and exhibit that three main zones of the transistor: *emitter, base* and *collector*. The emitter emits charge carriers. Since the emitter of the *npn* transistor is *n*-type, it emits electrons. The base controls the flow of charge carriers and is

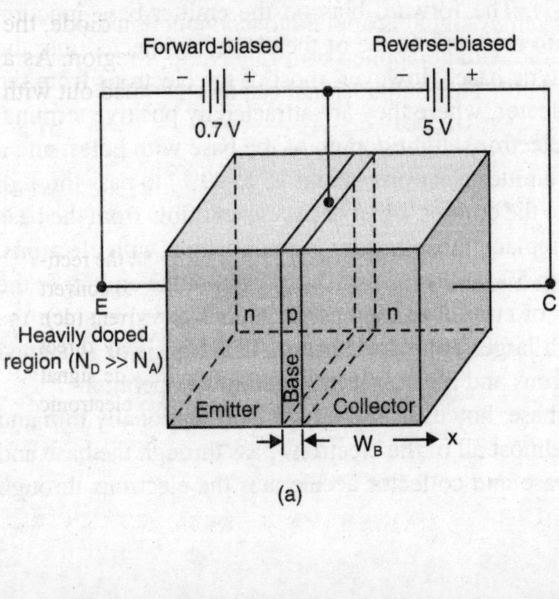

Fig. 3.37 *(a)* An *npn* bipolar junction transistor (BJT). The *n* region on the left is the emitter, the thin *p* region in the middle is the base, and the *n* region on the right is the collector. For normal operation the emitter-base junction is forward-biased and the collector-base junction is reverse-biased.

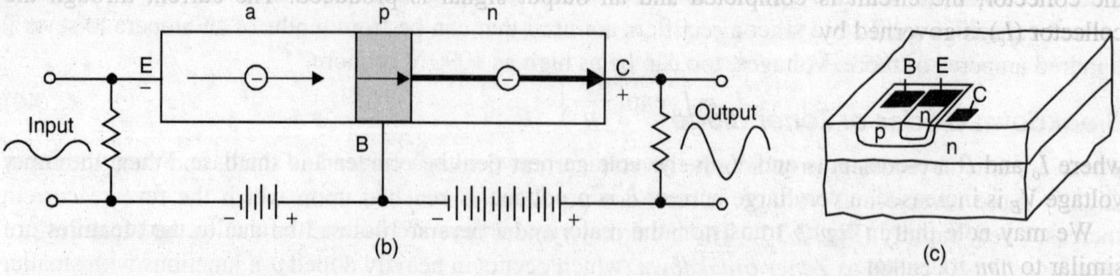

Fig. 3.37 (*b*) Biasing of an *n-p-n* bipolar transistor. (*c*) Schematic representation of an *n-p-n* bipolar transistor. The dark areas are the contact pads.

p-type for the *npn* transistor. The base is made very thin (about 10^{-3} cm thick) and is lightly doped so that only a small fraction of the charge carriers from the emitter will recombine with oppositely charge majority carriers of the base. The collector collects charge carriers mainly from the emitter. Since the collector zone of *npn* BJT is *n*-type, it collects mainly electrons from the emitter. A sketch of the cross-section of the *npn* BJT is shown in Fig. 3.38.

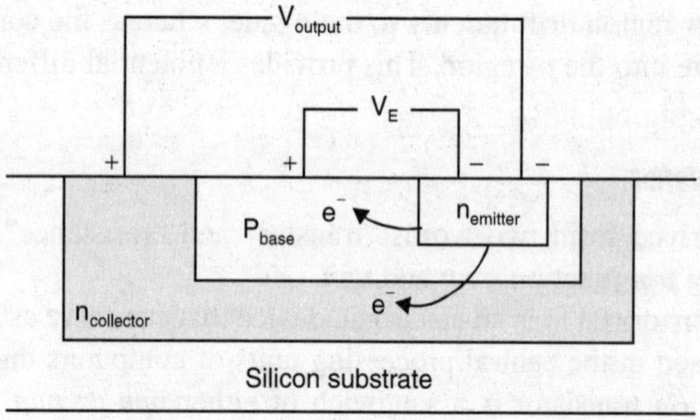

Fig. 3.38 Sketch of the cross-section of *npn* BJT.

Under normal operation of the *npn* transistor, the emitter-base junction is forward-biased and the collector-base junction is reverse-biased (Fig. 3.37). The forward-bias on the emitter-base junction causes an injection of electrons from the emitter into the base. Some of the electrons injected into the base are lost by recombination with holes in the *p*-type base. However, most of the electrons from the emitter pass right through the thin base into the collector, where they are attracted by positive terminal of the collector. Heavy doping of the emitter with electrons, light doping of the base with holes, and a very thin base are all factors that cause most of the emitter electrons (about 95 to 99%) to pass through to the collector. Very few holes flow from the base to the emitter. Most of the current flow from the base terminal to the base region is the flow of holes to replace those lost by recombination with electrons. The current flow to the base is small and is about 1 to 5% of the electron current from the emitter to the collector. In some respects, one can think the flow of current to the base as a control valve since the small base current can be used to control the much larger collector current. This transistor is named bipolar because both types of charge carriers (electrons and holes) are involved in its operation.

Electrons and holes attempt to recombine in the base; however, if the base is exceptionally thin and lightly doped, or if the combination time τ is long, almost all of the electrons pass through the base and enter the collector. The reverse-bias between the base and collector accelerates the electrons through the collector, the circuit is completed and an output signal is produced. The current through the collector (I_c) is governed by

$$I_c = I_o \exp\left(\frac{V_E}{B}\right) \qquad \ldots(40)$$

where I_o and B are constants and V_E is the voltage between the emitter and the base. When the input voltage V_E is increased, a very large current I_c is produced.

We may note that in a *pnp* transistor, the majority carriers are holes. Its function and features are similar to *npn* transistor.

Field effect transistor (FET): This is a second type of transistor, which is more often used for storing data in computer memories. This transistor behaves in a somewhat different manner than the bipolar junction transistors. A *metal oxide semiconductor field effect transistor* (MOSFET), in which two *n*-regions are formed within a *p*-type substrate. One of the *n*-type regions is called the source (S); the second is called the drain (D). The third component of the transistor is a conductor called a gate (G), *i.e.* separated from the semiconductor by a thin insulating layer of SiO_2 or Si_3N_4 (Fig. 3.39). This layer is about 10 nm thick. On the top of SiO_2 another layer of polysilicon (or metal) is deposited to form the third contact for the transistor, called the gate (G). Since SiO_2 is an excellent insulator, the gate connection is not in direct electrical contact with the *p*-type material below the oxide.

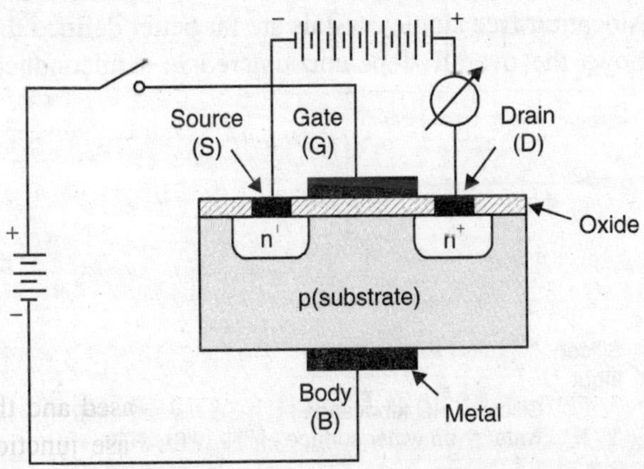

Fig. 3.39 Normally-off-type *n*-channel MOSFET. The dark areas symbolize the (aluminium) metallizations. The "oxide" layer consists of SiO_2 or Si_3N_4. This layer is about 10 nm thick. Quite often, the *B* and *S* terminals are interconnected.

For a simplified type of MOSFET or NMOS when no voltage is applied to the gate, the *p*-type material under the gate contains majority carriers that are holes and only a few electrons are attracted to the drain. However, when a positive voltage is applied to the gate, its electric field attracts electrons from the nearby n^+ source and drain regions to the thin layer beneath the surface of SiO_2 just under the gate so that this region becomes *n*-type silicon. With electrons being the majority-carriers. When electrons are present in this channel, a conducting path exists between the source and the drain. Obviously electrons will flow between the source and the drain if there is a positive voltage difference between them, producing an amplified signal ("on" state). By changing, the input voltage between the gate and the source, the number of electrons in the conductive path changes, thus also changing the output signal. When no voltage is applied to the gate, no electrons are attracted to the region between the source and the drain and there is no current flow from the source to the drain ("off"-state).

The MOSFET, like the bipolar transistor, is also capable of current amplification. The gain in MOSFET devices is usually measured in terms of a voltage ratio instead of current ratio as for the bipolar transistor. *p*-type MOSFETs with holes for majority carriers can be made in a similar way, using *p*-type islands for the source and the drain in an *n*-type substrate. Since the current carriers are electrons in NMOS and holes in PMOS ones, they are called as *majority carrier devices*.

FETs are generally less expensive to produce than the bipolar junction transistors. Since FETs occupy less space, they are preferred in microelectronic integrated circuits. MOSFETs are utilized in memories, microcomputers, logic circuits, amplifiers, analog switches and operational amplifiers. In addition to these, many other devices such as solar cells, photovoltaic panels, lasers, photodiodes, etc. are prepared using different semiconductors. We may note that only little power is expanded for the operation of MOSFET, which is vital to prevent overheating of the device. However, there are some applications for which bipolar transistors are necessary.

Manufacture and Fabrication of Microelectronic Integrated Circuits

To produce electronic components that require little power, operate at high speeds, yet are inexpensive, microelectronic integrated circuits formed on silicon chips may contain as many as 1,000,000 transistors or other devices, each having dimensions smaller than 1 μm (10^{-6} cm). Special technologies are required to produce these circuits. The first requirement for the most common devices is pure, single crystal silicon (Si). Usually, these crystals are grown using the Czochralski growth technique (Fig. 3.40). A seed crystal is used to grow large (up to 12″ diameter) silicon single crystals. Float zone and liquid encapsulated Czochralski techniques are also used. Usually single crystals are preferred because electrical properties of uniformly doped and essentially dislocation-free single crystals are far better defined than those of polycrystalline silicon. Figure 3.41 shows the overall steps encountered in semiconductor processing.

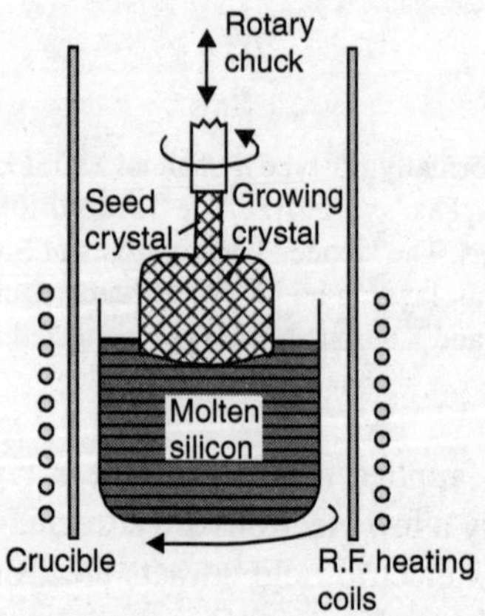

Fig. 3.40 Czochralski growth technique used for the growth of single crystals of silicon.

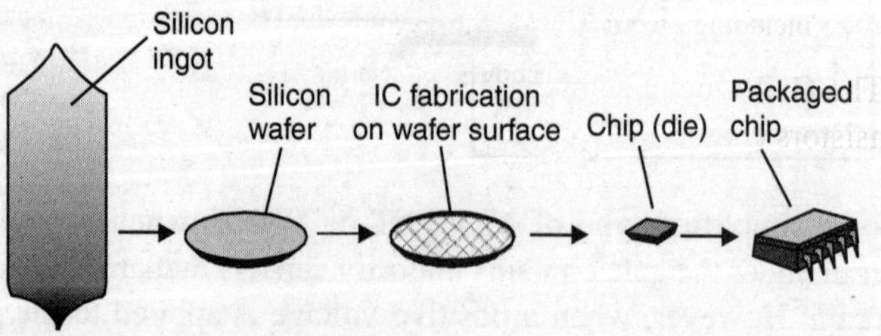

Fig. 3.41 Various step encountered in the processing of semiconductors.

For the production of an FET transistor, the essential steps are as shown in Fig. 3.42. In selected areas to create *p*- and *n*- type regions, diffusion and ion implantation are used. We may note that almost all processing of semiconductors is conducted in clean rooms to eliminate contamination due to air- and liquid-borne particles. This is why, the highest purity chemicals and solvents are used in semiconductor processing.

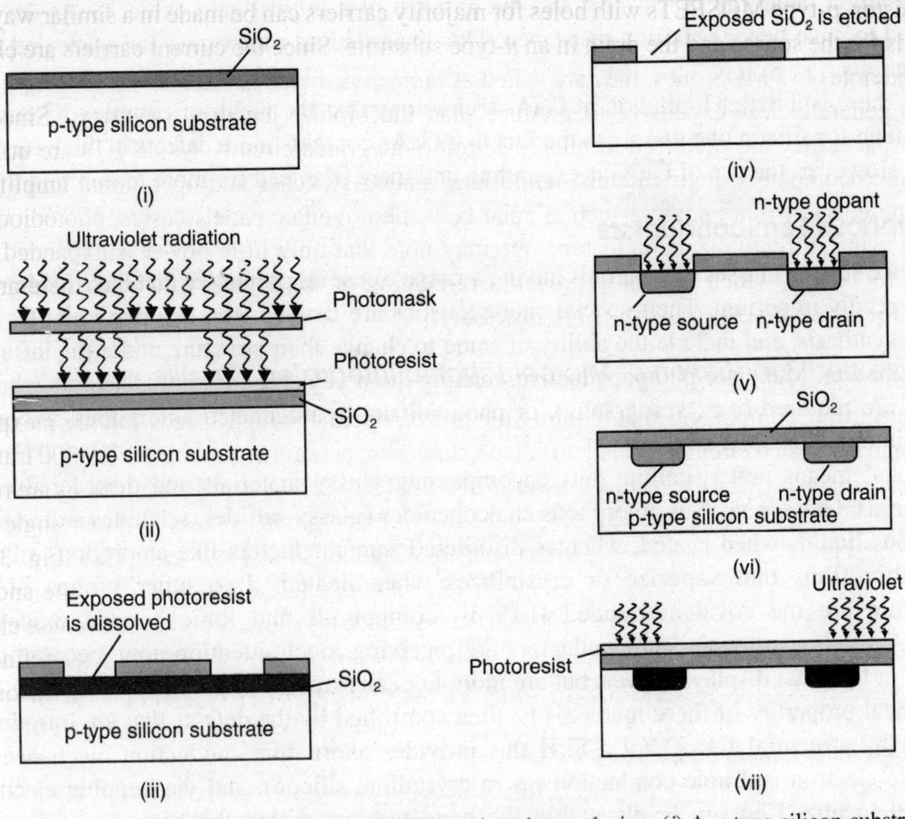

Fig. 3.42 Various steps for the production of an FET semiconductor device: (*i*) A *p*-type silicon substrate is oxidized. (*ii*) photolithography, ultraviolet radiation passing through a photomask, exposes a photoresist layer, (*iii*) the exposed photoresist is dissolved, (*iv*) by etching the exposed silica is removed, (*v*) to produce the source and drain, an *n*-type dopant is introduced, (*vi*) the silicon is again oxidized, (*vii*) to introduce other components including electrical levels, for the device, photolithography is repeated.

The GaAs metal-semiconductor field-effect transistors (MESFETs) are the most widely used GaAs transistors (Fig. 3.43).

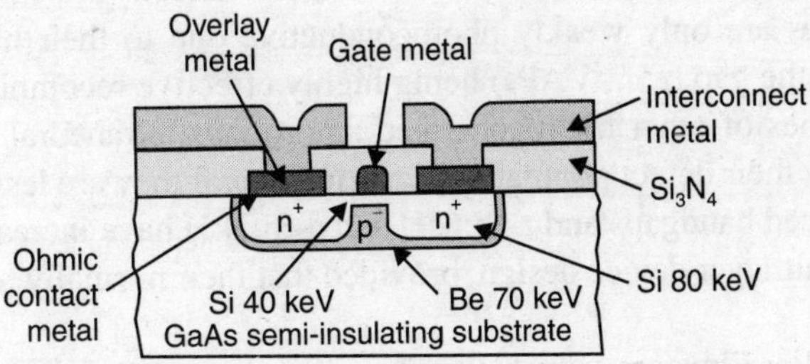

Fig. 3.43 A GaAs MESFET.

There are some advantages of GaAs MESFETs over silicon as devices for use in high-speed digital integrated circuits. Few of these are:

(*i*) Electrons travel faster in *n*-GaAs as is indicated by their high mobility in GaAs than in Si [$\mu_n = 0.720\ m^2/(V.s)$ for GaAs *vs.* $0.135 m^2/(V.s)$ for Si].

(*ii*) Due to larger band gap of 1.47 eV and the absence of a critical-gate oxide, GaAs devices are reported to have better radiation resistance. This is particularly important for space and military applications.

However, there is a major limitation of GaAs technology that the yield for complex IC circuitry is much lower than for silicon due mainly to the fact that GaAs contains more defects in the base material than silicon. However, the use of GaAs is expanding and there is a need for more research in this field.

3.9.7 Amorphous Semiconductors

These comprise several classes of materials having a range of useful properties that have made many of them commercially important. Their special characteristics are derived from the particular techniques used in their synthesis, and include the ability of some to change their structure under the influence of heat or illumination. Many are photoconductive, making them valuable for photodetectors when large area devices are required (*e.g.*; xerography), or photovoltaic, hydrogenated amorphous silicon being widely used in thin-film solar cells.

'Amorphous' means noncrystalline thus encompassing glassy materials and disordered covalent bonded materials. Glasses, such as amorphous chalcogenides (glassy sulfides, selenides and tellurides), melt to viscous liquids when heated, whereas disordered semiconductors like amorphous silicon are metastable thin films that vaporize or crystallizes when heated. Two other groups of glassy semiconductors are the covalent bonded II-IV-V_2 compounds and ionic bonded oxides (*e.g.*, $V_2O_5P_2O_5$). Amorphous organic semiconductors are receiving much attention now, because of their potential use in low-cost display devices, but are more logically treated as polymer materials.

The electrical properties of these materials is often controlled by the defects that are introduced by impurities or by structural flaws. For a-Si:H this provides alternative conduction mechanisms, via extended state electron and hole conduction (as in crystalline silicon), and via hopping electrons or holes through localized energy levels within the band tails or within the band gap. The relative importance of each route depends on the temperature and on the density of states (DOS) in each energy range, as well as on the mobility of charge carriers.

When a-Si: H is illuminated it may show strong photoconductivity, depending on the preparative conditions. Free carriers are generated by the illumination but may fall into defect states (energies within the band gap) where they can be released or can recombine with the oppositely charged carriers. The detailed response to light (wavelength, intensity, duration) and temperature thus depends on the distribution and concentration of defects. Luminescence effects are similarly influenced. Most amorphous chalcogenides are only weakly photoconductive due to their high density of gap states, those near the middle of the gap (*e.g.*, VAPs) being highly effective recombination centers.

The electrical properties of other hydrogenerated amorphous tetrahedral semiconductors based on silicon are determined by their defect populations, and in general they are less well behaved. a-Si:Ge:H and a-Si:Sn:H have reduced bandgaps and a-Si:C:H and a-Si:N:H have increased bandgaps which offer many possibilities for multilayer device design, provided that their normally high defect contents can be reduced or controlled.

Amorphous silicon alloy films are valuable as the active layers in thin-film photovoltaic cells, two dimensional optical position detectors, linear image sensors (optical scanners), and thin film transistors used in liquid crystal display panels. They also have used as antireflection coatings and planar optical waveguides. Amorphous chalcogenide films have been used as electrical switches and memory devices, for optical data storage, and as dry photoresistors and electron beam resists. Both types of amorphous material have been used in xerography and medical X-ray imaging detectors.

Amorphous silicon solar cells have power conversion efficiencies of ~12% for the most complicated structures. These are tandem cells that use different alloys (including a-Si:C:H) for the various layers, in order to enhance effective absorption of the solar spectrum. A serious drawback of using amorphous semiconductors in electronic devices is their short carrier diffusion length. In solar cells this is circumvented by ensuring that there is a built-in field to separate photogenerated electron-hole pairs before they can be recombine. Because doped amorphous silicon alloys have high defect densities, it is difficult to make the effective *pn* junctions that provide such built-in fields in conventional crystalline silicon cells. Instead, amorphous silicon cells use pin structures, where the *i*-layer is effectively undoped and provides an extended electric field between the *p-i* and *i-n* junctions. Long periods of illumination increase the dark current in these devices, as additional defects are generated. This '*Staebler-Wronski effect*' is believed to be associated with hydrogen movement at a passivated dangling bond (Si–H) and a nearby weak Si–Si bond. It is a cause of slow degradation in unstabilized solar cells although the change may be reversed by thermal annealing.

In contrast to '*minority carrier*' photovoltaic devices, thin-film transistors necessitate the control of majority carriers alone. To address each pixel in a liquid crystal display, a switching element is needed and simple thin-film transistors fulfill this role. Although low carrier mobilities in amorphous semiconductors restrict the current passed and switching speeds, the requirement for coating large area glass panels at low temperatures make a-Si:H with a-Si:N:H gate dielectric almost the only options. Improved performance is possible by using pulsed laser irradiation to change the amorphous semiconductor to polycrystalline material.

3.10 INSULATORS

We have read that in an ideal insulator, all valence electrons are occupied in bond formation and almost no free electrons are available for electrical conduction. Thus an insulator (electrical) is a non-metallic material that has a filled valence band at 0 K and relatively wide energy band gap. Consequently, the room temperature electrical conductivity is very low, less than about 10^{-10} $(\Omega\text{-m})^{-1}$. A good insulator may have a resistivity as high as 10^{14} (Ω-m). Common electrical insulating materials are polyethylene, bakelite, lucite, mica, PVC, rubber, porcelain, etc. Most polymers and ionic ceramics are insulating materials at room temperatures. Thermal agitation and imperfections adversely affect the insulating properties of a material but, as is real materials, a few free electrons are always there to conduct electricity.

Non-metallic crystals are held together by ionic and covalent bonds and valence electrons are much more closely associated with their atoms than in the metallic bond. Generally, insulators are those materials which have the electrons completely filled in the *Brillouin Zone* and also at the same time have wide energy gaps.

The conductivity of insulators increases with the rise in temperature because more and more conducting electrons are released. However, the materials which are insulators at 0 K but develop significant conductivities at room temperature become semiconductors. Room temperature conductivities of few selected non-metallic materials are given in Table 3.12. Of course, many insulating materials are used on the basis of their ability to insulate, and thus a high electrical resistivity is desirable.

One can classify the requirements of good insulating materials as electrical, mechanical, thermal and chemical. Electrically the insulating material should have high resistivity to reduce the leakage current and high dielectric strength so that it may withstand higher voltage without being punctured or broken down. Moreover, the insulator should have small dielectric loss.

Table 3.12 Room temperature conductivities for some selected non-metallic materials

Material	*Electrical conductivity* $[(\Omega\text{-}m)^{-1}]$	*Material*	*Electrical conductivity* $[(\Omega\text{-}m)^{-1}]$
Graphite	$3 \times 10^4 - 2 \times 10^5$	POLYMERS	
CERAMICS		Phenol-formal-dehyde	10^{-9}–10^{-10}
Concrete (Dry)	10^{-9}	Polymethyl-methacrylate	$< 10^{-12}$
Soda lime glass	10^{-10}–10^{-11}	Nylon 6, 6	10^{-12}–10^{-13}
Porcelain	10^{-10}–10^{-12}	Polystyrene	$< 10^{-14}$
Borosilicate glass	$\sim 10^{-13}$	Polyethylene	10^{-15}–10^{-17}
Aluminium oxide	$< 10^{-13}$	Polytetra fluoroethylene	$< 10^{-17}$
Fused silica	$< 10^{-13}$		

Insulators are used on the basis of volume and not weight and hence a low density is preferred. A uniform viscosity and for liquid insulators ensures uniform thermal and electrical properties.

Liquid and gaseous insulators are used also as coolants, *e.g.* transformer oil, hydrogen and helium are used for insulation and cooling purposes. Obviously, for such materials good thermal conductivity is a desirable property. To prevent mechanical damage, the insulator should also have small thermal expansion. Moreover, it should be non-ignitable or if ignitable it should be self-extinguishable.

Chemically, the insulating materials should be resistant to oils, liquid gas fumes, acids and alkalies. Moreover, the insulator should not absorb water particles as water lowers the insulation resistance and the dielectric strength.

Insulators should have certain mechanical properties depending on the use to which they are put. Thus when insulators are used for electric machine insulation, they should have sufficient mechanical strength to withstand vibration. In such cases, good heat conducting property is also desirable.

In making dielectric capacitors insulating materials with large electronic and ionic polarizabilities and therefore large permittivity are used. Titanium oxide with a permittivity ~100 is a good example of such a material. We may note that the use of molecules with a permanent dipole moment is not desirable because of possibility of large dielectric losses at high frequencies.

Properties of the some of the insulating materials are given in Table 3.13.

Table 3.13 Properties of Few Insulating Materials at Room Temperature

Material	*Dielectric constant* (ε_r)	*Dielectric strength (kV/mm)*	*Density* ($\times 10^{-3}$) *(kg/m³)*	*Resistivity* (ρ) *(Ω-m)*
Mica	4–5.5	60–125	2.5–2.7	10^{14}–10^{19}
Glass	4–10	20–30	2.2–4.0	10^{13}–10^{16}
Ebonite (RP)	4–4.5	25	1.3	1×10^{20}
Bakelite	4–4.6	10–40	1.2	—
Asbestos	—	2	2.3–2.6	2×10^7
Rubber (soft)	2.6–3.0	15–25	1.7–2.9	4×10^{15}
Silk (natural)	4.5	—	—	—
Polystyrene	2.2–2.8	25–50	1.05–1.65	$5 \times 10^{13} - 5 \times 10^{19}$
Polyvinyl chloride (PVC) resin	3.1–3.5	50	1.38	—

3.11 DIELECTRICS

These are the materials or insulators which have the unique characteristic of being able to store electric charge. The electrons in these substances are localized in the process of bonding the atoms together.

Obviously, covalent or ionic bonds, a mixture of both, or van der Waals bonding between closed shell atoms give rise to solids or gases which exhibit dielectric or insulating properties. Dielectric materials may be gases, liquids or solids with the exception of air which is the insulating material between the bare conductors of the overhead electric grid system. Liquid dielectrics are used mainly as impregnants for high voltage paper insulated cables and capacitors as filling and cooling media for transformers and circuit breakers. Most common properties of dielectric materials are: (*i*) dielectric constant, (*ii*) dielectric strength, (*iii*) insulation resistance, (*iv*) surface resistivity, (*v*) loss factor and (*vi*) tangent of loss factor in terms of a capacitor or phase difference (*vii*) polar and non-polar materials.

Materials, which are capable of separating electrical conductors circuit breakers, *e.g.* silicon, oils, liquid dielectrics have high dielectric constant, high resistance, high dielectric strength when moisture and impurities are removed from them. They have high temperature dissipation capacity and least dielectric losses. Their dielectric constant is greater than one. Following materials are important from engineering point of view:

Mica is the widely used insulating material in switch gears armature windings, electrical heating devices like iron, hot plates, etc. It is also used in capacitors for high frequency application. Mica is an inorganic compound of silicates of aluminium, soda potash and magnesia. It is crystalline in nature and can be easily split into very thin flat sheets. The two important types of mica are: (*i*) muscovite and (*ii*) phlogopite. Mica has a good dielectric strength and mechanical strength. Its dielectric constant varies between 5 and 7.5, loss tangent between 0.0003 and 0.015 and dielectric strength between 700 and 1000 kV/mm.

Asbestos is also used as an insulator in the form of paper, tape, cloth and board. Asbestos is widely used in panel boards, insulating tubes and cylinders in the construction of air cooled transformers. Asbestos is an inorganic material, which is used to designate a group of naturally occurring fibre material. Asbestos has good dielectric and mechanical properties.

Ceramics are generally non-metallic inorganic compounds, *e.g.* silicates, aluminates, oxides, carbides, borides, nitrides and hydroxides. Ceramics used as dielectrics may be broadly described as alumina, porcelains, ceramics, titanates, etc. These have excellent dielectric and mechanical properties. The dielectric constant of most commonly used ceramics varies between 4 and 10. These are used in switches in plug holders, thermocouples, cathode heaters, vacuum type ceramic metal seals, etc. Ceramic capacitors may be operated at high temperatures and can be moulded into any shape and size.

Electric grade ceramics are used for the manufacturing of insulators, terminal blocks, plates, frames, coils, etc. They must have low losses, good insulating properties and high strength.

Insulators for operation at low frequencies are made of electric-grade porcelain which possesses fairly good electric properties. A drawback is that it has high losses which increase sharply on heating above 200°C, and a low strength.

Insulating parts for operation at high frequencies are mostly made of steatite, a talc base material. Steatites contain no harmful impurities and retain their properties at temperatures up to 100°C. They can be pressed easily, shrink on burning by only 1–2%, and are suitable for making parts in which dense but porous structure and accurate dimensions are essential. In contrast to other kinds of ceramics, steatite can be cut quite easily (after burning). Among their drawbacks are cracking under the action of rapidly varying temperatures and some difficulties the burning involves.

Glass is an inorganic insulating material, which comprises of complex system of oxides. Silica (SiO_2) is the most essential constituent of many commercially used glasses. It is fused with alkali (like potash, soda, etc.) and some base (like lime, lead oxide, etc.). The silica glass (having 100% SiO_2) is the best insulating material. The dielectric constant of glass varies between 3.7 and 10 and loss tangent between 0.0003 and 0.01 and dielectric strength between 2.5 and 50 kV/mm. Glass is used in electric bulbs, X-ray tubes, mercury switches, electronic valves as insulating material. It is also used in capacitors as dielectric material.

Resins are organic polymers and may be natural or synthetic. The synthetic resins are produced artificially. The commonly used synthetic resins are polyethylene, polystyrene, polyvinyl chloride, acrylic resins, teflon, nylon, etc. These have good dielectric and mechanical properties. The dielectric constant of resins varies between 2 and 4.5, the loss tangent between 0.0002 to 0.04 and dielectric strength is quite high. These are used in transformers, high frequency capacitors. These are also used as a dielectric material in d.c. capacitors.

Rubber

These are organic polymers and may be natural or synthetic. The natural rubber obtained from rubber tree has limited applications due to its resistance to low and high temperatures. The synthetic rubbers are produced artificially by *copolymerisation* of isobutylene and isoprene. These have good electrical and thermal properties. The dielectric constant of rubber varies between 2.5 and 5, and loss tangent between 0.01 and 0.03. Rubber's are used as an insulating material for electric wires, tapes, cables, coatings, motor windings, transformers, etc.

Gaseous dielectrics, *e.g.* nitrogen, hydrogen, etc. have dielectric constant 1.0.

The best dielectrics are polystyrene, polycarbonate, polyethylene, polymide, mineral oil, pure alumina and pure silica.

3.11.1 Dielectric Constant

Consider a simple parallel-plate capacitor (a capacitor is a device that stores electric energy) with metal plates of area *A* separated by distance *d*, as shown in Fig. 3.44. Consider first the case in which the space between the plates is a vacuum. If a voltage *V* is applied across the plates, one plate will acquire a net charge of $+q$ and the other a net charge of $-q$. The charge q is found to be directly proportional to the applied voltage *V* as

$$q = CV \text{ or } C = \frac{q}{V}$$

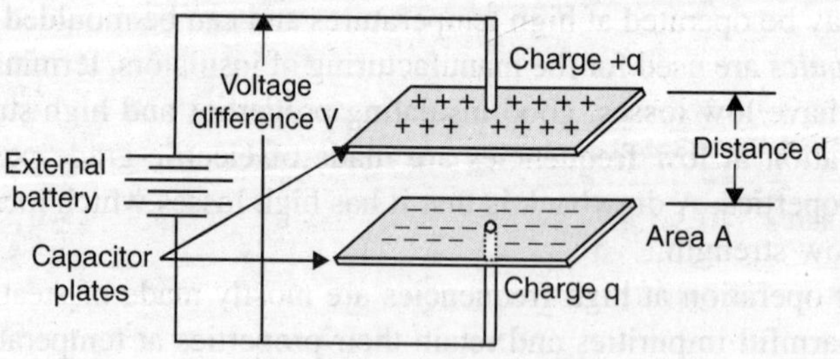

Fig. 3.44 Simple parallel-plate capacitor.

where *C* is a proportionality constant called the *capacitance* of the capacitor. The SI unit of capacitance is coulombs per volt (C/V), or the *farad* (F). Thus

$$1 \text{ farad} = \frac{1 \text{ coulomb}}{\text{volt}}$$

Since the farad is a much larger unit of capacitance than is normally encountered in electrical circuitry, the commonly used units are the *picofarad* (1 pF = 10^{-12} F) and the *microfarad* (1 μF = 10^{-6} F).

The capacitance of a capacitor is a measure of its ability to store electric charge. The more charge stored at the upper and lower plates of a capacitor, the higher is its capacitance.

The capacitance C for a parallel-plate capacitor whose area dimensions are much greater than the separation distance of the plates is given by

$$C = \varepsilon_0 \frac{A}{d}$$

where ε_0 = permittivity of free space = 8.854×10^{-12} F/m.

When a dielectric (electrical insulator) fills the space between the plates (Fig. 3.45), the capacitance of the capacitor is increased by a factor κ, which is called the *dielectric constant* of the dielectric material. For a parallel-plate capacitor with a dielectric between the capacitor plates,

$$C = \frac{\kappa \varepsilon_0 A}{d}$$

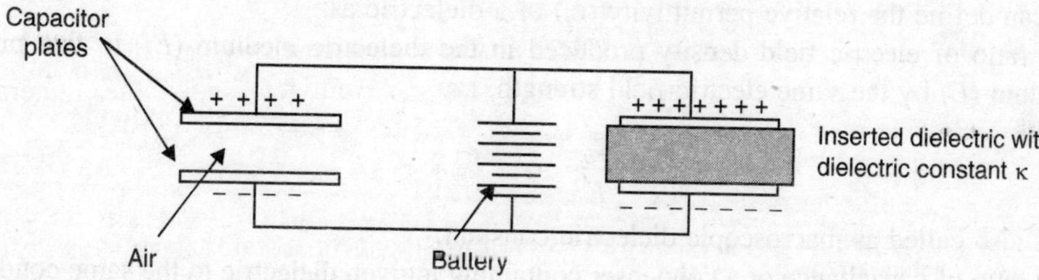

Fig. 3.45 Two parallel-plate capacitors under the same applied voltage. The capacitor on the right has a dielectric (insulator inserted between the plates), and as a result the charge on the plates is increased by a factor of κ above that on the plates of the capacitor without the dielectric.

Table 3.14(*a*) lists the dielectric constants for some ceramic insulator materials.

Table 3.14(a) Electrical Properties of Some Ceramic Insulator Materials

Material	*Volume* (Ω-*m*)	*Dielectric strength*		*Dielectric constant κ*		*Loss factor*	
		V/mil	*kV/mm*	*60 Hz*	*10^6 Hz*	*60 Hz*	*10^6 Hz*
Electrical porcelain insulators	10^{11}–10^{13}	55–300	2–12	6	...	0.06	
Streatite insulators	>10^{12}	145–280	6–11	6	6	0.008–0.090	0.007–0.025
Fosterite insulators	>10^{12}	250	9.8	...	6	...	0.001–0.002
Alumina insulators	>10^{12}	250	9.8	...	9	...	0.0008–0.009
Soda-lime glass	...	...	...	...	7.2	...	0.009
Fused silica	...	8	...	...	3.8	...	0.00004

The energy stored in a capacitor of a given volume at a given voltage is increased by the factor of the dielectric constant when the dielectric material is present. By using a material with a very high dielectric constant, very small capacitors with high capacitances can be produced.

Dielectric Strength: Another property besides the dielectric constant that is important in evaluating dielectrics is the *dielectric strength*. This quantity is a measure of the ability of the material to hold energy at high voltages. Dielectric strength is defined as the voltage per unit length (electric field or voltage gradient) at which failure occurs and thus is the maximum electric field than the dielectric can maintain without electrical breakdown.

Dielectric strength is most commonly measured in volts per mil (1 mil = 0.001 in) or kilovolts per millimeter. If the dielectric is subjected to a voltage gradient that is too intense, the strain of the electrons or ions in trying to pass through the dielectric may exceed its dielectric strength. If the dielectric strength is exceeded, the dielectric material begins to break down and the passage of current (electrons) occur.

Relative Permeability

In an electric field, the charge density D is directly proportional to the applied field, *i.e.*

$$D = \varepsilon E \qquad ...(41)$$

where ε is the dielectric constant or permittivity of the material placed between the electrodes. For vacuum,

$$D = \varepsilon_o E \qquad ...(41(a))$$

where $\varepsilon_o = 8.854 \times 10^{-12}$ farad/m is called the absolute permittivity.

One can define the relative permittivity (ε_r) of a dielectric as:

(*i*) The ratio of electric field density produced in the dielectric medium (E_o) to that produced in vacuum (E) by the same electric field strength, *i.e.*

$$\varepsilon_r = \frac{E_o}{E}$$

ε_r is also called as macroscopic dielectric constant.

(*ii*) The ratio of capacitance of a condenser containing a given dielectric to the same condenser with vacuum as dielectric.

Thus relative permittivity can be express as

$$\varepsilon_r = \frac{\varepsilon}{\varepsilon_o} \qquad ...(42)$$

Now, charge density = $\varepsilon_o \varepsilon_r E = \varepsilon_o E + \varepsilon_o (\varepsilon_r + 1) E$

where $\varepsilon_r - 1 = \chi$ is called electric susceptibility. χ = Bound charge density/free charge density.

Permittivity is influenced by permeability (μ), stress distribution, temperature and frequency.

3.11.2 Phase Difference and Dielectric Loss

A dipole tends to align itself along the direction of applied electric field and for ac fields tends to follow the field and be in a phase with it. However, the interaction of this dipole with other dipoles in the medium prevents this and this leads to dielectric loss. This loss appears as heat. This dielectric loss is connected with ε_r'', the imaginary part of the dielectric constant. The rate of loss of energy in unit volume of the material is obtained as

$$W = \frac{1}{2} \varepsilon_o \varepsilon_r'' \omega E_o^2 \qquad ...(43)$$

Obviously, the energy loss is proportional to ε_r''. Usually the dielectric loss is expressed in terms of a quantity called *'the loss tangent'*, tan δ. This is defined as $\tan \delta = \varepsilon_r''/\varepsilon_r'$. The angle δ is the compliment of angle ϕ, *i.e.* $90 - \phi = \delta$, where ϕ is the angle between the applied field and the resultant current vector (Fig. 3.10).

From alternating current studies, we have

$$\text{Power} = V_R I_R \cos \phi$$

where V_R and I_R are rms values of voltage and current, and for a circuit containing capacitance C and resistance R. We have

$$\tan\delta = \frac{1}{\omega CR} \qquad \text{...(44)}$$

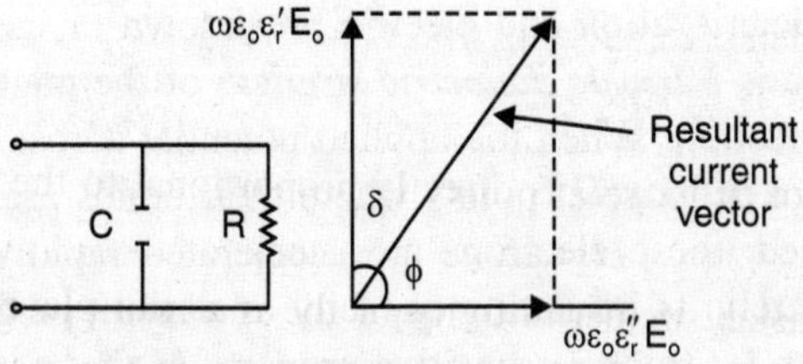

Fig. 3.46 Vector relationship between the field vector E_o, the current vector $\omega\,\varepsilon_o\,\varepsilon_r'\,E_o$ which leads the field by 90° and the current vector $\omega\,\varepsilon_o\,\varepsilon_r''\,E_o$ which is in phase with the field. The angle δ is shown.

Using (42), one obtains for a dielectric

$$W = \frac{1}{2}\varepsilon_o\varepsilon' \tan\delta\,\omega E_o^2 = \frac{1}{2}\frac{E_o^2}{\rho} \qquad \text{...(45)}$$

where ρ is the resistivity. Thus

$$\tan\delta - \frac{1}{\omega\varepsilon_o\varepsilon'\rho} \qquad \text{...(46)}$$

The energy losses in a dielectric material are due to: (*i*) ionization, (*ii*) leakage current, (*iii*) polarization, and (*iv*) structural inhomogeneity.

Ionization losses occurs in gases and solids having pores with entrapped gases. With the rise in field strength applied to a gas, a stage is reached when the gas molecules gets ionized due collisions. This leads to enhanced conduction leading to dielectric losses. At low or ordinary electric fields, the conductivity is low and hence the loss is also small (Eqn. (45)). For example, for $\rho \approx 10^{16}$ Ω-m, $\varepsilon_r' \approx 1$ and at 100 Hz, $\tan\delta < 2 \times 10^{-8}$. In case of solids with gas inclusions the loss tangent increases with voltage from the voltage required for the ionization of gas molecules to the value of voltage at which ionization is complete. When such inclusions are high, there are chances of the failure of the dielectric.

Due to the leakage current, the losses in highly conducting liquids and solids are high (Eqn. (44)). In case of neutral liquids the losses will be small as their conductivity is low. Purity of oil is quite important as otherwise dielectric losses increase due to conduction. Pure transformer oil has $\tan\delta \approx 0.0005$.

Dipole losses in the radiofrequency region are usually due to dipole rotation or ions jumping from one equilibrium position to another. The dielectric losses associated with ions, the frequencies of which fall in the infrared region, are usually referred to as *infrared absorption*. Similarly, the losses in the optical region, associated with the electrons, are referred to as *optical absorption*.

3.11.3 Dielectric Strength

As the voltage applied across a dielectric is increased, dielectric loses its insulation property at a certain voltage called the *breakdown voltage* (V_{br}). The corresponding field strength (E_{br}) is given by

$$E_{br} = \frac{V_{br}}{t} \qquad \text{...(47)}$$

Here t is the thickness of the material. This voltage, V_{br} per unit thickness of the material is called the *dielectric strength* and usually expressed in Volts/mil or kV/millimeter or mega Volt/m. The breakdown of a dielectric is due to the collision of accelerated electrons or ions with molecules. At relatively high fields, the electrons in the dielectric gain enough energy to knock other charged particles and make them available for conduction.

Dielectric failure involves deterioration and electric breakdown or cascading. Resistance to electronic breakdown in a dielectric is *intrinsic dielectric strength* of the material. Intrinsic breakdown in a dielectric is caused by imperfection. When the applied potential across the dielectric becomes high enough, a few electrons are broken loose at points where their bonds are strained by the presence of imperfections. Upon being freed, these electrons are accelerated rapidly through the material. The dielectric material loses completely its insulating capacity as a result of this process.

Obviously, dielectric strength is structure sensitive property of the material. We may note that the actual thickness of the material affects the breakdown potential (V_{br}) per unit thickness, *i.e.* dielectric strength. Usually thicker materials possess lower dielectric strength than thinner materials.

The fundamental breakdown mechanisms in a solid are: (*i*) *intrinsic breakdown,* (*ii*) *thermal breakdown,* (*iii*) *discharge breakdown* and (*iv*) *electrochemical breakdown.*

3.11.4 Power Factor

The ratio of power loss in a material to the product of applied voltage and current is called power factor. The power factor and loss angle depend upon the nature of the material, applied voltage, humidity, temperature and frequency.

3.11.5 Polar and Non-Polar Materials

Permanent dipole moments exist in some molecules by virtue of an asymmetrical arrangement of positively and negatively charged regions. Such molecules are termed polar molecules. For example, in the HCl molecule (Fig. 3.47(*a*)), the electron of hydrogen atom spends more time moving around the Cl atom than around the H atom. Therefore, the centre of negative charges does not coincide with that of positive charges, and the molecule has a dipole moment directed from the Cl atom to the H atom. That is, we may write H^+Cl^-. The electric dipole of HCl molecule is $p = 3.43 \times 10^{-30}$ C-m. In the CO molecule Fig. 3.47(*b*), the charge distribution is only slightly asymmetric and the electric dipole moment is relatively small, about 0.4×10^{-30} C-m, with the carbon atom corresponding to the positive and the oxygen atom to the negative end of the molecule.

Polar molecules can also induce dipoles in adjacent non-polar molecules, and a bond will form as a result of attractive forces between the two molecules. Moreover, the magnitude of this bond will be greater than for fluctuating induced dipoles.

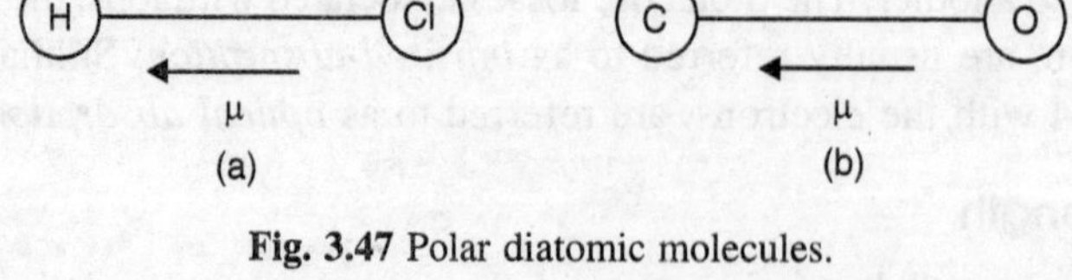

Fig. 3.47 Polar diatomic molecules.

van der Waals forces will also exist between adjacent polar molecules. We may note that the associated bonding energies are significantly greater than for bonds involving induced dipoles.

The strongest secondary bonding type, the hydrogen bond, is a special case of polar molecule bonding. We find that it occurs between molecules in which hydrogen is covalently bonded to fluorine (as in HF molecule), oxygen (as in H_2O molecule), and nitrogen (as in NH_3 molecule). For each H–F, H–O or H–N bond, the single hydrogen electron is shared with the other atom. Obviously, the hydrogen end of the bond is essentially a positively charged bare proton that is unscreened by any electrons. This highly positively charged end of the molecule is capable of a strong attractive force with the negative end of an adjacent molecule (Fig. 3.48). Clearly, this single proton forms a bridge between two negatively charged atoms. We may note that the magnitude of the hydrogen bond is generally greater than that of the other types of secondary bonds and may be as high as 0.52 eV/molecule (= 51 kJ/mol).

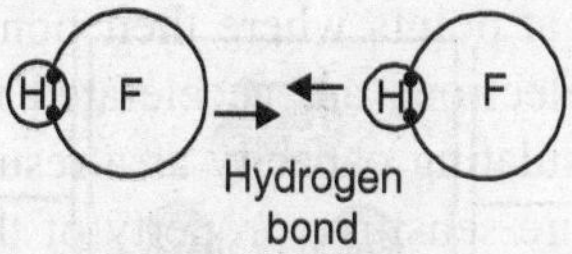

Fig. 3.48 Representation of hydrogen bonding in HF molecule.

We find that the melting and boiling temperatures for hydrogen fluoride and H_2O are abnormally high in light of their low molecular weights, as a consequence of hydrogen bonding.

Polarization

There are four primary mechanisms causing polarization: (*i*) electronic polarization, (*ii*) ionic polarization, (*iii*) molecular polarization and (*iv*) space charge (Fig. 3.49). However, their occurrence depends upon the electrical frequency of the applied field, just like the mechanical behaviour of materials depends on the strain rate. When we apply a very rapid rate of strain, such mechanisms as plastic deformations do not get a chance to let the material deform. Similarly, if we apply a rapidly alternating electric field, some polarization mechanisms may not get a chance to induce polarization in the material.

When a material is placed in an electric field (Fig. 3.49), *e.g.* between the plates of a condenser, the field strength of charged particles within the material interact with the electric field. If the material is a conductor, some of the free electrons simply move to the side nearest the positive electrode until they totally counteract the applied electric field. Obviously, no field is left within the material. The displacement of charged particles occurs almost simultaneously, bringing about equilibrium.

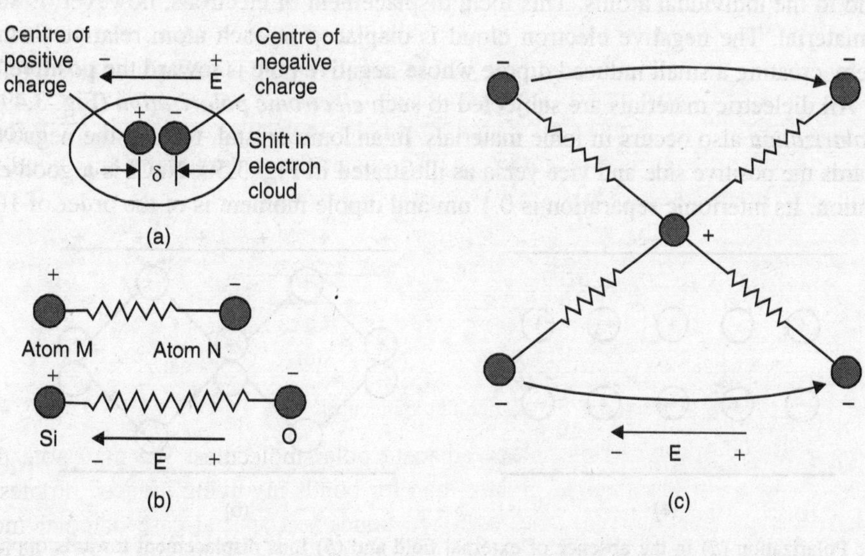

(Contd.)

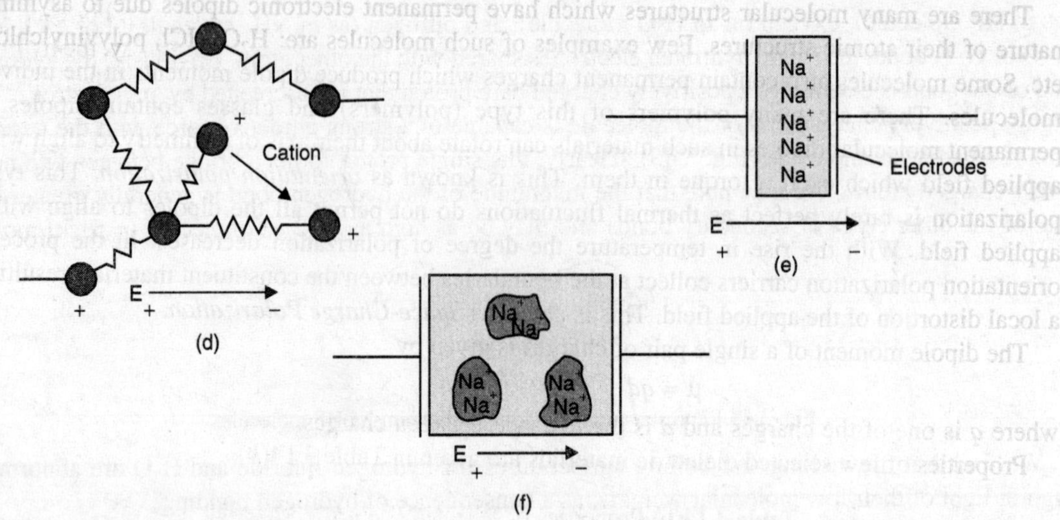

Fig. 3.49 Polarization mechanisms in materials: (*a*) electronic, (*b*) atomic or ionic, (*c*) high-frequency dipolar or orientation (present in ferroelectrics), (*d*) Low-frequency dipolar (present in linear dielectrics and glasses), (*e*) interfacial-space charge at electrodes, and (*f*) interfacial-space charge at heterogeneities such as grain boundaries.

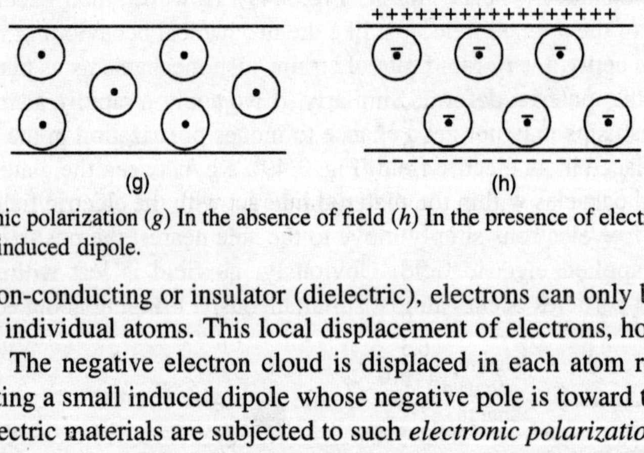

Fig. 3.49 Electronic polarization (*g*) In the absence of field (*h*) In the presence of electric field, the formation of induced dipole.

If the material is non-conducting or insulator (dielectric), electrons can only be displaced locally as they are bound to the individual atoms. This local displacement of electrons, however, is sufficient to polarize the material. The negative electron cloud is displaced in each atom relative to the positive nucleus, thereby creating a small induced dipole whose negative pole is toward the positive side of the electric field. All dielectric materials are subjected to such *electronic polarization* (Fig. 3.49).

Induced polarization also occurs in ionic materials. In an ionic crystal, for *e.g.* the negative ions are attracted towards the positive side and vice versa as illustrated in Fig. 3.50. NaCl is a good example of ionic polarization. Its interionic separation is 0.1 nm and dipole moment is of the order of 10^{-29} C-m.

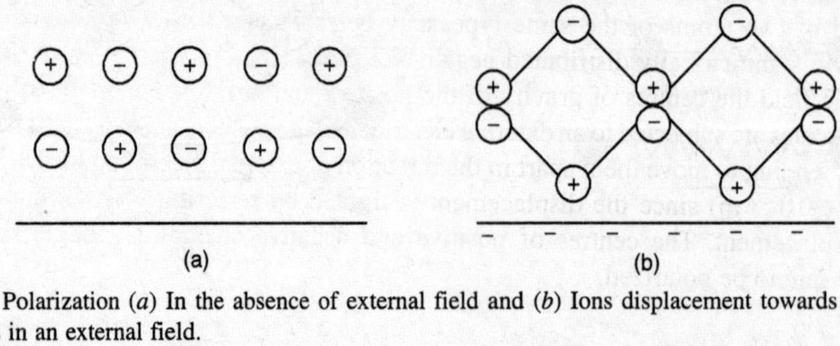

Fig. 3.50 Ionic Polarization (*a*) In the absence of external field and (*b*) Ions displacement towards opposite charged plates in an external field.

There are many molecular structures which have permanent electronic dipoles due to asymmetric nature of their atomic structures. Few examples of such molecules are: H_2O, HCl, polyvinylchloride, etc. Some molecules also contain permanent charges which produce dipole moments in the individual molecules. There are many polymers of this type (polymers) and glasses contain dipoles. The permanent molecular dipoles in such materials can rotate about their axis of symmetry to align with an applied field which exert a torque in them. This is known as *orientation polarization.* This type of polarization is rarely perfect as thermal fluctuations do not permit all the dipoles to align with the applied field. With the rise in temperature the degree of polarization decreases. In the process of orientation polarization carriers collect at the boundaries between the constituent materials resulting in a local distortion of the applied field. This is called as *Space-Charge Polarization.*

The dipole moment of a single pair of charges is given by

$$\mu = qd \qquad ...(48)$$

where q is one of the charges and d is the distance between charges.

Properties of few selected dielectric materials are given in Table 3.14(*b*).

Table 3.14(b) Properties of Selected Dielectric Materials

Material	*Dielectric constant* (at 60 Hz)	*Dielectric constant* (at 10^6 Hz)	*Dielectric strength* (10^6 V/m)	tanδ (at 10^6 Hz)	*Resistivity* (ohm · cm)
Polyethylene	2.3	2.3	20	0.00010	$>10^{16}$
Teflon	2.1	2.1	20	0.00007	10^{18}
Polystyrene	2.5	2.5	20	0.00020	10^{18}
PVC	3.5	3.2	40	0.05000	10^{12}
Nylon	4.0	3.6	20	0.04000	10^{15}
Rubber	4.0	3.2	24		
Phenolic	7.0	4.9	12	0.05000	10^{12}
Epoxy	4.0	3.6	18		10^{15}
Paraffin wax		2.3	10		10^{13}–10^{19}
Fused silica	3.8	3.8	10	0.00004	10^{11}–10^{12}
Soda-lime glass	7.0	7.0	10	0.00900	10^{15}
Al_2O_3	9.0	6.5	6	0.00100	10^{11}–10^{13}
TiO_2		14–110	8	0.00020	10^{13}–10^{18}
Mica		7.0	40		10^{13}
$BaTiO_3$		2000–5000	12	~0.0001	10^{8}–10^{15}
Water		78.3			10^{14}

Non-polar materials

Most of the hydrocarbons are non-polar. In non-polar materials the molecules, which are usually diatomic and composed of two atoms of the same type, may be represented as a positive nuclei of charge q surrounded by a symmetrically distributed negative electron cloud of charge $-q$. In the absence of an applied electric field the centres of gravity of the positive and negative charge distributions coincide. When the molecules are subjected to an external electric field the positive and negative charges experience electric forces tending to move them apart in the direction of external electric field. The distance moved is very small (~10^{-10} m) since the displacement is limited by restoring forces which increases with increasing displacement. The centres of positive and negative charges no longer coincide and the molecules are said to be polarized.

3.11.6 Insulation Resistance

This separates a number of a conductors at different electrical potentials and does not allow a large flow of electric current between them. The difference of potential may cause leakage of current along two paths, *i.e.* (*i*) over the surface of insulation and (*ii*) through the solid material of the insulators.

We may note that the resistance offered along the said two paths is not the same due to the nature of the material. The former is due to the surface resistivity of the material, whereas the latter is the result of the volume resistivity of the material. The combined effect of two is called the insulation resistance of the material.

We may define the surface resistivity as the resistance between the two opposite edges of a square of unit area of insulation surface. The volume resistivity is defined as the resistance presented to the flow of an electric current by a material of unit cross-section and of unit length at 0°C.

The resistive properties of real dielectrics are:

(*i*) Under the action of a potential, mobility with which any conducting species moves.

(*ii*) Temperature variation of conductivity to state why the conductivity of a dielectric varies in exponential manner with temperature.

(*iii*) To analyse the time response of dielectrics, frequency or time variation of conductivity.

(*iv*) While designing an equipment, the knowledge of breakdown or the highest voltage a dielectric will withstand is of interest. Usually insulators breakdown at 10^6 V/m to 10^9 V/m dc at 20°C.

3.11.7 Classification of Dielectrics

One can classify dielectric as under following three groups:

Class	*Effect of electric field*
Simple dielectrics	Creates dipoles
Paraelectrics	Orients dipoles
Ferroelectrics	Orients domains of aligned permanent dipoles

3.11.8 Ferroelectricity

The group of dielectric materials called *ferroelectrics* exhibit spontaneous polarization *i.e.*, polarization in the absence of an electric field. In a sense, ferroelectrics are the electric analog of the ferromagnets, which may display permanent magnetic behaviour. In ferroelectrics, the polarization can be changed and even reversed by an external electric field. The reversibility of the spontaneous polarization is due to the fact that the structure of a ferroelectric crystal can be derived from a non-polarized structure by small displacement of ions. In most ferroelectric crystals, this non-polarized structure becomes stable if the crystal is heated above a critical temperature, the *ferroelectric Curie temperature* (T_c); *i.e.* the crystal undergoes a phase transition from the polarized phase (ferroelectric phase) into an unpolarized phase (paraelectric phase). The change of the spontaneous polarization at T_c can be continuous or discontinuous. The T_c of different types of ferroelectric crystals range from a few degrees absolute to a few hundred degrees absolute.

From a practical standpoint ferroelectrics can be divided into two classes: (*i*) In this *first class* of ferroelectrics, polarization can occur only one crystal axis, *e.g.* Rochelle salt, KH_2PO_4, $(NH_4)_2\ SO_4$, guanidine aluminium sulphate hexahydrate, glycine sulphate, colemanite and thiourea. (*ii*) In this *second class* of ferroelectrics, spontaneous polarization can occur along several axes that are equivalent in the paraelectric phase, *e.g.* $BaTiO_3$-type (or perovskite type) ferroelectrics, $Cd_2Nb_2O_7$, PbN_2O_6, certain alums, such as methyl ammonium alum and $(NH_4)_2 - Cd_3(SO_4)_3$.

One can also place ferroelectrics as *proper ferroelectrics* and *improper ferroelectrics*. In proper ferroelectrics, *e.g.*, $BaTiO_3$, KH_3PO_4, and Rochelle salt, the spontaneous polarization is the order parameter. In improper ferroelectrics, the spontaneous polarization can be considered a by-product of another structural phase transition. The properties of some representative ferroelectrics are given in Table 3.15.

Table 3.15 Properties of Some Representative Ferroelectric Crystals

Group	*Crystal*	*Curie temperature* T_c (°K)	P_s $C/m^{-2} \times 10^{-2}$	*At T (°K)*
Ilmenites and	GeTe	670	—	—
Perovskites	$LiNbO_3$	1480	71	296
	$KNbO_3$	710	30	600
	$BaTiO_3$	393	26	300
	$SrTiO_3$	32	3	4.2
KDP type	KH_2PO_4 (KDP)	123	4.7	100
	KD_2PO_4	213	5.5	100
	$Rb_2H_2PO_4$	147	5.6	90
	KH_2AsO_4	97	5.0	78
TGS type	$(NH_2CH_2COOH)_3 \cdot H_2SO_4$ (Triglycine sulphate)	322	2.8	275
Rochelle salt type	$NaKC_4H_4O_6 \cdot 4H_2O$ ...(14.Rochelle salt)	296 (upper) 255 (lower)	0.25	275

The spontaneous polarization in ferroelectrics can occur in at least two equivalent crystal direction, thus a ferroelectric crystal consists in general of regions of homogeneous *domains*. Ferroelectrics of the first class consist of domains with parallel and antiparallel polarization [Fig. 3.51(*a*)], whereas ferroelectrics of the second class can assume much more complicated domain configurations [Fig. 3.51(*b*)]. The region between two adjacent domains is called a *domain wall*. Within this wall, the spontaneous polarization changes its direction.

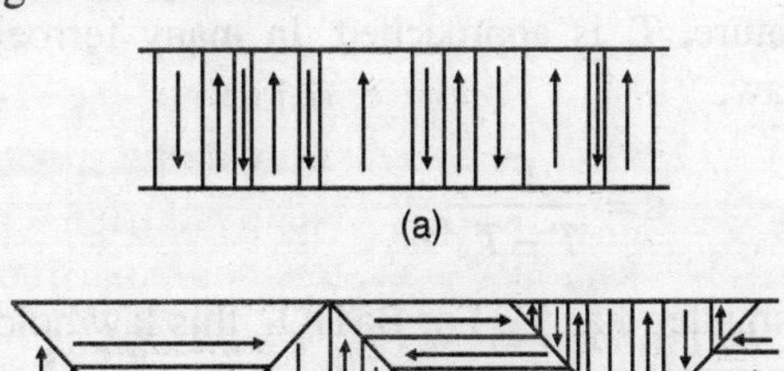

Fig. 3.51 Domain configurations (simplified) encountered in ferroelectric: (*a*) First class and (*b*) Second class.

When an electric field is applied to a specimen of a ferroelectric crystal, the polarization first rises rapidly with applied field to a value above which the dependence is linear. Linear extrapolation to zero field gives, the *saturation* or *spontaneous polarization*. On subsequently reducing the field to zero, remanent (residual) polarization remains. The negative field to reduce the polarization to zero is called the *coercive field*. Obviously, a ferroelectric crystal can be switched by the application of an electric field and a *hysteresis loop* is associated with the switching. The existence of a dielectric hysteresis loop in a dielectric material implies that the substance possesses a spontaneous polarization, P_s and the value of spontaneous polarization (depending upon the shape of hysteresis loop) depends upon a number of factors such as the dimensions of the specimen, the temperature, the texture of the crystal, the thermal and electrical properties of the crystal. A typical hysteresis loop is shown in Fig. 3.52. For most ferroelectrics, the values of P_s are between 10^{-7} and 10^{-4} C/cm^2. In non-ferroelectric dielectrics, electric fields between 10^5 and 10^8 V/cm would be necessary in order to achieve such large polarizations.

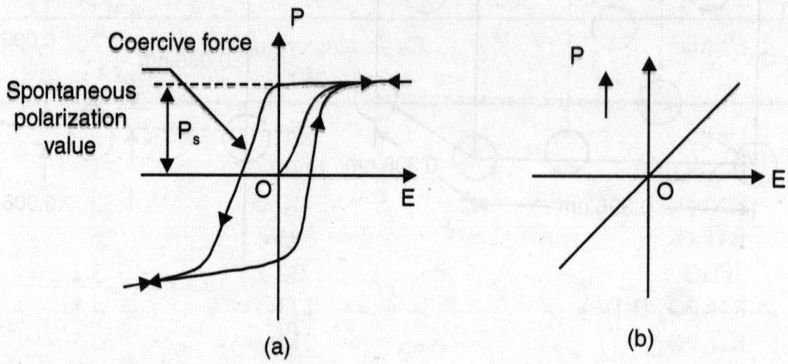

Fig. 3.52 Net polarization *P* of a ferroelectric crystal versus externally applied electric field: (*a*) Hysteresis, (*b*) *P – E* relation above curie temperature.

The hysteresis loop of a ferroelectric material changes its shape as the temperature is increased. The height and width decrease with increase of temperature. At a certain temperature known as *ferroelectric Curie temperature* (T_c), the loop merges to a straight line and the ferroelectric behaviour of the material disappears. The electric susceptibility χ_E in the *paraelectric phase* (above curie temperature, T_c) is related to the temperature by the Curie-Weiss law

$$\chi_E = \frac{C}{T - T_C} \qquad ...(49)$$

where C is Curie constant and $T_c \rightarrow$ *Curie-Weiss temperature*.

As a rule, the dielectric constant ε measured along a ferroelectric axis increases in the paraelectric phase when the Curie Temperature, T_c is approached. In many ferroelectrics, this increase can be approximated by Curie-Weiss law,

$$\varepsilon = \frac{C}{T - T_o} \qquad ...(50)$$

where T_o is equal or somewhat smaller than T_c. For $BaTiO_3$, this law holds unaltered up to frequencies of 2.4×10^{10} Hz. Ferroelectric crystals will have an extremely large dielectric constant.

One of the most common ferroelectrics is the barium titanate ($BaTiO_3$). The spontaneous polarization in this crystal is a consequence of the positioning of the Ba^{2+}, Ti^{4+}, and O^{2-} ions within the unit cell (Fig. 3.53). The Ba^{2+}, ions are located at the corners of the unit cell, which is of tetragonal symmetry (a cube that has been elongated slightly in one direction). The dipole moment results from the relative displacements of the O^{2-} and Ti^{4+} ions from their symmetrical positions as shown in the side view of the unit cell. We can see that O^{2-} ions are located near, but slightly below, the centers of each of six faces, whereas the Ti^{4+} ion is displaced upward from the unit cell corner. Obviously, a permanent ionic dipole moment is associated with each unit cell. However, when $BaTiO_3$ is heated above its *ferroelectric* T_c (= 120°C), the unit cell becomes cubic, and all ions assume symmetric positions within the cubic unit cell; the material now has a perovskite crystal structure, and ferroelectric behaviour

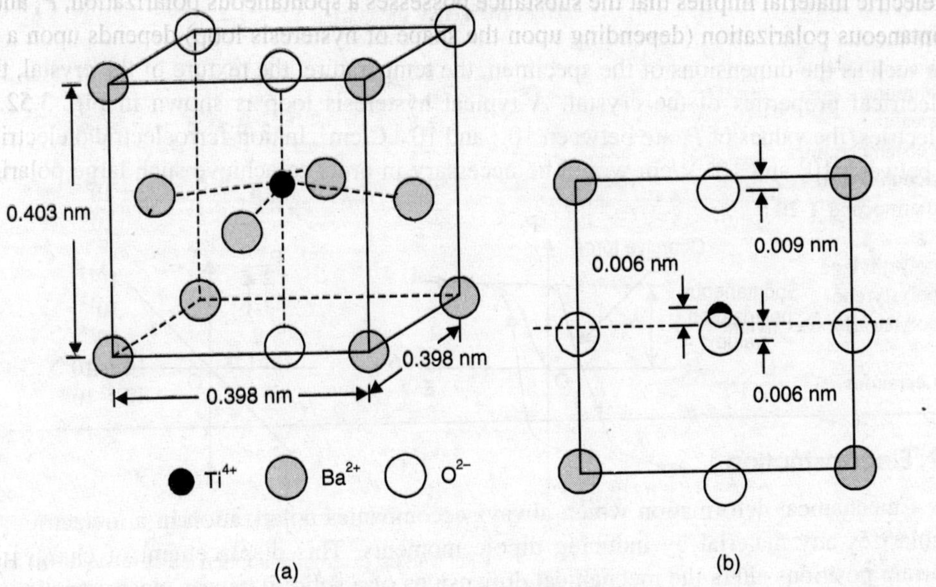

Fig. 3.53 $BaTiO_3$ unit cell (*a*) in an isomeric projection and (*b*) looking at one face, exhibiting the displacement of Ti^{4+} and O^{2-} ions from the center of the face.

ceases. Ceramic materials and especially ferroelectrics are the most important ones among dielectrics. Ceramics may have very diversified electric properties (Table 3.16) and are almost insusceptible to ageing and heating.

Spontaneous polarization of this group of materials results as a consequence of interactions between adjacent permanent dipoles where in they mutually align, all in the same direction. For example, with $BaTiO_3$, the relative displacements of O^{2-} and Ti^{4+} ions are in the same direction for all the unit cells within some volume region of the specimen. At room temperature, ε for $BaTiO_3$ may be as high as 5000. Consequently, capacitors made from these ferroelectric materials can be significantly smaller than capacitors made from other dielectric materials. This is why that these materials are widely used in the manufacture of miniatured capacitors.

Table 3.16 Properties of Dielectrics

Material	ε	*tan* $\delta \times 10^4$	ρ_v, (Ω-*m*)	E_{br} *(MV/m)*
Low frequency dielectrics				
Ceramics				
• electric grade porcelain	6.5	200	10^{11}	25
• ferroelectric ceramics T-7500	7500	300	10^7	2.5
• ferroelectric ceramics VK-1 for variconds	$(20–100) \times 10^3$	1800	10^6	2.5
• piezoelectric ceramics T-1700	1700	300	10^6	2.5
Plastics				
• polyvinyl chloride	4–13	200	10^{12}	16–30
• powdered phenolic plastic with wood dust	11–5	1100–2700	$10^5–10^9$	10–11
• same with minear filler	10–6	500–1000	$10^{10}–10^{11}$	18
• epoxy plastics	4–2.6	400–500	$10^9–10^{12}$	16–20

(Contd.)

	High frequency dielectrics			
Ceramics				
• ultraporcelain	8.5	2–10	10^{10}	20
• alumina	9.5	2	10^{13}	15
• steatite	6.5	3–9	10^{9}	20
• celsian ceramics	7.5	3	10^{10}	35
• ticond T-150	150	3	10^{8}	10
• thermocond T-20	20	3	10^{8}	10
Plastics				
• polyethylene	2.4	2–5	10^{13}	20
• polystyrene	2.5	3–5	10^{13}	30
• polytetrafluoroet hylene	2	2	10^{15}	20
Glass	3.5–16	18–175	10^{11}–10^{16}	30–100
Glass ceramics	5–7	3–20	10^{12}–10^{16}	28–48

3.11.9 Electrostriction

This is a mechanical deformation which always accompanies polarization in a dielectric. An electric field polarizes any material by inducing dipole moments. This displacement of charges from their equilibrium positions alters the mechanical dimensions of a solid; it causes electrostriction. However, mechanical stress applied to a neutral material cannot induce dipole moments; *i.e., electrostriction has no inverse*. If a mechanical distortion creates a voltage, the effect must be caused by permanent dipole moments anchored in the structure without a centre of symmetry.

3.11.10 Piezoelectricity or Pressure Electricity

Electricity, or electric polarity, resulting from the application of mechanical pressure on a dielectric crystal is called piezoelectricity or pressure electricity. The application of a mechanical stress produces in certain dielectric (electrically non-conducting) crystals an electric polarization (electric dipole moment per cubic meter) which is proportional to this stress. If the crystal is isolated, this polarization manifests itself as a voltage across the crystal, and if the crystal is short-circuited, a flow of charge can be observed during loading. Conversely, application of a voltage between certain faces of the crystal produces a mechanical distortion of the material. This reciprocal relationship is referred to as the piezoelectric effect. The phenomenon of generation of a voltage under mechanical stress is referred to as the *direct piezoelectric effect*, and the mechanical strain produced in the crystal under electric stress is called the *converse piezoelectric effect*. The piezoelectric effect is demonstrated in Fig. 3.54.

Piezoelectricity occurs only in insulating materials. Only few ceramic materials exhibit this property. Piezoelectric materials are used extensively in *transducers* for converting a mechanical strain into an electrical signal. Such devices include microphones, phonograph pickups, vibration-sensing elements, and the like. The converse effect, in which a mechanical output is derived from an electrical signal input, is also widely used in such devices as sonic and ultrasonic transducers, headphones, loudspeakers, and cutting heads for disk recording. Both the direct and converse effects are employed in devices in which the mechanical resonance frequency of the crystal is of importance. Such devices include electric wave filters and frequency control elements in electronic oscillator circuits.

The necessary condition for the piezoelectric effect is the absence of symmetry in the crystal structure. Of the 32 crystal classes, 21 lack a centre of symmetry, and with the exception of one class, all of these are piezoelectric. Piezoelectric materials include titanates of barium and lead, zirconate ($PbZrO_3$), ammonium dihydrogen phosphate ($NH_4H_2PO_4$), natural quartz. As stated earlier, this property is a characteristic of materials having complicated crystal structure with a low degree of symmetry. One may improve the piezoelectric behaviour of a polycrystalline specimen by heating above its Curie temperature and then cooling to room temperature in strong electric field.

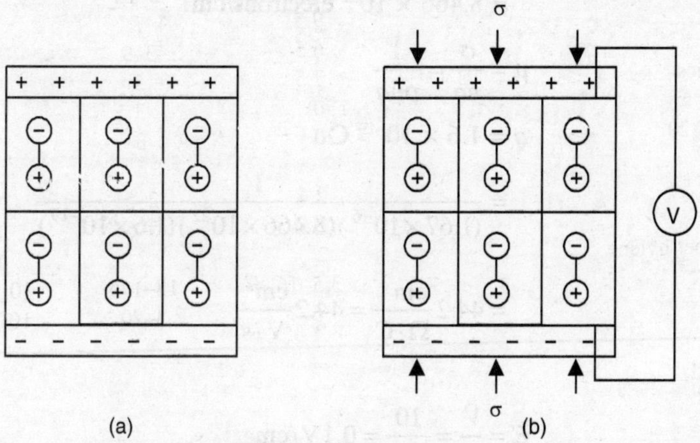

Fig. 3.54 Piezoelectric effect (*a*) dipoles within a piezoelectric material and (*b*) a voltage is generated when the material is subjected to a compressive stress.

Piezoelectrics are required to have high piezoelectric modulus and low losses. Ferroelectric ceramics have the piezoelectric modulus around 6×10^{-10} C/N, which is one or two order of magnitude higher than that of quartz. The structure of piezoelectric ceramics is a solid solution of barium titanate, barium niobate, lead niobate or lead titanate.

Piezoelectric materials have the following important applications: (*i*) frequency resonators, (*ii*) gramo-phone pickups, (*iii*) filters, (*iv*) ultrasonic flaw detectors, (*v*) underwater sonar transducers and (*vi*) air transducers (ear-phones, microphones, hearing aids, etc.).

3.11.11 Uses of Dielectrics

We have seen that dielectric materials are electrically insulative, yet susceptible to polarization in the presence of an electric field. This polarization phenomenon accounts for the ability of the dielectrics to increase the charge storing capability of capacitors. Now, we can summarize the main uses of dielectrics as follows: (*i*) piezoelectric and electro-optic devices, (*ii*) in capacitors, resistors and strain gauges, (*iii*) thermionic valves, radiation detectors, electric devices, dielectric amplifier, (*iv*) dielectrics are usually used as ordinary insulators in power cables, signal cables, electric motors, etc. and (*v*) dielectrics are used in transformers and various form of switchgear and generators where the dissipation problem of heat is active, and a common way of getting rid of it is to insulate with a transformer oil, *i.e.* mineral oil.

Example 1: Considering that all of the valence electrons contribute to the flow of current, (*a*) Calculate the mobility of an electron in copper and (*b*) also calculate the average drift velocity for electrons in a 100 cm copper wire when 10 V are applied. Given, the lattice parameter of Cu = 3.6151×10^{-10} cm and $\sigma = 5.98 \times 10^5\ \Omega^{-1}\ \text{cm}^{-1}$. [B. Tech.]

Solution: The copper has one valency. The number of valence electrons equals the number of copper atoms in the material. Since copper is FCC and hence there are 4 atoms/unit cell. Thus, the resistivity

$$\frac{1}{\sigma} = \frac{1}{5.98 \times 10^5} = 1.67 \times 10^{-6}\ \Omega\text{-cm.}$$

Now, $$n = \frac{(4 \text{ atoms/cell})(1 \text{ electron/atom})}{(3.6151 \times 10^{-8} \text{ cm})^3}$$

$$= 8.466 \times 10^{22} \text{ electrons/cm}^3$$

Now, $$\mu = \frac{\sigma}{nq} = \frac{1}{\rho nq}$$

where $$q = 1.6 \times 10^{-19} \text{ C}$$

$$= \frac{1}{(1.67 \times 10^{-6})(8.466 \times 10^{22})(1.6 \times 10^{-19})}$$

$$= 44.2 \frac{\text{cm}^2}{\Omega \cdot \text{C}} = 44.2 \frac{\text{cm}^2}{\text{V} \cdot \text{s}}$$

(*b*) Electric field,

$$E = \frac{V}{l} = \frac{10}{100} = 0.1 \text{V/cm}$$

$$\bar{v} = \mu E$$

$$= 44.2 \times 0.1 = 4.42 \text{ cm/s.}$$

Example 2: Calculate the resistance of an aluminium wire 100 m long and having cross-sectional area of 3 sq. mm at 20° C. Given, the resistivity of Al at 20°C = 2.66 × 10^{-8} Ω-m. [AMIE]

Solution: Given

$$l = 100 \text{ m}$$

$$\rho = 2.66 \times 10^{-8} \text{ } \Omega\text{-m.}$$

$$A = 3 \times 10^{-6} \text{ sq.m}$$

$$R = \frac{\rho l}{A}$$

$$= \frac{2.66 \times 10^{-8} \times 100}{3 \times 10^{-6}}$$

$$= 88.67 \times 10^{-2} \text{ } \Omega.$$

Example 3: The resistivity of pure copper is 1.56 micro-ohm-cm. An alloy of copper containing 2 atomic per cent nickel has a resistivity of 4.06 micro-ohm-cm. An alloy of copper, containing 1 atomic per cent silver, has a resistivity of 1.7 micro-ohm-cm. Find the resistivity of a copper alloy containing 1 atomic per cent nickel and 3 atomic per cent silver. [Diploma]

Solution: Resistivity of pure Cu = 1.56 micro-ohm-cm. Resistivity of Cu containing two atomic per cent Ni = 4.06 micro-ohm-cm.

∴ Increase due to one atomic % Ni $= \frac{4.06 - 1.56}{2}$

= 1.25 micro-ohm-cm

Now, resistivity of copper, containing one atomic per cent silver = 1.7 micro-ohm-cm.

∴ Increase in resistivity due to one atomic % Ag = 1.7 – 1.56 micro-ohm-cm.

= 0.14 micro-ohm-cm.

Resistivity of copper alloy containing one atomic per cent Ni and 3 atomic per cent.

$$\text{Ag} = 1.56 + 1.25 + 3 \times 0.14$$
$$= 1.56 + 1.25 + 0.42$$
$$= 3.23 \text{ micro-ohm-cm.}$$

Example 4: The resistivity of pure copper at room temperature is 1.8×10^{-8} Ω-m. The resistivity of Cu-4% Ni alloy at room temperature is 7×10^{-8} Ω-m. Show that the resistivity due to impurity scattering per % of Ni in the Cu lattice is 1.3×10^{-8} Ω-m. [Diploma]

Solution: Let us take that the resistivity of copper at 0 K is negligible. Now, 1.8×10^{-8} Ω-m is the thermal part of resistivity of copper as well as of the alloy between 0 K and room temperature.

$$\therefore \text{ Impurity scattering per \% of Ni is then} = \frac{(7.0 \times 1.8) \times 10^{-8}}{4} = 1.3 \times 10^{-8} \ \Omega\text{-m}.$$

Example 5: Calculate the mobility of an electron in copper and also the average drift velocity for electrons in 100 cm - Copper when 10 V is applied. Assume that all the valence electrons contribute to current flow. Given: lattice parameters of Cu = 3.1651×10^{-8} cm and conductivity (σ) = 5.98×10^5 Ω^{-1} cm^{-1}. [B.Tech.]

Solution: The valency of Cu is 1. Thus, the number of valence electrons equals the number of copper atoms in the material. Copper is FCC and there are 4 atoms per unit cell.

$$\text{The resistivity } (\rho) = \frac{1}{\sigma} = \frac{1}{5.98 \times 10^5}$$
$$= 1.67 \times 10^{-6} \ \Omega\text{-cm.}$$
$$n = \frac{(4 \text{ atoms/cell})(1 \text{ electron/atom})}{(3.6151 \times 10^{-8} \text{ cm})^3}$$
$$= 8.466 \times 10^{22} \text{ electrons/cm}^3$$

Now,
$$\mu = \frac{\sigma}{nq} = \frac{1}{\rho n q} \quad (q = 1.6 \times 10^{-19} \text{ C})$$
$$= \frac{1}{(1.67 \times 10^{-6})(8.466 \times 10^{22})(1.6 \times 10^{-19})}$$
$$= 44.2 \frac{\text{cm}^2}{\Omega \cdot \text{C}} = 44.2 \frac{\text{cm}^2}{\text{V} \cdot \text{s}}$$

Electric field,
$$E = \frac{V}{l} = \frac{10}{100} = 0.1 \text{ V/cm}$$

Since the mobility is 44.2 cm^2/V.s, and therefore
$$\bar{v} = \mu E = 44.2 \times 0.1 = 4.42 \text{ cm/s}.$$

Example 6: Calculate the electrical conductivity of pure copper at 400° C. Given: Conductivity of pure copper = 5.98×10^5 Ω^{-1} cm^{-1}, resistivity of Cu at room temperature = 1.67×10^{-6} Ω.cm and the temperature resistivity coefficient (α) = 0.0068 Ω.cm/° C.

Solution: We have

$$\rho = \rho_0 (1 + \alpha \Delta T)$$
$$= 1.67 \times 10^{-6} [1 + 0.0068(400 - 25)]$$
$$\rho = 5.929 \times 10^{-6} \ \Omega\text{-cm}$$

$$\therefore \quad \sigma = \frac{1}{\rho} = \frac{1}{5.929 \times 10^{-6}} = 1.69 \times 10^5 \ \Omega^{-1}\text{cm}^{-1}.$$

Example 7: Considering that electrical conductivity of MgO is determined primarily by the diffusion of the Mg^{2+} ions, find the mobility of the Mg^{2+} ions and also calculate the electrical conductivity of MgO at 1800° C. Given the diffusion coefficient of Mg^{2+} ions in MgO at 1800° C is 10^{-10} cm²/s. For MgO, z = 2/ion and $k_B = 1.38 \times 10^{-23}$ J/K, $q = 1.6 \times 10^{-19}$ C.

Solution: $\mu = \dfrac{zqD}{k_B T}$ $(T = 1800 + 273 = 2073 \text{ K})$

$$= \frac{2 \times 1.6 \times 10^{-19} \times 10^{-10}}{1.38 \times 10^{-23} \times 2073}$$
$$= 1.12 \times 10^{-9} \ \text{C} \cdot \text{cm}^2/\text{J} \cdot \text{s}$$

We know that charge in Coulomb is equivalent to Amp/sec and 1 J = Amp.sec.volts

We have

$$\mu = 1.12 \times 10^{-9} \ \text{cm}^2/\text{V.s}$$

MgO has the similar structure to NaCl with 4 magnesium ions/unit cell. The lattice parameter is 3.96×10^{-8} cm, so the number of Mg^{2+} ions/cm³, n

$$n = \frac{4 \ Mg^{2+} \text{ ions/cell}}{(3.96 \times 10^{-8} \text{ cm})^3}$$
$$= 6.4 \times 10^{22} \text{ ions/cm}^3$$

Now,
$$\sigma = nzqu$$
$$= (6.4 \times 10^{22}) \times 2 \times 1.6 \times 10^{-19} \times 1.12 \times 10^{-9}$$
$$= 22.94 \times 10^{-6} \ \text{c.cm}^2/\text{cm}^3.\text{V.s}$$
$$= 2.294 \times 10^{-5} \ \Omega^{-1}.\text{cm}^{-1}.$$

Example 8: The critical temperature of mercury is 4.2 K.

(*a*) Calculate the energy gap in electron volts at $T = 0$?

(*b*) Calculate the wavelength of a photon whose energy is just sufficient to break up cooper pairs in mercury at $T = 0$. In what region of the electromagnetic spectrum are such photons found?

Solution: (*a*) The Cooper pair binding energy, or gap energy, is

$$E_g = 3 \, kT_c$$
$$= 3 \times 1.4 \times 10^{-23} \ \text{J/°K} \times 4.2 \text{ K}$$
$$= 1.8 \times 10^{-22} \text{ J} = 1.1 \times 10^{-3} \ eV$$

(*b*) $E_g = hv = hc/\lambda$

$$\therefore \quad \lambda = \frac{hc}{E_g} = \frac{6.6 \times 10^{-34} \text{ J-s} \times 3 \times 10^8 \text{ m/s}}{1.8 \times 10^{-22} \text{ J}} = 1.1 \times 10^{-3} \text{ m}..$$

Example 9: Find the limiting magnetic field that will permit niobium to serve as a superconductor at liquid helium temperature. Given: H_o = 1970 Oersteds, T_c = 9.25 K [B.Tech., M.Sc.]

Solution: Temperature of liquid helium 4K and as such this will be operating temperature.

We have
$$H = H_o\left[1-\left(\frac{T}{T_c}\right)^2\right]$$

$$=1970\left[1-\left(\frac{4}{9.25}\right)^2\right]$$

$$= 1970/0.813 = 1602 \text{ oersted.}$$

Obviously, magnetic field (H) must remain below 1602 oersted so that niobium may serve as super-conductor at 4 K.

Example 10: Diameter of a wire is 0.20 cm and current flowing through it is 20 A. The maximum power dissipation along the wire is 4 W/m. Show that the minimum allowable conductivity of wire for this application is about 3.18×10^7 $(\Omega.\text{m})^{-1}$. [M.Sc., B.Tech.]

Solution: We have

$$P = iV = i^2R$$

and
$$R = \rho\frac{l}{A}$$

∴
$$P = i^2\rho\frac{l}{A} = \frac{i^2 l}{\sigma A}$$

Here
$$\rho = 4 \text{ W (in 1 m)}, i = 20 \text{ A}, l = 1 \text{ m}$$

$$A = \frac{\pi}{4}(0.0020 \text{ m})^2 = 3.14\times10^{-6}\text{m}^2$$

Thus
$$\sigma = \frac{i^2 l}{\rho A} = \frac{(20A)^2\times(1\text{m})}{(4\text{W})\times(3.14\times10^{-6}\text{m})}$$

$$= 3.18 \times 10^7\ \Omega^{-1}\text{-m}^{-1}$$

Obviously, for the given application, the conductivity (σ) of the wire be equal to or greater than 3.18 $\times 10^7\ \Omega^{-1}\ \text{m}^{-1}$.

Example 11: The density of Si is 2.33 g/cm³ and its atomic mass is 28.08 g/mol. Calculate the number of Si atoms per cubic meter.

Solution:

$$\text{Number of Si atoms per m}^3 = \frac{6.023\times10^{23}\text{ atoms}}{\text{mol}}$$

$$= \frac{1}{28.08 \text{ g/mol}}$$

$$= \frac{2.33\times10^{6}\ \text{g}}{\text{m}^3}$$

$$= 5.00 \times 10^{28}\ \text{atoms/m}^3.$$

Example 12: For Si at 300 K, $n_i = 1.5 \times 10^{16}$ carriers/m³, $q = 1.60 \times 10^{-19}$ C, $\mu_n = 0.135$ m²/(V.s), and $\mu_p = 0.048$ m²/(V.s). Calculate the electrical resistivity of Si. [B.Tech., B.Sc.]

Solution:

$$\rho = \frac{1}{\sigma} = \frac{1}{n_i q(\mu_n + \mu_p)}$$

$$= \frac{1}{\left(\dfrac{1.5\times10^{16}}{\text{m}^3}\right)(1.60\times10^{-19}\ \text{C})\left(\dfrac{0.135\text{m}^3}{\text{V}\cdot\text{s}} + \dfrac{0.048\,\text{m}^2}{\text{V}\cdot\text{s}}\right)}$$

$$= 2.28 \times 10^3\ \Omega.\text{m}.$$

Example 13: Estimate for Ge at 25° C: (*a*) the number of charge carriers (*b*) the fraction of total electrons in the valence band that are excited into the conduction band, and (*c*) $n_o \left(= \dfrac{n}{\exp(-E_g/2k_BT)}\right)$.

Given $\rho = 43$ Ω-cm, $E_g = 0.67$ eV, $\mu_n = 3900 \dfrac{\text{cm}^2}{\text{V}\cdot\text{s}}$ and $\mu_p = 1900$ cm²/V.s.

Solution: $2k_BT = 2 \times 8.63 \times 10^{-5}\ \text{eV} \times (273 + 25)$
$= 0.0514$ eV at 25° C.

$$\sigma = \frac{1}{\rho} = 0.023\ \Omega^{-1}\ \text{cm}^{-1}$$

$$(a)\ n = \frac{\sigma}{\mu_n + \mu_p} = \frac{0.023}{(1.6\times10^{-19})(3900+1900)}$$

$$= 2.5 \times 10^{13}\ \frac{\text{electrons}}{\text{cm}^3}$$

Obviously, there are 2.5×10^{13} electrons/cm³ and $2.5 \times 10^{13}\ \dfrac{\text{holes}}{\text{cm}^3}$. These charge carriers are helping to conduct a charge in Ge at 25° C.

(*b*) The lattice parameter of diamond cubic Ge is 5.6575×10^{-8} cm. The total number of electrons in the valence band of Ge is

$$\text{Total electrons} = \frac{\left(\dfrac{8\ \text{atoms}}{\text{cell}}\right)\left(\dfrac{4\ \text{electrons}}{\text{atom}}\right)}{(6.6575\times10^{-8}\ \text{cm})^3}$$

$$= 1.77 \times 10^{23}$$

Fraction, that is excited $= \dfrac{2.5 \times 10^{13}}{1.77 \times 10^{23}}$

$= 1.41 \times 10^{-10}$

(c) Now, $n_o = \dfrac{n}{\exp(-E_g/2k_BT)}$

$= \dfrac{2.5 \times 10^{13}}{\exp(-0.67/0.514)}$

$= 1.14 \times 10^{19}$ carriers/cm^3.

Example 14: You have to design a semiconductor based on Si, which can provide a constant conductivity of 100 Ω^{-1} cm^{-1} over a range of temperatures. Find the level of purity needed? [B.Tech.]

Solution: Si will have to be doped with atoms having a valence of +3. If we assume that the number of intrinsic carriers is small, then

$$\sigma = N_a q \mu_p$$

where $\sigma = 100\ \Omega^{-1}.\text{cm}^{-1}$, $\mu_p = 480$ cm^2/V.s

The number of charge carriers required

$$N_a = \frac{\sigma}{q\mu_p} = \frac{100}{1.6 \times 10^{-19} \times 480}$$

$= 1.30 \times 10^{18}$ acceptor atoms/cm^3.

Assuming that the lattice constant of Si remains unchanged as a result of doping, we compare the concentration of donor atoms in Si with the concentration of Si atoms.

$$N_a = \frac{\left(\dfrac{1 \text{ electron}}{\text{dopant atom}}\right)\left(\dfrac{x \text{ dopant atom}}{\text{Si atom}}\right)\left(\dfrac{8 \text{ Si atoms}}{\text{unit cell}}\right)}{(5.4307 \times 10^{-8}\text{ cm})^3}$$

$$= \frac{1.3 \times 10^{18} \times (5.4307 \times 10^{-8})^3}{8}$$

$= 26 \times 10^{-6}$ dopant atom/ Si atom

i.e. 26 dopant atoms/10^6 Si atoms.

We may note that possible dopant include *B*, Al, gallium and indium. For processing high purity chemicals and clean room conditions are essential for processing.

Example 15: A Si wafer is doped with 10^{21} phosphorous atoms/m^3. Calculate the majority carrier concentration, (*b*) the minority carrier concentration and (*c*) the electrical resistivity of the doped Si at room temperature, *i.e.* 300 K. You may assume complete ionization of the dopant atoms, $n_i(S_i) = 1.5 \times 10^{16}$ m^{-3}, $\mu_n = 0.135$ m^2/(V.s), and $\mu_p = 0.048$ m^2/(V.s).

Solution. Si is doped with *P* (group V element). Obviously, the doped silicon is *n*-type.

(*a*) $n_n = N_d = 10^{21}$ electrons/m^3

(*b*) $\dfrac{n_i^2}{N_d} = \dfrac{(1.5 \times 10^{16}\text{ m}^{-3})^2}{10^{21}\text{ m}^{-3}} = 2.25 \times 10^{11}\text{ holes/m}^3$

(c) $$\rho = \frac{1}{q\mu_n n_n} = \frac{1}{(1.6\times10^{-19}\text{ C})\left(0.135\frac{\text{m}^2}{\text{V}\cdot\text{s}}\right)\left(\frac{10^{21}}{\text{m}^3}\right)}$$

$$= 0.0463\ \Omega\text{.m.}$$

Example 16: A silicon semiconductor at 300 K is doped with 1.4×10^{16} boron atoms/cm³ plus 1.0×10^{16} phosphorous atoms/cm³. Calculate (*a*) the equilibrium electron and hole concentrations, (*b*) the mobilities of electrons and holes, and (*c*) the electrical resistivity. You may assume complete ionization of the dopant atoms. Given n_i(Si) = 1.5×10^{10} cm⁻³. [B.Tech.]

Solution: (*a*) *Majority carrier concentration*: The net concentration of immobile ions is equal to the acceptor ion concentration minus the donor ion concentration, *i.e.*

$$p_p \approx N_o - N_d = 1.4 \times 10^{16}\ B \text{ atoms/cm}^3 - 1.0 \times 10^{16}\ \rho \text{ atoms/cm}^3$$
$$= N_a = 4.0 \times 10^{15} \text{ holes/cm}^3$$

Minority carrier (electrons) concentration:

$$n_p = \frac{n_i^2}{n_a} = \frac{(1.50\times10^{10}\text{ cm}^{-3})^2}{4\times10^{15}\text{ cm}^{-3}}$$
$$= 5.6 \times 10^4 \text{ electrons/cm}^3$$

(*b*) *Mobilities of electrons and holes*:

For electrons, we have total impurity concentration

$$C_T = 2.4 \times 10^{16} \text{ ions/cm}^3 \text{ one obtains}$$
$$\mu_n = 900 \text{ cm}^2/(\text{V.s})$$

For holes,

$$C_T = 2.4\times10^{16} \text{ ions/cm}^3\text{, one obtains}$$
$$\mu_p = 300 \text{ cm}^2/(\text{V.s})$$

(*c*) *Electrical resistivity*: The doped semiconductor is *p*-type:

$$\rho = \frac{1}{q\mu_p p_p}$$

$$= \frac{1}{(1.60\times10^{-19}\text{ C})(300\text{ cm}^2/(\text{V}\cdot\text{s}))(4.0\times10^{15}/\text{cm}^3)}$$
$$= 5.2\ \Omega\text{.cm.}$$

Example 17: Find the intrinsic electrical conductivity of GaAs at 300 K and at 343 K. Also calculate the fraction of the current carried by the electrons in the intrinsic GaAs at 300 K. [B.Tech.]

Solution: (*i*) σ at 300 K

We have $\sigma = n_i q\,(\mu_n + \mu_p)$

$$= (1.4 \times 10^{12}\text{ m}^{-3})(1.60 \times 10^{-19}\text{ C}) \times [0.720\text{ m}^2/(\text{V.s}) + 0.020\text{ m}^2/(\text{V.s})]$$
$$= 1.66 \times 10^{-7}\ (\Omega\text{.m})^{-1}$$

(*ii*) σ at 343 K

$$\sigma = \sigma_o e^{-E_g/2kT}$$

$$\frac{\sigma_{343}}{\sigma_{300}} = \frac{\exp\left[-1.47\text{ eV}/\left[2\times 8.62\times 10^{-5}\text{ eV/K}\right](343\text{K})\right]}{\exp\left[-1.47\text{eV}/\left[2\times\left(8.62\times 10^{-5}\text{ eV/K}\right)(300\text{ K})\right]\right]}$$

$$\sigma_{343} = \sigma_{300}e^{3.56} = 1.66\times 10^{-7}\,(\Omega\cdot\text{m})^{-1}\times 35.2$$

$$= 5.84\times 10^{-6}\,(\Omega\cdot\text{m})^{-1}$$

(b) $$\frac{\sigma_n}{\sigma_n+\sigma_p} = \frac{n_i q\mu_n}{n_i q(\mu_n+\mu_p)}$$

$$= \frac{0.720\text{ m}^2/(\text{V}\cdot\text{s})}{0.720\text{ m}^2/(\text{V}\cdot\text{s})+0.020\text{ m}^2(\text{V}\cdot\text{s})}$$

$$= 0.973.$$

Example 18: A light emitting diode display made using a GaAs-GaP solid solution of composition 0.4 GaP – 0.6 GaAs has a direct bandgap of 1.9 eV. Show that the colour display by this LED is red.

[B. Tech.]

Solution: We know that in direct bandgap materials, a forward bias, *i.e.* an external voltage is used to inject carriers that recombine via radiative recombination and give off light. The relation between the emitted wavelength of light and bandgap is

$$E_g = h\nu = hc/\lambda$$

Here $h = 6.63 \times 10^{-3.4}$ J.s = 4.41×10^{-15} eV.s

$c = 3 \times 10^8$ m/s, λ = ? Putting the values in above relation, one obtains

$\therefore$ 1.9 eV = $1.24/\lambda$ (in μm)

$\therefore$ λ = 0.652 μm = 652 nm

This wavelength corresponds to red light. Obviously, LED displays red in colour.

Example 19: The saturation current density of a P-N junction Germanium diode is 250 mA/m² at 300° K. Find the voltage that would have to be applied across the junction to cause a forward current density of 10^5 A/m² to flow.

Solution: $$I = I_s\left[\exp\left(\frac{eV}{kT}-1\right)\right]$$

Dividing throughout by volume, one obtains the equation in the form of current density as

$$J = J_s\left[\exp\left(\frac{eV}{kT}-1\right)\right]$$

$$\exp\left(\frac{eV}{kT}\right)-1 = J/J_s$$

$$\exp\left(\frac{eV}{kT}\right)-1 = \frac{10^5}{250\times 10^{-3}} = 4\times 10^5 \qquad J_s = 250\text{ mA/m}^3$$

$$J = 10^5\text{ A/m}^2$$

$$\frac{eV}{kT} \simeq \log_e\left(4\times10^5\right) = 12.9$$

$$V = \frac{12.9\times1.38\times10^{-23}\times300}{1.6\times10^{-19}} = 0.33\text{ V.}$$

Example 20: Show that the relative dielectric constant of a barium titanate crystal, which, when inserted in a parallel plate condenser of area 10 mm × 10 mm and distance of separation of 2 mm, gives a capacitance of 10^{-9} F is 2259. [M.Sc.(M.S.)]

Solution: We have

$$C = 10^{-9}\text{ F}$$

$$d = 2\times10^{-3}\text{ m}$$

$$\varepsilon_o = 8.854\times10^{-12}$$

$$A = 10\times10^{-3}\times10\times10^{-3} = 10^{-4}\text{ m}^2$$

$$C = \frac{\varepsilon_r\varepsilon_o A}{d}$$

$$\therefore \quad \varepsilon_r = \frac{Cd}{\varepsilon_o A}$$

$$= \frac{10^{-19}\times2\times10^{-3}}{8.854\times10^{-12}\times10^{-4}}$$

$$= 2259.$$

Example 21: A simple parallel-plate capacitor is to be made to store 5.0×10^{-6} C at a potential of 8000 V. The separation distance between the plates is to be 0.30 mm. Calculate the area (in square meters) that the plates must have if the dielectric between the plates is (*a*) a vacuum ($\kappa = 1$) and (*b*) alumina ($\kappa = 9$). ($\varepsilon_o = 8.85\times10^{-12}$ F/m.) [B.Tech.]

Solution: $C = \dfrac{q}{V} = \dfrac{5.0\times10^{-6}\text{ C}}{8000\text{ V}} = 6.25\times10^{-10}\text{ F}$

$$A = \frac{Cd}{\varepsilon_o\kappa} = \frac{(6.25\times10^{-10}\text{ F})(0.30\times10^{-3}\text{ m})}{(8.85\times10^{-12}\text{ F/m})(\kappa)}$$

(*a*) For vacuum, $\kappa = 1$: $A = 0.021\text{ m}^2$

(*b*) For alumina, $\kappa = 9$: $A = 2.35\times10^{-3}\text{ m}^2$

As can be seen by these calculations, the insertion of a material with a high dielectric constant can appreciably reduce the area of the plates required.

Example 22: Using the given data for $BaTiO_3$ crystal, calculate the polarization. The shift of the titanium ion from the body centre is 0.06 Å. The oxygen anions of the side faces shift by 0.06 Å, while the oxygen anions of the top and bottom faces shift by 0.08 Å, all in a direction opposite to that of the titanium ion. [B.Tech.]

Solution: Let us calculate the dipole moments due to the effective number of each type of ion in the unit cell of $BaTiO_3$ as:

(*i*) dipole moment due to two O^{2-} ions on the four side faces

$$\mu_1 = Qd = 2 \times 2 \times 1.6 \times 10^{-19} \times 0.06 \times 10^{-10} \text{ C-m}$$

(*ii*) dipole moment due to one O^{2-} on top and bottom

$$\mu_2 = Qd = 1 \times 2 \times 1.6 \times 10^{-19} \times 0.08 \times 10^{-10} \text{ C-m}$$

(*iii*) dipole moment due to one Ti^{4+} ion at body centre

$$\mu_3 = Qd = 1 \times 4 \times 1.6 \times 10^{-19} \times 0.06 \times 10^{-10} \text{ C-m}$$

$\therefore$ Total dipole moment $= 1.02 \times 10^{-29}$ C-m

Now, the polarization is the total dipole moments per unit volume. Now, ignoring the effect due to barium ions, one obtains

$$P = \frac{\mu}{A} = \frac{1.02 \times 10^{-29}}{4.03 \times (3.98)^2 \times 10^{-30}}$$

$$= 0.16 \text{ C-m}^2.$$

SUMMARY

1. On the basis of electrical conductivity (σ) or its reciprocal, resistivity (ρ), materials are classified as superconductors, metals, semiconductors and insulators.
2. In the energy band model for electrical conduction in metals, the valence electrons of the metal atoms interact and interpenetrate each other to form energy bands. Since the energy bands of the valence electrons of metal atoms overlap, producing partially filled composite energy bands, very little energy is required to excite the highest energy electrons so that they become free to be conductive.
3. In pure metals, the resistivity increases with temperature. Resistivity is very sensitive to impurities and microstructural defects such as grain boundaries.
4. Resistivity of alloys is typically higher and dominated by alloying elements and is relatively stable with respect to temperature.
5. The phenomenon of flow of direct electric current through a material that has no resistance to that flow is called superconductivity and the material that exhibit this phenomenon is called superconductor.
6. Superconductors are materials that offer zero resistance to the flow of direct electrical current (d.c.) and exclude magnetic field, *i.e.* $\rho = 0$, $B = 0$ inside a superconductor.
7. The superconductive effect is seen at very low temperatures. The highest known T_c is 134 K in mercury cuprates.
8. In most Type-I superconductors, the superconductive effect is lost even under very small magnetic flux densities, whereas in Type-II superconductors the flux can penetrate while the material remains superconducting and above a certain critical field (H_{C_2}), the superconductivity is lost.
9. The energy band model for an insulator consists of a lower filled valence band and a higher empty conduction band. The valence band is separated from the conduction band by a large energy gap (about 6 to 7 eV). Obviously, for insulators to be conductive, a large amount of energy have to be supplied so that the valence electrons to "jump" the gap.
10. Semiconductors have conductivities between insulators and conductors. Intrinsic semiconductors have a relatively small energy gap (*i.e.* about 0.7 to 1.1 eV) between their valence and conduction bands.

11. Resistivity, a dielectric constant, the mobility of electrons, and the holes characterize semiconductors.
12. By doping the intrinsic semiconductors with impurity atoms to make them extrinsic, the amount of energy required to cause semiconductors to be conductive is greatly reduced.
13. A semiconductor prepared by dopants, which determine the number and type of charge carriers (electrons and holes) is called extrinsic semiconductor. Extrinsic behaviour can also be seen due to impurities.
14. Semiconductors can be classified as elemental (Si, Ge) or compound (SiC, GaAs). Both of these can be intrinsic or extrinsic (*n*- or *p*- type). Some semiconductors are direct (*e.g.* Ga As) or indirect (*e.g.* Si) bandgaps.
15. The *n*-type (negative) semiconductors have electrons for their majority carriers. The *p*-type (positive) semiconductors have holes (missing electrons) for their majority charge carriers.
16. By fabricating *pn* junctions in a single crystal of semiconductor such as Si, various types of semiconductor devices can be fabricated, *e.g. pn* function diodes, and *npn* transistors can be produced by using these junctions.
17. Modern microelectronic technology has developed to such an extent that thousands of transistors can be placed on a "*chip*" of semiconducting Si less than about 0.5 cm square and about 0.2 mm thick. Complex microelectronic technology has made possible highly sophisticated microprocessors and computer memories.
18. Dielectric materials have large bandgaps and do not conduct electricity. Although these materials are electrically insulator, yet susceptible to the presence of electric field. This polarization phenomenon accounts for the ability of dielectrics to increase the charge storing capabilities of capacitors, the efficiency of which is expressed in terms of a dielectric constant.
19. Polarization in dielectrics results from the inducement by or orientation with the electric field of atomic or molecular dipoles; a dipole is said to exist when there is a net spatial separation of positively and negatively charged entities. Possible polarization types include electronic, ionic and orientation; not all types need be present in a particular dielectric.
20. Materials in which dielectric polarization is linearly related to the electric field; the dielectric constant is not dependent on the electric field.
21. For alternating electric fields, whether a specific polarization type contributes to the total polarization and the dielectric constant depends on frequency, each polarization mechanism ceases to function when the applied field frequency exceeds to relaxation frequency.
22. Piezoelectrics are the materials that develop voltage upon the application of a stress and develop strain when an electric field is applied. In pyroelectrics, temperature change causes polarization. Polarization in these materials is induced by the imposition of external forces.
23. Ferroelectrics are materials that shows spontaneous and reversible dielectric polarization. $BaTiO_3$, PZT, PVDF are examples of ferroelectrics. These materials exhibit a large dielectric constant, even in a paraelectric state, and are often used to make capacitors. All ferroelectrics are piezoelectrics but converse is not always true.

SUGGESTED READINGS

1. S. L. Kakani and Amit Kakani, 'Material Science' (2nd edn. 2010), New Age Publishers, New Delhi-2.
2. S. L. Kakani and C. Hemjarani, 'Solid State Physics', Sultan Chand, New Delhi-2 (4th edn., 2004).
3. L. Solymar and D. Walsh, 'Electrical Properties of Materials', Oxford University Press, New York (6th edn. 1998).

4. R. E. Hummel, 'Understanding Materials Science', Springer (2004).
5. R. E. Hummel, 'Electronic Properties of Materials', 3rd edn., Springer (2001).
6. S. M. Sze, 'Physics of Semiconductor Devices', 2nd edn., Wiley (1981).
7. S. L. Kakani and K. C. Bhandari, 'Electronics', (4th edn. 2011), New Age Publishers, New Delhi-2.
8. S. L. Kakani and K. C. Bhandari, 'A Text Book of Electronics', (4th edn. 2010), New Age Publishers, New Delhi-2.
9. S. L. Kakani and S. Kakani, 'Superconductivity', (2nd edn. 2012), New Age Publishers, New Delhi-2.

REVIEW QUESTIONS

1. Distinguish between a metal, a semiconductor and an insulator on the basis of their energy band structure.
2. (*a*) Explain electrical conductivity on the basis of energy band diagram.
   (*b*) What is the importance of a energy band diagram and how does it compare the conductor and insulator? [A.M.I.E.]
3. What is an insulator? What is the reason that a material behaves an insulator? [A.M.I.E.]
4. Describe an expression for the conductivity of conductors from the first principle. [A.M.I.E.]
5. Discuss factors that affect the electrical resistance of materials. [A.M.I.E.]
6. Explain what insulators, conductors and semiconductors are? Give examples of each. [A.M.I.E.]
7. Define a dielectric. Explain the common properties and uses of dielectrics. [A.M.I.E.]
8. Describe with examples of three categories of dielectrics.
9. What are semiconductors? How they differ from metals and insulators? What are their characteristic properties? [B.E.]
10. Why an increase in temperature of semiconductor increases its conductivity? [B.E.]
11. Explain the terms: (*i*) Intrinsic semiconductors and (*ii*) Extrinsic semiconductors. [A.M.I.E.]
12. What is impurity conduction in semiconductors? Explain how the presence of a small impurity in a semiconductor modifies its conduction properties.
13. Define the terms: (*i*) Electron-hole pair, (*ii*) Donor and (*iii*) Acceptor. Give examples of each with suitable materials. [A.M.I.E.]
14. Describe *N*-type and *P*-type of semiconductors. [A.M.I.E.]
15. Describe a *P-N* junction and transistor with neat sketches. [B.E.]
16. Explain the process of recombination of electrons and holes in a semiconductor. If the electron-hole pair recombine, how can their concentration remain constant at a particular temperature?
17. What are extrinsic semiconductors? How can they be formed?
18. At a high temperature an extrinsic semiconductors like an intrinsic one, why?
19. Is it possible for compound semiconductors to exhibit intrinsic behaviour? Explain.
20. For each of the following pairs of semiconductors decide which will have the smaller bandgap energy, E_g, and then give reason: (*b*) Carbon (C), and Ge, (*b*) AlP and InSb (*c*) GaAs and ZnSe, (*d*) ZnSe and CdTe and (*e*) CdS and NaCl.
21. (*a*) Explain why no hole is generated by the electron excitation involving a donor impurity atom?
    (*b*) Explain why no free electron is generated by the electron excitation involving an acceptor impurity atom?
22. Will each of the following elements act as a donor or an acceptor when added to the indicated semiconducting material? Assume that the impurity elements are substitutional.

Impurity	*Semiconductor*
N	Si
B	Ge
Zn	GaAs
S	InSb
In	CdS
As	ZnTe

23. What is superconductivity? Mention some important property changes that occur in materials when they undergo phase change from normal to the superconducting state.
24. What are Type-I and Type-II superconducting materials? Give three examples of each why Type-II materials are preferred for applications of superconductivity? [B.E.]
25. Define critical temperature, critical field and critical current density for a superconductor. [A.M.I.E.]
26. Explain the behaviour of a superconductor in a magnetic field. What is Meissner effect? How does the critical magnetic field varies with temperature? [M.Sc. (M.S.)]
27. What is isotope effect in superconductors? How it provided a clue to the microscopic BCS theory of superconductivity? [B.E.]
28. Describe the effects of (*i*) magnetic field, (*ii*) frequency and (*iii*) isotopes on superconductors.
29. What are the characteristic lengths of a superconductor? How penetration depth varies with temperature? [B.E.]
30. Give a brief account of thermodynamics of a superconductor.
31. What is BCS theory of superconductivity? Write its salient features.
32. Write a short note on potential applications of superconductors. [B.E.]
33. What are d.c. and a.c. Josephson effects? Explain their importance.
34. What do you understand by quantum tunnelling? Explain its significance. [B.E.]
35. What are HTSC oxides? How they are different from conventional superconductors? What is their future prospects from applications of superconductivity point of view?
36. What is radiative and non-radiative combination? What type of materials are used to make LEDs?

PROBLEMS

1. The resistance of a given sample of copper of diameter 0.1 mm and length 0.2 m is 0.439 Ω. Calculate its resistivity and conductivity. [**Ans.** 1.72×10^{-8} Ω-m, 5.81×10^7/ Ω-m] [B.E.]
2. At room temperature the electrical conductivity and the electron mobility for Cu are $6 \times 10^7/(\Omega\text{-m})$ and 0.0030 $m^2/(V\text{-}s)$, respectively. (*a*) Find the number of free electrons per m^3 for Cu at room temperature. (*b*) What is the number of free electrons per Cu atom? Given density 8.9 gm/cm^3.

$$\left[(a)\ n = 1.25\times10^{29}/\text{m}^3,\ (b)\ 1.48\ \frac{\text{free electrons}}{\text{atom}}\right] \quad \text{[A.M.I.E., B.E.]}$$

3. Show that the percentage increase in resistivity of (*i*) Cu, (*ii*) nichrome on heating from 300 K to 1000 K will be 300% and 7% respectively. [A.M.I.E.]
4. For intrinsic gallium arsenide, the room temperature electrical conductivity is 10^{-6} $(\Omega\text{-m})^{-1}$, the electron and hole mobilities are, 0.85 and 0.04 m^2/V-s respectively. Compute the intrinsic concentration n_i at room temperature. [**Ans.** 7.0×10^{12} m^{-3}]

5. Calculate the electrical conductivity of intrinsic silicon at 150° C (423 K). [**Ans.** 0.52 (Ω-m)$^{-1}$]
6. The high-purity silicon is added 10^{23}/m^3 arsenic atoms: (*a*) Is this material *N*-type or *P*-type? (*b*) Calculate the room temperature electrical conductivity of this material. (*c*) Compute the conductivity at 100° C (373 K). [**Ans.** (*a*) *N*-type, (*b*) 1120 (Ω-m)$^{-1}$, (*c*) 640 (Ω-m)$^{-1}$]
7. The following electrical characteristics have been found for both intrinsic and *P*-type extrinsic indium phosphide (InP) at room temperature:

	$\sigma(\Omega\text{-}m)^{-1}$	$N(m^{-3})$	$P(m^{-3})$
Intrinsic	2.5×10^{-6}	3.0×10^{13}	3.0×10^{13}
Extrinsic (*N*-type)	3.6×10^{-5}	4.5×10^{14}	2.0×10^{12}

Calculate electrons and hole mobilities. [**Ans.** $\mu_e = 0.50$ m^2/V-s, $\mu_h = 0.02$ m^2/V-s]

8. Mobilities of electrons and holes in a sample of intrinsic germanium at room temperature are 3600 cm^2/V-s and 1700 cm^2/V-s respectively. If the electron and hole densities are each equal to 2.5×10^{13} per cc, calculate its conductivity. [**Ans.** 2.12 mho/m]
9. A current density of 100,000 A/cm^2 is applied to a gold wire 50 m in length. The resistance of wire is 2 Ω. Calculate the diameter of the wire and the voltage applied to the wire. [B.Tech.] [**Ans.** $d = 0.0865$ cm, 1174 V]
10. After determining the electrical conductivity of cobalt at 0° C, one decide to double that conductivity. To what temperature cobalt be cooled? [B.Tech.] [**Ans.** –70.8°C]
11. GaV_3 is to operate as a superconductor in liquid helium (4 K). For GaV_3, $T_c = 16.8$ K and $\mu_0 = 350{,}000$ oersted. Show that the maximum magnetic field that can be applied to the material is 330160 Oe. [M.Sc.]
12. Compare the number of charge carriers per cubic centimeter, the fraction of the total electrons in the valence band that are excited into the conduction band, and constant n_0 for Ge, Si and Sn. [$n_{Ge} = 1.767 \times 10^{23}$ per cm^3, f(Ge) $= 1.259 \times 10^{-10}$, n_o (Ge) $= 1.017 \times 10^{19}$ per cm^3]
13. Suppose one is able to produce a polarization of 5×10^{-8} C/m^2 in a 5 mm side cube of barium titanate. Show that the voltage produced will be 9.4 V. [B.Tech.]
14. A current of 2A flows through a slab of Sn, 1 mm thick and 2 mm wide, whose conductivity is 10^6 1/(Ω-m). Show that the voltage drop between two potential probes that are 10 cm apart along the slab is 0.1 V. [B.Tech.]
15. What current flows through a *pn* diode to which a voltage of 0.3 V is applied and a saturation current of 1μA is assumed at 300 K? [B.Sc(Hons)] [**Ans.** 0.11 A]
16. Calculate the mobility of electrons in a metal when the time between two electron/atom collisions is 2.5×10^{-12}s. [B.Sc(Hons)] [**Ans.** 0.44 m^2/V-s]
17. When a voltage of 5m V is applied to the emitter of a transistor, a current of 2m A is produced. When the voltage is increased to 8 mV, the current through the collector rises to 6m A. When the emitter voltage is doubled from 9 mV to 18 mV, show that the collector current increase is 2600%. [B.Tech.]

SHORT ANSWER QUESTIONS

1. What are electronic materials?

Ans. Electronic materials include conductors, semiconductors, superconductors, insulators and dielectrics.

2. What is the basis for the classification of electronic materials?

Ans. Electronic materials can be classified on the basis of their band structure. These materials can also be classified on the basis of the magnitude of their electrical conductivity. Materials with conductivity less than $10^{-12}\ \Omega^{-1}\ cm^{-1}$ are insulating, conductivity greater than $10^3\ \Omega^{-1}\ cm^{-1}$ are considered conductors and conductivity less than $10^3\ \Omega^{-1}\ cm^{-1}$ are considered semiconductors.

3. What is energy gap or bandgap (E_g)?

Ans. The energy between the top of the valence band and the bottom of the conduction band that a charge carrier must obtain before it can transfer a charge.

4. How resistivity of conductors depend on temperature and impurities?

Ans. In pure metals, the resistivity increases with temperature. Resistivity is very sensitive to impurities and microstructural defects such as grain boundaries.

5. What is the resistivity of alloys affected by alloying elements?

Ans. Resistivity of alloys is typically higher and dominated by alloying elements and is relatively stable with respect to temperature.

6. What is the classical model for electrical conduction in metals?

Ans. In the classical model for electrical conduction in metals, the outer valence electrons of the atoms of the metal are assumed to be free to move between the positive-ion cores (atoms without valence electrons) of the metal lattice. In the presence of applied electric potential the free electrons attain a *drift velocity.* The movement of electrons and their associated electric charge in a metal constitute an electric current. By convention, electric current is considered to be positive charge flow, which is in the opposite direction to electron flow.

7. What is the energy band model for electrical conduction in metals?

Ans. In this model, the valence electrons of the metal atoms interact and interpenetrate each other to form energy bands. Since the energy bands of the valence electrons of metal atoms overlap, producing partially filled composite energy bands, very little energy is required to excite the highest-energy electrons so that they become free to be conductive.

8. Give an idea of the energy band model for an insulator.

Ans. The energy band model for an insulator consists of a lower filled valence band and a higher empty conduction band. The valence band is separated from the conduction band by a large energy gap (about 6 to 7 eV).

9. What are superconductors?

Ans. Superconductors are materials that offer zero resistance to the flow of electric current and exclude a magnetic field ($R = 0$, $B = 0$ inside superconductor), *i.e.* superconductors are perfect conductors.

10. What are Type-I and Type-II superconductors?

Ans. In Type-I superconductors, the superconductive effect is lost even under very small magnetic flux densities. In Type-II superconductors, the flux can penetrate while the material remains superconducting and, above a certain critical field (H_{C_2}), the superconductivity is lost.

11. What are intrinsic semiconductors?

Ans. Intrinsic semiconductors have a relatively small energy gap (*i.e.* about 0.7 to 1.1 eV) between their valence and conduction bands. The properties of these semiconductors are controlled by the element or compound that makes the semiconductor and not by impurities or dopants.

12. What are extrinsic superconductors?

Ans. A semiconductor prepared by adding dopants, which determine the number of charge carriers is called extrinsic. Extrinsic behaviour can also be seen due to impurities. By adding the impurity atoms, the amount of energy required to cause semiconductors to be conductive is greatly reduced.

13. How many types of extrinsic semiconductors are there?

Ans. Extrinsic semiconductors can be *n*-type or *p*-type. The *n*-type (negative) semiconductors have electrons for their majority carriers. The *p*-type (positive) semiconductors have hole (missing electrons) for their majority charge carriers. Some semiconductors are direct (*e.g.*, GaAs) or indirect (*e.g.* Si) bandgaps.

14. Give examples of elemental and compound semiconductors.

Ans. Si and Ge are elemental whereas SiC and GaAs are compound semiconductors.

15. What is a *p-n* junction?

Ans. *p-n* junction is a device made by creating an *n*-type region in a *p*-type material (or vice versa). A *p-n* junction behaves as a diode and multiple *p-n* junctions function as transistors. *p-n* junction is also the basis of LEDs and solar cells. A *p-n* junction device permits current to flow in only one direction in a circuit, *i.e.* it act as a rectifier.

16. What is a reverse biasing?

Ans. When a junction device is connected so that the *p*-side is connected to a negative terminal; very little current flows through a *p-n* junction under reverse bias.

17. What is a zener diode?

Ans. Zener diode is a heavily doped *p-n* junction device which breaks down via a tunnelling mechanism, used as a device to limit current.

18. What is the impact of modern microelectronic technology?

Ans. Modern microelectronic technology has developed to such an extent that thousands of transistors can be placed on a "chip" of semiconducting silicon less than about 0.5 cm square and about 0.2 mm thick. Complex microelectronic technology has made possible highly sophisticated microprocessors and computer memories.

19. What is the impact of electronic materials on technology advancement?

Ans. Electronic materials have enabled many technologies ranging from spark plugs to high voltage line insulators to solar cells, computer chips, and several other sensors and actuators.

20. What are dielectrics?

Ans. Dielectrics have large bandgaps and do not conduct electricity. With insulators, the focus is on breakdown voltage or field. With other dielectrics, the emphasis is on the dielectric constant, tan δ, frequency and temperature dependence. Polarization mechanisms in materials dictate this dependence.

21. What are piezoelectrics?

Ans. These are the materials in which, the application of stress results in the development of a voltage; the application of a voltage causes strain. Obviously, these are the materials that develop voltage upon the application of a stress and develop strain when an electric field is applied. In pyroelectrics, temperature change causes polarization.

22. What are ferroelectrics?

Ans. These are the materials that exhibit a reversible and spontaneous polarization, *e.g.* $BaTiO_3$, PZT, PVDT. These materials exhibit a large dielectric constant, even in paraelectric state, and are often used to make capacitors. We may note that all ferroelectrics are piezoelectrics but the converse is not always true.

23. What are linear dielectrics?

Ans. Materials in which the dielectric polarization is linearly related to the electric field are called linear dielectrics. In these materials the dielectric constant is not dependent on the electric field.

24. What is microstructure sensitive property?

Ans. Property that depend on the microstructure of a material, *e.g.* conductivity, dielectric constant or yield strength is called microstructure sensitive property.

25. What is nonradiative recombination?

Ans. Combining of electrons and holes leading to generation of heat is called nonradiative recombination. This occurs mainly in indirect bandgap materials such as Si.

26. What is radiative recombination?

Ans. Recombination of holes and electrons that leads to emission of light is called radiative recombination. This occurs in direct bandgap materials.

27. What is a thermistor?

Ans. This is a semiconductor device that is particularly sensitive to changes in temperature, permitting it to serve as an accurate measure of temperature.

OBJECTIVE QUESTIONS

1. The resistance R of a conducting material of length L, resistivity ρ and cross-sectional area A is

(*a*) $R = \dfrac{L}{A\rho}$ (*b*) $R = \dfrac{L\rho}{A}$

(*c*) $R = LA\rho$ (*d*) $R = \dfrac{1}{LA\rho}$ [ ]

2. Current density J(A/cm^2) is defined as

(*a*) $J = \dfrac{I}{A}$ (*b*) $J = \dfrac{I^2}{A}$

(*c*) $J = \dfrac{I}{A^2}$ (*d*) $J = IA$ [ ]

3. If n and p present the concentrations of free electrons and holes and μ_n and μ_p the mobilities of electrons and holes, respectively, then the conductivity of semiconductors can be expressed as

(*a*) $\sigma = n\mu_n + p\mu_p$ (*b*) $\sigma = nq + pq$

(*c*) $\sigma = nq\mu_n + pq\mu_p$ (*d*) $\sigma = npq\,(\mu_n + \mu_p)$

4. In sodium atom the valence band and conduction band are represented by

(*a*) 3*s*, 3*p* energy levels (*b*) 3*s*, 4*s* energy levels

(*c*) 3*p*, 4*s* energy levels (*d*) 4*s*, 4*p* energy levels [ ]

5. The Fermi energy (E_F) is the energy level at which half of the possible energy levels in the band are occupied by
(a) protons (b) neutrons
(c) electrons (d) none of these

6. The covalent bonding in semiconductors, *e.g.* Si and Ge produces a complex change in the band structure called
(a) excitation (b) deexcitation
(c) hybridization (d) none of these

7. With the increase in temperature, the conductivity of semiconductors
(a) increases (b) decreases
(c) remains constant (d) sometimes increases and sometimes decreases

8. The bandgap in diamond is
(a) ~1.1 eV (b) ~2.5 eV
(c) ~ 3.7 eV (d) ~5.5 eV

9. The unit of electrical conductivity is
(a) ohm-cm (b) $ohm^{-1}\ cm^{-1}$
(c) ohm/cm (d) ohm/cm^2

10. The change in resistivity of a pure metal as a function of temperature can be estimated from the equation
(a) $\rho = \rho_{RT}(1 + \alpha_R)$ (b) $\rho = \rho_{RT}(1 + \alpha_R \Delta T)$
(c) $\rho = \rho_{RT}(1 + \alpha_R^2 \Delta T)$ (d) $\rho = \rho_{RT}(1 + \alpha_R/\Delta T)$

11. Imperfections in crystal structure of a metal scatter electrons, reducing its
(a) mobility and conductivity (b) mobility only
(c) conductivity only (d) none of these

12. A superconductor is a
(a) perfect diamagnet and perfect conductor
(b) perfect diamagnet only
(c) perfect conductor only
(d) none of the above

13. The effect of the temperature on the critical field (H_c) of a superconductor is given by
(a) $H_c = H_o\left[1 - \frac{T}{T_c}\right]$ (b) $H_c = H_o\left[1 - \left(\frac{T}{T_c}\right)^{1/2}\right]$
(c) $H_c = H_o\left[1 - \left(\frac{T}{T_c}\right)^{2}\right]$ (d) $H_c = H_o\left[1 - \left(\frac{T}{T_c}\right)^{-2}\right]$

14. The motion of magnetic flux lines limits the current transport in a superconductor within the
(a) grains (b) surface
(c) vortices (d) none of these

15. A carbon nanotube is a small single or multiwalled structure made from
(a) silicon (b) graphite
(c) carbon (d) aluminium

16. An extrinsic semiconductor (*n* or *p*-type) is preferred for devices since its properties are temperature stable and can be controlled using
    (*a*) ion implantation
    (*b*) diffusion of dopants
    (*c*) ion implantation or diffusion of dopants
    (*d*) none of the above
17. The temperature dependence conductivity in intrinsic semiconductors is nearly
    (*a*) linear (*b*) exponential
    (*c*) parabolic (*d*) none of these
18. When tetravalent impurity such as gallium or boron is added in Si or Ge, the semiconductor is called
    (*a*) intrinsic (*b*) *p*-type extrinsic
    (*c*) *n*-type extrinsic (*d*) none of these
19. A measure of how much electrical energy is lost due to motion of charge entities that respond to the electric field via different polarization mechanisms is called
    (*a*) heat loss (*b*) dielectric loss
    (*c*) heat as well as dielectric loss (*d*) none of above [ ]
20. A heavily doped *p-n* junction device which breaks down via a tunnelling mechanism, used as a device to limit current is called
    (*a*) zener diode (*b*) LED
    (*c*) varactor diode (*d*) none of these

ANSWERS

1. (*b*)	2. (*a*)	3. (*c*)	4. (*a*)	5. (*c*)	6. (*c*)
7. (*c*)	8. (*d*)	9. (*b*)	10. (*b*)	11. (*a*)	12. (*a*)
13. (*c*)	14. (*a*)	15. (*c*)	16. (*a, b, c*)	17. (*b*)	18. (*b*)
19. (*b*)	20. (*a*)				

CHAPTER

4

Ceramics

4.1 INTRODUCTION

Ceramics are inorganic, non-metallic materials that consist of metallic and non-metallic elements bonded together primarily by ionic and/or covalent bonds. The chemical compositions of these materials vary considerably, from simple compounds to mixtures of many complex phases bonded together.

The properties of these materials also vary greatly due to differences in bonding. In general, these materials are typically hard and brittle with low toughness and ductility. Due to their desirable characteristics such as high hardness, wear resistance, chemical stability, high temperature strength, and low coefficient of thermal expansion, advanced ceramics are being selected as the preferred material for many applications. These materials are usually good electrical and thermal insulators due to absence of conduction electrons. Applications of advanced ceramics are not limited to mineral processing, seals, valves, heat exchangers, metal-forming dies, adiabatic diesel engines, gas turbines, medical products and cutting tools. Ceramic cutting tools have many advantages over their conventional metal counterparts, including chemical stability, higher resistance to wear, higher not hardness, and superior heat dispersal during the chip removal process. Few examples of ceramic materials that are used to manufacture cutting tools are metal oxide composites (70% Al_2O_3 – 30% TiC), silicon-aluminium-oxynitride (sialons), and cubic boron nitride. These tools are manufactured through powder metallurgy processes in which ceramic particles are densified into a final shape through compaction and sintering. The chapter-opening images are examples of various types of metal removal products made from advanced ceramics.

Ceramic materials normally have relatively high melting temperatures and high chemical stability in many hostile environments are indispensable for many engineering designs.

In general, we can divide ceramic materials used for engineering applications into two groups: (*i*) traditional ceramic materials and (*ii*) engineering ceramic materials. Typically, traditional ceramics are made from three basic components: clay, silica (flint), and feldspar. Bricks and tiles are familiar examples of traditional ceramics. The engineering ceramics, in contrast, typically consist of pure or nearly pure compounds such as aluminium oxide (Al_2O_3), silicon carbide (SiC), and silicon nitride (Si_3N_4). Examples of the use of the engineering ceramics in high technology are SiC in high temperature areas of the experimental AGT-100 automotive gas turbine engine and aluminium oxide in

Table 4.1 Functional Classification of Ceramics

Function	*Application*	*Examples*
Electrical	Capacitor dielectrics	$BaTiO_3$, $SrTiO_3$, Ta_2O_5
	Microwave dielectrics	$Ba(Mg_{1/3}Ta_{2/3})O_3$, $Ba(Zn_{1/3}Ta_{2/3})O_3$
		$BaTi_4O_9$, $Ba_2Ti_9O_{20}$, $Zr_xSn_{1-x}TiO_4$, Al_2O_3
	Conductive oxides	In-doped SnO_2(*ITO*)
	Superconductors	$YBa_2Cu_3O_{7-x}$(*YBCO*)
	Electronic packaging	Al_2O_3
	Insulators	Porcelain
	Solid-oxide fuel cells	ZrO_2, $LaCrO_3$
	Piezoelectric	$Pb(Zr_xTi_{1-x})O_3$(*PZT*), $Pb(Mg_{1/3}Nb_{2/3})O_3$
	Electro-optical	*PLZT*, $LiNbO_3$
Magnetic	Recording media	γ-Fe_2O_3, CrO_2 ("chrome" cassettes)
	Ferrofluids, credit cards	Fe_3O_4
	Circulators, isolators,	Nickel zinc ferrite
	Inductors, magnets	Manganese zinc ferrite
Optical	Fibre optics	Doped SiO_2
	Glasses	SiO_2 based
	Lasers	Al_2O_3, yttrium aluminum garnate (*YAG*)
	Lighting	Al_2O_3, glasses
Automotive	Oxygen sensors, fuel cells	ZrO_2
	Catalyst support	Cordierite
	Spark plugs	Al_2O_3
	Tires	SiO_2
	Windshields/windows	SiO_2 based glasses
Mechanical/Structural	Cutting tools	WC-Co cermets
		Silicon-aluminum-oxynitride (*Sialon*)
		Al_2O_3
	Composites	SiC, Al_2O_3, silica glass fibres
	Abrasives	SiC, Al_2O_3, diamond, BN, $ZrSiO_4$
Biomedical	Implants	Hydroxyapatite
	Dentistry	Porcelain, Al_2O_3
	Ultrasound imaging	*PZT*
Construction	Buildings	Concrete
		Glass
		Sanitaryware
Others	Defense applications	*PZT*, B_4C
	Armor materials	
	Sensors	SnO_2
	Nuclear	UO_2
		Glasses for waste disposal
	Metals processing	Alumina and silica-based refractories, oxygen sensors, casting molds, etc.
Chemical	Catalysis	Various oxides (Al_2O_3, ZrO_2, ZnO, TiO_2)
	Air, liquid filtration	
	Sensors	
	Paints, rubber	
Domestic	Tiles, sanitaryware, whiteware, kitchenware, pottery, art, jewelry	Clay, alumina, and silica-based ceramics, glass-ceramics, diamond, ruby, cubic zirconia and other crystals

**Acronyms are indicated in itatics.*

the support base for integrated circuit chips in a thermal-conduction module. The functional classification of ceramics is given in Table 4.1 and properties of commonly used polycrystalline ceramics are given in Table 4.2.

Table 4.2 Properties of Commonly Encountered Polycrystalline Ceramics

Material	*Melting point (°C)*	*Thermal expansion coefficient ($\times 10^{-6}$ cm/cm)/°C*	*Knoop hardness (HK) (100 g)*
Al_2O_3	2000	~6.8	2100
BN	2732	0.57^a, -0.46^b	5000
SiC	2700	~3.7	2500
Diamond		1.02	7000
Mullite	1810	4.5	–
TiO_2	1840	8.8	–
Cubic ZrO_2	2700	10.5	–

[a] Perpendicular to pressing direction.
[b] Parallel to pressing direction.

4.2 CRYSTAL STRUCTURES OF SIMPLE CERAMICS

Table 4.3 shows some ceramic compounds with relatively simple crystal structures with their melting points. The atomic bonding in these ceramic compounds is a mixture of ionic and covalent types. One can obtain the approximate values for the percentages of ionic and covalent character for the bonds between these compounds by considering the electronegativity differences between the different types of atoms in the compounds and by using Pauling's equation for per cent ionic character (Eqn. 1) of bonding in compound *AB*

$$\%\ \text{ionic character} = \left(1 - e^{(-1/4)(X_A - X_B)^2}\right)(100\%) \qquad ...(1)$$

Table 4.3 Some Simple Ceramic Compounds with their Melting Points

Ceramic compound	*Formula*	*Melting point (°C)*
Hafnium carbide	HfC	4150
Titanium carbide	TiC	3120
Tungsten carbide	WC	2850
Magnesium oxide	MgO	2798
Zirconium dioxide	ZrO_2*	2750
Silicon carbide	SiC	2500
Boron carbide	B_4C_3	2450
Aluminum oxide	Al_2O_3	2050
Silicon dioxide†	SiO_2	1715
Silicon nitride	Si_3N_4	1700
Titanium dioxide	TiO_2	1605

* Is believed to have a monoclinic fluorite (distorted) crystal structure when melted.
† Cristobalite.

where X_A and X_B are the electronegativities of the atoms *A* and *B* in the compound. Table 4.4 shows that the per cent ionic or covalent character varies considerably in simple ceramic compounds. The amount of ionic or covalent bonding between the atoms of ceramic compounds is important since it determines to some extent what type of crystal structure will form in bulk ceramic compound.

Table 4.4 Per Cent Ionic and Covalent Bonding in Some Ceramic Compounds

Ceramic compound	*Bonding atoms*	*Electronegativity difference*	*% ionic character*	*% covalent character*
Zirconium dioxide, ZrO_2	Zr–O	2.3	73	27
Magnesium oxide, MgO	Mg–O	2.2	69	31
Aluminum oxide, Al_2O_3	Al–O	2.0	63	37
Silicon dioxide, SiO_2	Si–O	1.7	51	49
Silicon nitride, Si_3N_4	Si–N	1.3	34.5	65.5
Silicon carbide, SiC	Si–C	0.7	11	89

In ionic (ceramic) solids the packing of the ions is basically determined from the following factors:

(*i*) The relative size of the ions in the ionic solid (ions are assumed to be hard spheres with definite radii).

(*ii*) The need to balance electrostatic charges to maintain electrical neutrality in the ionic solid.

When ionic bonding between atoms takes place in the solid state, the energies of the atoms are lowered by the formation of the ions and their bonding into an ionic solid. Ionic solids tend to have their ions packed together as densely as possible to lower the overall energy of the solid as much as possible. The limitations to dense packing are the relative sizes of the ions and the necessity for maintaining the charge neutrality.

4.2.1 Ionic Solid: Size Limitations for the Dense Packing of Ions

Ionic solids consist of cations and anions. In ionic bonding some atoms lose their outer electrons to become *cations* and others gain outer electrons to become *anions*. Thus, cations are normally smaller than the anions they bond with. The number of anions that surround a central cation in an ionic solid is called the *coordination number* (CN) and corresponds to the number of nearest neighbours surrounding a central cation. For stability, as many anions as possible surround a central cation. However, the anions must make contact with the central cation, and charge neutrality must be maintained.

Examples of Ionically bonded crystals

(*i*) *CsCl* (Cesium Chloride): The crystal structure of CsCl is principally ionically bonded and there are equal numbers of Cs^+ and Cl^- ions. The eight chloride ions surround a central cesium cation at the $\left(\frac{1}{2}, \frac{1}{2}, \frac{1}{2}\right)$ position in the CsCl unit cell. Ionic crystals CsBr, TlCl and TlBr are examples that also have CsCl crystal structure. The intermetallic compounds AgMg, LiMg, AlNi, and β–Cu–Zn also have this structure. However, the CsCl structure is not of much importance for ceramic materials but does illustrate how higher radius ratios lead to higher coordination numbers in ionic crystal structures.

Sodium Chloride (NaCl) Crystal Structure

NaCl or rock salt crystal structure is highly ionically bonded and has the chemical formula NaCl. Obviously, there are an equal number of Na^+ and Cl^- ions to maintain charge neutrality. Figure 4.1(*a*) has negative Cl^- anions occupying regular FCC atom lattice sites and positive Na^+ cations occupying the interstitial sites between the FCC atom sites. The centres of the Na^+ and Cl^- ions occupy the following lattice positions, which are marked in Fig. 4.1(*a*).

$$Na^+: \quad (\tfrac{1}{2}, 0, 0) \quad (0, \tfrac{1}{2}, 0) \quad (0, 0, \tfrac{1}{2}) \quad (\tfrac{1}{2}, \tfrac{1}{2}, \tfrac{1}{2})$$

$$Cl^-: \quad (0, 0, 0) \quad (\tfrac{1}{2}, \tfrac{1}{2}, 0) \quad (\tfrac{1}{2}, 0, \tfrac{1}{2}) \quad (0, \tfrac{1}{2}, \tfrac{1}{2})$$

Since each central Na^+ cation is surrounded by six Cl^- anions, the structure has octahedral coordination (*i.e.*, CN = 6) as shown in Fig. 4.1(*b*). NaCl unit cell truncated is shown in Fig. 4.1(*c*).

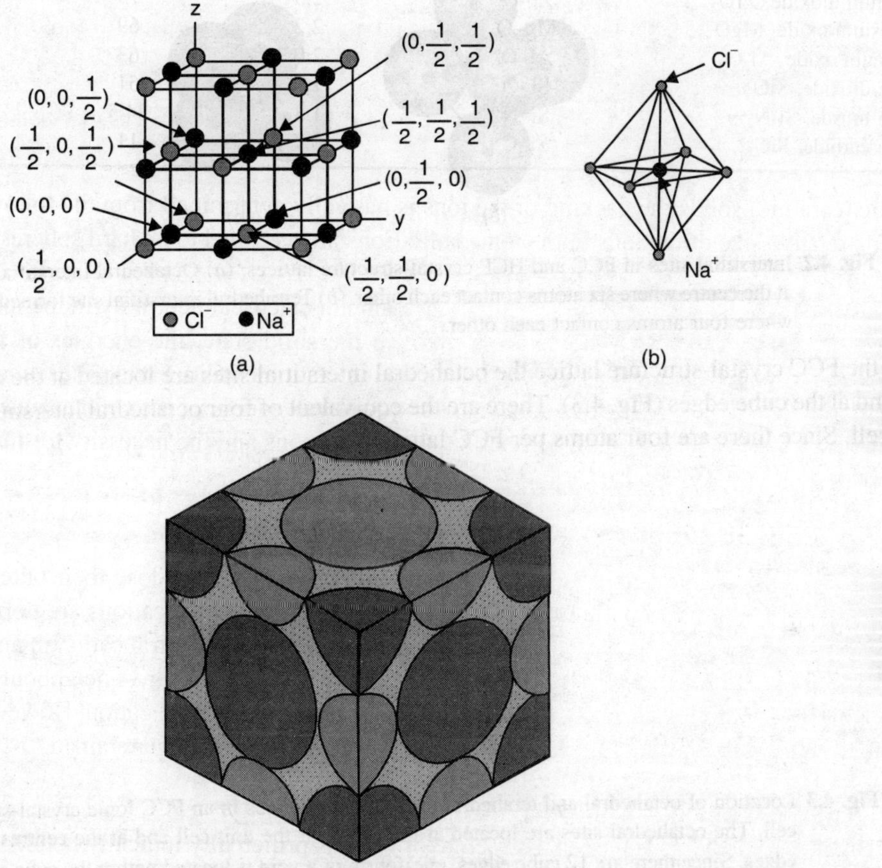

Fig. 4.1 (*a*) NaCl lattice-point unit cell indicating positions of the Na^+ (radii = 0.102 nm) and Cl^- (radii = 0.181 nm) ions. (*b*) Octahedron showing octahedral coordination of six Cl^- anions around a central Na^+ cation. (*c*) NaCl unit cell truncated.

4.2.2 Interstitial Sites in FCC and HCP Crystal Lattices

There are empty spaces or voids among the atoms or ions that are packed into a crystal structure lattice. These voids are *interstitial sites* in which atoms or ions other than those of the parent lattice can be fitted in. In the FCC and HCP crystal structures, which are close-packed structures, there are two types of interstitial sites: *Octahedral* and *tetrahedral*. In the octahedral site there are six nearest atoms or ions equidistant from the centre of the void Fig. 4.2(*a*). This site is called octahedral as the atoms or ions surrounding the centre of the site from an eight-sided octahedron. In the tetrahedral site there are four nearest atoms or ions equidistant from the centre of the tetrahedral site Fig. 4.2(*b*). A regular tetrahedron is formed when the centers of the four atoms surrounding the void are joined.

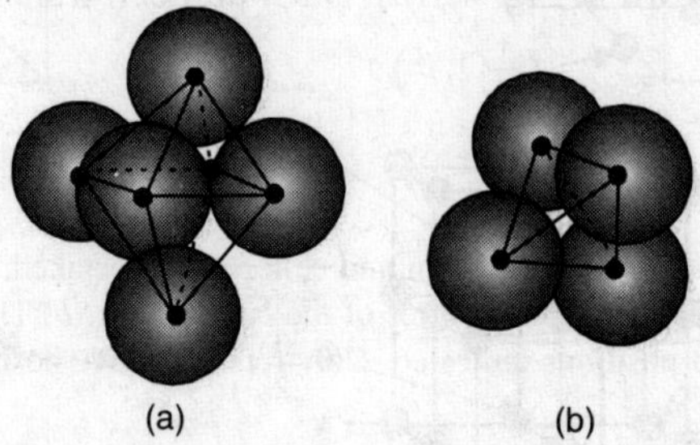

Fig. 4.2 Interstitial sites in FCC and HCP crystal-structure lattices. (*a*) Octahedral interstitial site formed at the centre where six atoms contact each other. (*b*) Tetrahedral interstitial site formed at the centre where four atoms contact each other.

In the FCC crystal-structure lattice the octahedral interstitial sites are located at the center of the unit cell and at the cube edges (Fig. 4.3). There are the equivalent of four octahedral interstitial sites per FCC unit cell. Since there are four atoms per FCC lattice.

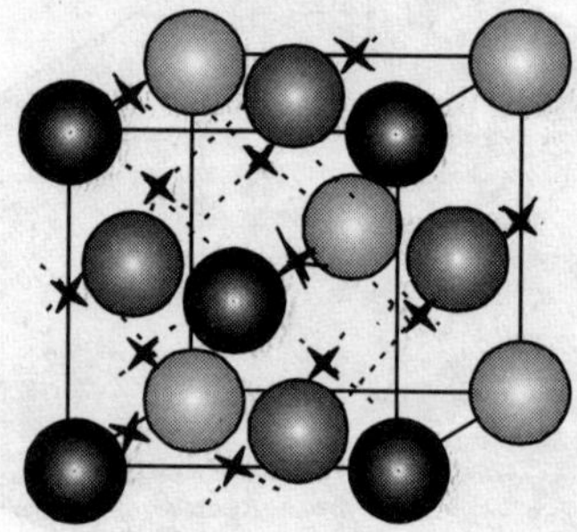

Fig. 4.3 Location of octahedral and terahedral interstitial void sites in an FCC ionic crystal-structure unit cell. The octahedral sites are located at the centre of the unit cell and at the centres of the cube edges. Since there are 12 cube edges, one-fourth of a void is located within the cube at each edge. Thus, there is the equivalent of $12 \times \frac{1}{4} = 3$ voids within the FCC unit cell at the cube edges. Therefore, there is the equivalent of four octahedral voids per FCC unit cell (one at the centre and the equivalent of three at the cube edges). The tetrahedral voids are located at the (1/4, 1/4, 1/4)-type sites, which are indicated by points with tetrahedrally directed rays. Thus there are a total of eight tetrahedral void sites located within the FCC unit cell.

Figure 4.4(*a*) indicates the lattice positions for octahedral interstitial sites in an FCC unit cell.

The tetrahedral sites in the FCC lattice are located at the (1/4, 1/4, 1/4)-type positions (Figs. 4.3 and 4.4(*b*)). In the FCC unit cell there are eight tetrahedral sites per unit cell or two per atom of the parent FCC unit cell. In the HCP crystal structure, because of similar close packing to the FCC structure, there is also the same number of octahedral interstitial sites as atoms in the HCP unit cell and twice as many tetrahedral sites as atoms.

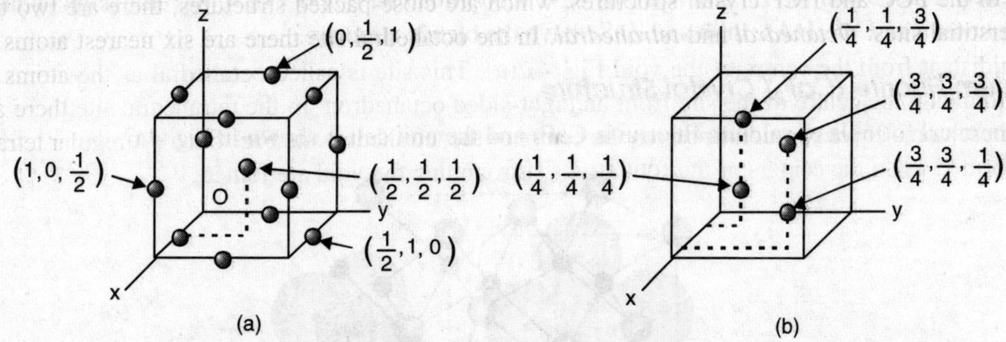

Fig. 4.4 Location of interstitial sites in the FCC atom unit cell. (*a*) The octahedral sites in the FCC unit cell are located at the centre of the unit cell and at the centres of the cube edges. (*b*) The tetrahedral sites in the FCC unit cell are located at the unit cell positions indicated. Only representative positions are located in the figure.

Zinc Blende (ZnS) Crystal Structure

The unit cell of ZnS is shown in Fig. 4.5, which has the equivalent of four zinc and four sulphur atoms. One type of atom (either S or Zn) occupies the lattice points of an FCC unit cell, and the other type (either S or Zn) occupies half the tetrahedral interstitial sites of the FCC unit cell. In the ZnS crystal-structure unit cell is shown in Fig. 4.4. Sulfur atoms occupy the FCC unit-cell atom positions and Zn atoms occupy half the tetrahedral interstitial positions of the FCC unit cell. The position coordinates of the S and Zn atoms in the ZnS crystal structure can be indicated as:

S atoms:	(0, 0, 0)	$\left(\frac{1}{2}, \frac{1}{2}, 0\right)$	$\left(\frac{1}{2}, 0, \frac{1}{2}\right)$	$\left(0, \frac{1}{2}, \frac{1}{2}\right)$
Zn atoms:	$\left(\frac{3}{4}, \frac{1}{4}, \frac{1}{4}\right)$	$\left(\frac{1}{4}, \frac{1}{4}, \frac{3}{4}\right)$	$\left(\frac{1}{4}, \frac{3}{4}, \frac{1}{4}\right)$	$\left(\frac{3}{4}, \frac{3}{4}, \frac{3}{4}\right)$

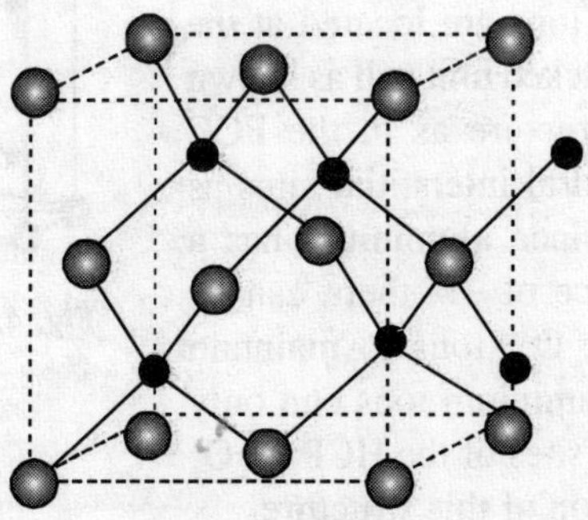

Fig. 4.5 Zinc blende (ZnS) crystal structure. In this unit cell the sulfur atoms occupy the FCC atom unit-cell sites (equivalent of four atoms). The zinc atoms occupy half the tetrahederal interstitial sites (four atoms). Each Zn or S atom has a coordination number of 4 and is tetrahedrally covalently bonded to other atoms.

According to Pauling's equation (1), the ZnS bond has 87% covalent character, and so the ZnS crystal structure must be essentially covalently bonded. As a result, the ZnS structure is tetrahedrally covalently bonded and the Zn and S atoms have a coordination number of 4. Several semiconducting compounds, *e.g.*, CdS, InAs, InSb and ZnSe have the blende structure.

Calcium Fluorite (CaF_2) Crystal Structure

The chemical formula of calcium fluorite is CaF_2 and the unit cell is shown in Fig. 4.6.

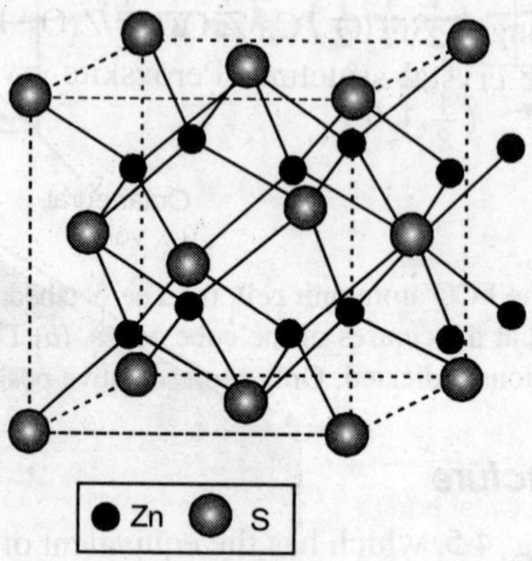

Fig. 4.6 CaF_2 (Calcium fluoride or fluorite) crystal structure. In this unit cell the Ca^{2+} ions are located at the FCC unit-cell sites (four ions). Eight fluorite ions occupy all the tetrahedral interstitial sites.

Antifluorite Crystal Structure

The antifluorite structure consists of an FCC unit cell with anions, *e.g.*, O^{2-} ions, occupying the FCC lattice points. Cations, *e.g.*, Li^+ occupy the eight tetrahed sites in the FCC lattice, *e.g.*, compounds with this structure are Li_2O, Na_2O, K_2O and Mg_2Si.

Corundum (Al_2O_3) Crystal Structure

Figure 4.7 shows the corundum (Al_2O_3) crystal structure. In the crystal structure the oxygen ions are located at the lattice sites of a hexagonal closed-packed unit cell as shown in Fig. 4.7. In the HCP crystal structure as in the FCC structure there are as many octahedral interstitial sites as there are atoms in the unit cell. Since aluminium has a valence of +3 and oxygen a valence of –2, there can be only two Al^{3+} ions for every three O^{2-} ions to maintain electrical neutrality. Clearly, the aluminium ions can only occupy two-thirds of the octahedral sites of the HCP Al_2O_3 lattice, which leads to some distortion of this structure.

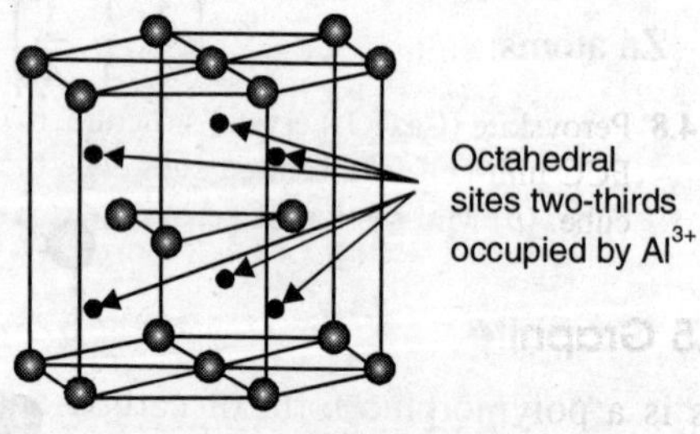

Fig. 4.7 Corundum (Al_2O_3) crystal structure. Oxygen ions (O^{2-}) occupy the HCP unit-cell sites. Aluminum ions (Al^{3+}) occupy only two-thirds of the octahedral interstitital sites to maintain electrical neutrality.

4.2.3 Spinel Crystal Structure

There are number of oxides having spinel ($MgAl_2O_4$) structure, which has the general formula AB_2O_4, where A is metal ion with *a* + 2 valence and B is a metal ion with *a* + 3 valence. The oxygen ions in the spinel structure form an FCC lattice and the A and B ions occupy tetrahedral and octahedral interstitial sites, depending on the particular type of spinel. Compounds having spinel structure are widely used for non-metallic magnetic materials for electronic applications.

4.2.4 Perovskite Crystal Structure

Ca^{2+} and O^{2-} ions in the perovskite ($CaTiO_3$) structure form an FCC unit cell with the Ca^{2+} ions at the corners of the unit cell and the O^{2-} ions in the centres of the faces of the unit cell as shown in Fig. 4.8. The highly charged Ti^{4+} ion is located at the octahedral interstitial site at the centre of the unit cell and is coordinated to six O^{2-} ions. Above 120°C, $BaTiO_3$ has the perovskite structure but below 120°C temperature its structure is slightly changed. $SrTiO_3$, Ca ZrO_3, Sr ZrO_3, La AlO_3, etc. are the examples of other compounds having perovskite crystal structure. Perovskite crystal structure is important for piezoelectric materials.

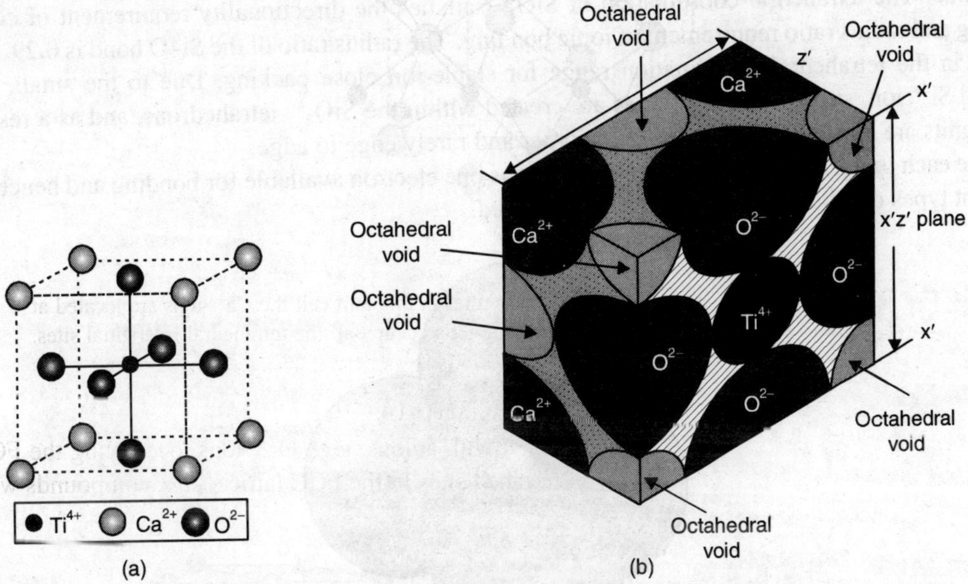

Fig. 4.8 Perovskite ($CaTiO_3$) crystal structure. (*a*) Calcium ions occupy FCC unit-cell corners, and oxygen ions occupy FCC unit-cell face-centred sites. The titanium ion occupies the octahedral interstitial site at the centre of the cube. (*b*) Mid-section of perovskite ($CaTiO_3$) crystal structure (truncated).

4.2.5 Graphite

This is a polymorphic form of carbon and is not a compound of a metal and a non-metal. However, this is sometimes considered a ceramic material. Graphite has a layered structure in which the carbon atoms in the layers are strongly covalently bonded in hexagonal arrays (Fig. 4.9). The layers are bonded together by weak secondary bonds and can thus slide past each other easily. The ease of sliding of the layers gives graphite in lubricating properties.

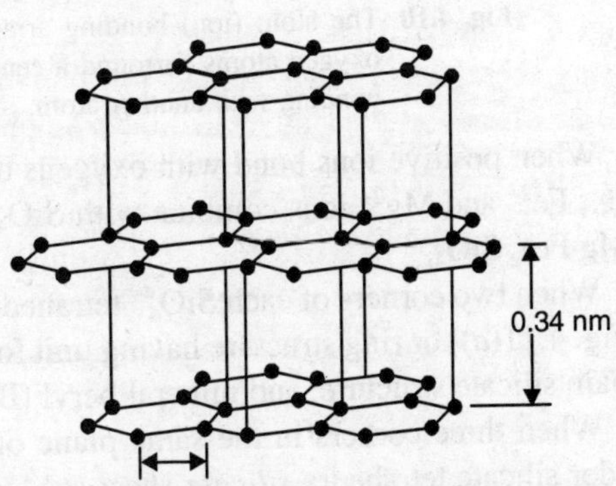

Fig. 4.9 The structure of crystalline graphite. Carbon atoms form layers of strongly covalently bonded hexagonal arrays. There are weak secondary bonds between the layers.

4.3 SILICATE STRUCTURES

There are many ceramic materials which contain silicate structures, which consist of silicon and oxygen atoms (ions) bonded together in various arrangements. In addition, a large number of naturally occurring minerals, *e.g.*, clays, feldspars, and micas are silicates since silicon and oxygen are the two most adundant elements in the earth's crust. Due to low cost, availability and special properties, many silicates are useful for engineering materials. Silicate structures are specially important for the engineering construction materials glass, brick and portland cement. Silicates are also used for making many important electrical insulative materials.

The basic building block of silicates SiO_4^{4-} tetrahedron as shown in Fig. 4.10. According to calculations from Pauling equation, the Si–O bond in the SiO_4^{4-} structure is about 50% covalent and 50% ionic. The tetrahedral coordination of SiO_4^{4-} satisfies the directionality requirement of covalent bonding and radius ratio requirement of ionic bonding. The radius ratio of the Si–O bond is 0.29. This is clearly in the tetrahedral coordination range for stable-ion close packing. Due to the small, highly charged Si^{4+} ion, strong bonding forces are created within the $SiO_4{}^{4-}$ tetrahedrons, and as a result the SiO_4^{4-} units are normally joined corner to corner and rarely edge to edge.

Since each oxygen of the silicate tetrahedron has one electron available for bonding and hence many different types of silicate structures can be produced.

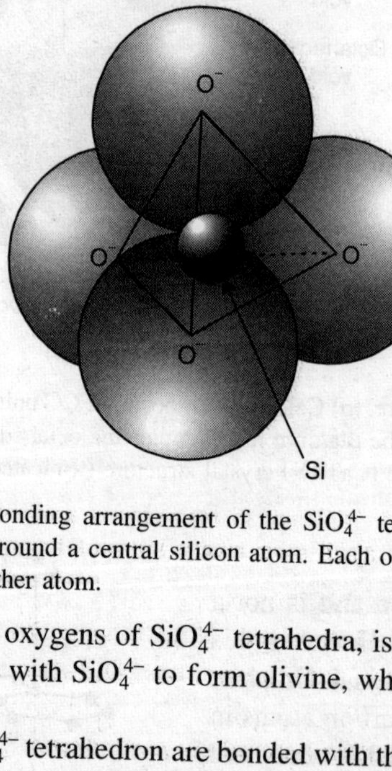

Fig. 4.10 The atom (ion) bonding arrangement of the SiO_4^{4-} tetrahedron. In this structure four oxygen atoms surround a central silicon atom. Each oxygen atom has one electron for bonding with another atom.

When positive ions bond with oxygens of SiO_4^{4-} tetrahedra, island silicate structures are produced, *e.g.*, Fe^{2+} and Mg^{2+} ions combine with SiO_4^{4-} to form olivine, which has the basic formula (chemical) $(Mg{\cdot}Fe)_2\ SiO_4$.

When two corners of each SiO_4^{4-} tetrahedron are bonded with the corners of other tetrahedra, a chain (Fig. 4.11(*a*)) or ring structure having unit formula SiO_3^{2-} results. The mineral enstatite ($Mg\ SiO_3$) has a chain silicate structure, and mineral beryl [$Be_3Al_2(SiO_3)_6$] has a ring silicate structure.

When three corners in the same plane of a silicate tetrahedron are bonded to the corners of three order silicate tetrahedra *silicate sheet structures* form as shown in Fig. 4.11(*a*). This structure has the unit formula $Si_2O_5^{2-}$. These silicate sheets can bond with other types of structural sheets because there is still one unbonded oxygen on silicate tetrahedron (Fig. 4.11(*b*)), *e.g.*, the negatively charged silicate sheet can bond with a positively charged sheet of $Al_2(OH)_4^{2+}$ to form a composite sheet of kaolinite (Fig. 4.12). The mineral kaolinite in its pure form consists of very small, flat plates roughly hexagonal in shape having average size being about 0.7 μm in diameter and 0.05 μm thick. The crystal plates are

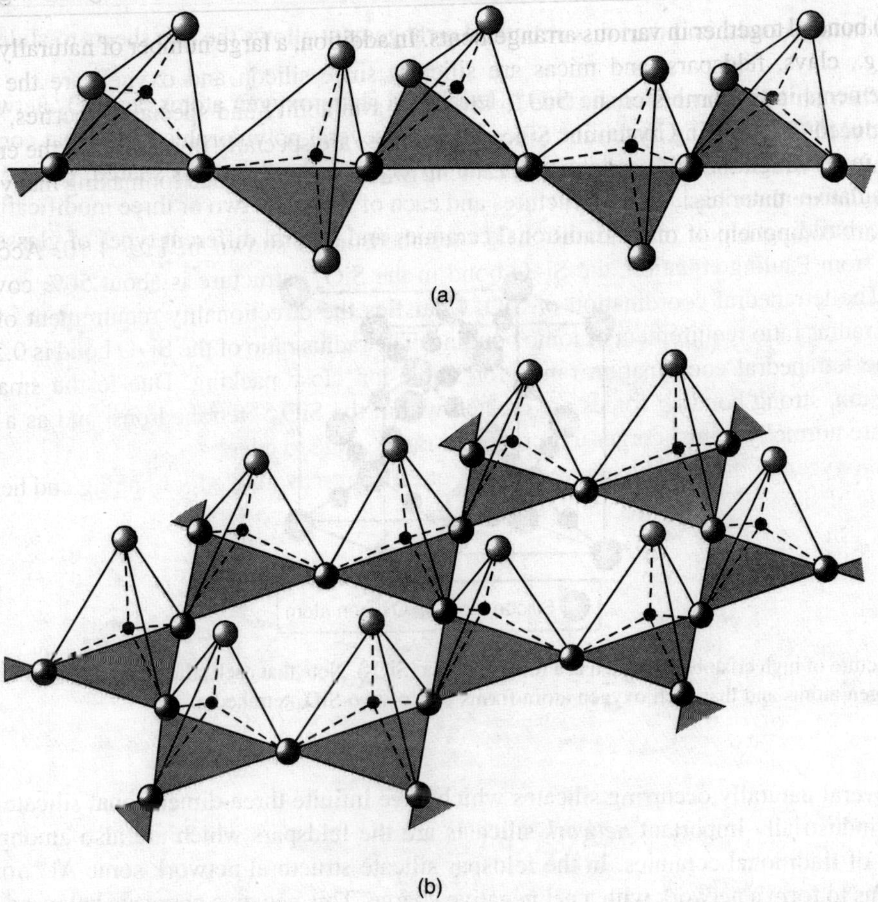

Fig. 4.11 (*a*) Silicate chain structure. Two of the four oxygen atoms of the SiO_4^{4-} tetrahedra are bonded to other tetrahedra to form silicate chains. (*b*) Silicate sheet structure. Three of the four oxygen atoms of the SiO_4^{4-} tetrahedra are bonded to other tetrahedra to form silicate sheets. The unbonded oxygen atoms are shown as lighter spheres.

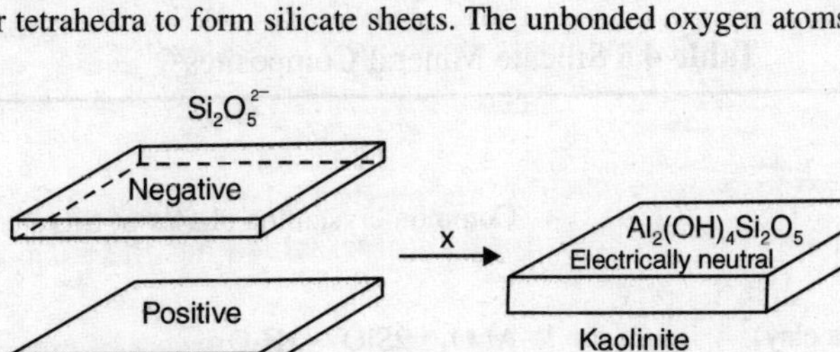

Fig. 4.12 Schematic diagram of the formation of kaolinite from sheets of $Al_2(OH)_4^{2+}$ and $Si_2O_5^{2-}$. All the primary bonds of the atoms in the kaolinite sheet are satisfied.

made of a series (~50) of parallel sheets bonded together by weak secondary bonds. Several high-grade clays consist mainly of mineral kaolinite. Mineral talk, in which a sheet of $Mg_3(OH)_2^{4+}$ bonds with two outer-layer $Si_2O_5^{2-}$ sheets (one on each side) to form a composite sheet with the unit chemical formula $Mg_3(OH)_2(Si_2\ O_5)_2$ is another example of a *sheet silicate.* The composite talc sheets are bonded together by weak secondary bonds, and thus this structural arrangement allows the talc sheets to slide over each other easily.

Silica: When all four corners of the SiO_4^{4-} tetrahedra share oxygen atoms, an SiO_2 network called *silica* is produced (Fig. 4.13). Crystalline silica exists in several polymorphic forms that correspond to different ways in which the silicate tetrahedra are arranged with all corners shared. *Quartz, tridymite* and *Cristobalite* are three basic silica structures and each of these has two or three modifications. Silica is an important component of many traditional ceramics and several different types of glasses.

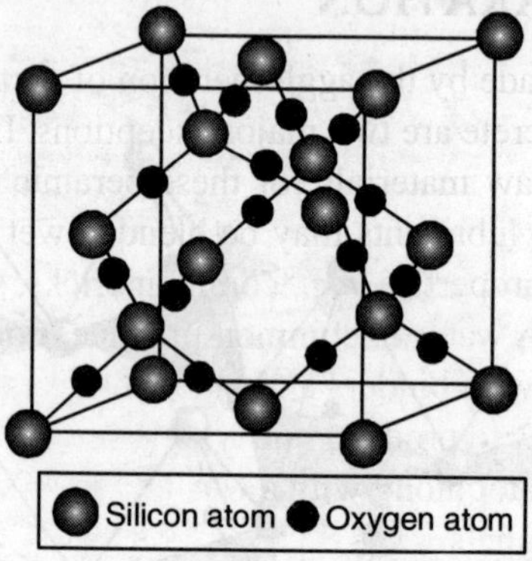

Fig. 4.13 Structure of high cristobalite which is a form of silica (SiO_2). Note that each silicon atom is surrounded by four oxygen atoms and that each oxygen atom forms part of two SiO_4 tetrahedra.

Feldspars

There are several naturally occurring silicates which have infinite three-dimensional silicate networks. Among the industrially important *network* silicates are the feldspars which are also among the main components of traditional ceramics. In the feldspar silicate structural network some Al^{3+} ions replace some Si^{4+} ions to form a network with a net negative charge. This negative charge is balanced with large ions of alkali and alkaline earth ions such as Na^+, K^+, Ca^{2+} and Ba^{2+}, which fit into interstitial positions. The ideal components of some silicate minerals are summarized in Table 4.5.

Table 4.5 Silicate Mineral Composites

Silica:	
Quartz Tridymite Cristobalite	Common crystalline phases of SiO_2
Alumina silicate:	
Kaolinite (china clay)	$Al_2O_3 \cdot 2SiO_2 \cdot 2H_2O$
Pyrophyllite	$Al_2O_3 \cdot 4SiO_2 \cdot H_2O$
Metakaolinite	$Al_2O_3 \cdot 2\ SiO_2$
Sillimanite	$Al_2O_3 \cdot SiO_2$
Mullite	$3Al_2O_3 \cdot 2SiO_2$
Alkali alumina silicate:	
Potash feldspar	$K_2O \cdot Al_2O_3\ 6SiO_2$
Soda feldspar	$Na_2O\ Al_2O_3\ 6SiO_2$
(Muscovite) mica	$K_2O \cdot 3Al_2O_3 \cdot 6SiO_2 \cdot 2H_2O$
Montmorillonite	$Na_2O \cdot 2MgO \cdot 5Al_2O_3 \cdot 24SiO_2 \cdot (6+n)H_2O$
Leucite	$K_2O \cdot Al_2O_3 \cdot 4SiO_2$

Magnesium silicate:	
Cordierite	$2MgO \cdot 5SiO_2 \cdot 2Al_2O_3$
Steatite	$3MgO \cdot 4SiO_2$
Talc	$3MgO \cdot 4SiO_2 \cdot H_2O$
Chrysotile (asbestos)	$3MgO \cdot 2SiO_2 \cdot 2H_2O$
Forsterite	$2MgO \cdot SiO_2$

4.4 MATERIALS PREPARATION

Several ceramic products are made by the agglomeration of particles. However, the production of glass products and the casting of concrete are two major exceptions. Depending on the required properties of the finished ceramic part, the raw materials for these ceramic products vary. The particles and other ingradients such as binders and lubricants may be blended wet or dry. However, for ceramic products that do not have very "critical" properties, *e.g.,* common bricks, sewer pipe, and other clay products, the blending of the ingradients with water is common practice. For some other ceramic products the raw materials are ground dry along with binders and other additives. Sometimes wet and dry processing of raw materials are combined, *e.g.,* to produce one type of high-alumina (Al_2O_3) insulator, the particulate raw materials are milled with water along with a wax binder to form a slurry that is subsequently spray-dried to form small, spherical pallets.

Ceramic products prepared by agglomerating particles may be formed by a variety of methods in the dry, plastic, or liquid conditions. In ceramic industry, the cold-forming processes are predominant, but hot forming processes are also used to some extent. Commonly used methods for ceramic formation are processing, slip casting and extrusion.

Ceramic particulate raw materials can be *pressed* in the dry, plastic, or wet condition into a die to form shaped products.

Dry pressing method is used commonly for products such as structural refactories (high-heat resistance materials) and electronic ceramic components. This method may be defined as the simultaneous uniaxial compaction and shaping of a granular powder along with small amounts of water and/or organic binder in a die. A series of operations for the dry pressing of ceramic powders into a simple shape is shown in Fig. 4.14. After cold processing, the parts are usually fixed (sintered) to achieve the required strength and microstructural properties. Dry pressing is extensively used due to the fact that it can form a wide variety of shapes rapidly with uniformity and close tolerances, *e.g.,* aluminas, titanates, and ferrites can be dry-pressed into sizes from a few mils to several inches in linear dimensions at a rate of up to about 500/minute.

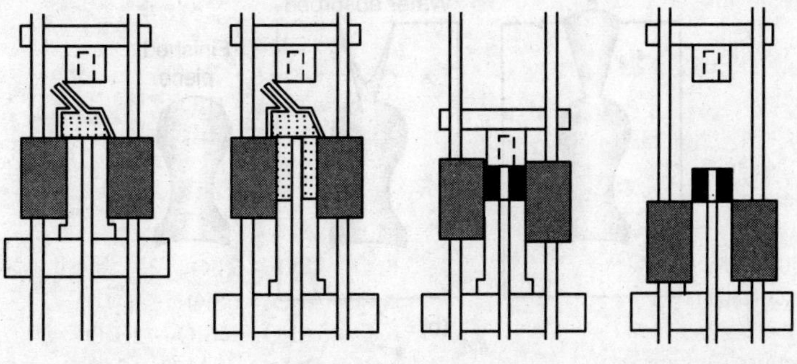

Fig. 4.14 Dry pressing of ceramic particles: (*a*) and (*b*) filling, (*c*) pressing and (*d*) ejection.

Isostatic pressing: In isostatic pressing the ceramic powder is loaded into a flexible (usually rubber), airtight container (called a bag), *i.e.* in inside a chamber of hydraulic fluid to which pressure is applied. The force of the applied pressure compacts the powder uniformly in all directions, with the final product taking the shape of the flexible container. After cold isostatic pressing the part must be fired (sintered) to achieve the required properties and microstructure. Ceramic parts manfactured by this process include refractories, bricks and shapes, spark plug insulators, radomes, carbide tools, crucibles and bearings.

Hot pressing: Ceramic parts of high density and improved mechanical properties are produced by combining the pressing and firing operations in this process. Both uniaxial and isostatic methods are used.

Slip casting: By using slip casting process, ceramic shapes can be cast as illustrated in Fig. 4.15. In this process the main steps are:

(*i*) Preparation of a powdered ceramic material and a liquid (usually clay and water) into a stable suspension called a slip.

(*ii*) Pouring the slip into a porous mold that is usually made of plaster of paris and permitting the liquid portion of the slip to partially absorbed by the mold. As the liquid is removed from the slip, a layer of semihard material is formed against the mold surface.

(*iii*) When a sufficient wall thickness has been formed, the casting process is interruped and the excess slip is poured out of the cavity as shown in Fig. 4.15(*a*). This is termed as *drain casting.* Alternatively, a solid shape may be made by allowing the casting to continue until the whole mold cavity is filled Fig. 4.15(*b*). This type of slip casting is termed as *solid casting.*

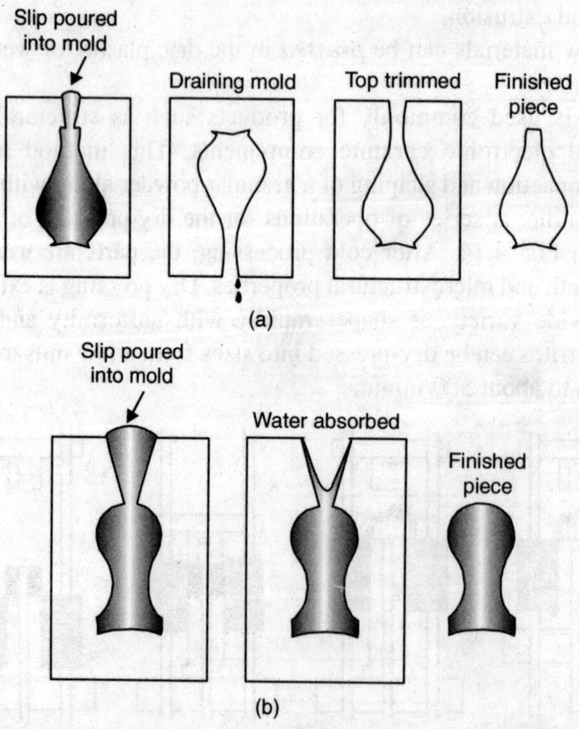

Fig. 4.15 Slip casting of ceramic shapes. (*a*) Drain casting in porous plaster of paris mold. (*b*) Solid casting.

(*iv*) The material in the mold is allowed to dry to provide adequate strength for handling and the subsequent removal of the part of the mold.

(*v*) Ultimately the cast part is fired to attain the required microstructure and properties.

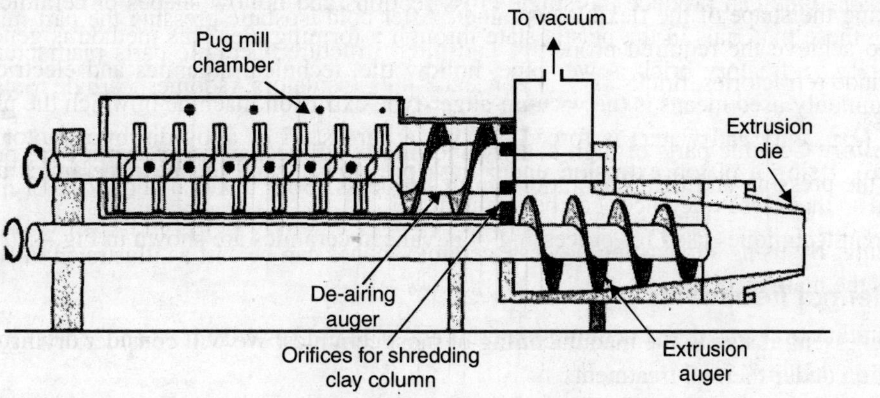

Fig. 4.16 Cross-section of combination mixing mill (pug mill) for ceramic materials and vacuum-auger extrusion machine.

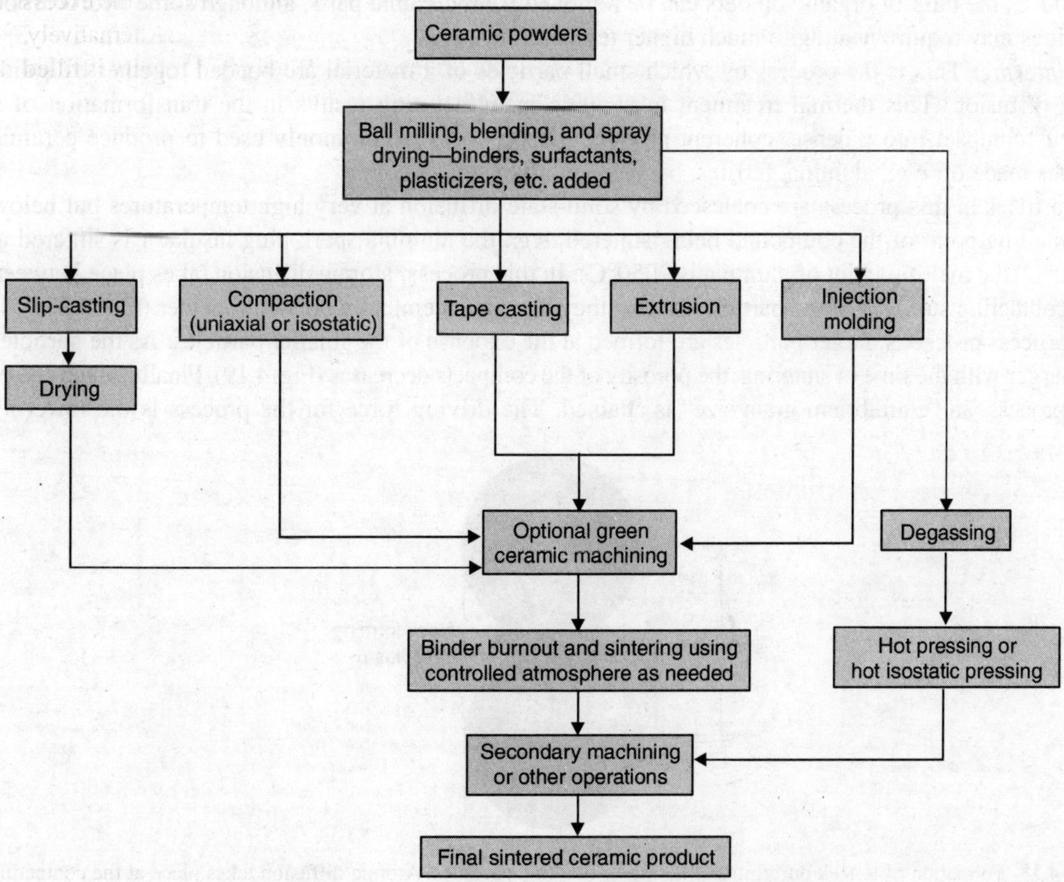

Fig. 4.17 Different techniques for processing of advanced ceramics.

This process (slip-casting) is advantageous for forming thin-walled and complex shapes of uniform thickness. This process is especially economical for development parts and short production runs. Pressure and vacuum casting are amongst the several new variations of the slip-casting process.

Extrusion: One can produce the single cross-sections and hollow shapes of ceramic materials by extruding these materials in the plastic state through a forming die. This method is generally used to produce, *e.g.*, refractory brick, sewer pipe, hollow tile, technical ceramics and electrical insulators. Most commonly used means is the vacuum-auger-type extrusion machine in which the plastic ceramic material (*e.g.*, clay and water) is forced through a hard steel or alloy die by a motor driven auger (Fig. 4.16). Using a piston extrusion under high pressure special technical ceramics are frequently produced so that close tolerances can be attained.

Different techniques used for processing of advanced ceramics are shown in Fig. 4.17.

4.4.1 Thermal Treatments

This is an essential step in the manufacturing of most ceramics. We will consider drying, sintering and vitrification under thermal treatments.

Drying and Binder Removal: The main purpose of drying ceramics is to remove water from the plastic ceramic body prior it is fixed at higher temperatures. Usually, drying to remove water is carried out at or below 100°C and can take about 24 hours for a large ceramic plant. By heating in the range 200 to 300°C, the bulk of organic binders can be removed from ceramic parts, although some hydrocarbon residues may require heating to much higher temperatures.

Sintering: This is the process by which small particles of a material are bonded together by solid-state diffusion. This thermal treatment in ceramic manufacturing results in the transformation of a porous compact into a dense, coherent product. This process is commonly used to produce ceramic shapes made of, *e.g.*, alumina, ferrites, beryllia and titanates.

Particles in this process are coalesced by solid-state diffusion at very high temperatures but below the melting point of the compound being sintered, *e.g.*, the alumina spark plug insulator is sintered at 1600°C (the melting point of alumina is 2050°C). In this process, atomic diffusion takes place between the contacting surfaces of the particles so that they become chemically bonded together (Fig. 4.18). As the process proceeds, larger particles are formed at the expense of the smaller particles. As the particles get larger with the time of sintering, the porosity of the compacts decreases (Fig. 4.19). Finally, at the end of the process, an "equilibrium grain size" is attained. The driving force for the process is the lowering

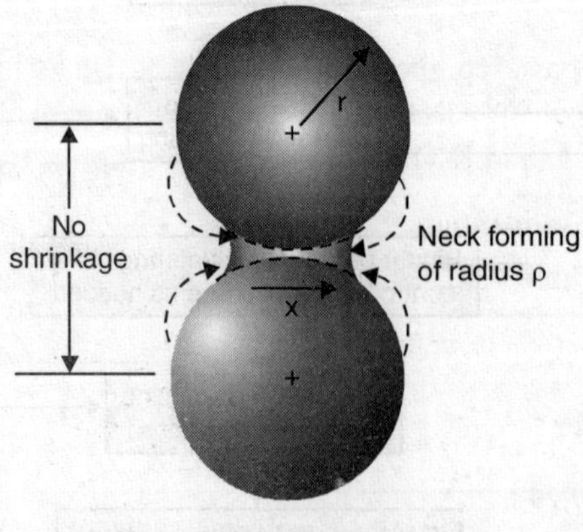

Fig. 4.18 Formation of a neck during the sintering of two fine particles. Atomic diffusion takes place at the contacting surfaces and enlarges the contact area to form a neck.

of the energy of the system. The high surface energy associated with the original individual small particles is replaced by the lower overall energy of the grain boundary surfaces of the sintered product.

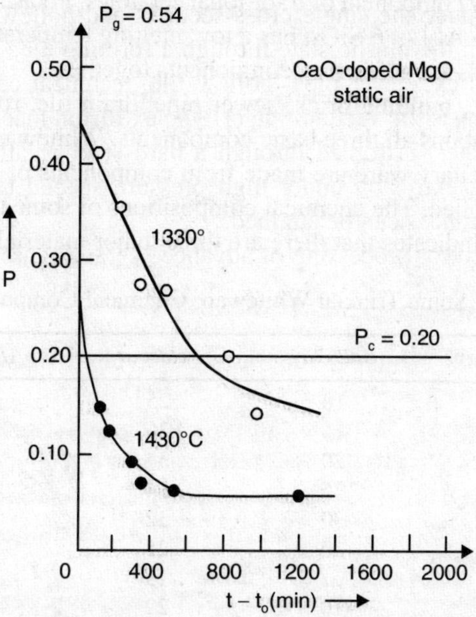

Fig. 4.19 Porosity versus time for MgO compacts doped with 0.2 wt % CaO and sintered in static air at 1330 and 1430°C. Note that the higher sintering temperature produces a more rapid decrease in porosity and a lower porosity level.

Vitrification: There are some ceramic products, *e.g.*, porcelain, structural clay products, and some electronic components contain a glass phase. This glass phase serves as a reaction medium by which diffusion can take place at a lower temperature than in the rest of the ceramic solid material. During the firing of these types of ceramic materials, a process termed as *vitrification* takes place whereby the glass phase liquefies and fills the pore spaces in the material. This liquid glass phase may also react with some of the remaining solid refractory matrix that bonds the unmelted particles together.

Traditional Ceramics

These are made from three basic components: clay, silica (flint), and feldspar. Clay consists mainly of hydrated aluminium silicates ($Al_2O_3 \cdot SiO_2 \cdot H_2O$) with small amounts of other oxides such as $TiO_2 \cdot Fe_2O_3$, MgO, CaO, Na_2O and K_2O. Chemical compositions of several industrial clays are given in Table 4.6.

Table 4.6 Chemical Compositions of Some Clays

	Weight percentages of major oxides								
Type of clay	*Al_2O_3*	*SiO_2*	*Fe_2O_3*	*TiO_2*	*CaO*	*MgO*	*Na_2O*	*K_2O*	*H_2O*	*Ignition loss*
Kaolin	37.4	45.5	1.68	1.30	0.004	0.03	0.011	0.005	13.9	
Tenn ball clay	30.9	54.0	0.74	1.50	0.14	0.20	0.45	0.72	...	11.4
Ky. ball clay	32.0	51.7	0.90	1.52	0.21	0.19	0.38	0.89	...	12.3

In traditional ceramics, the clay provides workability of the material before firing hardens it and constitutes the major body material. The silica (SiO_2) (also known as flint or quart) has a high melting temperature and is the refractory component of traditional ceramics. Potash (potassium) feldspar, which has the basic composition $K_2O \cdot Al_2O_3 . 6\ SiO_2$, has a low melting temperature and makes a glass when the ceramic mix is fired. It bonds the refractory components together.

Structural clay products, *e.g.*, building brick, sewer pipe, drain tile, roofing tile, and floor tile are made of natural clay, which contains all three basic components. Whiteware products such as electrical porcelain, dinner china, and sanitary ware are made from components of clay, silica, and feldspar for which the composition is controlled. The chemical compositions of some triaxial whitewares are given in Table 4.7. The term *triaxial* indicates that there are three major materials in their composition.

Table 4.7 Some Triaxial Whiteware Chemical Compositions

Type body	*China clay*	*Ball clay*	*Feldspar*	*Flint*	*Other*
Hard porcelain	40	10	25	25	
Electrical insulation ware	27	14	26	33	
Vitreous sanitary ware	30	20	34	18	
Electrical insulation	23	25	34	18	
Vitreous tile	26	30	32	12	
Semivitreous whiteware	23	30	25	21	
Bone china	25	...	15	22	38 bone ash
Hotel china	31	10	22	35	$2CaCO_3$
Dental porcelain	5	...	95		

The phase diagram 4.20 shows the typical composition ranges for different types of whitewares in the leucite mullite ternary. The circled areas indicates the composition ranges of some whitewares.

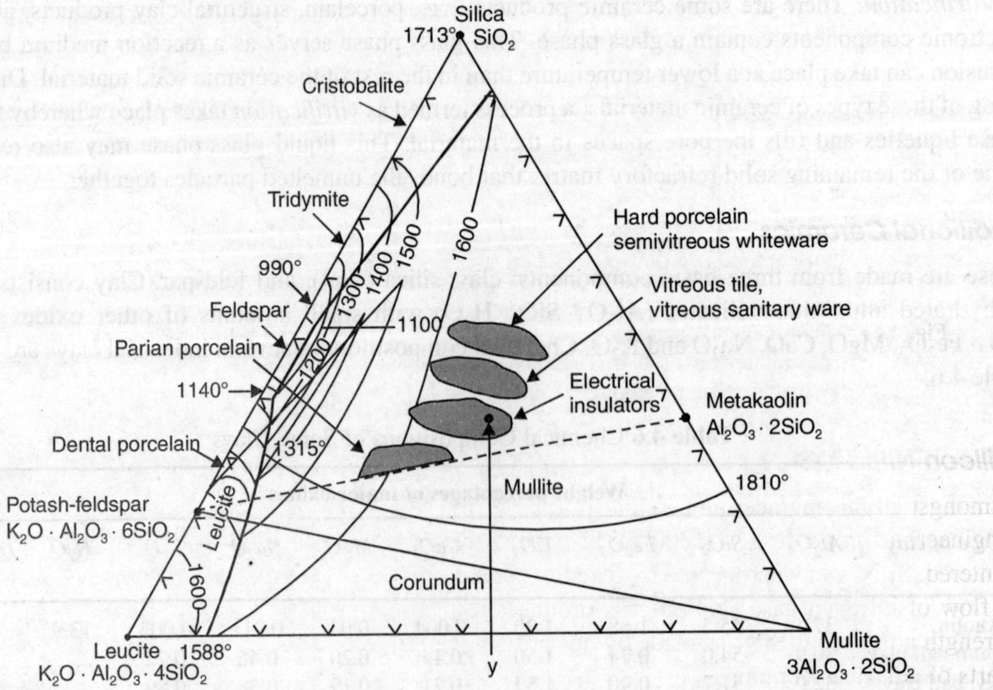

Fig. 4.20 Areas of triaxial whiteware compositions shown on the silica-leucite-mullite phase-equilibrium diagram.

The changes that takes place in the structure of triaxial bodies during firing are not completely due to their complexity.

Triaxial porcelains are satisfactory as insulators for 60-cycle use, but at high frequencies dielectric losses become too high. The considerable amounts of alkalies derived from the feldspar used as flux increase the electrical conductivity and dielectric losses of triaxial porcelains.

Engineering or Technical Ceramics

In contrast to the traditional ceramics, which are mainly based on clay, engineering or technical ceramics are basically pure compounds or nearly pure compounds of chiefly oxides, carbides, or nitrides, *e.g.*, alumina (Al_2O_3), silicon nitride (Si_3N_4), silicon carbide (SiC), and Zirconia (ZrO_2) combined with some other refractory oxides. The important properties, processes and applications of few of these engineering ceramics are as follows:

(*i*) *Alumina* (Al_2O_3): This was originally developed for refractory tubing and high-purity crucibles for high-temperature use and now finds wide applications, *e.g.*, in spark plug insulator material.

Aluminium oxide is commonly doped with magnesium oxide, cold pressed, and sintered, producing the type of microstructure as shown in Fig. 4.21. We can note the uniformity of alumina grain structure. Alumina is generally used for high-quality electrical applications where low dielectric loss and high resistivity are needed.

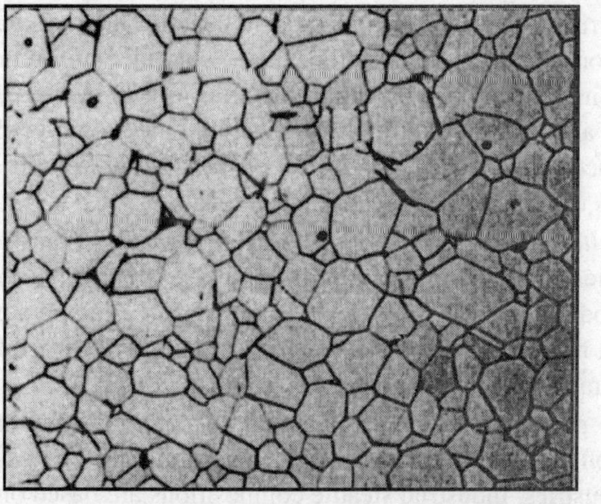

Fig. 4.21 Microstructure of sintered, powdered aluminium oxide doped with magnesium oxide. The sintering temperature was 1700°C. The microstructure is nearly pore-free, with only a few pores within the grains. (Magnification 500 ×).

Silicon Nitride (Si_3N_4)

Amongst all the engineering ceramics, probably the silicon nitride has the most useful combination of engineering properties. At 1800°C, Si_3N_4 dissociates significantly and therefore cannot be directly sintered. Si_3N_4 can be processed by reaction bonding in which a compact of silicon powder is nitrided in a flow of nitrogen gas. This process produces a microporous Si_3N_4 with moderate strength. Higher-strength non-porous Si_3N_4 is made by hot-pressing with 1–5% MgO. Si_3N_4 is being explored for use for parts of advanced ceramics.

Silicon carbide (SiC): This is a hard, refractory carbide with outstanding resistance to oxidation at high temperatures. Although a nonoxide, SiC at high temperatures forms a skin of SiO_2 that protects the main body of material SiC can be sintered at 2100°C with 0.5–1% B at a sintered aid. Silicon carbide is commonly used as a fibrous reinforcement for metal-matrix and ceramic-matrix composite materials.

Zirconia (ZrO_2): Pure zirconia is polymorphic and transforms from the tetragonal to monoclinic structure at about 1170°C with an accompanying volume expansion and so is subject to cracking. However, if we combine ZrO_2 with other refractory oxides, *e.g.*, CaO, MgO, and Y_2O_3, the cubic structure can be stabilized at room temperature and one finds some applications. One can produce a partially stabilized zirconia (PSZ) by combining ZrO_2 with 9% MgO, having especially high fracture toughness, which has led to several new ceramic applications.

4.5 ELECTRICAL PROPERTIES OF CERAMICS

There are many electrical and electronic applications of ceramic materials. Many types of ceramics are used for electrical insulators (dielectrics) for low-and high voltage electric currents. Other types of ceramics called *piezoelectrics* can convert weak pressure signals into electrical signals, and vice versa.

4.5.1 Ceramic Insulator Materials

The electrical and mechanical properties of ceramics make them especially suitable for many insulator applications in the electrical and electronic industries. The ionic and covalent bonding in these materials restricts electron and ion mobility and therefore these materials are quite suitable as good electrical insulators. These bondings make most ceramic materials strong but relatively brittle. However, the chemical compositions and microstructure of electrical-and-electronic-grade ceramics be more closely controlled than for structural ceramics, *e.g.*, bricks or tiles. Now, we will discuss some aspects of the structure and properties of few insulator ceramic materials:

(*i*) *Electrical porcelain*: Typical electrical porcelain consists approximately 50% clay ($Al_2O_3 \cdot 2SiO_2 \cdot 2H_2O$), 25% silica, and 25% feldspar ($K_2O \cdot Al_2O_3 \cdot 6\ SiO_2$). This composition makes a material that has good green-body plasticity and a wide firing range of temperature at a relatively low cost. However, this material has a high power-loss factor compared to other electrical insulator materials, which is due to highly mobile alkali ions.

(*ii*) *Steatite*: Steatite porcelains are good electrical insulators as they have low power-loss factors, low moisture absorption and good impact strength. These are used extensively by the electronic and electrical appliance industries. Industrial steatite compositions are based on about 90% talc ($3\ MgO \cdot 4\ SiO_2 \cdot H_2O$) and 10% clay. The microstructure of fired steatite consists of enstatite ($Mg \cdot SiO_3$) crystals bonded together by a glassy matrix.

Fosterite ($Mg_2\ SiO_4$): This has no alkali ions in a vitreous phase and thus it has a higher resistivity and lower electrical loss with increasing temperature than the steatite insulators. At higher frequencies, fosterite has lower loss dielectric properties.

Alumina: These ceramics have aluminium oxide (Al_2O_3) as the crystalline phase bonded with a glassy matrix. The glass phase, which is normally alkali-free, is compounded from mixtures of clay, talc, and alkaline earth fluxes and is usually alkali-free. These ceramics have relatively high dielectric strengths and low dielectric losses along with relatively high-strengths. Sintered alumina (99% Al_2O_3) is widely used as a substrate for electronic-device applications because of its low dielectric losses and smooth surface. Alumina is also used for ultralow-loss applications where a large energy transfer through a ceramic window is necessary, *e.g.* for radomes.

4.5.2 Ceramic Materials for Capacitors

Ceramic materials are generally used as dielectric materials for capacitors, with disc ceramic capacitors being by far the most common type of ceramic capacitor (Fig. 4.22). These very small flat-disk ceramic capacitors consists mainly of barium titanate ($BaTiO_3$) alongwith other additives (Table 4.8). $BaTiO_3$ is used as its very high dielectric constant of 1200 to 1500. One can raise its dielectric constant to values of many thousands with additives.

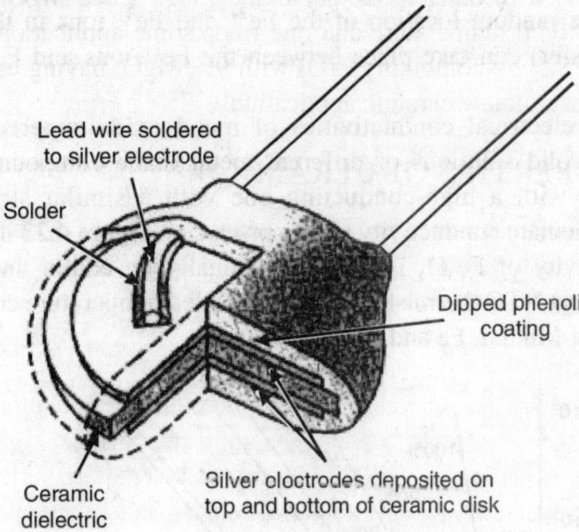

Fig. 4.22 Ceramic capacitors.

Table 4.8 Representative Formulations for Some Ceramic Dielectric Materials for Capacitors

Dielectric constant κ	*Formulation*
325	$BaTiO_3 + CaTiO_3$ + low % $Bi_2Sn_3O_9$
2100	$BaTiO_3$ + low % $CaZrO_3$ and Nb_2O_5
6500	$BaTiO_3$ + low % $CaZrO_3$ or $CaTiO_3 + BaZrO_3$

Ceramic chip capacitors are used in some ceramic-based thick-film hybrid electronic circuits. Chip capacitors can provide appreciably higher capacitance per unit area values and can be added to the thick-film circuit by a simple soldering or bonding operation.

4.5.3 Ceramic Semiconductors

There are some ceramic compounds having semiconducting properties and these are important for the operation of some electrical devices, *e.g.*, *thermistor* or thermally sensitive resistor, which is used for temperature measurement and control. We shall restrict to the discussion of negative temperature coefficient (NTC) type of thermistor whose resistance decreases with increasing temperature, *i.e.* as the temperature increases, the thermistor becomes more conductive, as in the case of a silicon semiconductor.

Most commonly used ceramic semiconducting materials for NTC thermistors are sintered oxides of elements Mn, Ni, Fe, Co and Cu. Solid-solution combinations of the oxides of these elements are used to obtain the necessary range of electrical conductivities with temperature changes.

We first consider the ceramic compound magnetite, Fe_3O_4, which has a relatively low resistivity of about 10^{-5} Ω-m as compared to a value of about 10^8 Ω-m for most regular transition metal oxides. Fe_3O_4 has the inverse spinel structure with the composition $FeO \cdot Fe_2O_3$ that can be expressed as

$$Fe^{2+}\,(Fe^{3+}, Fe^{3+})\,O_4$$

In this structure the oxygen ions occupy FCC lattice sites, with the Fe^{2+} ions in octahedral sites and the Fe^{3+}ions half in octahedral sites and half in tetrahedral sites. The good electrical conductivity of Fe_3O_4 is attributed to the random location of the Fe^{2+} and Fe^{3+} ions in the octahedral sites so that electron "hopping" (transfer) can take place between the Fe^{2+} ions and Fe^{3+} ions while maintaining charge neutrality.

One can control the electrical conductivities of metal oxide superconducting compounds for thermistors by forming solid solutions of different metal oxide compounds. By combining a low-conducting metal oxide with a high-conducting one with a similar structure, a semiconducting compound with an intermediate conductivity can be produced. Figure 4.23 illustrates this effect. Figure shows how the conductivity of Fe_3O_4 is reduced gradually by adding increasing amounts in solid solution of Mg Cr_2O_4. Most NTC thermistors with controlled temperature coefficients of resistivity are made of solid solutions of Mn, Ni, Fe and Co oxides.

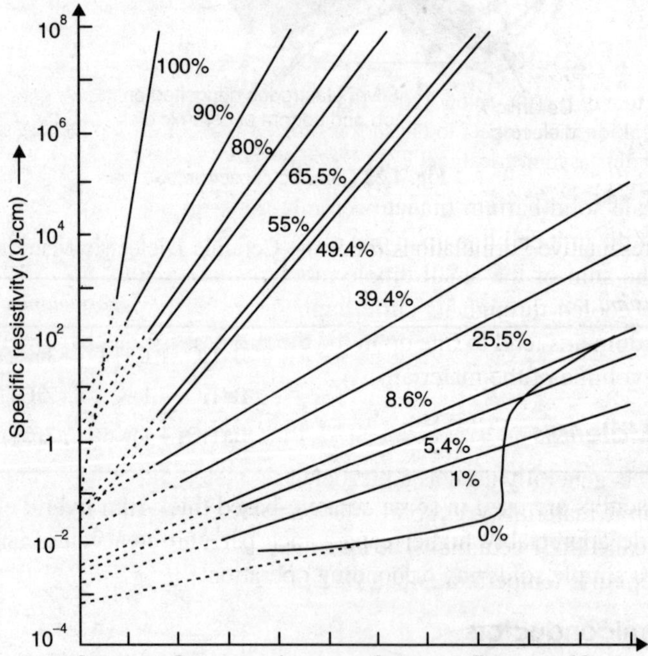

Fig. 4.23 Specific resistivity of solid solution of Fe_3O_4 and $MgCr_2O_4$. Mole per cent $MgCr_2O_4$ is indicated on the curves.

4.5.4 Ferroelectric Ceramics

There are some ceramic ionic crystalline materials having unit cells that do not have a centre of symmetry and as a result their unit cells contain a small electric dipole, *e.g.*, $BaTiO_3$. Above 120°C, $BaTiO_3$ has the regular cubic symmetrical perovskite structure (Fig. 4.15). Below 120°C, the central Ti^{4+} ion and the surrounding O^{2-} ions of the $BaTiO_3$ unit cell shift slightly in opposite directions to create a small electric dipole moment (Fig. 4.24). This shifting of the ion positions at the critical temperature of 120°C, called the *Curie temperature*, changes the crystal structure of $BaTiO_3$ from cubic to slightly tetragonal.

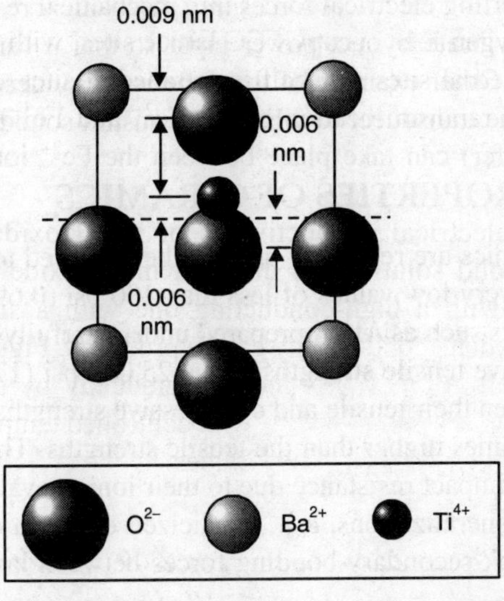

Fig. 4.24 The structure of $BaTiO_3$ below 120°C (its Curie temperature) is slightly tetragonal due to a slight shift of the Ti^{4+} central ion with respect to the surrounding O^{2-} ions of the unit cell. A small electric dipole moment is present in this asymmetrical unit cell.

On a larger scale solid barium titanate ceramic material has a domain structure in which the small electric dipoles of the unit cell line up in one direction. The resultant dipole moment of a unit volume of this material is the sum of the small dipole moments of the unit cells. When polycrystalline barium titanate is slowly cooled through its curie temperature in the presence of a strong electric field, the dipoles of all the domains tend to line up in the direction of the electric field to produce a strong dipole moment per unit volume of the material.

Piezoelectric Materials

Although $BaTiO_3$ is generally used as a piezoelectric material, but it has largely been replaced by other piezoelectric ceramic materials, *e.g.*, solid solutions of lead zirconate ($PbZrO_3$) and lead titanate ($PbTiO_3$) to make what are called PZT ceramics. The PZT materials have a broader range of piezoelectric properties, including a higher curie temperature, than $BaTiO_3$.

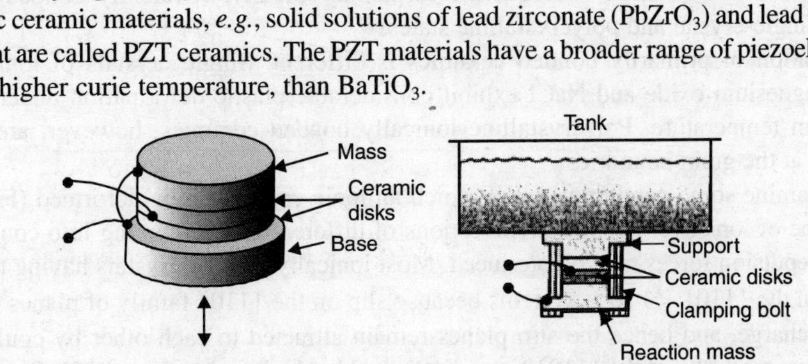

Fig. 4.25 (*a*) Piezoelectric compression accelerometer. (*b*) Piezoelectric ceramic elements in an ultrasonic cleaning apparatus.

Piezoelectric ceramics finds many industrial applications, *e.g.*, for the case of converting mechanical forces into electrical responses are the piezoelectric compression accelerator (Fig. 4.25*a*), which can measure vibratory accelerations occurring over a wide range of frequencies, and the phonograph catridge in which electrical responses are "picked up" from a stylus vibrating in record grooves. An example for the case of converting electrical forces into mechanical responses is the ultrasonic cleaning transducer that is caused to vibrate by a.c. power input so that it can induce violent agitation of the liquid in a tank (Fig. 4.25*b*). Other example of this type is the underwater sound transducer in which electric power input causes the transducer to vibrate to transmit sound waves.

4.6 MECHANICAL PROPERTIES OF CERAMICS

As a class of materials, ceramics are relatively brittle. The observed tensile strength of these materials varies greatly, ranging from very low values of less than 100 psi (0.69 MPa) to about 10^6 psi (7×10^3 MPa) for whiskers of ceramics such as Al_2O_3 prepared under carefully controlled conditions. As a class of materials, few ceramics have tensile strengths above 25,000 psi (172 MPa). Ceramic materials also have a large difference between their tensile and compressive strengths, with the compressive strengths usually being about 5 to 10 times higher than the tensile strengths. There are many ceramic materials, which are hard and have low impact resistance due to their ionic-covalent bindings. However, there are several exceptions to these generalizations, *e.g.*, plasticized clay is a ceramic material that is soft and easily deformable due to weak secondary bonding forces between layers of strongly ionic-covalently bonded ions.

Mechanisms for the Deformation of Ceramic Materials

The lack of plasticity in crystalline ceramics is due to their ionic and covalent bonds. In metals, plastic flow takes place mainly by the movement of line faults (dislocations) in the crystal structure over special crystal slip planes. In metals, dislocations move relatively low stresses due to the non-directional nature of the metallic bond and because all atoms involved in the bonding have an equally distributed negative charge at their surfaces, *i.e.*, there are no positive or negatively charged ions involved in the metallic bonding process.

In covalent crystals and covalently bonded ceramics, the bonding between atoms is specific and directional, involving the exchange of electron charge between pairs of electrons. Obviously, when covalent crystals are stressed to a sufficient extent, they exhibit brittle fracture to a separation of electron-pair bonds without their subsequent reformation. Clearly, covalently bonded ceramics are brittle in both single-crystal and polycrystalline states.

The deformation of primarily bonded ceramics is different. Single crystals of ionically bonded solids, *e.g.*, magnesium oxide and NaCl exhibit considerable plastic deformation under compressive stresses at room temperature. Polycrystalline ionically bonded ceramics, however, are brittle, with cracks forming at the grain boundries.

Now, we examine some conditions under which an ionic crystal can be deformed (Fig. 4.25). The slip of one plane of ions over another involves ions of different charge coming into contact, and thus attractive and repulsion forces may be produced. Most ionically bonded crystals having the NaCl-type structure slip on the $\{110\} < 1\bar{1}0 >$ systems because slip on the {110} family of planes involves only ions of unlike charge, and hence the slip planes remain attracted to each other by coulombic forces during the slip process. Slip of the {110} type is indicated by the line *AA*′ (Fig. 4.25). On the other hand slip on the {110} family of planes is rarely observed because ions of the same charge come into contact, which will tend to separate the planes of ions slipping over each other. This {110}-type slip is indicated by the plane *BB′* (Fig. 4.26). Several ceramic materials in the single-crystal form show considerable plasticity. However, in polycrystalline ceramics adjacent grains must change shape during deformation. Since there are limited slip systems in ionically bonded solids, cracking occurs at the grain boundaries and subsequent brittle fracture occurs. As most industrially important ceramics are polycrystalline, most ceramic materials tend to be brittle.

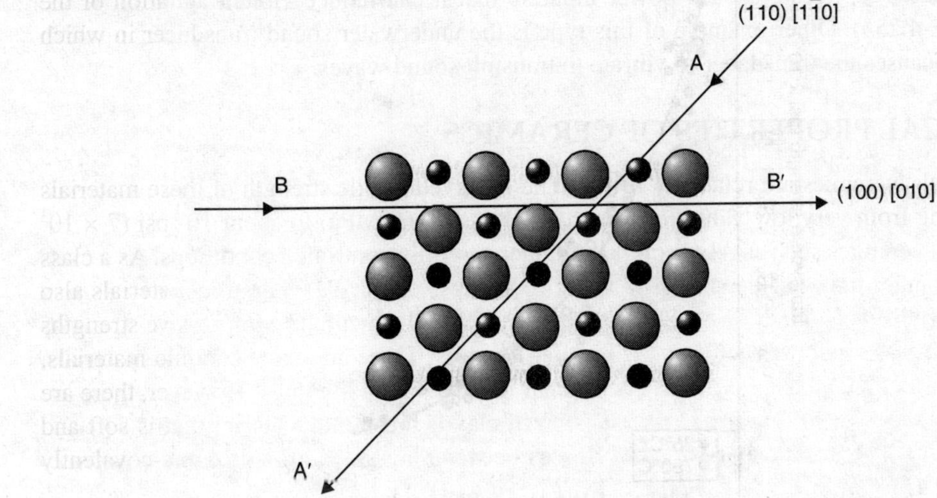

Fig. 4.26 Top view of NaCl crystal structure indicating (*a*) slip on the (110) plane and in the [1$\bar{1}$0] direction (line *AA′*) and (*b*) slip on the (100) plane in the [010] direction (line *BB′*).

4.6.1 Factors Affecting the Strength of Ceramic Materials

The mechanical failure of ceramic materials occurs mainly from structural defects. In ceramic materials the principal sources of fracture include surface cracks produced during surface finishing, voids (porosity), inclusions and large grains produced during processing.

Pores in brittle ceramic materials are regions where stress concentrates, and when the stress at a pore reaches a critical value, a crack forms and propagates since there are no large, energy-absorbing processes in these materials, *e.g.*, those that operate in ductile metals during deformation. Obviously, once cracks start to propagate, they continue to grow until fracture occurs. Pores are also detrimental to the strength of ceramic materials due to the fact that they decrease the cross-sectional area over which a load is applied and hence lower the stress a material can support. Obviously, the size and volume fraction of pores in ceramic materials are important factors affecting their strength. How an increasing volume fraction of pores decreases the transverse tensile strength of alumina is shown in Fig. 4.27.

Flaws in processed ceramics may also be critical in determining the fracture strength of a ceramic material. A large flaw may be the major factor affecting the strength of a ceramic. In fully dense ceramic materials in which there are no large pores, the flaw size is usually related to the grain size. For ceramics free from porosity, the strength of a pure ceramic material is a function of its grain size, with finer-grain-size ceramics having smaller-size flaws at their grain boundaries and hence being stronger than large-grain-size ones.

Obviously, the strength of a polycrystalline ceramic material is determined by several factors that include chemical composition, microstructure, and surface conditions as major factors. Temperature and environment also are important as well as the type of stress and how it is applied. However, the failure of most ceramic materials at room temperature usually originates at the largest flaw.

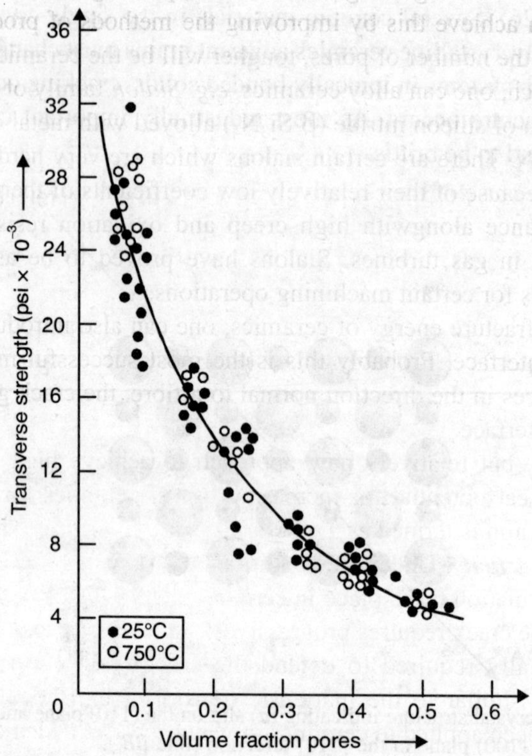

Fig. 4.27 The effect of porosity on the transverse strength of pure alumina.

4.6.2 Toughness of Ceramics

Ceramic materials find limited use in structural applications due to their brittleness. Due to their combination of covalent-ionic bonding, ceramics have inherently low toughness. However, by the use of processes, *e.g.*, hot pressing ceramics with additives and reaction bonding, engineering ceramics with improved toughness have been obtained.

The fracture toughness equation is

$$K_{1C} = Y\sigma_f \sqrt{\pi a} \qquad \text{...(2)}$$

where K_{IC} is fracture toughness measured in $MPa - \sqrt{m}\left(ksi - \sqrt{in}\right)$, the fracture stress σ_f in MPa (*ksi*), and a (half the size of the largest internal flaw) in meters (inches). *Y* is a dimensionless constant equal to about 1.

The stress at which ceramics fracture is significantly lower than their flow stress except at high temperatures. Similar to other materials, the critical fracture stress (σ_{fr}) of ceramic materials is directly related to the elastic modulus (*E*), and the fracture surface energy (γ), and varies inversely with the flaw size (*c*), in accordance with *Griffith Orowan relation.*

$$\sigma_{fr} \propto \left(\frac{E\gamma}{c}\right)^{1/2} \qquad \text{...(3)}$$

For materials which have inherently high elastic moduli, it is very difficult to increase elastic modulus. Thus, to enhance the fracture strength of ceramics, most efforts have been directed towards reducing the flaw size by decreasing the grain size or improving surface finish through the use of ultrafine powders. One can achieve this by improving the methods of processing which may reduce porosity. Obviously, lesser the number of pores, tougher will be the ceramic.

As an alternative approach, one can alloy ceramics, *e.g. Sialon* family of ceramics. Sialon family of ceramics is based on a form of silicon nitride (β-Si_3N_4) alloyed with metal oxides and nitrides such as MgO, Y_2O_3, Al_2O_3 and AlN. There are certain sialons which are very hard and wear resistance, they also resist thermal shock because of their relatively low coefficients of thermal expansion (CTE). The high thermal shock resistance alongwith high creep and oxidation resistance make sialons quite valuable materials for use in gas turbines. Sialons have proved to be as outstanding material for manufacture of cutting tools for certain machining operations.

In order to enhance the fracture energy of ceramics, one can also introduce fibres with high elastic modulus and with weak interface. Probably this is the most successful method of producing tough materials. If a crack advances in the direction normal to a fibre, the crack gets branched off along the fibre length due to weak interface.

There is more complex, but relatively new approach to achieve high toughness and strength in ceramics and which has great potential for increasing use of ceramics for wear resistance and other applications. This phenomenon is termed as *Transformation toughening*. This phenomenon involves a special form of crack tip *plasticity*. Under the conditions of high stress developed at a crack tip, strain induced martensite transformation takes place in certain materials. This transformation absorbs elastic energy and consequently the crack requires progressively more energy to propagate the just the surface energy (γ) which is normally required to extend the crack walls. Obviously, the transformation toughening of ceramics is similar to the martensic transformation thoughening that is achieved in metals. This principle has been applied to ceramics in which ZrO_2 is added or ceramics based on ZrO_2.

We now look into the mechanisms that produce transformation toughening in a ZrO_2 – 9 mol % MgO ceramic material. Pure zirconia ZrO_2 exists in three different crystal structures: *monoclinic* from room temperature to 1170°C, tetragonal from 1170 to 2370°C, and cubic (the fluorite structure) above 2370°C.

The transformation of pure ZrO_2 from tetragonal to monoclinic structure is martensitic and cannot be suppressed by rapid cooling. Moreover, this transformation is accompanied by a volume increase of about 9%, so it is quite impossible to fabricate articles from pure ZrO_2. When we add 10 mol% of other refractory oxides such as CaO, MgO and Y_2O_3, the cubic form of ZrO_2 is stabilized so that it can exist at room temperature in the metastable state, and one can fabricate articles from this material. Cubic ZrO_2 combined with stabilizing oxides so that it retains the cubic structure at room temperature is termed to as *fully stabilized zirconia*.

Very recently zirconia-refractory oxide ceramic materials with enhanced toughness and strength have been produced by taking advantage of their phase transformation. One of the very important zirconia compound ceramics is partially stabilized zirconia (PSZ), which contains 9 mol % MgO. When a mixture of ZrO_2 – 9 mol % MgO is sintered at about 1800°C (Fig. 4.28), and then rapidly cooled to room temperature, it will be in the *all-metastable cubic structure.* However, when we reheat this material to 1400°C and held for a sufficient time, a fine metastable submicroscopic precipitate with the tettragonal structure is precipitated. This material is termed as *partially stabilized zirconia* (PSZ). Figure 4.28(*b*) under the action of stresses that cause small cracks in the ceramic material, the tetragonal phase transforms to the monoclinic phase, causing a volume expansion of the precipitate that retards the crack propagation by a kind of crack-closing mechanism. By impending the advances of cracks, the ceramic is "toughened" Fig. 4.28(*c*). We may note that partially stabilized zirconia has a fracture toughness of 8 + MPa$-\sqrt{m}$, which is higher than the fracture toughness of all the other engineering ceramics.

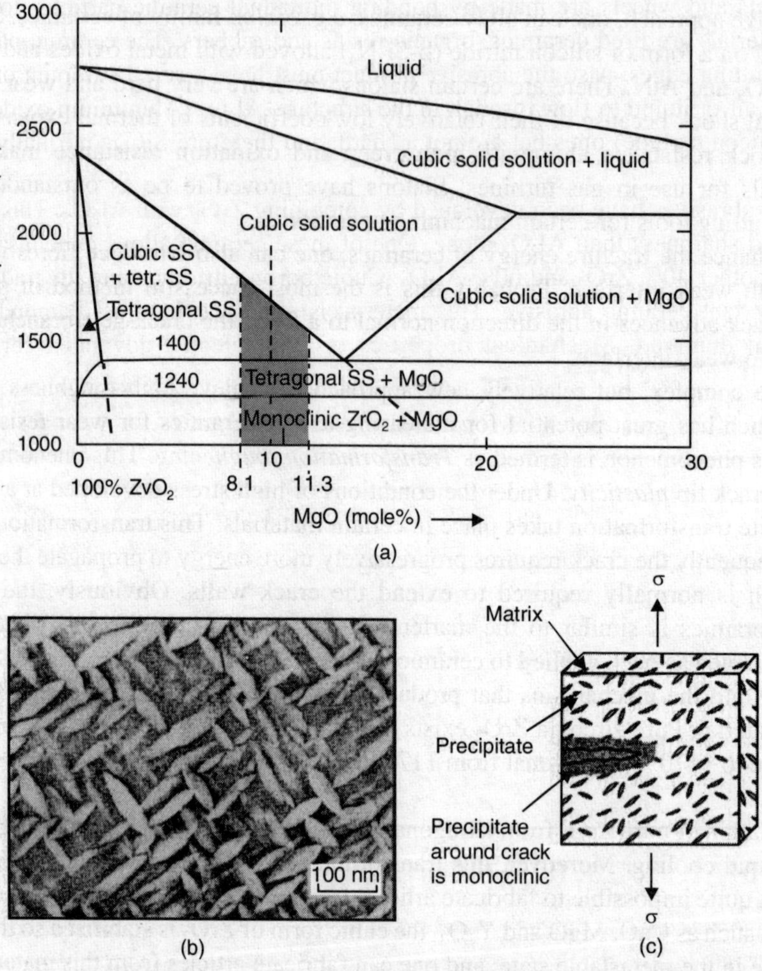

Fig. 4.28 (*a*) Phase diagram of high-ZrO_2 part of the ZrO_2–MgO binary phase diagram. The shaded area represents the region used for combining MgO with ZrO_2 to produce partially stabilized zirconia. (*b*) Transmission electron micrograph of optimally aged MgO–partially stabilized ZrO_2 showing the tetragonal oblate spheroid precipitate. Upon the application of sufficient stress, these particles transform to the monoclinic phase with a volume expansion. (*c*) Schematic diagram illustrating the transformation of the tetragonal precipitate to the monoclinic phase around a crack in a partially stabilized ZrO_2–9 mol % MgO ceramic specimen.

Fatigue failure in metals occurs under repeated cyclic stresses due to the nucleation and growth of cracks within a work-hardened area of a specimen. Due to the ionic-covalent bonding of the atoms in a ceramic materials, there is an absence of plasticity in ceramics during cyclic stressing. As a consequence, fatigue fracture in ceramic is rare. Very recently, results of stable fatigue-crack growth at room temperature under compression-compression stress cycle in notched plates of polycrystalline alumina have been reported. A straight fatigue crack was produced after 79,000 compression cycle. Microcrack propagation along grain boundaries led to final intergranular fatigue failure.

The high hardness of some ceramic materials makes them useful as *abrasive* materials for cutting, grinding and polishing other materials having lower hardness. Examples of two most commonly used manufactured ceramic abrasives are fused alumina (Al_2O_3) and SiC (silicon carbide). Abrasive products such as sheets and wheels are made by bonding individual ceramic particles together. Examples of bonding materials are fired ceramics, organic resins and rubbers. The ceramic particles must be hard with sharp cutting edges. Also the abrasive product must have a certain amount of porosity to provide channels for air or liquid to flow through in the structure. Al_2O_3 (Aluminium oxide) grains are tougher than SiC (silicon carbide) ones but are not as hard, and therefore SiC is normally used for the harder materials.

Improved abrasives have been developed by combining ZrO_2 with Al_2O_3. This has higher strength, hardness, and sharpness than Al_2O_3 alone. One of these ceramic alloys contains 25% ZrO_2 and 75% Al_2O_3 and another 40% ZrO_2 and 60% Al_2O_3. Cubic boron nitride having the trade name '*Borazon*' is another important ceramic abrasive. This material is almost as hard as diamond but has better heat stability than diamond. Mechanical properties of few selected advanced ceramics are given in Table 4.9.

Table 4.9 Mechanical Properties of Selected Advanced Ceramics

Material	*Density* (g/cm^3)	*Tensile strength* (*psi*)	*Flexural strength* (*psi*)	*Compressive strength* (*psi*)	*Young's modulus* (*psi*)	*Fracture toughness* ($psi\sqrt{in.}$)
Al_2O_3	3.98	30,000	80,000	400,000	56×10^6	5,000
SiC (sintered)	3.1	25,000	80,000	560,000	60×10^6	4,000
Si_3N_4 (reaction bonded)	2.5	20,000	35,000	150,000	30×10^6	3,000
Si_3N_4 (hot pressed)	3.2	80,000	130,000	500,000	45×10^6	5,000
Sialon	3.24	60,000	140,000	500,000	45×10^6	9,000
ZrO_2 (partially stabilized)	5.8	65,000	100,000	270,000	30×10^6	10,000
ZrO_2 (transformation toughened)	5.8	50,000	115,000	250.000	29×10^6	11,000

4.7 THERMAL PROPERTIES OF CERAMICS

Due to strong ionic-covalent bonding, most ceramic materials have low thermal conductivities. Ceramics are good thermal insulators. A comparison of thermal conductivities of many ceramic materials is shown in Fig. 4.29. Ceramic materials due to their high heat resistance are used as *refractories*, which are materials that resist the action of hot environments, both liquid and gaseous. Refractories are used widely by the metals, chemical, ceramic and glass industries.

There are many pure ceramic compounds with high melting points, *e.g.*, Al_2O_3 and magnesium oxide, which could be used as industrial refractory materials. However, these materials are expensive and difficult to form into shapes. This is why, most industrial refractories are made of mixtures of ceramic compounds. The compositions of some refractory brick compositions and some of their applications are listed in Table 4.10.

Important properties of ceramic refractory materials are low-and high-temperature strengths, bulk density and porosity. Most ceramic refractories have bulk densities that range from 2.1 to 3.3 g/cm^3 (132 to 206 lb/ft^3). Dense refractories with low porosity have higher resistance to corrosion and erosion and to penetration by liquids and gases. However, high amount of porosity is desirable for insulating refractories. Insulating refractories are generally used as backing for brick or refractory material of higher density and refractoriness.

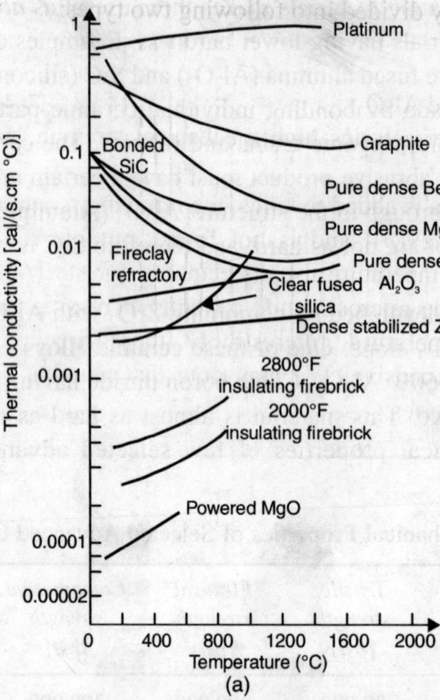

Fig. 4.29 (*a*) Thermal conductivity (logarithmic scale) of ceramic materials over a wide temperature range.

Table 4.10(a) Compositions and Applications for Some Refractory Brick Materials

	Composition (wt %)			
	SiO_2	Al_2O_3	MgO	Other
Acidic types:				
Silica brick	95–99			
Superduty fireclay brick	53	42		
High-duty fireclay brick	51–54	37–41		
High-alumina brick	0–50	45–99+		
Basic types:				
Magnesite	0.5–5		91–98	0.6–4 CaO
Magnesite-chrome	2–7	6-13	50–82	18–24 Cr_2O_3
Dolomite (burned)			38–50	38–58 CaO
Special types:				
Zircon	32			66 ZrO_2
Silicon carbide	6	2		91 SiC

Applications for some refractories:

Superduty fireclay bricks: linings for aluminium-melting furnaces, rotary kilns, blast furnaces, and hot-metal transfer ladles.

High-duty fireclay brick: linings for cement and lime kilns, blast furnaces, and incinerators.

High-alumina brick: boiler furnaces, spent-acid regenerating furnaces, phosphate furnaces, glass-tank refiner walls, carbon black furnaces, continuous-casting tundish linings, coal gasification reactor linings, and petroleum coke kilns.

Silica brick: chemical reactor linings, glass tank parts, ceramic kilns, and coke ovens.

Magnesite brick: basic-oxygen-process furnace linings for steelmaking.

Zircon brick: glass-tank bottom paving and continuous-casting nozzles.

Industrial ceramic are commonly divided into following two types:

(i) Acidic Refractories

These are based mainly on SiO_2 and Al_2O_3.

Silica refractories have high refractories high mechanical strength, and rigidity at temperatures almost to their melting points.

When a small amount of alumina is added to silica, the refractory contains a very low melting-point eutectic microconstituent [Fig. 4.29(*b*)] and is not found suitable for refractory applications at temperatures above ~1600°C, a temperature usually required for steel making. However, when one adds larger amounts of alumina, the microstructure contains increasing amounts of mullite, $3Al_2O_3 \cdot 2SiO_2$, having high melting temperature. Interestingly, these fireclay refractories are generally relatively weak, but they are inexpensive. We may note that alumina concentrations above ~50% constitute the high-alumina refractories.

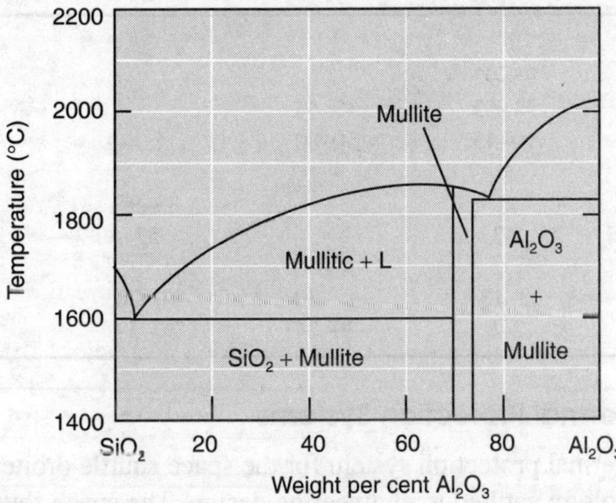

(b)

Fig. 4.29 (*b*) SiO_2-Al_2O_3 phase diagram, the basis for alumina silicate refractories.

Fireclays are based on a mixture of plastic fireclay, fine clay, and clay (coarse-particle) grog. In the unfired (green) condition these refractories consist of a mixture of particles varying from coarse to extremely fine. After firing, the fine particles form a ceramic bond between the larger particles.

High-alumina refractories contain from 50 to 99 per cent alumina and have higher fusion temperatures than fireclay bricks. They can be used for more severe furnace conditions and the higher temperatures that fireclay bricks but are more expensive.

(ii) Basic Refractories

Basic refractories are based mainly on magnesia (MgO), lime (CaO), and Cr_2O_3 (chrome ore) or mixture of two or more of these materials. As a group, basic refractories have high bulk densities, high melting temperatures and good resistance to chemical attack by basic slags and oxides but are quite expensive. Basic refractories containing a high percentage (92 to 95%) of magnesia are used extensively for linings in the basic-oxygen steelmaking process.

(iii) Neutral Refractories

These refractories include chromite and chromite-magnesite. These might be used to separate acid and basic refractories, preventing them from attacking one another.

(iv) Special Refractories

In several refractory applications, particularly when oxygen is not present, carbon or graphite is used. Other refractory materials are zirconia (ZrO_2), Zircon ($ZrO_2 \times SiO_2$) and a variety of nitrides, carbides, and borides. We may note that most of the carbides, *e.g.* TiC and ZrC do not resist oxidation well, and their high-temperature applications are particularly suited to reducing conditions. SiC is an exceptional, *i.e.* when SiC is oxidized at high-temperatures, a thin layer of SiO_2 forms at the surface, protecting the SiC from further oxidation upto about 1500°C. Nitrides and borides also have high melting temperatures and are less susceptible to oxidation. Interestingly, some of the oxides and nitrides are candidates for use in jet engines.

A comparison of refractories based on their chemical behaviour is given in Table 4.10(*b*)

Table 4.10(b) Compositions of typical refractories (weight percents)

Refractory	*SiO_2*	*Al_2O_3*	*MgO*	*Fe_2O_3*	*Cr_2O_3*
Acidica					
Silica	95–97				
Superduty firebrick	51–53	43–44			
High-alumina firebrick	10–45	50–80			
Basic					
Magnesite			83–93	2–7	
Olivine	43		57		
Neutral					
Chromite	3–13	12–30	10–20	12–25	30–50
Chromite-magnesite	2–8	20–24	30–39	9–12	30–50

4.7.1 Space Shuttle Thermal Protection Systems

The development of the thermal protection system for the space shuttle orbiter is an excellent example of modern materials technology applied to engineering design. The space shuttle orbiter could be used for at least 100 missions, new insulative ceramic tile materials were used.

About 70% of the orbiter's external surface is protected from heat by approximately 24,000 individual ceramic tiles made from a silica-fibre compound. Figure 4.30 shows space shuttle thermal protection systems. This material has a density of only 4 kg/ft^3 (9 lb/ft^3) and can withstand temperatures as high as 1260°C.

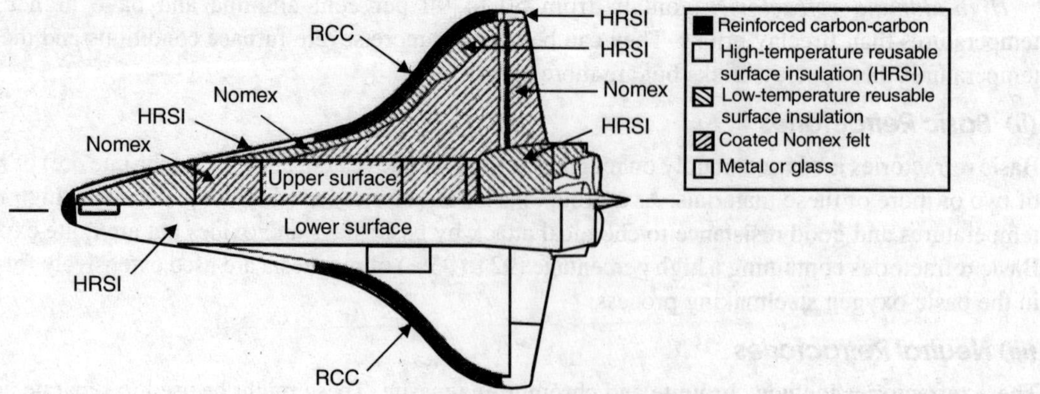

Fig. 4.30 Space shuttle thermal protection systems.

4.8 GLASSES

A glass is a ceramic material in that it is made from inorganic materials at high temperatures. However, glass is distinguished from other ceramic materials in that its constituents are heated to fusion and then cooled to a rigid state without crystallization. Obviously, a glass is an *inorganic product of fusion that has cooled to a rigid condition without crystallisation.* A characteristic of a glass is that it has a non-crystalline or amorphous structure, *i.e.* the molecules in a glass are not arranged in a regular repetitive long-range order as exists in a crystalline solid. In a glass the molecules change their orientation in a random manner throughout the material.

The combination of transparency and hardness at room temperature along with sufficient strength and excellent corrosion resistance to most environments make these materials indispensable for several engineering applications, *e.g.* construction and vehicle glazing. In the electrical industry, glass is essential for various types of lamps because of its insulative properties and ability to provide a vacuum tight enclosure. In the electronics industry electron tubes also require the vacuum-tight enclosure provided by glass along with its insulative properties for lead-in connectors. The high chemical resistance of glass makes them useful for laboratory equipments and apparatus for corrosion-resistant liners for pipes and reaction vessels in the chemical industry.

4.8.1 Glass Transition Temperature

Figure 4.31 illustrates that the solidification behaviour of a glass is different from that of a crystalline solid. Figure 4.31 shows a plot of specific volume (reciprocal of density) versus temperature for two types of materials. A liquid that forms a crystalline solid upon solidifying (*i.e.*, a pure metal) will normally crystallize at its melting point with a significant decrease in specific volume, as illustrated by the path *ABC* in Fig. 4.31. In contrast, a liquid that forms a glass upon cooling does not crystallize but follows a path like *AD* (Fig. 4.31). Liquid of this type becomes more viscous as its temperature is lowered and transforms from a rubbery, soft plastic state to a rigid, brittle glassy state in a narrow temperature range where the slope of the specific volume versus temperature curve is markedly decreased. The point of intersection of the two slopes of this curve define a transformation point called the *glass transition temperature* T_g. This point is structure sensitive, with faster cooling rates producing higher values of T_g.

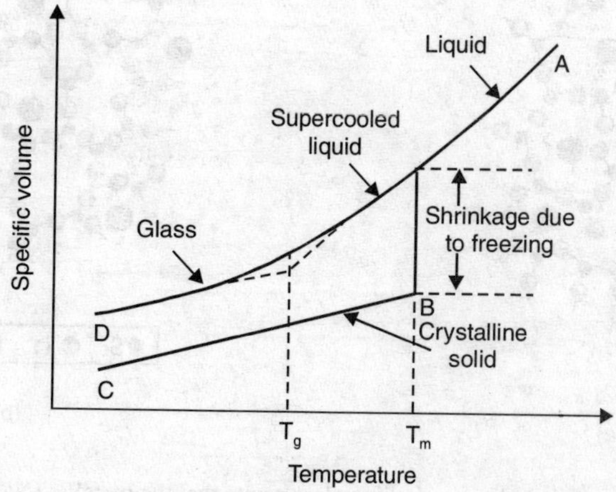

Fig. 4.31 Solidification of crystalline and glassy (amorphous) materials showing changes in specific volume. T_g is the glass transition temperature of the glassy material. T_m is the melting temperature of the crystalline material.

4.8.2 Structure of Glasses

Most inorganic glasses are based on the *glass-forming oxide* silica (SiO_2). The fundamental subunit in silica-based glasses is SiO_4^{4-} tetrahedron in which (Si^{4+}) atom (ion) in the tetrahedron is covalently ionically bonded to four oxygen atoms (ions) as illustrated in Fig. 4.32(*a*). In crystalline silica, *e.g.*, cristobalite, the Si–O tetrahedra are joined corner to corner in a regular arrangement, producing long-range order as idealized in Fig. 4.32(*b*). In a simple silica glass the tetrahedra are joined corner to corner to form a *loose network* with no long-range order (Fig. 4.32(*c*)).

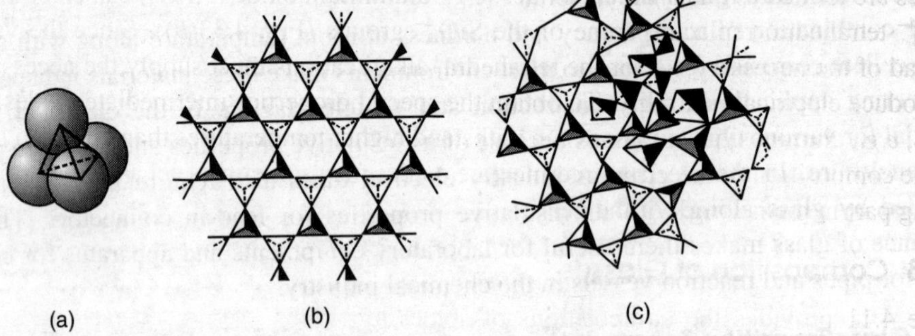

Fig. 4.32 Schematic representation of (*a*) a silicon-oxygen tetrahedron, (*b*) ideal crystalline silica (cristobalite) in which the tetrahedra have long-range order, and (*c*) a simple silica glass in which the tetrahedra have no long-range order.

Boron oxide (B_2O_3) is also a glass-forming oxide and by itself forms subunits that are flat triangles with the boron atom slightly out of the plane of the oxygen atoms. However, in borosilicate glasses that have additions of alkali and alkaline earth oxides, BO_3^{3-} triangles can be converted to BO_4^{4-} tetrahedra, with the alkali or alkaline earth cations providing the necessary electroneutrality. We may note that B_2O_3 is an important addition to many types of commercial glasses such as borosilicate and aluminoborosilicate glasses.

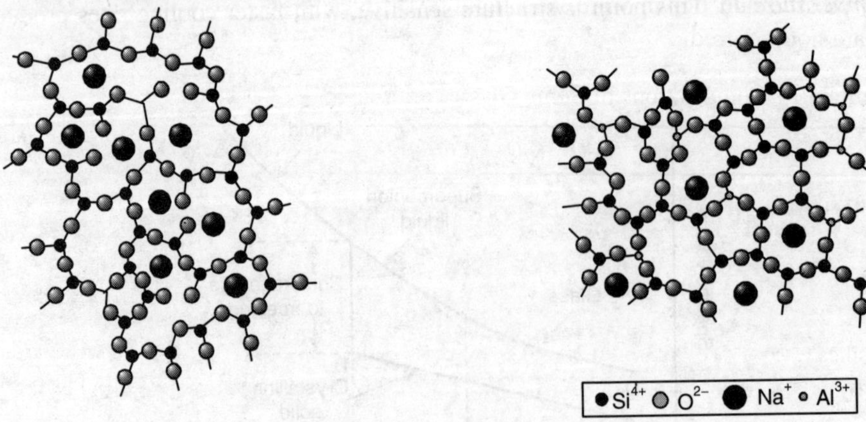

Fig. 4.33 (*a*) Network-modified glass (soda-lime glass); note that the metallic (Na^+) ions do not form part of the network. (*b*) Intermediate-oxide glass (alumina-silica) glass; note that the small metallic (Al^{3+}) ions form part of the network.

Oxides that break up the glass network are termed as *network modifiers* or *Glass-modifying oxides* Alkali oxides, *e.g.*, Na_2O and K_2O and alkaline earth oxides, *e.g.*, CaO and MgO are added to silica glass to lower its viscosity so that it can be worked and formed more easily. The oxygen atoms from these oxides enter the silica network at points joining the tetrahedra and break up the network producing oxygen atoms with an unshared electron (Fig. 4.33(*a*)). The Na^+ and K^+ ions from the alkali oxides Na_2O and K_2O do not enter the network but remain as metal ions ionically bonded in the interstices of the network. By filling some of the interstices, these ions promote crystallisation of the glass.

Some oxides cannot form a glass network by themselves but can join into an existing network. These oxides are termed as *intermediate oxides*, *e.g.*, aluminium oxide (Al_2O_3) can enter the silica network as AlO_4^{4-} tetrahedra, replacing some of the SiO_4^{4-} groups (Fig. 4.33(*b*)). Since the valence of Al is +3 instead of the necessary +4 for the tetrahedra, alkali cations must supply the necessary other electrons to produce electrical neutrality. To obtain the special properties intermediate oxides are added to silica glass, *e.g.*, aluminosilicate glasses can withstand higher temperatures than common glasses. Depending on the composition of the glass, intermediate oxides may sometimes act as network modifiers as well as taking part in the network of the glass.

4.8.3 Composition of Glasses

Table 4.11 provides the compositions of some important types of glasses along with their special properties and applications. *Fused silica glass* is the most important single-component glass. It has a high spectral transmission and is not subjected to radiation damage, which causes browning of other glasses. This is why the fused silica glass is an ideal glass for space vehicle windows, wind tunnel windows, and optical systems in spectrophotometric devices. However, it is difficult to process the silica glass and moreover it is expensive.

Soda-lime glass: This is most commonly produced glass which accounts for about 90% of all the glass produced. In this glass the basic composition is 71 to 73% SiO_2, 12 to 14% Na_2O and 10 to 12% CaO. The Na_2O and CaO decrease the softening point of this glass from 1600 to about 730°C so that the soda-lime glass is easier to form. An addition of 1 to 4% MgO is used for preventing devitrification, and an addition of 0.5 to 1.5% Al_2O_3 is used to increase durability. This glass is used for flat glass, containers, pressed and blown ware, and lighting products where high chemical durability and heat resistance are not required.

Table 4.11 Compositions of Some Glasses with their Special Properties and Applications

Glass	SiO_2	Na_2O	K_2O	CaO	B_2O_3	Al_2O_3	*Other*	*Remarks*
1. (Fused) silica	99.5+							Difficult to melt and fabricate but usable to 1000°C. Very low expansion and high thermal shock resistance.
2. 96% silica	96.3	<0.2	<0.2		2.9	0.4		Fabricate from relatively soft borosilicate glass; heat to separate SiO_2 and B_2O_3 phase; acid leach B_2O_3 phase; heat to consolidate pores.

Glass	SiO_2	Na_2O	K_2O	CaO	B_2O_3	Al_2O_3	Other	Remarks
3. Soda-lime:plate glass	71–73	12–14		10–12		0.5–1.5	MgO, 1–4	Easily fabricated. Widely used in slightly varying grades, for windows, containers, and electric bulbs.
4. Lead silicate: Electrical	63	7.6	6	0.3	0.2	0.6	PbO, 21 MgO, 0.2	Readily melted and fabricated with good electrical properties.
5. High-lead	35		7.2				PbO, 58	High lead absorbs x-rays; high refractive used in achromatic lenses. Decorative crystal glass.
6. Borosilicate: Low expansion	80.5	3.8	0.4		12.9	2.2		Low expansion, good thermal shock resistance, and chemical stability. Widely used in chemical industry.
7. Low electrical loss	70.0		0.5		28.0	1.1	PbO, 1.2	Low dielectric loss.
8. Aluminoborosilicate: Standard apparatus	74.7	6.4	0.5	0.9	9.6	5.6	B_2O, 2.2	Increased alumina, lower boric oxide improves chemical durability.
9. Low alkali (*E-glass*)	54.5	0.5		22	8.5	14.5		Widely used for fibres in glass resin composites.
10. Aluminosilicate	57	1.0		5.5	4	20.5	MgO, 12	High-temperature strength, low expansion.
11. Glass-ceramic	40–70					10–35	MgO, 10–30 TiO_2, 7–15	Crystalline ceramic made by devitrifying glass. Easy fabrication (as glass), good properties. Various glasses and catalysts.

Borosilicate glasses: The replacement of alkali oxides by boric oxide in the silica glassy network produces a lower-expansion glass. When B_2O_3 enters the silica network, it weakens its structure and considerably lowers the softening point of the silica glass. The weakening effect is caused due to the presence of planar three-coordinate borons. This type of glass also known as *Pyrex glass* is used for laboratory equipment, piping, ovenware and sealed-bean headlights.

Lead glasses: Lead oxide is usually a modifier in the silica network but can also act as a network former. Lead glasses with high lead oxide contents are low melting and are useful for solder-sealing glasses. High lead glasses are also used for shielding from high energy radiation, and they find application for radiation windows, fluorescent lamp envelopes, and television bulbs. However, due to their high refractive indexes, lead glasses are used for some optical glasses and for decorative purpose glasses.

Silicate glasses: These glasses are most widely used. *Fused silica,* formed from pure SiO_2, has a high melting point, and the dimensional changes during heating and cooling are small (Fig. 4.34). However, the silicate glasses contain additional oxides (Table 4.12). The oxides such as silica behave as *glass formers,* an intermediate oxide (*e.g.* lead oxide or aluminium oxide) does not form a glass by itself but is incorporated into the network structure of the glass formers. A third group of oxides, the *modifiers,* break up the network structure and eventually cause the glass to devitrify, or crystallize.

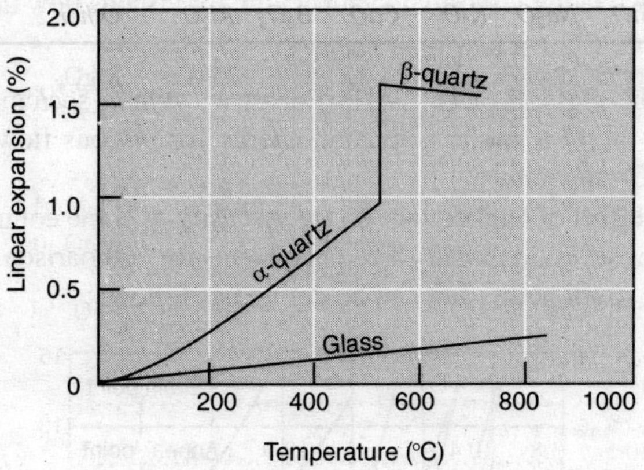

Fig. 4.34 The expansion of quartz. In addition to the regular—almost linear— expansion, a large, abrupt expansion accompanies the α- to β-quartz transformation. However, glasses expand uniformly.

Table 4.12 Division of the Oxides into Glass Formers, Intermediates and Modifiers

Glass formers	*Intermediates*	*Modifiers*
B_2O_3	TiO_2	Y_2O_3
SiO_2	ZnO	MgO
GeO_2	PbO_2	CaO
P_2O_5	Al_2O_3	PbO
V_2O_3	BeO	Na_2O

Modifiers break up the silica network if the oxygen-to-silicate ratio (O:Si) increases significantly. When Na_2O is added, the sodium ion enter holes within the network rather than getting part of the network. However, the oxygen ion that enters with the Na_2O dees become part of the network. When this happens, there are not enough silicon ions to combine with extra oxygen ions and keep the network intact. This means, a high O:Si ratio causes the remaining silica tetrahedra to form chains, rings, or compounds, and the silica no longer transforms to a glass. It is found that when O:Si ratio is above about 2.5, silica glasses are difficult to form; above a ratio of 3, a glass forms only when special precautions are observed, *e.g.* the use of rapid cooling rates.

It is observed that modification lowers the melting point and viscosity of silica and this makes it possible to produce glass at lowers temperatures. One finds that the addition of Na_2O produces eutectics with low melting temperatures. Adding CaO, which reduces the solubility of the glass in water, further modifies these glasses.

4.8.4 Viscous Deformation of Glasses

A glass is found to behave as a viscous (super cooled) liquid above its glass transition temperature (t_g). Under stress, groups of silicate atoms (ions) can slide past each other, allowing permanent deformation of the glass. Above t_g, interatomic bonding forces resist deformation, but cannot prevent viscous flow of the glass if the applied stress is sufficiently high. As the temperature of the glass is progressively increased above its t_g, the viscosity of the glass decreases and viscous flow becomes easier. The effect of temperature on the viscosity of a glass follows an *Arrhenius-type equation* except that the sign of the exponential term is positive instead of negative as is usually the case, *i.e.* for diffusivity, the Arrhenius-type equation is $D = D_0$

exp (−Q/RT). The equation relating viscosity to temperature for viscous flow in a glass is

$$\eta = \eta_0 \exp [+Q/RT] \quad ...(4)$$

where η is viscosity of the glass, P or Pa · s[1P (Poise) = 1 dyne · sec/cm^2; 1 Pa · s (Pascal-second) = 1N.s/m^2; 1P = 0.1 Pa · s], Q is molar activation energy for viscous flow, R is universal molar gas constant and T is absolute temperature.

Figure 4.35 shows the effect of temperature on the viscosity of some commercial glasses. Horizontal lines in the figure indicates several viscosity reference points for comparison. These points are working, softening, annealing, and strain points and can be defined as follows:

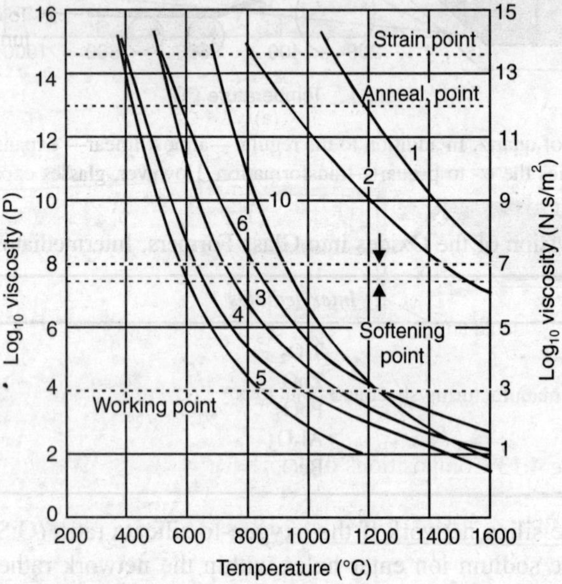

Fig. 4.35 Effect of temperature on the viscosities of various types of glasses. Numbers on curves refer to compositions in Table 4.8.

(*i*) *Working point:* Viscosity = 10^4 Poises (10^3 Pa · s). At this temperature operations of glass fabrication can be carried out.

(*ii*) *Softening point* = 10^8 Poises (10^7 Pa · s). At this temperature the glass will flow at an appreciable rate under its own weight. However, this point cannot be defined by a precise viscosity because it depends on the density and surface tension of the glass.

(*iii*) *Annealing point* = $10^{14.5}$ Poises ($10^{13.5}$ Pa · s). Below this temperature the glass is rigid and stress relaxation occurs only at a slow rate. The interval between the annealing and strain points is commonly considered the *annealing range of a glass.*

Usually the glasses are melted at a temperature that corresponds to a viscosity of about 10^2 Poises (10 Pa · s). During forming, the viscosities of glasses are compared qualitatively. We may note that a *hard glass* has a high softening point, whereas a *soft glass* has a lower softening point. A *long glass* has a large temperature difference between its softening and strain points, *i.e.* the long glass solidifies slower than a *short glass* as the temperature is decreased.

4.8.5 Forming Methods for Glasses

Glass products are formed by first heating the glass to a high temperature to produce a viscous liquid and then molding, drawing, or rolling it into a desired shape.

We may note that sheet and plate glass are produced when the glass is in the molten state. Techniques include the molten glass through water-cooled rolls or floating the molten glass over a pool of liquid tin (Fig. 4.36). The liquid tin process produces an exceptionally smooth surface on the glass. We may note that the development of the float-glass process was a genuine break through in the area of glass processing. Table 4.13 shows the composition of typical glasses.

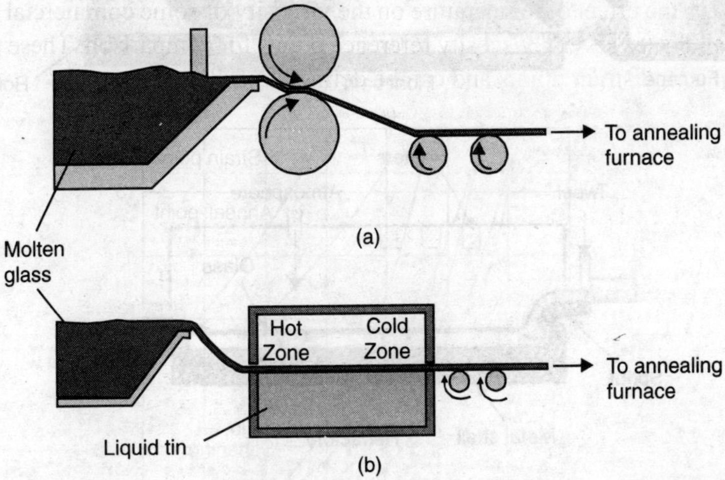

Fig. 4.36 Techniques for manufacturing sheet and plate glass: (*a*) rolling and (*b*) floating the glass on molten tin.

Table 4.13 Compositions of Typical Glasses (in Weight per cent)

Glass	SiO_2	Al_2O_3	CaO	Na_2O	B_2O_3	MgO	PbO	*Others*
Fused silica	99							
Vycor™	96				4			
Pyrex™	81	2		4	12			
Glass jars	74	1	5	15		4		
Window glass	72	1	10	14		2		
Plate glass/Float glass	73	1	13	13				
Light bulbs	74	1	5	16		4		
Fibres	54	14	16		10	4		
Thermometer	73	6		10	10			
Lead glass	67			6			17	10% K_2O
Optical flint	50			1			19	13% BaO, 8% K_2O, Z_nO
Optical crown	70			8		10		2% BaO, 8% K_2O
E-glass fibres	55	15	20		10			
S-glass fibres	65	25				10		

About 85% of the flat glass produced is made by *float process* in which a ribbon of glass moves out of the melting furnace and floats on the surface of a bath of molten tin (Fig. 4.37). While moving across the molten tin the glass ribbon is cooled and while under a chemically controlled atmosphere (Fig. 4.37).

When its surfaces are sufficiently hard, the glass sheet is removed from the furnace without being marked by rollers and passed through a long annealing furnace called a *lehr*, where residual stresses are removed. This process is used for forming sheet and plate glass.

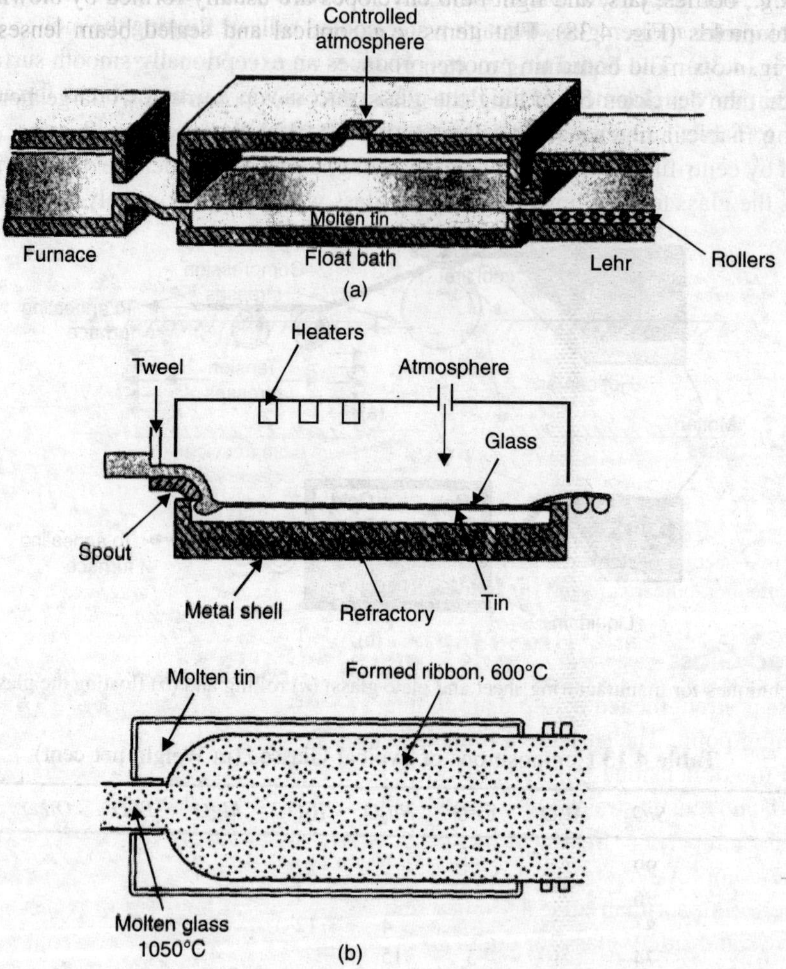

Fig. 4.37 (*a*) Diagram of the float-glass process. (*b*) Side and top schematic views of the float-glass process.

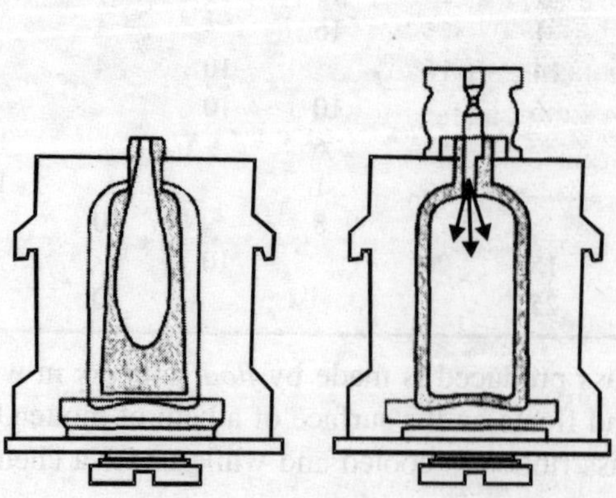

Fig. 4.38 (*a*) Reheat and (*b*) final blow stages of a glass blowing machine process.

Deep items, *e.g.*, bottles, jars, and light bulb envelopes are usually formed by blowing air to force molten glass into molds (Fig. 4.38). Flat items, *e.g.*, optical and sealed beam lenses are made by pressing a plunger into a mold containing molten glass.

One can make many articles by casting the glass into an open mold. A large borosilicate glass telescope mirror 6 m in diameter was made by casting the glass. Funnel shaped items, *e.g.*, television tubes are formed by centrifugal casting. Gobs of molten glass from a feeder are dropped into a spinning mold that causes the glass to flow upward to form a glass wall of approximately uniform thickness.

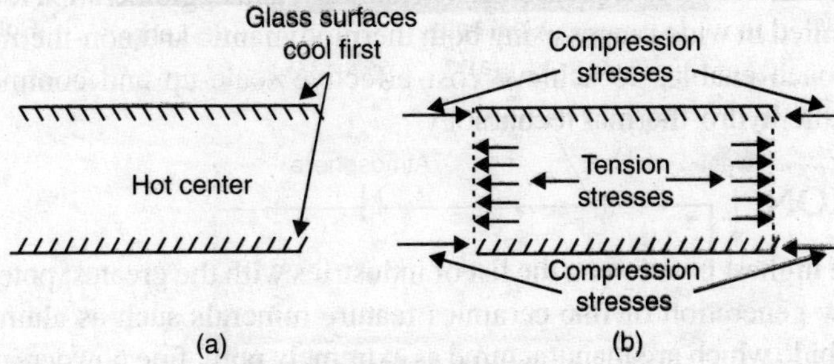

Fig. 4.39 Cross-section of tempered glass (*a*) after surface has cooled from high temperature near glass-softening temperature and (*b*) after centre has cooled.

4.8.6 Tempered Glass

This type of glass is strengthened by rapid air cooling of the surface of the glass after it has been heated to near its softening point. The surface of the glass cools first and contracts, while the interior is warm and readjusts to the dimensional change with little stress. When the interior cools and contracts, the surfaces are rigid, and tensile stresses are created in the interior of the glass and compressive stresses on the surfaces. This "tempering" treatment increases the strength of the glass as applied tensile stresses must surpass the compressive stresses on the surface prior to fracture occurs. Tempered glass has a higher resistance to impact than annealed glass and is about four times stronger than annealed glass. Auto side windows and safety glass for doors are items that are thermally tempered.

4.8.7 Chemically Strengthened Glass

One can increase the strength of glass by special chemical treatments, *e.g.*, if a sodium aluminosilicate glass is immersed in a bath of potassium nitrate at a temperature ~50°C below its stress point (~500°C) for 6 to 10 hours, the smaller sodium ions near the surface of the glass are replaced by larger potassium ions. The introduction of the larger potassium ions into the surface of the glass produces compressive stresses at the surface and corresponding tensile stresses at the centre. This process can be used on thinner cross-sections than can thermal tempering since the compressive layer is much thinner.

4.9 SMART CERAMIC MATERIALS

There is a growing demand from various industries to synthesize smart materials with more and more sophisticated features requires the use of smart starting materials, with very well-defined and controlled specific properties, such as size, shape (morphology), chemical composition and defect structure, surfrace functionalization, etc. This in turn requires use of advanced synthetic methods, which, in addition to producing superior product, should be also inexpensive and environmentally friendly. A large family of smart ceramics, primarily powders and coatings has emerged that can be prepared under very mild hydrothermal conditions (T < 200°C, P < 1.5 MPa). This has generated considerable interest in hydrothermal technology. Hydrothermal synthesis is environmentally friendly, energy conserving low-temperature crystallisation of all forms of smart materials directly from aqueous solutions. Hydrothermal crystallization affords excellent control of chemical composition (oxides, nanoxides, native elements, dopants, solid solutions), as well as shape, size, and morphology of ceramics. Understanding the physicochemical processes occurring in the aqueous solution is the key for intelligent engineering of hydrothermal synthesis of smart ceramics. Thermodynamic modelling is used to predict formation conditions for phases of interest. Subsequent size, morphology, and agglomeration level of the synthesized ceramics can be controlled in wide ranges using both thermodynamic and non-thermodynamic variables. Such intelligent approach enables to achieve cost effective scale-up and commercial production of smart ceramics using the hydro-thermal technology.

4.10 CONCLUSION

Ceramics occupies the highest position on the list of industries with the greatest potential for commercial development. The new generation of fine ceramics feature minerals such as aluminium oxide, silicon nitride, and silicon carbide which are manufactured as extremely pure, fine powders prior to consolidation at high temperatures to yield a durable dense structure. The control of these new ceramics provides the design engineer with a class of ceramics which are stronger, lighter, and harder than most metallic competitions. Moreover, ceramics can operate at much higher temperatures and they do not degrade significantly due to oxidation as do commercial metals, *e.g.* irons and steels.

These materials do, however, have *Achilles heel:* brittleness. This brittleness is associated with the number of extremely small flaws such as cracks, voids, and impurities present in the material, and also the amount of energy required to fracture the material in the presence of these flaws. This means that the manufacturing processes must be extremely carefully controlled, since flaws as small as 10 to 50 micrometers can reduce the strength of a ceramic part to only a few per cent of its ideal strength thereby rendering the part useless. While material toughness is a major limitation of the current generation of advanced structural ceramics, the many significant advantages of these materials have not deterred further research on these compounds while other research efforts are focused on this limiting material characteristic. These latter efforts are typically focused on synthesizing tougher materials through the prosecution of research on microstructural design, transformation toughening, and ceramic composites. Where the incorporation of ceramic particulates, whiskers, or continuous fibres in a ceramic matrix can yield a composite material that absorbs more energy during fracture, than a geometrically identical part fabricated in the matrix material alone.

The highly desirable characteristics offered by these new ceramics will impact a broad segment of the industrial sector such as automotive, electronics, aerospace, medical and telecommunications segments. Ceramics technologies have developed a high profile in the advanced ceramic consumer goods industries through the marketing of pliers, scissors and ball-point pen tips, etc. Structural ceramic parts featuring alumina, SiC, and silicon nitride are commercially available for wear resistant parts, turbo charger, rocket arms, and cutting tools. Coatings and bearings for orthopedic or dental implants are also available. Some of these materials function as resorbable materials to provide temporary support until the human body can gradually replace it, while others are surface-active materials which form a bond while the surrounding tissue in order to stimulate growth.

One of the foci of ceramic materials research and development is concerned with the automobile industry. Research engineers are projecting the development of engineers which are 30% to 40% more fuel efficient than the current generation of engines, and moreover, these engines run at higher temperatures without lubrication or cooling. The electronics industry currently consumes the vast majority of technical ceramic production, but it too will benefit from this new generation of ceramics through the development of superior computer chips, photonic devices, capacitors, and the like, which will trigger the evolution of the *optical computer* where information pulses travel at the speed of light. Obviously, this new generation of computers will be able to process information much more quickly than the current generation of electron-based machines, and they will also be capable of accommodating a greater volume of data.

Example 1: Calculate the ionic packing factor for CsCl. Ionic radii are $Cs^+ = 0.170$ nm and $Cl^- = 0.181$ nm. [B.Tech.]

Solution: The ions touch each other across the cube diagonal of the CsCl unit cell, as shown in Fig. 4.40. Let $r = Cs^+$ ion and $R = Cl^-$ ion. Thus

$$\sqrt{3}\,a = 2r + 2R$$

$$= 2(0.170 \text{ nm} + 0.181 \text{ nm})$$

$$a = 0.405 \text{ nm}$$

$$\text{CsCl ionic packing factor} = \frac{\frac{4}{3}\pi r^3 (1\,Cs^+ \text{ ion}) + \frac{4}{3}\pi R^3 (1\,Cl^- \text{ ion})}{a^3}$$

$$= \frac{\frac{4}{3}\pi(0.170 \text{ nm})^3 + \frac{4}{3}\pi(0.181 \text{ nm})^3}{(0.405 \text{ nm})^3}$$

$$= 0.68$$

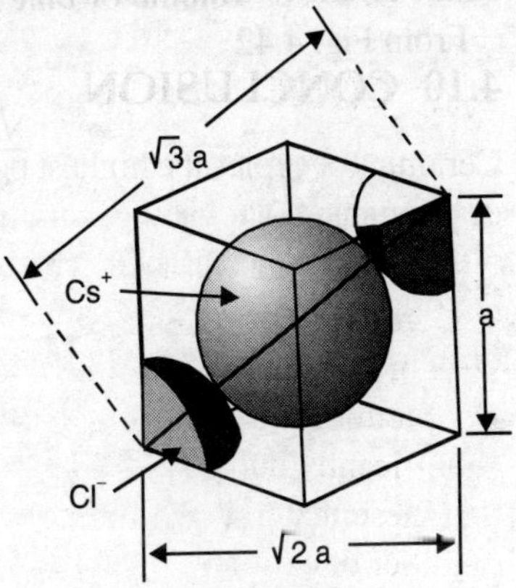

Fig. 4.40

Example 2: Calculate the linear density of Ca^{2+} and O^{2-} ions in ions per nanometer in the [110] direction of CaO, which has the NaCl structure, (Ionic radii: $Ca^{2+} = 0.106$ nm and $O^{2-} = 0.132$ nm.) [B.Tech.]

Solution: From Fig. 4.41, we see that the [110] direction passes through $2O^{2-}$ ion diameters in traversing from (0, 0, 0) to (1, 1, 0) ion positions. The length of the [110] distance across the base face of a unit cube is $\sqrt{2}\,a$, where a is the length of a side of the cube, or lattice constant. From Fig. 4.41 of the cube face of the NaCl unit cell, we see that $a = 2r + 2R$. Thus, for CaO,

$$a = 2(r_{Ca^{2+}} + R_{O^{2-}})$$

$$= 2(0.106 \text{ nm} + 0.132 \text{ nm}) = 0.476 \text{ nm}$$

The linear density of the O^{2-} ions in the [110] direction is

$$\rho_L = \frac{2O^{2-}}{\sqrt{2}a} = \frac{2O^{2-}}{\sqrt{2}(0.476 \text{ nm})} = 2.97\ O^{2-}/\text{nm}$$

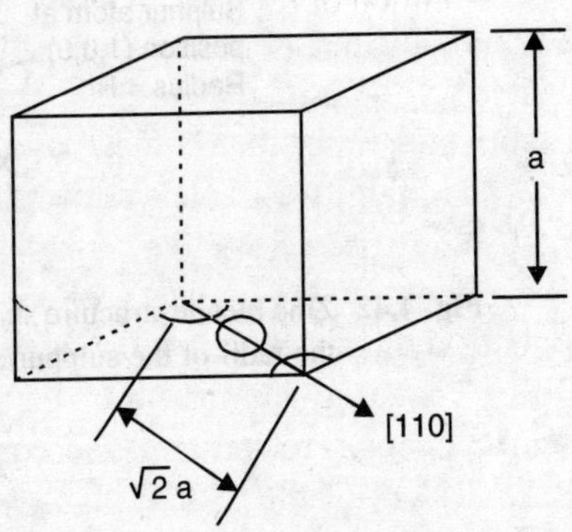

Fig. 4.41

The linear density of Ca^{2+} ions in the [110] direction is also 2.97 Ca^{2+}/nm if we shift the origin of the [110] direction from (0, 0, 0) to $\left(0, \frac{1}{2}, 0\right)$. Thus, the solution to the problem is that there are 2.97 (Ca^{2+} or O^{2-})/nm in the [110] direction.

Example 3: Calculate the density of zinc blende (ZnS). Assume the structure to consist of ions and that the ionic radius of Zn^{2+} = 0.060 nm and that of S^{2-} = 0.174 nm. [B.Tech.]

Solution: $\text{Density} = \dfrac{\text{mass of unit cell}}{\text{volume of unit cell}}$

There are four zinc ions and four sulfur ions per unit cell. Thus

$$\text{Mass of unit cell} = \frac{\left(4Zn^{2+} \times 65.37\,\text{g/mol}\right) + \left(4S^{2-} \times 32.06\,\text{g/mol}\right)}{6.02 \times 10^{23}\ \text{atoms/mol}}$$

$$= 6.47 \times 10^{-22}\,\text{g}$$

$$\text{Volume of unit cell} = a^3$$

From Fig. 4.42

$$\frac{\sqrt{3}}{4}a = r_{Zn^{2+}} + R_{S^{2-}} = 0.060\text{ nm} + 0.174\text{ nm} = 0.234\text{ nm}$$

$$a = 5.40 \times 10^{-8}\text{ cm}$$

$$a^3 = 1.57 \times 10^{-22}\text{ cm}^3$$

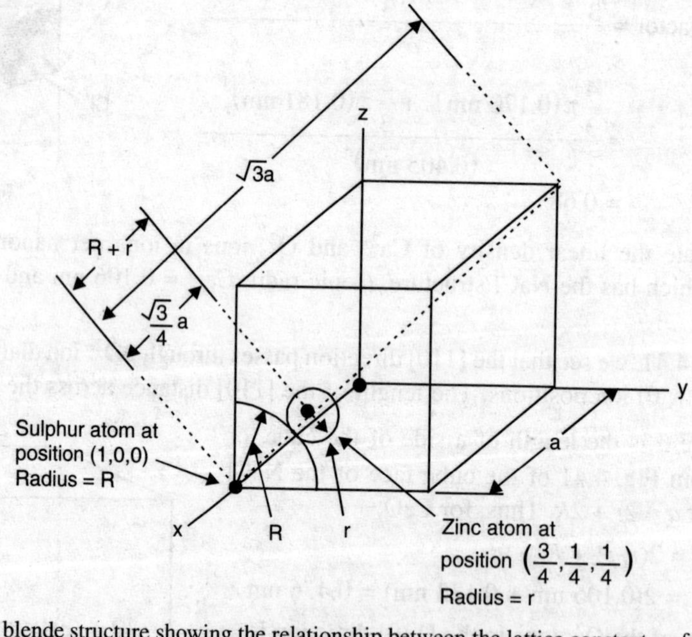

Fig. 4.42 Zinc blende structure showing the relationship between the lattice constant a of the unit cell and the radii of the sulphur and zinc atoms (ions):

$$\frac{\sqrt{3}}{4}a = r_{Zn^{2+}} + R_{S^{2-}}$$

or

$$a = \frac{4}{\sqrt{3}}(r + R)$$

Thus

$$\text{Density} = \frac{\text{mass}}{\text{volume}} = \frac{6.47 \times 10^{-22}\text{ g}}{1.57 \times 10^{-22}\text{ cm}^3} = 4.12\,\text{g/cm}^3$$

The handbook value for the density of ZnS (cubic) is 4.10 g/cm^3.

Example 4: A reaction-bonded silicon nitride ceramic has a strength of 300 MPa and a fracture toughness of 3.6 MPa-$\sqrt{m}$. What is the largest-size internal crack that this material can support without fracturing? Use Y = 1 in the fracture-toughness equation. [B.Tech.]

Solution:

$$\sigma_f = 300 \text{ MPa} \qquad K_{IC} = 3.6 \text{ MPa-}\sqrt{m} \qquad a = ? \quad Y = 1$$

$$K_{IC} = Y_{\sigma_f} \sqrt{\pi a}$$

or

$$a = \frac{K_{IC}^2}{\pi \sigma_f^2} = \frac{\left(3.6 \text{ MPa-}\sqrt{\text{m}}\right)^2}{\pi (300 \text{ MPa})^2}$$

$$= 4.58 \times 10^{-5}\text{m} = 45.8 \text{ μm}$$

Thus, the largest internal crack = $2a$ = 2(45.8 μm) = 91.6 μm.

Example 5: A 96 per cent silica glass has a viscosity of 10^{13} P at its annealing point of 940°C and a viscosity of 10^8 P at its softening point of 1470°C. Calculate the activation energy in kilojoules per mole for the viscous flow of this glass in this tempe..ature range. [B.Tech.]

Solution:

Annealing point of glass = T_{ap} = 940°C + 273 = 1213 K $\qquad \eta_{ap} = 10^{13}$ P

Softening point of glass = T_{sp} = 1470°C + 273 = 1743 K $\qquad \eta_{sp} = 10^8$ P

R = gas constant
= 8.314 J/(mol.K)

Q = ? J/mol.

Using equation $\eta = \eta_0 e^{Q/RT}$,

$$\eta_{ap} = \eta_0 e^{Q/RT_{ap}} \qquad \text{or} \qquad \frac{\eta_{ap}}{\eta_{sp}} = \exp\left[\frac{Q}{R}\left(\frac{1}{T_{ap}} - \frac{1}{T_{sp}}\right)\right] = \frac{10^{13}\text{P}}{10^8\text{P}} = 10^5$$

$$\eta_{sp} = \eta_0 e^{Q/RT_{sp}}$$

$$10^5 = \exp\left[\frac{Q}{8.314}\left(\frac{1}{1213\text{K}} - \frac{1}{1743\text{ K}}\right)\right]$$

$$\ln 10^5 = \frac{Q}{8.314}\left(8.244 \times 10^{-4} - 5.737 \times 10^{-4}\right) = \frac{Q}{8.314}\left(2.507 \times 10^{-4}\right)$$

$$11.51 = Q(3.01 \times 10^{-5})$$

$$Q = 3.82 \times 10^5 \text{ J/mol} = 382 \text{ kJ/mol.}$$

Example 6: SiC particles are compacted and fired at high temperature to produce a strong ceramic shape. Subsequently the ceramic shape is weighted when dry-(360 g), after soaking in water (385 g), and while suspended in water (224 g). Given the specific gravity of SiC = 3.2 g/cm^3, find the apparent porosity, the true porosity, and the fraction of the pore volume that is closed. [B.Tech.]

Solution: We have

(*i*) Apparent porosity $= \dfrac{W_w - W_d}{W_w - W_s} \times 100$

$$= \frac{385-360}{385-224} \times 100 = 15.5\%$$

(*ii*) Bulk density $= \dfrac{W_d}{W_w - W_s} = \dfrac{360}{385-224}$

$$= 2.2 \text{ gm/cm}^3$$

We may note that the denominator is equal numerically to the volume of the ceramic, since density of water is 1 gm/cm³.

(*iii*) True porocity $= \dfrac{\rho - B}{\rho} \times 100 = \dfrac{3.2 \times 2.24}{3.2} \times 100$

$$= 30\%$$

The closed pore percentage = true porosity – apparent porosity = 30 – 15.5 = 14.5%

Fraction closed pores $= \dfrac{14.5}{30} = 0.48.$

Example 7: When B_2O_3 is introduced in silica, it produces good chemical resistance. The O : Si ratio is not more than 2.5, *i.e.* the glassware to have a low melting temperature to make the glass-forming process easier and more economical. Design such a glass. [B.Tech.]

Solution: We know that B_2O_3 reduces the melting temperature of silica and hence we would like to add as much B_2O_3 in silica as possible. Since O : Si ratio is no more than 2.5 and hence B_2O_3 amount is limited. We determine the amount of B_2O_3 for getting exactly an O : Si ratio of 2.5. Let f_b denotes the mole fraction of B_2O_3 added to the glass and $(1 - f_B)$ be the mole fraction of SiO_2. Thus

$$\frac{\text{O}}{\text{Si}} = \frac{\left(3\dfrac{\text{O ions}}{\text{B}_2\text{O}_3}\right)(f_B) + \left(2\dfrac{\text{O ions}}{2}\right)(1-f_B)}{1\left(\dfrac{\text{Si ion}}{\text{SiO}_2}\right)(1-f_B)} = 2.5$$

$\therefore$ $$3f_B + 2 - 2f_B = 2.5 - 2.5\, f_B$$

or $$f_B = 0.143$$

Thus, we must produce a glass containing no more than 14.3 mol % B_2O_3. In wt per cent, we have

$$\text{Wt \% } \text{B}_2\text{O}_3 = \frac{f_B(69.62 \text{ g/mol})}{(f_B)(69.62 \text{ g/mol}) + (1-f_B)(60.08 \text{ g/mol})} \times 100$$

$$= \frac{0.143 \times 69.62}{0.143 \times 69.62 + 0.857 \times 60.08} \times 100$$

$$= 16.2$$

SUMMARY

- *Ceramic materials* are inorganic, non-metallic, materials consisting of metallic and non-metallic elements bonded together primarily by ionic and/or covalent bonds. Due to this the chemical compositions and structures of ceramic materials vary considerably. Ceramic materials may consist of a single compound, *e.g.*, pure aluminium oxide, or they may be composed of a mixture of many complex phases such as the mixture of clay, silica, and feldspar in electrical porcelain.
- Due to differences in bonding, the properties of ceramic materials also vary greatly. In general, most ceramic materials are typically hard and brittle with low impact resistance and ductility. Consequently, in most designs of engineering, high stresses in these materials are normally avoided, especially if ceramic materials are tensile stresses.
- Due to absence of conduction electrons, ceramic materials are normally good electrical and thermal insulators and thus several ceramics are used for electrical insulation and refractories.
- Some ceramic materials can be highly polarised with electric charge and are used for dielectric materials for capacitors.
- Polycrystalline ceramics exhibit a brittle behaviour, partly because of porosity. Since most polycrystalline ceramics cannot plastically deform (unless special conditions with respect to temperature and strain rates are crated), the porosity limits the ability of the material to withstand a tensity load.
- Ceramics play a critical role in a wide array of technologies related to electronic, optical, magnetic and energy, Several advanced ceramics play a very important role in providing thermal insulation and high temperature properties. Applications of advanced ceramics range from credit cards, houses for silicon chips, tiles for the space energy efficient glasses.
- Traditional ceramics find a very important role as refractories for metals processing and consumer applications.
- Ceramic processing is commonly conducted with the help of compaction and sintering. For specialized applications, isostatic compaction, not pressing, and hot isostatic pressing are used to achieve higher-densification levels.
- Permanent polarization of some ceramic materials produce piezoelectric properties that permit these materials to be used as electromechanical transducers. Other ceramic materials, *e.g.*, Fe_3O_4 are semiconductors and find application for thermistors for temperature measurement.
- The processing of ceramic materials usually involves the agglomeration of small particles by a variety of methods in the dry, plastic or liquid states. Cold forming processes predominate in the ceramic industry, however hot-forming processes are also used.
- Pressing, slip-casting, and extrusion are commonly used ceramic-forming processes. After forming, ceramic materials are usually given a thermal treatment such as sintering or vitrification. During sintering, the small particles of a formed article are bonded together by solid-state diffusion at high temperatures.
- Glass-ceramics are formed using controlled crystallisation of inorganic glasses. These materials find wide use for kitchenware and several other applications.
- Many silicates and other ceramics form glasses rather easily, since the kinetics of crystallisation are sluggish. Glasses are formed in the form of sheets using float-glass or fibres and other shapes. Silicate glasses find use in significant number of applications, *e.g.* window glasses, fibre optics, windshields and fibre glass.
- The melting or formation of a glass is called vitrification. In vitrification, a glass phase serves as a reaction medium to bond the unmelted particles together.

- A glass is a ceramic material that is made from inorganic materials at high temperatures and is distinguished from other ceramics in that its constituents are heated to fusion and then cooled to the rigid condition without crystallization. Most inorganic glasses are based on a network of ionically covalently bonded silica (SiO_2) tetrahedra. Additions of other oxides such as Na_2O and CaO modify the silica network to provide a more workable glass. Other additions to glasses create a spectrum to properties.
- The centre of the temperature range in which a non-crystalline solid changes from being glass-brittle to being viscous is called glass transition temperature.
- Glasses have special properties such as transparency, hardness at room temperature, and excellent resistance to most environments that make them important for several engineering designs.
- Ceramics in the form of fibres, thin films, coatings, and single crystals, find several different applications.

REVIEW QUESTIONS

1. What is a ceramic material? Mention some properties common to most materials.
2. What is the difference between traditional and engineering ceramic materials? Give examples of each.
3. Using Pauling equation (% ionic character = $1 - e^{(-1/4)(X^A - X^B)^2}$ (100%)], Compare the per cent covalent character of the following compounds: hafnium carbide, titanium carbide, tantalum carbide, silicon carbide and boron carbide.
4. Why are only two-thirds of the octahedral interstitial sites filled by Al^{3+} ions when the oxygen ions occupy the HCP lattice sites in Al_2O_3?
5. Describe the perovskite structure. What fraction of the octahedral interstitial sites are occupied by the tetravalent cation?
6. What is the antifluorite structure? What ionic components have this structure? What fraction of tetrahedral interstitial sites are occupied by cations?
7. Why are the layers of graphite able to slide past each other easily?
8. Mention the basic steps in the processing of ceramic products by the agglomeration for processing.
9. Describe the dry-processing method for producing such ceramic products as technical ceramic compounds and structural refractories. Mention the advantages of dry-pressing ceramic materials.
10. Mention the advantages of hot-pressing ceramic materials.
11. Mention the advantages of slip-casting.
12. What types of ceramic products are produced by extrusion? Mention the advantages and limitations of this process.
13. Describe vitrification process. In what type of ceramics does vitrification take place?
14. Describe the approximate composition of kaolin clay.
15. What is Feldspar? Mention the role it have in traditional ceramics.
16. Mention the kinds of ions which cause an increase in the conductivity of electrical porcelain.
17. Mention three major applications for ceramic materials in the electrical-electronics industries.
18. Define dielectric loss angle and dielectric loss factor for a dielectric material: why is a high dielectric loss factor undesirable?
19. Write approximate composition of electrical porcelain. What is a major disadvantage of electrical porcelain as electrical insulative material?

20. Write approximate composition of steatite. Mention the desirable electrical properties of steatite as electrical insulative material.
21. Write composition of fosterite and explain why fosterite is an excellent insulator material?
22. Explain, why sintered alumina is widely used as a substance for electronic device applications.
23. What do you understand by a thermistor and NTC thermistory? What materials are used to make NTC thermistors?
24. Explain the mechanism for electrical conduction in Fe_3O_4.
25. What are the PZT piezoelectric materials? Explain, why are considered superior to $BaTiO_3$ piezoelectric materials?
26. Mention some of the properties of glasses that make them indispensable for several engineering designs.
27. What do you understand by tempered glass? How it is produced? Why tempered glass is considerably stronger in tension than annealed glass? Mention some applications of tempered glass.

PROBLEMS

1. Calculate the density of CsBr in grams per cubic centimeter. CsBr has the CsCl structure. Ionic radii are $Cs^+ = 0.165$ nm and $Br^- = 0.196$ nm. [B.Tech.]
2. Calculate the density of ZrO_2 in gms/cubic centimeter, which has the CaF_2 crystal structure. Ionic radii are $Zr^{4+} = 0.087$ nm and $O^{2-} = 0.132$ nm. [B.Tech.]
3. Calculate the ionic packing factor for $CaTiO_3$, which has the perovskite structure. Given ionic radii of $Sr^{2+} = 0.127$ nm, $Sn^{4+} = 0.074$ nm, and $O^{2-} = 0.132$ nm. You may assume $a = 2\left(r_{sn^{4+}} + r_{0^{2-}}\right)$. [B.Tech.]
4. Calculate the planar density in ions per square nanometer in the (111) and (110) planes of ThO_2, which has the fluorite structure. Ionic radii are $Th^{4+} - 0.110$ nm and $O^{2-} - 0.132$ nm. [B.Tech.]

SHORT ANSWER QUESTIONS

1. What do you understand by ceramic materials?

Ans. These are the inorganic, non-metallic materials that consist of metallic and non-metallic elements bonded together primarily by ionic and/or covalent bonds.

2. What do you understand by dry pressing?

Ans. The simultaneous uniaxial compaction and shaping of ceramic granular particles (and binder) in a die is called dry pressing.

3. What is isostatic pressing?

Ans. The simultaneous uniaxial compaction and shaping of a ceramic powder (and binder) by pressure applied uniformly in all directions is called isostatic pressing.

4. Explain vitrification.

Ans. Melting or formation of a glass is termed as vitrification. This process is used to produce a viscous liquid glass in a ceramic mixture upon firing. Upon cooling, the liquid phase solidifies and forms a vitreous or glassy matrix that bonds the unmelted particles of the ceramic material together.

5. What do you understand by slip-casting?

Ans. This is a process used for ceramic shape formation in which a suspension of ceramic particles and water are poured into porous mold and then some of the water from the cast material diffuses into the mold, leaving a solid shape in the mold.

6. What is a thermistor?

Ans. A thermistor is a ceramic semiconductor device that changes in resistivity as the temperature changes and is used to measure and control temperature.

7. What is a transducer?

Ans. Transducer is a device that is actuated by power from one source and transmits power in another form to a second system, *e.g.*, a transducer can convert input sound energy into an output electrical response.

8. What do you understand by refractory (ceramic) material?

Ans. This is a material that can withstand the action of a hot environment.

9. What is a glass?

Ans. A glass is a ceramic material that is made from other ceramics in that its constituents are heated to fusion and then cooled to the rigid condition without crystallisation.

10. What is glass transition temperature?

Ans. The centre of the temperature range in which a non-crystalline solid changes from being glass-brittle to being viscous is called glass transition temperature (t_g).

11. Mention glass reference points.

Ans. Working point, softening point, Annealing point and strain point.

12. What is a flat glass?

Ans. Flat glass that is produced by having a ribbon of molten glass cool to the glass brittle state while floating on the top of a flat bath of molten tin and under a reducing temperature.

13. What are thermally and chemically tempered glasses?

Ans. Glass that has been reheated to near its softening temperature and then rapidly cooled in air to introduce compressive stresses near its surface are called thermally tempered glass. Glass that has been given a chemical treatment to introduce large ions into its surface to cause compressive stresses at its surface is called chemically tempered glass.

14. What is injection molding?

Ans. This is a processing technique in which a thermoplastic mass (loaded with ceramic powder) is mixed in an extruder—like set up and then injected into a die to form complex parts. In the case of ceramics, the thermoplastics is burnt off. Also find wide use for thermoplastics.

15. What do you understand by cold isostatic pressing (CIP)?

Ans. CIP is a powder-shaping technique in which hydrostatic pressure is applied during compaction. This is used to achieve a higher green ceramic density or compaction of more complex shape.

16. What do you understand by hot isostatic pressing (HIP)?

Ans. HIP is a powder-processing technique in which large pieces of metals, alloys and ceramics can be produced using sintering under a hydrostatic pressure generated by a gas.

17. What is injection molding?

Ans. Injection molding is a processing technique in which a thermoplastic mass (loaded with ceramic powder) is mixed in an extruder-like set up and then injected into a die to form complex parts. In the case of ceramics, the thermoplastic is burnt off. Also, widely used for thermoplastics.

18. What do you understand by tape casting?

Ans. Tape casting is a process for making thin sheets of ceramics using a ceramic slurry consisting of binders, plasticizers, etc. The slurry is cast with the help of a blade onto a plastic substrate. The resultant green tape is then dried, cut, and machined and used to make electronic ceramic and other devices.

19. How glass-ceramics are formed?

Ans. Glass-ceramics are formed using controlled crystallisation of inorganic glasses. These materials find wide use for kitchenware and several other applications.

20. What are silicate glasses?

Ans. Generally, the silicate glasses contain additional oxides. While oxides such as silica behave as *glass formers,* an intermediate oxide (such as lead oxide or aluminium oxide) does not form a glass by itself but is incorporated into the network structure of the glass formers. A third of oxides, the modifiers, bread up the network structure and eventually cause the glass to diversify, or crystallize.

21. What is tempered glass?

Ans. This is produced by quenching the surface of plate glass with air, causing the surface layers to cool and contract. When the centre cools, its contraction is restrained by the already rigid surface, which is placed in compression. This type of glass is capable of withstanding much higher tensile stresses and impact blows than untampered glass.

22. Mention, where tempered glass is used?

Ans. Tempered glass is used in car and home windows, shelving for refrigerators, ovens, furniture, and many other applications where safety is important.

23. What is laminated glass?

Ans. Laminated glass consists of two annealed glass pieces with a polymer (such as polyvinyl butyral or PVB) in between, is used to make car windshields.

24. What are refractories?

Ans. Refractory materials are important components of equipment used in the production, refining, and handling of metals and glasses, for constructing heat-treating furnaces, and for other high-temperature processing equipment. However, the refractories must survive at high temperatures without being corroded or weakened by the surrounding environment. Typical refractories are composed of coarse oxide particles bonded by a fine refractory material. The finer material melts during firing, providing bonding. In some cases, refractory bricks contain about 20% to 25% apparent porosity to provide improved thermal insulation.

25. What is the effect of temperature and composition on the viscosity of glass?

Ans. Figure 4.35 shows the effect of temperature and composition on the viscosity of glass.

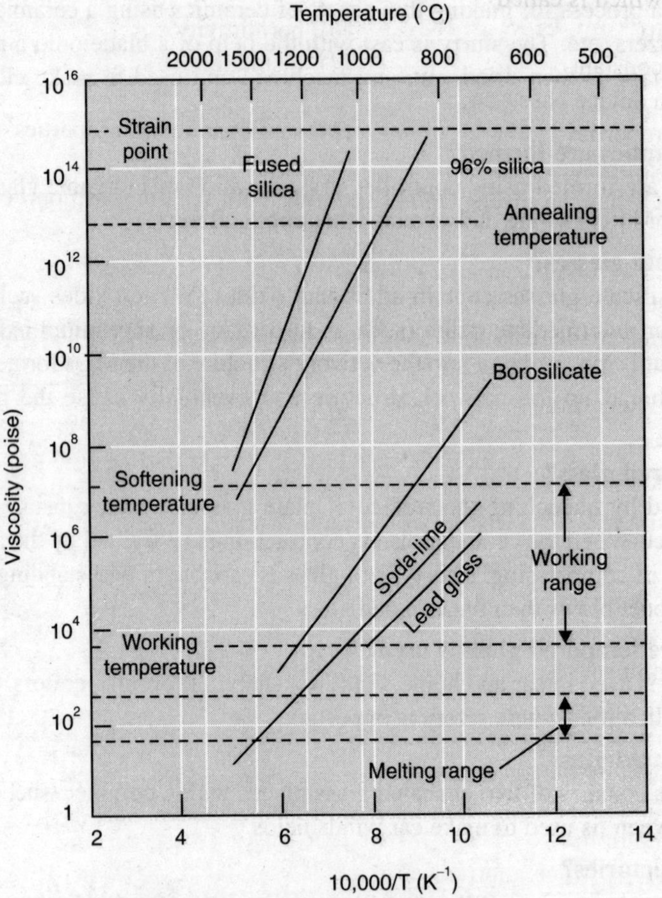

Fig. 4.43 The effect of temperature and composition on the viscosity of glass.

OBJECTIVE QUESTIONS

1. The radius ratio for CsCl is 0.94. This has
   (*a*) cubic coordination (CN = 8) (*b*) octahedral coordination (CN = 6)
   (*c*) cubic as well as octahedral coordination
   (*d*) none of the above is correct
2. $MgAl_2O_4$ has
   (*a*) spinel structure (*b*) corundum structure
   (*c*) perovskite structure (*d*) none of these
3. Above 1710°C silica is
   (*a*) liquid (*b*) gas
   (*c*) solid (*d*) plasma
4. The process by which small particles of a material are bonded together by solid-state diffusion is called
   (*a*) fusion (*b*) vitrification
   (*c*) sintering (*d*) extrusion

5. When a dielectric fills the space between the plates, the capacitance of the capacitor is increased by a factor K, which is called
   (*a*) permeability (*b*) permittivity
   (*c*) dielectric constant (*d*) insulation
6. Thermistor is a device based on
   (*a*) metallic properties (*b*) semiconducting properties
   (*c*) insulating properties (*d*) none of these
7. Magnetite Fe_3O_4 has the inverse spinel structure with the composition $FeO \cdot Fe_2O_3$ that can be written as
   (*a*) $Fe^{2+} (Fe^{3+}, Fe^{3+}) O_4$ (*b*) $Fe_2O \cdot FeO_3$
   (*c*) $FeO_2 \cdot Fe_2O_2$ (*d*) $Fe \cdot Fe \cdot O_2 FeO_2$
8. Ceramic materials made from solid solutions of lead zirconate ($PbZrO_3$) and lead titanate ($PbTiO_3$) is called
   (*a*) ferroelectric ceramics (*b*) PZT ceramics
   (*c*) steatite (*d*) fosterite
9. The fracture-toughness relation is
   (*a*) $K_{IC} = Y\sigma_f \pi a$ (*b*) $K_{IC} = Y\sigma_f \sqrt{\pi a}$
   (*c*) $K_{IC} = Y/\sigma_f \sqrt{\pi a}$ (*d*) $K_{IC} = Y\sigma_f/\sqrt{\pi a}$
   (The symbols have the usual meaning)
10. Because of high heat resistance, ceramic materials are used as
    (*a*) abrasive materials (*b*) refractories
    (*c*) semiconducting materials (*d*) None of these
11. Oxides that break up the glass network are known as
    (*a*) network modifiers (*b*) network insulators
    (*c*) network multiplier (*d*) network broker
12. The temperature at which a ferroelectric material on cooling undergoes a crystal structure change that produces spontaneous polarization in the material is called
    (*a*) neel temperature (*b*) curie temperature
    (*c*) weiss temperature (*d*) None of these

ANSWERS

1. (*a*)	2. (*a*)	3. (*a*)	4. (*c*)	5. (*c*)	6. (*b*)
7. (*a*)	8. (*b*)	9. (*b*)	10. (*b*)	11. (*a*)	12. (*b*)

CHAPTER

5

Polymers

5.1 INTRODUCTION

The word 'polymer' literally means "many parts", *i.e.* the basic make up of polymers consists of smaller units (mers) joined together either naturally or synthetically to produce polymers or macromolecules. A polymeric solid material may be considered to be one that contains many chemically bounded parts or units that themselves are bonded together to form a solid. Polymers are materials consisting of giant or macromolecules, chain-like molecules having average molecular weights from 10,000 to more than 1,000,000 g/mole built by joining many mers or units through chemical bonding. Molecular weight is defined as the sum of atomic masses in each molecule. Nature has given us a number of polymers like proteins, carbohydrates, silk, wool, cotton, rubber, leather, etc. Man has made a variety of polymers with wide ranging properties such as softness like silk and wool and strong like steel. Most polymers, solids or liquids, are carbon-based organic; however, they can be inorganic (*e.g.* silicones based on Si–O network).

The first semi-synthetic (modified) polymer was cellur nitrate made available in market in 1860. In 1910, purely bakelite was introduced. Lot of developments in the science and technology of polymers have taken place since 1920. The plastic industry has grown very fast and surpassed the growth of almost all other industries. Plastics are materials that are composed principally of naturally occurring and modified or artificially made polymers usually containing additives such as fibres, fillers, pigments, and the like that further enhance their properties. Plastics include thermoplastics, thermoset materials, and elastometers (natural or synthetic). Plastics are slowly replacing almost all our conventional materials. Scientists have made replacement for natural materials. Currently, research is being carried out on *conducting polymers, i.e.* synthetic metals. Although, plastics can be recycled but most of them are non-biodegradable and hence plastic waste poses a major threat to the environment.

5.2 BASIC DEFINITIONS

(*i*) **Polymer:** Polymers are materials made from giant (or macromolecular), chain-like molecules having average molecular weights from 10,000 to more than 1,000,000 g/mol built by the joining of many mers or units by chemical bonds, *i.e.* a polymer is a large molecule built by the repetition of small, simple chemical unit called monomer. The repeating chemical units are covalently linked to each other in a macromolecule. Polymers are usually, but not always, carbon based. The properties of a polymer are entirely different from those of the chemicals (or monomers) from which it is formed.

(*ii*) **Mer:** A unit group of atoms and molecules that defines a characteristic arrangement for a polymer.

(*iii*) **Monomer:** Monomer is defined as a simple molecule with two or more binding sites through which it forms covalent linkages with other monomer molecules to form the macromolecule. Obviously, monomers are building blocks of polymers. Interestingly, all simple molecules cannot behave as monomers, *e.g.* molecules like ammonia, water, ethanol, etc. are not monomers. Only molecules with two or more bonding sites can act as monomers, *e.g.* alkenes, vinyl chloride, adipic acid, glycol with two bonding sites act as monomers.

The repeat unit of polymer is usually equivalent or nearly equivalent to the *monomer,* or the starting material from which the polymer is formed (Table 5.1).

Table 5.1 Some Linear High Polymers, their Monomers and their Repeat Units

Polymer	*Monomer*	*Repeat units*
Polyethylene	$CH_2{=}CH_2$	$-CH_2CH_2-$
Poly (vinyl chloride)	$CH_2{=}CHCl$	$-CH_2CHCl$
Polyisobutylene	CH_3 $CH_2{=}C$ CH_3	CH_3 CH_2-C- CH_3
Polystyrene[1]	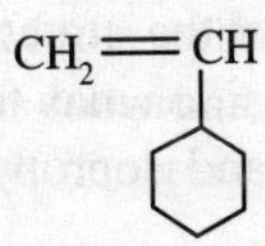	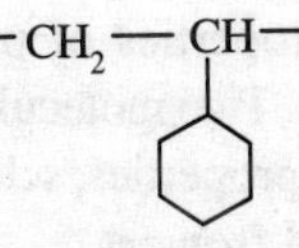
Poly caprolactan (6-nylon)	$H-N(CH_2)_5C-OH$ $H \quad O$	$-N(CH_2)_5C-$ $H \quad O$
Polyisoprene (natural rubber)	$CH_2{=}CH-C{=}CH_2$ CH_3	$CH_2CH{=}C-CH_2-$ CH_3

1.  represents benezene ring, double bonds being omitted.

(*iv*) **Polymerisation:** This is defined as the chemical reaction in which a monomer is converted to the polymer under specific conditions. However, monomer alone cannot undergo polymerisation, but requires the presence of chemical called *initiator.*

$$\text{Monomer} + \xrightarrow[\text{Pressure}]{\text{Initiator required}} \text{Polymer}$$

$$n\left[\mathrm{CH_2{=}\overset{\displaystyle Cl}{\overset{|}{C}}H}\right] \xrightarrow[\text{Initiator}]{\text{Peroxide}} \underset{\substack{\text{PVC}\\ \text{(Polymer)}}}{\left(\mathrm{-CH_2-\overset{\displaystyle Cl}{\overset{|}{C}}H}\right)_n}$$

(*v*) **Polymerisation Process:** Polymerisation, or the joining of large unit molecules called monomers, utilizes the valence of the partially filled outer shell of the carbon atom (carbon has a valence of 4) to join smaller units together to form larger chains of molecules. Oxygen, sulphur, silicon, or nitrogen can be used to replace the carbon atom.

Two conditions must be met for polymerisation to occur: (*i*) A molecule must have at least two locations that have unsatisfied bonds, which will easily join with other molecules. This requirement means starting with a molecule that has a double bond, such as carbon. Because the carbon molecule has a double bond, each bond is a pair of shared electrons. If one of the bonds between the carbon atoms opens up, a single bond exists, leaving the other two electrons to share with other atoms. If another carbon atom passes by that has opened up its double bond, the two can join to form a chain. This procedure continues, producing a polymer chain, and is called polymerisation. The process continues as long as the second condition is met. (*ii*) The second condition necessary for polymerisation is that after each reaction, at least two open locations must remain. Many chains form within polymers. These chains form threads that entwine themselves around each other for strength.

Polymers can be strengthened by *cross-linking.* Cross linking occurs when the double bonds between atoms within a chain are broken and these atoms or molecules form, or link up, with neighbouring atoms. This linkage provides additional strength to the chain and reduces the *slippage* that occurs between molecules. Slippage occurs when the polymer threads slide past each other when subjected to a load.

The properties of polymers also depend on the structure as well as the composition of the molecule. Two molecules with the same composition may form two different-configurations having different properties, sch as propyl (1-propanol) and isopropyl (2-propanol) alcohol. These variations are called *isomers.*

Polymerisation takes place through *addition polymerisation, copolymerisation,* or *condensation polymerisation.* In the polymerisation process, a large unit molecule, the *monomer,* is added to another monomer to forms a large chain, the polymer (referring to many parts), which has a number of repeated units, mers. Mers are the smallest units recongnizable in the chain. The degree of polymerisation is the number of repeating units that have identical structures within the chain formed by the polymer. Addition polymerisation involves only one type of mer. Figure 5.1 shows polymerisation by addition.

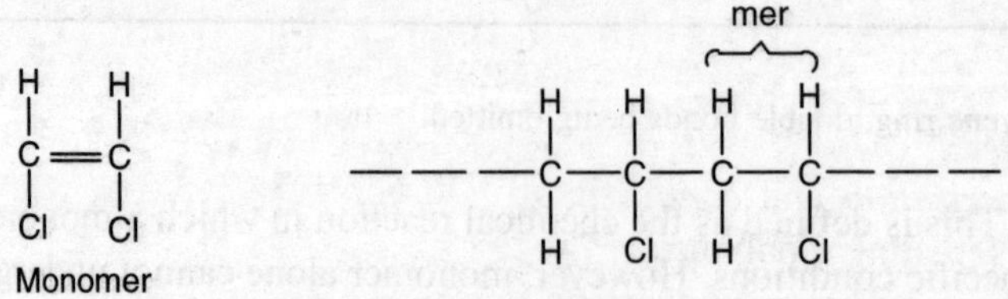

Fig. 5.1 Addition polymerisation.

In copolymerisation, more than one molecule makes up the mer. Acrylonitride-butadiene-styrene (ABS) is an example of a copolymer. Figure 5.2 shows the copolymerisation process for ABS polymers.

Condensation polymerisation involves the chemical reaction of two or more chemicals to form a new molecule. This chemical reaction produces a condensate or nonpolymerisable by product,

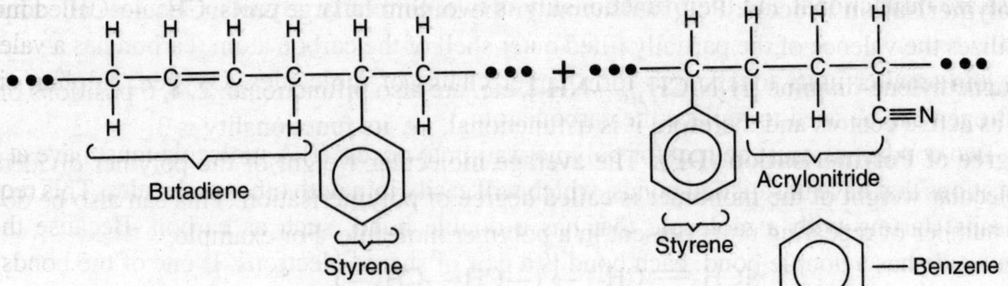

Fig. 5.2 Copolymerisation.

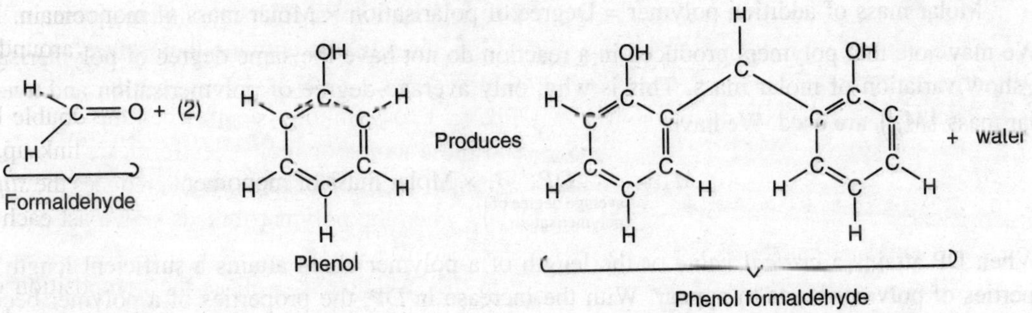

Fig. 5.3 Condensation polymerisation.

usually water. A catalyst is often required to start and maintain the reaction. It can also be used to control the reaction rate. The process of condensation polymerisation is shown in Fig. 5.3.

With some exceptions, polymers made in chain reactions often contain only carbon atoms in the main chain (*homochain polymers*), whereas polymers made in step reactions may have other atoms, originating in the monomer functional groups, as part of the chain (heterochain polymers). For example

	Monomer	**Polymer**
1.	Vinyl chloride	Polyvinyl chloride
2.	Vinyl acetate	Polyvinyl acetate
3.	Ethylene	Polyethylene
4.	Styrene	Polystyrene

Addition polymerisation and *condensation polymerisation* are the two main ways to conduct polymerisation (creating a polymer).

(*vi*) **Elastomers:** These are polymers thermoplastics or lightly crossed linked thermosets) that have an elastic deformation > 200%.

(*vii*) **Functionality:** This is defined as the number of bonding sites present in the molecule of the monomer. For example, all double bonded compounds $CH_2 = CH_2$ (ethane), $CH_3—CH = CH_2$ (propene), $CH_2 = CHCl$ (vinyl chloride) have two bonding sites, the double bond. Obviously, these are bifunctional and their functionality is two. Similarly, glycols $\begin{array}{l}CH_2—OH\\ |\\ CH_2—OH\end{array}$, adipic acid, hexamethylene-diamine $[H_2N(CH_2)_6—NH_2]$, etc. are also bifunctional. 2, 4, 6 positions of phenol act as active centres and therefore it is trifunctional, *i.e.* its functionality is 3.

(*viii*) **Degree of Polymerisation (DP):** The average molecular weight of the polymer divided by the molecular weight of the monomer is called degree of polymerisation. This can also be defined as the number of repeating units present in a polymer molecule. For example,

$$n[CH_2 = CH_2] \rightarrow (—CH_2—CH_2—)_n$$

where n is a whole number and it is called the degree of polymerisation. In case of addition polymerisation, the molar mass of the polymer can be calculated with the help of following formula:

Molar mass of addition polymer = Degree of polarisation × Molar mass of monomer

We may note that polymers produced in a reaction do not have the same degree of polymerisation and show variation of molar mass. This is why, only average degree of polymerisation and average molar mass $(\bar{M}_n)$ are used. We have

$$\bar{M}_n = \underset{\left(\begin{array}{c}\text{average degree of}\\ \text{polymerisation}\end{array}\right)}{\overline{DP}} \times \text{Molar mass of monomer}$$

When DP attains a *critical value* or the length of a polymer chain attains a sufficient length, the properties of polymer starts to appear. With the increase in DP, the properties of a polymer become more and more prominents, *i.e.* polymer size increases. Although, the molecular weight for most of the polymers ranges between 8,000 to 150,000 but higher molecular weights in the range of 10^8 are also in existence. For example, C_2H_4 (ethylene) is a gas, its oligomers having DP in the range 3 to 20 are

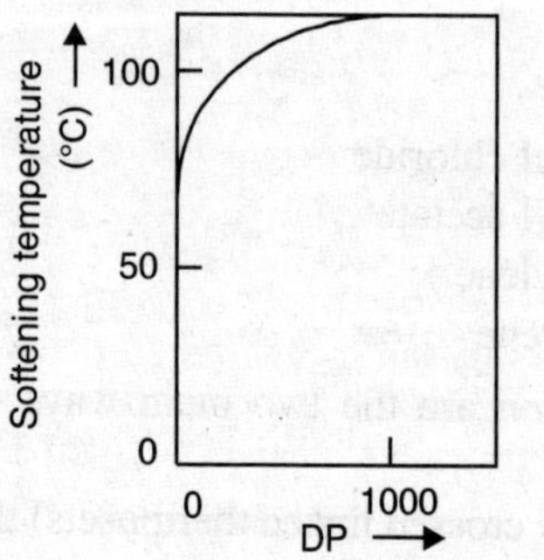

Fig. 5.4

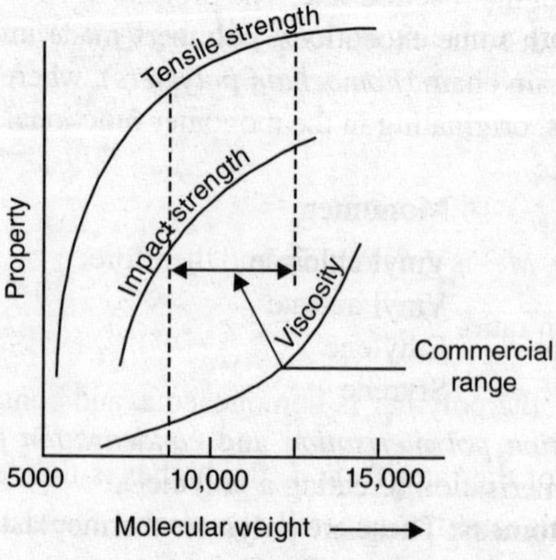

Fig. 5.5

liquids. With the increase in DP their viscosity goes on increasing. The polymers of ethylene (C_2H_4) with DP in the range 25 to 40 are like gases and with DP in the range 50 to 60, they are like waxes. When DP increases beyond 350, ethylene polymers become resins. The effect of DP on softening temperature is shown in Fig. 5.4. From figure it is obvious that with the increase in DP, the softening temperature also increases, but when DP increases beyond 1000, the rise in softening temperature is small.

The tensile strength increases with the molecular weight or DP. The viscosity also increases tensile strength rapidly in the initial stages but viscosity increases when molecular weight is large. Impact strength also increases with the increase in molecular weight (Fig. 5.5). Polymers with higher molecular weights possess superior mechanical properties but the fabrication becomes difficult due to higher viscosities. One has to make a compromise between properties and processability for commercial range.

(*ix*) **Molar Masses of Polymers:** Usually, polymers contain a mixture of chains of different lengths (or different degree of polymerisation). This means, it is not possible to assign a definite molar mass to a polymer. It is quite usual to adopt statistical methods in assigning average molar masses of polymers. Two important averaging are adopted: (*i*) number average molar mass, and (*ii*) weight average molar mass.

Number average molar: $\bar{M}_n$ is the ratio of the sum of molar masses of individual molecules to the total number of molecules in the mixture.

Thus, if in a sample n_1 molecules have molar mass M_1, n_2 molecules with molar mass M_2, etc. till n_i with M_i then,

$$\bar{M}_n = \frac{n_1M_1 + n_2M_2 + \ldots + n_iM_i}{n_1 + n_2 + \ldots + n_i} = \frac{\Sigma n_iM_i}{\Sigma n_i} \quad \ldots(1)$$

Weight average molar mass (M_w): In a mixture, if w is the mass of polymers with molar mass M_1, W_2 with molar mass M_2.... upto W_i with molar mass M_i then,

$$\bar{M}_w = \frac{w_1M_1 + w_2M_2 + \ldots + w_iM_i}{w_1 + w_2 + \ldots + w_i} = \frac{\Sigma w_iM_i}{\Sigma w_i} \quad \ldots(2)$$

But number of moles $n = \dfrac{w}{M}$ and $w = nM$

Therefore replacing w_1 by n_1, M_1, w_2 by n_2 M_2... w_i *by* $n_i M_i$ *in Eqn.* (2), we have

$$\bar{M}_w = \frac{n_1M_1^2 + n_2M_2^2 + \ldots + n_iM_i^2}{n_1M_1 + n_2M_2 + \ldots + n_iM_i} = \frac{\Sigma n_iM_i}{\Sigma n_i}$$

Usually $\bar{M}_w$ is greater than and therefore $\dfrac{M_w}{M_n}$ is greater than or equal to one. This ratio is called the distribution ratio.

If $\dfrac{\bar{M}_w}{\bar{M}_n}$ the polymer is homogeneous and contains polymers of same chain length. The higher the deviation of the ratio from 1, the higher is the degree of heterogeneity of the polymer. However, in practice this ratio lies generally between 1.5 and 3 but in some cases it may exceed 30.

(*x*) **Molecular Weight and its Distribution:** In both chain and stepwise polymerisation, the length of a chain is determined by purely random events. In step reactions, the chain length is determined by the local availability of reactive groups at the ends of the growing chains. In radical polymerisation, chain length is determined by the time during which the chain grows before it diffuses into the vicinity of a second free radical and the two react. In either case, the polymeric product contains molecules having many different chain lengths.

Extremely, large molecular weights (sometimes 'molecular mass' 'molar mass' and 'relative molar mass' are also used) are reported in polymers with very long chains. During the polymerisation process in which these large macromolecules are synthesized from smaller molecules, not all polymer chains will grow to the same length; this results in a distribution of chain lengths or molecular weights. Normally, an average molecular weight is specified, which may be determined by the measurement of various physical properties, *e.g.* viscosity and osmotic pressure.

Average molecular weight can be defined through several ways. One can obtain the *number average molecular weight* $\overline{M}_n$ by dividing the chains into a series of size ranges and thus determining the number of fraction of chains within each size range. One can express this number average molecular weight as

$$\overline{M}_n = \sum_i x_i M_i \qquad ...(3)$$

Here M_i represents the mean (middle) molecular weight of size range i, and x_i is the fraction of the total number of chains within the corresponding size range.

Another average, a *weight average* molecular weight $\overline{M}_w$ is based on the weight fraction of molecules within the various size ranges is defined as

$$\overline{M}_w = \sum_i w_i M_i \qquad ...(4)$$

Here w_i is the weight fraction concentration of the molecules having molecular weight M_i. A typical molecular weight distribution along with these molecular weight averages for a polymer is shown in Fig. 5.6.

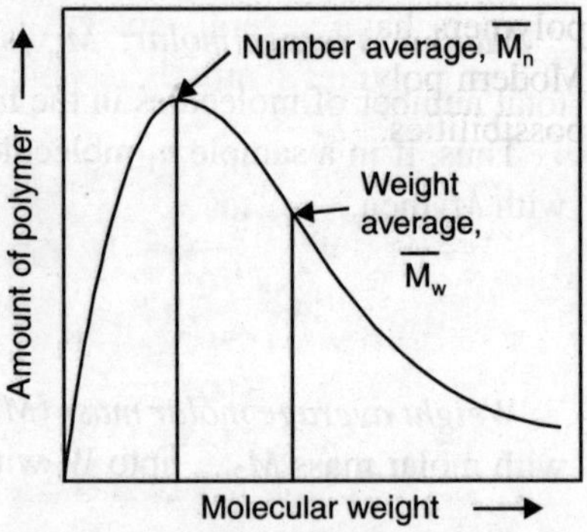

Fig. 5.6 Distribution of molecular weights in a typical polymer.

For typical polymers the number average lies near the peak of the weight-distribution curve or the most probable molecular weight. Both the averages represented by relations (3) and (4) are used in the interpretation of the properties of polymers. The ration $\overline{M}_w / \overline{M}_n$ indicates the degree of distribution of the molecular weights. When all the molecules are identical, the said ratio is unity. However, in practice this ratio lies generally between 1.5 and 3 but in some cases it may exceed 30.

Another way of expressing average chain size of a polymer is the *degree of polymerisation n*, which represents the average number of mer units in a chain. Both number-average (n_n) and weight-average (n_w) degrees are possible. These can be expressed as

$$n_n = \frac{\overline{M}_n}{\overline{m}} \qquad ...(5)$$

and

$$n_w = \frac{\overline{M}_w}{\overline{m}} \qquad ...(6)$$

Here $\overline{M}_n$ and $\overline{M}_w$ represents the number-average and weight-average molecular weights as defined above, while $\overline{m}$ is the mer molecular weight. One can determine $\overline{m}$ for a copolymer (having two or more different mer units) from

$$\bar{m} = \sum f_j m_j \quad \text{...(7)}$$

where f_j and m_j are the chain fraction and molecular weight of mer j respectively.

(*xi*) *Molecular Structure*: Polymers consist of large number of molecular chains, each of which may bend, coil and kink (Fig. 5.7*a*). This leads to extensive interwining and entanglement of neighbouring chain molecules. These random coils and molecular entanglements are responsible for a number of important characteristics of polymers, to include the large elastic extensions displayed by the rubber materials.

There are some mechanical and thermal characteristics of polymers which are a function of the ability of chain segments to experience rotation in response to applied stresses or thermal vibrations. We may note that rotational flexibility is dependent on mer structure and chemistry, *e.g.*, the region of a chain segment that has a double bond (C=C) is rotationally rigid. Moreover, introduction of a bulky or large side group of atoms restricts rotational movement, *e.g.*, polystyrene molecules, which have a phenyl side group are more resistant to rotational motion than are polyethylene chains.

The physical characteristics of a polymer is found to depend not only its molecular weight and shape but also on differences in structures of molecular chains. In addition to various isomeric configurations, polymers have several molecular structures including linear, branched, crosslinked, and network. Modern polymer synthesis techniques permit one to have considerable control over various structural possibilities.

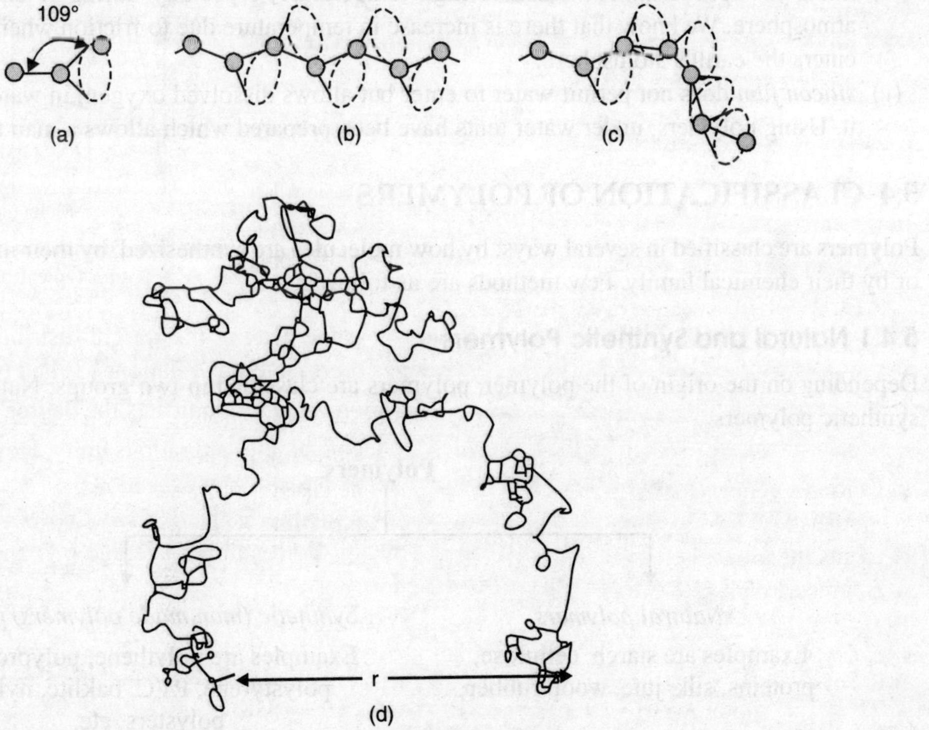

Fig. 5.7 How polymer chain shape is influenced by the positioning of backbone carbon atom (circle). (*a*) the rightmost atom may lie anywhere on the dashed circle and still subtend a 109° angle with the bond between the other two atoms (*b*) and (*c*) straight and twisted chain segments generation (*d*) single polymer chain molecule that has numerous random kinks and coils produced by chain body rotations.

5.3 GENERAL CHARACTERISTICS OF POLYMERS

General characteristics of polymers can be summarized as follows:

(*i*) Polymers have long chain structures. The individual molecule of a polymer is very large, *i.e.,* it may consists of thousands of similar small molecules, all bonded together covalently.

(*ii*) All polymers have one thing common, *i.e.,* carbon, which further bonds with hydrogen, nitrogen, halogens or other organic or inorganic substances.

(*iii*) Although, polymer's structure may be crystalline in simple materials but generally they are non-crystalline solids at room temperatures. No doubt, polymers pass through a viscous stage during formation.

(*iv*) Polymers have light weight and they can be easily fabricated and shaped. They are poor conductors of electricity and their thermal conductivity is also low. Moreover, the polymers are resistant to chemical attack and decay.

Polymers are widely used in different fields. Polymers are used in indefinite number of forms, *e.g.* as telephone sets, paints, radio and television cabinets, coatings, adhesive and other countless objects. Few notable developments in polymer materials are:

(*i*) polymer materials far strong than steel according to weight to weight ratio.

(*ii*) polymer materials suitable for repairing damaged kidney or heart.

(*iii*) materials act as *adhesive* so strong that the use of nails in wood and leather can be avoided.

(*iv*) *lining for spacecraft* to withstand high temperature, especially during re-entering into earth's atmosphere. We know that there is increase in temperature due to friction when the spacecraft re-enters the earth's atomsphere.

(*v*) *silicon film* does not permit water to enter but allows dissolved oxygen, in water, to pass through it. Using polymers, under water tents have been prepared which allows a man to live and work.

5.4 CLASSIFICATION OF POLYMERS

Polymers are classified in several ways: by how molecules are synthesized, by their molecular structure, or by their chemical family. Few methods are as follows:

5.4.1 Natural and Synthetic Polymers

Depending on the origin of the polymer, polymers are classified in two groups: Natural polymers and synthetic polymers.

Polymers

- *Natural polymers*
  Examples are starch, cellulose, proteins, silk, jute, wool, rubber
- *Synthetic (man made polymers) polymers*
  Examples are polythene, polypropene, polystyrene, PVC, baklite, nylon, polysters, etc.

There are some semi-synthetic polymers which are modified natural polymers like cellulose acetate, cellulose nitrate, etc.

5.4.2 Addition and Condensation Polymers

Based on the nature of chemical reaction involved in the formation of the polymers, polymers are classified into (*a*) addition polymers and (*b*) condensation polymers.

Polymers

Addition polymers

They are formed by the addition of olifinic monomers without the elimination of bye-products. Molar mass of the homo polymer will be the product of the degree of poly-merization and the molar mass of the monomer.

Examples: Polythene, polypropylene, Plexiglass PVC, etc.

Condensation polymers

They are formed by the intermolecular condensation of two different monomers with the elimination of byeproducts like Ammonia water HCl etc. Molar mass of the polymer will be the product of the degree of polymerization and the molar mass of the repeating unit.

Examples: Bakelite, nylon polyester

(*c*) According to structure and end use, polymers are classified into following four types:

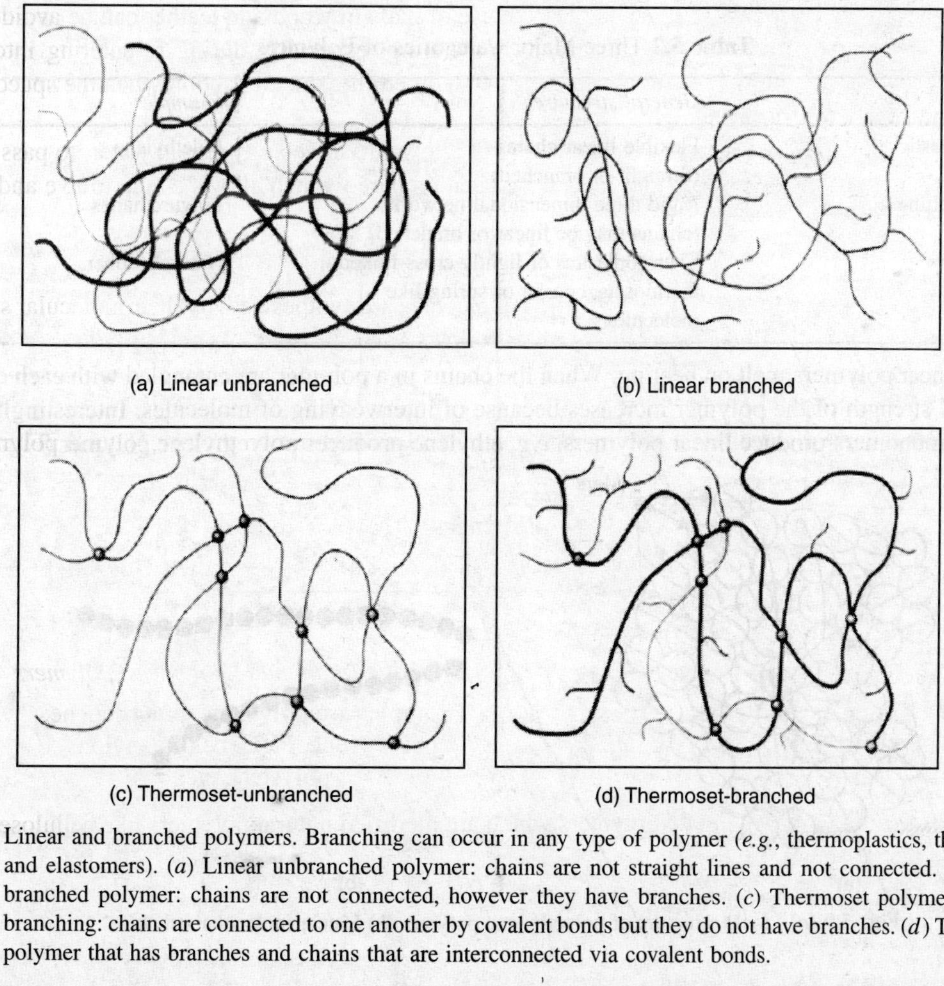

Fig. 5.8 Linear and branched polymers. Branching can occur in any type of polymer (*e.g.*, thermoplastics, thermosets and elastomers). (*a*) Linear unbranched polymer: chains are not straight lines and not connected. (*b*) linear branched polymer: chains are not connected, however they have branches. (*c*) Thermoset polymer without branching: chains are connected to one another by covalent bonds but they do not have branches. (*d*) Thermoset polymer that has branches and chains that are interconnected via covalent bonds.

(*i*) ***Elastomers:*** When stretched, these polymers undergo large elongation and retain their shape on releasing the force. The polymer chain is long and coiled and in between chains only weak forces of attraction exists. Few examples of elastomers are: natural rubber, Buna-s rubber, etc.

(*ii*) ***Fibres:*** These are thin thread like polymers. These polymers do not undergo deformation. These contain long chain molecular strains held together through hydrogen bonding. Few examples of fibres are: Silk, wool, nylon, cotton, polysters, terylene, etc.

(*iii*) ***Resins:*** These are polymers of low molar mass in liquid or semisolid form. These polymers may undergo further polymerisation to give plastics (high molar mass). Examples of resins are: Epoxy resins, phenol-formaldehyde resin, urea formaldehyde resin, etc.

(*iv*) ***Plastics:*** These are polymers which can be moulded into desired articles by application of pressure and temperature. Few examples of plastic polymers are: PVC, plexiglass.

(*d*) ***Linear or Branched Polymers:*** One can also classify polymers is to state if the polymer is a *linear polymer* or a *branched polymer* (Fig. 5.8).

Linear Polymer: This consists of spaghetti—like molecular chains. In a branched polymer, there are primary polymer chains and secondary offshoots of smaller chains that stem from these main chains. Although, we say that chains are "linear", the chains are actually not in the form of straight lines. A better way to describe polymers is in terms of their mechanical and thermal behaviours. Three major categories of polymers are compared in Table 5.2.

Table 5.2 Three Major Categories of Polymers

Category	*General structure*	*Example*
Thermoplastic	Flexible linear chains (Straight or branched)	Polethylene
Thermosetting	Rigid three dimensional network (chains may be linear or branched)	Polyurethanes
Elastomers	Thermoplastics or lightly cross-linked thermosets, consist of spring like molecules	Natural rubber

These linear polymers melt on heating. When the chains in a polymer are entangled with each other, mechanical strength of the polymer increases because of interweaving of molecules. Interestingly, bi-functional monomers produce linear polymers, *e.g.* ethylene produces polyethylene polymer (Fig. 5.9).

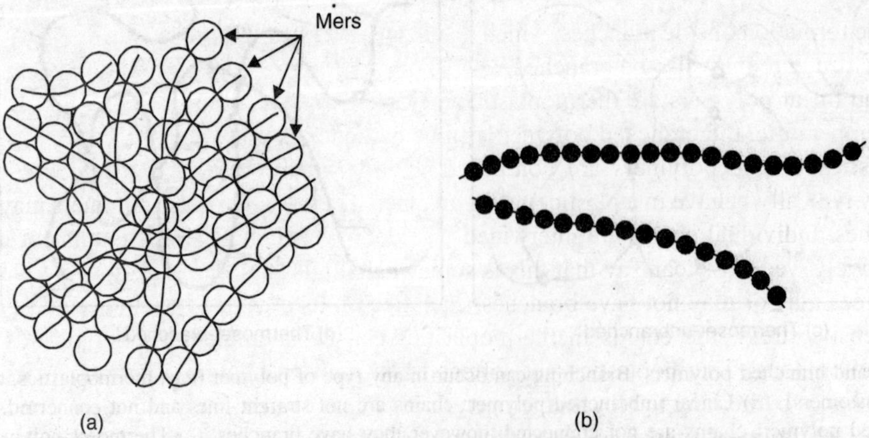

Fig. 5.9 (*a*) Linear structure of polyethylene polymer (*b*) Linear molecular structure.

The polymer fibres used in textile industry are the linear polymers. These polymers have high molecular weight and fixed orientation of the molecules parallel to fibre axis.

Branched Polymers

In contrast to the linear chain molecules, some polymers have branched chains, often as a result of side reactions during polymerisation (Fig. 5.10*a*). Branched polymers may be synthesized in which side branch chains are connected to the main ones as illustrated in Fig. 5.11. The chain packing efficiency is

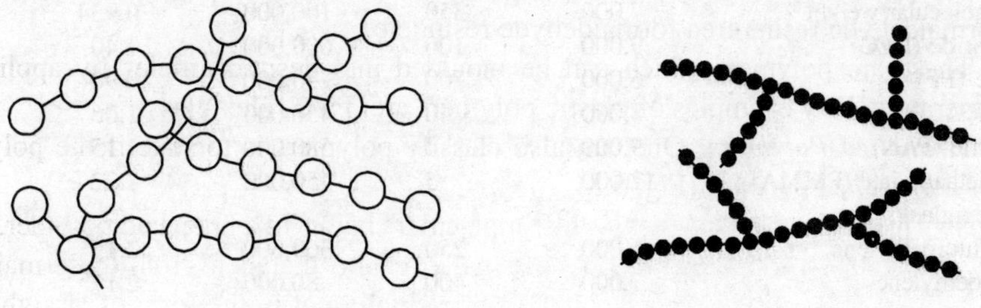

(a) (b)

Fig. 5.10 (*a*) Branching (*b*) Alternative representation.

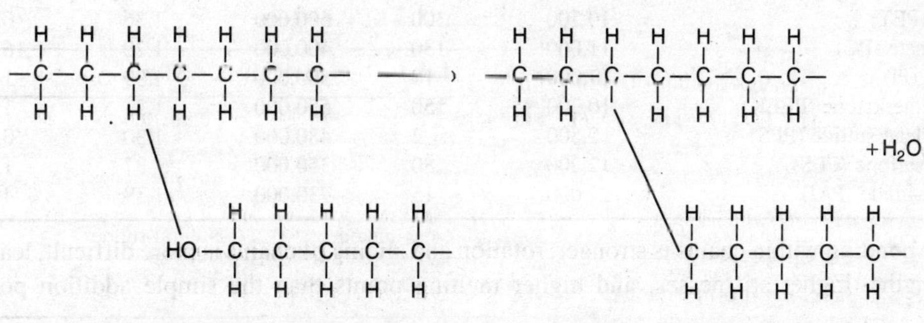

Polyethylene chains Branch linking of chain Polyethylene

Fig. 5.11 Illustration of branching mechanism. This is achieved by removing a side atom from the main chain and attaching another C—C bond.

reduced with the formation of side branches, which results in lowering of polymer density. The polymers that form linear structure may also be branched.

Branched and linear polymers are thermoplastic and have gained a great industrial importance. By changing the temperature, the branched polymers can be hardened or softened.

Thermoplastics: These polymers are composed of long chains produced by joining together monomers; they typically behave in a plastic, ductile manner. The chains in these polymers may or may not have branches. Individual chains are interwined. van der Waals bonds between atoms of different chains are relatively weak. We can say that this is somewhat similar to a few trees that are tangled up together. The trees may or may not have branches, each tree on its own and not connected to another. By applying tensile stress, the chains in thermoplastics can be untangled. Thermoplastics can be crystalline or amorphous. Thermoplastics get softened and melt on heating. Thermoplastics can be processed into shapes by heating to elevated temperatures. These can be easily recycled. These can be said to be a special group of polymers having the processing ease of thermoplastics and the elastic behaviour of elastomers. Some of the mechanical properties of thermoplastics are given in Table 5.3.

Table 5.3 Properties of Selected Thermoplastics

	Tensile strength (psi)	*% Elongation*	*Elastic modulus (psi)*	*Density (g/cm³)*	*Izod impact (ft lb/in.)*
Polyethylene (PE):					
Low-density	3,000	800	40,000	0.92	9.0
High-density	5,500	130	180,000	0.96	4.0
Ultrahigh molecular weight	7,000	350	100,000	0.934	30.0
Polyvinyl chloride (PVC)	9,000	100	600,000	1.40	
Polypropylene (PP)	6,000	700	220,000	0.90	1.0
Polystyrene (PS)	8,000	60	450,000	1.06	0.4
Polyacrylonitrile (PAN)	9,000	4	580,000	1.15	4.8
Polymethyl methacrylate (PMMA) (acrylic, Plexiglas)	12,000	5	450,000	1.22	0.5
Polychlorotrifluoroethylene	6,000	250	300,000	2.15	2.6
Polytetrafluoroethylene (PTFE, Teflon)	7,000	400	80,000	2.17	3.0
Polyoxymethylene (POM) (acetal)	12,000	75	520,000	1.42	2.3
Polyamide (PA) (nylon)	12,000	300	500,000	1.14	2.1
Polyester (PET)	10,500	300	600,000	1.36	0.6
Polycarbonate (PC)	11,000	130	400,000	1.20	16.0
Polyimide (PI)	17,000	10	300,000	1.39	1.5
Polyetheretherketone (PEEK)	10,200	150	550,000	1.31	1.6
Polyphenylene sulfide (PPS)	9,500	2	480,000	1.30	0.5
Polyether sulfone (PES)	12,200	80	350,000	1.37	1.6
Polyamide-imide (PAI)	27,000	15	730,000	1.39	4.0

Since the bonding within chains is stronger, rotation and sliding of chains is more difficult, leading to higher strengths, higher stiffnesses, and higher melting points than the simple addition polymers (Table 5.4).

There are some cases, where good impact properties can be gained from the complex chains with polycarbonates being particularly remarkable. Interestingly, polycarbonates (Lexan™, Merlon™ and Sparlux™) are found useful for making bulletproof windows, compact disks for data storage, etc.

There are large number of polymers, which typically are used for special applications and in relatively small quantities. These are formed from complex monomers, often by the condensation mechanism.

Table 5.4 Repeat Units and Applications for Selected Addition Thermoplastics

Polymer	*Repeat unit*	*Application*	*Polymer*	*Repeat unit*	*Application*
Polyethylene (PE)	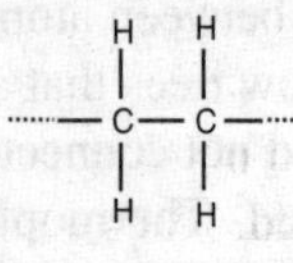	Packing films, wire insulation, squeeze bottles, tubing, household items	Polyacrylonitrile (PAN)	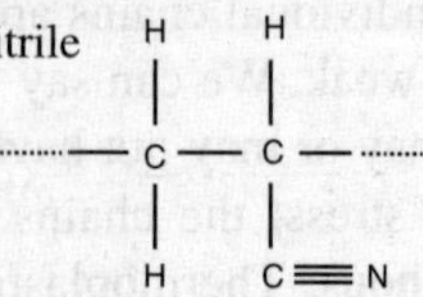	Textile fibres, precursor for carbon fibres, food container

Polymer	Repeat unit	Applications
Polyvinyl chloride (PVC)	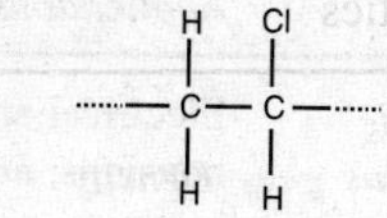	Pipe, valves, fittings, floor tile, wire insulation, vinyl automobile roofs
Polymethyl methacrylate (PMMA) (acrylic-plexiglas)	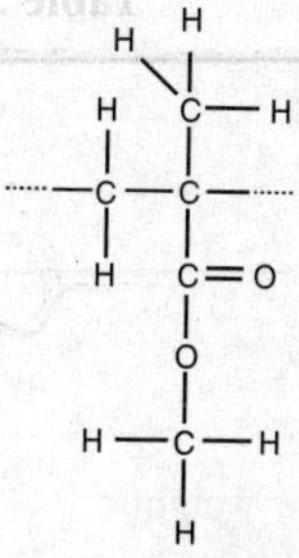	Windows, windshields coatings, hard contact lenses, lighted signs
Polypropylene (PP)	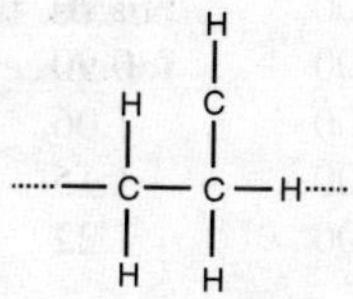	Tanks, carpet fibres, rope, packaging
Polystyrene (PS)	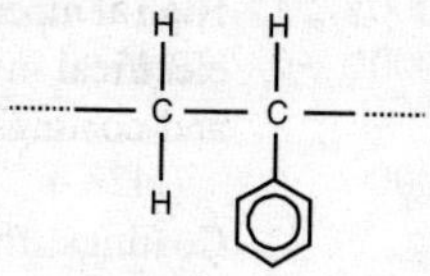	Packaging and insulation foams, lighting panels, appliance components, egg cartons
Polychlorotri-fluoroethylene	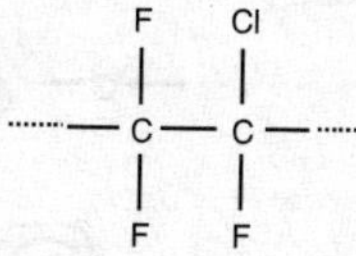	Valve components, gaskets, tubing electrical insulation
Polytetra-fluoroethylene (Teflon) (PTFE)	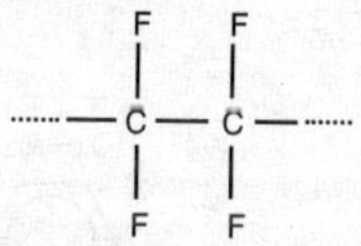	Seals, valves, nonstick coatings

Oxygen, nitrogen, sulphur, and benezene rings (or aromatic groups) may be incorporated into the chain. The repeat units and typical applications for these complex polymers are shown in Table 5.5.

Table 5.5 Repeat Units and Applications for Complex Thermoplastics

Polymer	*Repeat unit*	*Applications*
Polyoxymethylene (acetal) (POM)	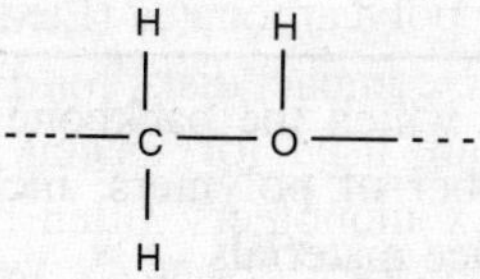	Plumbing fixtures, pens, bearings, gears, fan blades
Polyamide (nylon) (PA)	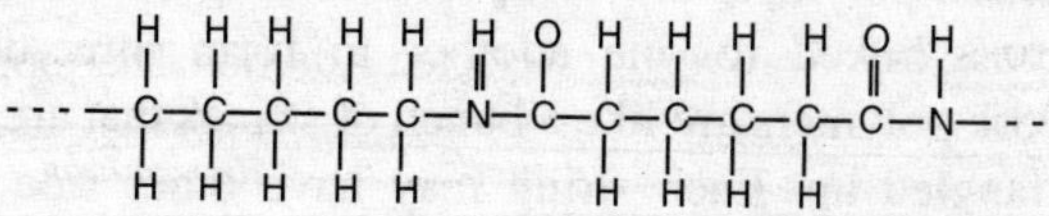	Bearings, gears, fibres, rope, automotive components, electrical components
Polyester (PET)	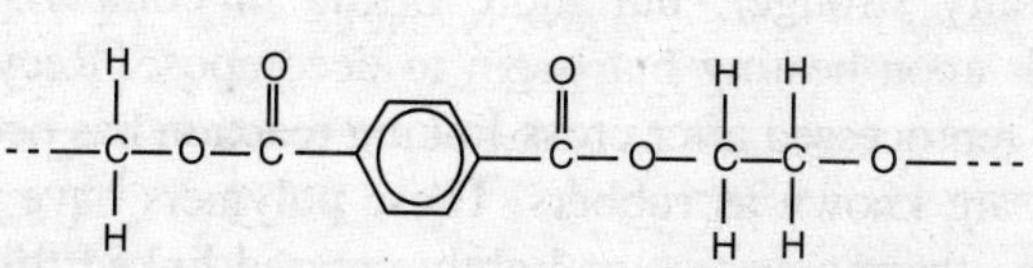	Fibres, photographic film, recording tape, boil-inbag containers beverage containers

(Contd.)

Polymer	*Repeat unit*	*Applications*
Polycarbonate (PC)	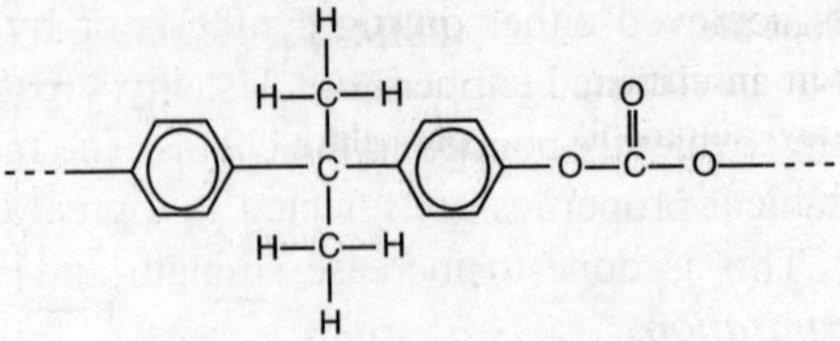	Electrical and appliance housings, automotive components, football helmets, returnable bottles, compact discs (CD's)
Polyimide (PI)	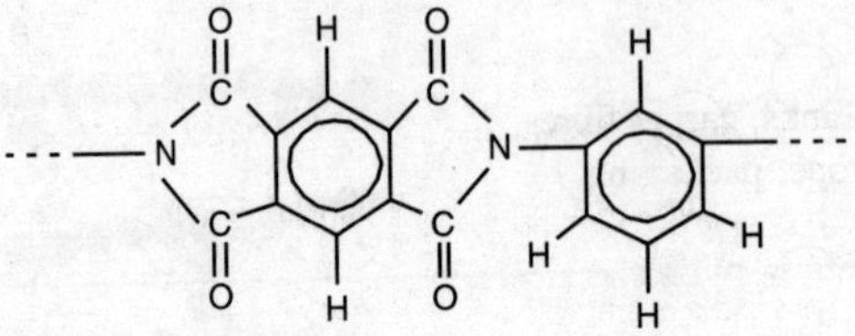	Adhesives, circuit boards, fibres for space shuttle
Polyetheretherketone (PEEK)	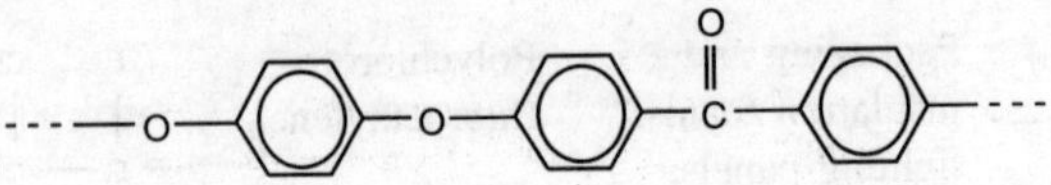	High-temperature electrical insulation and coatings
Polyphenylene sulfide (PPS)	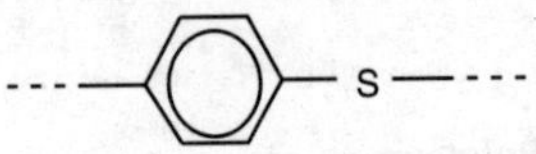	Coatings, fluid-handling components, electronic components, hair dryer components
Polyether sulfone (PES)	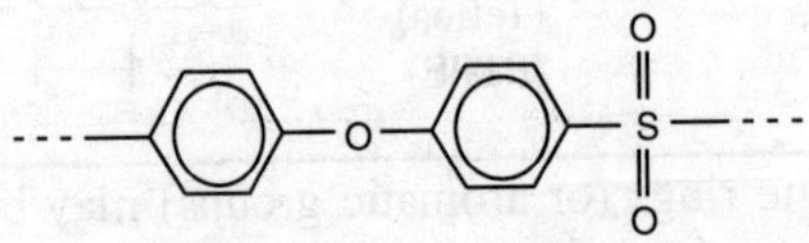	Electrical components, coffeemakers, hair dryers, microwave oven components
Polyamide-imide (PAI)	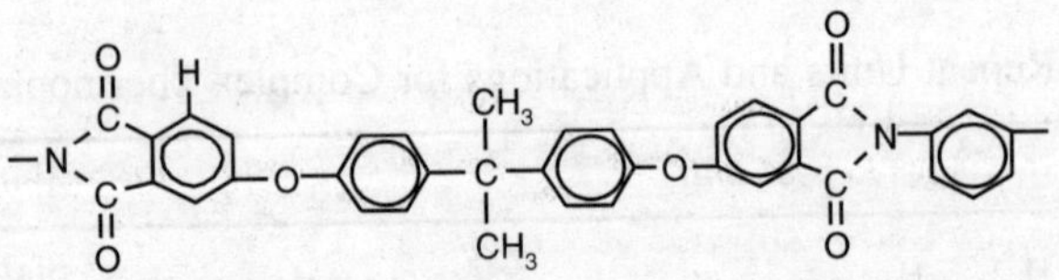	Electronic components, aerospace and automotive applications

Polyoxymethylene, or acetal, is a simple example in which the backbone of the polymer chain contains alternating carbon and oxygen atoms. A number of polymers, including polymides and polyetheretherketone (PEEK), are quite important aerospace materials.

Thermosetting Polymers: These polymers are composed of long chains (linear or branched) of molecules that are strongly cross-linked to one another to form three-dimensional networks. Thermosetting polymers or network polymers are like a bunch of strings that are knotted to one another in several places and not just tangled up. Each string may have other side strings attached to it. Thermoset polymers are generally stronger, but more brittle in comparison to thermoplastics. Thermoset polymers do not melt upon heating but begin to decompose. Recycling of thermosets is difficult as they cannot easily be reprocessed after cross-linking reaction has occurred.

Elastomers: These polymers are known as rubbers. These polymers have an elastic deformation > 200%. These polymers may be thermoplastics or lightly crossed-linked thermosets. The polymer chains consist of coil-like molecules that can reversibly stretch by applying a force.

Cross-linked Polymers

In these polymers, adjusted linear chains are joined one to another at various positions by covalent bonds (Fig. 5.12). Crosslinking is achieved either during synthesis or by a non-reversible chemical reaction that is usually carried out at an elevated temperature. Usually, crosslinking is accomplished by additive atoms or molecules that are covalently bonded to the chains. The movement of adjacent chains is greatly restricted and the mechanical properties are affected to a great extent. Many of the rubber elastic materials are cross-linked. This is done to increase strength and reduce plasticity in natural rubber. This process is called *vulcanization.*

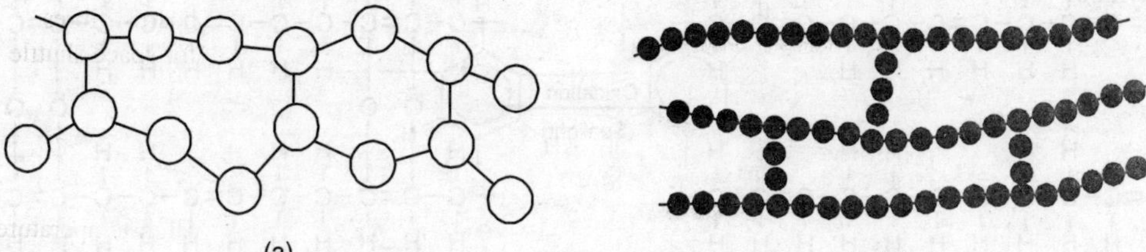

Fig. 5.12 (*a*) Crosslinked structure (*b*) Alternative representation.

Vulcanization is the crosslinking process in elastomers, which is achieved by a non-reversible chemical reaction, ordinarily carried out at an elevated temperature. In several vulcanizing reactions, sulphur compounds are added to the heated elastomer; chains of sulphur atoms bond with adjusted polymer backbone chains crosslink them, which is accomplished according the reaction given below (Fig. 5.13) in which two crosslinks shows consist of *m* and *n* sulphur atoms. Main chainsites of crosslink are carbon atoms that were doubly bonded prior to vulcanization but, after vulcanization, have become singly bonded.

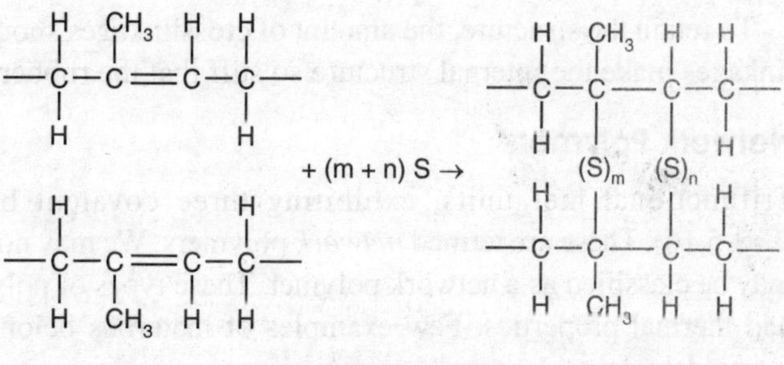

Fig. 5.13 Vulcanization.

Prior to vulcanization, *i.e.* unvulcanized rubber is soft and tacky and also has poor resistance to abrasion. Modulus of elasticity, tensile strength, and resistance to degradation by oxidation all are enhanced by this technique. The sulphur content controls the number of anchoring points. The magnitude of the modulus of elasticity is directly proportional to the density of the crosslinks. Figure 5.14 shows the stress-strain curves for both vulcanized and unvulcanized natural rubber.

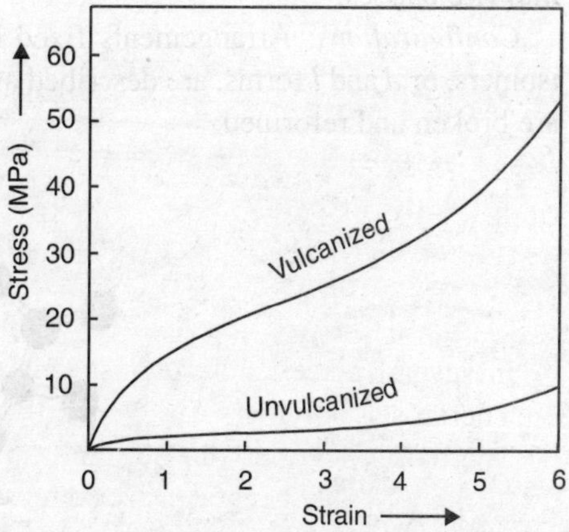

Fig. 5.14 Stress-strain curves to 600% elongation for unvulcanized and vulcanized natural rubber.

In order to produce a rubber that is capable of large extensions without rupture of the primary chain bonds, there must be relatively few crosslinks, and these must be widely separated. One obtains the useful rubber when about 1 to 5 parts (by weight) of sulphur are added to 100 parts of rubber. However, increasing further the sulphur content hardens the rubber and also reduces its extensibility.

Ageing of Polymers

We may note that crosslinking is sometimes harmful too. Under the catalytic action of sunlight, the rubber gets cross-linked with oxygen of air over a long period of time. This is termed as ageing of polymers (Fig. 5.15). The crosslinked rubber loses its flexibility and elasticity.

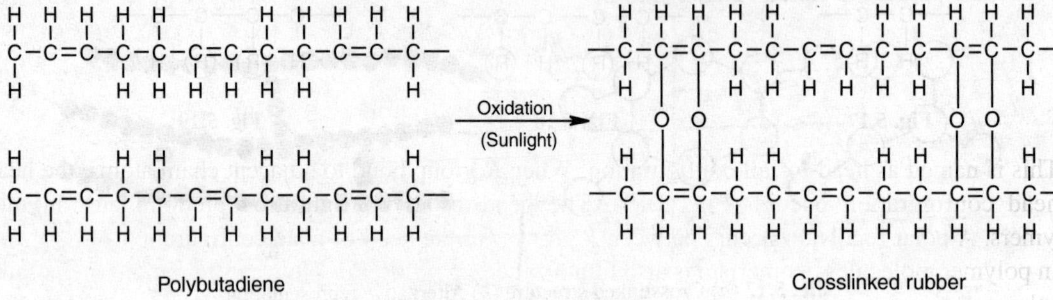

Fig. 5.15 Ageing of polymers.

To retain the structure, the amount of crosslinkages should be as small as sufficient. Excessive cross-linkages make the internal structure so stiff that the rubbery state will be destroyed.

Network Polymers

Trifunctional mer units, exhibiting three covalent bonds, form three-dimensional networks (Fig. 5.16). These are termed *network* polymers. We may note that a polymer which is highly crosslinked may be classified as a network polymer. These types of polymeric materials have distinctive mechanical and thermal properties. Few examples of materials belonging to this group are epoxies and phenol-formaldehyde.

The Texture of Polymers: One can divide the geometrical arrangements of atoms in a polymer chain into two classes:

Configurations: Arrangements fixed by chemical bonding in the molecule, such as cis and trans isomers, or *d* and *l* forms, are described as configurations. This cannot be altered unless chemical bonds are broken and reformed.

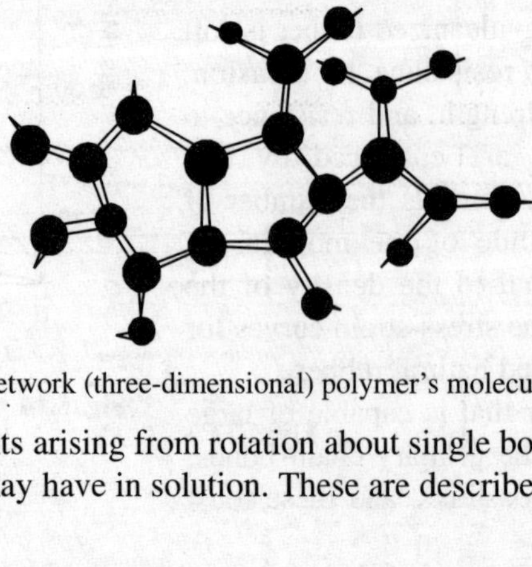

Fig. 5.16 Network (three-dimensional) polymer's molecular structure.

Conformations: Arrangements arising from rotation about single bonds, which include the mainfold forms that the polymer chain may have in solution. These are described as *conformations.*

5.5 MOLECULAR CONFIGURATIONS

The polymers, which have one side atom or group of atom bonded to the main chain, the regularity and symmetry of the side group arrangement can significantly influence the properties. Let us consider the mer unit (Fig. 5.16) in which *R* represents an atom or the side group other than hydrogen, *e.g.* Cl, CH_3. When the *R* side groups of successive mer units bond to alternate carbon atom, one arrangement is possible as shown in Fig. 5.17.

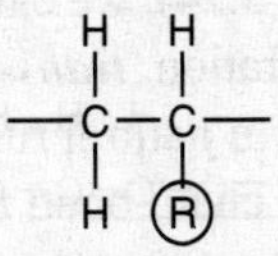

Fig. 5.17

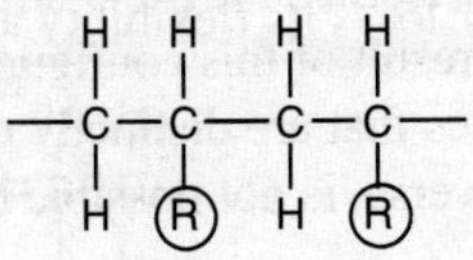

Fig. 5.18

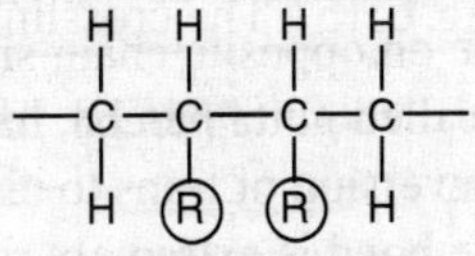

Fig. 5.19

This is named as head-to-tail configuration. When *R* group bond to adjacent chain atoms, the head-to-head configuration occurs (Fig. 5.19). The head-to-tail configuration predominates in most polymers. A polar repulsion occurs between *R* groups for the head-to-head configuration.

In polymer molecules, isomerism is also found, where in different atomic configurations are possible for the same composition. Stereoisomerism and geometrical isomerism are two subclasses of isomerism.

5.5.1 Stereoisomerism

This refers to the situation in which atoms are linked together in the same order (head-to-tail) but differ in their spatial arrangement. In an *isotactic configuration,* all the *R* groups are situated on the same side of the chain (Fig. 5.20). One stereoisomer is shown.

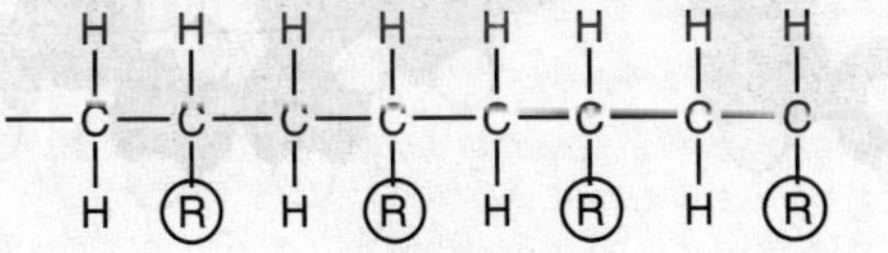

Fig. 5.20

The *R* groups alternate sides of the chain in a *syndiotactic configuration* (Fig. 5.21).

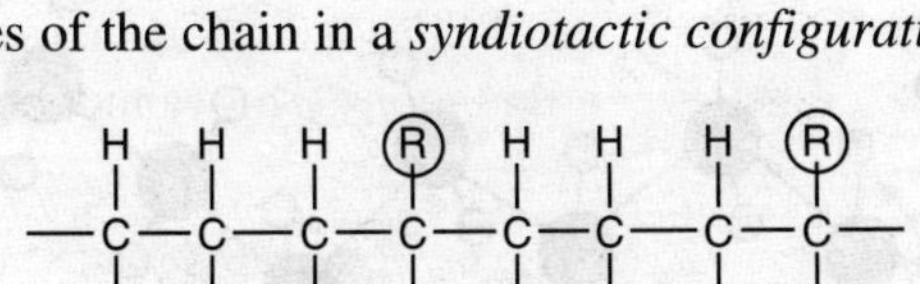

Fig. 5.21

The term *atactic configuration* is used for random positioning (Fig. 5.22).

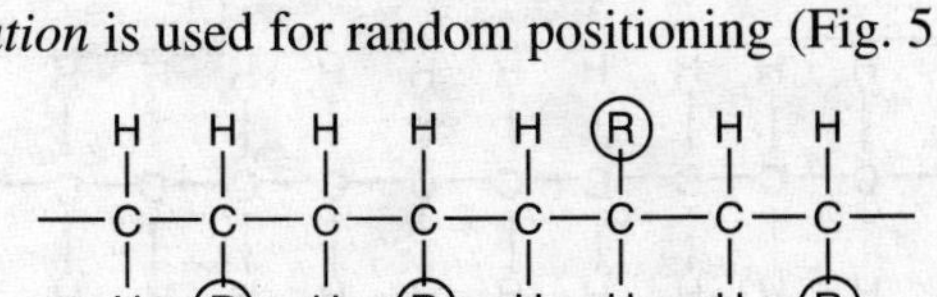

Fig. 5.22

5.5.2 Geometrical Isomerism

Geometrical isomers or other important chain configurations are possible within mer units having a double bond between chain carbon atoms. Bonded to each of the carbon atoms participating in the double bond is a single-bonded atom or radical, which may be situated on one side of the chain or its opposite. Let us consider the isoprene mer having the structure (Fig. 5.23) in which the CH_3 and the H are positioned on the same side of the chain. This is called *cis* structure and the resulting polymer, cis-polyisoprene, is natural rubber. The alternate isomer, the *trans*-structure (Fig. 5.24), we can see that CH_3 and H reside on opposite chain sides. As a result of this configurational alteration, *trans*-polyisoprene, sometimes called gutta percha, has properties that are distinctly different from natural rubber. We may note that conversion of trans to cis, or vice versa, is not possible by a simple chain bond rotation as the chain double bond is extremely rigid.

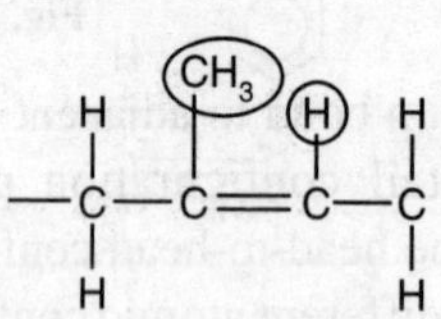

Fig. 5.23

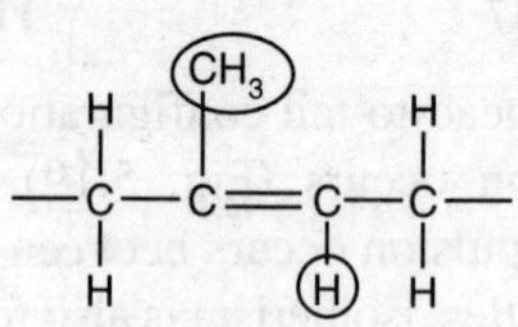

Fig. 5.24

5.6 REPRESENTATIVE STRUCTURES

One could represent a polyethylene, the simplest of the thermoplastics as shown in Fig. 5.25.

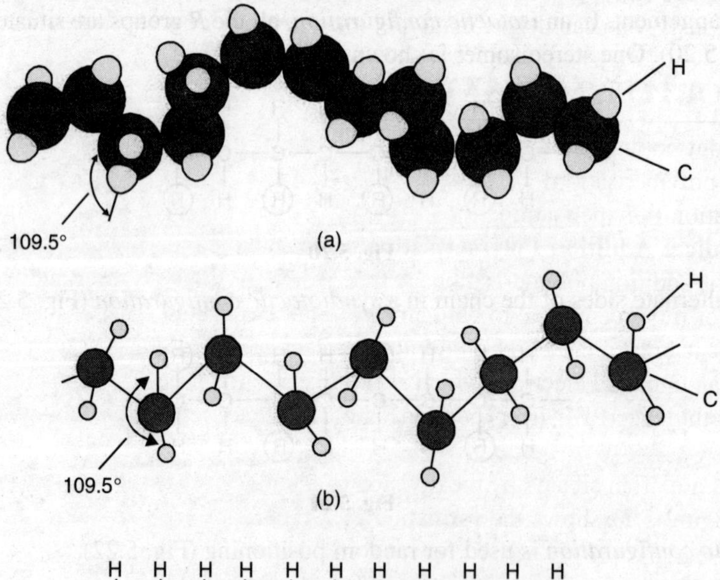

Fig. 5.25 Three different ways to represent the structure of polyethylene: (*a*) three-dimensional 'solid' model, (*b*) three-dimensional 'space' model, and (*c*) a simple 'two-dimensional model.

We can see that the polymer chain consists of a backbone of carbon atoms; two hydrogen atoms are bonded to each carbon atom in the chain. The chain twists and turns throughout space. Polyethylene has no branches (Fig. 5.25) and hence is a linear thermoplastic. The essential elements of the polymer structure are included in Fig. 5.25(*c*) and hence used to describe various polymers. A number of polymers include ring structures, such as benzene ring found in polystyrene and other polymers is shown in Fig. 5.26.

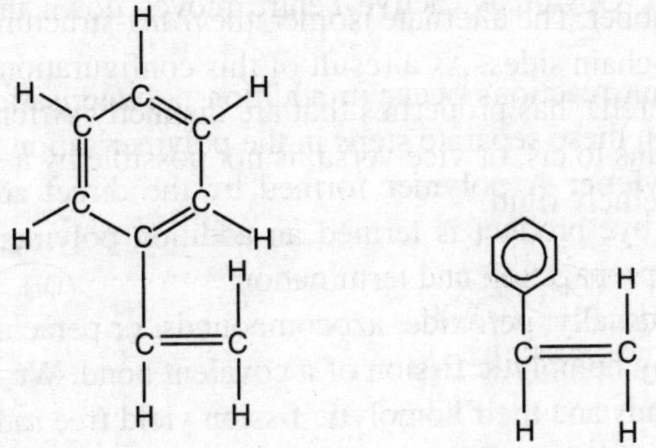

Fig. 5.26 Two ways to represent the benzene ring. In this case, the benzene ring is shown attached to a pair of carbon atoms, producing styrene.

If in Fig. 5.25(*c*), we replace one of the hydrogen atoms in CH_2 with (CH_3), a benezene ring, or chlorine, one obtains the structure of polypropylene, polystyrene, and polyvinyl chloride (PVC). If we replace all II in CH_2 groups with fluorine (F), one obtains the structure of polytetrafluoroethylene or Teflon™. Obviously, one can derive many polymer structures from the structure of polyethylene.

5.7 TYPES OF POLYMERISATION

One can carry out polymerisation reaction through one of the following two ways:

(*i*) Addition polymerisation

(*ii*) Condensation polymerisation.

(*i*) **Free Radical Addition Polymerisation Mechanism:** Free radical polymerisation is the most common type of addition polymerisation. A free radical is simply a molecule with unpaired electron. The tendency for this free radical to gain an additional electron in order to form a pair makes it highly reactive so that it breaks the bond on another molecule by stealing an electron, leaving behind that molecule with an unpaired electron (which is another free radical). The formation of the most common polymer, polyethylene (PE), from ethylene molecules is an example of addition or chain-growth polymerisation.

Free radicals are often created by the division of a molecule (an initiator) into two fragments along a single bond. Figure 5.27 shows the formation of a radical from its *initiator,* in this case, *e.g.* Benzol peroxide.

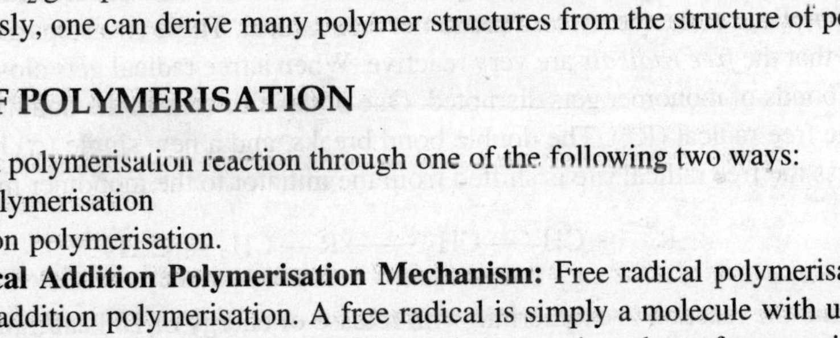

Fig. 5.27 Formation of a radical from its initiator.

In presence of an appropriate combination of heat, pressure, and catalysts, the double bond between the carbon atoms is broken and replaced with a single covalent bond. The ends of the monomer are now *free radicals,* each carbon atom has an unpaired electron that it may share with other free radicals. We may note that addition polymerisation occurs because the original monomer contains a double covalent bond between the carbon atoms. The double bond is unsaturated bond. After changing to a single bond, the carbon atoms are still joined, but they become active; other *repeat units* or *mers* can be added to produce the polymer chain. Obviously, active centre moves down the chain as the polymerisation occurs.

There are three significant reactions occur in addition polymerisation: *initiation, propagation* and *termination.* We will explain these separate steps in the polymerisation of ethylene.

Polymerisation of Ethylene: A polymer formed by the direct addition of repeated monomers without the elimination of bye-product is termed an addition polymer. The addition polymerisation follows through: initiation, propagation and termination.

Initiation: An initiator (usually, peroxide, azocompounds, or peracids) in a reaction process break-down to give free radicals by homolytic fission of a covalent bond. We may note that the initiators are thermally unstable compounds and their homolytic fission yield free radicals.

$$\underset{\text{Initiator molecule}}{R-R} \rightarrow \underset{\text{Free radicals}}{2R^*}$$

Here R stands for peroxides, azocompounds, or peracids.

The endothermic energy required to obtain free radicals is supplied by the absorption of heat (thermal initiation) or by absorption of light (photochemical). The initiator free radical may combine with a monomer olefinic molecule to form monomer free radical. These two steps are termed *initiation.*

We may note that the *free radicals* are very reactive. When a free radical gets close to a double bond, one of the bond/bonds of monomer gets disrupted. One of the electrons in the double bond of monomer is attracted to the free radical (R*). The double bond breaks, and a new single (σ) bond is formed and during the process the free radical site is shifted from the initiator to the monomer molecule as follows:

$$\underset{\text{Free radical}}{R^*} + \underset{\text{Ethylene}}{CH_2 - CH_2} \longrightarrow \underset{\text{Monomer free radical}}{R - CH_2 - C^*H_2}$$

We may not that the reaction is exothermic with release of energy of 20 kcal/mol.

Preparation: The monomer free radical obtained above attacks the double bond of a fresh monomer, (CH_2—CH_2) molecule to form a dimer free radical with a shift of free radical to the second monomer molecule. With another molecule of monomer, the dimmer free radicals add on to form a trimer-free radical. The chain length goes on increasing and polymer free radical (a growing polymer) is formed. However, each step involves growth of polymer by one unit. In each process, a monomer unit is added, there is an energy release of about 20 kcal/mole. The process is as follows:

$$\underset{\text{Monomer free radical}}{R - CH_2 - C^*H_2} + \underset{\text{Ethylene}}{CH_2 - CH_2} \rightarrow \underset{\text{Dimer free radical}}{R - CH_2 - CH_2 - CH_2 - C^*H_2}$$

$$\rightarrow \underset{\text{Dimer-free radical}}{R - CH_2 - CH_2 - CH_2 - C^*H_2} + (n-2)(CH_2 - CH_2) \rightarrow \underset{\text{(Polymer)}}{R - (CH_2)_n - CH_2 - C^*H_2}$$

where n = number of molecules/(moles) of monomer, (CH_2 — CH_2) added to the chain during polymerisation.

Termination: We need polymers that have a controlled average molecular weight and a molecular weight distribution. This means the polymerisation reactions must have an "off" switch as well. Chain length during the process of polymerisation does not increase indefinitely. Depending on the conditions, chain propagation comes to an end at any stage when the polymer-free radical is converted to a stable polymer molecule. Termination of a chains may take place through two different mechanisms:

(i) *Combination or Coupling:* By coupling of one polymer free radical with another free radical or initiator free radical to form stable dead polymer.

$$R-(CH_2)_{n1}-CH_2-C^*H_2+R-(CH_2)_{n2}-CH_2-C^*H_2$$

$$\rightarrow R-(CH_2)_{n1}-CH_2-CH_2-(CH_2)_{n2}-CH_2-CH_2-R$$

(Dead heavy polymer)

(ii) *By disproportion of two polymer free radicals:* The active end of one chain may remove a hydrogen atom from a second chain through the process of *disproportionation* and as a result, two dead polymers, one unsaturated and other saturated are formed. This reaction terminates two chains, rather than combining two chains into one larger chain. Sometimes, compounds known as terminators are added to end the polymerisation reactions. In general, for thermoplastics, the higher the average molecular weight the higher will be the melting temperature and higher will be the Young's modulus of the polymer. Termination by disproportion is illustrated here. During the termination by disproportion, one H-atom from a growing polymer chain is shifted to the other growing polymer chain as a result of two dead polymers, one saturated and one unsaturated, are formed.

$$R-(CH_2)_{n1}-CH_2-C^*H_2+R-(CH_2)_{n2}-CH_2-C^*H_2$$

$$\rightarrow R-(CH_2)_{n1}-CH_2-CH_3+R-(CH_2)_{n2}-CH=CH_2$$

The following example illustrates an addition polymerisation reaction:

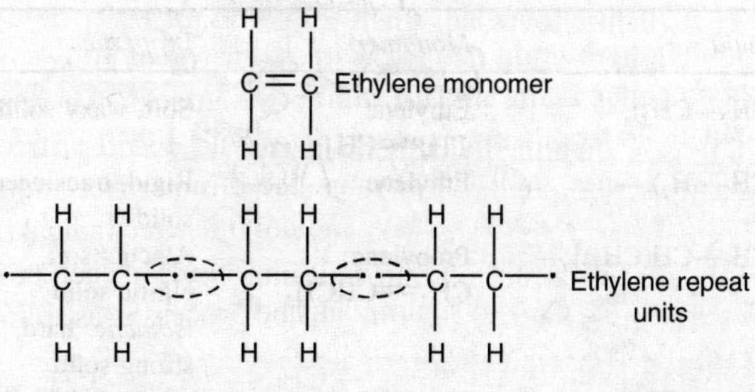

Fig. 5.28 The addition reaction for producing polyethylene from ethylene molecules. The unsaturated double bond in the monomer is broken to produce active sites, which then attract additional repeat units to either end to produce a chain.

The termination of polyethylene chain growth is shown in Fig. 5.29.

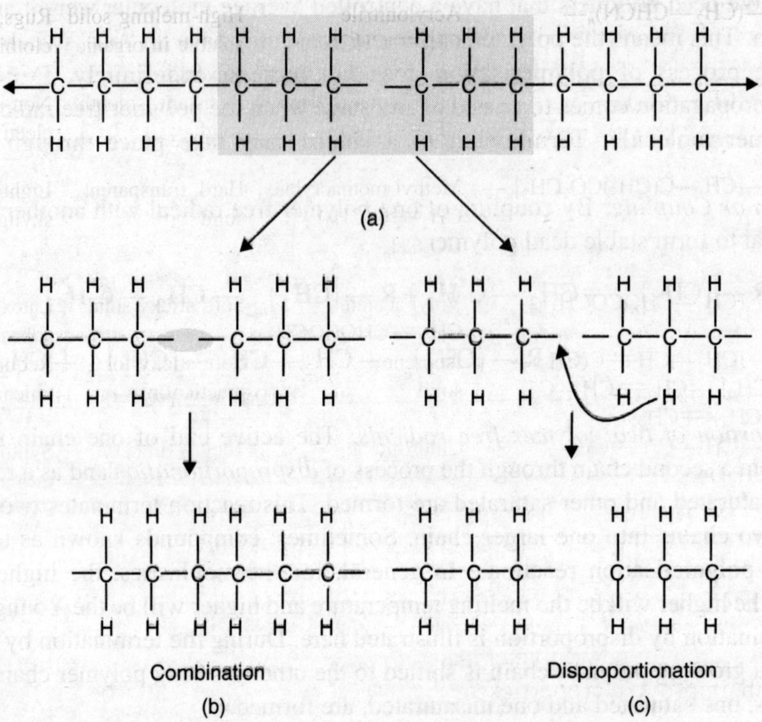

Fig. 5.29 Termination of polyethylene chain growth: (*a*) the active ends of two chains come into close proximity, (*b*) the two chains undergo combination and become one large chain, and (*c*) rearrangement of a hydrogen atom and creation of a double covalent bond by disproportionation cause termination of two chains.

Some common addition polymers are listed in Table 5.6.

Table 5.6 Some Common Addition Polymers

Name(s)	*Formula*	*Monomer*	*Properties*	*Uses*
Polyethylene low density (LDPE)	$—(CH_2—CH_2)_n—$	Ethylene $CH_2{=}CH_2$	Soft, waxy solid	Film wrap, plastic bags
Polyethylene high density (HDPE)	$—(CH_2—H_2)_n—$	Ethylene	Rigid, translucent solid	Electrical insulation bottles, toys
Polypropylene (PP) different grades	$—[CH_2—CH(CH_3)]_n—$	Propylene $CH_2{=}CHCH_3$	Atactic: soft, elastic solid *isotactic:* hard, strong solid	Similar to LDPE carpet, upholstery
Poly(vinyl chloride) (PVC)	$—(CH_2—CHCl)_n—$	Vinyl chloride $CH_2{=}CHCl$	strong rigid solid	Pipes, siding, flooring
Poly(vinylidene chloride) (Saran A)	$—(CH_2—CCl_2)_n—$	vinylidene chloride $CH_2{=}CCl_2$	Dense, high-melting solid	Seat covers, films
Polystyrene (PS)	$—[CH_2—CH(C_6H_5)]_n—$	Styrene $CH_2{=}CHC_6H_5$	Hard, rigid, clear solid soluble in organic solvents	Toys, cabinets packaging (foamed)

Polyacrylonitrile (PAN, Orlon, Acrilan)	$—(CH_2—CHCN)_n—$	Acrylonitrile $CH_2{=}CHCN$	High-melting solid soluble in organic solvents	Rugs, blankets clothing
Polytetrafluoro-ethylene (PTFE. Teflon)	$—(CF_2—CF_2)_n—$	Tetrafluoroethylene $CF_2{=}CF_2$	Resistant, smooth solid	Non-stick surfaces electrical insulation
Poly(methyl methacrylate) (PMMA, Lucite, Plexiglas)	$—[CH,—C(CH_3)CO_2CH_3]_n—$	Methyl methacrylate $CH_2{=}C(CH_3)$ CO_2CH_3	Hard, transparent solid	Lighting covers, skylights signs
Poly(vinyl acetate) (PVAc)	$—(CH_2—CHOCOCH_3)_n—$	Vinyl acetate $CH_2{=}CHOCOCH_3$	Soft, sticky solid	Latex paints, adhesives
cis-Polyisoprene natural rubber	$—[CH_2—CH{=}C(CH_3)—CH_2]_n—CH_2{=}CH—C(CH_3){=}CH_2$	Isoprene solid	Soft, sticky for practical use	Requires vulcanization
Polychloroprene (cis + trans) (Neoprene)	$—[CH_2—CH{=}CCl—CH_2]_n—$	Chloroprene $CH_2{=}CH—CCl{=}CH_2$	Tough, rubbery solid	Synthetic rubber oil resistant

5.8 COPOLYMERS

There is a continuous search for new polymer materials that can be easily and economically synthesized and fabricated, with improved properties or better property combinations that are offered by homopolymers.

One can prepare polymers with more than one kind of mer in the chain, *i.e.* different types of mers react to form a chain. Such resulting polymers are named as *copolymers*. The technique used for copolymers is named as *copolymerisation*. A specific type of addition polymerisation process is known as *Co-polymerisation*. There are different types of copolymers which can be produced depending upon the type of monomers and the method of synthesis. Copolymers are of following types:

(*i*) ***Random Copolymer:*** This type has random distribution of various mers in its chain. If M and R represents two different mers, then the random copolymer can be depicted as shown in Fig. 5.30. Obviously, two different units are randomly dispersed along the chain.

MMRMRRRMMRMRRMMMRMMRRRRMM ....

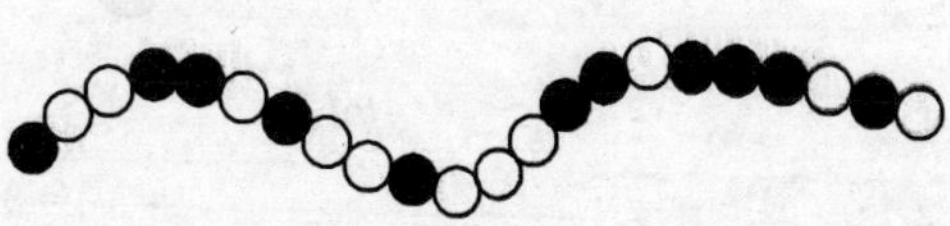

Fig. 5.30 Random copolymer (with two different type of mers)

(*ii*) ***Alternating Copolymer:*** As the name suggests, the two mer units alternate chain positions as shown in Fig. 5.31.

(*iii*) ***Block Copolymer:*** Identical mers are clustered in blocks along the chain (Fig. 5.32).

MRMRMRMRMRMRMR . . .

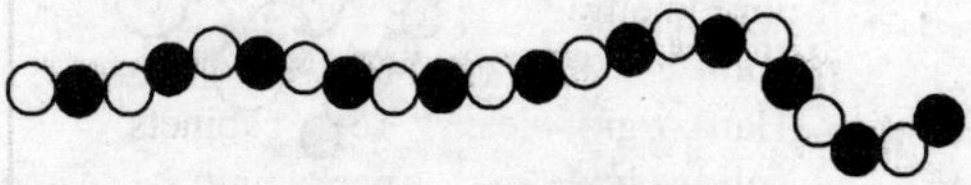

Fig. 5.31 Alternative copolymer (mers are arranged alternatively)

MMMMMMMMMMMMMRRRRRRRRRRRRRMMMMM . . .

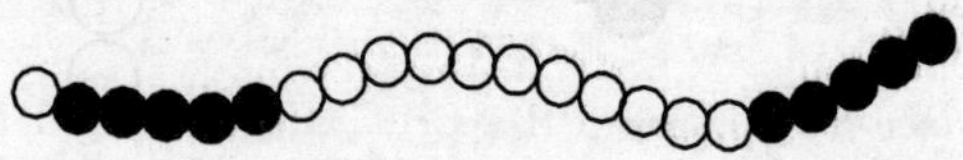

Fig. 5.32 Block copolymer (mers arranged in blocks)

(*iv*) ***Grafted Copolymer:*** It is the copolymer in which homopolymer side branches of one type may be grafted to homopolymer main chains that are composed of a different mer (Fig. 5.33).

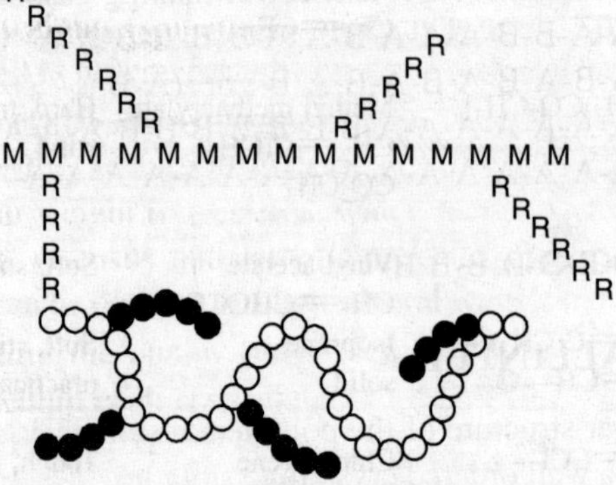

Fig. 5.33 Grafted copolymer (one types of mers are attached to one or more sides of the main chain formed by other type of mers).

Copolymers have better physical and mechanical properties. Copolymerisation is just similar to the alloying process in metallurgy. Similar to alloys, one can synthesize various copolymers with desired

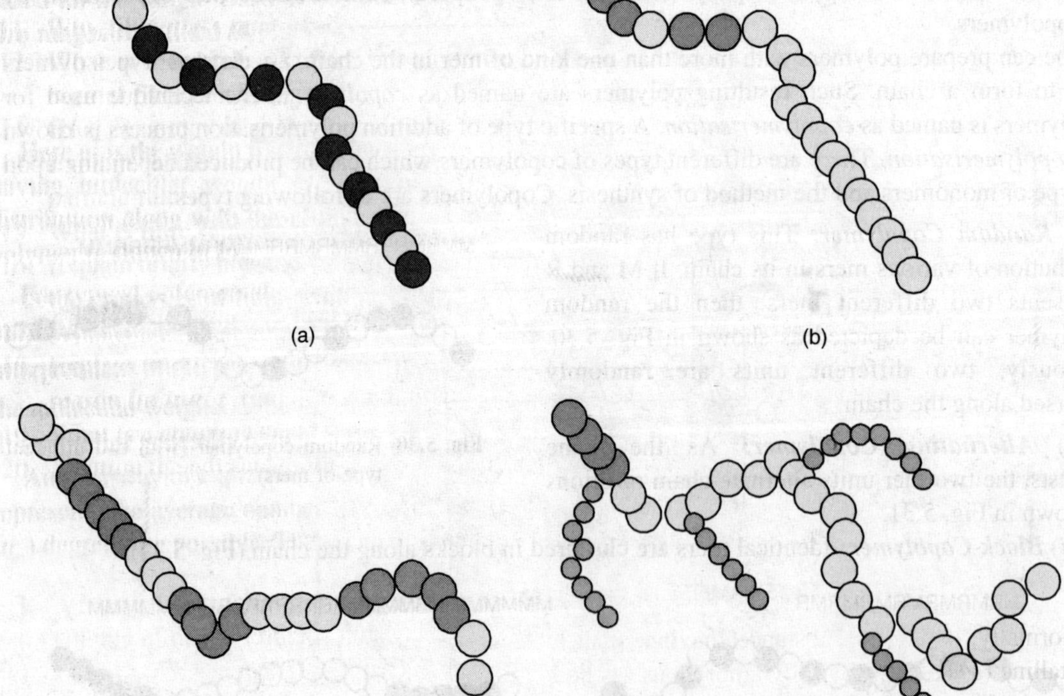

Fig. 5.34 Four types of copolymers: (*a*) alternating monomers, (*b*) random monomers, (*c*) block copolymers, and (*d*) grafted copolymers. Circles of different sizes represent different monomers.

combination of properties. Synthetic rubbers are often copolymers. Styrene-butadiene rubber (SBR) is a common random polymer from which automobile tyres are prepared. Nitride butadiene rubber (NBR) is another random copolymer composed of acrylonitrile and butadene. NBR is highly elastic and also resistant to swelling in organic solvents. NBR is used for making gasoline hoses.

- *Random copolymers:* -A-A-B-B-A-A-A-B-A-A-B-B-A-B-A-A-A-B-A-A-A-B-B-
- *Alternate copolymers:* -A-B-A-B-A-B-A-B-A-B-, or -(A-B-)$_n$-
- *Block copolymers:* -A-A-A-A-A-A-A-B-B-B-B-B-B-B-A-A-A-A-A-A-A-B-B-...
- *Graft copolymers:* -A-A-A-A-A-A-A-A-A-A-A-A-A-A-A-A-A-A-A-A-A-

  |

  -B-B-B-B-B-B-B-

5.9 POLYMER CRYSTALLINITY

The complexity of the molecular structure of the polymers results in a large variety of supermolecular structures and can extend from a well ordered crystalline structure to a completely amorphous structure. One can think of *polymer crystallinity* as the packing of molecular chains so as to produce an ordered atomic array. We know that crystal structure may be specified in terms of unit cells, which are often quite complex. The unit cell for polyethylene and its relationship to the molecular structure is shown in Fig. 5.35. This unit cell for polyethylene has orthorhombic geometry. However, the chain molecules also extend beyond the unit cell shown in Fig. 5.35.

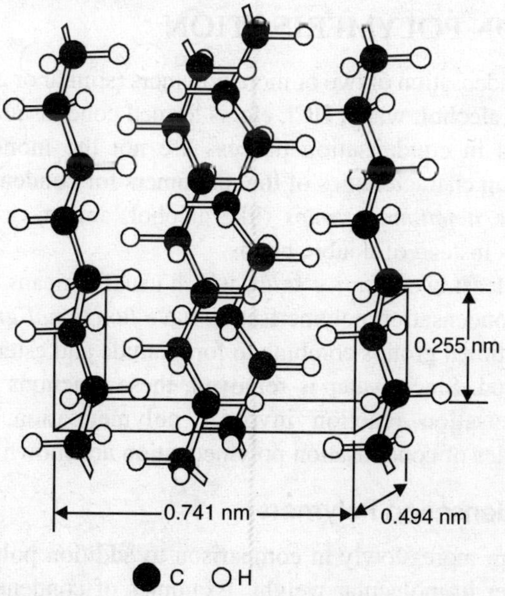

Fig. 5.35 Molecular chains in a unit cell for polyethylene.

Normally, molecular substances having small molecules, *e.g.* water and methane are either totally crystalline (as solids) or totally amorphous (as liquids). Polymer molecules, as a consequence of their size and often complexity are often only partially crystalline (or semicrystalline), having crystalline regions dispersed within the remaining amorphous material.

The density of a crystalline polymer is found to be greater than an amorphous one of the same material and molecular weight, since the chains are more closely packed together for the crystalline structure. One may determine the degree of crystallinity by weight from accurate density measurements, according to

$$\% \text{ crystallinity} = \frac{\rho_c(\rho_s - \rho_a)}{\rho_s(\rho_c - \rho_a)} \times 100 \quad ...(5)$$

where ρ_s is the density of a specimen for which we have to determine the per cent crystallinity, ρ_a is the density of the totally amorphous polymer, and ρ_c is the density of the perfectly crystalline polymer. One will have to determine the values of ρ_a and ρ_c by other experimental methods.

For a polymer, the degree of crystallinity depends on the rate of cooling during solidification and also on the chain configuration. The molecular chemistry also influence the ability of a polymer to crystallize. Crystallization is also not favoured in polymers that are composed of chemically complex mer structures, *e.g.* polyisoprene.

Crystallization is easily accomplished for linear polymers as there as virtually no restrictions to prevent chain alignment. However, branched polymers are never high crystalline as the side branches interfere with crystallization. Most network and crosslinked polymers are almost totally amorphous. However, a few crosslinked polymers are partially crystalline. There is a greater tendency for the development of non-crystallinity for coploymers due to irregular and random arrangement for the mer.

Obviously, the physical properties of polymeric materials to some extent are influenced by the degree of crystallinity. We may note that crystalline polymers are usually stronger and more resistant to dissolution and softening by heat.

5.10 CONDENSATION POLYMERISATION

A polymer formed by the condensation of two or more polymers (similar or different) with the elimination of molecules like ammonia, alcohol, water, HCl, etc. is termed condensation polymer.

The monomers involved in condensation process are not the monomers as those in addition polymerisation. The two main characteristics of the monomers for condensation polymerisation are:

(*i*) These monomers have *functional groups* (like alcohol, amine, or carboxylic acid groups) or carboxylic acid groups instead of double bonds.

(*ii*) Each monomer has at least two *reactive sides,* which usually means two functional groups.

Monomers involved in condensation polymerisation have *functional groups* (carboxylic amino and alcohol groups). These functional groups combine to form amide and ester linkages. When this occurs, a water molecule is removed. Since water is removed, these reactions are termed as condensation reactions. When a condensation reaction involves polymerisation, we term it condensation polymerisation. Few examples of condensation polymerisation are shown in Fig. 5.36.

Characteristics of Condensation Polymers

Condensation polymers form more slowly in comparison to addition polymers, often requiring heat, and they are generally lower in molecular weight. Examples of condensation polymers are given in Table 5.7.

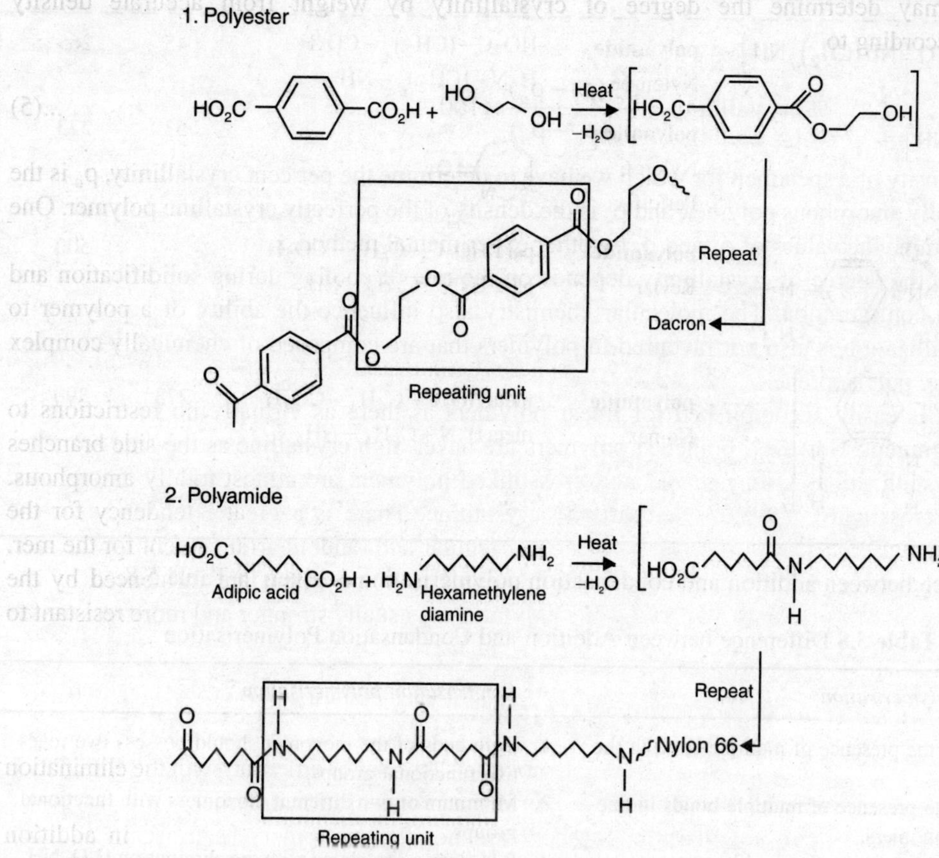

Fig. 5.36 Examples of condensation polymerisation.

Table 5.7 Condensation Polymer Type and their t_g Data

Formula	*Type*	*Components*	T_g°C	T_m°C
$\sim[CO(CH_2)_4\,CO-OCH_2CH_2O]_n\sim$	**polyester**	$HO_2C-(CH_2)_4-CO_2H$ $HO-CH_2CH_2-OH$	< 0	50
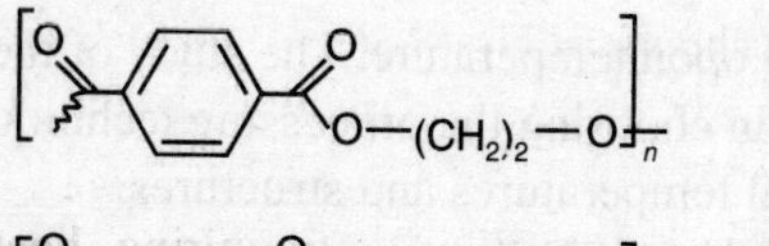	**polyester** Dacron Mylar	para $HO_2C-C_6H_4-CO_2H$ $HO-CH_2CH_2-OH$	70	265
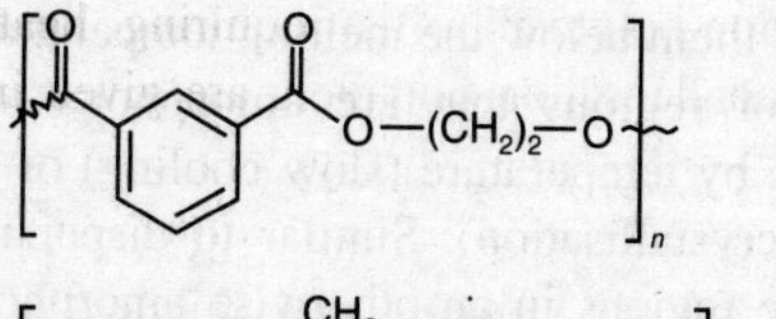	**polyester**	meta $HO_2C-C_6H_4-CO_2H$ $HO-CH_2CH_2-OH$	50	240
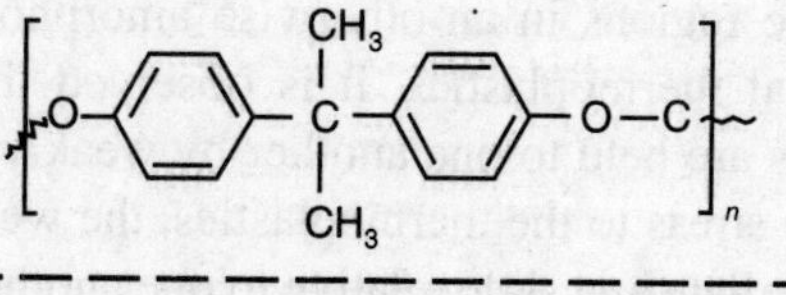	**polycarbonate** Lexan	$(HO-C_6H_4-)_2\,C(CH_3)_2$ (Bisphenol A) $X_2C{=}O$ $(X = OCH_3$ or Cl)	150	267

(*Contd.*)

$\sim\left[CO(CH_2)_4CO-NH(CH_2)_6NH\right]_n\sim$	**polyamide** Nylon 66	$HO_2C-(CH_2)_4-CO_2H$ $H_2N-(CH_2)_6-NH_2$	45	265
$\sim\left[CO(CH_2)_5NH\right]_n\sim$	**polyamide** Nylon 6 Perlon	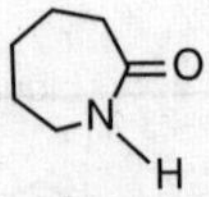	53	223
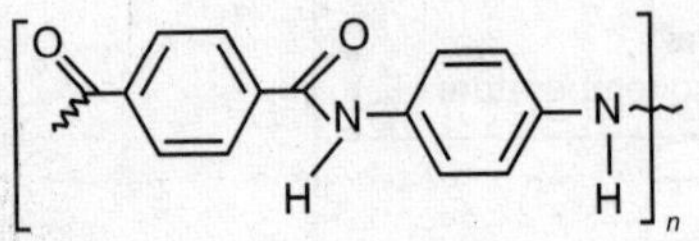	**polyamide** Kevlar	para $HO_2C-C_6H_4-CO_2H$ para $H_2N-C_6H_4-NH_2$	-	500
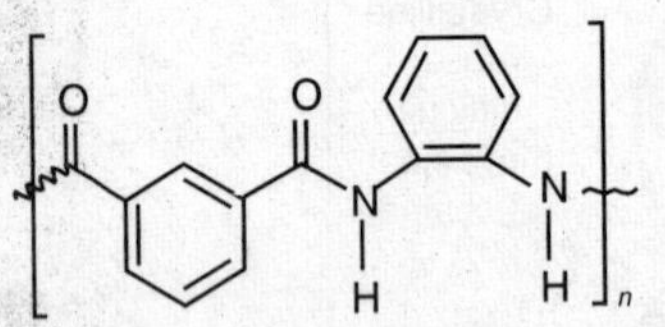	**polyamide** Nomex	meta $HO_2C-C_6H_4-CO_2H$ meta $H_2N-C_6H_4-NH_2$	273	390

The difference between addition and condensation polymerisation is given in Table 5.8.

Table 5.8 Difference between Addition and Condensation Polymerisation

Addition polymerisation	*Condensation polymerisation*
1. It demands the presence of multiple bond in the monomer.	1. Both ends of the monomer should possess two reactive functional groups.
2. It prefers the presence of multiple bonds in one or more monomers.	2. Minimum of two different monomers with functional groups.
3. No elimination of bye-product like water.	3. It is always associated with the elimination H_2O, NH_3 etc.
4. A thermoplastic is obtained which is a homo chain polymer.	4. A thermosetting or thermoplastic plastic (hetero chain) is obtained.
5. It is observed that chain growth is at one active centre.	5. Chain growth is observed at two active centres.

5.11 EFFECT OF TEMPERATURE ON THERMOPLASTICS

It is found that the properties of thermoplastics change depending upon temperature. The study of these changes can help us (*a*) better design components, and (*b*) guide in choosing the processing techniques that need to be used. Figures 5.37 and 5.38 exhibit several critical temperatures and structures.

Thermoplastics can be amorphous or crystalline once we cool them below the melting temperature, T_m (Fig. 5.37). Usually, engineered thermoplastics consist of regions that are amorphous and crystalline. One can introduce the crystallinity in thermoplastics by temperature (slow cooling) or by application of stress that can untangle chains (stress-induced crystallisation). Similar to dispersion strengthening of metallic materials, the formation of crystalline regions in an otherwise amorphous matrix helps increase the strength of thermoplastics. In typical thermoplastics, it is observed that bonding within the chains is covalent, but the long coiled chains are held to one another by weak van der Waals bonds and by entanglement. When we apply a tensile stress to the thermoplastics, the weak bonding between the chains can be overcome and the chains can rotate and slide relative to one another.

We may note that the ease with which the chains slide depends on both temperature and polymer structure.

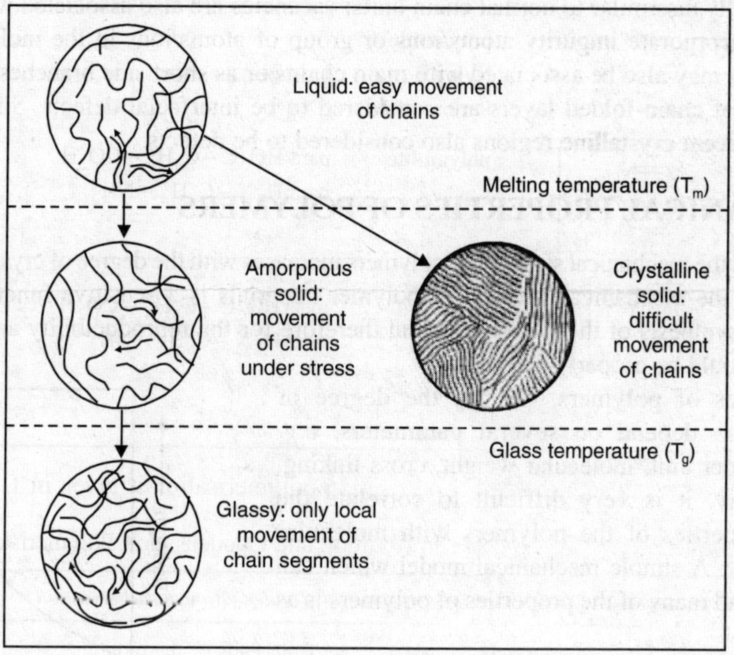

Fig. 5.37 The effect of temperature on the structure and behaviour of thermoplastics.

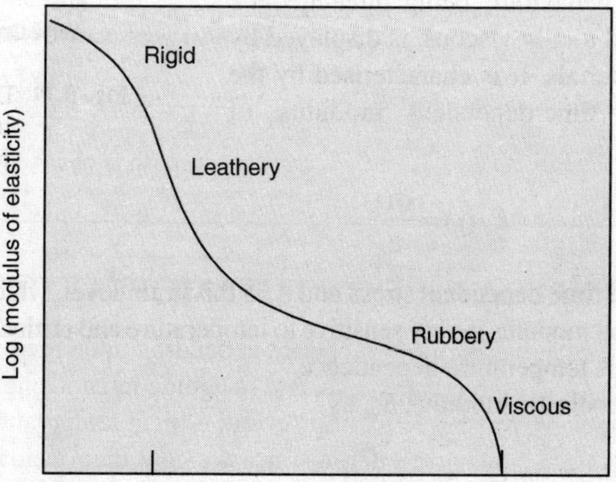

Fig. 5.38 The effect of temperature on the modulus of elasticity for an amorphous thermoplastic. Note that T_g and T_m are not fixed.

5.12 DEFECTS IN POLYMERS

The point defect concept in polymers is different than in metals and ceramics due to the chain-like macromolecules and the nature of the crystalline state for polymers. However, point defects similar to those found in metals have been observed in crystalline regions of polymeric materials; these include vacancies and interstitial atoms and ions. Chain ends are also considered to be defects in as much as they are chemically dissimilar to normal chain units; vacancies are also associated with the chain ends. One may also incorporate impurity atoms/ions or group of atoms/ions in the molecular structure as interstitials. They may also be associated with main chains or as short side branches.

The surfaces of chain-folded layers are considered to be interfacial defects. Similarly boundaries between two adjacent crystalline regions also considered to be defects.

5.13 MECHANICAL PROPERTIES OF POLYMERS

It is observed that the mechanical strength of polymers increases with the degree of crystallinity (Fig. 5.39). This shows that the mechanical strength of polymer materials is a sensitive function of the process variables in the synthesis of these materials and therefore for the reproducibility and reliability of the products they should be properly controlled.

The properties of polymers, besides the degree of crystallinity, also depend on several parameters, *e.g.* nature of monomer unit, molecular weight, cross-linking, etc. This is why, it is very difficult to correlate the mechanical properties of the polymers with molecular structure directly. A simple mechanical model which can help to understand many of the properties of polymers is as follows:

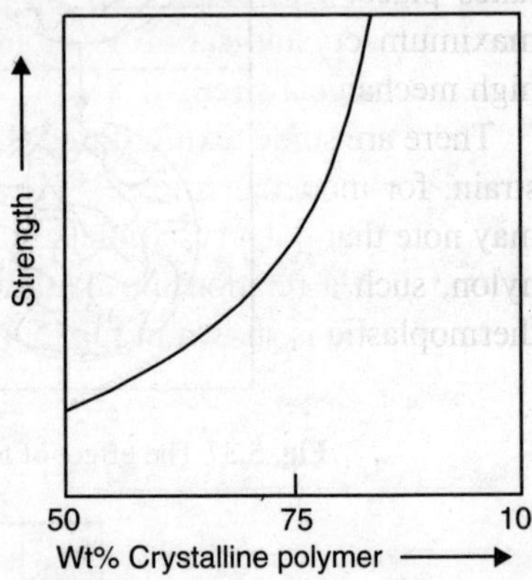

Fig. 5.39 The increase of the strength of the polyethylene with crystallinity of the polymer.

When subjected to external stresses, many polymers undergo not only elastic deformation but also *viscous flow.* Viscoelastic mechanical behaviour, being inter-mediate between totally elastic and totally viscous, is displayed by a number of polymer materials. It is characterised by the *relaxation modulus,* a time-dependent modulus of elasticity,

$$E_r(t) = \frac{\sigma(t)}{\varepsilon_o} \qquad \text{...(6)}$$

where $\sigma(t)$ is the measured time dependent stress and ε_o is the strain level, which is maintained constant. The magnitude of relaxation modulus is very sensitive to temperature and critical to in-service temperature range for elastomers in this temperature dependence.

One can define the viscoelastic modulus E_{ve} as

$$E_{ve} = \frac{\sigma}{(e_{ela} + e_{vis})} \qquad \text{...(7)}$$

The subscripts in Eqn. (7) indicate the contribution from elastic and viscous flow deformations. Figure 5.39 shows a plot of viscoelastic modulus as a function of temperature for different types of polymers.

For $E_{ve} > E_{ve}^3$, the properties of polymers are glassy and for $E_{ve}^3 > E > E_{ve}^2$ leather like and for $E_{ve}^2 > E > E_{ve}^1$ rubber like and for $E < E_{ve}^1$ viscous.

The polymers behave as a glassy material for $T < T_g$. For $T > T_g$, one enters into the viscoelastic region in which the behaviour is leather like and then rubber like (Fig. 5.40). The modulus of elasticity in this region is low and is a function of time depending on the viscosity of the medium. We may note that in a crystalline polymer the rubber like region is not present.

The mechanical properties are considerably affected by crystallisation. Crystallisation inhibit viscous flow. Although the composition of natural rubber and gutta percha is same but they have different molecular structure. Gutta percha being crystalline is hard and brittle, whereas rubber is soft and can easily be elongated. During the process of deformation, crystallisation is also produced. Rubber can be elongated by 500–600% and also the strain is reversible. During such a stretching, crystallisation takes place. However, with further extensions, when maximum crystallisation is attained, rubber achieves high mechanical strength.

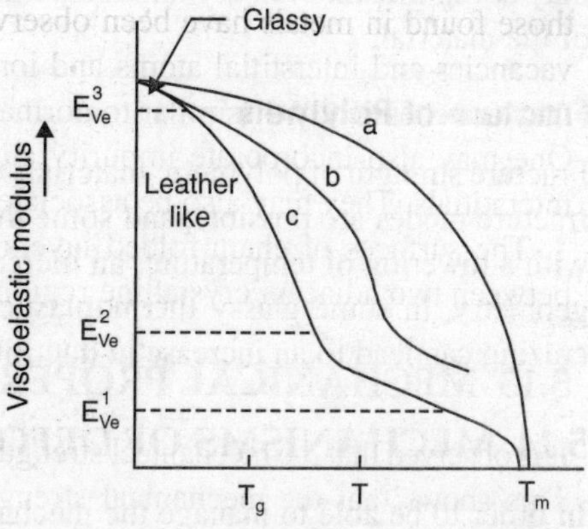

Fig. 5.40 Viscoelastic modulus as a function of temperature: (*a*) crystalline, (*b*) cross-linked, and (*c*) amorphous polymers.

There are some textile fibres which show a linear relationship between stress and strain, *i.e.* stress ∝ strain, for moderate stresses. However, beyond a certain point, the elongation increases rapidly. We may note that polycrystalline structure does not change on the application of stress, *e.g.*, in the case of nylon, such a relationship is almost entirely linear. The stress-strain curve for 6, 6-nylon, a typical thermoplastic is shown in Fig. 5.41.

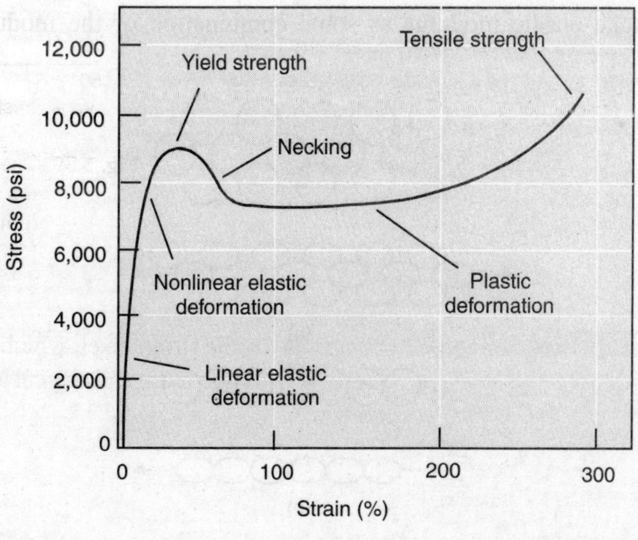

Fig. 5.41 The stress-strain curve for 6, 6-nylon a typical thermoplastic polymer.

Concluding, we can say that in polymers, molecular structure which is controlled by synthesis parameters influences the properties to a great extent.

Viscoelastic Creep

When the stress level is maintained constant, many polymeric materials are found to be susceptible to time-dependent deformation. Such deformation is termed as *viscoelastic creep*. This type of deformation may be significant even at room temperature and under modest stresses that lie below the yield strength of the material.

Fracture of Polymers

Fracture strength of polymeric materials are low relative to metals and ceramics. Both brittle and ductile fracture modes are possible, and some thermoplastic materials experience a ductile-to-brittle transition with a lowering of temperature, an increase in strain rate, and/or an alteration of specimen thickness or geometry. In some glassy thermoplastics, the crack-formation process may be preceded by crazing; crazing can lead to an increase in ductility and toughness of the material.

5.14 MECHANISMS OF DEFORMATION

In order to be able to manage the mechanical characteristics of polymeric materials, an understanding of deformation mechanisms of these materials is important. In this regard, deformation mechanisms for two different types of polymers *semicrystalline* and *elastomeric* are of importance. The deformation mechanisms are:

(*i*) *Elastic Deformation Mechanism*: The elastic deformation depends upon the amount of bond-straightening (Fig. 5.42*a*) and bond lengthening (Fig. 5.42*b*). The mechanism of elastic deformation in semicrystalline polymers in response to tensile stresses is the elongation of the chain molecules from their stable conformations, in the direction of the applied stress, by the bending and stretching of the strong chain covalent bonds. There may also be some slight displacement of adjacent molecules, which is resisted by relatively weak secondary or van der Waals bonds. Semicrystalline materials may be considered as composite materials since they are composed of both crystalline and amorphous regions. Obviously, one may take elastic modulus as some combination of the moduli of crystalline and amorphous phases.

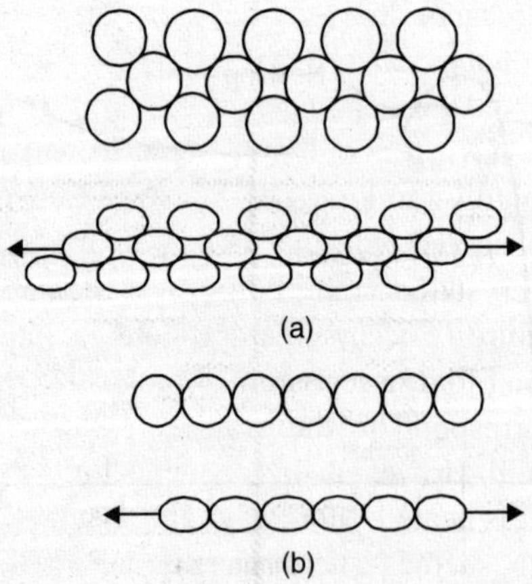

Fig. 5.42 Elastic deformation: (*a*) bond straightening, (*b*) bond lengthening.

(*ii*) *Plastic Deformation Mechanism*: Plastic deformation occurs due to slip between adjacent molecules of polymer material. Since the molecules are bonded by weak attractive forces and hence the slip occurs very easily. Plastic deformation is more prominent where alignment of molecules is linear. Plastic deformation occurs due to the slippage at the weaker points between the molecules and not due to the breaking of intermolecular bonds. The linear polymers with crosslinking show high degree of plastic deformation, whereas network polymers do not show high deformation and they are thus brittle.

The mechanism of plastic deformation can be best described by the interactions between lamellar and intervening amorphous regions in response to an applied tensile load. This process occurs in several stages.

5.15 CRYSTALLISATION, MELTING AND GLASS TRANSITION PHENOMENA IN POLYMERS

With respect to the design and processing of polymeric materials, three phenomena, namely, crystallisation, melting and glass transition are important. As mentioned earlier, crystallisation is the process by which upon cooling, an ordered (*i.e.*, crystalline) solid phase is produced from a liquid melt having a highly random molecular structure. When a polymer is heated, the occurrence of melting transformation is the reverse process. The glass-transition phenomenon occurs with amorphous or non-crystallisable polymers that, when cooled from a liquid melt, become rigid solids yet retain the disordered molecular structure, a characteristic of liquid state. The melting and glass transition temperatures of a polymer, *i.e.* T_m and T_g respectively, are important for knowing the temperature range over which a particular polymer may be utilized and processed. The magnitude of T_m and T_g increase with increasing chain stiffness; stiffness is enhanced by the presence of chain double bonds and side groups that are either bulky or polar. Molecular weight and degree of branching also affect T_m and T_g.

We may note that alterations of physical and mechanical properties attend crystallisation, melting, and glass transition. Interestingly, for semicrystalline polymers, crystalline regions will experience melting (and crystallisation), while non-crystalline areas pass through the glass transition.

The time dependence of crystallisation is the same as for many solid-state transformations. The fraction crystallised y is a function of time t according to the Avrami equation,

$$y = 1 - \exp\,[-kt^n] \qquad \text{...(8)}$$

where k and n are time-independent constants, whose values depend on the crystallising system. Rate of crystallisation is equal to the reciprocal of time required for crystallisation to proceed to 50% completion. This rate depends on crystallisation temperature and also on molecular weight.

The melting of a polymer crystal on heating, at the *melting temperature*, T_m corresponds to the transformation of a solid material, having an ordered structure of aligned molecular chains, to a viscous liquid in which the structure is highly random. There are many features distinctive to the melting of polymers that are not normally observed with metals and ceramics, *e.g.* consequences of the polymer molecular structure and lamellar crystalline morphology.

The glass transition at temperature T_g occurs in amorphous (or glassy) and semicrystalline polymers, and is due to a reduction in motion of large segments of molecular chains with decreasing temperature. Upon cooling, this transition correspond to the gradual transformation from a liquid to a rubbery material, and finally to a rigid solid. However, this sequence of events occurs in reverse order when a rigid glass a temperature below T_g is heated. Moreover, abrupt changes in other physical properties also observed with glass transition; *e.g.* stiffness, heat capacity and coefficient of thermal expansion.

Below the *glass temperature* T_g, the linear amorphous polymers gets hard, brittle and glass like.

$$\underset{\text{(Hard, brittle plastic)}}{\text{Glass state}} \xrightarrow{T_g} \underset{\text{(Soft flexible)}}{\text{Rubbery state}}$$

This is again not a fixed temperature but a range of temperatures. When the polymer cools below t_g, certain properties, *e.g.* density or modulus of rigicity. Change at a different rate (Fig. 5.43).

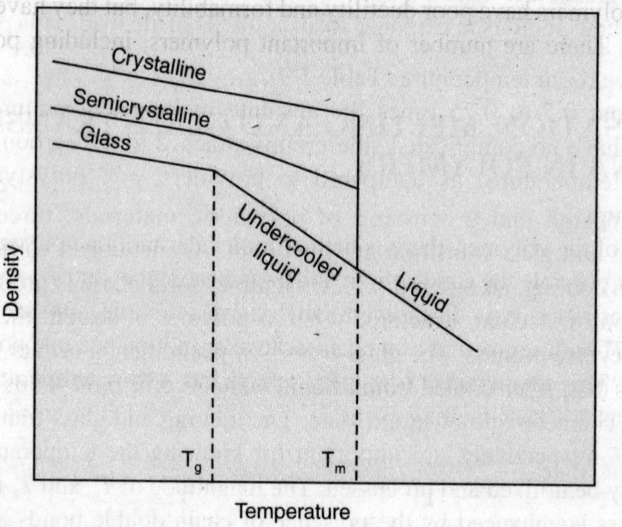

Fig. 5.43 The relationship between the density and the temperature of the polymer shows the melting and glass temperatures. However, T_g and T_m are not fixed; rather, they are ranges of temperatures.

Table 5.9 Melting and Glass Temperature-ranges (°C) for Selected Thermoplastics and Elastomers

Polymer	*Melting temperature range*	*Glass temperature range (T_g)*	*Processing temperature range*
Addition polymers			
Low-density (LD) polyethylene	98–115	–90 to –25	149–232
High-density (HD) polyethylene	130–137	–110	177–260
Polyvinyl chloride	175–212	87	
Polypropylene	160–180	–25 to –20	190–288
Polystyrene	240	85–125	
Polyacrylonitrile	320	107	
Polytetrafluoroethylene (Teflon)	327		
Polychlorotrifluoroethylene	220		
Polymethyl methacrylate (acrylic)		90–105	
Acrylonitrile butadiene styrene (ABS)	110–125	100	177–260
Condensation polymers			
Acetal	181	–85	
6,6-nylon	243–260	49	260–327
Cellulose acetate	230		
Polycarbonate	230	149	271–300
Polyester	255	75	
Polyethylene terephthalate (PET)	212–265	66–80	227–349
Elastomers			
Silicone		–123	
Polybutadiene	120	–90	
Polychloroprene	80	–50	
Polyisoprene	30	–73	

Although glassy polymers have poor ductility and formability, but they have good strength, stiffness, and creep resistance. There are number of important polymers, including polystyrene and polyvinyl chloride, have t_g above room temperature (Table 5.9).

T_g is typically about 0.5 to 0.75 times the absolute melting temperature T_m. Polymers such as polyethylene, which have no complicated side groups attached to the carbon backbone, have low t_g's (even below room temperature) as compared to polymers, *e.g.* polystyrene, which have more complicated side groups.

In the glassy state of the polymer, there is neither molecular motion nor segmental motion. When all chain motions are not possible the rigid solid results. On heating beyond T_g, segmental motion becomes possible but molecular mobility is disallowed. Obviously, a flexible, soft rubbery state is reached. On heating much above T_g both segmental as well as molecular motion become possible and polymer flows like a viscous liquid. This temperature is usually referred as a flow temperature since any polymer is only a mixture and has no sharp melting point.

$$\underset{\text{(Frozen state)}}{\text{Glass state}} \xrightarrow{\text{Above } T_g} \underset{\text{(Viscoelastic state)}}{\text{Rubbery state}} \xrightarrow{\text{Above } T_f} \underset{\text{(Viscofluid state)}}{\text{Viscous liquid}}$$

A number of factors affect the value of T_g.

1. *Crystallinity:* In crystalline polymers the polymer chains are arranged in a regular parallel fashion. Each chain is bound to the other by strong forces like H-bonding. Hence crystalline polymers have higher T_g than amorphous polymers.
2. *Molecular mass:* Generally T_g of a polymer increases with molar mass up to a particular value and beyond that there is no change.
3. The presence of side chain which hinder free rotation also increase T_g.
4. The presence of plasticizers, reduce T_g of a polymer (like impunities on melting point of pure solids).
5. Stereo regularity of polymers increase T_g. Thus T_g of an isolactic polymer is greater than that of syndiotactic polymer whose T_g is greater than atactic polymer.

Importance of T_g Value: T_g value is not only a measure of flexibility of a polymer but also gives us an idea of the thermal expansion, heat capacity, electrical and mechanical properties of the polymer. Thus the workability and usefulness of a polymer over a range of temperature can be obtained from its T_g.

5.16 POLYMER CHARACTERISTICS

As stated in Section 5.3, polymers are broadly classified in three categories:

(*i*) *Plastics*: These are the organic materials that can be easily shaped or moulded by mechanical or chemical action, with or without the application of heat. Obviously, a plastic is a permanently deformable or mouldable material. Probably, the largest number of different polymeric materials fall under this category. Polyethylene, polypropylene, polyvinyl chloride, polystyrene, and fluorocarbons, epoxies, phenolics and polyesters may be put under this category. Plastics have a wide variety of combination of properties. There are some plastics which are very rigid and brittle, whereas others are flexible, exhibiting both elastic and plastic deformations when stressed, and some time experiencing sufficient deformation prior to fracture.

Polymers belonging to this class of materials may have any degree of crystallinity, and all molecular structures and configurations (*i.e.*, linear, branched, isotactic, etc.) are possible. Plastic materials may further be classified as thermoplastic or thermosetting. Table 5.10 summarizes the characteristics and typical applications for a number of plastics. Plastic have following properties:

(*i*) Plastics are light weight materials that is their densities and specific gravities are low.

Table 5.10 Characteristics and Typical Applications of Some Plastic Materials

Plastic material (Type) (1)	*Trade names (2)*	*Characteristics (3)*	*Typical applications (4)*
		Thermoplastics	
Acrylonitrile-butadiene-styrene (ABS)	Abson Cycolac Kralastic Lustran Novodur Tybrene	Outstanding strength and toughness, resistant to heat distortion; good electrical properties; flammable and soluble in some organic solvents	Refrigerator linings, lawn and garden equipment, toys, highway safety devices
Acrylics (polymethyl methacrylate)	Acrylite Diakon Lucite Plexiglas	Outstanding light transmission and resistance to weathering; only fair mechanical properties	Lenses, transparent aircraft enclosures, drafting equipment, outdoor signs
Fluorocarbons (PTFE or TFE)	Teflon Fluon Halar Halon Hostaflon TF	Chemically inert in almost all environments, excellent electrical properties; low coefficient of friction; may be used to 260°C (500°F); relatively weak and poor cold-flow properties	Anticorrosive seals, chemical pipes and valves, bearings, antiadhesive coatings, high-temperature electronic parts
Polyamides (nylons)	Nylon Durethan Herox Nomex Ultramid Zytel	Good mechanical strength, abrasion resistance, and toughness; low coefficient of friction; absorbs water and some other liquids	Bearings, gears, cams, bushings, handles, and jacketing for wires and cables
Polycarbonates	Baylon Iupilon Lexan Makrolon Merlon Nuclon	Dimensionally stable; low water absorption; transparent; very good impact resistance and ductility; chemical resistance not outstanding	Safety helmets, lenses, light globes, base for photographic film
Polyethylene	Alathon Alkathene Ethron Fortiflex Hi-fax Petrothene Rigidex Zendel	Chemically resistant and electrically insulating; tough and relatively low coefficient of friction; low strength and poor resistance to weathering	Flexible bottles, toys, tumblers, battery parts, ice trays, film wrapping materials
Polypropylene	Bexphane Herculon Meraklon Moplen Poly-pro Pro-fax Propathene	Resistant to heat distortion; excellent electrical properties and fatigue strength; chemically inert; relatively inexpensive; poor resistance to UV light	Sterilizable bottles, packaging film, TV cabinets, luggage

(Contd.)

Polystyrene	Carinex Celatron Hostyren Lustrex Styron Vestyron	Excellent electrical properties and optical clarity; good thermal and dimensional stability; relatively inexpensive	Wall tile, battery cases, toys, indoor lighting panels, appliance housings
Vinyls	Darvic Exon Geon Pee Vee Cee Pliovic Saran Tygon	Good low-cost, general-purpose materials; ordinarily rigid, but may be made flexible with plasticizers; often copolymerized; susceptible to heat distortion	Floor coverings, pipe, electrical wire insulation, garden hose, phonograph records
Polyester (PET or PETE)	Celanar Crastin Dacron Hylar Melinex Mylar Terylem	One of the toughest of plastic films; excellent fatigue and tear strength, and resistance to humidity, acids, greases, oils, and solvents	Magnetic recording tapes, clothing, automotive tire cords, beverage containers
		Thermosetting Polymers	
Epoxies	Araldite Epikote Epon Epi-rez Lekutherm Nepoxide	Excellent combination of mechanical properties and corrosion resistance; dimensionally stable; good adhesion; relatively inexpensive; good electrical properties	Electrical moldings, sinks, adhesives, protective coatings, used with fibreglass laminates
Phenolics	Bakelite Amberol Arofene Durite Resinox	Excellent thermal stability to over 150°C (300°F); may be compounded with a large number of resins, fillers, etc.; inexpensive	Motor housings, telephones, auto distributors, electrical fixtures
Polyesters	Aropol Baygal Derakane Laguval Laminac Selectron	Excellent electrical properties and low cost; can be formulated for room- or high-temperature use; often fibre reinforced	Helnets, fibreglass boats, auto body components, chairs, fans

(*ii*) They have high resistance to chemical attack, *i.e.* corrosion.
(*iii*) They have very good surface finish when obtained from the dies.
(*iv*) They have high thermal and electrical insulation.
(*v*) As compared to metals their strength is low and also low elastic modulus value.
(*vi*) As compared to metals, plastic have low softening and thermal degradation temperature.
(*vii*) Plastics are poor to fair dimensional stability, particularly in moist conditions.
(*viii*) Plastics are available in different colours as well as transparent form.
(*ix*) With respect to metals, plastics have high coefficient of thermal expansion.
(*x*) These are unsuitable for high temperature applications.
(*xi*) Plastics can be easily fabricated.

Table 5.11 Important Characteristics and Typical Applications of Few Common Elastomers

Material (chemical type)	*Trade/ common name*	*Elongation (%)*	*Useful temperature range [°C]*	*Characteristics*	*Applications (typical)*
Natural poly-isoprene	Natural rubber (NR)	500–760	–60 to 120	Excellent physical properties; good resistance to cutting, gouging, and abrasion; low heat, ozone, and oil resistance; good electrical properties	Pneumatic tyres and tubes; heels and soles; gaskets
Styrene-butadiene copolymer	GRS, Buna S (SBR)	450–500	–60 to 120	Good physical properties; excellent abrasion resistance; not oil, ozone, or weather resistant; electrical properties good, but not outstanding	Same as natural rubber
Acrylonitrile-butadiene copolymer	Buna A, Nitrile (NBR)	400–600	–50 to 150	Excellent resistance to vegetable, animal, and petroleum oils; poor low-temperature properties; electrical properties not outstanding	Gasoline, chemical, and oil hose; seals and O-rings; heels and soles
Chloroprene	Neoprene (CR)	100–800	–50 to 105	Excellent ozone, heat, and weathering resistance; good oil resistance; excellent flame resistance; not as good in electrical applications as natural rubber	Wire and cable; chem. tank linings; belts, seals, hoses, and gaskets
Polysiloxane	Silicone (VMQ)	100–800	–115 to 315	Excellent resistance to high and low temperatures; low strength; excellent electrical properties	High-and low-temperature insulation; seals, diaphragms; tubing for food and medical uses

There are many plastics with outstanding properties. Polystyrene and polymethyl methacrylate are especially useful for applications in which optical transparency is critical, however, it is imperative that the material be highly amorphous or, if semicrystalline, have very small crystallites. The fluorocarbons have a very low coefficient of friction and moreover they are extremely resistant to attack by a host of chemicals, even at relatively high temperatures. They find wide range of uses as coatings. On nonstick cookware, in bearings and bushings, and for high temperature electronic components.

(*ii*) *Elastomers*: These materials have properties similar to polymers. These materials may be repeated stretched or elongated and will return to their original condition upon release of the force producing the elongation. They exhibit elastic behaviour, as compared to polymers, which exhibit greater plastic properties. Included in this category are rubber and rubber like materials. A distinction is made between rubber and elastomers. Rubber must withstand a 200% elongation and rapidly return to its original dimensions. This property of a material to recover from elastic deformation is termed as *resilience.* Table 5.11 summarizes properties and applications of few common elastomers.

There are several criteria which have to be met for a polymer to be elastomeric: (*i*) It must be easily crystallise. We may note that elastomeric materials are amorphous, having molecular chains that are naturally coiled and kinked in the unstressed state. (*ii*) For the coiled chains, chain bond rotations must be relatively free so that they may readily respond to an applied force. (*iii*) To experience relatively large elastic deformations for elastomers, the onset of the plastic deformation must be delayed. This objective is achieved by restricting the motions of chains past one another by crosslinking. The crosslinks acts as anchor points between the chains and prevent chain slippage to occur. In many elastomers, crosslinking is carried out in a process called vulcanization. (*iv*) The elastomer must be above its glass transition temperature. The lowest temperatures at which rubber like behaviour persists for good number of common elastomers is between –50 and –90°C. An elastomer becomes brittle below its glass transition temperature.

The properties of common elastomers are typical and, of course, depend on the degree of vulcanization and on whether any reinforcement used. Natural rubber has an outstanding combination of desirable properties and still utilized to a large degree. However, the most important synthetic elastomer is SBR (styrene butadiene copolymer) which is used especially in automobile tyres, reinforced with carbon black. Acrylonitrile butadiene copolymer (NBR) is another common synthetic elastomer, which is highly resistant to degradation and swelling.

The mechanical properties of even vulcanized rubbers for many applications, *e.g.* automobile tyres, are not satisfactory in terms of tensile strength, abrasion and tear resistance, and stiffness. One may improve these characteristics by additives such as carbon black.

Nowadays, silicon rubbers are widely used for several applications. For these materials the backbone carbon chain is replaced by a chain that alternates Si and oxygen atoms (Fig. 5.44) where *R* and *R′* represent the side-bonded atoms such as hydrogen or group of atoms, *e.g.* CH_3. As an example, the mer structure of polydimethylsiloxane is (Fig. 5.45). No doubt, as elastomers these materials are crosslinked.

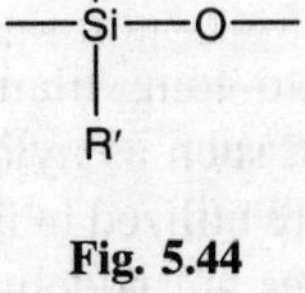

Fig. 5.44

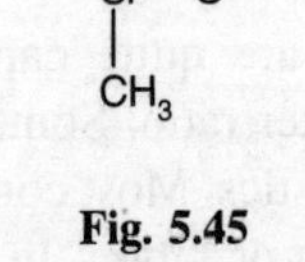

Fig. 5.45

At low temperatures, the silicone elastomers possess a high degree of flexibility (–90°C) and yet these are stable to temperatures as high as 250°C. Moreover, they are resistant to weathering and lubricating oils. A more attractive characteristic is that some silicone rubbers vulcanize at room temperature (RTV rubbers).

Thermoplastic elastomers, sometimes referred to as *elastoplastics,* result from the copolymerisation of two or more monomers. One monomer is used to provide the hard, crystalline features, whereas the other monomer produces the soft, amorphous features. When combined, these form a thermoplastic material that exhibits properties similar to the harder, vulcanised elastomers. Thermoplastic elastomers can be either molded or extruded into finished shapes. Urethanes were the first thermoplastic elastomers to be produced. These products are used in such applications as gaskets, gears and fuel lines. Other thermoplastic elastomers include copolyester, styrene, styrene-butadiene, and the olefins. Copolyester elastomers have more properties but are more expensive than urethanes. They are used as hydraulic hoses, couplings, and cable insulation. Styrene copolymers are less expensive than either urethanes or copolyesters. They have a lower tensile strength but exhibit greater elongation. SBR is used in medical products, tubing, packaging materials, adhesives and sealants. Thermoplastic olefins (TPO) are used in tubing, seals, gaskets, and electrical products. The molecular structure for a thermoplastic elastomer is shown in Fig. 5.46.

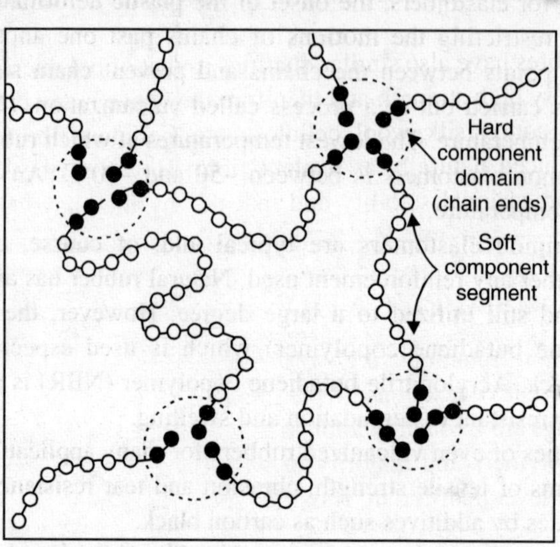

Fig. 5.46 Molecular structure for a thermoplastic elastomer. It consists of 'soft' (*i.e.* butadiene or isoprene) mer centre-chain segments and hard (*i.e.* styrene) domains (chain ends), which act as physical cross-links at room temperature.

(*iii*) *Fibres*: These consists of long molecular chains and all the chains are aligned in the direction of length of the fibre. The directional properties are improved by alignment, *i.e.* the strength of fibres are greatly enhanced in the direction parallel to the fibre length. These provide strength to the molecular units.

The fibre polymers are quite capable of being drawn into long filaments having at least a 100 : 1 length to diameter ratio. Some polymers used as fibres, such as nylon and cellulose acetate, serve equally well as plastics. Most commercial fibre polymers are utilized in the textile industry, being woven or knit into cloth or fabric. In addition, the aramid fibres are widely employed in *composite* materials. A fibre polymer, to be used as a textile material must have some restrictive physical and chemical properties. Fibres, while is used may be subjected to a variety of mechanical deformations, *e.g.* stretching, twisting, shearing and abrasion. Obviously, they must have a high tensile strength over a relatively wide temperature range and a high modulus of elasticity and also abrasion resistances. We may note that these properties are governed by the chemistry of polymer chains and also by fibre drawing process.

Fibres materials should have relatively high molecular weight. The structure and configuration of the chains should allow the production of a highly crystalline polymer as the tensile strength increases with degree of crystallinity. This translates into a requirement for linear and unbranched chains that are symmetrical and have regularly repeating mer units.

Synthetic fibres are produced by the following methods: (*i*) Melt spinning (*ii*) Dry spinning and wet spinning. To improve fibres properties, *i.e.*, to increase/decrease resistance, water or fire proofs, etc., chemical treatment is done. The choice of method mainly depends upon the properties of polymers, *i.e.*, heat stability, melting point and solubility in suitable solvents.

The thermal properties of fibre polymers are important from the point of view of washing and maintaining clothes. These properties include melting and glass transition temperatures. Also, fibre polymers must show chemical stability to a rather extensive variety of enviornments, including acids, bases, dry cleaning solvents, bleaches and sunlight. Moreover, they must be relatively non-flammable and amenable to drying.

Metallic and ceramic fibres have also similar qualities as that of polymeric fibres. However, metallic and ceramic fibres have very high strength and therefore they have many industrial applications. There are also natural fibres, *e.g.* cotton, silk, wool, cellulose, etc. Cellulose fibres are flexible and are strong in tension. Cotton is produced naturally as a plant fibre. However, the synthetic fibres are cheap, durable and have more dimensional stability and that is why they have almost replaced the natural fibres.

5.17 MISCELLANEOUS APPLICATIONS OF POLYMERS

In order to serve one or more functions, *e.g.* (*i*) to protect the item from the environment that may produce corrosive or deteriorative reactions, (*ii*) to improve the appearance of item, and (*iii*) to provide electrical insulation, coatings are frequently applied to the surface of materials. Many of the ingredients in coating materials are polymers and majority of which are organic in origin. One finds that these organic coatings fall into several different classifications, *e.g.* paint, varnish, enamel, lacquer, and shellac.

Adhesives: This is a substance which is used to join together surfaces of two solid materials (usually termed as "adherends") to produce a joint with a high shear strength. The inherent strength of the adhesive may be much less than that of the adherend materials, nevertheless, a strong joint may be produced if the adhesive layer is thin and continuous. If a good joint is formed, the adherend material may fracture or rupture before the adhesive.

Polymeric materials belonging to the groups: thermoplastics, thermosetting resigns, elastomeric compounds, and natural adhesives (animal glue, casein, starch, and rosin) may serve adhesive functions. Polymers adhesives are widely used to join a large variety of material combinations; *e.g.* metal-metal, metal-plastic, metal-ceramic, etc. The service temperature limitation is a major primary drawback. The strength of organic polymers decreases rapidly with rise in temperature but they maintain their mechanical integrity only at relatively low temperatures.

Films: Polymeric materials have recently found widespread use in the form of *thin films.* Films, whose thicknesses lies between 0.025 and 0.125 mm are fabricated and widely used as bags for packaging food products and other items, as textile products, and a host of other uses. The main characteristics of the materials produced and used as films include low density, a high degree of flexibility, high tensile and tear strengths, resistance to attack by moisture and other chemicals, and low permeability to some gases, especially water vapour. The polymers which satisfy these criteria and are manufactured in film form include polyethylene, polypropylene, cellophane and cellulose acetate, etc.

Foams: These are the plastic materials which contain a relatively high volume percentage of small pores. Thermoplastic as well as thermosetting materials are used as foams. These materials include polyurethane, rubber, polystyrene, and polyvinyl chloride. Foams are widely used as cushions in automobiles and furniture as well as in packaging and thermal insulations. The process of foaming is carried out by incorporating into the batch of material a blowing agent that upon heating, decomposes with the liberation of a gas. Gas bubbles are created throughout the now fluid-mass, which remains as pores upon cooling and yield to a sponge-like structure. One can obtain the same effect by bubbling an inert gas through a material while it is in molten state.

5.18 ADVANCED POLYMERIC MATERIALS

In the past few years a number of new polymer materials having unique and desirable combinations of properties have been developed. Many of these polymer materials have found niches in new technologies and also replaced several other materials. Some of these materials include: *ultrahigh molecular weight polyethylene, liquid crystal polymers,* and *thermoplastic elastomers.*

(*i*) *Ultrahigh Molecular Weight Polyethylene* (*UHMWPE*): Most linear polyethylenes have weight-average molecular weights in the range 100,000–200,000, commercially two higher-molecular-weight grades are widely used.

High-Molecular Weight (HMW) polyethylene has $\overline{M}_w$ in the range 300,000–500,000. This can be processed by usual techniques and moreover, it has improved environmental stress crack resistance, impact and tensile strength, and long-term strength retention. HMWPE is used for pipe, film and large blow-molded containers where these properties are important.

UHMWPE has $\overline{M}_w$ in the range 3,000,000–6,000,000 g/mol, which an order magnitude greater than that of high density polyethylene. In fibre form, the trade name of UHMWPE is *spectra.* These materials exhibit following extraordinary characteristics: (*a*) exceptionally high impact-resistance, (*b*) outstanding resistance to wear and abrasion, (*c*) very low-coefficient of friction, (*d*) extremely good chemical resistance to normally encountered solvents, (*e*) self-lubricating and nonstick surface, (*f*) high sound damping and energy absorption characteristics, (*g*) excellent low-temperature properties and (*h*) excellent dielectric properties and electrically insulating.

However, the mechanical properties of this material diminish rapidly with decreasing temperature as it has relatively low melting temperature.

This material does not melt or flow in the normal thermoplastic manner because of its extremely high molecular weight. Therefore, most of its fabrication is based on modifications of the *compression-molding technique*.

This material finds numerous and diverse applications due to its unusual combination of properties, *e.g.* bullet proof vests, composite military helmets, ski-bottom surfaces, fishing line, bowling alley, golf ball cores, bearings, sprockets, gaskets, value seats, conveyor-belt parts and ice-skating rink surfaces, biomedical prostheses, blood filters, marking pen nibs, pump impellers, bushings, material handling equipment (for coal, cement, gravel, grain, etc.).

(*ii*) *Liquid Crystal Polymers* (*LCPs*): These polymer materials are a group of chemically complex and structurally distinct materials. These materials have unique properties and used in diverse applications.

These materials are composed of extended, rod-shaped, and rigid molecules. In view of their molecular arrangement, LCPs do not fall within any of conventional liquid, amorphous, crystalline, semicrystalline classifications, but these may be considered as a new state of matter, *i.e.*, the *liquid crystalline state,* being neither crystalline nor liquid. In the melt (or liquid) condition, other polymer molecules are randomly oriented, whereas LCPs molecules can become aligned in highly ordered configurations. Based on orientation and positional orderings, there are three types of liquid crystals: *smectic, nematic* and *cholesteric*.

LCPs finds principal use in liquid crystal displays (LCDs), on digital watches, laptop computers, and also for other digital displays. Here cholesteric types of LCPs are used, which, at room temperature, are fluid liquids, transparent, and optically anisotropic. The displays consist of two sheets of glass between them is sandwitched the LCP material. The outer face of each glass is coated with a transparent and electrically conductive film and also, in addition, into this film on the side that is viewed or etched the character-forming number/letter elements. When a voltage is applied through the conductive films, *i.e.* between these two glass sheets, over one of these character-forming regions causes a disruption of the orientation of the LCP molecules in this region, a darkening of LCP material, and, in turn, the formation of a visible character.

There are some nematic type of LCPs, which are rigid solids at room temperature. The outstanding combination properties and processing characteristics of these LCPs have found uses in variety of commercial applications. These LCPs materials exhibit the following behaviours:

(*a*) They exhibit excellent thermal stability and may be used up to 230°C.
(*b*) They are stiff and strong. Their tensile moduli range between 10 and 24 GPa and tensile strengths between 125 to 255 MPa.
(*c*) They have high impact strengths, which are retained upon cooling to relatively low temperatures.
(*d*) They exhibit chemical inertness to a wide variety of acids, bleaches, solvents, etc.
(*e*) They show inherent flame resistance and combustion products that are relatively nontoxic.

As regards the processing and fabrication characteristics of these materials, we have

(*a*) One may use all conventional processing techniques available for thermoplastic materials.
(*b*) These materials have extremely low shrinkage and warpage take place during molding.
(*c*) These materials have exceptional dimensional repeatability from part to part.
(*d*) The low melt viscosity of these materials permits molding of thin sections and/or complex shapes.
(*e*) The low heats of fusion of these materials result in rapid melting and subsequent cooling. This helps to shorten molding cycle times.
(*f*) The anisotropric finished part properties of these materials mean that molecular orientation effects are produced from melt flow during molding.

LCPs are widely used by electronics industry (interconnect devices, relay and capacitor housings, brackets, etc.), by the medical equipment industry (in components to be repeatedly sterilized), and in photocopiers and fibre optic components.

(*iii*) *Thermoplastic Elastomers*: See Section 15.

5.19 POLYMER ADDITIVES

The properties of the polymers discussed so far are intrinsic ones, *i.e.*, characteristic of or fundamental of or fundamental to the specific polymer. Many times, one find it necessary to modify the mechanical, chemical, and physical properties to a greater degree than is possible by the simple alteration of this fundamental molecular structure. To enhance or modify many of these properties, foreign substances called *additives* are knowingly introduces so that a polymer material may found more serviceable. Following are the typical additives:

(*i*) *Fillers*: Many plastics are virtually useless alone but are converted into highly serviceable products by combining them with particulate or fibrous solids. This improves their tensile and compressive strengths, abrasion resistance, toughness, dimensional and thermal stability, and other properties. Materials used as particulate fillers include silica flour and sand, wood flour (finely powered sawdust), glass, clay, talc, limestone and also some synthetic polymers. The principal fillers used can be divided into two types: *particulate* and *fibrous*. Among the particulate fillers are silica products, including sand, quartz and diatomaceous earth; silicates, including clay, mica, talc, asbestors, etc. Fibrous fillers, some quite old (cellulosics) and some new "space-age" products (metallic fibres, whiskers) include cellulosic fibres, such as alpha-cellulose and cotton flock; synthetic fibres, etc. Particle sizes range all the way from 10 nm to macroscopic dimensions. In recent years the use of mineral fillers has increased markedly, with emphasis on calcium carbonate. As these inexpensive materials replace some volume of the more expensive polymer, the cost of final product is reduced.

(*ii*) *Plasticizers*: These are added to plastic materials to improve flow and, therefore, processability, and to reduce the brittleness of the product. Thus the addition of plasticizers to plastics improves their flexibility, ductility, and toughness of polymers. Moreover, their presence also produces reductions in hardness and stiffness. Usually, plasticizers are liquids having low vapour pressures and low molecular weights. The small plasticizer molecules occupy positions between the large polymer chains thereby effectively increasing the interchain distance with a reduction in the secondary intermolecular bonding. Plasticizers are commonly used in polymers to make them soft, flexible and tough materials from those of a hard, brittle, glass like solid. An example is the plasticisation of poly (vinyl chloride) and vinylchloride acetate copolymers. One can also achieve similar changes in properties, of course, by alterating the molecular structure of polymer (*e.g.* by copolymerisation, sometimes called *internal plasticisation*). We may note that the plasticizer lowers the glass transition temperature, so that at ambient conditions one may use the polymers in applications requiring some degree of pilability and ductility. These applications are thin sheets or films, raincoats, tubing and curtains.

(*iii*) *Stabilisers*: Under normal environmental conditions, some polymeric materials are subject to rapid deteriorian, usually in terms of mechanical integrity. Usually, this deterioration is a result of exposure to light, in particular ultraviolet (UV) radiation, and also oxidation. UV radiation interacts with and causes a severance of some of the covalent bonds along the molecular chain, which may also result in some crosslinking. Oxidation deterioration is a consequence of the chemical interaction between oxygen atoms and the polymer molecules. Additives that counteract these deteriorative processes are named as *stabilisers*. An important example is vinyl resins. Other applications of stabilizers include the use of carbon black to prevent photochemical degradation (by excluding light), as in polyethylene, and of ultraviolet light absorbers, such as hydrobenzophenones, to improve the light stability of both plastics and their colorants.

(*iv*) *Colorants*: Colorants for plastics include a wide variety of inorganic and organic materials. These impart a specific colour to a polymer. Colorants may be added in the form of dyes or pigments. A few are molecularly dispersed (oil-soluble dyes) or have small particle size and a refractive index near that of the plastic (*organic pigments* such as the phthalocyanines) and lead to transparent coloured products when incorporated into transparent plastics. Others, including *inorganic pigments*, impart opacity to the plastic. Common colorants for plastics include, among many others, titanium dioxide and barium sulphate (white), phthalocyanine blues and greens, ultramarine blues, chrome greens, quinacridone reds and magnets, molybdate orange, cadmium reds and yellows, iron oxide and chrome yellow, carbon black, flake aluminium for a silver metallic effect, and lead carbonate or mica for pearlescence.

The molecules in a dye dissolve and become part of the molecular structure of the polymer, whereas pigments are filler materials which do not dissolve and remain as a separate phase, *i.e.*, usually they have a small particle size, are transparent, and have a refractive index just equal to that of the parent polymer. Other may impart *opacity* as well as *colour* to the polymer as mentioned above.

(*v*) *Flame retardants*: The flammability is a major concern for polymeric materials, especially in the manufacturing of textile goods and children's toys. In pure form, most of the polymers are flammable;

exceptions are those polymers which contain significant contents of chlorine and/or fluorine, such as polyvinyl chloride and polytetrafluoroethylene. One can enhance the flammability resistance of the remaining combustible polymers by additives called *flame retardants.* These flame retardants may function by initiating a chemical reaction that causes a cooling of the combustion region and a cessation of burning. The most useful material imparting flame retardance to plastics is antimony trioxide.

5.20 MANUFACTURING PROCESSES INVOLVING POLYMERS

Polymers are produced from a variety of naturally occurring materials, including wood, air, water, petroleum, natural gas, and salt, etc. Chemists separate these naturally occurring materials into their basic elements and then reassemble these basic elements into different, distinct polymers using heat, pressure and chemical action. For example, polystyrene is produced from coal and petroleum or natural gas. The chemist begins by extracting benzene from coal and ethylene, gas from the natural gas or petroleum. The basic elements benzene and ethylene, are combined to form ethylbenzene. Through the application of heat and pressure, the ethylbenzene is formed and ground to produce polystyrene beads. Several polymers are available in powder, pellet, or premixed form.

Many polymers are also available as liquid casting system. Epoxies are an example of a liquid casting system. Another type of casting polymer is *plastisol.* A plastisol is a liquid vinyl polymer that solidifies when it contacts a heated surface. Plastisols are widely used in dipping processes for insulated coatings.

Polymers are available in expanded or expandable forms. Expanded polymers are low-density cellular polymers available in two varieties, open- and closed cell polymers. These expanded, or *foamed, polymers* come in two forms, rigid and flexible. An example of a rigid, foamed polymer is polystyrene, or styrofoam, which is widely used in packaging. One type of flexible foamed polymer is urethane, which is used in seat cushions and craft work. Foam insulation is an example of expandable polymers. Expandable polymers are used to fill voids and provide thermal and acoustic insulation.

Several methods are used to fashion products from polymers. The method used for the specific polymer depends on several factors: (*a*) whether the polymer material is thermoplastic or thermosetting; (*b*) if material is thermoplastic, the temperature at which if softens; (*c*) the atmospheric stability of the material being formed; and (*d*) the geometry and size of the finished product.

Several methods are used to fashion products from polymers. Some of the more common production methods are casting, blow moulding, compression moulding, transfer moulding, injection moulding, extrusion, lamination, vacuum forming, cold forming, filament winding, calendaring and foaming. Each of these methods has advantages and disadvantages, depending on the type of material used, the product design, and the cost involved.

(*i*) *Castings*: This is the simplest of the manufacturing methods, because no fillers are used and no pressure is applied. However, a mould is required. There are four basic type of casting methods: *simple casting, dip casting, slush casting* and *rotational casting*. In the simple casting process, a prototype, or model, of the material is made first. Dip casting requires a preheated mandrel, a dipping station, a curing oven, a cooling station, and a method to remove the finished casting. Slush casting uses a split method, which has a internal cavity formed in the shape of the finished part. Rotational casting or rotational moulding utilizes a predetermined amount of plastisol, which is placed inside a preheated split mould.

(*ii*) *Moulding*: This is the widely used common method for forming plastic polymers. There are several moulding techniques used, which include compression, transfer, blow, injection and extrusion moulding. For each of the said technique, a finely pelletized for granulized plastic is forced, at an elevated temperature and by pressure, to flow into, fill, and assume the shape of the mould cavity.

5.20.1 Compression and Transfer Moulding

In compression moulding, the moulding powder is compressed between two parts of heated metal mould (Fig. 5.47). Both mould pieces are heated, however, only one is movable. The mould is closed, and heat and pressure are applied, causing the plastic material to become viscous and conform to the mould shape. Prior to moulding, raw material may be mixed and cold pressed into a disc, which is called a preform. Preheating of the preform reduces moulding time and pressure, extend the die lifetime, and produces more uniform finished piece.

This technique is suitable for large and small parts of thermosetting plastics as well as large and bulky parts of thermoplastics. However, it is used with thermoplastics, is more time consuming and expensive.

Transfer moulding is generally used for thermosetting plastics, specially for manufacturing small parts having small deep holes, cavities, metal inserts, etc., because compressive moulding is not suitable for these applications. In this technique, a variation of compression moulding, the solid ingradients are first melted in a heat transfer chamber. As the molten material is injected into the mould chamber, the pressure is distributed more uniformly overall surfaces (Fig. 5.48). This procedure is more suitable for thermosetting polymers and for pieces having complex geometries.

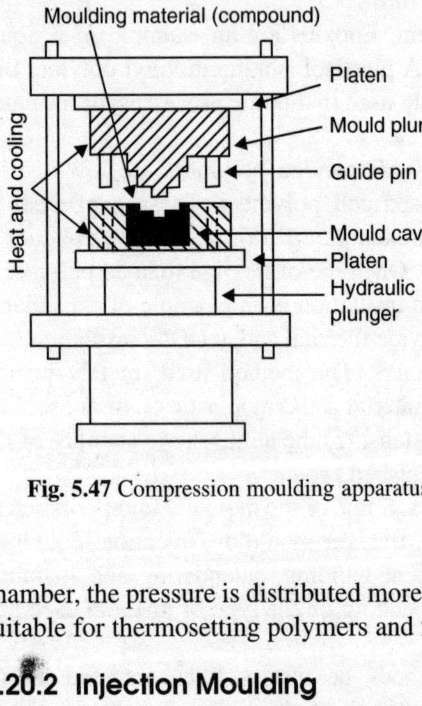

Fig. 5.47 Compression moulding apparatus.

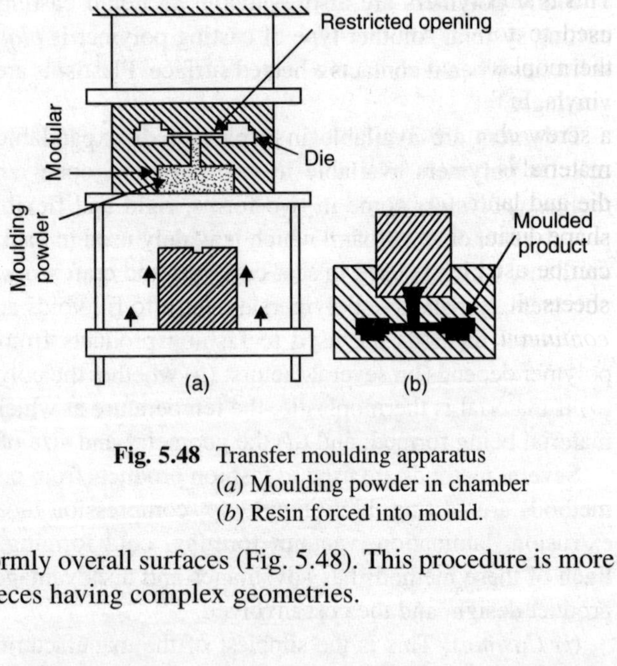

Fig. 5.48 Transfer moulding apparatus
(*a*) Moulding powder in chamber
(*b*) Resin forced into mould.

5.20.2 Injection Moulding

This is used to form thermoplastic materials. Commonly used thermoplastic polymers include acrylics, fluorocarbons, nylon, polyethylene, polystyrene and vinyls. It is an extremely fast method of production. A schematic cross-section of the apparatus is shown in Fig. 5.49. Raw material is fed into a hopper, which is screw-fed into the pressure chamber. As the screw turns, the material is fed into the heating chamber, where it is melted. The melted material continues through a nozzle that is seated against the mould. The nozzle allows the molten material to flow into the cavity through a system of gates and runners. The die remains relatively cool; thus, the product solidifies almost upon filling the cavity. To prevent the premature cooling (and resultant solidifying) of the molten material, the material is forced through the barrel into the cavity under pressure. The mould is then opened and the part is ejected. The method is used to produce such items as toys, plastic models and parts, tile, food containers, battery cases, and radio cabinets.

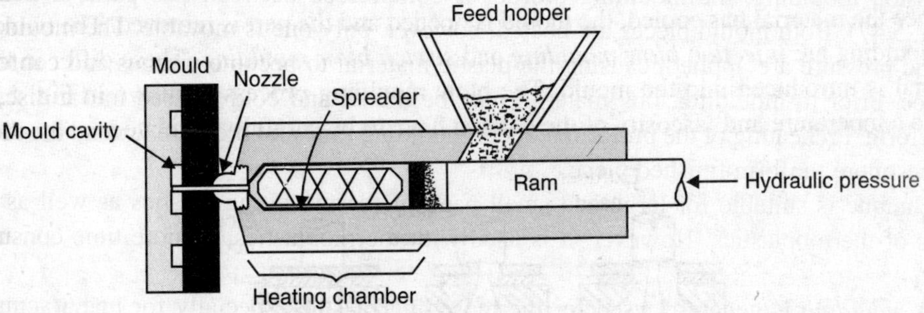

Fig. 5.49 Injection moulding apparatus.

5.20.3 Extrusion Moulding

This is simply injection moulding of a viscous thermoplastic through an open-ended die. This process is used to form materials that have long, uniform-cross-section. It is typically involved in forming thermoplastic polymers, such as acrylics, cellulosics, fluorocarbons, nylon, styrene, polyethylene and vinyls. In this process raw material is fed into a hopper, where it is gravity-fed into a screw chamber. In a screw chamber, a rotating screw forces the material into and through a preheated section. Here the material is compressed and then forced through a heated die onto a conveyor. As the material leaves the die and lands on the belt, it is cooled by an air or water spray to harden it sufficiently, preserving its shape during transportation (Fig. 5.50). Usually, the material is then cut to length or coiled. This process can be used to produce any material that has a uniform cross-section, such as pipe, tubing, rods and sheets. It is also used to form pellets for injection moulding. Extrusion is very fast and allows for *continuous production*.

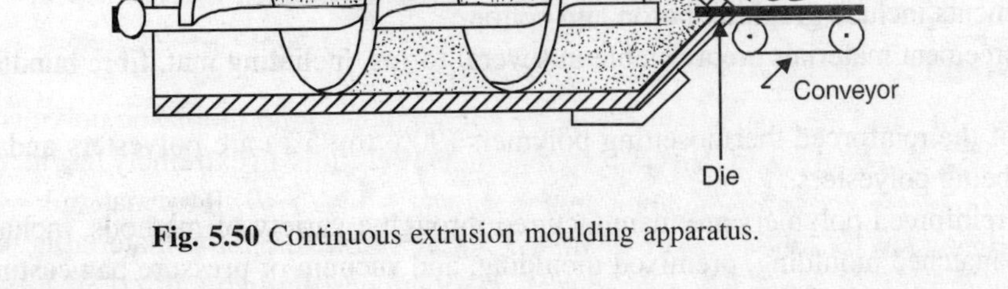

Fig. 5.50 Continuous extrusion moulding apparatus.

5.20.4 Extrusion Blow Moulding

This process is used to form hollow shapes, such as bottles and containers. A thermoplastic tube called a *parison* is fed through the opening formed by the two halves of split mould. Once the necessary length of the parison has entered the mould, the two halves close, sealing one end of the parison, forcing the parison to expand, following the contour of the inner mould cavity. This mould is kept cool during the blow moulding process, so that the thermoplastic parison solidifies as it comes into contact with the mould. Once the material has cooled, the mould is opened and the part is removed. The two other types of blow moulding are *injection blow moulding* and *stretch blow moulding*. These differ in the way the raw material is introduced into the mould. The blow moulding process is shown in Fig. 5.51. In this process the temperature and viscosity of the parison have to be carefully regulated.

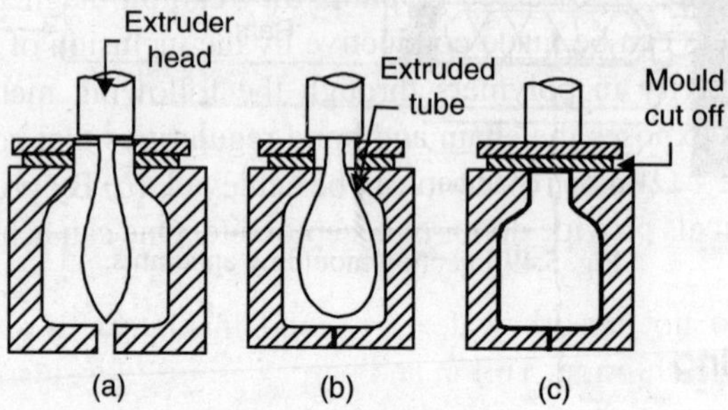

Fig. 5.51 Extrusion blow moulding (*a*) mould open (*b*) mould closed, tube expanding under pressure and (*c*) tube fully blows

5.21 REINFORCED POLYMERS

These are polymers whose properties have been altered through the use of reinforcement material. Glass, mineral and carbon fibres can be added to polymers to alter their properties. Most polymers are available as glass fibre-reinforced products. The glass fibres, whose diameters range from 0.0002 to 0.001 inch in diameter, are coated with a *resin* and a coupling agent. Property increase of 200% or better can be obtained through the glass fibre-reinforcement.

The strength of these reinforced polymers depends on the type and quantity of *filler* used and the degree to which the resin has wetted the reinforcement material. One of the more common filler materials is glass fibre.

Carbon fibre-reinforced materials are more expensive than fibre-reinforced materials. Carbon fibre-reinforced materials have high tensile strength, stiffness, and greater related mechanical properties. They also produce lower coefficients of expansion, improved creep resistance, better wear resistance, greater toughness, and higher strength-to-weight ratios than glass fibre-reinforced materials.

Glass and carbon are the most common fibre-reinforcements. Other materials used in fibre-reinforcements include graphite, boron and cotton.

Reinforcement materials are available in several forms, including mat, fibre bundles, chopper fibres and cloth.

Most of the reinforced thermosetting polymers (90% to 95%) are polyesters and epoxies, with the majority being polyesters.

These reinforced polymers are manufactured through a variety of methods, including hand lay-up, spray-up, matched moulding, premixed moulding, and vacuum or pressure bag casting.

These materials are replacing metals in a variety of applications. They offer similar or better strength, characteristics at lower weight and often, at lower cost. They are frequently easier to produce, are resistant to corrosion and chemicals, and offer a wide variety of manufacturing processes, from the very simple and inexpensive to the higher volume and more intricate processes.

5.22 BEHAVIOUR OF POLYMERS

(*i*) *Electrical Behaviour*: Polymer materials are very good insulators. They can be easily applied on the wires as polymerised liquid as well as partially polymerised liquid to provide a *uniform coating*. This coating gets polymerised around the wire. There are some plastics which allow the formation of film easily and therefore such plastics are used for wire insulation.

Polymers do not exhibit any magnetic properties. However, when polymers are combined with filler materials, *e.g.* the addition of finely powdered graphite, they exhibit magnetic properties and conduct electricity. Obviously, polymers can be made conductive by the inclusion of second conductive phase. One can achieve the conductivity in polymers through the following methods: (*a*) *Mistreating by gamma ray irradiations*. This exposes the chain and bond regularities that contain donor and acceptor sites. Resistivities in the range 1 Ω-m to 10^{10} Ω-m can be achieved. (*b*) By producing a molecular chain which has radicals. These radicals provide donor or acceptor site. One can achieve resistivities as low as 0.12 Ω-m.

Those polymers which do not absorb water are good insulators, *e.g.* polyethylene, polytetrafluorethylene PTFE, polyvinyle chloride. Tuflon and perpex are rigid insulators.

The polymers which are sensitive to polarisation of structure exhibit *dielectric* properties. The polymers which contain natural dipoles, the value of dielectric constant is more, *e.g.* the polymers containing polar groups, *e.g.* chloride, hydroxal or sulphur atom readily become electrically polarised when placed in a electric field, resulting in high dielectric constant.

The dipoles at low frequencies respond to the changes in electric field. With the rise in frequency, the response of polar groups decreases and only electronic polarisation occurs. Figure 5.52 shows the behaviour of dielectric constant with frequency. The dielectric constant decreases with the increase of frequency and also with temperature.

The dielectric constant of plastics which contain plasticisers or fillers, increases artificially. This phenomenon is known as *interfacial polarisation*.

(*ii*) *Mechanical properties*: See Section 5.13.

(*iii*) *Chemical properties*: Although the organic materials are covalently bonded but they can be ionized to some extent, *e.g.* acetic acid produces negative acetate radical and positive hydrogen ion, as follows:

$$CH_3COOH \rightarrow CH_3COO^- + H^+$$

Obviously, due to ionisation of organic salts and when the reaction proceeds backward, a weak acid is generated.

The polymers which are capable of only limited ionisation are termed as *polyelectrolytes*. These polyelectrolyte materials possess both covalent and ionic bonds. They are called ionic conductors of electricity. Since their diffusion rate is slow and hence conduction through them is lesser.

During chemical reaction, the chemical bonds break and recombine. However, for polymerisation process, the chemical reactions are essential.

Most of the organic materials are flammable materials. Organic compounds react with oxygen and produce CO_2 and H_2O, *i.e.* complete combustion of hydrocarbon,

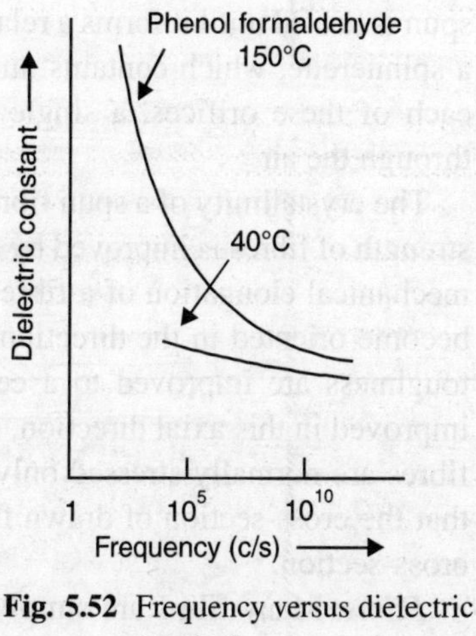

Fig. 5.52 Frequency versus dielectric constant.

$$C_nH_m + \left(n + \frac{m}{4}\right) O_2 \rightarrow nCO_2 + \frac{m}{2} H_2O$$

During this process due to formation of CO_2 energy is released but energy is also required to dissociate the hydrocarbons into their constituent elements. Hydrocarbons dissociate into carbon and hydrogen and then again combine with oxygen to form CO_2 and H_2O. If hydrogen in the chain is replaced by some other element in the chain, the combustibility reduces.

The explosives have sufficient oxygen and therefore combustion in them is easy and quick. Such materials require negligible energy for their dissociation. The dissociation of cellulose nitrate and nitroglycerine are as follows:

$$\underset{\text{(Cellulose nitrate)}}{2C_6H_7\,(ONO_2)_3\,O_2 + 2O_2} \rightarrow \underset{\text{(Combustion products)}}{CO_2 + 11CO + 7H_2O + 6NO}$$

$$\underset{\text{(Nitroglycerine)}}{2C_3H_5\,(ONO_2)_3} \rightarrow \underset{\text{(Combustion products)}}{CO_2 + 5CO + 5H_2O + 6NO}$$

Nitroglycerine for its complete combustion does not require any external source of oxygen, whereas cellulose nitrate requires very little oxygen from air.

(iv) Hardness: Most plastic materials have low scratch resistance. One can improve the hardness of these materials by lamination with glass fibre.

(v) Thermal Properties: Due to depolymerisation, the polymerised thermosetting polymers lose their strength at high temperature. However, one can expose thermoplastic polymers to high temperatures.

Polymers become unsuitable at subzero temperature as they get embrittled at low temperature.

These materials show low thermal diffusivity, poor heat resistance, high flammability and poor ignition and burning capacity in comparison to metals and ceramics.

(vi) Optical Properties: Clear reflectivity, permanent colouring and lustre are some advantages of polymers. One can make them fully transparent to fully opaque as per requirement.

5.23 FABRICATION OF FIBRES AND FILMS

(i) Fibres: The process by which fibres are formed from bulk polymer material is called *spinning*. Most often, fibres are spun from the molten state in a process called melt spinning. First, the material to be spun is heated until it forms a relatively viscous liquid and then it is pumped down through a plate called a spinnerette, which contains numerous small round holes. When the molten material passes through each of these orifices, a single fibre is formed, which solidifies almost immediately upon passing through the air.

The crystallinity of a spun fibre will depend on its rate of cooling during the process of spinning. The strength of fibres is improved by a *postforming process* called *drawing*. However, drawing is simply the mechanical elongation of a fibre in the direction of its axis. The molecular chains during this process become oriented in the direction of drawing, such that the tensile strength, modulus of elasticity, and toughness are improved to a certain extent. Although the mechanical strength of a drawn fibre is improved in this axial direction, strength is reduced in a transverse or radial direction. Moreover, since fibres are normally stressed only along the axis, this strength differential is not critical. We may note that the cross-section of drawn fibres is nearly circular, and the properties are uniform throughout the cross-section.

Films: Many films are simply extruded through a thin die slit and then this may be followed by a rolling operation that serves to reduce thickness and improve strength. Alternatively, film may be blown: continuous tubing is extruded through an annular die; then, by maintaining a very carefully controlled positive gas pressure inside the tube, one may reduce wall thickness continuously to produce a thin cylindrical film, which may be cut and laid flat. Some of the newer films are produced by coextrusion; that is, multilayers of more than one polymer type are extruded simultaneously.

5.24 WOOD

Wood is a natural, renewable, organic substance with a wide variety of uses. The main constituent of wood is cellulose. Of particular interest are trees that grow to sufficient height with diameters large enough to be useful in the production of lumber and wood products. Other plants, such as bushes, vines, and shrubs, are also wood, but these plants have little use in structural application.

Wood is widely available and easy to work, although it burns easily, has a variety of defects, and is subject to attack by insects and fungi. Therefore, special precautions must be taken to preserve it. When wood has been prepared for subsequent use in building and construction, it is called *lumber*.

Wood is generally divided into two easily recognizable areas, *heatwood* and *sapwood.* Very young trees are sapwood. Heatwood is usually called as *truewood*. Heatwood cells are dead and no longer contribute to the life processes of a tree.

Wood typically consists of tiny, hollow, fibrous cell that are composed of *cellulose* and *lignin*. The typical composition of wood is 60% cellulose, 28% lignin and 12% other materials. Lignin acts primarily as a cement to hold the wood cells together, thus supporting the tree. Cellulose is made up primarily of carbohydrates produced by photosynthesis.

Trees are often classified as *hardwoods* or *softwoods*. This distinction is based on their cell structure and not on their mechanical properties. However, most hardwoods are harder and more durable than softwoods, but this relationship is not an absolute. In general, hardwoods are *deciduous*, or broad-leaves, trees, which drop their leaves in the fall. Softwoods are *conifers*, which bear needles instead of leaves and produce seed cones. The softest and hardest of all woods are found in the tropics; the lightest is balsa and the heaviest is iron wood, or lignum vitae. The principal distinction between the two types of wood is that hardwoods contain very large cells, known as *vessels*, or *pores*, whereas softwoods do not. In addition, hardwoods have wood *parenchyma* and ray cells. Wood parenchyma are thin, soft-walled cells that often surround the vessels and contribute to the grain pattern of the wood. The principal woods used are listed in Table 5.12.

Table 5.12 Principal Types of Wood

Wood	*Uses*
(*i*) *Softwoods*	
Cedar	Clothes, closets, pencils
Cypress	Tanks, silos, storage containers
Douglas fir	Plywood
Sitka spruce	Aircraft applications
Southern pine	Railroad ties, boxes, trim
Western white pine	Matches, boxes, crating
(*ii*) *Hardwoods*	
Ash	Tool handles, baseball bats
Basswood	Boxes, crating
Hickory	Tool handles
Maple	Flooring, boxes, crating
Oak	Posts, flooring, pallets, trim

Table 5.13 Woods Used for Construction

Use	*Wood*
Structural	• Yellow pine, douglas fir, hemlock, red wood, cedar
Light framing	• Spruce, hemlock, yellow pine, fir, cedar
Siding	• Yellow pine, cypress, red wood, spruce, cedar, hemlock, fir
Exterior finish	• White pine, cypress, red wood, spruce, cedar, hemlock, fir
Interior finish	• *Painted*: White and yellow pine, birch, gum, red wood, popular, spruce, cedar, fir *Natural or Stained*: Oak, white pine, birch, red wood, yellow pine, walnut, ash, cherry, gum, maple
Sashes, doors, frames	• White pine, fir, birch, red wood, yellow pine
Sheathing, roofing	• Fir, hemlock, pine, red wood, spruce
Flooring	• Yellow pine, maple, oak, hemlock, birch, beech, spruce, gum, fir

Wood, for construction purposes, is primarily used as structural timber; light framing, siding; exterior finish; interior finish, sashes, doors, and frames; sheathing, roofing, and subflooring; and flooring. The principal woods used for each of these purposes are shown in Table 5.13.

The physical properties of wood, including the grain pattern, physical characteristics, weight and moisture content, relate to the structure of the wood. Mechanical properties, such as compressive strength, flexural strength, and modulus of elasticity are associated with the ability of the wood to withstand applied forces.

Lumber grading is performed to allow the user to select the best quality product suited to the application. In almost all grades, for all types of lumber, grading is based on the number, character, and location of various features and defects that may lower the strength, durability or visual value of the product. Common features include knots, checks, splits, pitch pockets and strains.

Example 1: A copolymer consists of 15 wt % polyvinyl acetate (PVA) and 85 wt % polyvinyl chloride (PVC). Determine the mole fraction of each component. [B.Tech. 2005]

Solution: Let the basis be 100 g of copolymer; therefore we have 15 g of PVA and 85 g of PVC. First we determine the number of moles of each component that we have, and then we calculate the mole fractions of each.

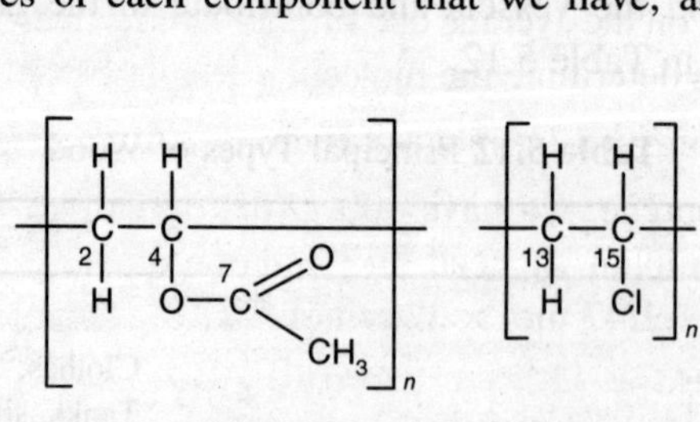

Fig. 5.53 Structural formulas for the mers of (*a*) polyvinyl acetate and (*b*) polyvinyl chloride.

Moles of polyvinyl acetate. The molecular weight of the PVA mer is obtained by adding up the atomic masses of the atoms in the structural formula for the PVA mer (Fig. 5.53*a*):

4 C atoms × 12 g/mol + 6 H atoms × 1 g/mol + 2 O atoms × 16 g/mol = 86 g/mol

$$\text{No. of moles of PVA in 100 g of copolymer} = \frac{15\,\text{g}}{86\,\text{g/mol}} = 0.174$$

Moles of polyvinyl chloride. The molecular weight of the PVC mer is obtained from Fig. 5.53*b*
2 C atoms × 12 g/mol + 3 H atoms × 1 g/mol + 1 Cl; atoms × 35.5 g/mol = 62.5 g/mol

$$\text{No. of moles of PVC in 100 g of copolymer} = \frac{85\,\text{g}}{62.5\,\text{g/mol}} = 1.36$$

$$\text{Mole fraction of PVA} = \frac{0.174}{0.174 + 1.36} = 0.113$$

$$\text{Mole fraction of PVC} = \frac{1.36}{0.174 + 1.36} = 0.887.$$

Example 2: Determine the mole fractions of vinyl chloride and vinyl acetate in a copolymer having a molecular weight of 10,520 g/mol and a degree of polymerisation (DP) of 160. [B.Tech. 2006]

Solution: From Example 1, the molecular weight of the PVC mer is 62.5 g/mol and that of the PVA mer is 86 g/mol.

Since the sum of the mole fractions of polyvinyl chloride, f_{vc}, and polyvinyl acetate, f_{va} = 1, $f_{va} = 1 - f_{vc}$. Thus, the average molecular weight of the copolymer mer is

$$MW_{av}(\text{mer}) = f_{vc}\,MW_{vc} + f_{va}\,MW_{va} = f_{vc}\,MW_{vc} + (1 - f_{vc})MW_{va}$$

The average molecular weight of the copolymer mer is also

$$MW_{av}\,(\text{mer}) = \frac{MW_{av}\,(\text{polymer})}{DP} = \frac{10{,}520\ \text{g/mol}}{160\ \text{mers}} = 65.75\ \text{g/(mol.mer)}$$

The value of f_{vc} can be obtained by equating the two equations of MW_{av} (mer).

$$f_{vc}\,(62.5) + (1 - f_{vc})(86) = 65.75 \text{ or } f_{vc} = 0.86$$

$$f_{va} = (1 - f_{vc}) - 1 \quad 0.86 = 0.14.$$

Example 3: How much sulfur must be added to 100 g of polyisoprene rubber to crosslink 5 per cent of the mers? Assume all available sulphur is used and that only one sulphur atom is involved in each cross-linking bond. [B.Tech.]

Solution: As shown in Fig. 5.54, on the average one sulphur atom will be involved with one polyisoprene mer in the crosslinking. First we determine the molecular weight of the polyisoprene mer.

MW (polyisoprene) = 5 C atoms × 12 g/mol + 8 H atoms × 1 g/mol = 68.0 g/mol

Thus, with 100 g of polyisoprene, we have 100 g/(68.0 g/mol) = 1.47 mol of polyisoprene. For 100 per cent crosslinking with sulphur we need 1.47 mol of *S* or

1.47 mol × 32 g/mol = 47.0 g sulphur

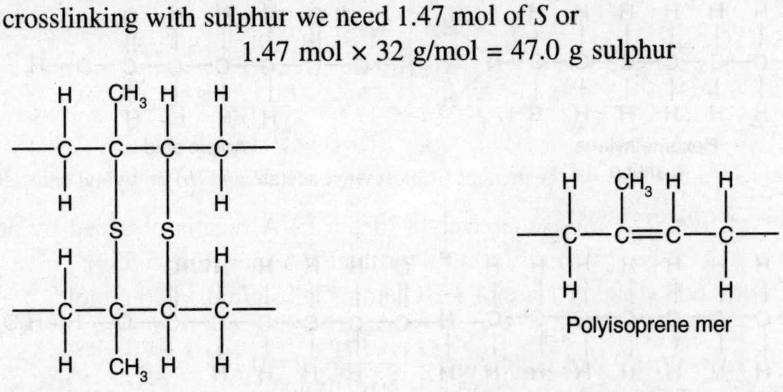

Fig. 5.54 cis-1, 4 polyisoprene chain after crosslinking with sulphur at the active double-bond sites.

For crosslinking 5 per cent of the bonds we need only

$$0.05 \times 47.0 \text{ g } S = 2.35 \text{ g } S$$

Example 4: A butadiene-styrene rubber is made by polymerising one monomer of styrene with eight monomers of butadiene. If 20 per cent of the crosslink sites are to be bonded with sulphur, what weight per cent sulphur is required? [B.Tech.]

Solution: Basis: 100 g of copolymer.

In the copolymer we have one mole of styrene combined with eight moles of polybutadiene. Thus on a weight basis we have:

8 moles of polybutadiene × 54 g/mol = 432 g
1 mole of polystyrene × 104 g/mol = 104 g
Total weight of copolymer = 536 g

The weight ratio of polybutadiene to copolymer = 432 /536 = 0.806

Thus, in 100 g of copolymer we have 100 g × 0.806 = 80.6 g of butadiene or 80.6 g/54 g = 1.493 moles of polybutadiene.

Grams of S for 20 per cent crosslinking = (1.493 moles) (32 g/mole) (0.20) = 9.55 g S.

$$\text{Weight } \% S = \left(\frac{9.55\text{g}}{100\text{g} + 9.55\text{g}}\right) 100\% = 8.72\%.$$

Example 5: A butadiene-acrylonitrile rubber is made by polymerising one acrylonitrile monomer with three butadiene monomers. How much sulfur is required to react with 100 kg of this rubber to crosslink 20 per cent of the crosslink sites? [B.Tech.]

Solution: Basis: 100 g of copolymer

3 moles of polybutadiene × 54 g = 162 g
1 mole of polyacrylonitrile × 53 g = 53 g
Total weight of copolymer = 215 g

The weight ratio of polybutadiene to copolymer = 162 g/215 g = 0.7535

In 100 g of copolymer we have 100 g × 0.7535 = 75.35 g or 75.35 g/54 g/mol = 1.395 mol

Weight of S for 20 per cent crosslinking = (1.395 mol) (32 g/mol) (0.20) = 8.93 g S or 8.93 kg.

Example 6: Calculate the degree of polymerisation if 6, 6-nylon has a molecular weight of 120,000 g/mol. [B. Tech.]

Solution: The linear nylon chain is produced when a hydrogen atom from hexamethylene diamine combines with an OH group from adipic acid to form a water molecule as shown below:

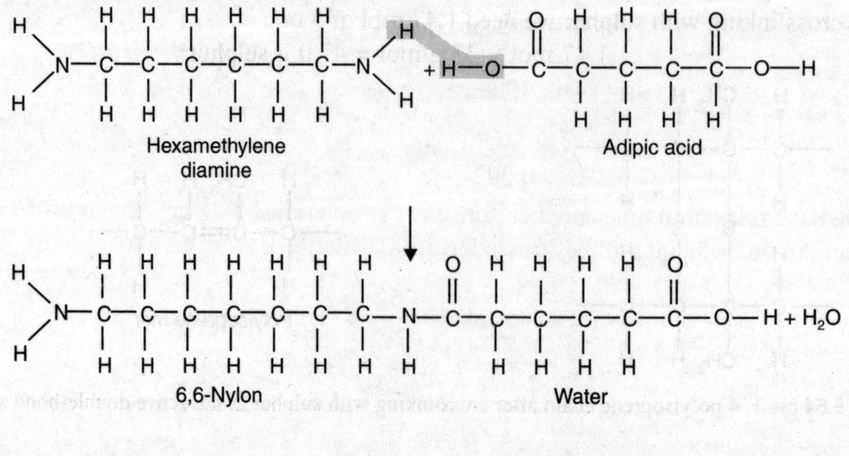

We may note that the reaction can continue at both ends of new molecule and consequently, long chains may form. This polymer is called 6, 6-nylon polymer as both monomers contain six carbon atoms.

When a long chain forms, there is, on an average, one water molecule released for each reacting molecule. The molecular weights are 116 g/mol for hexamethylene diamine, 146 g/mol for adipic acid, and 18 g/mol for water. The repeat unit for 6, 6-nylon is

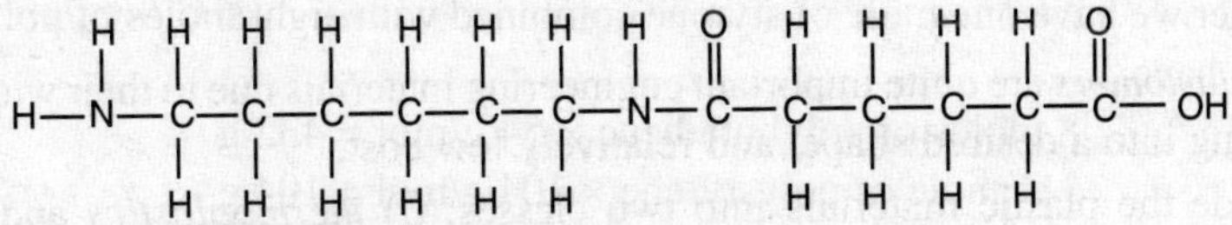

The molecular weight of the repeat unit is the sum of the molecular weights of the two monomers, minus that of two water molecules that are produced:

$$M_{\text{repeat unit}} = 116 + 146 - 2(18) = 226 \text{ g/mol}$$

$$\text{Degree of polymerisation} = \frac{1,20,000}{226} = 531$$

We may note that the degree of polymerisation refers to the total number of repeat units in the chain. The chain contains 531 hexamethylene diamine and 531 adipic acid.

Example 7: A band of polyisoprene is to hold together a bundle of steel rods for up to one year. If the stress on the band is less than 1500 psi, the band will not hold the rods tightly. Design the initial stress that must be applied to a polyisoprene band when it is slipped over the steel. Experiments revealed that an initial stress of 1000 psi decreased to 980 psi after six weeks. [B.Tech.]

Solution: Although the strain of the elastomer band may be constant, but the stress will decrease over time due to stress relaxation. We have

$$\sigma = \sigma_0 \exp\left(-\frac{t}{\lambda}\right)$$

where σ_0 is original stress and λ is relaxation time given by

$$\lambda = \lambda_0 \exp\left(\frac{Q}{RT}\right)$$

Here σ_0 is constant, Q is the activation energy related to the ease with which polymer chains slide past each other.

Here $\sigma = 980$ psi, $\sigma_0 = 1000$ psi

and $t = 6$ weeks

$\therefore$ $980 = 1000 \exp(-6/\lambda)$

or $\lambda = \dfrac{6}{0.202} = 297$ weeks.

Now, we have relaxation time and therefore, we can determine the stress that must be initially placed onto the band in order that it will be stressed to 1500 psi after 1 year (52 weeks). Thus

$$1500 = \sigma_0 \exp\left(-\frac{52}{297}\right)$$

or $$\sigma_0 = \frac{1500}{0.839} = 1788 \text{ psi.}$$

Obviously, the polyisoprene band must be made significantly undersized to it can slip over the materials it is holding together with a tension of 1788 psi. After one year, the stress will be 1500 psi.

SUMMARY

- *Plastics* and *elastomers* are quite important engineering materials due to their wide range properties, ease of forming into a desired shape, and relatively low cost.
- One can divide the plastic materials into two classes: (*i*) *thermoplastics* and (*ii*) *thermosetting plastics* (thermosets).
- A plastic material that requires heat to make it formable (plastic) and upon cooling, retain its shape is termed *thermoplastic*. These are composed of chain polymers with the bonds between the chains being of the secondary permanent dipole type. These materials can be repeatedly softened when heated and harden when cooled. Polyethylenes, vinyls, acrylics, cellulosics, and nylons are typical examples of thermoplastics.
- A plastic material that has undergone a chemical reaction by the action of heat, catalysis, etc., leading to a crosslinked network macromolecular structure is termed *thermosetting plastic* (*thermoset*). These cannot be remelted and reprocessed since when they are heated they degrade and decompose. There are some thermosetting reactions which take place at room temperature without the use of heat and pressure. Phenolics, unsaturated polysters, and epoxies are examples of typical thermoset plastics.
- The chemicals required for producing plastics are derived mainly from petroleum, natural gas and coal.
- Plastic materials are produced by the polymerising of many small molecules called *monomers*. Monomer is a simple molecular compound that can be covalently bonded together to form long molecular chains (polymers). Ethylene is an example of monomer.
- Thermoplastics are composed of long-molecular-chain polymers, with the bonding forces between the chains being of the secondary permanent dipole type.
- A high-molecular-mass compound whose structure consists of a large number of small repeating units called *mers* is termed *chain polymer*. In most polymers, carbon atoms make up most of the main chain atoms.
- A repeating unit in chain polymer molecule is called *mer*.
- Thermosetting plastics are covalently bonded throughout with strong covalent bonding between all the atoms.
- The chemical reaction in which high-molecular mass molecules are formed from monomers is termed *polymerisation.*
- The chemical reaction in which high-molecular-mass molecules are formed from two or more monomers are termed *copolymerisation.*
- The polymerisation mechanism whereby each polymer molecule increases in size at a rapid rate once growth has started is called *chain polymerisation*. This type of reaction takes place in three steps: (*i*) chain initiation, (*ii*) chain propagation, and (*iii*) chain termination. The name implies a chain reaction and usually initiated by some external source, *e.g.* the chain polymerisation of ethylene into polyethylene.

- The molecular mass of a polymer chain divided by the molecular mass of its mer is called *degree of polymerisation*.
- The number of active bonding sites in a monomer is called *functionality*. If the monomer has two bonding sites, it is said to be *bifunctional*.
- A polymer consisting of only one type of monomeric unit is called *homopolymer*.
- A polymer chain consisting of two or more types of monomeric units is called *copolymer*.
- The formation of primary valence bonds between polymer chain molecules is called *crosslinking*. When extensive crosslinking occurs as in the case of thermosetting resins, crosslinking makes one supermolecule of all the atoms.
- The centre of the temperature range where a heated thermoplastic upon cooling changes from a rubbery, leathery state to that of brittle glass is called *glass transition temperature*.
- Chemical agents added to plastic compounds to improve flow and processibility and to reduce brittleness is called *plasticizers*. Example is plasticized polyvinyl chloride.
- A low-cost inert substance added to plastics to make them less costly is termed *filler*. Fillers may also improve some physical properties such as tensile strength, impact strength, hardness, wear resistance, etc.
- Thermosetting polymers are highly crossedlinked into a three-dimensional network structure. Typically, high glass temperatures, good strength, and brittle behaviour are found. Once cross-linking occurs, these polymers cannot be easily recycled.
- *Elastomers* or *rubbers* form a large subdivision of polymeric materials and are of great engineering importance. A material that at room temperature stretches under a low stress to at least twicc its length and then quickly returns to almost its original length upon removal of the stress.
- The isomer of 1, 4 polyisoprene that has the methyl group and hydrogen on the same side of the central double bond of its mer is *cis-1, 4 polyisoprene*. Natural rubber consists mainly of this isomer.
- *Natural rubber* is obtained from tree plantations and is still much in demand (about 30% of world's rubber supply) because of its superior elastic properties. Synthetic rubbers account for about 70% of the world's rubber supply, with styrene-butadiene being the most commonly used type. Other synthetic rubbers such as nitrile and polychloroprene (neoprene) are used for applications where special properties such as resistance to oils and solvents are needed.
- The isomer of 1, 4 polyisoprene that has the methyl group and hydrogen on opposite sides of the central double bond of its mer is expressed as trans *1, 4 polyisoprene*.
- A chemical reaction that causes crosslinking of polymer chains is called *vulcanisation.* Vulcanisation usually refers to the crosslinking of rubber molecular chains with sulphur. However, the word vulcanisation is also used for other crosslinking reactions of polymers such as those that occurs in some silicone rubbers.
- Manufacturing processes depend on the behaviour of the polymers. Processes such as extrusion, injection molding, thermoforming, casting, drawing, and spinning are made possible by the viscoelastic behaviour of the thermoplastics. The non-reversible behaviours of bonding in thermosetting polymers limits their processing to fewer techniques, *e.g.* as compression molding, transfer molding and reaction-injection molding.

SUGGESTED READINGS

1. L.H. Sperling, *Introduction to Physical Polymer Science*, Wiley, New York (2001).

2. A. Rudin, *The Elements of Polymer Science and Engineering*, 2nd ed., Academic Press, San Diego (1998).
3. C.A. Harper (ed.), *Handbook of Plastics, Elastomers and Composites*, 3rd ed., McGraw-Hill, New York (1996).

REVIEW QUESTIONS

1. What is a polymer? Write their main characteristics? What are their uses?
2. What are meant by the terms *saturated* and *unsaturated* in relation to polymers?
3. Differentiate between *polymorphism* and *isomerism*.
4. Explain three types of polymerisation.
5. What do you understand by polymerisation? What is the difference between addition polymerisation and condensation polymerisation? [AMIE]
6. What is meant by polymerisation? What are the two broad classifications of polymers? Discuss them. [AMIE]
7. Define the term 'polymerisation'. Explain the different types of polymers with examples and applications. [AMIE]
8. What is polymerisation? With the help of suitable examples, compare and contrast and processes of addition polymerisation and condensation polymerisation. [AMIE, M.Sc. (M.S.)]
9. What are the basic differences between the thermosetting and thermoplastic polymers? [B.E.]
10. Name three commonly used thermoplastic polymers and their applications. [Diploma]
11. Name three commonly used thermosetting polymers and their applications. [Diploma]
12. (*a*) Discuss briefly the different polymerisation mechanisms.
    (*b*) What are the general properties of polymeric materials? [AMIE]
13. (*a*) What is the difference between *configuration* and conformation in relation to polymer chains? [B.E.]
14. Make comparisons of thermoplastic and thermosetting polymers (*i*) on the basis of mechanical characteristics upon heating, and (*ii*) according to possible molecular structure. [AMIE, B.E.]
15. (*a*) What feature is necessary in a monomer for addition polymerisation to be possible? Is this same feature necessity for condensation polymerisation?
    (*b*) Describe the difference between thermoplastics and thermosetting polymers in terms of applied stress, increased temperature and atomic structure.
    (*c*) Describe briefly how each of the following modifies the strength of a polymer: (*i*) increased degree of polymerisation, (*ii*) increased branching, (*iii*) increased crosslinking and (*iv*) increased crystallinity. [AMIE, M.Sc.(M.S.)]
16. What is the effect of the use of plasticizer or thermoplastic materials? [AMIE]
17. Explain the types of bonding that exist in polymers? [B.E.]
18. Explain the lamination process and the purpose for laminating plastics. [Diploma]
19. List and describe the six major production techniques that are used in the plastics industry. [M.Sc.]
20. (*a*) List the principal additives used in the manufacture of plastic articles. (*b*) What are the special properties of plastics that make them useful engineering materials? [AMIE]
21. What is the difference between compression moulding and transfer moulding of plastic parts? For what type of work is the latter used? [AMIE]
22. What are linear polymers? Explain the difference between addition and condensation polymerisation. [AMIE]

23. Explain briefly why the tendency of a polymer to crystallise decreases with increasing molecular weight. [B.E.]
24. Describe the phenomenon of viscoelasticity. [B.E.]
25. For thermoplastic polymers, cite five factors that favour brittle fracture. [B.E.]
26. Which of the following would you expect to be elastomers and which thermosetting polymers at room temperatures? Justify your answer:
    (*i*) linear and crystalline polyethylene
    (*ii*) phenolformaldehyde
    (*iii*) heavily crosslinked polyisoprene having a glass-transition temperature of 50°C
    (*iv*) lightly crosslinked polyisoprene having a glass-transition temperature of –60°C
    (*v*) linear and partially amorphous polyvinyl chloride. [B.E.]
27. What is the distinction between dye and pigment colorants? [B.E.]
28. Cite four factors that determine what fabrication technique is used to form polymeric materials. [B.E.]
29. Why must fibre materials that are melt spun and then drawn by thermoplastic? Mention two reasons. [B.E.]
30. Write short notes:
    (*a*) Fillers for plastics (*b*) Plasticizers (*c*) Bakelite (*d*) Polyvinyl chloride (*e*) Plastic as a insulating material (*f*) Wood as a natural polymer (*g*) Polymerisation (*h*) Difference between thermosetting and thermoplastic resins (*i*) Protective coatings (*j*) Natural rubber (*k*) Copolymerisation (*l*) Vulcanization (*m*) Die and pigment colorants (*n*) Lamination processes and the purpose for laminating plastics (*o*) Rheology (*p*) Foam (*q*) Elastomer (*r*) Cis- and transforms of elastomers.
31. In what ways do elastomers differs from polymers?
32. Why are elastomers and not polymers used in applications such as tyres?

PROBLEMS

1. The number average molecular weight of a polypropylene is 1,000,000 gm/mol. Compute the number average degree of polymerisation. [B.E.]
   [**Ans.** 23,700]
2. For a linear polymer molecule, the total chain length L depends on the bond length between chain atoms d, the total number of bonds in the molecule N, and the angle between adjacent backbone chain atoms θ, as follows: $L = Nd \sin(\theta/2)$. Moreover, the average end to end distance for a serier of polymer molecule r is equal to, $r = d\sqrt{N}$. A linear polyethylene has a number-average molecular weight of 300,000 gm/mol. Determine the average values of L and r for this material.
   [B.E., REC, M.Sc. (M.S.)] [**Ans.** L = 2682 nm, r = 22.5 nm]
3. The tensile strength and number-average molecular weight for two polymethyl methacrylate materials are given below

Tensile strength MPa	*Number average molecular weight (gm/mol)*
107	40,000
170	60,000

Find the tensile strength at a number average molecular weight of 30,000 gm/mol. [B.E.]
[**Ans.** 44 MPa]

SHORT ANSWER QUESTIONS

1. What are polymers?

Ans. Most of the polymeric materials are composed of very large molecules-chains of carbon atoms, to which are side-bounded various atoms or radicals.

2. What do you understand by mers?

Ans. The macromolecules of polymers may be though of as being composed of mers, smaller structural entities, which are repeated along the chain.

3. When molecular entanglements occur?

Ans. When the chain assume twisted, coiled, and kinked shapes or contours.

4. What are the structures possible with regard to molecular structure of polymers?

Ans. With regard to molecular structure, linear, branched, crosslinked, and network structures are possible, in addition to isotactic, syndiotactic, and atactic, stereoisomers, and cis and trans geometrical isomers.

5. How polymers are classified with regard to their behaviour at elevated temperature?

Ans. Thermoplasts and thermosets. The thermoplasts comprise of plastics, fibres and elastomers.

6. What are thermoplastic polymers?

Ans. These materials have linear and branched structure and they soften when heated and harden when cooled.

7. What are thermoset polymers?

Ans. In contrast to thermoplast materials, thermosets, once having hardened, will not soften upon heating. Their structures are crosslinked and network.

8. When the crystallinity exist?

Ans. When the packing of molecular chains is such as to produce an ordered atomic arrangement, the condition of crystallinity is said to exist. In addition to being entirely amorphous, polymers may also exhibit varying degrees of crystallinity; *i.e.*, crystalline regions are interdispered within amorphous areas. Crystallinity is facilitated for polymers that are chemically simple and that have regular and symmetrical chain structures. The crystallinity of long chain polymers is influenced by a number of factors, *e.g.* chain length, branching, presence of large random side groups and plasticizers.

9. How you will grow polymer single crystals?

Ans. One can grow this from dilute solutions as thin platelets and having chain-folded structures.

10. What are spherulites?

Ans. Many semicrystalline polymers form spherulites. Each spherulite consists of a collection of ribbon-like chain-folded lamellar crystallites that radiate outward from its centre.

11. On the basis of stress-strain behaviour, how you will classify polymers?

Ans. Brittle, plastic and highly elastic.

12. What do you understand by viscoelastic mechanical behaviour of polymers?

Ans. This type of behaviour, being intermediate between totally elastic and totally viscous, is displayed by a number of polymeric materials. It is characterised by the relaxation modulus, a time-dependent modulus of elasticity. The magnitude of the relaxation modulus is very sensitive to temperature; critical to the in-service temperature range for elastomers in this temperature dependence.

13. What is the importance of melting and glass transition temperatures for polymers?

Ans. These parameters are important relative to the temperature range over which a particular polymer may be utilised and processed. The magnitude of T_m and T_g increase with increasing chain stiffness; stiffness is enhanced by the presence of chain double bonds and side groups that are either bulky or polar. Molecular weight and degree of branching also affect T_m and T_g.

14. What do you understand by *aging*?

Ans. Slow degradation of polymers as a result of exposure to low levels of heat, oxygen, bacteria, or ultraviolet rays is called *aging*.

15. What do you understand by blushing?

Ans. A thermoplastic bent repeatedly leads to small volumes of the material crystallising: this leads to voids that ultimately cause the material to fail.

16. What is deflection or distortion temperature?

Ans. The temperature at which a polymer will a given amount under a standard lead is called deflection or distortion temperature.

17. What is *Diene*?

Ans. A group of monomers that contain two double-covalent bonds is called *diene*. These monomers are often used in producing elastomers.

18. What do you understand by interpenetrating polymer networks?

Ans. Polymer structures produced by intertwinning two separate polymer structures or networks is called interpenetrating polymer networks.

19. What is spinnerette?

Ans. Spinnerette is an extrusion die containing many small openings through which hot or molten polymer is forced to produced filaments. Rotation of the spinnerette twists the filaments into a yarn.

20. What do you understand by tacticity of polymers?

Ans. Polymer molecules have different arrangements of adjacent centres within molecules which can be referred to as tacticity of polymers. On the basis of this, they are classified as

- isotactic polymers,
- atactic polymers, and
- syndiotactic polymers

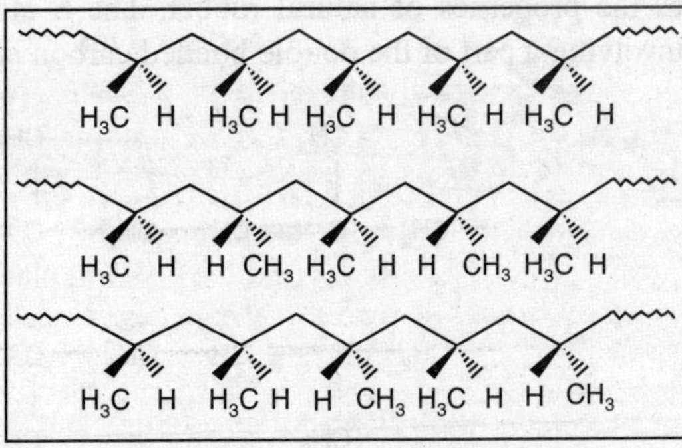

Example: Vinyl polymer.

21. What is a natural rubber?

Ans. Natural rubber is an addition polymer formed from the monomer $CH_2{=}C(CH_3){-}CH{=}CH_2$ called isoprene (2 methyl 1:3 butadiene). The average degree of polymerisation of rubber is around 5000. Addition between molecules of isoprene takes place by 1 : 4 addition and one double bond shifts between 2 and 3 positions

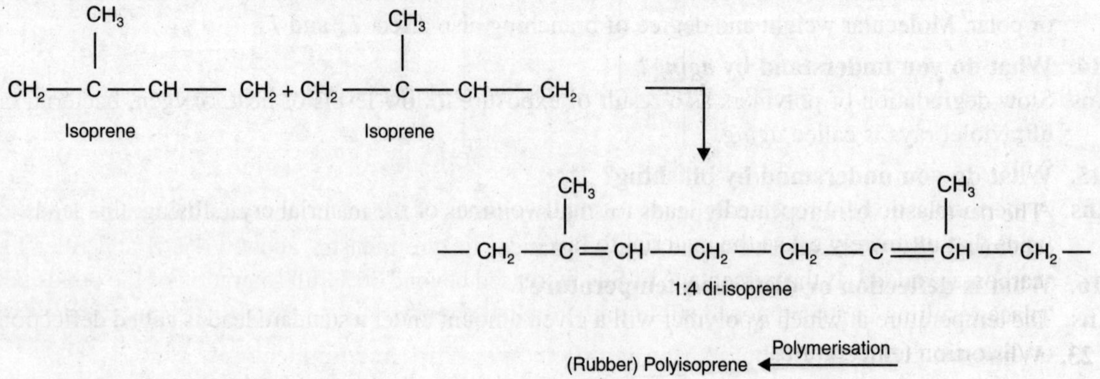

Since each isoprene unit even now contains a C = C bond, polyisoprene can exist in 2 isomeric form *cis* and *trans.*

Natural rubber is the *cis* isomer while the *trans* isomer is called *gutta-percha.*

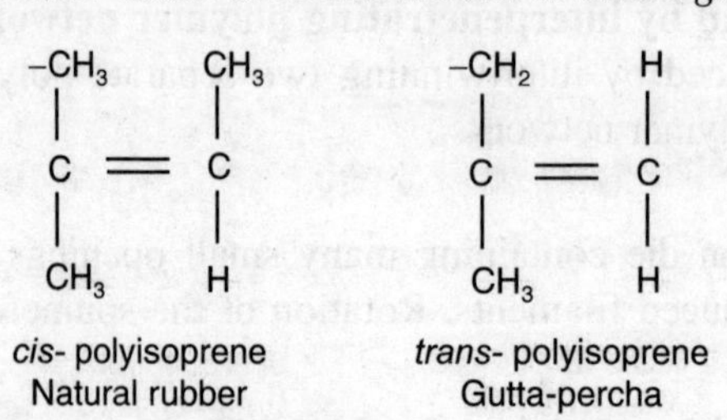

22. What do you understand by *vulcanisation*?

Ans. This is the process involving natural rubber with sulphur around 135°C for 1 to 4 hours. This improves the properties of natural rubber. The S atoms form bridges between poly-neoprene chains involving a part of the double bonded carbon atoms.

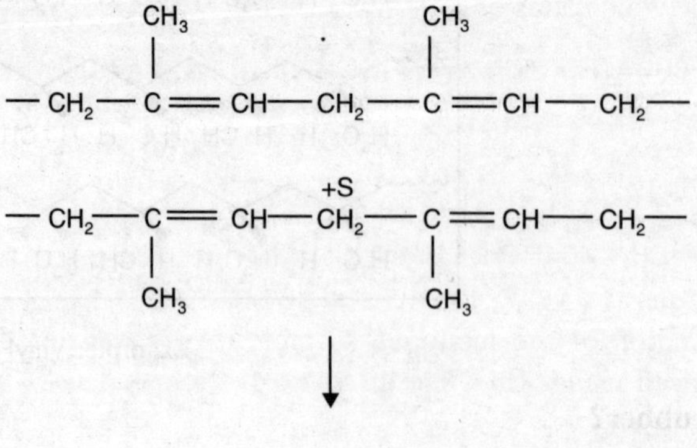

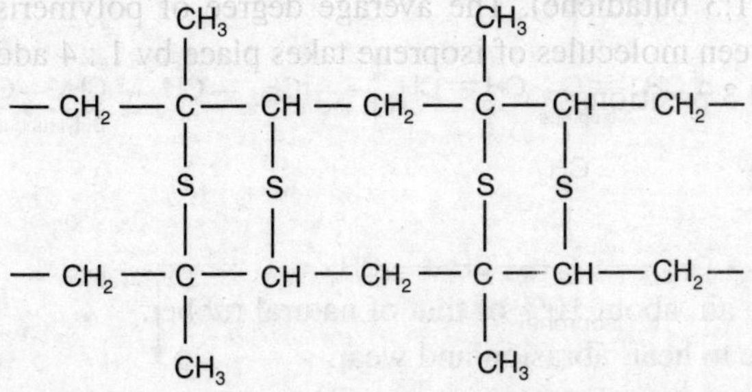

The degree of vulcanisation depends on the amount of *S* added. It increases the tensile strength, chemical resistance and wear and tear resistance. In tyre industry about 3-4% S and filler like carbon are added. If the percentage of S is increased beyond 30% full saturation of the bonds take place resulting in a rigid non-flexible rubber called ebonite.

23. What are synthetic rubbers? Mention their properties and applications.

Ans. Synthetic rubber has slightly modified structures from natural rubber—and exhibit properties that are more conductive for their technical uses. The following are some of the commonly used synthetic rubbers:

1. *Buna-S rubber:* It is a copolymer formed by the emulsion polymerisation of a mixture of 1 : 3 butadiene and styrene in the presence of peroxide catalyst at 5°C. The rubber obtained is called *Buna-S* rubber or SBR [Styrene butadiene rubber] or *cold rubber.* In buna-S, Bu stands for butadiene, Na for sodium (used as catalyst in earlier days), and *S* for styrene.

$$n\underset{\displaystyle C_6H_5}{CH}{=}CH_2 + nCH_2{=}CH{-}CH{=}CH_2 \rightarrow (-\overset{\displaystyle C_6H_5}{CH}-CH_2-CH{=}CH-CH_2-)_n$$

SBR contains unsaturation and is vulcanised with sulphur. It is used in tyre treads and shoe soles because of its high resistance to abrasion.

2. *Neoprene rubbers (chloroprene rubbers):* Neoprene rubber or chloroprene rubber is polymer of 2 chloro 1 : 3 butadiene. The monomer is prepared from acetylene. Catalytic dimerisation of acetylene gives vinyl acetylene which adds on a molecule of HCl to form 2 chloro 1 : 3 butadiene.

$$\underset{\text{(Acetylene)}}{CH{\equiv}CH + CH{\equiv}CH} \rightarrow \underset{\text{(Catalytic additions)}}{CH_2{=}CH{-}C{\equiv}CH}$$

$$CH_2{=}CH_2-C{\equiv}CH + HCl \rightarrow \underset{\text{2, 1:3 chlorobutadiene}}{-[CH_2-\overset{\displaystyle Cl}{CH}-CH-CH_2]_n}$$

Emulsion polymerisation of the monomer is presence of peroxide catalyst gives chloropene. Neoprene is vulcanised using ZnO. The product so formed is used in making conveyor belts, hoses, etc.

3. *Butyl rubber:* It is a copolymer formed from isobutylene with small amounts of isoprene (1 to 5%) at 90°C.

$$\underset{\text{Isobutylene}}{m[CH_2{=}C]-CH_3} + \underset{\text{Isoprene}}{n[CH_2{=}\overset{\overset{\large CH_3}{|}}{C}-CH{=}CH_2]} \rightarrow \underset{\text{Butyl rubber}}{-[CH_2-CH_2-CH_2-C{=}CH-\dot{C}H_2]_n-}$$

Properties

1. It is chemically inert.
2. It has permeability to air, about 10% of that of natural rubber.
3. It has good resistance to heat, abrasion and wear.
4. It is soluble in benzene.
5. It is not easily attacked by ozone.
6. It has good electrical insulating characteristics.

Applications

- It is widely used in automotive inner tubes for the tyres.
- It is used for making conveyor belts.
- It is also used as insulating material in electrical appliances.

24. Compare natural and synthetic rubbers.

Ans. Comparison of properties of natural and synthetic rubbers.

Property	*Natural rubber*	*Synthetic rubber*
1. Tensile strength	Low	High
2. Chemical resistivity	Low—gets oxidized even in air	High—not oxidized in air.
3. Action of heat	Cold condition it is hard and brittle, at higher temperatures soft and sticky.	Can withstand effect of heat over a range of temperature.
4. With organic solvents	Swells and dissolves	Does not swell and dissolve.
5. Ageing	Undergoes ageing quickly	Resists ageing.
6. Elasticity	On increased stress undergoes permanent deformation.	Has high elasticity.

25. How thermoplastic chains can be made more rigid and stronger?

Ans. The thermoplastic chains can be made more rigid and stronger by using nonsymmetrical monomers that increase the bonding strength between the chains and make it more difficult for the chains to disentangle when stress is applied. Further, there are several monomers produce more rigid chains containing atoms or groups of atoms other than carbon; this structure also produces high-strength thermoplastics.

26. What are thermoplastic elastomers?

Ans. These have combine features of both thermoplastics and elastomers. At high temperatures, these polymers behave as thermoplastics and are plastically formed into shapes; at low temperatures, they behave as elastomers.

27. How crystallinity in polymers is measured?

Ans. The weight percentage of the structure that is crystalline can be calculated from the density of the polymer

$$\% \text{ Crystalline} = \frac{\rho_c(\rho-\rho_c)}{\rho(\rho_c-\rho_a)}\times 100$$

where ρ is the measured density of the polymer, ρ_a is the density of amorphous polymer, and ρ_c is the density of completely crystalline polymer.

OBJECTIVE QUESTIONS

1. The side group in polyvinyl chloride is
(*a*) H, Cl, Cl, Cl (*b*) H, H, H, Cl
(*c*) H, H, H, H (*d*) H, H, CH_3, Cl
2. The factors that promote non-crystallinity in polymers are
(*a*) branching and addition of plasticisers (*b*) copolymerisation
(*c*) large random side groups (*d*) all of these
3. The time average molecular weight is expressed as (symbols have usual meanings)
(*a*) $\overline{M}_n = \sum x_i M_i$ (*b*) $\overline{M}_n = \Sigma M_i/x_i$
(*c*) $\overline{M}_n = \sum \sqrt{x_i M_i}$ (*d*) None of these
4. A weight-average molecular weight $\overline{M}_w$ is calculated according to (symbols have usual meanings)
(*a*) $\overline{M}_w = \sum w_i M_i$ (*b*) $\overline{M}_w = \sum \sqrt{w_i M_i}$
(*c*) $\overline{M}_w = \sum M_i/w_i$ (*d*) $\overline{M}_w = \sum (W_i M_i)^2$
5. Relaxation modulus $E_r(t)$ for viscoelastic polymers is given by (symbols have usual meanings)
(*a*) $E_r(t) = \sqrt{\sigma(t)}/\varepsilon_0$ (*b*) $E_r(t) = \sigma(t)/\varepsilon_0$
(*c*) $E_r(t) = \varepsilon_0\sqrt{\sigma(t)}$ (*d*) $E_r(t) = \sqrt{\varepsilon_0}\,\sigma(t)$
6. Apart from three *H* atoms, the fourth side group in the monomer of polyproplene is
(*a*) C_6H_6 (*b*) CH_3 (*c*) C_6H_5 (*d*) $COOCH_3$
7. If C-C bond length is 1.54 Å, the length of uncoiled polyethylene chain with 500 mers is
(*a*) 650 Å (*b*) 1258 Å (*c*) 1628 Å (*d*) 1836 Å
8. The degree of polymerisation in polyethylene is 50,000. The average molecular weight is
(*a*) 650,000 amu (*b*) 1,300,000 amu
(*c*) 1,400,000 amu (*d*) 1,720,000 amu
9. The number of double bonds required in the monomer for crosslinking is
(*a*) 2 (*b*) 3 (*c*) 4 (*d*) 1
10. Plasticisers are added in plastics to
(*a*) reduce hardness (*b*) improve flexibility
(*c*) improve their strength (*d*) reduce glass transition temperature (t_g)
11. The windows of aeroplane are made in
(*a*) PVC (*b*) PTFE (*c*) PMMA (*d*) none of these
12. Which one is not a characteristic of plastics?
(*a*) Low density (*b*) Machinability (*c*) High strength (*d*) Large plastic deformation

13. In teflon, *i.e.*, PTFE monomer, the four side groups are
    (*a*) H, H, H, CH_3 (*b*) F, F, F, F
    (*c*) H, H, H, H (*d*) H, H, Cl, Cl
14. The weight of sulfur required to achieve 5% cross-linking of 68 kg of polyisoprene is
    (*a*) 0.8 kg (*b*) 1.6 kg
    (*c*) 3.2 kg (*d*) 5.6 kg
15. An ebonite is
    (*a*) copolymer (*b*) natural wood
    (*c*) natural rubber (*d*) none of these
16. Plastics are used in automobiles because of their
    (*a*) low melting point (*b*) high strength
    (*c*) capacity to be moulded with good finish
    (*d*) none of these
17. Choose the correct statement:
    (*a*) Thermoplastics are either amorphous or crystalline but thermosets are amorphous
    (*b*) Thermoplastics are crystalline
    (*c*) Thermosetting and thermoplastic polymers are essentially amorphous
    (*d*) Thermosetting plastics are crystalline

ANSWERS

1. (*b*) 2. (*d*) 3. (*a*) 4. (*a*) 5. (*b*) 6. (*b*) 7. (*b*) 8. (*c*) 9. (*a*)
10. (*b*) 11. (*c*) 12. (*c*) 13. (*b*) 14. (*b*) 15. (*c*) 16. (*c*) 17. (*a*)

CHAPTER

6

Composites

6.1 INTRODUCTION

Many of modern technologies require materials with unusual combinations of properties that cannot be met by the metals, conventional metallic alloys, ceramics, and polymeric materials, *e.g.* materials needed for aerospace, underwater, and transportation applications. For example engineers working in aircraft industry are looking and searching for structural materials that have low densities, are strong, stiff, and abrasion and impact resistant, and are not easily corroded. Obviously, this is a rather formidable combination of characteristics. Usually strong materials are relatively dense; also, increasing the strength or stiffness generally results in a decrease in impact strength. Therefore, in composites the whole is greater than the sum of the parts (synergy). Wood, celery, bamboo, and corn are all examples of nature's composites, where two materials combine to reinforce and bind together. We have already discussed in previous chapters, composites of sort which include multiphase metal alloys, ceramics and polymers. A composite is considered to be any multiphase material that exhibits a significant proportion of the properties of both constituent phases such that a better combination of properties is realized. This is termed as the *principle of combined action*. According to this principle, better combinations are fashioned by the judicious combination of two or more distinct materials.

All composites generally have one thing in common: a *matrix* or *binder* combined with a reinforcing material. Obviously, a composite consists of a matrix material, dispered within which is a dispersion of one or more phases of another material. If the fibres are directionally oriented and continuous, the material is termed an *advanced composite*. Reinforced concrete is a good example of composite material. When concrete is reinforced with steel rebar, the concrete becomes the matrix, which surrounds the reinforcing fibre, the rebar. Another example is reinforced fibre glass products such as fishing rods. Glass fibres are set in a thermosetting resin matrix. This produces a strong, lightweight, flexible fishing rod. Other fibres are produced from aramid (Kevlar and Nomex), boron, carbon, graphite, and ultrahigh-molecular-weight polyethylene (spectra). The matrix for these materials is typically a thermosetting epoxy resin. These materials provide some exceptional increase in mechanical properties, sometimes three to six times greater than steel. Another example of a composite material is pearlitic steels. The microstructure of this material consists of alternating layers of α-ferrite and cementite. The ferrite phase is soft and ductile, whereas cementite is hard and very brittle. The combined mechanical characteristics of this composite, *i.e.* pearlite (reasonably high ductility and strength) are superior to those of either of the constituent phases. Other common varieties of composites include combinations: fibre-resin, fibre-ceramic, carbon-metal, metal concrete, metal resin and wood plastic. Most of the contemporary advanced composites use glass, kevlar (an aramid), or one of the various types of graphite fibres. *Fibre composites* form an important subset of this class of engineering materials. Obviously, a composite is a multiphase material that is artificially made, as to one that occurs naturally. In addition, the constituent phases in a composite must be chemically dissimilar and separated by a distinct interface. This is why most metallic alloys and many ceramics do not fit this definition because their multiple phases are formed as a consequence of natural phenomena.

There are many reasons for making composites: the incorporation of fibres into brittle ceramics produces a composite of enhanced toughness. Fillers, such as the presence of aggregate in concrete, reduce the overall cost of the product, and additionally improve the compressive strength. The second phase may furthermore be a gas, as in the manufacture of foamed products of low density.

On the basis of strength and stiffness alone fibre reinforced composite materials may not be superior to metals of comparable strength, but when the specific modulus (*i.e.* modulus per unit weight) and specific strength are considered, then their use implies that the weight of components can be reduced. This is an important factor in all forms of transport, where reductions of weight result in greater energy savings.

In order to produce a new generation of extraordinary materials, scientists and engineers while designing composite materials, have ingeniously combined various metals, ceramics and polymers.

One can classify composite materials as per simple scheme shown in Fig. 6.1. There are three main divisions: particle reinforced, fibre-reinforced and structural composites. We note that there exist atleast two subdivisions for each main division.

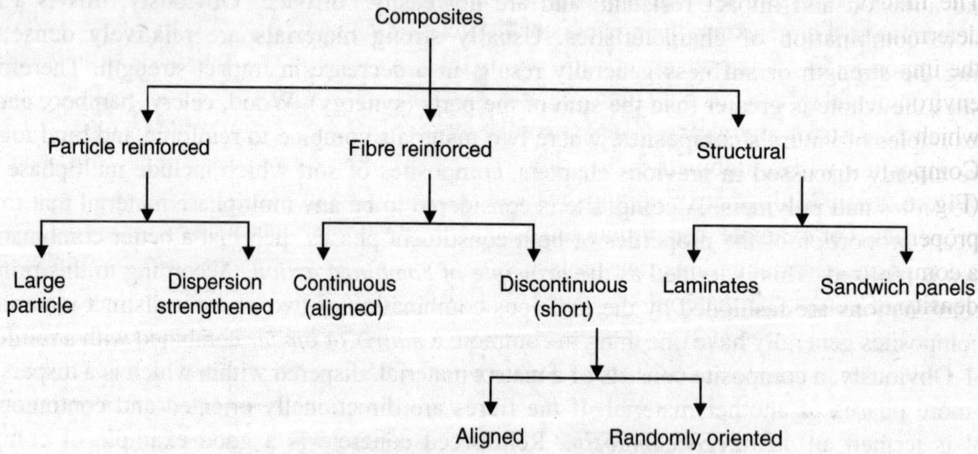

Fig. 6.1 A simple classification schemes for the various composite types.

The dispersed phase for particle-reinforced composites is equiaxed (*i.e.*, particle dimensions are approximately the same in all directions) and the dispersed phase for fibre-reinforced composites has the geometry of a fibre-reinforced composites has the geometry of a fibre (*i.e.*, a large-length-to-diameter ratio). Structural composites are combinations of composites and homogeneous materials.

Examples of these three groups include concrete, a mixture of cement and aggregate, which is a particulate composite; fibre glass, a mixture of glass fibres imbedded in a resin matrix, which is a fibre composite; and plywood, alternating layers of laminate veneers, which is a laminate composite.

6.2 GENERAL CHARACTERISTICS OF COMPOSITES

(*i*) Composite materials are superior to all other known structural materials in specific strength and stiffness, high temperature strength, fatigue strength and other properties. The desired combination of properties can be tailored in advance and realized in the manufacture of a particular material. Moreover, the material can be shaped in this process as close as possible to the form of final products or even structural units.

(*ii*) Composite materials are complex materials whose components differ strongly from each other in the properties, are mutually insoluble or only slightly soluble and divided by distinct boundaries.

(*iii*) The principle of manufacture of composites has been borrowed from nature. Trunks and stems of plants and bones of man and animals are examples of natural composites. In wood, cellulose fibres are bonded by plastic lignin, in bones, thin and strong fibres of phosphates are bonded by plastic collagen.

(*iv*) The properties of composites mainly depend on the physico-mechanical properties of their components and the strength of bonds between them. A characteristic feature of composite materials is that the merits of their components are fully utilized. Composite materials may acquire certain valuable properties not found in the components. For obtaining the optimal properties in composites, their components are chosen so as to have sharply different, but complementary properties.

(*v*) The base, or matrix, of composites may consist of metals or alloys (metallic composites), polymers, carbon and ceramic materials (non-metallic composites).

(*vi*) The matrix is essentially the binding and shaping component in composites. Its properties determine to a large extent the process conditions for the manufacture of composite materials and the important operating characteristics, *e.g.*, working temperature, fatigue strength, resistance to environmental effects, density, and specific strength. Some composites have a combined matrix which consists of alternating layers (two or more) of different composition.

(*vii*) Composites with combined matrix may be called *multi-matrix*, or *multi-layer* composites (Fig. 6.2(*a*)). Multi-matrix composites can be characterized by a wider spectrum of useful properties. For example, use of titanium as an addition to aluminium may increase the strength of a composite material in directions transverse to fibres. Aluminium layers in a matrix diminish the density of composite material.

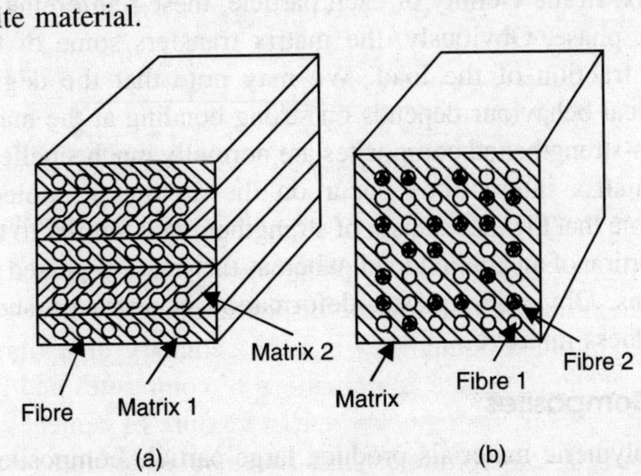

Fig. 6.2 Schemes of (*a*) multi-matrix and (*b*) polyfibre composite materials.

(*viii*) Fillers, *i.e.*, other components are uniformly distributed in a matrix. These plays the major part in strengthening of composites and thus they are called *strengtheners*. Fillers should possess high values of strength, hardness and elastic modulus. These characteristics should be substantially higher than those of the matrix. With an increase of the elastic modulus and ultimate strength of a filler, the corresponding properties of a composite material also increase, but do not reach the value of the filler. Fillers are alternatively called reinforcing components. This is a broader term than 'strengthener', it does not specify the particular strengthening role of filler which may used for improving other properties of a composite.

The properties of a composite material can also depend on the shape (geometry), dimensions, concentration and distribution of filler (reinforcement pattern).

As regards their shape, fillers are divided into three main groups: (*i*) zero-dimensional (*ii*) one-dimensional, and (*iii*) two-dimensional.

By the reinforcement pattern, composite materials are divided into three groups: with uniaxial, biaxial, and triaxial reinforcement.

(*ix*) Fillers of different shape may be used for obtaining a wider complex of properties or enhancing a particular property of a composite material. For example, the strength of bond between one-dimensional filler elements (glass or carbon fibres) and a polymer matrix can be increased by introducing a zero-dimensional filler (particles of asbestos, silicon carbide, etc.). The same purpose can be achieved by reinforcing a composite material with fillers of the same shape, but different composition. For example, the modulus of elasticity of composite materials with a polymer matrix reinforced by glass fibres can be increased by additional reinforcement with boron fibres.

Composite materials containing two or more different fillers are termed *complex-reinforced composites*. (Fig. 6.2(*b*)).

6.3 PARTICLE-REINFORCED COMPOSITES

These can be further classified under two subgroups: (*i*) *large particle* and (*ii*) *dispersion-strengthened composites*. The distinction between these is based upon reinforcement or strengthening mechanism. The term 'large' indicate that particle-matrix interactions cannot be treated on the atomic or molecular level; rather continum mechanics is used. The particulate phase for most of these composites is harder and stiffer than the matrix. In the vicinity of each particle, these reinforcing particles tend to restrain movement of the matrix phase. Obviously, the matrix transfers some of the applied stress to the particles, which bear a fraction of the load. We may note that the degree of reinforcement or improvement of mechanical behaviour depends on strong bonding at the matrix particle interface.

Particles for dispersion-strengthened composites are normally much smaller (diameter between 0.01 and 0.1 μm). Particle-matrix interactions occur on the atomic or molecular level and lead to strengthening. We may note that the mechanism of strengthening is similar to that for precipitation. The matrix bears the major portion of an applied load, whereas the small dispersed particle hinder or impede the motion of dislocations. Obviously, plastic deformation is restricted such that yield and tensile strengths, as well as hardness improve.

6.3.1 Large-particle Composites

Fillers added to some polymeric materials produce large-particle composites. The fillers modify or improve the properties of the material and/or replace some of the polymer volume with less expensive filler material.

Concrete is another familiar example of large-particle composite. Concrete is composed of cement (the matrix), and sand and gravel (the particulates).

We may note that particles can have quite a variety of geometries, but they should be approximately the same dimension in all directions (equiaxed). Particules should be small and evenly distributed throughout the matrix for effective reinforcement. Moreover, the volume fraction of the two phases influences the behaviour; mechanical properties are enhanced with increasing particulate content. The *rule of mixtures* equations reveal that the elastic modulus should fall between an upper bound given by

$$E_c(u) = E_m V_m + E_p V_p \quad ...(1)$$

and a lower bound, or limit,

$$E_c(t) = \frac{E_m E_p}{V_m E_p + V_p E_m} \quad ...(2)$$

where E and V denote the elastic modulus and volume fraction, respectively and the subscript c, m and p represent composite, matrix and particulate phases respectively.

Large-particle composites are utilized with all three material types, *i.e.* metals, polymers and ceramics. Examples of ceramic-metal composite are *cermets*. Cemented carbide, which is composed of extremely hard particles of a refractory carbide ceramic such as tungsten carbide (WC) or titanium carbide (TiC), embedded in a matrix of a metal such as cobalt or nickel is the most common cermet. These composites are widely used as cutting tools for hardened steels.

Interestingly, both elastomers and plastics are frequently reinforced with various particulate materials. The use of many of the modern rubbers would have been severely restricted without reinforcing particulate materials, *e.g.* as carbon black. Carbon black consists of very small and essentially spherical particles of carbon. Carbon black is produced by the combustion of natural or oil in an atmosphere that has only a limited air supply. When carbon black is added to vulcanised rubber, this material which is very cheap, enhances tensile strength, toughness, and tear and abrasion resistance. Automobile tire contain on the order of 15 to 30 volume percentage of carbon black.

In order that carbon black may provide significant reinforcement, the particle size must be evenly small, with diameters between 20 and 50 nm. Moreover, the particle must be evenly distributed throughout the rubber and must form a strong adhesive bond with the rubber matrix. Particle reinforcement using other materials, *e.g.* silica, is much less effective as this special interaction between the rubber molecules and particle surfaces does not exist.

Concrete

This is a common large-particle composite in which both matrix and disperse phases are ceramic materials. Broadly speaking, concrete implies a composite material consisting of an aggregate of the particles that are bound together in a solid body by some type of binding medium, *i.e.*, a cement. The two most familiar concretes are those made with portland and asphaltic cements, where the aggregate is gravel and sand. Asphaltic cement is used primarily as a paving material on a wider scale, whereas portland cement concrete is used extensively as a structural building material.

Portland cement concrete

The ingredients for portland cement concrete are portland cement, a fine aggregate (sand), a coarse aggregate (gravel), and water. The aggregate particles act as a filler material to reduce the overall cost of concrete product as they are cheap, whereas cement is relatively costly. The ingredients have to be added in correct proportions so that one may achieve the optimum strength and workability of concrete mixture. One can achieve dense packing of the aggregate and good interfacial contact by having particles of two different sizes. The fine particles of sand should fill be void spaces between the gravel particles. Normally, these aggregrates comprise between 60% and 80% of the total volume. However, the amount of cement-water paste should be sufficient to coat all the sand and gravel particles, otherwise the cementitious bonds will be incomplete.

This is a major material for construction. One can pour portland cement concrete in place and it hardens at room temperature, and even when submerged in water. However, as a structural material, it is found to be relatively weak and extremely brittle; the tensile strength is approximately 10 to 15 times smaller than its compressive strength. Moreover, large concrete structure can experience considerable thermal expansion and also contraction with temperature fluctuations. Water also penetrates into external pores, which can cause severe cracking in cold weather (as a consequence of freeze-thaw cycles).

Reinforced concrete

One can increase the strength of portland cement concrete by additional reinforcement. One can achieve this by means of steel rods, wires, bars (rebar), or mesh, which are embedded into the fresh and uncured concrete. Obviously, the reinforcement renders the hardened structure capable of supporting greater tensile, compressive and shear stresses. Considerable reinforcement is maintained even when cracks develop in the concrete.

The coefficient of thermal expansion for steel is nearly the same as that of concrete and thus it serves as a suitable reinforcement. Moreover, steel is not rapidly corroded in the cement environment and also a relatively strong adhesive bond is formed between it and the curved concrete. One can enhance this adhesion by the incorporation of contours into the surface of the steel member. This permits a greater degree of interlocking.

One can also reinforce portland cement concrete by mixing into the fresh concrete fibres of a high-modulus material, *e.g.* glass, nylon, steel and polyethylene. However, when exposed to cement environment some fibre materials experience rapid deterioration and therefore care must be exercised in using this type of reinforcement.

There is another reinforcement technique of strengthening concrete. This involves the introduction of residual compressive stresses into the structural member. The resulting material is called *prestressed concrete*. The characteristic of brittle ceramics, *i.e.* they are stronger in compression than in tension is utilized in this method. Obviously, to fracture a prestressed concrete member, the magnitude of the precompressive stress must be exceeded by an applied tensile stress.

In one prestressed technique, inside the empty moulds high-strength steel wires are positioned and stretched with a high tensile force, which is maintained constant. After placing the concrete and allowing them to harden, the tension is released. As the wires contract, they put the structure in a state of compression because the stress is transmitted to the concrete through the concrete-wire bond that is formed.

There is also another technique known as *posttensioning*. In this technique stresses are applied after the concrete hardens. Sheet metal or rubber tubes are situated inside and pass through the concrete forms, around which the concrete is cast. Steel wires are fed through the resulting holes after the cement has hardened and tension is applied to the wires through jacks attached and abutted to the faces of the structure. Also, a compressive stress is imposed on the concrete piece, this time by the jacks. To protect the wire from corrosion, the empty spaces inside the tubing are filled finally with a grout.

We must note that the concrete that is prestressed should be of a high quality and there must be a low shrinkage and low creep rate. Usually, prestressed concretes are prefabricated and used mostly for railway bridges and highway.

6.3.2 Dispersion-strengthened Composites

This type of composite contains small particulates or dispersions, which increase the strengths of the composite by blocking the movement of dislocations. The *dispersoid* is typically a stable oxide of the original material. A common example is sintered aluminium powder (SAP). SAP has an aluminium matrix, which contains up to 14% aluminium oxide (Al_2O_3). This composite is produced with the *powder metallurgy process*, where the powders are mixed, compacted at high pressures, and sintered together. Sintering involves heating a material until the particles of the material fuse together. Only the edges of the particles are generally bonded together; the whole particle does not melt. In terms of sintered ceramics, the product is a strong, rigid, brittle product that exhibit good compressive strength, high melting points, and good heat resistance. Examples of dispersion-strengthened composites include Ag–CdO, used as an electrical contact material, Pb–PbO, used in battery plates, and Be–BeO, used in nuclear reactor and aerospace components. The high-temperature strength of nickel alloy may be enhanced significantly by the addition of about 3 volume percentage of thoria (ThO_2) as finely dispersed particles. This material is known as thoria-dispersed (or TD) nickel.

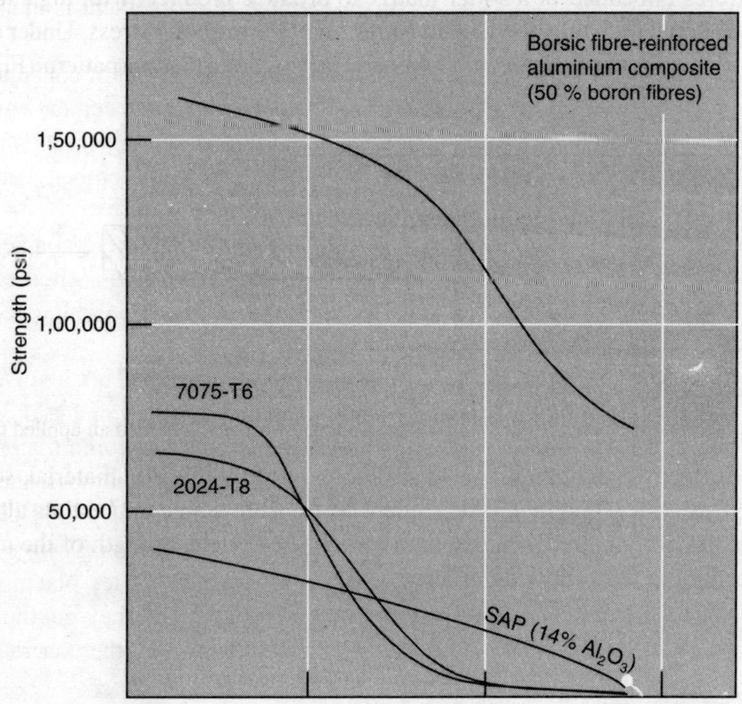

Fig. 6.2 (*c*) A comparison of the yield strength of dispersion-strengthened sintered Al powder (SAP) composite with two conventional high-phase high strength Al alloys. The composite has benefits almost above 300°C. For comparison, a fibre-reinforced Al composite is shown.

Dispersion-strengthened *nanocomposite* materials containing particles 10 to 250 nm in diameter is classified as *particulate composites.* The *dispersoids,* usually a metallic oxide, are introduced into the matrix by means other than traditional phase transformations. No doubt, the small particles are not coherent with the matrix, they block the movement of dislocations and produce a pronounced strengthening effect.

The dispersion-strengthened composites may be weaker at room temperature in comparison to the traditional age—hardened alloys, which contain a coherent precipitate. However, the composites do not catastrophically soften by averaging, overtempering, grain growth, or coarsening of the dispersed phase, the strength of the composite decreases only gradually with rise in temperature as shown in Fig. 6.2(*c*).

We may note that the dispersoid must have a low solubility in the matrix and should not chemically react with the matrix. However, a small amount of solubility may help improve the bonding between the dispersant and the matrix. Cu_2O dissolves in copper at high temperatures; thus, the Cu_2O–Cu system would not be effective. On the other hand, Al_2O_3 does not dissolve in aluminium; the Al_2O_3–Al system does give an effective dispersion-strengthened material.

6.4 FIBRE-REINFORCED COMPOSITES

These are strong fibres imbedded in a softer matrix to produce products with high strength-to-weight ratios. The matrix material transmits the load to fibres, which absorb the stress. Under an applied stress, fibre-matrix bond ceases at the fibre ends, yielding a matrix deformation pattern (Fig. 6.3).

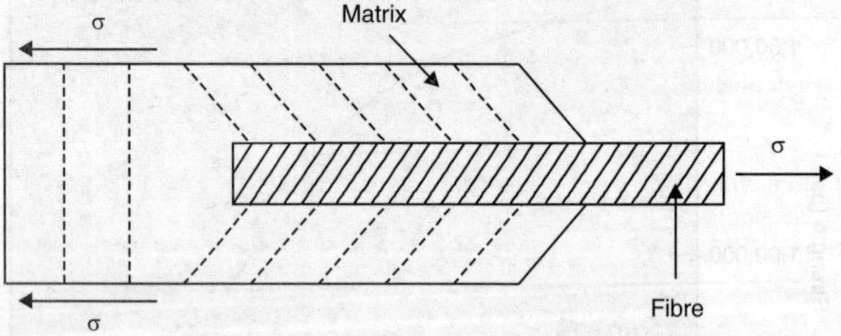

Fig. 6.3 The deformation pattern in the matrix surrounding a fibre, subjected to an applied tensile load.

In order to have effective strengthening and stiffening of the composite material, some critical fibre length is esential. This critical length l_c is dependent on the fibre diameter *d* and its ultimate (or tensile) strength σ_f, and on the fibre-matrix bond strength (or the shear yield strength of the matrix, whichever is smaller) τ_c according to the following relation

$$l_c = \frac{\sigma_f^* d}{2\tau_c} \quad ...3(a)$$

The critical length, l_c is on the order of 1 mm for a number of glass and carbon fibre-matrix combinations and ranges between 20 and 150 times the fibre diameter.

Obviously, the strength of these composites comes from the bonding between the reinforcement fibres and the matrix. The length-to-diameter, or *aspect,* ratio of the fibres used as reinforcement influences the properties of the composite. The higher the aspect ratio, the stronger the composite. Therefore, long, continuous fibres are better than short ones for composite construction. However, continuous fibres are more difficult to produce and place in the matrix. Shorter fibres are easier to place in the matrix but offer poor reinforcement. Some trade-off is made when shorter, discontinuous fibres are used with aspect ratios greater than a specified minimum value. The greater the number of fibres, the stronger the composite. This holds true up to about 80% of the volume of the composite, where the matrix can no longer completely surround the fibres.

The strength of the composite can be estimated from

$$\sigma_c = E_f \sigma_f^* \left(1 - \frac{l_c}{2l}\right) + E_m \sigma_m \qquad \text{...3(b)}$$

where σ_m is the stress on the matrix when the fibre breaks. $\sigma_f^* = TS_f$ is the tensile strength of the fibre.

Fibres for which $l >> l_c$ (normally $l > 15\ l_c$) are called *continuous*, whereas fibres which have lengths shorter than this are termed *discontinuous*. The matrix deforms around the discontinuous fibre having length less than l_c, such that there is virtually no stress transference and little reinforcement by the fibre. These are essentially the particulale composites as discussed earlier. In order to affect a significant improvement in strength of the composite, the fibres must be continuous. Mechanical properties of some reinforcing fibres are given in Table 6.1.

Table 6.1 Mechanical Properties of Some Reinforcing Fibres

Material	*Density, ρ (kg × m⁻³ × 10³)*	*Young's modulus, E (GPa)*	*Tensile strength σ* (GPa)*	*Fibre radius, r (μm)*
• E-glass fibres	2.56	76	1.4–2.5	10
• Carbon fibres (High modulus)	1.75	390	2.2	8.0
• Carbon fibres (High strength)	1.95	250	2.7	8.0
• Kevlar fibres	1.45	125	3.2	12
• Silicon carbide (Monofilament)	3.00	410	8.6	140
• Silicon carbide (Nicalon)	2.50	180	5.9	14
• Alumina (Saffil)	2.80	100	1.0	3

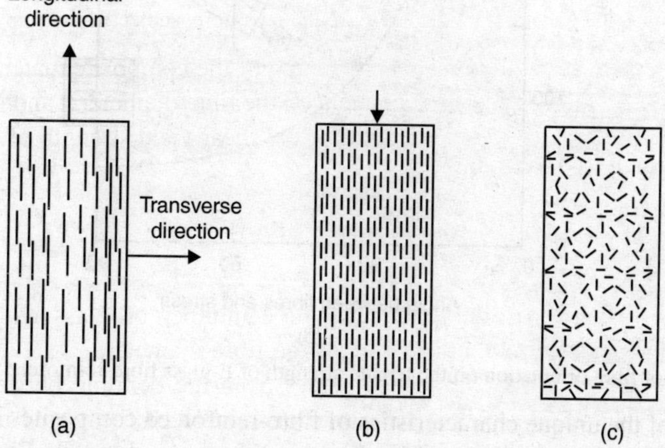

Fig. 6.4 Representation of (*a*) continuous and aligned, (*b*) discontinuous and aligned, and (*c*) discontinuous and randomly oriented fibre-reinforced composites.

The arrangement or *orientation* of the fibres, relative to each other, the fibre *concentration,* and *distribution* all have a significant influence on the strength and other properties of fibre-reinforced composites. There are two possible extremes with respect to orientation: (*i*) a parallel alignment of the longitudinal axis of the fibres in the single direction, and (*ii*) a totally random alignment. Continuous fibres are normally aligned as shown in Fig. 6.4(*a*), discontinuous fibres may be aligned shown in Fig. 6.4(*b*) and randomly oriented, as or partially oriented as shown in Fig. 6.4(*c*). One can realize overall better composite properties when fibre distribution is uniform.

One may introduce the reinforcing fibres into the matrix in a number of orientations. We may note that short, randomly oriented fibres having a small aspect ratio—typical of fibre glass—are easily introduced into the matrix and exhibit relatively isotropic behaviour in the composite.

Long, or even continuous, unidirectional arrangements of fibres yield anisotropic properties, with particularly good strength and stiffness parallel to the fibres. Often, these fibres are designated as 0° piles, indicating that all of the fibres are aligned with the direction of the applied stress. However, unidirectional orientations provide poor properties, when the load is perpendicular to the fibres [Fig. 6.4(*d*)].

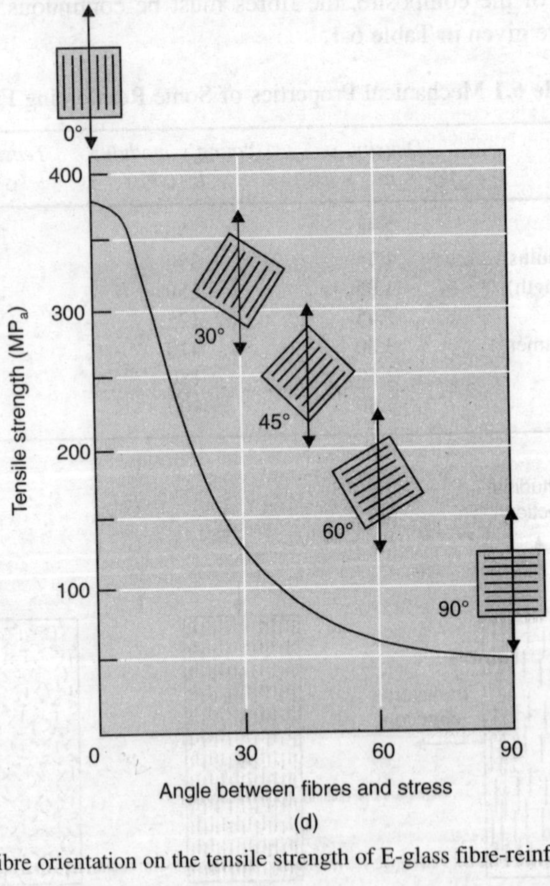

(d)

Fig. 6.4 (*d*) Effect of fibre orientation on the tensile strength of E-glass fibre-reinforced epoxy composites.

Interestingly, one of the unique characteristics of fibre-reinforced composites is that their properties can be tailored to achieve different types of loading conditions. As shown in Fig. 6.4(*e*) and (*f*), long, continuous fibres can be introduced in several directions; in orthogonal arrangements (0°/90° piles), good strength is achieved in two perpendicular directions. We may note that more complicated arrangements (such as 0°/± 45°/90° piles) provide reinforce-ment in multiple directions.

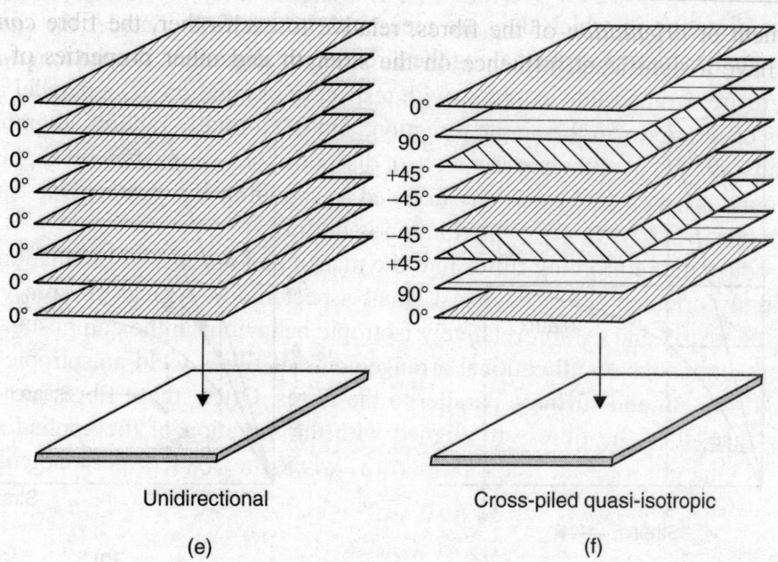

Fig. 6.4 (*e*) Tapes containing aligned fibres can be joined to produce a multi-layered unidirectional composite structure. (*f*) Tapes containing aligned fibres can be joined with different orientations to produce a quasi-isotropic composite. In this case, a 0°/± 45°/90° composite is formed.

One can also arrange fibres in three dimensional patterns. In even the simplest of fabric weaves, the fibres in each individual layer of fabric have some small degree of orientation in a third direction When the fabric layers are knitted or stiched together, better three dimensional reinforcement occurs. One can also use more complicated three-dimensional weaves [Fig. 6.4(*g*)].

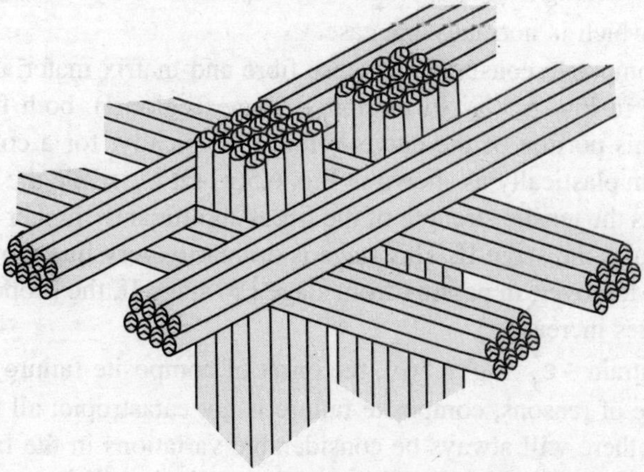

Fig. 6.4 (*g*) A three-dimensional weave for fibre-reinforced composites.

6.4.1 Continuous and Aligned Fibre Composites

Mechanical behaviour of this type of composite depend on: (*i*) stress-strain behaviours of fibre and matrix phases, (*ii*) the phase volume fractions, and (*iii*) the direction in which the stress or load is applied. Moreover, the properties of a composite having its fibres aligned are highly anisotropic, *i.e.*, dependent on the direction in which they are measured. Let us first consider the situation shown in Fig. 6.5(*a*), wherein the stress is applied along the direction of alignment, *i.e.* the *longitudinal direction.*

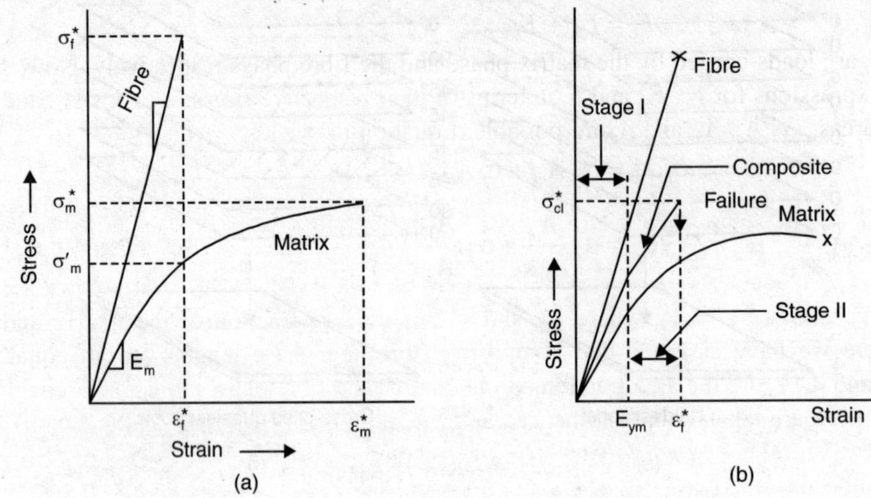

Fig. 6.5 (*a*) Stress-strain curves for brittle fibre and ductile matrix materials. For both the materials stresses and strains are noted. (*b*) Stress-strain curve for an aligned fibre-reinforced composite, which is exposed to a uniaxial stress applied in the direction of alignment. We may note curve for fibre and matrix materials of part (*a*) are also superimposed.

Let us assume the stress versus strain behaviours for fibre and matrix phases shown in Fig. 6.5(*a*). Here we consider the fibre to be totally brittle and the matrix phase to be reasonably ductile. The fracture strengths in tension for fibre and matrix are indicated by σ_f^* and σ_m^* respectively, in Fig. 6.5(*a*). The corresponding fracture strains are represented by ε_f^* and ε_m^* respectively. It is assumed that $\varepsilon_m^* > \varepsilon_f^*$, which is normally the case.

A fibre-reinforced composite consisting of these fibre and matrix materials will exhibit the stress-strain behaviour shown in Fig. 6.5(*b*). In the initial stage (region I), both fibres and matrix deform elastically. Normally, this portion of the curve is linear. Typically, for a composite of this type, the matrix yields and deform plastically as shown in Fig. 6.5(*b*) (at ε_{ym} while the fibres continue to stretch elastically, in as much as the tensile strength of the fibres significantly higher than the yield strength of the matrix. This constitutes the stage II. This stage is ordinarily very linear but of diminished slope in comparison to stage I. Moreover, in passing from stage I to stage II, the proportion of the applied load that is borne by the fibres increases.

Corresponding to a strain ~ ε_f^* Fig. 6.5(*b*), the onset of composite failure begins as the fibres start to fracture. For a couple of reasons, composite failure is not catastropic: all the fibres do not fracture at the same time, since there will always be considerable variations in the fracture strength of brittle fibre materials. Moreover, even after fibre failure, the matrix is still intact in as much as $\varepsilon_f^* < \varepsilon_m^*$ Fig. 6.5(*a*). Obviously, these fractured fibres, which are shorter than the original ones, are still embedded within the intact matrix, and consequently are capable of sustaining a diminished load as the matrix continues to plastically deform.

Elastic Behaviour: Longitudinal Loading

Let us study the elastic behaviour of a continuous and oriented fibrous composite that is loaded in the direction of fibre alignment direction (*i.e., longitudinal loading*). Assuming that the fibre-matrix interfacial bond is very good and the situation is that of *isostrain*, *i.e.* deformation of both matrix and fibres is same. Thus the total load sustained by the composite, F_c is given by

$$F_c = F_m + F_f \quad \text{...(4)}$$

where F_m and F_f are loads carried by the matrix phase and the fibre phase respectively. Now, stress $F = \sigma A$, thus the expressions for F_c, F_m and F_f in terms of their respective stresses, *i.e.* σ_c, σ_m and σ_f and cross-sectional areas, *i.e.* A_c, A_m and A_f are possible. Equation (4) yields

$$\sigma_c A_c = \sigma_m A_m + \sigma_f A_f \quad \text{...(5)}$$

or

$$\sigma_c = \sigma_m \frac{A_m}{A_c} + \sigma_f \frac{A_f}{A_c} \quad \text{...(6)}$$

Now, $A_m/A_c = V_m$ and $A_f/A_c = V_f$, where V_m and V_f are volume fraction of the matrix and fibres respectively. Here we have assumed that composite, fibre and phase lengths are all equal. Truly speaking A_m/A_c and A_f/A_c are the area fractions of the matrix and fibre phases. Since phase lengths for both matrix and fibre are taken equal and hence we have taken A_m/A_c and A_f/A_c equivalent to V_m and V_f respectively.

Thus, Eqn. (6) becomes

$$\sigma_c = \sigma_m V_m + \sigma_f V_f \quad \text{...(7)}$$

In the light of assumption of isostrain state, we have

$$\varepsilon_c = \varepsilon_m = \varepsilon_f \quad \text{...(8)}$$

Dividing each term in Eqn. (7) by its respective strain, we obtain

$$\frac{\sigma_c}{\varepsilon_c} = \frac{\sigma_m}{\varepsilon_m} V_m + \frac{\sigma_f}{\varepsilon_f} V_f \quad \text{...(9)}$$

If composite, matrix, and fibre deformations are all elastic, then $\sigma_c/\varepsilon_c = E_c$, $\sigma_m/\varepsilon_m = E_m$ and $\sigma_f/\varepsilon_f = E_f$, where E's being the moduli of elasticity for the respective phase. Substitution in (9), we obtain an expression for the modulus of elasticity of a continuous and aligned fibrous composite *in the direction of alignment*, E_{cl}, as

$$E_{cl} = E_m V_m + E_f V_f \quad \text{...(10}a\text{)}$$

or

$$E_{cl} = E_m (1 - V_f) + E_f V_f \quad \text{...(10}b\text{)}$$

Since the composite consists of only matrix and fibre phase, *i.e.*, $V_m + V_f = 1$.

Obviously, E_{cl} is equal to the volume-fraction weighted average of the moduli of the elasticity of the fibre and matrix phases. We may note that other properties, including density, also have this dependence on volume fractions.

We can also show that for longitudinal loading, the ratio of the load carried by the fibres to that carried by the matrix is

$$\frac{F_f}{F_m} = \frac{E_f V_f}{E_m V_m} \quad \text{...(11)}$$

Elastic Behaviour—Transverse Loading

A continuous and oriented fibre composite may be loaded in transverse direction (*i.e.*, *transverse loading*). Obviously, load is applied at a 90° to the direction of fibre alignment. For this situation the stress σ to which the composite as well as both phases are exposed is the same, *i.e.*

$$\sigma_c = \sigma_m = \sigma_f = \sigma \quad \text{...(12)}$$

This state is called as *isostress state.* Now, the strain or deformation of entire composite ε_c is

$$\varepsilon_c = \varepsilon_m V_m + \varepsilon_f V_f \quad \text{...(13)}$$

or

$$\frac{\sigma}{E_{ct}} = \frac{\sigma}{E_m} V_m + \frac{\sigma}{E_f} V_f \qquad (\because \varepsilon = \sigma/E) \quad \text{...(14)}$$

where E_{ct} is the modulus of elasticity in the transverse direction. Dividing (14) through σ, yields

$$\frac{1}{E_{ct}} = \frac{V_m}{E_m} + \frac{V_f}{E_f} \quad \text{...(15)}$$

or

$$E_{ct} = \frac{E_m E_f}{V_m E_f + V_f E_m} = \frac{E_m E_f}{(1 - V_f) E_f + V_f E_m} \quad \text{...(16)}$$

Strength of composites: The tensile strength of fibre-reinforced composite (TS_c) depends on the bonding between the fibres and the matrix. However, sometimes one may use the rule of mixtures to approximate the tensile strength of a composite containing continuous, parallel fibres:

$$TS_c = E_f TS_f + E_f \sigma_m$$

where Ts_f is the tensile strength of the fibre and σ_m is the stress acting on the matrix. We may note that other properties, such as ductility, impact properties, fatigue properties, and creep properties are difficult to predict even for unidirectionally aligned fibres.

Longitudinal Tensile Strength

Let us study the strength characteristics of continuous and aligned fibre-reinforced composites that are loaded in the *longitudinal direction.* In this case, strength is usually taken as the maximum stress on the stress-strain curve Fig. 6.5(*b*). Usually, this point corresponds to fibre fracture, and indicates the onset of composite failure. Longitudinal tensile strength values for three common fibrous composites are given in Table 6.2. We may note that the failure of this type of composite material is relatively complex process and there are several different failure modes possible. Again, the mode that operates for a specific composite will depend on fibre and matrix properties, and the nature and strength of the fibre-matrix interfacial bond.

Table 6.2 Longitudinal and Transverse Tensile Strengths (Typical) for Unidirectional Fibre-reinforced Composites. The Fibre Content in each is about 50 Vol%

Material	*Longitudinal tensile strength (MPa)*	*Transverse tensile strength (MPa)*
Glass-polyester	700	20
Carbon (high modulus)–epoxy	1000	35
Kevlar-epoxy	1200	20

For $\varepsilon^*_f < \varepsilon^*_m$ Fig. 6.5(*a*), we note that fibres will fail before the matrix. Once the fibres have fractured, the majority of the load that was borne by fibres is now gets transferred to the matrix. One obtains the expression for the longitudinal strength of the composite, σ^*_{cl} as

$$\sigma^*_{cl} = \sigma^*_m (1 - V_f) + \sigma^*_f V_f \quad \text{...(17)}$$

where σ^*_m is the stress in the matrix at fibre failure Fig. 6.4(*a*) and σ^*_f is the fibre tensile strength.

Transverse Tensile Strength

The strength of continuous and unidirectional fibrous composites are highly anisotropic. Normally, such composites are designed to be loaded along the high strength, longitudinal direction. However, transverse tensile loads may also be present during in service applications. Under these prevailing circumstances, premature failure may result in as much as transverse strength is usually extremely low. Sometimes, it lies below the tensile strength of the matrix. Obviously, the reinforcing effect of the fibres is negative one. Transverse tensile strengths for few unidirectional composites are given in Table 6.2.

Properties of both the fibre and matrix, the fibre-matrix bond strength, and the presence of voids have significant influence on the transverse strength. In order to improve the transverse strength of these composites, one will have to modify the properties of matrix.

6.4.2 Discontinuous and Aligned Fibre Composites

Although reinforcement efficiency is lower for discontinuous than for continuous fibres, discontinuous and aligned fibre composites Fig. 6.4(*b*) are more in demand. Chopped glass fibres are used mostly, however, carbon and aramid discontinuous fibres are also used. One can produce these short fibre composites with moduli of elasticity and tensile strengths approaching 90% and 50%, respectively, of their continuous fibre counterparts.

The longitudinal strength (σ^*_{cd}) for a discontinuous and aligned fibre composite having a uniform distribution of fibres and in which $l > l_c$, is given by

$$\sigma^*_{cd} = \sigma^*_f V_f \left(1 - \frac{l_c}{2l}\right) + \sigma'_m (1 - V_f) \qquad \text{...(18)}$$

Here $^*_f$ and $'_m$ represent, respectively, the fracture strength of the fibre and the stress in the matrix when the composite fails Fig. 6.5(*a*).

For $l < l_c$, *i.e.*, fibre length is less than critical, then the longitudinal strength ($\sigma^*_{cd'}$) is given by

$$\sigma^*_{cd'} = \frac{l\tau_c}{d} V_f + \sigma'_m (1 - V_f) \qquad \text{...(19)}$$

where *d* is the diameter of the fibre and τ_c is the smaller of the either the fibre-matrix bond strength or the matrix shear yield strength.

6.4.3 Discontinuous and Randomly Oriented Fibre Composites

Normally, fibres are used when the fibre orientation is random, short and discontinuous. Reinforcement of this type is shown in Fig. 6.4(*c*). One can use a 'rule-of mixture' expression for the elastic modulus similar to Eqn. 10(*a*), as follows:

$$E_{cd} = K E_f V_f + E_m V_m \qquad \text{...(20)}$$

Here *K* is a fibre efficiency parameter, which depends on V_f and the ratio E_f/E_m. However, its magnitude will be less than unity in the range 0.1 to 0.6. Thus, the modulus increases in some proportion to the volume fraction of fibre for random fibre reinforcement (as with oriented).

We have seen that aligned fibrous components are inherently anisotropic in that the maximum strength and reinforcement are achieved along the alignment (longitudinal) direction. Fibre reinforcement is virtually non-existent in the transverse direction and fractures usually occurs at relatively low tensile stresses. The composite strength lies between these extremes for other stress orientations.

6.4.4 Laminar Composites

When multidirectional stresses are imposed within a single plane, aligned layers that are fastened together one on top of another at different orientations are frequently utilized. These are called *laminar composites*. These are generally designed to provide high strength and low cost at a lighter weight. A familiar laminar composite is plywood, where the veneers are joined by adhesives, typically phenolic or amine resins. The individual odd number of piles are staked so that the grain in each layer runs perpendicular to that of the layers above and below it (Fig. 6.6). This technique offers plywood that is strong and yet cheaper. Some safety glass is a laminated structure, where an adhesive such as polyvinyl butyral is used between two outer layers of glass to keep the glass from flying when broken. Formica is another common laminate used for countertops. Laminates require two or more layers be bonded together. Laminations may also be constructed using fabric material such as cotton, paper, or woven glass fibres embedded in a plastic matrix. Obviously, a laminar composite has relatively high strength in a number of directions in the two-dimensional plane; however, the strength in any given direction is, of course, lower than it would be if all the fibres were oriented in that direction. Modern ski is one example of a relatively complex *laminated structure*.

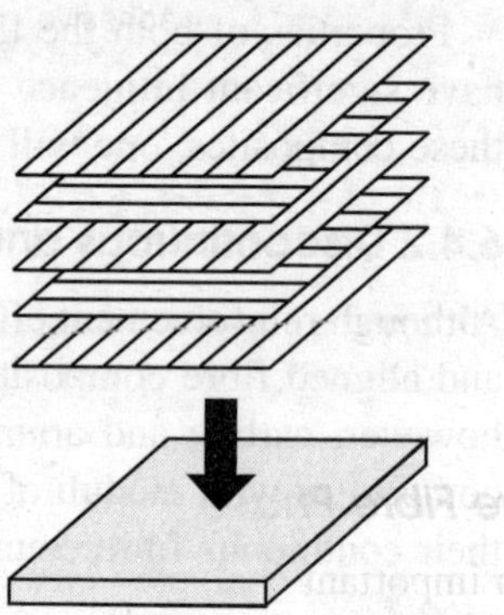

Fig. 6.6 Laminar composite. The stacking of successive oriented, fibre-reinforced layers.

Applications involving totally multidirectional applied stresses generally use discontinuous fibres, which are randomly oriented in the matrix material. We may note that the reinforcement efficiency is found only 1/5th that of an aligned composite in the longitudinal direction; however, the mechanical characteristics are isotropic.

For a particular composite, consideration of orientation and fibre length will depend on the level and nature of the applied stress as well as fabrication cost. Short-fibre composites (both aligned and randomly oriented) production rates are rapid, and intricate shapes can be formed that are not possible with continuous fibre-reinforcement. Moreover, fabrication costs are considerably lower than for continuous and aligned.

Sandwich Structures

These have thin layers of facing materials over a low density material, or combcore, such as a polymer foam or expanded metal structure. A familiar sandwich-structured composite is corrugated cardboard. The corrugated paper core is covered by two faces of thin paper. In structures of this type, the facing material serves to fix the inner core in place. The core provides the strength. Typical face materials include aluminium alloys, fibre-reinforced plastics, titanium, steel and plywood.

Structurally, the core serves two purposes: (*i*) it separates the faces and resists deformations perpendicular to face plane, and (*ii*) it provides a certain degree of sheer rigidity along planes that are perpendicular to the faces. Foamed polymers, synthetic rubbers, inorganic cements, balsa wood, etc. materials and structures are used for cores. Core has lower stiffness and lower strength.

Another popular core consists of a '*honeycomb*' structure, which finds wide use in industries such as the aircraft industry, where higher strength and lower weight are important factors. The honeycomb structure consists of thin foils that have been formed into interlocking hexagonal cells, with axis oriented perpendicular to the face panels. The material used may be similar to the face material. These structures are light weight, stiff, and strong and can be filled to provide sound and vibration damping. The honeycomb structure is shown in Fig. 6.7.

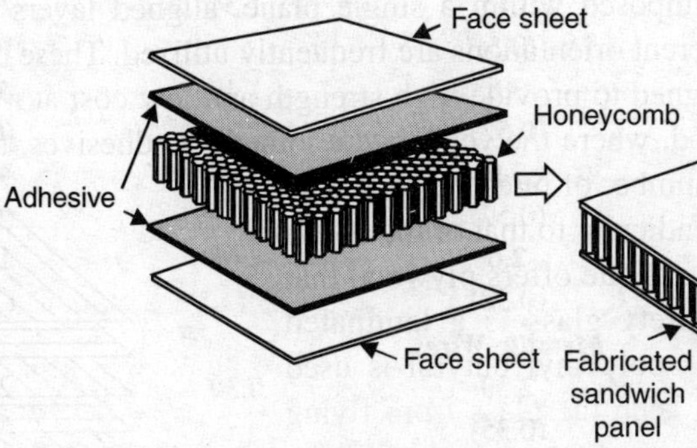

Fig. 6.7 Honeycomb structure.

The Fibre Phase

An important characteristics of most materials, especially of brittle ones, is that a small diameter fibre is much stronger than the bulk material. We have read that the probability of the presence of a critical surface flaw that can lead to fracture diminishes with decreasing specimen volume. This feature of materials is used to advantage in the fibre-reinforced composites. Moreover, the materials used for reinforcing fibres have high tensile strengths.

Fibres can be grouped into three different classed based on their *diameter* and *character*: *whiskers, fibres* and *wires*. Whiskers are not utilized extensively as a reinforcement medium because they are extremely expensive and moreover, it is difficult and impractical to incorporate whiskers into a matrix. Graphite silicon carbide, silicon nitride, and aluminium oxide are few whisker materials. Some mechanical characteristics of whiskers, fibres and metallic wires are given in Table 6.3.

Table 6.3 (a) Some Mechanical Characteristics of Few Fibre-reinforcement Materials

Material	*Specific gravity*	*Tensile strength [GPa (10^6 psi)]*	*Specific strength (GPa)*	*Modulus of elasticity [GPa (10^6 psi)]*	*Specific modulus (GPa)*
		Whiskers			
Graphite	2.2	20 (*c*)	9.1	700 (100)	318
Silicon nitride	3.2	5–7 (0.75–1.0)	1.56–2.2	350–380 (50–55)	109–118
Aluminium oxide	4.0	10–20 (1–3)	2.5–5.0	700–1500 (100–220)	175–375
Silicon carbide	3.2	20 (*c*)	6.25	480 (70)	150
		Fibres			
Aluminium oxide	3.95	1.38 (0.2)	0.35	379 (55)	96

(*Contd.*)

Aramid (Kevlar 49)	1.44	3.6–4.1 (0.525–0.600)	2.5–2.85	131 (19)	91
Carbon[a]	1.78–2.15	1.5–4.8 (0.22–0.70)	0.70–2.70	228–724 (32–100)	106–407
E-Glass	2.58	3.45 (0.5)	1.34	72.5 (10.5)	28.1
Boron	2.57	3.6 (0.52)	1.40	400 (60)	156
Silicon carbide	3.0	3.9 (0.57)	1.30	400 (60)	133
UHMWPE (Spectra 900)	0.97	2.6 (0.38)	2.68	117 (17)	121
Metallic Wires					
High-strength steel	7.9	2.39 (0.35)	0.30	210 (30)	26.6
Molybdenum	10.2	2.2 (0.32)	0.22	324 (47)	31.8
Tungsten	19.3	2.89 (0.42)	0.15	407 (59)	21.1

[a] The term "carbon" instead of "graphite" is used to denote these fibres, since they are composed of crystalline graphite regions, and also of non-crystalline material and areas of crystal misalignment.

Materials, which fall under the head *fibres* are either polycrystalline or amorphous and have small diameters. Fibrous materials are usually either polymers or ceramics (*e.g.*, the polymer aramids, glass, carbon, boron, aluminium oxide, and silicon carbide). Fibre-reinforced composites have been used for centuries, *e.g.* straw has been used in the making of mud bricks since the Pharaohs. Steel rebar has been used for many years in reinforced concrete structure. Fibre glass cloth and resin have been used to repair many automobiles fenders. These applications are among the more familiar.

Fine wires have relatively large diameters. Typical materials under this group include steel, molybdenum and tungsten. Wires are widely used as a radical steel reinforcement in automobile tyres, in filament-wound rocket castings, and in wire-wound high pressure houses.

It is interesting to note that in most fibre-reinforced composites, the fibres are strong, stiff, and light weight. We may note that if the composite is to be used at elevated temperatures, the fibre should also have a high melting temperature. Obviously, the *specific strength* and *specific modulus* of the fibre are important characteristics, we have

$$\text{Specific strength} = \frac{TS}{\rho}$$

and

$$\text{Specific modulus} = \frac{E}{\rho}$$

Here *TS* is the tensile strength, ρ is the density, and *E* is the modulus of elasticity. Table 6.3(*b*) shows the properties of typical fibres. We may remember that 1 gm/cm^3 = 0.0361 lb/in^3. Further, the highest specific modulus is usually found in materials having a low atomic number and covalent bonding.

Table 6.3 (*b*) Properties of Selected Reinforcing Materials

Material	*Density* (ρ) *(g/cm³)**	*Tensile strength (TS) (ksi)*	*Modulus of elasticity (E) (× 10⁶ psi)*	*Melting temperature (°C)*	*Specific modulus (× 10⁷ in)*	*Specific strength (× 10⁶ in)*
Polymers:						
Kevlar™	1.44	650	18.0	500	34.7	12.5
Nylon	1.14	120	0.4	249	1.0	2.9
Polyethylene	0.97	480	25.0	147	7.1	13.7
Metals:						
Be	1.83	185	44.0	1277	77.5	2.8
Boron	2.36	500	55.0	2030	64.7	4.7
W	19.40	580	59.0	3410	8.5	0.8
Glass:						
E-glass	2.55	500	10.5	<1725	11.4	5.6
S-glass	2.50	650	12.6	<1725	14.0	7.2
Carbon:						
HS (high strength)	1.75	820	40.0	3700	63.5	13.0
HM (high modulus)	1.90	270	77.0	3700	112.0	3.9
Ceramics:						
Al_2O_3	3.95	300	55.0	2015	38.8	2.1
B_4C	2.36	330	70.0	2450	82.4	3.9
SiC	3.00	570	70.0	2700	47.3	5.3
ZrO_2	4.84	300	50.0	2677	28.6	1.7
Whiskers:						
Al_2O_3	3.96	3000	62.0	1982	43.4	21.0
Cr	7.20	1290	35.0	1890	13.4	4.9
Graphite	1.66	3000	102.0	3700	170.0	50.2
SiC	3.18	3000	70.0	2700	60.8	26.2
Si_3N_4	3.18	2000	55.0	–	47.8	17.5

$*1\frac{gm}{cm^3} = 0.0361\frac{lb}{in.^3}$.

The Matrix Phase

The matrix phase of a fibrous composites may be a polymer, or ceramic and a metal. Metals and polymers are generally used as matrix materials because some ductility is desirable. To improve fracture toughness for ceramic-matrix composites, the reinforcing component is added.

The matrix phase serves several functions for fibre-reinforced composites: (*i*) it binds the fibres together and acts as a medium by which an externally applied stress is transmitted and distributed to the fibres. However, only a very small proportion of an applied load is sustained by the matrix phase. Also, the matrix material should be ductile. The elastic modulus of the fibre should be much higher than that of the matrix, (*ii*) Matrix has to protect the individual fibres from surface damage as a result of mechanical abrasion or chemical reactions with the environment. These interactions may cause surface flaws capable of forming cracks, which may lead to failure at low tensile stress levels. Lastly, the matrix separates the fibres and, by virtue of its relative softness and plasticity, prevents the propagation of brittle cracks from fibre to fibre, which may result in catastrophic failure. Obviously, matrix phase serves as a barrier to crack propagation. It is possible that some of the individual fibres fail, total composite fracture will not occur until large number of adjacent fibres, one having failed, form a cluster of critical size.

To minimize fibre pull-out, it is essential that adhesive bonding forces between fibre and matrix be high. In reality, in the choice of the matrix-fibre combination bonding strength is an important factor. The magnitude of this bond plays vital role in the ultimate strength of the composite. To maximize the stress transmittance from the weak matrix to the strong fibres, adequate bonding is essential.

6.4.5 Polymer-Matrix Composites (PMCs)

These materials consist of a polymer resin (here resin denote a high-molecular weight reinforcing plastic) as the matrix, with fibres as the reinforcement medium. In light of their room-temperature properties, ease of fabrication, and cost, these PMCs are used in the greatest diversity of composite applications, as well as in huge quantities. In accordance with reinforcement type (*i.e.*, glass, carbon and aramid), various classifications of PMCs, along with their applications and the various polymer resins that are used, are as follows:

(a) Glass Fibre-Reinforced Polymer (GFRP) Composites: Fibre glass is a composite consisting of glass fibres, which may be continuous or discontinuous and contained within a polymer-matrix. These are commonly produced in '*E*' glass (*E* is for electrical), because it draws well and has good strength and stiffness. A typical composition (Wt%) would be $52SiO_2$, $17CaO$, $14Al_2O_3$, $10Ba_2O_3$ with some oxides of Mg, Na and K, and molten glass is gravity fed into a series of platinum bushings each of which has several hundred holes in its base. Fine glass filaments are drawn mechanically as the glass exudes from the holes, then wound on to drums at speeds of several thousand metres per minute.

The strength of the glass fibres is dependent upon the surface damage arising when they rub against each other during processing. The application of a size coating early at an early stage during manufacture minimises this degradation in properties, by reducing the propensity for forming these 'Griffith' cracks. The size consists of an emulsified polymer in water, and also has the effect of binding the fibres together for ease of further processing.

We are familiar with many fibre glass applications; *e.g.* automotive and marine bodies, plastic pipes, industrial floorings and containers. In order to decrease vehicle weight and boost fuel efficiencies, the transport industries are utilizing to maximum extent of glass fibre-reinforced plastics. Many new applications are under investigation by the automative industry.

(b) Carbon Fibre-Reinforced Polymer (CFRP) Composites: These consist of small crystallites of graphite (crystal structure, see chapter 3). The atoms in the basal planes are held together by very strong covalent bonds, and there are weak van der Waals forces between the layers. To obtain high modulus and high strength the layer planes of the graphite have to the aligned parallel to the axis of the fibre, and the modulus of carbon fibres depends on the degree of perfection of alignment of atom planes. This varies considerably with the particular manufacturing route adopted, of which there are three main possibilities:

(*i*) Starting with the polymer PAN (polyacrylonitride), which closely resembles polyethylene in molecular confirmation, it is converted into a fibre and then stretched to produce alignment of the molecular chains along the fibre axis. While still under tension, it is heated in oxygen to form crosslinks between ladder-molecules and finally chemically reduced to give (at high temperatures) a graphic structure. The final graphitization temperature determines whether the fibres have maximum stiffness but a relatively low strength (Type-I fibres), or whether they develop maximum strength (Type-II).

(*ii*) Alternatively, fibres may be produced by melt-spinning molten pitch. During the spinning process, the orifice causes the planar molecules to become aligned. It is then treated, whilst held under tension in order to maintain its preferred orientation, at temperatures up to 2000 °C to form the requisite grains of graphite.

(*iii*) It is also possible to stretch either of the fibre types described above during the graphitization stage, giving further orientation of the layers parallel to the fibre axis.

Carbon fibres have the highest specific modulus and specific strength of all reinforcing fibre materials. Carbon fibres retain their high tensile modulus and high strength at elevated temperatures. However, high temperature oxidation may be a problem. Carbon fibres are not affected by moisture or a wide variety of solvents, acids and bases at room temperature. Carbon fibres exhibit a diversity of physical and mechanical characteristics, allowing composites incorporating these fibres to have specific engineering properties.

On the basis of tensile modulus, carbon fibres can have four classes: *standard, intermediate*, *high* and *ultrahigh* moduli. Fibre diameters generally range between 4 and 10 μm. Both *continuous* and *chopped* forms of these fibres are available. Moreover, carbon fibres are usually coated with a protective epoxy size that also improves adhesion with the polymer matrix.

Carbon-reinforced polymer composites are extensively used in sports and recreational equipment (fishing rods, golf clubs), filament-wound rocket motor cases, pressure vessel and aircraft structural components—both military and commercial, fixed wing and helicopters.

(c) Aramid Fibre-Reinforced Polymer Composites: Aramid fibres are high-strength, high modulus materials, especially desirable for their outstanding strength-to-weight ratios, which are superior to metals. Chemically, this group of materials is called as poly paraphenylene terephthalamide. There are a number of aramid materials. Trade names of two of these most common materials are Kevlar and Nomex. Kevlar has several grades (viz, Kevlar 29, 49 and 149) that have different mechanical behaviours. Mer chemistry and mode of chain alignment for aramid (Kevlar) are shown in Fig. 6.8(*a*). These fibres have longitudinal tensile strengths and tensile moduli (Table 6.3) and they are higher than for other polymeric fibre materials, but they are relatively weak in compression. This material is also known for its toughness, impact resistance, and resistance to creep and fatigue failure. Although aramids are thermoplastics, they are nevertheless, resistant to combustion and stable to relatively high temperatures. The temperatures range over which aramids retain their high mechanical properties is between –200 and 200°C. Aramids are susceptible to degradation by strong acids and bases, but they are relatively inert in other solvents and chemicals.

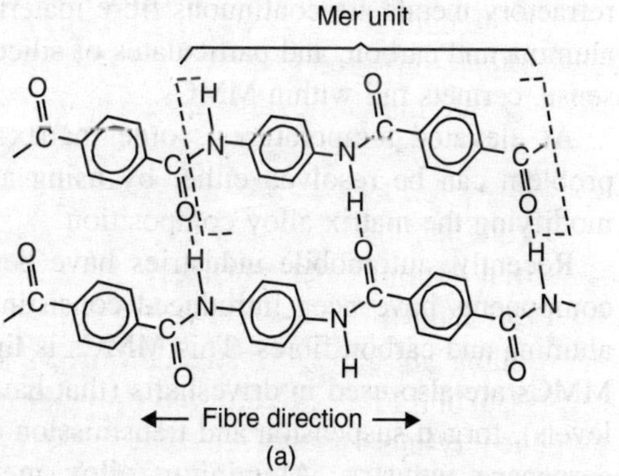

Fig. 6.8 (*a*) Mer and chain structures for aramid (Kevlar) fibres.

Usually, aramid fibres are used in composites having polymer matrices. Common matrix materials are the epoxies and polyesters. Aramid fibres are relatively flexible and somewhat flexible. Aramid composites are used in ballistic products (bullet-proof vests), sporting goods, tyres, ropes, missle cases, pressure levels, and as a replacement for asbestos in automotive brake and cluch linings, and gaskets.

6.4.6 Other Fibre Reinforcement Materials

Other fibre materials are boron, silicon carbide, and aluminium oxide. However, these fibre materials are used to much lesser degrees. Boron fibre-reinforced polymer composites find use in military aircraft components, helicopter robot blades, and some sporting goods. Silicon carbide and alumina fibres are used in tennis rackets, rocket nose cones and circuit boards.

Polymer-matrix materials

Polyesters and vinyl esters are the most widely used and least expensive polymer resins. These matrix materials are basically used for fibre glass reinforced composites. For mutations of a large number of resin provide a wide range of properties for these materials. The epoxies are more expensive and in addition to wide ranging commercial applications, also find extensive use in PMCs for aerospace applications. In comparison to polyesters and vinyl resins, epoxies have better mechanical properties and resistance to moisture. Polymide resins are used for high temperature applications (~230°C). High-temperature thermoplastic resins may find wider future aerospace application. These include polyetheretherketon (PEEK), polyphenylene sulphide (PPS) and polyetherimide (PEI).

Metal-matrix composites (MMCs)

The matrix in these composites is a ductile metal. These composites can be used at high service temperatures than their base metal counterparts. The reinforcement in these materials may improve specific stiffness, specific strength, abrasion resistance, creep resistance, thermal conductivity, and dimensional stability. In comparison to polymer-matrix composites, these materials have higher operating temperatures, non-flammability and greater resistance to degradation by organic fluids. In comparison to PMCs, MMCs are much more expensive and therefore, their use is somewhat restricted.

Alloys of aluminium, magnesium, titanium, and copper, and superalloys are used as matrix materials. The reinforcement may be in the form of particulates, both continuous and discontinuous fibres, and whiskers (concentrations range 10 to 60 Vol%) carbon, silicon carbide, boron, alumina, and refractory metals are continuous fibre materials, whereas silicon carbide whiskers, chopped fibres of alumina and carbon, and particulates of silicon carbide and alumina are discontinuous reinforcements sense, cermets fall within MMCs.

At elevated temperatures, some matrix-reinforcement combinations are highly reactive. This problem can be resolved either by using a protective surface coating to the reinforcement or by modifying the matrix alloy composition.

Recently, automobile industries have begun to use MMCs in their products, *e.g.* some engine components have been introduced consisting of an aluminium-alloy matrix that is reinforced with alumina and carbon fibres. This MMCs is light in weight and can resist wear and thermal distortion. MMCs are also used in driveshafts (that have higher rotational speeds and reduced vibrational noise levels), forged suspension and transmission components, and extruded stabilizer. MMCs are used by aerospace industry. Aluminium alloy metal-matrix composites, boron fibres are used as the reinforcement for the Space Shuttle Orbiter, and continuous graphite fibres for Hubble Telescope.

Using refractory metals such as tungsten, the high temperature creep and rupture properties of some of the superalloys (Ni-and Co based alloys) may be enhanced by fibre-reinforcement.

Ceramic-matrix composites (CMCs)

The fracture toughnesses of ceramics have been improved significantly by the development of CMCs-particulates, fibres, or whiskers of one ceramic material that have been embedded into a matrix of another ceramic. The fracture toughnesses of CMCs materials lies between 6 and 20 MPa$\sqrt{m}$. Increasing fibre content improves strength and fracture toughness. There is a considerable reduction in the scatter of fracture strengths for whisker-reinforced ceramics relative to their unreinforced counterparts. These CMCs exhibit improved high-temperature creep behaviour and resistance to thermal shock.

SiC whisker-reinforced aluminas are being utilized as cutting tool inserts for machining hard metal alloys. We may note that tool lives for these materials are greater than for cemented carbides.

Carbon-carbon composites

Carbon-carbon composite, *i.e.*, carbon fibre-reinforced carbon-matrix composite is one of the most advanced and promising engineering material. Obviously, in this material, both reinforcement and matrix are carbon. These materials have high-tensile modulii and tensile strengths that are retained to temperatures in excess of 2000°C, resistance to creep, and relatively large fracture toughness values. These materials have low coefficients of thermal expansion and relatively high thermal conductivities. These characteristics of these materials, coupled with high strengths, give rise to a relatively low susceptibility to thermal shock. The major drawback of these materials is a propensity to high-temperature oxidation. These materials are relatively new and expensive, and therefore, are not in wide use.

These materials are employed in rocket motors, as friction, materials in aircraft and high-performance automobiles, for hot-pressing molds, in components for advanced turbine engines, and as ablative shields for re-entry vehicles.

A comparison of specific strength of various carbon-carbon composites with that of other high-temperature materials relative to temperature is shown in Fig. 6.8(*b*).

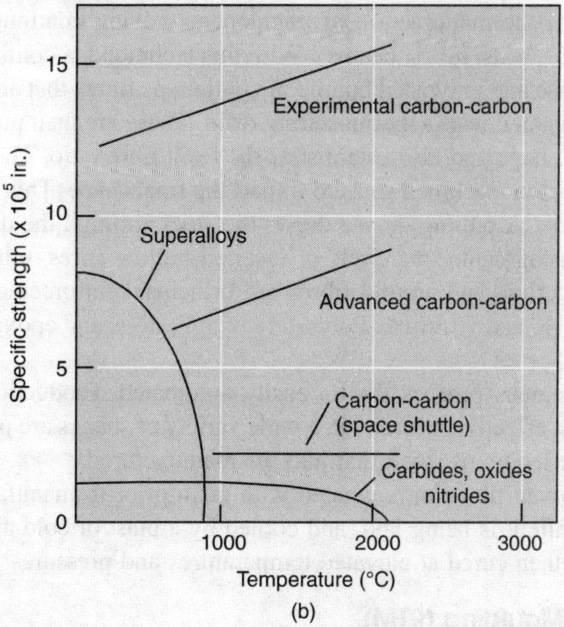

(b)

Fig. 6.8 (*b*) A comparison of the specific strength of various carbon-carbon composites with that of other high-temperature materials relative to temperature.

Hybrid composite materials (HCM)

The composites obtained by using two or more different kinds of fibres in a single matrix are termed as *hybrid*. Hybrid composites have a better all-round combination of properties than composites containing only a single fibre type. Although a variety of fibre combinations and matrix materials are used, but in the most common system, both carbon and glass fibres are incorporated into a polymeric resin. The carbon fibres are expensive, but they are strong and relatively stiff and provide a low density reinforcement. Glass fibres lack the stiffness of carbon, but they are inexpensive. The glass-carbon hybrid may be produced at a lower cost than either of the comparable all-carbon or all-glass reinforced plastics, and also it is stronger and tougher.

The properties of hybrid composites are anisotropic. When these are stressed in tension, failure is usually non-catastrophic, *i.e.*, does not occur suddenly. The carbon fibres are the first to fail, at which time the load is transferred to the glass fibres. Upon failures of the glass fibres, the matrix phase will have to sustain the applied load. We may note that eventual composite failure concurs with that of the matrix phase.

Hybrid composites find applications in lightweight land, water and air transport structural components, lightweight orthopedic components and sporting goods.

6.5 FABRICATION

Fabrication of composite materials can be accomplished through hand lay-up, injection moulding, lamination, filament winding, or pultrusion. The oldest method of composite manufacture is *hand lay-up*. In it, the resin is poured into a mould or applied to the surface and then the fabric is rolled into the resin or applied to the surface. The hand lay-up technique is used in boat building, auto repair, and home projects, because it requires minimal equipment and the result cures at room temperature.

Pultrusion is used for the manufacture of components having continuous lengths and a constant cross-sectional shape, (*i.e.* rods, tubes, beams). With this technique, a continuous fibre *rovings* or *tows* (A roving or tow is a loose and untwisted bundle of continuous fibres that are drawn together as parallel strands), are first impregnated with a thermosetting resin. These are then pulled through a steel die that performs to the desired shape and also establishes the resin/fibre ratio. Then the stock passes through a curing die that is precision machined so as to impart the final shape. This die is also heated to initiate curing of the resin matrix. A pulling device draws the stock through the dies and also determines the production speed. By using centre mandrels or inserted hollow cores, tubes and hollow sections are made possible. Glass, carbon and aramid fibres are principal reinforcements and normally added in concentrations between 40 and 70 Vol%. Polyesters, vinyl esters, and epoxy resins are commonly used matrix materials.

Pultrusion is a continuous process that is easily automated. Production rates are relatively high which makes it very cost effective. Moreover, a wide variety of shapes are possible. Also, there is really no practical limit to the length of stock that may be manufactured.

Prepag fibres are woven fibres impregnated with resin prior to manufacture. The prepag fabric is laid in place, heated while it is being laid, and cooled by a blast of cold air to set it in place initially. The formed product is then cured at elevated temperatures and pressures.

6.5.1 Resin Transfer Moulding (RTM)

This is a two-piece matching cavity-mould system used for moulding composites (Fig. 6.9).

The reinforcement material is placed in the mould cavity. A chopped mat is often draped along the inside of the mould. The mould halves are clamped together and the resin is introduced into the cavity. RTM offers fast curing times and is labour-intensive than other methods.

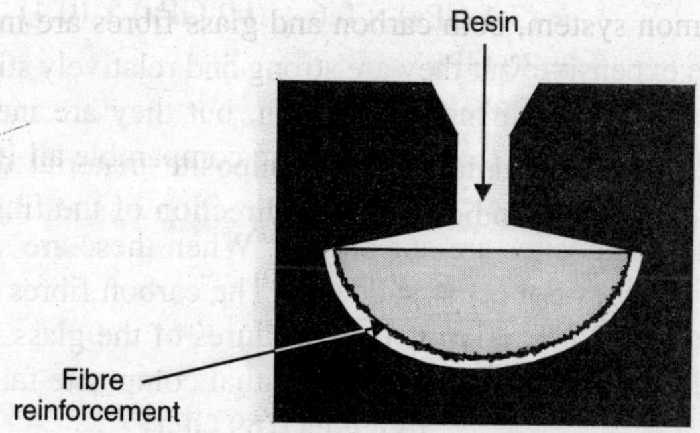

Fig. 6.9 Resin transfer moulding.

Ultimate Reinforced Thermoset Reaction Injection (URTRI)

This process allows the use of cores in the moulding process. A core is cast from a high-temperature epoxy foam. The core is wrapped with layers of graphite fabric and is positioned in the mould cavity. Epoxy resin is pumped into the heated mould and cured for about 5 min. The mould is then opened, and the part is extracted. This process is shown in Fig. 6.10.

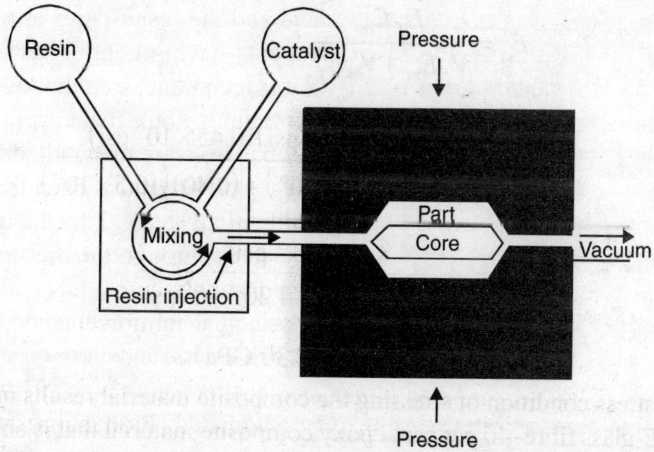

Fig. 6.10 Ultimately reinforced thermoset reaction injection process (URTRI).

The primary drawback of composites is the time it takes to manufacture them. Parts that are stamped out of metal every few seconds take several minutes to complete if made of composites. The increased time often increases the cost of manufacturing. However, with the rapid advances in technology, composite production has increased and applications have broadened, *e.g.* more and more automotive parts are being prepared from composites.

Example 1: A continuous and aligned glass fibre-reinforced composite consists of 40 Vol% of glass fibres having a modulus of elasticity 69 GPa and 60 Vol% of a polyester resin that, when hardened, displays a modulus of 3.4 GPa. Calculate the modulus of elasticity of this composite in the longitudinal directions. [B.E.]

Solution: We have $E_{cl} = E_m V_m + E_f V_f$

$= (3.4 \text{ GPa}) \times 0.6 + (69 \text{ GPa}) \times (0.4)$
$= 30 \text{ GPa}$
$= 4.3 \times 10^6 \text{ psi}.$

Example 2: Calculate the elastic modulus of the composite material described in Example 1, but assume that the stress is applied perpendicular to the direction of the fibre alignment.

Solution: We have

$$E_{cl} = \frac{E_m E_f}{V_m E_f + V_f E_m}$$

$$= \frac{(3.4 \text{ GPa})(69 \text{ GPa})}{(0.6)(69 \text{ GPa}) + (0.4)(3.4 \text{ GPa})}$$

$$= 5.5 \text{ GPa} = 0.81 \times 10^6 \text{ psi}.$$

Example 3: Calculate the modulus of elasticity for a composite material consisting of 60 per cent by volume of continuous E-glass fibre and 40 per cent epoxy resin for the matrix when stressed under **isostress conditions** (*i.e.*, the material is stressed perpendicular to the continuous fibre). The modulus of elasticity of the E-glass is 10.5 × 10⁶ psi and that of the epoxy resin in 0.45 × 10⁶ psi. [B. Tech.]

Solution:

$$E_c = \frac{E_f E_m}{V_f E_m + V_m E_f}$$

$$= \frac{(10.5 \times 10^6 \text{ psi})(0.45 \times 10^6 \text{ psi})}{(0.60)(0.45 \times 10^6) + (0.40)(10.5 \times 10^6)}$$

$$= \frac{4.72 \times 10^{12} \text{ psi}^2}{0.27 \times 10^6 \text{ psi} + 4.20 \times 10^6 \text{ psi}}$$

$$= 1.06 \times 10^6 \text{ psi} (7.30 \text{ GPa}).$$

Note that the isostress condition of stressing the composite material results in a modulus of elasticity for the 60 per cent *E*-glass fibre–40 per cent epoxy composite material that is about six times lower than that obtained by stressing under the isostrain condition.

Example 4: A unidirectional Kevlar 49 fibre-epoxy composite contains 60 per cent by volume of Kevlar 49 fibres and 40 per cent epoxy resin. The density of the Kevlar 49 fibres is 1.48 Mg/m³ and that of the epoxy resin is 1.20 Mg/m³. *(a)* What are the weight percentages of Kevlar 49 and epoxy resin in the composite material, and *(b)* what is the average density of the composite? [B. Tech.]

Solution: Basis is 1m³ of composite material. Therefore we have 0.60 m³ of Kevlar 49 and 0.40 m³ of epoxy resin. Density = mass/volume, or

$$\rho = \frac{m}{V} \quad \text{and} \quad m = \rho V$$

(*a*) Mass of Kevlar 49 = ρV = (1.48 Mg/m³) (0.60 m³) = 0.888 Mg

Mass of epoxy resin = ρV = (1.20 Mg/m³) (0.40 m³) = 0.480 Mg

Total mass = 1.368 Mg

$$\text{Wt. \% Kevlar 49} = \frac{0.888 \text{ Mg}}{1.368 \text{ Mg}} \times 100\% = 64.9\%$$

$$\text{Wt\% epoxy resin} = \frac{0.480 \text{ Mg}}{1.368 \text{ Mg}} \times 100\% = 35.1\%$$

(*b*) Average density of composite is

$$\rho_c = \frac{m}{V} = \frac{1.368 \text{ Mg}}{1 \text{ m}^3} = 1.37 \text{ Mg/m}^3.$$

Example 5: A piece of wood containing moisture weighs 165.3 g and after oven drying to a constant weight, weighs 147.5 g. What is its per cent moisture content? [B. Tech.]

Solution: The weight of water in the wood sample is equal to the weight of the wet wood sample minus its weight after oven drying to a constant weight. Thus,

$$\text{\% Moisture content} = \frac{\text{Weight of wet wood} - \text{Weight of dry wood}}{\text{Weight of dry wood}} \times 100\%$$

$$= \frac{165.3\text{g} - 147.5 \text{ g}}{147.5 \text{ g}} \times 100\% = 12.1\%.$$

Example 6: A metal-matrix composite is made with 80 per cent by volume of aluminium alloy 2124-T6 and 20 per cent by volume of SiC whiskers. The density of the 2124-T6 alloy is 2.77 g/cm^3 and that of the whiskers is 3.10 g/cm^3. Calculate the average density of the composite material. [B. Tech.]

Solution: Basis is 1 m^3 of material; thus we have 0.80 m^3 of 2124 alloy and 0.20 m^3 of SiC fibre in 1 m^3 of material.

Mass of 2124 alloy in 1 m^3 = (0.80 m^3) (2.77 Mg/m^3) = 2.22 Mg

Mass of SiC whiskers in 1 m^3 = (0.20 m^3)(3.10 Mg/m^3) = 0.62 Mg

Total mass in 1 m^3 of composite = 2.84 Mg

$$\text{Ave. density} = \frac{\text{mass}}{\text{unit volume}} = \frac{\text{Total mass of material in 1m}^3}{1\text{m}^3} = 2.84 \text{ Mg/m}^3.$$

Example 7: Consider that 2wt% ThO_2 is added to nickel. Each ThO_2 particle has a diameter of 1000 Å. Show that particles present in each cubic centimeter is 35.4 × 10^{12} particles/cm^3. [B.Tech.]

Solution: We have densities of ThO_2 and Ni as 9.69 and 8.9 g/cm^3, respectively. The volume fraction is:

$$f_{ThO_2} = \frac{\frac{2}{9.69}}{\frac{2}{9.69} + \frac{98}{8.9}} = 0.0184$$

Clearly, there is 0.0184 cm^3 of ThO_2 per cm^3 of composite. The volume of each ThO_2 sphere is:

$$V_{ThO_2} = \frac{4}{3}\pi r^3 = \frac{4}{3}\pi(0.5 \times 10^{-5} \text{ cm})^3$$

$$= 0.52 \times 10^{-15}\ \text{cm}^3$$

$\therefore$ Concentration of ThO_2 particles

$$= \frac{0.0184}{0.52 \times 10^{-15}}$$

$$= 35.4 \times 10^{12}\ \text{particles/cm}^3.$$

Example 8: A cemented carbide cutting tool used for machining contains 75 wt% WC, 15% TiC, 5 wt% TaC and 5 wt%. Co. Show that the density of the said composite is 11.50 g/cm³. [B. Tech.]

Solution: We have

$$f_{WC} = \frac{75/15.77}{75/15.77 + 15/4.94 + 5/14.5 + 5/8.9} = \frac{4.76}{8.70} = 0.547$$

$$f_{TiC} = \frac{15/4.94}{8.70} = 0.349$$

$$f_{TaC} = \frac{5/14.5}{8.70} = 0.040$$

$$f_{Co} = \frac{5/8.90}{8.70} = 0.064$$

Now, applying the rule of mixtures, we have the density of composite

$$\rho_c = \Sigma(f_i \rho_i) = \left[\frac{0.547}{15.77} + \frac{0.349}{4.94} + \frac{0.040}{14.5} + \frac{0.064}{8.9}\right]$$

$$= 11.50\ \text{g/cm}^3.$$

Example 9: When a stress (σ) is applied along the axis of the fibres, the rule of mixtures for the modulus of elasticity of a fibre-reinforced composite can be expressed as

$$E_{c,\,11} = f_m E_m + f_f E_f$$

where the notations have usual meanings. [B.Tech.]

Solution: We have, the total force acting on the composite is the sum of the forces carried by each constituent of the composite, *i.e.*

$$F_c = F_m + F_f$$

But $\quad F = \sigma A$

$$\therefore \quad \sigma_c A_c = \sigma_m A_m + \sigma_f A_f$$

or $$\sigma_c = \sigma_m \left(\frac{A_m}{A_c}\right) + \sigma_f \left(\frac{A_f}{A_c}\right)$$

Now, if the fibres have a uniform cross-section, the area fraction equals the volume fraction *f*, we have

$$\sigma_c = \sigma_m \cdot f_m + \sigma_f \cdot f_f$$

According to Hooke's law, $\sigma = \varepsilon E$, we have

$$E_{c,11}\,\varepsilon_c = E_m\,\varepsilon_m f_m + E_f\,\varepsilon_f f_f$$

Now, if the fibres are rigidly bonded to the matrix, both the fibres and the matrix must stretch equal amounts (iso-strain conditions):

$$\varepsilon_c = \varepsilon_m = \varepsilon_f$$

or
$$E_{c,11} = f_m E_m + f_f E_f$$

Example 10: Show that the equation for the modulus of elasticity of a fibre-reinforced composite is

$$\frac{1}{E_{c,\perp}} = \frac{f_m}{E_m} + \frac{f_f}{E_f}$$

where the stress is applied perpendicular to the axis of the fibre.

Solution: Obviously, the strains are no longer equal, but the weighted sum of the strains in each component equals the total strain in the composite, whereas the stresses in each component are equal (iso-stress conditions):

$$\varepsilon_c = f_m\,\varepsilon_m + f_f\,\varepsilon_f$$

or
$$\frac{\sigma_c}{E_c} = f_m\left(\frac{\sigma_m}{E_m}\right) + f_f\left(\frac{\sigma_f}{E_f}\right)$$

But $\sigma_c = \sigma_m = \sigma_f$. Thus, we have

$$\frac{1}{E_{c,\perp}} = \frac{f_m}{E_m} + \frac{f_f}{E_f}.$$

Example 11: Glass fibres in nylon provide reinforcement. If the nylon contains 30 Vol% E-glass, show that almost all of the load is carried by the glass fibres. [B.Tech.]

Solution: We have the modulus of elasticity for each component of the composite is

$$E_{\text{glass}} = 10.5 \times 10^6 \text{ psi},\ E_{\text{nylon}} = 0.4 \times 10^6 \text{ psi}$$

Now, both the nylon and the glass fibres have equal strain if bonding is good. Thus, we have

$$\varepsilon_c = \varepsilon_m = \varepsilon_f$$

or
$$\varepsilon_m = \frac{\sigma_m}{E_m} = \varepsilon_f = \frac{\sigma_f}{E_f}$$

or
$$\frac{\sigma_f}{\sigma_m} = \frac{E_f}{E_m} = \frac{10.5\times10^6}{0.4\times10^6} = 26.25$$

$$\text{Now, the fraction} = \frac{F_f}{F_f + F_m} = \frac{\sigma_f A_f}{\sigma_f A_f + \sigma_m A_m}$$

$$= \frac{\sigma_f\,(0.3)}{\sigma_m\,(0.3) + \sigma_m\,(0.7)}$$

$$= \frac{0.3}{0.3 + 0.7\left(\frac{\sigma_m}{\sigma_f}\right)}$$

$$= \frac{0.3}{0.3 + 0.7 \times \frac{1}{26.25}}$$

$$= 0.92$$

$$\approx 92\%.$$

Obviously, almost all of the load is carried by the glass fibres.

SUMMARY

- A materials system composed of a mixture or combination of two or more micro or macroconstituents that differ in form and chemical composition and are essentially insoluble in each other are termed *composite material.*
- Composite materials consisting of a mixture of a matrix of a plastic material such as a polyester or epoxy strengthened by fibres of high strength such as glass, carbon, or aramid are called *fibre-reinforced plastics*. Of these three fibres, glass fibres are the cheapest in cost and have intermediate strength and highest density in comparison to others. The fibres provide the high strength and stiffness and the plastic matrix bonds the fibres together and supports them.
- *Carbon fibres* have high strength, high modulus, and low density but are expensive and therefore used only for applications requiring their especially high strength to weight ratio.
- Fibres made from *E* (electrical) glass, which is a borosilicate glass and which is the most commonly used glass for fibres for fibre glass-reinforced plastics are *E-glass fibres*.
  Fibres made from *S*-glass, which is a magnesia-alumina-silicate glass and used for fibres for fibre glass-reinforced plastics when extra-high-strength are required is called *S-glass fibres*.
- *Carbon fibres* have high strength, high modulus and low density but are expensive. These are produced mainly from polyacrylonitride (PAN) or pitch that are stretched to align the fibrillar network structure within each carbon fibre and which are heated to remove oxygen, nitrogen, and hydrogen from the starting or precursor fibres. These are used only for applications requiring their especially high strength-to-weight ratio due to their high cost.
- Fibres produced by chemical synthesis and used for fibre-reinforced plastics are called *armid fibres*. These fibres have an aromatic (benzene ring type) polyamide linear structure.
  These fibres are relatively expensive and therefore used for applications where a high strength-to weight ratio is required alongwith better flexibility than carbon fibres.
- A collection of numerous fibres in a straight-laid bundle, specified according to the number of fibres it contains is called *tow* (of fibres) *e.g.* 6000 fibres/tow.
- A product made by bonding sheets of a material together, usually with heat and pressure is called *laminate*. One layer of multilayer laminate is called *laminate ply* or *lamina*.
- The most commonly used matrices for glass fibres for fibre-reinforced plastic composites are the *polyesters*, whereas the most commonly used matrices for *carbon-fibre reinforced plastics* are the epoxies.

- *Carbon-fibre-reinforced-epoxy* composite materials are used extensively for aircraft and aeroplane applications.
- *Glass fibre-reinforced polyester* composite materials have a much wider usage and find application in the building, transportation, marine and aircraft industries.
- A process for producing fibre-reinforced plastics by winding continuous reinforcement previously impregnated with a plastic resin on a rotating mandrel is called *filament winding*. When a sufficient number of layers have been applied, the wound form is cured and the mandrel removed.
- A compound of plastic resin, filler, and reinforcing fibre used to make fibre-reinforced plastic composite material is called *sheet-molding compound* (SMC).
- A mixture of portland cement, fine aggregate, coarse aggregate and water is called *concrete* (portland cement type). This as a construction material has advantages that includes unstable compressive strength, economy, castability on the job, durability, fine resistance, and aesthetic appearance. One can increase the low tensile strength of concrete significantly by reinforcement with steel rods.
- Concrete containing steel wires or bars to resist tensile forces is called *reinforced concrete.*
- Reinforced concrete in which internal compressive stresses have been introduced to counteract tensile stresses resulting from severe loads is termed as prestressed concrete. When concrete is poured over pretensioned steel wires or rods in prestressed concrete, then it is termed as *pretensioned (prestressed) concrete.*
- A bitumen consisting mainly of hydrocarbons having a wide range of molecular weights is called *asphalt.* Most asphalt is obtained from petroleum refining.
- Mixtures of asphalt and aggregate that are used mainly for road paying is termed as *asphalt mixes.*
- *Wood* is a natural composite material consisting essentially of cellulose fibres bonded together by a matrix of polymeric material made up mainly of lignin. The macrostructure of wood consists of *sapwood*, which is made up primarily of living cells and which carries the nutrients, and *heartwood,* which is composed of dead cells. The two main types of woods are *softwoods* and *heart woods.* Softwoods have exposed seeds and narrow (needle like) leaves, whereas hardwoods have covered seeds and broad leaves.
- Trees that have exposed seeds and arrow leaves (needles) are *softwood* trees, *e.g.* pine, fir and spruce.
- Trees that have covered seeds and board leaves are *hardwood trees, e.g.* oak, maple and ash.
- A very complex crosslinked three-dimensional polymeric material formed from phenolic units is called *Lignin.*
- The outer part of the tree stem of a living tree that contains some living cells that store for the tree is called *sapwood.*
- The innermost part of the tree stem that in the living tree contains only dead cells is called *heartwood.*
- The microstructure of wood consists of arrays of cells mainly in the longitudinal direction of the tree stem. Softwoods have long, thin-walled tubular cells called tracheids, whereas hardwoods have a dense cell structure that contains large vessels for the conduction of fluids.
- A ribbon-like aggregate of cells extending radially in the tree stem; the tissue of the ray is primarily composed of food-storing parenchyma cells is called *wood ray.*
- A tubular structure formed by the union of smaller cell elements in a longitudinal row is called wood vessel.
- Wood as a construction material has advantages that include usable strength, economy, ease of workability, and durability is properly protected.

SUGGESTED READINGS

1. K.K. Chawla, *Composite Materials Science and Engineering*, 2nd ed., Springer-Verlag, New York (1998).
2. D. Hull and T.W. Clyne, *An Introduction to Composite Materials*, 2nd ed., Cambridge University Press, New York (1996).
3. S.T. Peters, *Handbook of Composites*, 2nd ed. Kulwer Academic, Norwell, MA (1998).

REVIEW QUESTIONS

1. Define the term composite and explain what are advanced composites? Give few examples. [AMIE]
2. What factors influence the final properties of composites?
3. What is the difference between matrix and dispersed phases in a composite material?
4. Contrast the mechanical characteristics of matrix and dispersed phase for fibre-reinforced composite materials. [B.E.]
5. What is difference between cement and concrete?
6. Mention three important limitations that restrict the use of concrete as a structural material.
7. What are the functions of the matrix phase for a polymer-matrix fibre-reinforced composite.
8. For a polymer-matrix fibre-reinforced composite, compare the desired mechanical characteristics of matrix and fibre phases.
9. Why glass fibres are most commonly used for reinforcement?
10. What is the difference between carbon and graphite?
11. Why fibre glass reinforced composites are utilized extensively?
12. What is a hybride composite? Mention two important advantages of hybrid composites over normal fibre composites.
13. Give examples of applications where composites are used for load bearing applications.
14. Explain, what is a dispersion-strengthened composite? What is difference between this and a particle-reinforced composite?
15. What is fibre-reinforced composite? What fibre-reinforcing materials are commonly used?
16. Explain briefly how the volume of fibre, fibre-reorientation, and fibre strength and modulus affect the properties of fibre-reinforced composites.
17. What is a laminar composite?
18. Explain briefly in what supporting equipment composite materials are used? Mention the main reason for using composites in these applications.
19. What is a coupling agent? What is 'sizing' as it relates to the production of glass fibres?
20. Mention the advantages of using ceramic matrix composites.

PROBLEMS

1. A tensile load of 200 N is applied to a composite material of 1 sq. mm cross-sectional area. The volume of parallel fibre is 40%. Determine the stress in fibres, if axis of load is (*i*) parallel to the fibres, and (*ii*) perpendicular to the fibres. Given that the ratio of Young's moduli of fibre material and matrix material is 7. [B.E.]

   [**Ans.** (*i*) 58.8 N/mm^2 (*ii*) 411.6 N/mm^2]
2. Calculate the longitudinal tensile strength of an aligned glass fibre-epoxy matrix composite in which the average fibre diameter and length are 0.015 mm and 2.0 mm, respectively, and the

volume fraction of fibres is 0.25. Assume that (*i*) the fibre-matrix bond strength is 100 MPa (*ii*) the fracture strength of the fibres is 3500 MPa, and (*iii*) the matrix stress at composite failure is 5.5 MPa. [B.E.]

[Ans. σ^*_{cd} = 822 MPa]

3. **Boron coated with SiC (also known as Borsic) reinforced aluminium containing 40 Vol% fibres is an important high-temperature, light weight composite material. Estimate the density, modulus of elasticity, and tensile strength parallel to the fibre. The properties of individual components are:**

Material	*Density (ρ) (g/cm³)*	*Modulus of elasticity (E) (psi)*	*Tensile strength (TS) (psi)*
Fibres	2.36	55,000,000	400,000
Aluminium	2.70	10,000,000	5000

Also estimate the modulus of elasticity perpendicular to the fibres. [B.Tech.]

[Ans. $E_{c,11}$ 28 × 106 psi, $E_{c,\perp}$ = 14.9 × 10^6 psi]

Estimated modulus of elasticity = 163,000 psi.

4. Ni containing 2 wt% Th is produced in powered form, consolidated into a part, and sintered in the presence of oxygen, causing all of the Th to produce ThO_2 spheres 80 nm in diameter. Taking density of ThO_2 = 9.86 g/cm³, show that the number of spheres per cm³ is 7.65×10^{13}. [B.Tech.]

5. Calculate the density of a cemented carbide (or cermet), based on a titanium matrix if the composite contains 50 wt% WC, 22 wt% TaC, and 14 wt% TiC. Given ρ_{wc} = 15.77 g/cm³, ρ_{TiC} = 4.94 g/cm³, ρ_{TaC} = 14.5 g/cm³. [B.Tech.]

[Ans. 9.408 g/cm³]

6. To produce a low-cost composite having a modulus of elasticity greater than 120,000 psi and a tensile strength greater than 2000 psi, how much clay have to be added to 10 kg of polyethylene. Given, the density of clay = 2.4 g/cm³ and the density of polyethylene = 0.92 g/cm³. [B.Tech.]

[Ans. 11.12 to 22.2 kg]

7. An epoxy matrix is reinforced with 40 Vol % *E*-glass fibres to produce a 2 cm diameter composite that is to withstand a load of 25,000 N. Show that the stress acting on each fibre is 188 MPa. [B. Tech.]

SHORT ANSWER QUESTIONS

1. What are 'composites'?

Ans. These are artificially produced multiphase materials having a desirable combination of the best properties of the constituent phases. Usually, one phase (the matrix) is continuous and completely surrounds the other (the disperse phase).

2. What is 'concrete'?

Ans. This is a type of large particle, composite consists of an aggregate of particles bonded together with cement.

3. What is 'portland cement concrete'?

Ans. The aggregate consists of sand and gravel. The cementitious bond develops as a result of chemical reaction between the portland cement and water.

4. Out of the several composite types, for which the potential for reinforcement efficiency is greatest?

Ans. Fibre reinforced.

5. What types of fibres will be for short and discontinuous fibrous composites?

Ans. The fibres may be either aligned or randomly oriented.

6. Can polymer-matrix be reinforced with glass, carbon and aramid fibres?

Ans. Yes.

7. Which are the more advanced composites?

Ans. (*i*) Carbon-Carbon and (*ii*) Hybrids.

8. Write two kinds of structural composites.

Ans. (*i*) Laminar composites and (*ii*) Sandwich panels.

9. What do you understand by *brazing*?

Ans. Brazing is a process in which a liquid filler metal is introduced by capillary action between two solid based materials that are to be joined. Solidification of the brazing alloy provides the bond.

10. What are Aramid fibres?

Ans. These are polymer fibres, such as Kevlar™, formed from polyamides, which contain the benezene ring in the backbone of the polymer.

11. What are cemented carbides?

Ans. These are particulate composites containing hard ceramic particles bonded with a soft metallic matrix. We may note that the composite combines high hardness and cutting ability, yet still has good shock resistance.

12. What do you understand by cladding?

Ans. A laminar composite is produced when a corrosion-resistant or high-hardness layer of a laminar composite is formed onto a less expensive or high-strength backing.

13. What is a nanocomposite?

Ans. A nanocomposite is a material in which the dispersed phase is nano-sized and it distributed in the continuous matrix.

14. What are prepegs?

Ans. These are layers of fibres in unpolymerised resins. After the prepegs are stacked to form a desired structure, polymerisation joins the layers together.

15. State the rule of mixtures.

Ans. According to this rule, the properties of a composite material are a function of the volume fraction of each material in the composite.

16. What is a sandwich?

Ans. This is a composite material constructed to a light weight, low density material surrounded by dense, solid layers. The sandwich combines overall light weight with excellent stiffness.

17. What are whiskers?

Ans. Whiskers are very fine fibres grown in a manner that produces angle crystals with no mobile dislocations, thus giving nearly theoretical strengths.

18. Mention the fields in which composites have applications.

Ans. Composites have many applications in construction, automative, aerospace, sports, micro-electronics and several other industries.

19. What are dispersion-strengthened materials?

Ans. These materials contain exceptionally small oxide particles in a metal matrix. The small stable dispersoids interfere with slip, providing good mechanical properties at elevated temperatures.

20. What are particulate composites?

Ans. These contain particles that impart combinations of properties to the composite. Metal-matrix composites contain ceramic or metallic particles that provide improved strength and wear resistance and assure good electrical conductivity, toughness, or corrosion resistance. Polymer-matrix composites contain particles that enhance stiffness, heat resistance, or electrical conductivity while maintaining light weight, low cost or ease of fabrication.

21. Briefly mention the importance of fibre-reinforced composites.

Ans. These composites provide improvements in strength, stiffness, or high temperature performance in metals and polymers and impart toughness to ceramics. Fibres typically have low densities and give high specific strength and specific modules, but they often are very brittle. Fibres can be continuous or discontinuous. We may note that discontinuous fibres with a high aspect ratio *l/d* produce better reinforcement.

22. What are laminar composites?

Ans. These are built of layers of different materials. These layers may be sheets of different metals, with one metal providing strength, and the other providing hardness or corrosion resistance. Layers may also include sheets of fibre-reinforced material bonded to metal or polymer sheets or even to fibre-reinforced sheets having different fibre orientations. We may note that laminar composites are always anisotropic.

23. Write an expression for the density of a particulate composite.

Ans. $\rho_c = \sum f_i \cdot \rho_i = f_1\rho_1 + f_2\rho_2 + ... + f_n\rho_n$

where ρ_c is the density of composite, ρ_1, ρ_2 ..., ρ_n are the densities of each constituent in the composite and $f_1, f_2, f_3, ..., f_n$ are the volume fraction of each constituent.

24. What type of materials are used for electrical contacts and how they are produced?

Ans. Materials used for electrical contacts in switches and relays should have a good combination of wear resistance and electrical conductivity. Otherwise, the contacts erode, causing poor contact and arcing. These characteristics are found in tungsten-reinforced silver. A tungsten powder compact is made using conventional powder metallurgical processes as shown in Fig. 6.11 so that

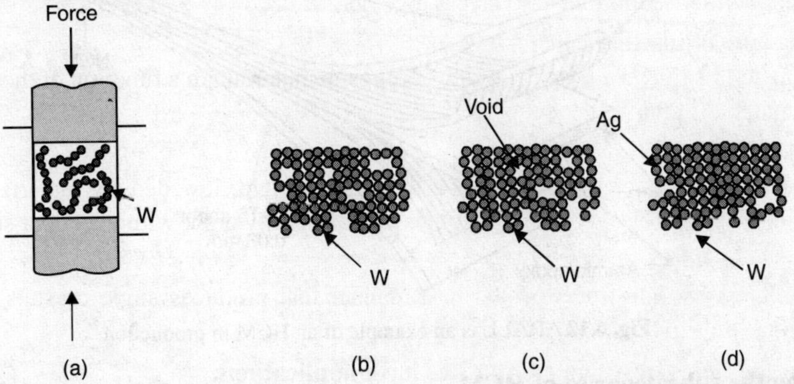

Fig. 6.11 Various steps in producing a silver-tungsten electrical composite: (*a*) Tungsten powders are pressed, (*b*) a low-density compact is produced, (*c*) sintering joins the tungsten powders, and (*d*) liquid silver is infiltrated into the pores between the particles.

high interconnected porosity is produced. Then, liquid silver is infiltrated to fill the interconnected voids. Both the silver and the tungsten are continuous. Obviously, the pure silver efficiently conducts current while the hard tungsten provides wear resistance.

25. What are matrix hybridized composites?

Ans. The matrix is one of the important constituents of the composites. Hybridising a matrix with another matrix also means toughening of the matrix. Such matrices when used to prepare the composites also can lead to a hybrid composite. Toughening of the matrix is carried out in order to reduce the brittleness of the binder matrix and as a consequence to improve the (mechanical and chemical resistance) properties of composites. An unmodified matrix usually consists of single-phase materials, while the addition of modifiers turns the toughened matrix into a multiphase system.

26. What do you understand by reinforcement and matrix hybridized composites?

Ans. Composites of this type consist of multiphase materials (four phases) in which both the matrix and the reinforcements have been hybridised. In one way, it can be said that it is the combination of the above two types.

The potential for the usage of the hybrid composites far out weighs any negative aspects since there are a vast number of plastic/plastic, plastic/metal armid-reinforced aluminium laminate (ARALL), metal/metal (MMC), metal/ceramic, ceramic/ceramic (space shuttle outer tile), and ceramic/plastic composite combination systems yet to be explored, investigated, developed, and in use.

27. What are hybrid composite materials (HCM)?

Ans. HCM are defined as a composite material system derived from the integrating of dissimilar materials, at least one of which is a basic composite material. A typical example of a HCM is a reinforced polymer composite combined with a conventional unreinforced homogeneous metal. The HCM blends the desirable properties of two or more types of materials into a single material system, which displays the beneficial characteristics of the separate constituents. An example of hybrid composite is ARALL (Fig. 6.12), which consists of high-strength aluminium alloy sheets interleafed with layers of aramid fibre-reinforced adhesive.

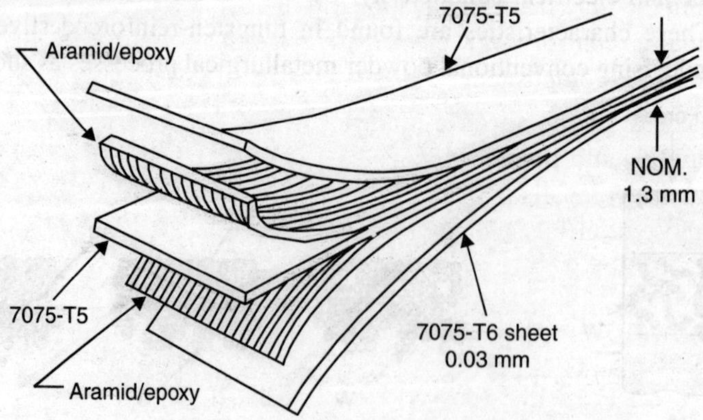

Fig. 6.12 ARALL is an example of an HCM in production.

28. Mention the subcategories of HCM.

Ans. (*i*) HCM, (*ii*) selective reinforcements, (*iii*) thermal management, (*iv*) selective reinforcements, and (*v*) ultralight weight materials.

29. Mention some of new composite materials.

Ans. New composite materials include nanocomposites, functionally gradient materials (FGMs), hybrid materials (Hy mats), IPNs, microinfiltrated macrolaminted composites (MIMLCs), and liquid crystal polymers (LPs).

30. What do you understand by smart skins and smart structures?

Ans. Smart skins and smart structures are related in that each contains embedded, nonstructural elements. A smart structure contains sensors that monitor the health of the structure itself, such as fibre sensors that monitor the health of the structure itself, such as fibre optics to determine temperature and structural deformations or cracks. A smart skin contains circuitry and electronic components that enable the skin to double as part of the electronic system of the parent vehicle, be it an aircraft or a missile.

31. What are *ultralight weight* materials?

Ans. These materials includes the emerging family of liquid crystal ordered polymers, which by virtue of their molecular structure, exhibit extremely high specific strength and specific stiffness. These highly directional materials are similar to composite-materials in that the long, ordered molecular chains within the polymer act very much like reinforcing fibres in a composite material. Examples of these materials are poly-p-phenylene benzobisthiazole (PB2T) and poly-ϕ-phenylene benzobisoxazole (PBO). Another example is gel-spun PE.

32. What are adaptive materials?

Ans. These materials are nowadays already reliable for industrial applications and can be easily introduced in systems newly designed, with their introduction already taken into consideration, or in older systems as a retrofit. They have the advantage of their application being easily achicved, independently of the base material, since for instance they can be bonded (surface glued or embedded) to metal, carbon, composite, etc. Adaptive systems can improve aircraft reliability, reducing failures through vibration control or reduction.

33. Explain matrix properties?

Ans. The matrix supports the fibres and keeps them in proper position, transfer the load to the strong fibres, protects the fibres from damage during manufacture and use of the composite, and prevents cracks in the fibre from propagating throughout the entire composite. The matrix usually provides the major control over electrical properties, chemical behaviour, and elevated-temperature use of the composite.

34. Make a comparison of the specific modulus and specific strength of composite materials with those of metals and polymers.

Ans. Figure 6.13 shows a comparison of specific modulus and specific strength of several composite materials with those of metals and polymers.

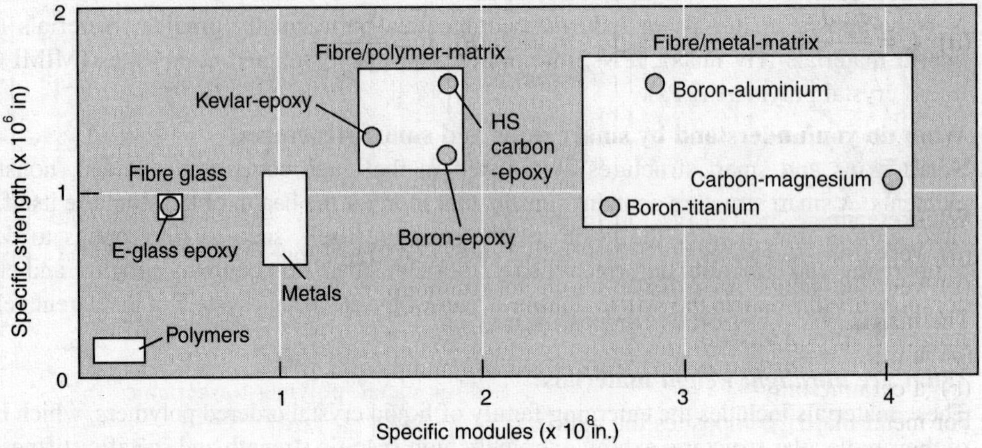

Fig. 6.13 A comparison of the specific modulus and specific strength of several composite materials with those of metals and polymers.

35. Give few examples of fibre-reinforced materials and mention their applications.

Ans. Table 6.5 below gives examples and applications of fibre-reinforced materials.

Table 6.5 Examples of Fibre-reinforced Materials and Their Applications

Material	*Applications*
Borsic aluminum	Fan blades in engines, other aircraft and aerospace applications
Kevlar™-epoxy and Kevlar™-polyester	Aircraft, aerospace applications (including space shuttle), boat hulls, sporting goods (including tennis rackets, golf club shafts, fishing rods), flak jackets
Graphite-polymer	Aerospace and automotive applications, sporting goods
Glass-polymer	Lightweight automotive applications, water and marine applications, corrosion-resistant applications, sporting goods equipment, aircraft and aeroscape components.

36. How laminar composites are built?

Ans. These are built of layers of different materials. These layers may be sheets of different metals, with one metal providing strength, and the other providing hardness or corrosion resistance. Layers may also include sheets of fibre-reinforced material bonded to metal or polymer sheets or even to fibre-reinforced sheets having different fibre orientations. We may note that the laminar composites are always anisotropic.

OBJECTIVE QUESTIONS

1. The cermets are examples of
   (*a*) ceramic-metal composites (*b*) carbon-carbon composites
   (*c*) metal-matrix composites (*d*) none of the above
2. The critical fibre length (l_c) is related to the fibre diameter d, ultimate (or tensile) strength σ_f^* and fibre-matrix bond strength τ_c as

(a) $l_c = \frac{\sigma_f^* d}{\tau_c^2}$ (b) $l_c = \frac{\sigma_f^* d}{2\tau_c}$

(c) $l_c = \frac{\sqrt{\sigma_f^* d}}{\tau_c}$ (d) $l_c = \sqrt{\sigma_f^* \, d\tau_c}$

3. Whiskers are
   (a) very thin steel wires (b) very thin copper wires
   (c) very thin single crystals (d) none of these
4. The matrix phase of fibrous composites may be
   (a) a metal only (b) a polymer only
   (c) a ceramic only (d) a metal, polymer, or ceramic
5. For metal-matrix composites the matrix is a
   (a) ductile metal (b) polymer
   (c) ceramic (d) none of these
6. A laminar composite is composed of
   (a) one-dimensional sheets or panels (b) two-dimensional sheets or panels
   (c) three-dimensional sheets or panels (d) none of these
7. Most advanced and promising composite is
   (a) carbon-carbon (b) metal-matrix
   (c) polymer-matrix (d) none of these
8. Trade names for two most common aramid materials are
   (a) silicon carbide, silicon nitride (b) e-glass, aluminium oxide
   (c) kevlar, nomex (d) high strength steel, aluminium oxide

ANSWERS

1. (a) 2. (b) 3. (c) 4. (d) 5. (a) 6. (b) 7. (a) 8. (c)

CHAPTER 7

Smart (or Intelligent) Materials

7.1 INTRODUCTION

Smart materials are a group of new and state of the art materials invented more than 30 years ago, but their development and improvement over the past three decades has led to new, more varied uses of these adaptable materials. These new classes have properties that can be altered by temperature, moisture, electric or magnetic fields, pH, and stress. They can change shape and colour, become stronger or produce voltage as a result of external stimuli. The adjective "smart" implies that these materials are able to sense changes in predetermined manners-traits that are also found in living organisms. Usually, 'smart' concept is also extended to rather sophisticated systems that consists of both smart and traditional materials.

Components of smart material (or system) include some type of sensor (that detects an input signal), and an actuator (that performs a responsive and adaptive function). Actuators may be called upon to change shape, position, natural frequency or mechanical characteristics to changes in temperature, electric fields and/or magnetic fields.

There are four types of materials commonly used for actuators: shape memory alloys, piezoelectric ceramics, magnetostrictive materials and electrorheological/magnetorheological fluids. *Shape memory alloys* (SMAs) such as a brass, nickel-titanium and gold-cadmium are materials, which alter their shape in response to changes in temperature and then return to their original shape. Piezoelectric ceramics expand and contract in response to an applied electric field (or voltage); conversely, they also generate an electric field when their dimensions are altered. We may note that the behaviour of magnetostrictive materials is analogous to that of the piezoelectrics, except that they are responsive to magnetic fields. Also, electrorheological and magnetorheological fluids are liquids that experience dramatic changes in viscosity upon the application of electric and magnetic fields, respectively.

Materials/devices employed as sensors include optical fibres, piezoelectric materials (including some polymers), and microelectromechanical (MEMS) devices.

In the past three decades, SMAs have found widespread use in applications such as miniature surgical tools that can twist and pull when a small amount of heat is applied, wires that expand and contract for use by dentists in straightening teeth, cell phone antennas that resist breaking, stents and bone plates that must fuse together or expand in order to stay in place, and flexible eyeglass lens.

SMAs, which find widespread applications today, should be recognized along with the familiar metals, plastics, ceramics, composites, powder metals and specialty-type multifunctional materials (SMAs, micromechanical systems (MEMs), functionally graded materials (FGMs), and nanomaterials).

By using smart materials instead of adding mass, engineers can endow structures with built-in responses to a myriad of contingencies. In their various forms, these materials can perform as actuators, which can adapt to their environments by changing characteristics such as shape and stiffness, or as sensors, which provide the actuators with information about structural and environmental changes. Sensors introduced into aircraft designs will help technicians and mechanics to provide a more sophisticated inspection technique for maintenance and repairs.

SMAs are beginning to play an important role in civil engineering designs for dams, bridges, highways and buildings. Sensors embedded throughout a concrete and composite structure can act like nerves, sensing when areas of structure begin to degrade and alerting maintenance engineers to the need for repairs.

The automative industry is also eager to incorporate intelligent materials technology. Currently, researchers are trying to develop smart car seats that can identify primary occupants and adapt to their preferences for height, leg room, back support and so forth. More profound changes are looming in automotive design based on smart materials, *e.g.* the technology to enable cars to tell owners how much air pressures tires have, when oil changes are needed, and other maintenance information exists as of today. The development of ultrahigh-fidelity stereo speakers using piezoelectric actuators, which expand and contract in thousands of a second in response to applied voltage, is aimed at turning whole house walls or car interiors into speakers by imbedding them with the tiny actuators. Obviously, in future, we would not need to install separate speakers in our homes and cars in an attempt to achieve maximum musical efforts. Obviously, cars and houses will offer built-in surround sound.

A new technology for implants that may improve construction or repair of bones in the face, skull and jaw has been developed. This provides a method for making scaffolds for bone tissue. The scaffold is seeded with the patient's own cells and is formed with a cement paste made of minerals also found in natural bone. The paste is mixed with beads of a natural polymer (made from seaweed) filled with bone cells. The paste is shaped or injected into a bone cavity and then allowed to harden with the encapsulated cells dispersed throughout the structure. The natural polymer beads gradually dissolve when exposed to the body's fluids, creating a scaffold that is filled by the now released bone cells.

The cement, a calcium phosphate material, is strengthened by adding chitosan, a bipolymer extracted from crustacean shells. The implant is further reinforced to about the same strength as spongy natural bone by covering it with several layers of a biodegradable fibre mesh already used in clinical practice.

Bone cells are very smart. They can tell us the difference between materials that are bioactive compared to bioinert polymers. The material is so designed to be similar to mineral in bone so that cells readily attach to the scaffold.

Recent advances in non-destructive evaluation (NDE) sensor technologies, health monitoring, and life prediction models are revolutionizing component inspections and life management, and will significantly improve the reliability and airworthiness of aerospace systems. Several key advances in NDE, health, monitoring and prognostics programmes include emerging structural health monitoring technologies for aerospace applications, high temperature health monitoring, advances in NDE technologies for measurement of subsurface residual stresses, computational methods and advanced NDE techniques, and materials damage assessment techniques and prognosis models for prediction of remaining useful life. Obviously, these technologies will help for lowering maintenance costs, integrated damage prognosis and life prediction, enhancing reliability and safety and improving the performance and operational efficiency of current and future aerospace systems.

Smart machines is another area of focus, *e.g.* moving composite manufactures out of hand lay-up and open molding into more environmentally acceptable closed-molding alternatives where a reliable source of performs is a necessity for volume production in closed-mold processes, such as resin molding (RTM), vacuum-assisted RTM (VARTM), and vacuum-assisted resin infusion molding (VARIM).

Composites are now used in monitoring systems where they are combined with other materials and sensors *e.g.* composite "shape sensing mat" used with a metallic riser system. The flexible mat, which incorporates fibre optics, wraps around a steel riser, and enable operators to excessive bonding and fatigue life during riser deployment on a dynamic positioned oil drilling ship as well as downhole coiled tubing complete with power and data transmission capabilities for drilling or workover applications and oilfield pipelines.

Efforts are there to equip machines with artificial intelligence capacity. Efforts are there to develop software that will enable machines to interpret and act on commands with near human common sense. In the present chapter, we will briefly discuss these materials.

Among several smart materials, shape memory alloys (SMAs), piezoelectric materials and fibre optic sensors are currently receiving wide applications.

7.2 SHAPE MEMORY ALLOYS

The shape memory phenomenon manifests itself when a shape memory alloy (SMA) is plastically deformed in the low-temperature martensitic condition, and upon removal of the external loads regains its original shape when heated. The exact mechanism by which the shape recovery takes place is not very well understood, however, the process of regaining the original shape is known to be associated with a reverse transformation of the deformed martensitic phase to the higher temperature austenitic phase. Typical examples exhibiting shape memory effect include the copper alloy systems of Cu-Zn, Cu-Zn-Al, Cu-Zn-Ga, Cu-Zn-Sn, Cu-Zn-Si, Cu-Zn-Ni, Cu-Sn and alloys of Au-Cd, Ni-Al and Fe-Pt. Nitinol, Nitinol, NiTi (Nickel-titanium alloy) is the most common of the SMAs or transformation metals. Obviously, SMAs are metallic compounds. SMAs exhibit a cubic high temperature crystal structure which transforms by martensite reaction into another crystal structure (tetragonal, orthorhombic, hexagonal, etc.) when temperature decreases. Martensite transformation enables large shape changes and the shape memory effect (SME), but heating and cooling are needed to complete the phase transformation. SME can produce large relative strains and high stress output, but usually small actuation frequency due to the slow cooling process. The SMA's have typically twined martensitic structure with mobile twinned boundaries. The multivalent twin structure can easily be deformed to the single-variant state to yield the maximum SME shape change.

SMAs are a unique class of materials, which remember their shape even after severe deformation, *i.e.* when a SMA is once deformed in the cold shape (martensite) these materials will stay deformed until heated, where, upon heating they will spontaneously return to their original pre-determined hot shape (austenite).

It is observed that, the structural changes at the atomic level contributes to this unique properties of the materials.

Martensite transformations fundamentally involve a lattice transformation featuring shear deformation and a coordinated atomic movement, which maintains a one-to-one lattice correspondence between the lattice points in the parent phase and the transformed phase. The martensitic phase is a substitutional or interstitial solid solution. The transformation is diffusion-free which yields the same concentration of solute atoms dissolved in the martensitic phase as in the parent phase. Martensitic transformations are typically characterized by well-defined shape changes or surface reliefs.

Martensitic transformations are characterized by a reverse transformation which is absent in the deformation fields associated with slip or twinning, therefore the deformation behaviour of alloys undergoing martensitic transformations is significantly different from those pertaining to ordinary metals and alloys. The fundamental characteristic of the reversible shape memory effect is that regardless of the severity of the deformation of the low-temperature martensitic phases, heating induces a reverse transformation which allows the specimen to return to the original shape corresponding to high temperature austentite phase. If this shape does not change upon cooling the specimen and transforming it to a martensitic phase, the phenomenon is called a *'one way' shape-memory effect*. One can induce a reversible shape-memory effect in SMAs by employing the following techniques:

(*i*) Subject the martensite to substantial deformation beyond a prescribed limit;

(*ii*) Subject the parent phase to deformation, and cool the specimen under suitable constraints, and maintain this state under stress for a prolonged period of time.

(*iii*) Deform the martensitic phase, and heat the specimen in order to induce the reverse transformation.

(*iv*) Deform the specimen after creating minute precipitation in the present phase.

All the above techniques create sites of internal stress through some mechanism within the high-temperature parent phase, which is essentially homogeneous and has the ability to recover its shape in the reverse transformation upon heating. Moreover, these sites of internal stress control the martensitic transformation which is initiated by cooling.

7.2.1 Types of Shape Memory Alloys

SMAs are of two types: The *one way shape memory alloy,* and *two way shape memory alloy.*

The materials which exhibit shape memory effect (*i.e.* taking their own shape) only upon heating are said to have one way shape memory.

In contrast, there are some materials which exhibit shape memory effect both during heating and cooling. Such materials are said to have two way shape memory.

7.2.2 Crystal Structure of Shape Memory Alloy

The shape memory alloys are found to have two distinct crystal structures or phases. The effect of temperature and internal stresses determines the crystal structure or the phase that the shape memory alloy will have at a particular instant.

The phase of shape memory alloy which exists at high temperature is called *austenite* phase (microscopically they possess small platelet structure) and, the other phase which is found to exist at low temperature is called the *martensite* phase (microscopically they possess needle like structure). In both the cases the material remains in solid form.

In austenite phase material possesses a highly symmetric structure, such as the one in cubic system.

However, in the martensitic phase the structure is of low symmetry, such as monoclinic. The basis for the shape memory effect is that, these materials can easily transform to and from martensite.

The shape memory materials could be broadly classified into *thermal induced shape memory, stress induced shape memory and ferro-magnetic shape memory materials*.

7.3 TEMPERATURE INDUCED TRANSFORMATION

Figure 7.1 shows the phase transformation taking place in SMAs not at a particular temperature but over a range of temperature. This transformation is called the temperature induced transformation.

The temperature induced transformation is characterised by four temperatures, M_s and M_f during cooling and A_s and A_f during heating.

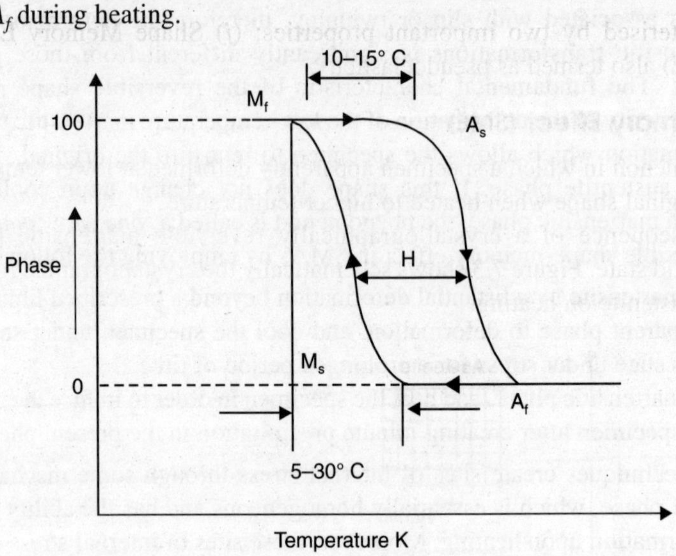

Fig. 7.1 Temperature induced transformation.

M_s and M_f indicate the temperature at which the transformation from the parent phase austenite into martensite starts and finishes respectively.

Similarly A_s and A_f indicate the temperature at which the reverse transformation from martensite to austenite starts and finishes upon heating.

It is observed that the overall transformation has a temperature range of 10–15° C depending on the chemical composition of the alloy. It is found that this overall transformation exhibit hysteresis in which, the transformations on heating and cooling do not overlap. However, this hysteresis is found to depend on the composition of the alloy system.

7.4 STRESS INDUCED TRANSFORMATION

When a stress is applied over the SMA, there is a possibility for phase transformation.

We may note that the stress induced transformation takes place at a constant temperature. At a tempera-ture above the A_f temperature, the martensite phase can be induced by applying stress over the austenite phase as shown in Fig. 7.2.

When stress exceeds above a certain value the austenite phase undergoes a large elastic deformation. Stressing beyond the elastic limit will result in permanent plastic strains. When the stress is removed, the material almost completely recovers to the parent austenite phase at a much lower value of stress as shown in Fig. 7.2.

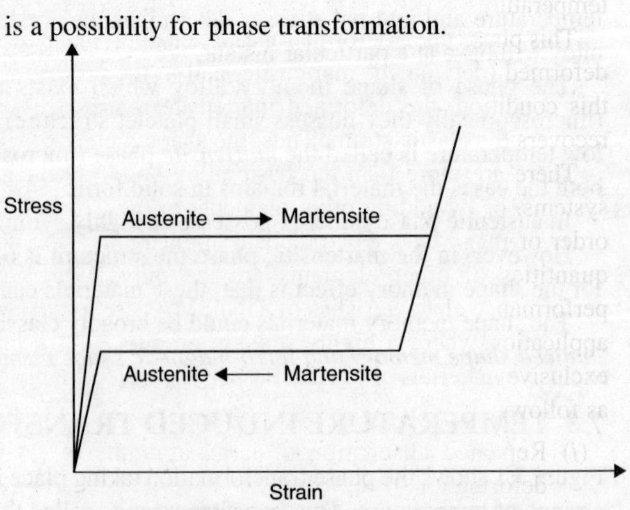

Fig. 7.2 Stress induced transformation.

7.5 FUNCTIONAL PROPERTIES

SMAs are characterised by two important properties: (*i*) Shape Memory Effect (SME) and (*ii*) Superelasticity (SE) also termed as pseudoelasticity.

7.5.1 Shape Memory Effect (SME)

SME is the phenomenon in which a specimen apparently deformed at lower temperature and reverts to its undeformed original shape when heated to higher temperature.

SME is a consequence of a crystallographically reversible martensitic phase transformation occurring in the solid state. Figure 7.3 shows schematically the crystallographic formation of martensite and reversion of austenite on heating.

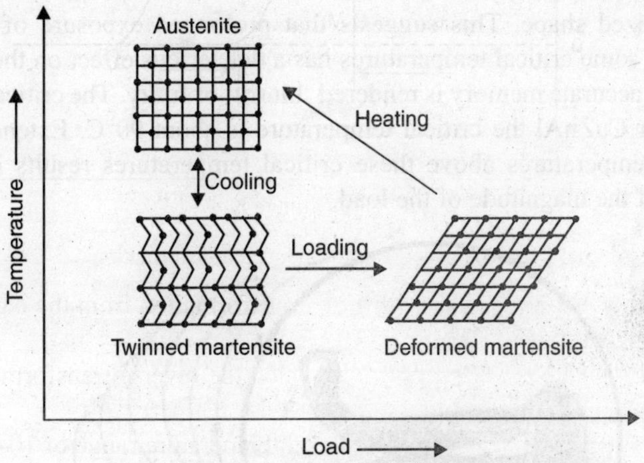

Fig. 7.3 Shape memory effect.

From Fig. 7.3, it is seen that the high temperature austenitic structure undergoes twinning as the temperature is lowered. This twinned structure having microscopically needle like structure is called martensite. This phase is relatively soft and easily deformed phase of SMA which exists at lower temperatures.

This phase upon deformation (*i.e.*, on applying external stress) takes on a particular shape called the deformed (detwinned) martensite, and in the process undergoes a large elastic strain. When heated in this condition, the deformed martensite returns to the stable austenite structure and in the process recovers the elastic strain. Thus, the shape memory phenomenon is observed.

There are more than ten basic alloy systems which exhibit this effect. By combining these alloy systems, or by adding other elements to these systems, the number of alloy systems increases by an order of magnitude. For practical applications TiNi and CuZnAl and their combinations with small quantitics of other elements are the only viable SMAs currently available in the market. The performance characteristics of TiNi SMAs are found superior to those of CuZnAl SMAs. Therefore, in application where a highly reliable product with a long fatigue life is desired, TiNi alloys are the exclusive materials of choice. Properties, which enable Ni–Ti wires for civil engineering application are as follows:

(*i*) Repeated absorption of large amounts of strain energy under loading without permanent deformation.

(*ii*) Possibility to obtain a wide range of cyclic behaviour from supplemental and fully recentering to highly dissipating by simply varying the characteristics of SMA components.
(*iii*) Usable strain range of 70%.
(*iv*) Extraordinary fatigue resistance under large strain cycles.

Currently, SMAs are mainly applied in medical sciences, electrical, aerospace and mechanical engineering and the recent studies indicate that they can open new applications in civil engineering specifically in seismic protection of buildings. It is reported that the fatigue behaviour of Cu-Zn-Al SMAs is comparable with steel. If larger diameter rods can be manufactured, it has a potential for use in civil engineering applications.

When SMA products are subjected to elevated temperatures for prolonged period of time, the shape at the high-temperatures is partially memorized, which inhibits the product from accurately reverting to a previously memorized shape. This suggests that prolonged exposure of SMA products to high temperatures beyond some critical temperatures has a deleterious effect on the memory of the alloy in the sense that a fairly accurate memory is rendered 'blurry' or fuzzy. The critical temperature for TiNi is about 250°C, and for CuZnAl the critical temperature is about 90°C. Extended exposure to thermal environments with temperatures above these critical temperatures results in an impaired memory function regardless of the magnitude of the load.

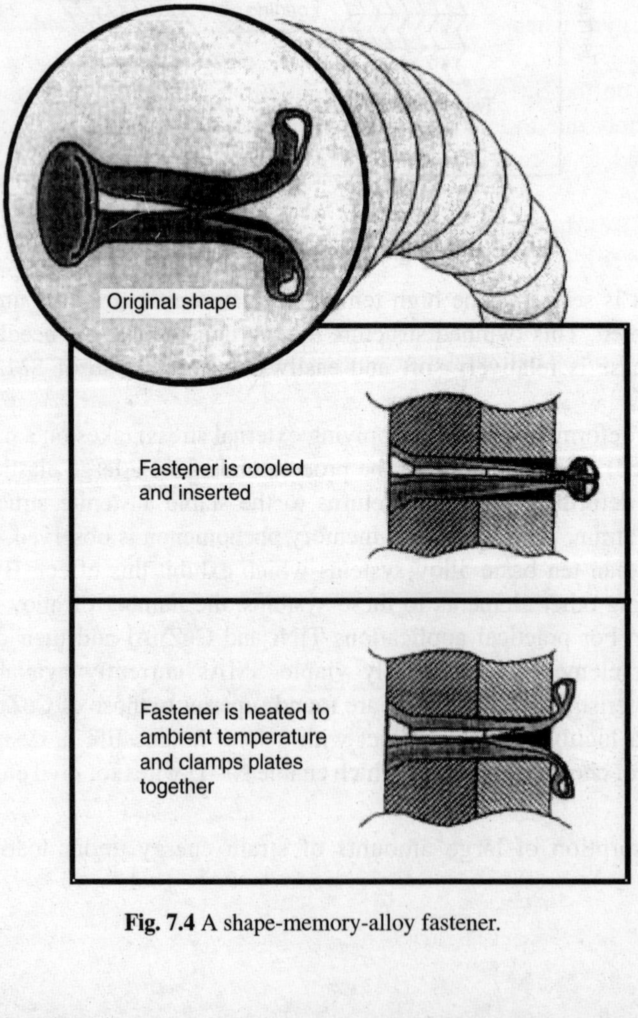

Fig. 7.4 A shape-memory-alloy fastener.

Another technical consideration in the practical application of SMAs pertains to fastening/joining protocols when SMAs are fastened and/or joined to other conventional materials. These significant issues arise due to the fact that SMAs undergo expansions and contractions which are as high as 2% to 6% under routine service scenarios. These changes in dimensions contrast sharply with the magnitude of expansions and contractions encountered in typical traditional materials which have an upper bound of approximately 0.5%. Obviously, if SMAs are welded or soldered to other materials they can easily fail at the joint when subjected to repeated loading. TiNi and CuZnAl can also be brazed using silver fillers, however, the brazed region can fail due to cyclic loading of only a few hundred cycles even when the strains are very small. This is why, it is desirable to devise some other mechanism for joining SMAs to other materials as shown in Fig. 7.4.

It is evident from the figure that the SMA is in contact with the conventional material over a very small region, and moreover, it is also free to move relative to other material. For similar reasons SMAs cannot be plated or painted. However, in some cases, CuZnAl is plated or painted to avoid corrosion or stress corrosion cracking.

7.5.2 Super Elasticity (SE) (Pseudoelasticity)

This refers to the ability of SMA to return to its original shape upon unloading after a substantial deformation. This SE is based on stress induced martensitic (SIM) transformation. SE in SMAs occurs at a constant temperature when the alloy is completely composed of austenite phase (temperature is greater than A_f).

The stress (load) on the SMA is increased until the austenite phase becomes transformed into a martensite phase simply due to loading. However, as soon as the loading is decreased the martensite starts to transform back to austenite since, the temperature is still above A_f. As a result, it returns back to its original shape.

Superelasticity causes the material to be extremely elastic is also known as pseudoelasticity. SE is nonlinear and moreover it is temperature and strain dependent. SE is a mechanical type of behaviour of SMAs.

Figure 7.5 shows the stress-strain behaviour of different phases at constant temperature for NiTi alloy, when tested in tension below, within and above its transformation range.

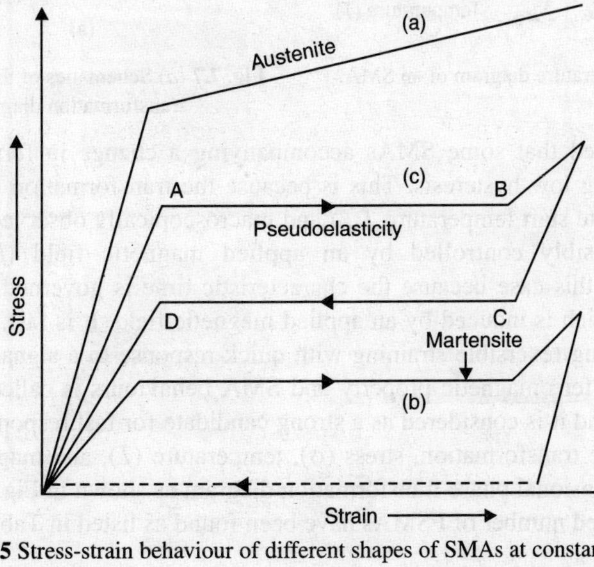

Fig. 7.5 Stress-strain behaviour of different shapes of SMAs at constant temperature.

At a temperature above A_f, it is found that the austenite phase have much higher yield and flow stresses (Fig. 7.5(*b*)). However, the martensite is easily deformed to several per cent strain at quite a low stress (Fig. 7.5(*b*)).

A shown by the dashed line in Fig. 7.5, this martensite phase is found to come back to the austenite phase upon heating after removing the stress. However, no such shape recovery is found in the austenite phase upon straining and heating because no phase change takes place.

The above two behaviours are termed as thermomechanical behaviours. Fig. 7.5(*c*) shows an interesting stress-strain behaviour where the material is completely composed of austenite phase ($t > A_f$). At this temperature, the martensite phase can be stressed induced (curve *AB*). Again, upon loading, the material is found to return to its original shape (curve CD) austenite without applying heat. This is basically a pure mechanical behaviour and is referred to as *superelasticity* or *pseudoelasticity*.

SE behaviour is applied in making eye glass frames, cellular phone antennae's, medical tools, and orthodontic arches.

7.6 FERROMAGNETIC SHAPE MEMORY ALLOY

The phase transformation of SMAs can be controlled by stress and temperature. Their relationship can be described by a stress-temperature phase diagram as shown in Fig. 7.6.

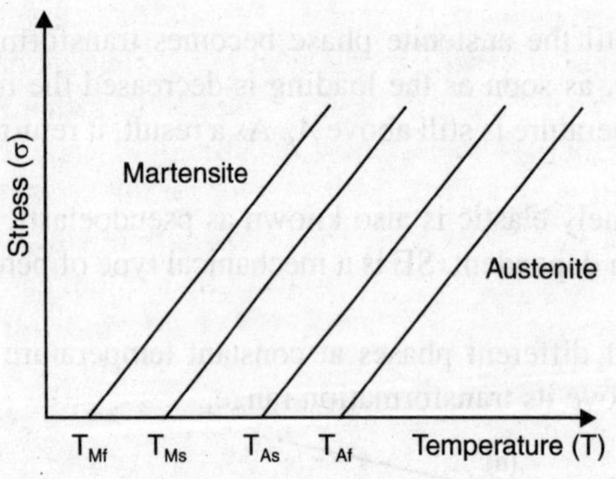

Fig. 7.6 Stress temperature diagram of an SMA.

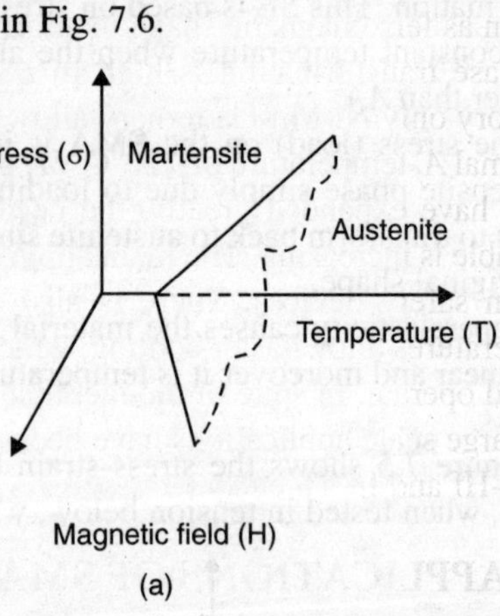

Fig. 7.7 (*a*) Schematics of three-dimensional phase transformation diagram of FSMAs.

Recently, it is observed that some SMAs accompanying a change in ferromagnetism at phase transformation and having low hysteresis. This is because the transformation characteristics such as transition point (martensite start temperature T_{Ms}) and macroscopically observed strain, caused by the transformation, are possibly controlled by an applied magnetic field (H). The response of transformation is fast in this case because the characteristic time is governed by the formation and growth of martensite, which is induced by an applied magnetic field. It is fast, thus, it is plausible to produce an actuator having reversible straining with quick response to a signal imposed or detected. Such an alloy with both ferromagnetic property and SMA behaviours is called ferromagnetic shape memory alloy (FSMA) and it is considered as a strong candidate for fast responsive actuator material. The relationship of phase transformation, stress (σ), temperature (T), and magnetic field (H) can be presented as a three-dimensional phase transformation diagram as shown in Fig. 7.7(*a*).

Currently, only a limited number of FSMAs have been found as listed in Table 7.1.

Table 7.1 List of FSMAs

Type	Alloy	T_{Ms}
Heusler	Ni_2MnGa	–150° C to 150° C
	Fe-Ni-Co-Ti	< –150° C
Iron-based alloys	Fe-23-25 Pt	< –48° C
	Fe-28-31Pd	–197° C to 80° C
	Fe-Pd-Ni	< R T
Cobalt based alloy	NiAlCo	

7.7 COMMERCIALLY IMPORTANT SMAs

Although a relatively wide variety of alloys are known to exhibit the SME, only some alloys find commercial applications. In fact only one major system became successful in industrial application: NiTi(X, Y) in which X, Y are elements replaced by Ni or Ti. This alloy has been found to be the most useful of all SMAs. The generic name for the family of NiTi alloy is *Nitinol* which stands for Nickel Titanium Naval Ordance Laboratory. Besides NiTi system, a lot of attention has been given in earlier times to Cu-based alloys, *i.e.* Cu-Al-Ni, Cu-Zn-Al and Co-Ni-Al. There is also a subfamily of SMAs known as ferromagnetic shape memory alloys (FSMAs) which are magnetically driven. The frequency of phase transformation is high in FSMAs compared to the cycles of heating and cooling. In this category only Ni-Mn-Ga ternary alloy has given satisfactory performance. Actual SMAs are limited to maximal A_f temperature of 120°C, M_s generally being below 100°C. However, since market demand on SMA have expanded greatly, the need for SMA transforming at higher temperatures than presently available is increasing. The main application area of interest are actuators in automobile and oil industry and in safety devices. There is also an interest in robotics since SMAs with high transformation temperatures allow faster cooling, which would significantly increase the bandwidth in which the robot should operate. In spite of the fact that many alloy systems show high transformation temperatures, no real large scale applications have been developed. Few examples of other SMAs are:Mn-Cu, Ti-Pd-Ni, Ni-Ti-Hf and In-Tl, etc.

7.8 APPLICATIONS OF SMAs

(i) Aircraft and Space Industry

SMAs can be used as fine-tuned helicopter blades and as lock ring electrical connectors. Moreover, they are also used in antenna opening, high-damping parts, hubble telescope and triggering devices.

Smart materials for active structure applications are mainly interesting because of their high energy density. On the other hand, their strain or stroke capacity is rather limited, as compared with other materials for aircraft structures, or with actuators. And they are rather heavy. SMAs have mainly been investigated in the form of wires or torque tubes for the active deformation of aerodynamic surfaces. To date, the achieved results are not very promising for aircraft control by active structures. SMAs could provide larger deformations and higher forces, but are not fast enough for flight control inputs, and their thermal energy supply issues are rather complex within the airframe and aircraft environment.

(ii) Automobile Industry

SMAs in this industry are used in making spring actuators, clutch systems, thermostats, oil pressure control unit and high pressure sealing plugs.

(iii) Smart Battery Materials

Batteries are major power sources for portable electronic devices as well as for automobile starting and ignition. Batteries are also intensively pursued for a widespread hybrid electric vehicle (HEV) and plug-in-hybrid electric vehicle (PHEV) applications to address the growing environmental concerns and increasing global energy demands. Moreover, batteries are critical as energy storage devices for effectively utilizing solar and wind energies, which are intermittent in nature. The performance factors of batteries are linked to the materials used as well as the engineering involved in fabricating them.

Lithium ion batteries have revolutionized the portable electronics market because of the development of smart cathode, anode and electrolyte compositions. Although only 50% of the theoretical capacity of the currently employed layered $LiCoO_2$ cathode can be used in practical cells, significant progress is being made in recent years and complex layer oxide solid solutions delivering two times higher capacity than $LiCoO_2$ have been identified. These high-energy density cathodes are attractive for portable applications. Despite the success in portable devices, lithium ion batteries are yet to enter the HEV and PHEV arena mainly due to the high cost, safety concerns and limited power capability of the layered $LiCoO_2$ cathode. In this regard, olivine $LiFePO_4$ and stabilized spinel $LiMn_2O_4$ (cation-and anion substituted) cathodes are appealing as they offer high power capability and better safety while both Fe and Mn are inexpensive and environmentally benign. Accordingly, these cathodes are being pursued for HEV and PHEV applications.

Although most of lithium ion cells are currently made with carbon anode, several new material systems such as nanostructured tin alloys and nanocomposites based on Si are being intensively pursued. Some major issues with some of these materials are large volume changes occurring during the charge-discharge processes and high irreversible capacity loss during the first cycle. Another attractive anode system is the spinel $Li_4Ti_5O_{12}$, but the drawback is its higher discharge voltage (−1.5 V *vs.* lithium) and the consequent decrease in cell voltage as well as lower capacity (−175 mAh/g) compared to carbon anode.

(iv) Medical Field

Probably the largest commercial application of SMAs are in the field of bioengineering and medical applications *i.e.*

(*a*) SMAs are used as dental arch wires. These wires make the misaligned teeth gradually to return to their original shape exerting a small and nearly constant force on the misaligned teeth. The use of this wire is less discomfort and more efficient in faster tooth movement.

(*b*) They are also used as blood clot filter.

(*c*) Nitimol needle wire localizers are used to locate and mark breast tumours so that subsequent surgery can be more exact and less invasive.

(*d*) Smart materials are used to make tweezers to remove foreign objects through small incisions.

(*e*) SMAs are used as guide wires for catheters through blood vessels.

(*f*) SMAs also find use in designing micro surgical instruments and micro grippers, etc.

(*g*) TiNi SMAs have been employed in the development of artificial joints including femoral heads and sockets. Fig. 7.7(*b*) shows a typical application of SMAs in artificial hip joints. TiNi SMAs are employed for bone plates and marrow pins for healing fractures in the femur, tibia and intestrocharter bones, and superelastic TiNi wires are employed for connecting broken bones.

(*h*) SMA blood clot filter has also been developed. Prototype filters employing TiNi wires for trapping blood clots moving through the Vena Cava have been developed for an exploratory procedure. These wires feature a memorized configuration which permits them to act as filters and trap blood

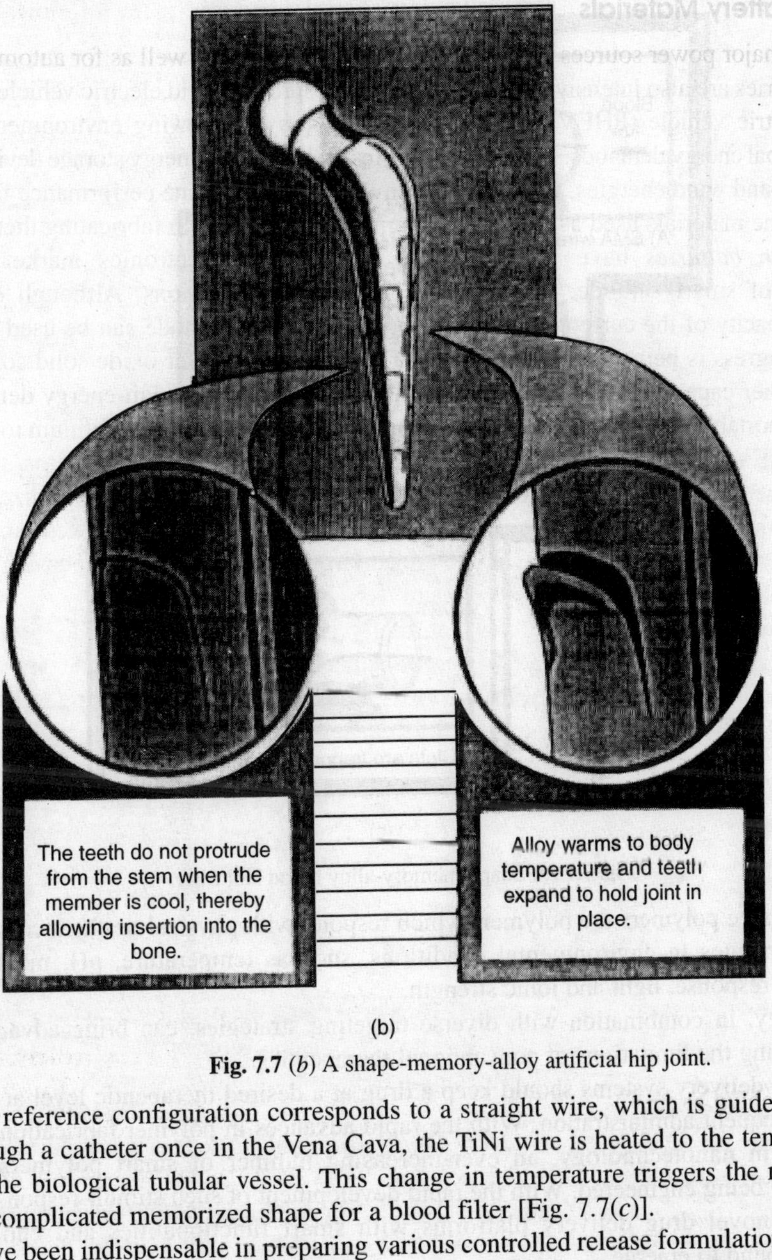

Fig. 7.7 (*b*) A shape-memory-alloy artificial hip joint.

clots. The reference configuration corresponds to a straight wire, which is guided into the Vena Cava through a catheter once in the Vena Cava, the TiNi wire is heated to the temperature of the blood in the biological tubular vessel. This change in temperature triggers the reversion of the wire to a complicated memorized shape for a blood filter [Fig. 7.7(*c*)].

Polymers have been indispensable in preparing various controlled release formulations. Of the many routes of drug administration, oral delivery is most attractive due to high patient convenience and compliance, but it is often very challenging. For example, oral delivery of drugs that are poorly soluble or are not stable in the gastrointestinal tract, or have high molecular weights has been difficult. Delivery of high molecular weight drugs, such as peptide and protein drugs and oligonucleotides, by parenteral administration has also been difficult due to their short circulation half-lives as well as lack of suitable delivery systems for long-term delivery. A number of approaches have been explored to increase the bioavailability of hydrophilic/hydrophobic drugs, peptides and proteins by conjugating drugs with natural or synthetic macromolecules or loading in polymeric drug delivery systems.

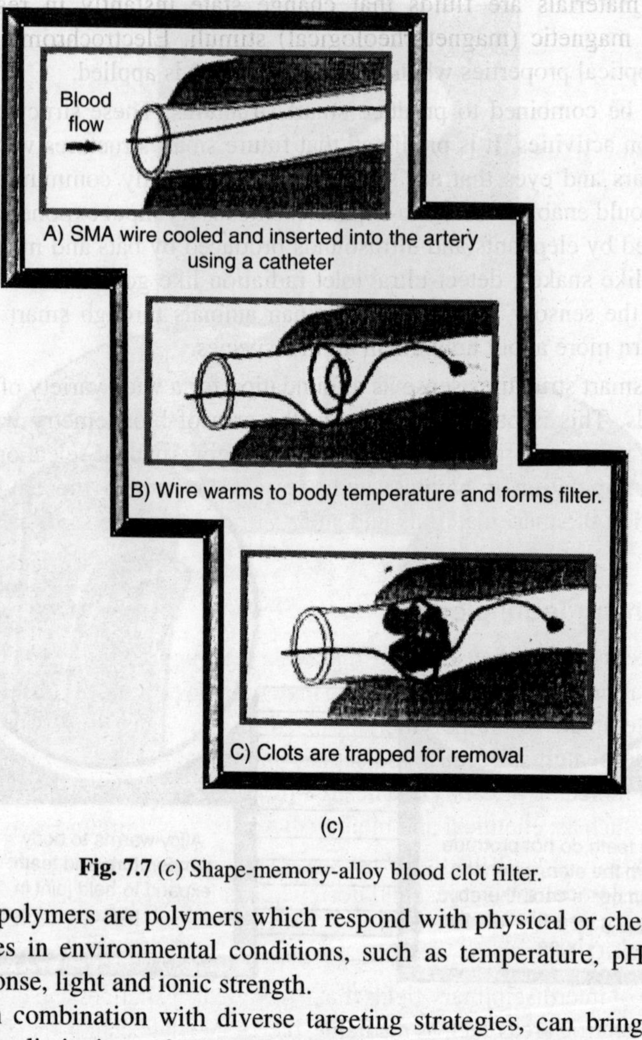

Fig. 7.7 (*c*) Shape-memory-alloy blood clot filter.

Stimuli-responsive polymers are polymers which respond with physical or chemical transformation in response to changes in environmental conditions, such as temperature, pH, mechanical stress, electric/magnetic response, light and ionic strength.

Nanotechnology, in combination with diverse targeting strategies, can bring advantages to drug delivery overcoming the limitations of conventional therapeutics.

The ideal drug delivery systems should keep a drug at a desired therapeutic level at the target site while avoiding frequent administration. With the rapid advances in polymer fabrication keeping pace with the strides in nanotechnology, an ever-increasing number of smart polymers with unique characteristics are being engineered. With the rapid development of such stimuli-responsive polymers, a whole line of novel drug delivery platforms with smart functionalities and enhanced patient compliance are bound to emerge.

(v) To Study of Aquatic Animals

Smart materials sense aspects of the environment and respond appropriately. They also reverse that response when conditions change, *e.g.* ferroelectric-based piezoelectric generate electric polarity when subjected to mechanical stress. Conversely, if one applies a voltage to piezoelectric crystals, the crystals respond by changing shape in a linear direction. Electrostrictive and magnetostrictive are also smart materials. These materials change size in response to electric or magnetic stimuli, respectively and produce a voltage when stretched.

Rheological smart materials are fluids that change state instantly in response to electrical (electrorheological) or magnetic (magnetorheological) stimuli. Electrochromic materials have the ability to change their optical properties when an electric current is applied.

Smart materials can be combined to produce smart structures. These structures perform sensing, controlling and actuation activities. It is predicted that future smart structures will be used to develop supersensitive noses, ears and eyes that are smart enough to directly communicate with the human brain. Such structures could enable humans to expand their sensory input exponentially. We could hear the infrasounds produced by elephants and ultrasounds produced by bats and many marine mammals. We could see infrared like snakes, detect ultraviolet radiation like goldfish and smell the world like dogs. By experiencing the sensory worlds of nonhuman animals through smart materials and smart structures, we could learn more about nonhuman animal species.

Smart materials and smart structures sense as a foundation for a wide variety of the equipment used to study aquatic animals. This is particularly true in the area of biotelemetry where smart tags and receivers are playing an increasing role in understanding foraging, habitat selection, mating behaviour, migration, navigation, population dynamics, and the ways in which the environment constraint behaviour. The application of smart materials and smart structures to the study of aquatic animals has increased significantly.

(vi) Molecular Imprinting Technology

Molecular imprinting technology creates 'smart materials' by integrating a template into the polymer formulation during the fabrication of these polymeric materials. In this way, molecular level information is transferred from the template to the polymers in the form of a three-dimensional "memory" of the template's shape and other features. Because molecularly, imprinted polymers (MIPs) can be tailored to specially bind a targeted molecule or catalyze a desired reaction, these smart materials hold tremendous promise for applications such as: chemical and biological separations, immunoassays, catalysts, sensors, nanotechnology, biomedical, *e.g.* drug delivery. Compounds currently utilized in these areas are nonspecific polymers, biological molecules, and designed small molecule receptors.

(vii) Biomedical Sensing

This is a typical example of interdisciplinary fields that poses great challenges and desires the cooperation of experts of numerous scientific areas such as medicine, bioengineering, chemistry, electrical engineering, material sciences, nanotechnology in developing biosensors and other sensitive biomedical constructions such as drug delivery systems. The development in the field of micro- and nanotechnology and material science enabled the preparation of miniaturized biosensors that are applicable in biomedical diagnostics and monitoring.

Biosensors are special sensors using a biologically active material for measurement or indication. They consists of two main parts: the receptor and the transducer (this is not a physical separation). The receptor interacts with the analyte selectively; the transducer produces a signal (electrical or optical) as a result of formal interaction. This signal carries information about the concentration of the analyte if certain conditions are ensured.

Biosensors can employ different transducers like calorimetric types, electrochemical types (potentiometric, amperometric or conductometric, electrochemical cells, as well as modified Ion-sensitive field effect transistor(ISFETs)) fibre optic transducers, or gravimetric type resonator transducers.

Latest development in the area of biomedical sensing are stimulated by the recent progress in smart materials designing. Smart materials are mostly polymers that respond to external stimuli with considerable property changes.

Biomedical applications cover not only research laboratory applications or hospital instruments but home diagnostic kits, *e.g.* glucose sensors for diabetic patients.

The development of biomedical sensing systems is still on the rise. Advances in material science and nanotechnology can be exploited in biosensor preparation in order to fabricating more and more sophisticated biosensing systems using smart materials.

(viii) Intelligent Chemical Indicators

Generally, chemical indicators may be defined as stimulus responsive materials that can provide useful information about changes in their environment. Organic dyes, hydrogel or "smart polymers", shape memory alloys, thermochromic or photochromic inks, and liquid crystals are some examples. They may function by forming structurally altered ionic or molecular complexes with species in their environment through chemical or physical interactions involving proton exchange, chelation, hydrogen bonding, dipole-dipole interactions or van der Waals forces. The resulting characteristic biochemical, chemical, optical, magnetic, thermal or mechanical changes can be tailored to provide the desired indication response.

Smart materials or devices may be defined as materials that produce strong visually perceptible changes in physical or chemical property in response to small physical or chemical stimuli in the medium. The material properties measured may include pH, concentration, composition, solubility, humidity, pressure, temperature, light intensity, electric and magnetic field, shape, air-velocity, heat capacity, thermal conductivity, melting point or reaction rates. Chemical-indicating devices can respond reversibly or irreversibly to small changes in the physical or chemical properties in their environment in a predictable manner. They may be regarded as smart materials because of the range of material properties they encompass. Typical materials include shape memory alloys, piezoelectric materials, magnetostrictive substances, electrorheological and magnetorheological fluids, hydrogen polymers, photo and thermoresponsive dyes. The application of chemical indicating devices has come to stay and now available. In many cases, they are living upto expectations in reducing the incidence of counterfeiting, forgery, tampering or piracy, and losses of perishable products. Temperature and time-temperature integrator or indicators (TTIs) are now being used in routing packaging of some critical food and chemical products. New indicators covering a large range of activation energies for many spoilage reactions need to be designed.

(ix) Discrete Applications of SMAs

SMA based devices have been discrete in nature with typical applications in switching and actuation. Figure 7.8 shows a typical application of this kind involving a temperature fuse.

SMA actuators have the distinct advantage of small size, because they are constructed as a single unit of SMA with a simple internal material-based mechanism for actuation. This is in sharp contrast with traditional actuators which employ more than one component and rely on more complex actuation mechanisms. Moreover, SMA actuators are not influenced by other environmental factors such as humidity and vacuum, for example. SMA actuators can be employed as positioning devices for scanning electron microscopes. Other applications area include reaction vessels, nuclear reactors, chemical plants, etc.

Two main types of SMA actuators, namely biased and differential have been developed for *robots*. Biasing uses a coil spring to generate the bias force that opposes the unidirectional force of SMA. In the

Fig. 7.8 A temperature fuse involving a shape-memory-alloy.

differential type, the spring is replaced with another SMA and the opposing forces control the actuation. Figure 7.9 shows the assembly of micro-robot actuated by SMAs. This micro-robot features five degrees

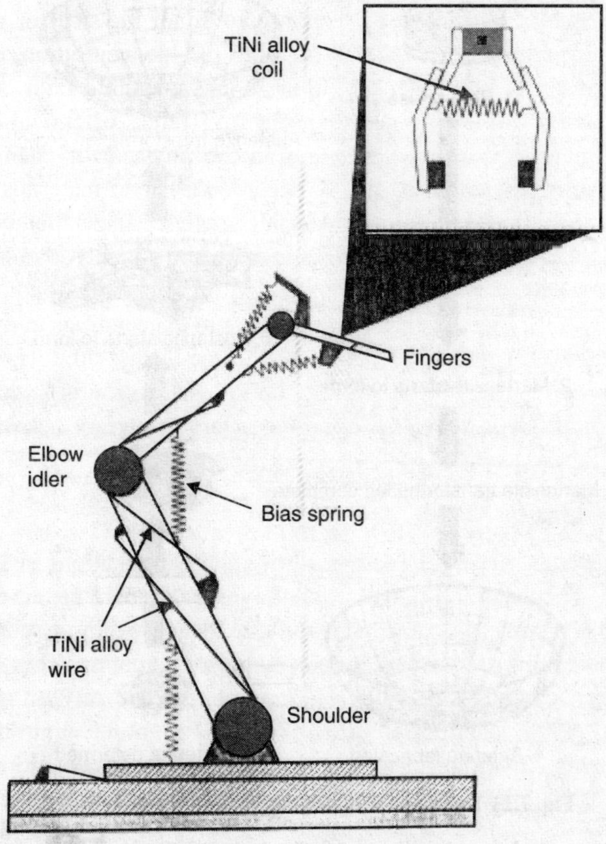

Fig. 7.9 A micro-robot actuated by shape-memory-alloys.

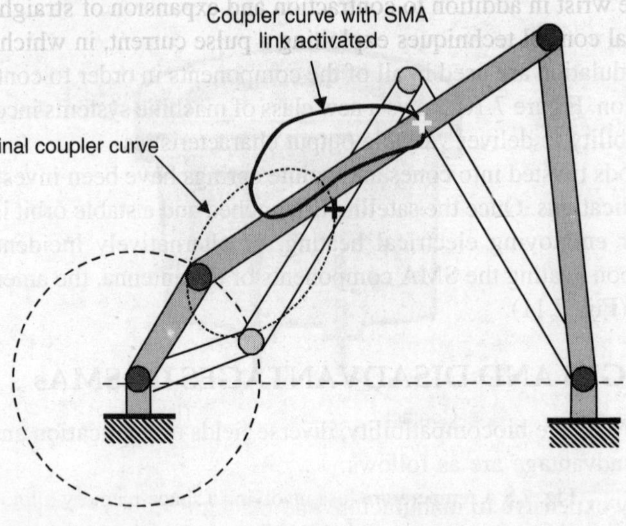

Fig. 7.10 Enhancing the coupler curve characteristics of a four-bar mechanism by utilizing shape-memory-alloys.

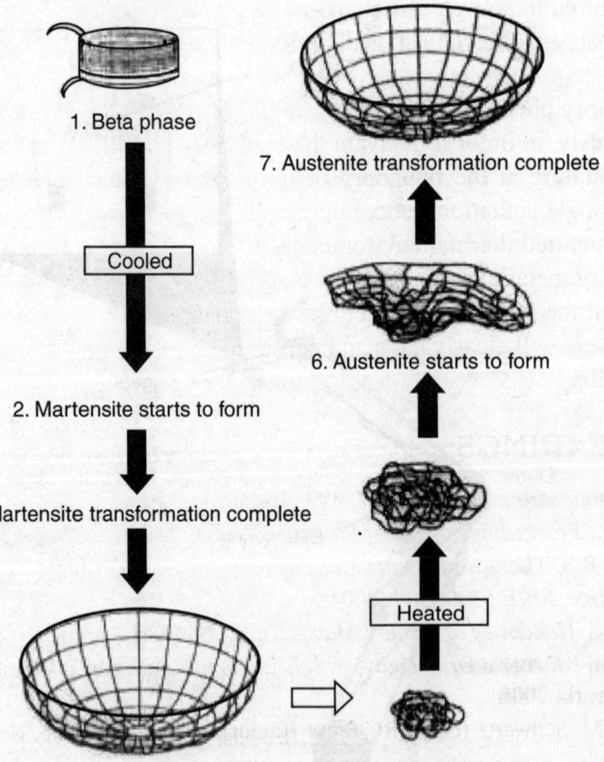

Fig. 7.11 The shape-memory-alloy memorization process.

of freedom corresponding to the capabilities of the human fingers, wrist, elbow and shoulder, for example. The variety of robotic maneuvers and operations are coordinated by activating the TiNi coils in the fingers and the wrist in addition to contraction and expansion of straight TiNi wires of the elbow and shoulders. Digital control techniques exploiting a pulse current, in which the current is modulated with pulse width modulation are used in all of the components in order to control their spatial positions and speeds of operation. Figure 7.10 shows a new class of machine systems incorporating SMA actuators which have the capability to deliver variable output characteristics.

TiNi sheets and rods twisted into cones and volute springs have been investigated for deployment in satellite antenna applications. Once the satellite is launched and a stable orbit is established, the antenna is warmed by either employing electrical heating, or alternatively incident sunlight is utilized for heating purposes. Upon heating the SMA components of the antenna, the antenna then unfolds in space as umbrella unfolds (Fig. 7.11).

7.9 ADVANTAGES AND DISADVANTAGES OF SMAs

The advantages of SMAs are biocompatibility, diverse fields of application and good mechanical properties. The major disadvantage are as follows:

(*a*) They are highly expensive to manufacture and machine it.

(*b*) The *response time* of smart materials featuring SMAs is greater than that for the competing technologies due to the inherent thermal time-constants associated with this class of material.

(*c*) The *energy* required to activate the shape-memory phenomenon is quite larger than that required with other approaches for certain class of problems because of the attendant thermal losses primarily by conduction.

(*d*) The shape-memory phenomenon occurs at a *single specific temperature* for a prescribed class of materials obviously, in order to activate the material, the SMA must be heated to the prescribed transition temperature at the boundary of the martensitic and austenitic domains of the phase diagram. This single actuation temperature could be problematical if the material is required to operate in unstructured thermal environments.

(*e*) The embedding of metallic wires in a host material for continuum applications will result in stress concentrations at the interface between the two constituents when the SMA wires are activated. These stress raisers will clearly have an undesirable effect upon the structural integrity of the part and its fatigue life.

SUGGESTED READINGS

1. G. Yang, et al., *Engineering Structures* 24, 309 (2002).
2. A. Suleman, et al., *Proceedings of Smart Structures and Materials*, 3674 (1999).
3. M.V. Gandhi and B.S. Thompson, *Smart Materials and Structures*, Chapman and Hall (1992).
4. V. Chem, et al., *Proc. SPIE* 5390, 309 (2004).
5. J.O. Besenhard, ed. *Handbook of Battery Materials*, Wiley-VCH, Weinheim, 1999.
6. I.K. Kwon, et al. in *Polymeric Drug Delivery Systems*, H.S. Nalwa and T. Webster, eds., American Scientific Publishers, New York, 2006.
7. Smart Materials, M. Schwartz (ed.) CRC Press: Taylor and Francis Group, New York, 2009.

SUMMARY

1. The ability of certain materials to develop microstructures that, after being deformed, can return the material to its initial shape after heating is termed shape-memory effect, *e.g.* Ni-Ti alloys.

2. Smart materials are the materials which can sense an external stimulus (*e.g.*, stress, pressure, temperature change, magnetic field, etc.) and initiate a response. Passive smart materials can sense external stimulus, actively smart materials have sensing and actuation capabilities.
3. Components of a smart material (or system) include some type of sensor (that detect an input signal), and an actuator (that performs a responsive and adaptive function).
4. Actuators may be called upon to change shape, position, natural frequency, or mechanical characteristics in response to changes in temperature, electric fields, and/or magnetic fields.
5. Four types of materials are commonly used for actuators: shape memory alloys (SMAs), piezoelectric ceramics, magneto-strictive materials and electrorheological/magnetorheological fluids.
6. Shape memory alloys are metals that, after having been deformed, revert back to their original shapes when temperature is changed.
7. Piezoelectric ceramics expand and contract in response to an applied electric field (or voltage): conversely, they also generate an electric field when their dimensions are altered.
8. The behaviour of magnetostrictive materials is analogous to that of the piezoelectrics except that they are responsive to magnetic fields.
9. Electrorheological and magnetorheological fluids are liquids that experience dramatic changes in viscosity upon the application of electric and magnetic fields, respectively.
10. Materials/devices employed as sensors include optical fibres, piezoelectric materials (including some polymers) and microelectromechanical devices, *e.g.* one type of smart system is used in helicopters to reduce aerodynamic cockpit noise that is created by the rotating rotor blades. Piezoelectric sensors inserted into blades, monitor blade stresses and deformations; feedback signals from these sensors are fed into a computer-controlled adaptive device, which generates noise-canceling antinoise.
11. Shape-memory alloys are of two types: (*i*) One way SMAs, which exhibit shape-memory effect only upon heating are said to have one way shape-memory; and (*ii*) two way shape-memory-alloys which exhibit shape-memory effect both during heating and cooling.
12. Shape-memory-alloys are said to have two distinct crystal structure or phases. The effect of temperature and internal stresses determines the crystal structure or the phase that the SMA will have at a particular instant. The phase which exists at high temperature is called the austenite phase and, the other phase which is found to exist at low temperature is called the martensite phase. In both the cases, material remains in solid form.
13. The shape memory materials could be broadly classified into thermal induced shape-memory, stress induced shape-memory and ferromagnetic shape memory materials.
14. If the phase transformation in SMAs takes place not at a particular temperature but over a range of temperature, the transformation is called the temperature induced transformation.
15. The stress induced transformation in SMAs takes place at constant temperature.
16. The SMAs are characterized by two important properties: (*i*) Shape memory effect (SME), and superelasticity (SE) also called pseudoelasticity.
17. The most common SMA is an alloy of Ni-Ti. Its generic name is Nitinol.
18. SMAs find applications in: (*i*) medical field, (*ii*) automobile industry, (*iii*) aircraft and space industry, (*iv*) consumer products, etc.
19. The advantages of SMAs are biocompatibility, diverse fields of application and good mechanical properties.
20. The major disadvantage of SMAs is that they are highly expensive to manufacture and machine it.

REVIEW QUESTIONS

1. What are shape-memory alloys? How many types of shape memory alloys are there? Explain the temperature induced and stress induced transformations in detail.
2. What are smart materials? How they differ from normal materials? Give examples of smart materials.
3. What are the features incorporate by current generation of smart materials and structures?
4. What are actuator materials? Give examples.
5. Mention primitive functions of intelligent materials.
6. Give examples of materials, which exhibit shape-memory effect.
   [**Hint:** Copper alloy systems of Cu-Zn, Cu-Zn-Al, Cu-Zn-Ga, Gu-Zn-Sn, Cu-Zn-Si, Cu-Zn-Ni, Cu-Au-Zn, Cu-Sn and alloys of Au-Cd, Ni-Al and Fe-Pt. Nitinol, a nickel-titanium alloy, is the most common of SMAs or transformation metals.]
7. What impediments are there to applications of shape memory alloys?
8. Give a brief account of functionally graded polymer blend materials. What prospects are their for application in the functionally graded blends?
   [Functionally graded materials made of metals and ceramics have characteristics of improved strength against thermal stress as well as excellent electromagnetic and optical properties. There have also been some reports on functionally graded polymeric materials. These materials are expected to be used in the future as a substitute for laminates because of their superiority in their strength and thermal shock resistance. The superiority seems to be due to the lack of interface, which suppresses the break at the interface and the thermal stress. Moreover, the excellent wide constitutional gradient for the graded structure results in improvement in several types of physical properties.]
9. Mention the properties of Ni-Ti wire which enable its use for civil engineering application.
10. Why Cu-Zn-Al SMA is considered as substitute for steel?
11. How cultural heritage structures can be restored by using smart materials?
12. What are hybrid composites? Give a brief account.
13. What are ferromagnetic shape memory alloy actuators? Give a brief account.
14. Give a brief account of aircraft applications of smart structures.
15. Give a brief account of smart battery materials with special emphasis on Lithium ion batteries.
16. Give a brief account of piezoelectric and electrostrictive ceramics transducers and actuators.
17. Give a brief account of optical fibre sensors.
18. What is intelligent processing of materials? How this is achieved?
19. What are shape-memory alloys? Explain their effects.
    [SMAs are metallic compounds. They exhibit a cubic high-temperature crystal structure, which transforms by martensitic reaction into another crystal structure (tetragonal, orthorhombic, hexagonal, etc.) when temperature decreases. Martensitic transformation enables large shape changes and the shape-memory effect (SME), but heating and cooling are needed to complete the phase transformation. SME can produce large relative strains and high stress output, but usually small actuation frequency due to the slow cooling process. The SMAs have typically twinned martensitic structure with mobile twin boundaries. The multivariant twin structure can easily be deformed to the single-variant state to yield the maximum SME shape change.
    To trigger the SMAs with a magnetic field, alloys must possess magnetic ordering, *i.e.* to be ferromagnetic or ferri/antiferro magnetic. These alloys may exhibit (*i*) magnetic field induced strain (MFIS) by magnetic shape memory effect (MSME) or by magnetic field-induced martensite

transformation (MF-SIM), (*ii*) Magnetic field-assisted superelasticity (MFAs), (*iii*) Conventional magnetostriction (MS), and (*iv*) giant magnetocaloric effect (GMCE).]

20. Explain smart coating technologies for corrosion protection. How the new developments in nanotechnology, biotechnology, biomimetic materials and functional materials will enhance the field?
21. Give a brief account of smart polymers for biotechnology and elastomers.
22. What are active truss structures?
    [The concept of active trusses is quite natural: it consists of replacing one or several passive bars by active struts. Piezoelectric transducers are ideally suited for this purpose, because of their high stiffness and virtually unlimited resolution, other types of transducer, based *e.g.* on electrostrictive materials can also be used. Integral force feedback (IFF) is one very effective control strategies. IFF is based on a collocated force sensor and has guaranteed stability as long as perfect amplifier and actuator dynamics are assumed].
23. What is molecular imprinting technology?
    [Molecular imprinting technology creates "smart materials" by integrating a template into the polymer formulation during the fabrication of these polymeric materials. In this way, molecular level information is transferred from the template to the polymers in the form of a three-dimensional "memory" of the template's shape and other features. Because molecularly imprinted polymers (MIPs) can be tailored to specifically bind a targeted molecule or catalyse a desired reaction, these smart materials hold tremendous promise for applications such as: (*i*) chemical and biological separations, (*ii*) immunoassays, (*iii*) catalysts, (*iv*) sensors, (*v*) nanotechnology, and (*vi*) biomedical, *e.g.* drug delivery].

SHORT ANSWER QUESTIONS

1. What is shape-memory effect?

Ans. The ability of certain materials to develop microstructures that after being deformed, can return the material to the initial shape when heated, *e.g.* Ni-Ti alloys.

2. What are smart materials?

Ans. These are the materials that can sense an external stimulus, *e.g.* stress, pressure, temperature change, magnetic field, etc. and initiate a response. Passively smart materials can sense external stimulus, actively smart materials have sensing and actuation capabilities.

3. What are ferromagnetic shape memory alloys?

Ans. Ferromagnetic shape-memory alloys show a shape-memory effect in response to a magnetic field. Few examples of these materials are Ni_2 MnGa, Fe-Pd and Fe_3Pt. Under the influence of magnetic field, the frequency of phase transformation is found to be high compared to the cycles of heating and cooling.

4. Can the memory of shape-memory alloys triggered by stress or temperature change?

Ans. Yes.

5. Does shape-memory alloys show a superelastic behaviour?

Ans. Yes, recoverable strains upto 10% are possible. This is the reason that shape-memory alloys have been used with a great success in applications as orthodontic wires, eye-glass frames, underwires for brassieres and antennas for cellular phones. In these applications use of superelastic effect is made and not the shape-memory effect.

6. What are the components of a smart material (or system)?

Ans. (*i*) Some type of sensor, which detects an input signal.

(*ii*) An actuator, which performs a responsive and adaptive function.

7. What type of materials are commonly used for actuators?

Ans. (*i*) Shape-memory alloys, (*ii*) piezoelectric ceramics, (*iii*) magnetorestrictive materials, and electrorheological/magnetorheological fluids.

8. What are the materials/devices employed as sensors?

Ans. Optical fibres, piezoelectric materials (including some polymers) and micromechanical devices.

9. What are advanced fabrics?

Ans. New fibrous materials often have smart properties, *i.e.* they can sense and respond to chemical or biological substances or react to physical stimuli such as blood pressure, body temperature or blood oxygen levels. It has also been proposed that computers or other electronic systems be integrated into textiles ("Wearable mother boards") among them are photovoltaics, energy conversion systems (mechanical to electrical), switches or storage elements.

In the medical field, antimicrobial, thrombonic and superabsorbing textiles are used. For the home or automobile, advanced filters that hold back pollen down to nanometer sizes and even remove odours have been developed.

Other important textiles are made from high-strength fibres (Kevlar, carbon strands, nylon, M5) for protection against impact (bullet proof vests). They possess a high stiffness in order to transfer that impact stress along the fibres.

There is also a strong interest in protective suits or overgarments that shield humans from biological contaminants or excessive heat, such as a fire. They should be light weight and provide comfort over long time periods. This is accomplished by utilizing semipermeable adsorptive carbon liners within the garment. Consequently, carbon layers adsorb vapours that may pass into the clothing, thus providing protection. Other suits consist of barrier laminates that are completely impermeable, *e.g.* Tyvek-reinforced polyproplene films with a rugged outer-shell-fabric such as aluminized kevlar and polybenzobisoxazole (PBO). Recently, membranes that permit the penetration of water vapour molecules to allow transpiration and cooling, but are impermeable for larger molecules have been developed. These *"perm-selective"* membranes have been used for water purification by reverse osmosis and are made of poly (vinyl alcohol), cellulose acetate, cellulosic cotton, or poly (allylamine).

Chameleon fibres is capable of its colour upon an electrical or biological stimulus from a bright orange to a camouflaging dark green. Among these smart materials are conducting polymers made from polypyrrole or polyaniline, which are coated on woven fabrics made of poly (ethylene terephthalate).

10. Why residual stress is undesirable and disadvantageous in the fabrication of almost all planar thin film electronic devices?

Ans. Residual stress could lead to failure of the device, or else can modify the device performance undesirably, *e.g.* microelectronic or microelectromechanical systems (MEMs).

11. What are the requirements to synthesise smart materials with more and more sophisticated features to meet growing demand from various industries?

Ans. To synthesise smart materials with more and more sophisticated features requires the use of smart starting materials, with very-well defined and controlled specific properties, such as size, shape (morphology), chemical composition and defect structure, surface functionalisation, etc.

12. What are the characteristics of functionally graded materials made of metals and ceramics?

Ans. These materials have characteristics of improved strength against thermal stress as well as excellent electromagnetic and optical properties.

13. What are shape-memory alloys?

Ans. SMA refers to the ability of certain alloys (Ni-Ti, Cu-Al-Zn, etc.) to undergo large strains, while recovering their initial configuration at the end of the deformation process spontaneously or by heating without any residual deformation. The particular properties of SMAs are strictly associated with a solid-solid phase transformation. Currently, SMAs are mainly applied in medical sciences, electrical, aerospace and mechanical engineering and recent studies indicate that they can open new applications in civil engineering specifically in seismic protection of buildings.

14. What are the properties which enable Ni-Ti wires for civil engineering applications?

Ans. (*i*) Repeated absorption of large amounts of strain energy under loading without permanent deformation.

(*ii*) Possibility to obtain a wide range of cyclic behaviour from supplemental and fully reentering to highly dissipating by simply varying the characteristics of SMA components.

(*iii*) Useable strain range of 70%.

(*iv*) Extraordinary fatigue resistance under large strain cycles.

15. Name the alloy whose fatigue behaviour is comparable with steel.

Ans. Cu-Zn-Al SMAs. If larger diameter rods can be manufactured, it has a potential for use in civil engineering applications.

16. What is smart concrete?

Ans. A mere addition of 0.5% specially treated carbon fibres enables the increase of electrical conductivity of concrete. Putting a load on this concrete reduces the effectiveness of the contact between each fibre and the surrounding matrix and thus slightly reduces its conductivity. Because of this peculiar property, the product is called "smart concrete". The concrete could serve both as a structural material as well as a sensor.

17. What are reinforcement hybridized composites?

Ans. These are the composites, which consists of two or more types of reinforcements embedded in a single matrix. The reinforcements can be of different types: (*i*) Synthetic (man-made) fibres, (*ii*) Natural fibres, (*iii*) Natural and synthetic fibres, and miscellaneous reinforcements.

18. What are matrix hybridized composites?

Ans. The matrix is one of the important constituents of the composites. Hybridizing a matrix with another matrix also means toughening of the matrix. Such matrices when used to prepare the composite also can lead to a hybrid composite.

19. How lithium ion batteries have revolutionized the portable electronics market?

Ans. Lithium ion batteries have revolutionized the portable electronics market because of the development of smart cathode, anode and electrolyte compositions. Although only 50% of the theoretical capacity of the currently employed layers $LiCoO_2$ cathode can be used in practical cells, significant progress is being made in recent years and complex layered oxide solid solutions delivering two times higher capacity than $LiCoO_2$ have been identified. Overall, continued focus on the development of new, smart materials and process has the potential to increase the energy density of the lithium ion batteries for portable as well as to make them viable for hybrid electric vehicle (HEV) and plug-in hybrid electric vehicle (PHEV) applications. Successful development of lithium ion technology for automobiles will have a profound impact in addressing one of the critical challenges facing humankind in the 21st century increasing energy needs, environmental concerns and global warming.

20. Mention two unique properties of SMAs.

Ans. (*i*) Shape-memory effect (SME), and

(*ii*) Superelasticity.

21. Classify the shape-memory alloys.

Ans. SMAs can broadly be classified into following three categories:

(*i*) Thermal induced shape-memory-alloy.

(*ii*) Stress induced shape-memory-alloy.

(*iii*) Ferromagnetic shape-memory-alloy.

22. How many types of crystal structures or phases are exhibited by SMAs?

Ans. SMAs exhibit two crystal structures or phases:

(*i*) Austenite phase at high temperature.

(*ii*) Martensite phase at lower temperature.

23. What makes SMAs to remember their shape?

Ans. The structural changes at the atomic level causes SMAs to remember their shape, *i.e.* the easy to and fro martensite transformation is the reason.

24. Explain, how phase transformation takes place in SMAs?

Ans. There are two types of phase transformations:

(*i*) Temperature induced phase transformation. This takes place over a range of temperature but not at a particular temperature.

(*ii*) Stress induced phase transformation. This takes place due to applied external stress.

25. What is superelasticity (SE) or pseudoelasticity in SMAs?

Ans. Superelasticity or pseudoelasticity refers to the ability of SMAs to return to their original shape upon unloading after a substantial deformation. This is based on Stress Induced Martensitic (SIM) transformation.

26. Write the generic name of Ni-Ti alloy.

Ans. Nitinol.

27. Write any two applications of SMAs.

Ans. (*i*) SMAs are used for making spring actuators, clutch systems and thermostats in the automobile industry.

(*ii*) SMAs are used in making surgical instruments and tweezers in medical field.

28. What are the advantages and disadvantages of SMAs?

Ans. Advantages of SMAs are biocompatibility, good mechanical properties and their applications in diverse fields.

The disadvantages of SMAs is that they are highly expensive to manufacture and machine it.

29. Why piezoelectric materials are most popularly used in smart structures and systems?

Ans. This is because the same piezoelectric material is applicable for both sensors and actuators in principle.

30. What is the natural advantage of the sol-gel process?

Ans. The natural advantage of the sol-gel process is the fact that a film of oxide can be placed on a substrate directly from solution. Other methods of depositing films of oxides are sputtering or chemical vapour deposition (CVD). These methods may succeed in some cases, but the sol-gel process is simpler than either one.

FILL IN THE BLANKS

1. Shape-memory-alloys (SMAs) are metals which exhibit ……… and pseudoelasticity or ……… . [shape-memory effect, superelasticity]
2. Shape-memory-alloys remember their shape even after severe ……… . [deformation]
3. Smart materials can sense an external ……… (*e.g.* stress, pressure, temperature, change, magnetic field, etc.). [stimulus]
4. Passively smart materials can sense external ……… . [stimulus]
5. Actively smart materials have ……… and ……… capabilities. [sensing, actuation]
6. Shape-memory-alloys deformed above a critical temperature show a large ……… elastic deformation as a result of a stress induced ……… transformation. [reversible, martensitic]
7. Shape-memory effect is the ability of certain materials to develop microstructures that, after being ………. can return the material to its ……… shape when heated. [deformed, initial]
8. Some of the materials exhibit shape-memory effect both during heating and cooling are said to have ……… shape-memory. [two way]
9. The shape-memory materials could be broadly classified into ……… shape-memory……… shape-memory and ……… shape-memory materials. [thermal induced, stress induced, ferromagnetic]
10. Ni-Ti alloy is most common smart-memory alloy. The generic name of this alloys is ……… [Nitinol]
11. Ferromagnetic shape-memory-alloys are ……… driven. [magnetically]
12. Growing demand from various industries to synthesise smart materials with more and more sophisticated features requires the use of ……… starting materials, with very well-defined and controlled specific ………, such as size, shape (morphology), chemical composition and defect structure, surface functionalisation, etc. [smart, properties]
13. Graded materials made of metals and ceramics have characteristics of improved ……… against thermal stress as well as ……… electromagnetic and optical properties. [strength, excellent]
14. Functionally graded polymer blends are expected to be used in the future as a substitute for ….. because of their ……… in their strength and thermal shock resistance. [laminates, superiority]
15. Currently, shape-memory-alloys (SMAs) are mainly applied in medical sciences, electrical ……… and mechanical engineering and recent studies indicate that they can open new applications in ……… engineering specifically in ……… of buildings. [aerospace, civil, seismic protection]
16. Shape-memory-alloys are scientifically attractive and can be used as ……… and actuators. [sensors]
17. An alloy with both ferromagnetic property and SMA behaviours is called ……… (FSMA) and it is considered as a strong candidate for fast responsive ……… . [ferromagnetic shape-memory-alloy, actuator]
18. Smart materials for active structure applications are mainly interesting because of their……… density. [high energy]
19. Lithium ion batteries have revolutionised the probable electronics market because of the development of smart ………, and electrolyte compositions. [cathode, anode]
20. Although most of the lithium ion cells are currently made with carbon anode, several new material systems such as nanostructured ……… and nanocomposites based on ……… are being intensively pursued. [tin alloys, Si]

CHAPTER

8

Nanomaterials

8.1 INTRODUCTION

Nanoscience and nanotechnology primarily study with the synthesis, characterisation, exploration and exploitation of nanomaterials. Nanomaterials are characterised by at least one dimension in the nanometer (1 nm = 10^{-9} m) range. Nanostructures constitute a bridge between molecules and infinite bulk systems. Individual nanostructures include clusters, nanocrystals, nanowires, quantum dots, and nanotubes, whereas collections of nanostructures involve arrays, assemblies, and super lattices of individual nanostructures. The physical and chemical properties of nanostructured materials can differ significantly of the atomic–molecular or bulk materials having same composition. The uniqueness of the structural characteristics, energetics, response, dynamics, and chemistry of nanostructured materials constitutes the basis of nanoscience. Suitable control of the properties and response of these materials can lead to new devices and technologies, *i.e.* nanotechnology.

Nanotechnology can be considered as a group of emerging technologies in which structure of matter is controlled at the nanometer scale, to produce new, interesting nanostructured materials and nano-devices that have useful and unique properties. Nanotechnology imposes only limited control of structure at the nano-metre scale producing useful products.

The impact of nanotechnology is already being felt in the form of new computer memories that provide rapid access to stored data, that can hold more of this data than the minidrives used in iPODs and do not need any external power source to retain the data. It is being felt in the form of prototypes for photovoltaic cells that can literally be sprayed onto buildings or computers to provide cheap power sources. And it is being felt in the form of "nanoengineered" gels that spread the recovery of damaged nerve cells.

Nanoscience and nanotechnology have grown explosively in the past two decades due to increasing availability of methods of synthesis of these materials as well as tools of characterisation and manipulation. Several innovative methods of synthesising nanoparticles and nanotubes and their assemblies are now available. The size – dependent electrical, optical and magnetic properties of individual nanostructures, metals and other materials are being understood in a better manner. These size dependent properties of these materials can be utilized to tailor them as per the requirements of a specific application.

There is great vitality in the area of nanoscience and nanotechnology and immense opportunities. Truly speaking, nanoscience is a interdisciplinary area covering physics, chemistry, biology, materials and engineering and technology.

Nanoscience and nanotechnology has tremendous scope for applications over a wide range. This is likely to benefit various industrial sectors, including chemical and electronic industries, as well as manufacturing. Health care, medical practice, space research and environmental protection will definitely benefit from nanoscience.

Studies are being conducted on the potential use of nanomaterials in diverse applications including hydrogen storage, ion-sensing and gas sensing, surface-modified nanoparticles for enhanced oil recovery adsorption of chemical and biological agent on to nanoparticles, active electrode materials for lithium-ion batteries, light emitting devices, dental composition, etc. Stronger and lighter metals could be used in space as well as implants in the human body. Drugs for immediate effect in a body can be produced and could be delivered to specific location in a very short time. Electronic devices of smaller size, *e.g.* transistors can be fabricated, helping computers operate faster. Nanotechnology also finds applications in producing quantum dot lasers or tiny lasers, high density storage devices, nanoelectro-mechanical systems (NEMS), etc. Nanoscience and nanotechnology has many applications in biological sciences too. The list of applications is increasing day-by-day with the progress of research in this field. One of the major and difficult problems facing the design of nanostructure based systems in understanding how to interconnect and address them. Obviously, the success of this field will depend on the development of new device and manufacturing technologies.

One can also see nanotechnology at work in nature. There are some species living in the sea produce shells surrounding them by making use of nanoscale calcium carbonate particles. These shells are quite strong and do not yield to pressure easily. Moreover, cracks produced on the surface of these shells do not propagate inside and thus the animals remain safe.

The field of nanoscience and nanotechnology has matured so rapidly and so fast that it is probably hard to find a segment of any technical subject where the implications of nanomaterials have not been explored at least to a preliminary extent. There is every reason to believe the field will make much progress in the coming decade.

8.2 NANOSCALE MIRACLES

The key element of nanoscience and nanotechnology is 'carbon', single sixth chemical element denoted with the symbol 'C', occupying a unique position in the periodic table of elements. Sometimes nanotechnology is also termed as 'carbon nanotechnology'. Carbon is a fundamental element in the living world, *i.e.* it is the basis of existence of life on the earth. We know that this element, as compound forms is present in the human body, food items, atmospheres, plants, fuels, etc. Carbon has four electrons in its outermost orbit, *i.e.* its valency is 4. Thus, carbon can bind itself or other atoms and form different compounds. Carbon atoms can form complicated network of molecules, which is the basis of organic chemistry. Carbon is the only element in the periodic table that has allotropes from zero dimensions to three dimensions. **Diamond, graphite** and **amorphous carbon** are considered to be three main allotropic forms of carbon. Out of these three allotropic forms two well known solid forms of carbon, *i.e.* diamond and graphite exhibit entirely different properties. Each of them is quite useful in its own way. Diamond is a hard material and known as a precious gem, largely used in jewelery. Diamond owes its sparkle partly to its high refractive index. Diamond is a poor conductor of electricity. On the other hand, graphite is remarkably soft and fair electrical conductor used in pencils. The difference in their properties is due to variation in the hybridisation of carbon atoms. Diamond is sp^3–bonded (hybridisation) where as graphite is sp^2–bonded (hybridisation) as shown in Figs. 8.1 and 8.2 respectively. In sp^2–bonding among the four valence electrons of carbon, three are covalently bonded by other three carbon atoms

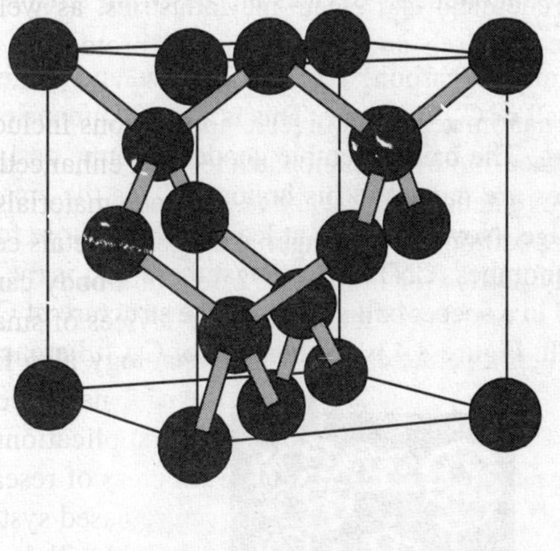

Fig. 8.1 Diamond structure.

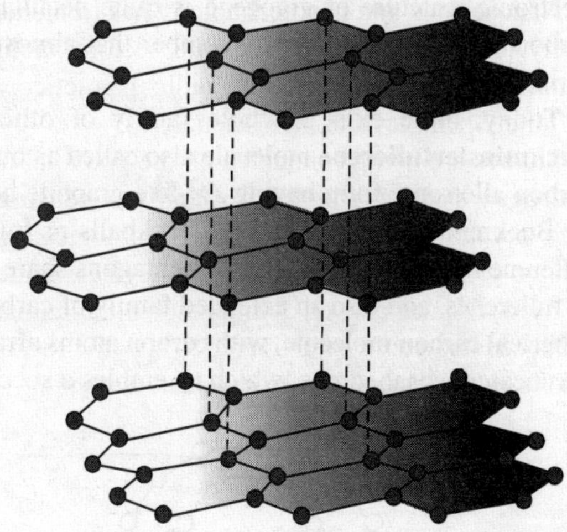

Fig. 8.2 Graphite structure.

and the fourth electron is weakly bounded to the carbon atom. This weakly bounded electron is mainly responsible for the interesting electron transport in graphene (Fig. 8.3).

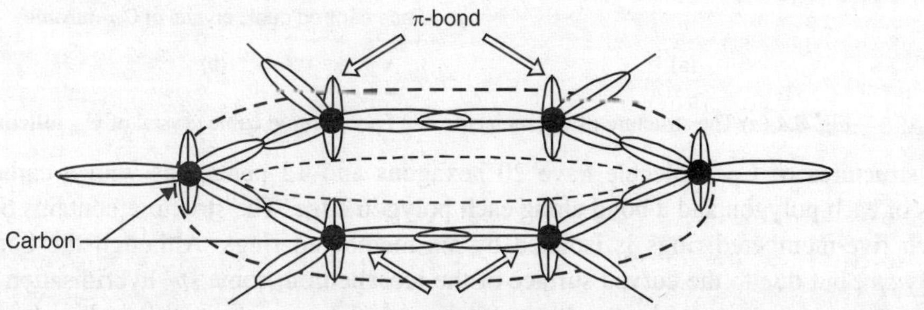

Fig. 8.3 Schematic of basic hexagonal bonding structure in grapheme sheet where carbons are denoted with circles, the σ-bonds are connected carbon nuclei in plane and π-bonds are out of the plane.

The electronic structure or the manner in which electrons are placed according to their energy are different in both diamond and graphite materials. This is why their physical properties are different. We may remember that the electronic structure of a solid depends on its atomic number and crystal structure or ordering of atoms in the lattice.

There are other three phases or allotropic forms of carbon exist only in nano form. "*Graphene*", refers to a monolayer of 2*d* lattice of sp^2-bonded carbon atoms with bond length approximately 0.14 nm arranged in a honey comb lattice (with bonding similar to graphite). Graphene could be produced only recently in 2004 by micromechanical cleavage of graphite. We know that graphite is a layered structure with strong 2*d* bonds and very weak (van der Waal type) interlayer coupling. This property of graphite is used to cleave graphite crystal and 2*d* graphene is produced. *Graphene* exhibit ballistic transport and quantum hall effect observed only in two dimensions. We may note that these effects are not observed in 3*D* bulk solids. Obviously, graphene is truly a 2*D* crystal and is a gapless semiconductor. Graphene is found to be a very promising candidate to make **ballistic field effect transistor**. Moreover, the electronic structure of graphene is quite peculiar and one can make its use for the development of carbon transistor. We may remember that almost all semiconductor devices are based on silicon and germanium.

Today, there exist a whole family of other forms of carbon, *e.g.* the hollow and **cagelike Buckminsterfullerene** molecule also called as buckyball or C_{60} fullerene. This is a zero "dimensional" carbon allotropic form having sp^2-like graphite bonding. The balls resemble geodesic dome, designed by Buckminister Fuller, and the buckballs or fullerenes are named in his honour. C_{60} is the smallest fullerene molecule in which two pentagons share an edge. Now, there are at least thirty or more forms of fullerenes, and also an extended family of carbon nanotubes (CNTS). Amongst these, C_{60} is the first spherical carbon molecule, with carbon atoms arranged in a soccer ball shape, *i.e.* the structure of C_{60} is a truncated icosahedrons, which resembles a soccer ball. Figure 8.4 is a schematic of C_{60} fullerene.

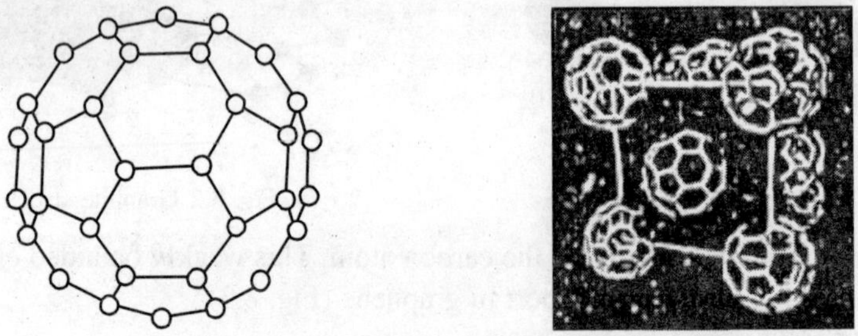

Face centred cubic crystal of C_{60} (fullerite)

(a) (b)

Fig. 8.4 (*a*) The structure of C_{60} molecule, (*b*) Face centred cubic crystal of C_{60} fullerite.

The structures of C_{60} molecule have 20 hexagons and 12 pentagons with a carbon atom at the vertices of each polygon and a bond along each polygon edge. The structure contains 60 carbon atoms in which five-membered rings is isolated by six-membered rings. Although the bonding in C_{60} is basically sp^2, but due to the curved surface of the icosahedron, some sp^3 hybridisation is also present. C_{60} has a unique property of being able to carry something inside it. When doped with alkali metal atom, C_{60} fullerene molecule exhibit superconductivity. We may note that C_{60} molecule show wave-particle duality and hence it is a truly quantum mechanical system.

The main advantage gained upon functionalisation of fullerenes is a substantial increase in their solubility. The existence of a great diversity of synthetic protocols combined with the high number of chemical reactions that have been mainly applied to C_{60} has led to the formation of a wide variety of functionalised fullerenes. Minute quantities of fullerenes in the form of C_{60}, C_{70}, C_{76} and C_{84} molecules are produced in nature. They have been observed in carbon root and also found to be formed by lightning discharges in the atmosphere.

Figure 8.5 shows a slightly elongated and the spherical carbon molecule, has seventy carbon atoms and is known as C_{70}. The structure of C_{70} has extra six-membered carbon rings. Similarly there are also a good number of other potential structures containing the same number of carbon atoms.

However, their peculiar shape depend on whether five-membered rings are isolated or not, or whether seven-membered rings are present or not. So far no significant applications of fullerenes have

Fig. 8.5 Structure of roughly ball shaped C_{70} fullerene.

been developed, yet, scientists are vey excited about the one new allotropes of carbon which is **carbon nanotube** (CNT). The current interest of researchers on CNTs is a direct consequence of the synthesis of **Buckminster – fullerene,** C_{60} and other fullerenes. They are molecules composed entirely of carbon, in the form of a tube (Fig. 8.6).

Fig. 8.6 Carbon nanotube structure.

We have seen that carbon is the most amasing element in the periodic table. In the past 30 years, we have experienced the renaissance in three materials families: **graphite intercalation compounds** (GICs), **fullerene solids** and **nanotubes.** GICs provided a wealth of detailed physical and chemical information, while immediate applications were frustrated by cost, lack of air stability, and most importantly, the lack of new or greatly improved applications and properties. Though the foundations of physics and chemistry remain unshaken, the Li-ion battery industry has certainly benefitted greatly from this knowledge. Fullerene solids fared better, but no large volume application has yet emerged. However, the elucidation of electronic properties in this highly correlated system has made invaluable contributions to fundamental science. Nanotubes may offer the best prospects yet. Much of the current research and commercialisation of nanotechnology rely on tubes, balls and wires prepared from carbon atoms/or other materials. The carbon based tubes can be CNTs, bucky tubes and very long tubes, often referred to as **nanowires.** The combination of enhanced properties encourages the drive to multifunctional materials. Extension to other tubular and nanowire-like materials provide ample scope for new discoveries.

The various parameters of carbon atoms in different dimensional carbon materials are given in Table 8.1.

Table 8.1 Various Parameters of Carbon Atom in Different Dimensional Carbon Materials

Parameters	*Fullerene*	*Carbon nanotubes*	*Graphite*	*Diamond*
Dimension	Zero	One	Two	Three
Hybridisation	sp^2	sp^2	sp^3	sp^3
Crystals structure	FCC	FCC	Hexagonal	Cubic
Lattice constant (Å)	14.15	–	a = 2.456 Å c = 6.696 Å	3.513 Å –
Colour	Black solid	Black	Grey	Colourless
Density (g/cm^3)	1.72	1.2–2.0	2.26	3.51
Specific gravity	1.7–1.9	2	2.2	3.52
Bond length (Å)	1.4	1.44	1.42	1.54
Bond energy (eV/mole)	> 25	> 25	25	15
Melting point (°C)	> 800 (sublimes)	3652	3652	3550
Thermal conductivity (Wm^{-1}K^{-1})	0.4		80–230	900–2320
Young's modulus (TPa)		1–1.2		
Tensile strength (GPa)		> 60		
Band gap (eV)		0.5 eV (d = 1.5 nm) 15 meV ($d \approx$ 1.5 nm)		5.47
Electrical conductivity (A/cm^2)		10^9		3.513

Nanomaterials have already shown a high potential that some of their applications in biology, medicine and drug delivery mechanisms have raised several ethical issues. We will discuss them in subsequent sections.

8.3 STUDY OF NANOSCALE MATERIALS

Transmission Electron Microscope (TEM)

Transmission electron microscope is very important tool in materials science for investigating the fine scale structure of materials.

TEM is a microscopy technique whereby a beam of electrons is transmitted through an ultra thin specimen, interacting with the specimen magnified and focussed by an objective lens and appears on an imaging screen, a fluorescent screen in most TEMs, plus a monitor, or on a layer of photographic film, or to be detected by a sensor such as a CCD camera. The first practical TEM was built by Arebus and Hillier at the University of Toronto in 1938.

Figure 8.7 shows schematically a TEM illustrating the basic principles of its operation. An electron beam is produced by a heated tungsten filament at the top of an evacuated column and is accelerated down the column by a high voltage (usually from 75 to 120 kV). Electromagnetic coils are used to condense the electron beam, and then it is passed through a very thin (*i.e.* about 100 nm thick or less) section of a thin metal specimen placed on the specimen stage (Fig. 8.7). The specimen area examined must be very thin so that some of the electrons entering are able to pass through it. As the electrons pass through the specimen, some are absorbed and some of them are scattered so that they change direction.

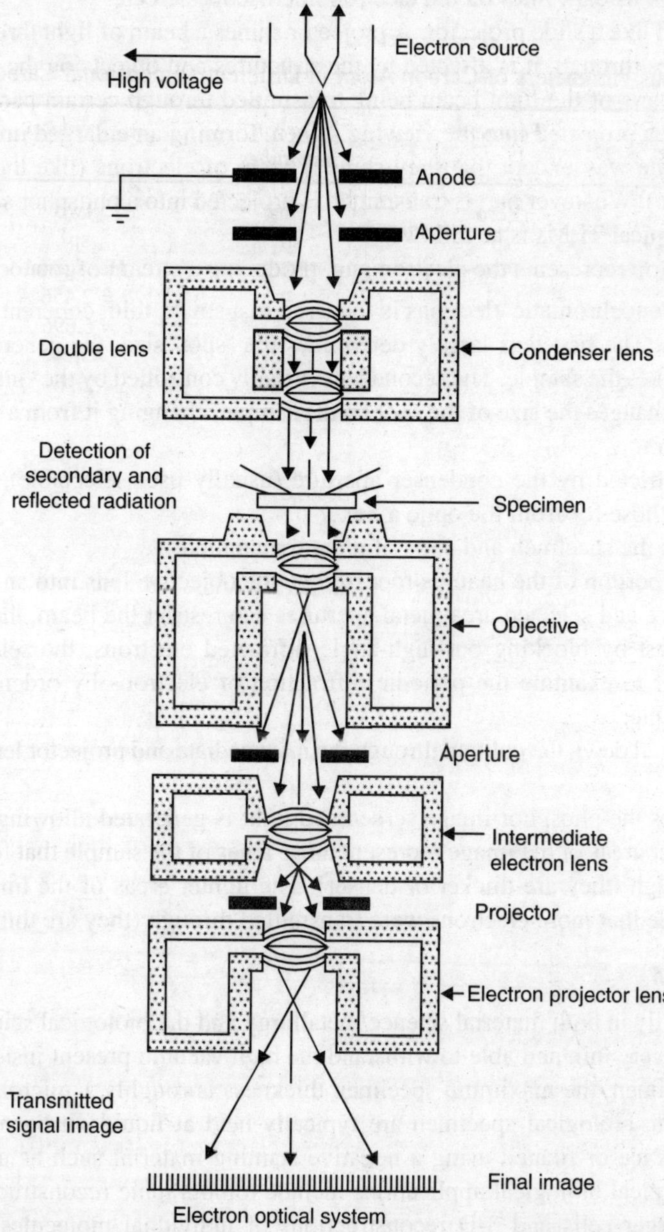

Fig. 8.7 Schematic arrangement of electron lens system in a transmission electron microscope (TEM). All the lenses are enclosed in a column that is evacuated during operation. The path of the electron beam from the electron source to the final projected transmitted image is indicated by arrows. A specimen thin enough to allow an electron beam to be transmitted though it is placed between the condenser and objective lenses as indicated.

Differences in crystal atomic arrangement will cause electron scattering. After the electron beam has passed through the specimen, it is focussed with the objective coil and then enlarged and projected on a fluorescent screen (Fig. 8.7). A region in a metal specimen that tends to scatter electrons to a high degree will appear dark on the viewing screen. Thus dislocations that have an irregular linear atomic arrangement will appear as dark lines on the electron microscope screen.

A TEM works much like a slide projector. A projector shines a beam of light through (transmits) the slide, as the light passes through, it is affected by the structures and objects on the slide. These effects result in only certain parts of the light beam being transmitted through certain parts of the slide. This transmitted beam is then projected into the viewing screen, forming an enlarged image of the slide.

TEMs work the same way except that they shine a beam of electrons (like the light) through the specimen (like the slide). Whatever part is transmitted is projected into a phosphor screen for the user to see. The working of typical TEMs is as follows:

The 'source' at the top represents the electron gun, producing a stream of monochromatic electrons.

- This stream of monochromatic electrons is focussed to a small, thin, coherent beam by the use of condenser lenses. The first lens largely determines the 'spot' size, the general size range of the final spot that strikes the sample. The second lens (usually controlled by the "intensity of brightness knob") actually changes the size of the spot on the sample, changing it from a wide dispersed spot to a pinpoint beam.
- The beam is restricted by the condenser aperture (usually user selectable), knocking out high angle electrons (those for from the optic axis).
- The beam strikes the specimen and parts of it are transmitted.
- This transmitted portion of the beam is focussed by the objective lens into an image.
- Optional objective and selected area metal apertures can restrict the beam, the objective aperture enhancing contrast by blocking out high-angle diffracted electrons, the selected area aperture enabling the user to examine the periodic diffraction of electrons by ordered arrangements of atoms in the sample.
- The image is passed down the column through the intermediate and projector lenses, being enlarged all the way.
- The image strikes the phosphor image screen and light is generated allowing the user to see the image. The darker areas of the image represent those areas of the sample that fewer electrons were transmitted through (they are thicker or denser). The lighter areas of the image represent those area of the sample that more electrons were transmitted through (they are thinner or less dense).

Applications of TEM

The TEM is used heavily in both material science/metallurgy and the biological sciences. In both cases the specimen must be very thin and able to withstand the high vacuum present inside the instrument.

For biological specimen, the maximum specimen thickness is roughly 1 micrometre. To withstand the instrument vacuum, biological specimen are typically held at liquid nitrogen temperatures after embedding in vitreous ice or fixated using a negative staining material such as uranyl acetate or by plastic embedding. Typical biological applications include tomographic reconstructions of small cells or thin sections of larger cells and 3-D reconstructions of individual molecules via single particle reconstruction.

Limitations

There are a number of drawbacks to the TEM technique. Many materials require extensive sample preparation to produce a sample thin enough to be electron transparent, which makes TEM analysis a relatively time consuming process with a low output of samples. The structure of the sample may also be changed during the preparation process. Also the field of view is relatively small, raising the possibility that the region analysed may not be characteristic of the whole sample. There is a possibility that the sample may be damaged by the electron beam, particularly in the case of biological materials.

Scanning Electron Microscope (SEM)

Figure 8.8 shows the schematic arrangement of SEM. Unlike the TEM, where electrons of the high voltage beam form the image of the specimen, the SEM produces images by detecting low-energy (<50 eV) secondary electrons that are emitted from the surface of the specimen due to excitation by the primary electron beam. In the SEM, the electron beam is allowed to move across the substance, with detectors generating an image by mapping the detected signals with beam position.

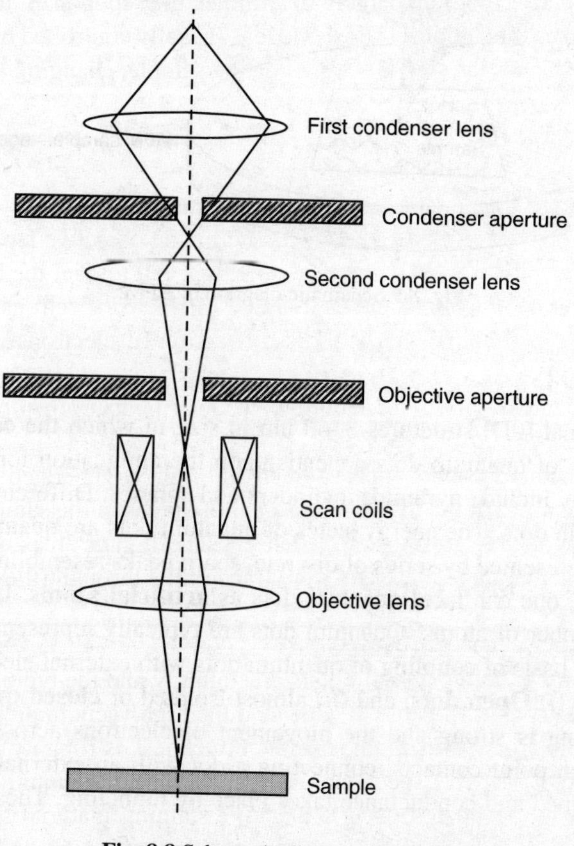

Fig. 8.8 Schematic diagram of SEM.

Usually, the TEM resolution is about an order of magnitude higher than the SEM resolution; however, because the SEM image relies on surface processes rather than transmission, it is possible to image bulk samples and also has a much greater depth of view. Thus SEM can produce images that are batter representation of the 3D structure of the sample.

Scanning probe microscopes are of two types: (*i*) the **atomic force microscope** (AFM) and (*ii*) the **scanning tunneling microscope** (STM). In the AFM technique, a tiny tip moves up and down in response to the electromagnetic forces between the atoms of the surface and those on the tip. The motion of this tip is recorded and used to generate an image of the atomic surface using a computer. Figure 8.9 shows a schematic arrangement of AFM.

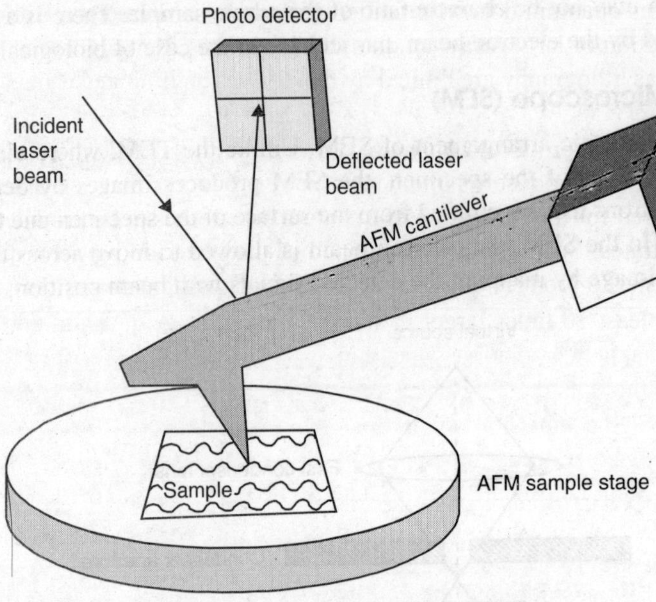

Fig. 8.9 Schematic diagram of AFM.

8.4 QUANTUM DOT

These are zero-dimensional (0D) structures 3–60 nm in size in which the carries are confined in all three directions. The size of quantum dots depends upon the application for which it is synthesised. Common shapes generally include pyramids, cylinders and spheres. Different synthesis route produce different kinds of quantum dots. The energy states of quantum dots are quantised in all directions and the density of states is represented by series of discrete, sharp peaks resembling that of an atom. On the basis of this comparison, one can label quantum dots as **artificial atoms.** Usually quantum dots are formed by a definite number of atoms. Quantum dots are typically represented by atomic clusters or **nanocrystallites**. On the basis of coupling of quantum dots with external electric circuits, one can put them under two classes : (*i*) **Open dots**, and (*ii*) almost isolated or **closed quantum dots**. In an open quantum dot, the coupling is strong and the movement of electrons across dot – lead junctions is allowed classically. When point contacts, connecting a dot with an external circuit, are pinched off, effective barriers are formed and conductance takes place by tunneling. These are almost –isolated or closed quantum dots.

An ensemble of **quantum dots** spatially arranged with a period (the distance between the nearest dots) comparable or shorter than the electron **de Broglie wavelength** is called a **quantum dot solid.** The basic properties of such dense quantum dot solid are expected to reproduce features inherent in solids, *i.e.* formation of energy bands in perfect lattice and existence of localised and delocalised electron states in disordered quantum dot structures. The quantum dot solid can be considered as a special form of **colloidal crystals.** Electronic and optical properties of quantum dot solids can be determined by electron confinement in each dot and collective effects arising form periodic special organisation of the dots.

Quantum dots can be fabricated with either a top-down technique or bottom-up technique. The top-down technique produces quantum dots with limited uniformity in diameters. Uniformity in diameters is crucial for the production of large arrays of dots that have consistent properties. Unfortunately, top-down approaches such as lithography are limited by the diffraction limit and cannot create dense arrays of quantum dots.

The most common way to produce a quantum dot is through the bottom-up approach. Bottom-up approach is a proper way to produce quantum dots in dense arrays that self-assemble in an orderly manner. The sites required for the nucleation of quantum dots seems to be provided by the high strain developed on the substrates. Such strain is developed at the interface between the substrate and the deposited layer. It is easier to produce quantum dots having consistent properties using approach.

A quantum dot is at least 10 times larger in size than an atom. A quantum dot emits only specific wavelength of electromagnetic radiation depending on its size. Quantum theory predicts that with decreasing size of quantum dots there will be a corresponding increase in energy of emitted light.

The ability to control the emission properties of a quantum dot has tremendous technological applications as well as of great academic interest to researchers. Unlike any other quantum structure, quantum dots have excellent confinement properties. This makes quantum dots extremely efficient in emitting light. Quantum dots have been the source of some of the world's most powerful lasers produced to date. **Quantum dot lasers** combines the length scales defining the confinement of photons (100 nm) and the length scales defining the confinement of electrons and holes (10 nm). Moreover, quantum dot lasers exhibit reduced temperature resistivity. For fabrication, it is necessary to form high quality, uniform quantum dots in the active layer. The practicality of a **quantum dot laser** is still being improved. In medical studies, quantum dots are already in practice as **tags** that can be inserted into a patient's bodies. One can see these tags using most medical scanning technologies can help pinpoint the biological processes inside the body as they take place.

8.5 QUANTUM WIRE OR NANOWIRE

A nanowire or a quantum wire is a one dimensional (1 D) structure that has a diameter of the order of a few nanometres. Such a wire is also known as a quantum wire. The length-to-width ratio, called the aspect ratio, for typical nanowires is about 1000 or more. As compared to quantum films, one more dimensions of the structure appears to be so small as to provide quantum confinement. Nanowires have many interesting properties that are not observed in bulk materials. This is because electrons in a nanowire occupy energy levels that are different from the energy levels or bands observed in bulk materials. Charge carriers are free to move only along the wire. Thus, only one kinetic component along with the confined energy contributes to the total energy. As a consequence, the density of states has $E^{1/2}$ dependence for each of the discrete pairs of states in the confined directions.

Quantum confinement exhibited by certain nanowires, *e.g.* carbon nanowires is reflected in the discrete values of the electrical conductivity. Such discrete values arise from a quantum mechanical restriction on the energy of electrons (and hence their number) travelling through nanowire. The difference in the successive values of conductivity is called *quantum conductivity.*

Nanowires have been grown out of metals, traditional semiconductors such as silicon and gallium, and a variety of polymers.

Conductivity of Nanowires

The conductivity of a nanowire is observed to be much less in comparison to the corresponding conductivity of the bulk material. This is due to a variety of reasons. One reason is the scattering of electrons from the boundaries of the wire. When the width of the wire is smaller than the mean free path of the free electrons in the bulk material the scattering of electrons from the boundaries increases. In copper, for example, the mean free path (at 300 K) is 40 nm. Copper nanowires less than 40 nm width will shorten the mean free path to the wire width, resulting in reduced conductivity. Silicon nanowires coated with nickel have shown improved conductivity.

Nanowire conductivity is found to be strongly influenced by the edge effects. The edge effects are due to atoms that are at the nanowire surface and are not fully bonded to neighbouring atoms as the atoms within the bulk of the nanowire. The unbounded atoms are often a source of defects within the nanowire and seems to cause the nanowire to conduct electricity more poorly in comparison to the bulk material. As a nanowire reduces in size, the surface atoms gets more numerous as compared to the atoms within the nanowire, and edge effects become more prominent.

The conductivity of quantum wire can undergo quantization. This can be explained on the basis of the quantization of energy of the electrons travelling through a nanowire. The conductivity of nanowire is thus the effect of the sum of the transport by separate *channels* of different quantized energy levels. The thinner the nanowire is, the smaller the number of channels available to the transport of electrons. One can verify this by measuring the conductivity of a nanowire while stretching it. Its conductivity decreases in a stepwise manner while its diameter shrinks.

Use of Nanowires

Applications of nanowires include sensors, interconnects, memory devices (potentially) logic devices and processors.

It is possible that semiconductor nanowire crossings may be important to the future of digital computing. No doubt, there are many other uses for nanowire, the only ones that actually take advantage of physics in the nanometre regime seems to be applications in fabrication of electronic devices.

Conducting nanowires offer the possibility of connecting molecular-scale devices in a **molecular computer.** Dispersions of conducing nanowires in different polymers are being investigated for use as transparent electrodes for flexible flat-screen displays.

Due to their high value of Young's moduli, application of nanowires in mechanically enhancing composites is also being investigated. Since nanowires appear in bundles, they may be used as tribological additives to improve friction characteristics and reliability of electronic transducers and actuators.

8.6 MODELLING OF QUANTUM SIZE EFFECT

The experimentally observed size effect can be explained with the help of following theoretical models.

1. ***Weak confinement region*** $(R >> a_B)$: The region wherein size of the cluster, R, is greater than the Bohr radius, a_B of the exciton in the bulk material is termed weak confinement region. In this region, due to size quantisation of exciton (exciton is a bound pair of electron and hole formed due to excitation of electron at an elevated energy level due to interaction with electromagnetic interaction) there is small increase, ΔE, in the exciton energy. This is given by $\Delta E = \hbar^2\pi^2/2MR^2$, where M is the mass of the exciton and is given by $M = m_e^* + m_h^*$, where m_e^* and m_h^*, are effective masses of electron and hole respectively, and R is radius of the cluster.

2. ***Medium confinement region*** ($R \approx a_B$)**:** The region wherein size of the cluster is comparable to the Bohr radius of the exciton in the bulk material is termed medium confinement region. This is an usual situation observed in certain semiconductor compounds. In this region due to size quantisation of exciton there is increase in the exciton energy. This increase in energy is given by

$$\Delta E = \frac{\hbar^2 \pi^2}{2m_e^* R^2} \qquad ...(1)$$

3. ***Strong confinement region*** ($R << a_B$)**:** The region wherein size of the cluster is smaller than the Bohr radius of the exciton in the bulk material is termed strong confinement region. There is an increase in the exciton energy due to size quantisation of exciton. This increase in energy is given by:

$$\Delta E = \frac{\hbar^2 \pi^2}{2\mu R^2} \qquad ...(2)$$

where μ is the reduced effective mass of the electron.

The **EMA model** was further improved by Brus in 1984 and called as **Brus model** or **single-band effective-mass model.** In 1986, Kayanuma further improved this model and found to be in agreement with experimental results.

The **empirical tight binding model** (ETBM) proposed by Wang and Herron in 1990 was successful in explaining many phenomena observed in optical spectra.

Empirical pseudo-potential model (EPM) developed by Ram Krishna and Friesner in 1991 is well established and found to be a reliable approach. **Effective bond order model** (EBOM) developed by Einevoll in 1992 helped in calculating energy levels for particular phase of certain semiconductor compounds. However, it may be noted that all the models are not applicable for cluster sizes smaller than or equal to 1 nm.

8.7 SURFACE AND INTERFACE EFFECTS

Significant changes are reported in the properties of nanomaterials due to the spatial confinement of charge carriers in three dimensions. The surface properties of nanomaterials also play an important role in determining the properties of these materials. The larger surface-to-volume ratio observed at the nanoscale is one of the reasons for the typical properties of these materials. The number of surface atoms in a nanomaterial is appreciably larger than those in the bulk. This causes the defects to be located at the surface. These surface atoms and defects seems to play a major role in deciding the properties of these materials.

In a small cluster, a large fraction of atoms occupy surface positions, *e.g.*, a 50-Å Cds cluster has about 15% of its atoms lying on its surface. The effect of relatively large proportion of atoms on or near the surface of clusters becomes prominent in the study of surface photons, absorbed species, and enhancement in the reactivity by using catalyst. The existence of this vast interface between the cluster and the surrounding medium seems to have a profound effect on the properties of the cluster. Whatever is the method of preparation, in most of the cases, the clusters have an imperfect surface. This leaves defects on the surface. These defects act as traps for the movement of the charge carriers after the photoexcitation. The presence of the trapped electrons or holes can modify the optical properties of these clusters and can give rise to **non-linear optical effects**. This can also alter the nature of the photochemical reactions which are of considerable interest in the field of photocatalysis. Controlling the cluster surface and in turn the cluster properties is a major challenge before researchers and provides a unique opportunity in this exciting and promising field.

The surface-to-volume ratio is important in the process of catalysis. The surface area of a catalyst is generally expressed in the units of square metre per gram and called as the **specific surface area** (S).

$$S = \frac{\text{Area}}{\text{Volume} \times \text{Density}} \qquad \text{...(3)}$$

The specific surface area per gram for geometrical shapes generally observed in nanomaterials are as follows:

$S = 6 \times 10^3/pd$ for a sphere with diameter d

$S = 6 \times 10^3/pa$ for a cube with side a

$S = 2 \times 10^3/pl$ for a thin disc whose length l is very small compared to its diameter d

$S = 4 \times 10^3/pd$ for a long cylinder with its diameter d very small compared to its length l.

8.8 PROPERTIES OF NANOMATERIALS

Research on nanomaterials has been stimulated by their technological applications. These materials exhibit entirely new and special properties. Technologically useful properties of these materials are not limited to their structural, chemical or mechanical behaviours. For example, multilayes represent examples of materials in which one can modify or tune a property for specific application by sensitively controlling the individual layer thickness.

Very different physical laws dominate the nanoscale materials. Properties of material change. Some of these materials are better conductors of electricity or heat, some are mechanically stronger, some have different magnetic properties, and some reflect light better or change colours as their size changes. Nanoscale materials have a large surface area available for interactions compared to bulk materials of equal volume. We know that nanomaterials are formed by grouping a small number of molecules together. These are called **nanoclusters** or **simply clusters**. The smaller number of molecules in a cluster also gives rise to some properties that are observed only in the nanomaterials.

8.8.1 Optical Properties

Confining electrons to small geometries give rise to **'particle in a box'** energy levels. This quantum confinement creates new energy states and results in the optoelectronic properties of semiconductors. Because of their role in quantum dots, nanoparticles made of elements, which are normal constituents of semiconductors, have been the subject of much study, with particular emphasis on their electronic properties.

Optical properties such as the colour of a material depends its electronic structure. As the size of the material is reduced, the number of molecules that form the material gets smaller. This results in smaller size of the clusters that form the nanomaterial. The energy difference between two energy levels is inversely proportional to the size of the box, the size of a cluster (clusters atomic) are large "molecules" containing typically from 10 to 2000 atoms. They have specific properties (mainly due to their large surface-to-volume ratio), which are size dependent and different from both those of atoms and the bulk materials) in this case. The energy levels of the electrons in the constituent molecules are no longer continuous as against the case of a solid. The energy levels get spread. This separation affects the energies required for the transitions of electrons to the excited states and the subsequent emission when the electron falls back to the ground state. This energy difference determines the colour of the light emitted. Clusters of different size therefore show different colours. The energy levels for electrons in clusters with different sizes are shown in Fig. 8.10.

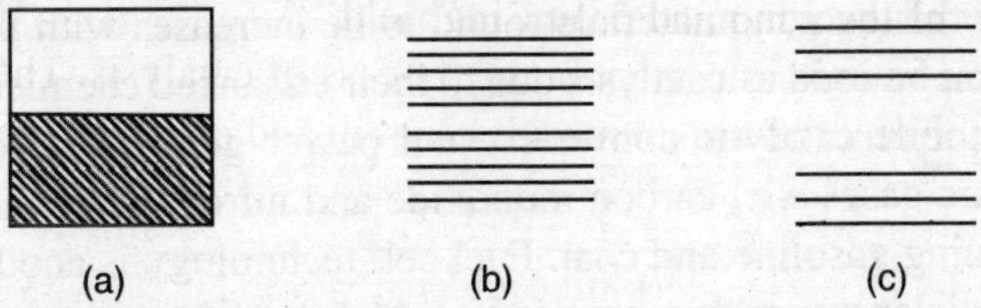

Fig. 8.10 Change in separation of energy level with size of cluster: (*a*) bulk metal, (*b*) large metal cluster and (*c*) small metal cluster.

For example, a piece of gold cut in smaller size is observed to change its colour in the nanosize regime. Depending on the size it can turn red, blue, yellow or some other colour. Materials having different size reflect and absorb light differently. Semiconductor cadmium selenide, in nanoscale size, can turn to any colour from red to blue depending on its size.

Similar changes in colours have been reported in the nature also. For example, leaves of a tree or plant appear green because the atomic arrangement in them is such that it absorbs all colours except green which is reflected. When leaves die, chlorophyll, the major constituent responsible for their green colour, breaks down changing the atomic arrangement and they appear brown. Colours observed on a peacock feather are also due to the nanosize particles on it.

8.8.2 Electrical Properties

Nanomaterials can store considerably more energy than the conventional materials due to their large surface area. This property of nanomaterials finds very interesting and useful applications, *e.g.* high-energy-density batteries.

We know that conventional and rechargeable batteries are used in almost all applications that require electric power. The energy density (storage capacity) of these batteries is quite low and they require frequent recharging. Nanocrystalline materials are found to be good as separator plates in batteries as they can hold considerably more energy than the separator plates made from conventional bulk materials, *e.g.*, nickel-metal hydride batteries made of nanocrystalline nickel and metal hydrides are likely to require far less frequent recharging and to last much longer.

Large electrochromic display devices

The *electrochromic* (a colour change of matter caused by an electric current) *device* consists of materials whose optical properties can be changed by passing electric current through them or by applying an electric field. These devices are similar to *liquid-crystal displays* (LCD) commonly used in calculators and watches and are primarily used in public hoardings and ticker boards to display information. The resolution, brightness, and contrast of the display of these devices depend on the grain size of the material used in them. Nanomaterials, such as tungstic oxide gel (tungsten oxide), are being explored for this purpose.

8.8.3 Chemical Properties

The electronic structure of nanomaterials depends on their size and therefore their ability to react chemically depends on their size. The higher surface-to-volume ratio is also responsible for their higher chemical reactivity. For example, gold is fairly inert when in the bulk form, but at the nanoscale, it is highly reactive.

Applications of chemical properties of nanomaterials

One of the major factors for the chemical applications of nanomaterials is the increase in their surface area. The chemical activity of the nanomaterials found to be increases with the increase in the surface area. Nanosize materials can be used as catalysts due to their enhanced chemical activity. Nanomaterials find applications in automobile catalytic converters and power generation equipment. Nanomaterials react with harmful and toxic gases *e.g.*, carbon monoxide and nitrogen oxide to prevent environmental pollution arising from burning gasoline and coal. Fuel cell technology is another important application of the noble metal nanoparticles where they are used as catalysts. The present-day fuel cell catalysts are based on the platinum (Pt) group metals. Pt and Pt-Ru alloys are some of the most frequently used catalysts from this group. Obviously, the use of these metals is one major factor for higher costs of these cells. One possibility to produce economical catalysts is the use of bimetallic nanoparticles.

Gold nanoparticles are widely studied as they have several potential applications. For *e.g.*, they have been used for the identification of the DNA sequences associated with disease, as well as viruses. New devices for use in computers can also be fabricated by using these materials. Obviously, these materials would therefore be excellent components in nanoelectronics, optics and biology.

8.8.4 Mechanical Properties

Scientific challenges in nanoscience and nanotechnology include the development of nanomaterials with novel mechanical properties. The need for scratch, mar, and/or abrasion resistance is well established in various markets, including fingernail polishes, flooring, plastic glazing, headlamp covers and other automotive parts, transportation windows, and optical lenses, where clear scratch-resistant coatings are used. Many of the mechanical properties of materials are modified at the nanoscale. Hardness and elastic modulus, fracture toughness, scratch resistance, and fatigue strength are some properties that are modified. Energy dissipation, mechanical coupling within arrays of components, and mechanical non-linearities are influenced by structuring components at the nanometre scale.

A number of important questions regarding the mechanical behaviour of nanocrystalline metals are yet to be answered; fundamentally the underlying mechanisms of deformation at both low and high temperatures are unknown. The major research challenge in this area currently lies in the areas of synthesis and processing. Until nanocrystalline materials can be produced which are dense, free of flaws, of high purity, and of sufficient quantity to carry out a variety of mechanical tests, measured values will be influenced by extrinsic rather than intrinsic properties. For high temperature applications, a great amount of research work needs to be performed in controlling grain sizes so that the correct balance of ductility, formability, and strength are achieved. Inspite of these challenging problems, nanocrystalline materials offer a wide variety of opportunities for fabricating new materials with precisely engineered microstructures where strength and ductility can be finely tuned. Moreover, the ability to structure materials on the nanoscale level makes it possible to test fundamental ideas about grain boundaries, dislocations and deformations

Mechanical applications of nanomaterials

Nanocrystalline materials possess unusual properties and their deformation behaviour has attracted intense activity in recent years.

Tougher and harder cutting tools can be developed using nanomaterials. Nanomaterials required for these purposes are made from tungsten carbide, tantalum carbide, and titanium carbide. Such tools are found harder, much more wear-resistant, erosion-resistant, and also last longer than their conventional (large-grained) counterparts. Moreover, for the miniaturisation of microelectronic circuits, the industry needs microdrills (drill bits with diameter less than the thickness of an average human hair or 100 vtm) with enhanced edge retention and far better wear resistance. Nanocrystalline carbides are quite stronger, harder, and wear-resistant and therefore they are currently being used in these microdrills.

Nanomaterials can be used to manufacture automobiles with greater fuel efficiency. Nanomaterials are quite stronger, harder, much more wear-resistant and hence they could be coated on spark plugs and other engine parts. Moreover, automobiles waste significant amounts of energy by losing the thermal energy generated by the engine. Nanocrystalline ceramics, *e.g.* zirconia and alumina retain heat much more efficiently. Engine cylinders coated with such materials are under consideration. This would result in complete and efficient combustion of the fuel.

One of the key properties required of aircraft components is their fatigue strength. Fatigue strength decreases with the age of the component. Moreover, the fatigue strength increases with a reduction in the grain size of the material. Nanomaterials provide such a significant reduction in the grain size over conventional materials that the fatigue life is increased by an average of 200–300%. In spacecraft, elevated-temperature strength of the material is crucial as the components (such as rocket engines, thrusters, and vectoring nozzles) operate at much higher temperatures than aircrafts and higher speeds. Nanomaterials are useful to enhance performance characteristics of aerospace components as well.

Ceramics are quite hard, brittle, and at same time difficult to machine even at high temperatures. However, with a reduction in grain size, the properties of ceramics change drastically. Ceramics made of nanocrystals are found to be ductile in nature. They can be pressed and sintered into various shapes at significantly lower temperatures. Zirconia in bulk form for example, is a hard, brittle ceramic. At nanoscale, it turns into a super plastic material, *i.e.*, it can be deformed to a great extent (up to 300% of its original length). These ceramics have nanocrystalline grains. Ceramics based on silicon nitride (Si_3N_4) and silicon carbide (SiC) have been used in automotive applications as high-strength springs, ball bearings, and valve lifters. They have also been used as components in high-temperature furnaces as they possess good formability and machinabilty combined with excellent physical, chemical and mechanical properties.

Nanocrystalline materials are found to provide better thermal insulation. Aerogels are porous and extremely light weight nanocrystalline materials. They can withstand 100 times their weight. They are currently being used for insulation in offices and homes. They are also being used as materials for "smart" windows, which darken when the sun is too bright and lighten otherwise.

8.5.5 Magnetic Properties

The magnetic properties of nanometer sized particles differ in several respects from the properties of their bulk counterparts. The large fraction of atoms located on surfaces or interfaces for example, whose local environments differ greatly, from those of the interior atoms leads to a blurring of the distinction between intrinsic and extrinsic properties since nano phase particles can be as much as 50% surface material, new magnetic properties characteristics of surfaces and interfaces become important from applications point of view.

Very interesting magnetic properties are observed in small **clusters**. The magnetic moment of each atom in cluster interacts with the moments of other atoms and forces all the other moments to orient in one direction with respect to some symmetry axis of the cluster. The cluster thus has a net magnetic moment and it is magnetised. As the cluster size decreases, it becomes easier for the atoms in a cluster to orient their magnetic moments in one direction and display ferromagnetic behaviour. In some cases, non-magnetic materials show a prominent increase in their magnetic moment when they contain less than 20 atoms.

The strength of a magnet is measured in, terms of **coercivity** and **saturation magnetisation**. It is observed that these values increase with a decrease in the grain size and increase in the specific surface area (surface area per unit volume) of the grains. Obviously, nanomaterials present magnetic properties that are not observed so far.

Magnetic applications of nanomaterials

Today information technologies ranging from personal computers to main frames use magnetic materials to store information on tapes, floppy diskettes, and hard disks. Our seemingly insatiable appetite for more computer memory will probably be met by a variety of magnetic recording technologies based on nanocrystalline thin-film media and magneto – optic materials. Personal computers and several of our consumer and industrial electronics components are now powered largely by lightweight switch mode supplies using new magnetic materials technology that was unavailable two decades ago. Magnetic materials touch many other aspects of our lives.

Magnetic properties of nanometer sized particle of both ferrimagnetic and ferromagnetic materials have attracted considerable attention in recent years. Such interest has been generated for practical as well as theoretical regions. Some of the application areas being explored are in high-density magnetic recording media, as ferrofluids, giant magnetoresistive systems. Some notable applications of nanostructured magnetic materials are: Storage applications, giant magneto resistance (GMR) recording head, high power magnets.

Magnets made of nanocrystalline yttrium-samarium-cobalt grains possess very unusual magnetic properties due to their extremely large surface area. Such magnets are observed to possess high magnetic retention. Typical applications for these high-power rare-earth magnets include quieter submarines, automobile alternators land-based power generators, and motors for ships, ultra-sensitive analytical instruments, and magnetic resonance imaging (MR1) in medical diagnostics.

8.8.6 Melting Temperature

The large surface-to-volume ratio of nanomaterials exhibit a very unusual effect on their thermal properties. They show substantially lower melting temperatures as compared to the solids. Negative heat capacities are also reported in some nanomaterials. The theoretical analysis and calorimetric investigations show that in the temperature range 10 K ≤ θD (Debye temperature) the heat capacity of nanopowers is from 1.2 to 2 times higher than that of coarse grain bulk materials. One can easily explain these observations. When a large system *e.g.* a solid is heated, the energy added is completely converted into potential energy of the atoms/ molecules. The solid is continuously converted to liquid form due to this added energy. The kinetic energy of the atoms/molecules and therefore the temperature of the solid thus remain constant. A small system like a nanomaterial on the other hand, tries to avoid partially molten states and converts some part of its kinetic energy into potential energy. Obviously, the kinetic energy of the nanomaterial decreases and it becomes cooler. Studies of the variation in the melting temperatures of gold reveal such behaviour. Variation in the melting temperature of gold with reduction in the particle size is shown in Fig. 8.11.

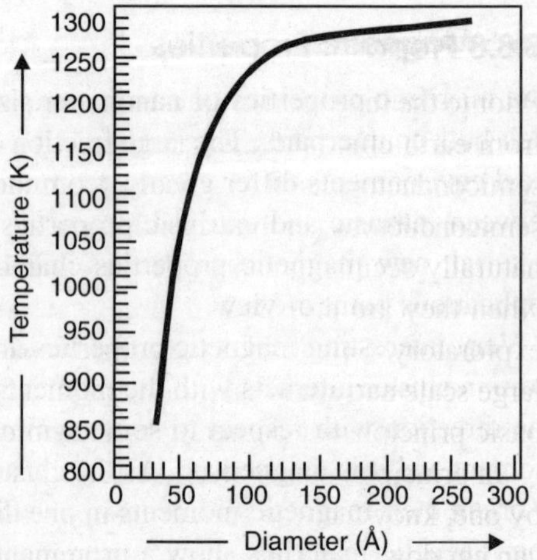

Fig. 8.11 Variation in the melting temperature of gold with reduction in the particle size.

8.8.7 Structural Properties

One can determine the size of a nanocluster by the number of molecules that form it. Naturally, the arrangement of these molecules is quite different when the number of molecules forming the cluster changes.

The size dependence of the structural properties leads to changes in many other properties of nanomaterials. Structural properties of metallic nanocluster are studied using the XRD analysis.

Aluminum is face-centred cubic in the bulk form. It is observed to be in the face-centred cubic hexagonal close packed, and the icosahedral structure at nanoscale. Similarly, indium is observed to undergo a transition from its tetragonal structure in the bulk form to the face-centred. The changes in the structure of aluminium is shown in Fig. 8.12.

The lattice structure of nanometer-sized crystallites in nanocrystalline samples synthesized by various methods showed that the lattice structure of the nanometer-sized crystallites evidently deviates from the equilibrium state. One may classify the deviation as: (*i*) distorted lattice structures and stoichiometric line compounds, and (*ii*) formation of metastable phase below a critical crystalline size.

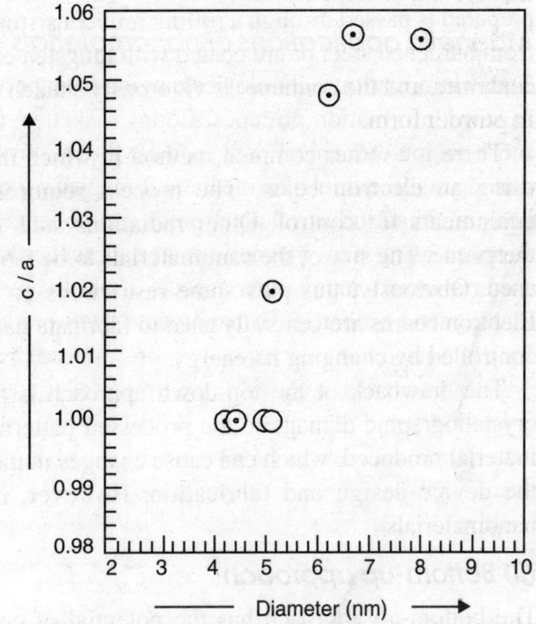

Fig. 8.12 Changes in the structure of aluminium.

Structural changes in the nanaomaterials also cause changes in the electronic structure of the clusters. The band levels and the valence energy change when the number of molecules in a cluster changes. The energy levels are in the form of a continuum when the material is in the bulk form. As the number of molecules is reduced, the energy levels become discrete and the material behaves in a different way. This change in the energy levels is shown in Fig. 8.12.

8.8.8 Preparation of Nanoscale Materials

Prior to the invention of transistor, metals were treated as the most important solids. Metals are obtained from earth crust and then purified. Usually, metals are used in the form of alloy. Similarly, elemental semiconductors such as silicon are also used in the form of either alloys or compounds. The compound semiconductors, ceramics, oxides, and insulators used in industry are prepared in the laboratory. Although naturally occurring nanoscale materials have been discovered, use of nanocrystals will be feasible only when they are prepared in large scale and at an affordable cost. As of today, research in this field is exploratory. Scientists are engaged in exploring various methods to produce nanoscale materials on a large scale and study their properties. A variety of methods are attempted. Here, we briefly describe the basic principles of preparation of nanoscale materials.

In principle, nanosize or nanostructured materials can be prepared by either arranging the atoms one by one, known as the *bottom-up approach,* using the techniques such as atom force microcopies or one can cut down the bulk or scrape the macroscopic solid to obtain nanosize solid known as the *top-down approach.*

(i) Top-down approach

This approach involves the breaking down of the bulk material into nanosize structures or particles. Truly speaking, these techniques are an extension of those that have been used for producing micron-sized particles. An example of such a technique is high energy wet ball milling. The material to be prepared is passed through a rolling mill consisting of two hard rollers. These rollers are generally made from hardened steel or are coated with tungsten carbide. The complete process is carried out in a sealed container, and the container is vigorously shaken continuously. This method produces the nanoparticles in powder form.

There is another common method in which the bulk material is cut to the required size and shape using an electron beam. The process requires a very high vacuum and involves sophisticated equipments for control. Other radiations such as lasers, X-rays or UV rays are also used in this technique. The size of the nanomaterials to be fabricated is governed by the wavelength of the radiation used. Obviously, this puts some restrictions on the radiations that can be used for nanofabrication. Electron beams are generally used to fabricate nanoscale devices as the wavelength of the beam can be controlled by changing its energy.

The drawback of the top-down approach is the imperfection of surface structure and significant crystallographic damage to the processed patterns. Further, it can also introduce internal stress in the material produced, which can cause changes in the properties of the material. This causes difficulties in the device design and fabrication. However, this approach is useful for the bulk production of nanomaterials.

(ii) Bottom-up approach

The bottom-up approach has the potential of creating less waste and hence more economical. This approach refers to the build up of a material from the bottom: atom-bottom molecule-by-molecule, or cluster-by-cluster. This is a method used for a long time to prepare small sized materials. There are several different methods involving the bottom-up approach. Wet chemical synthesis of nanomaterials or the sol-gel method is a very commonly used method. This method promises a better chance to obtain nanostructures with less defects, more homogeneous chemical composition.

(iii) Chemical methods

Chemical methods of preparing nanomaterials are quite simpler in principle, versatile, and are inexpensive. Large number of materials such as metals, semiconductors, insulators, ceramics, and other complicated solids are grown in nanostructured form by employing these methods. The chemical reactions can be carried at low or at high-temperatures. Low-temperature methods use reactions of aqueous solutions, while high temperature methods use different solvents. A salt containing cation and the one containing anion is mixed under controlled condition of concentration, rate of reaction, temperature, and pH of the solvents in presence of some organic molecules that work as binders. The large organic molecules bind to cations and/or anions and control the size of the precipitate. Depending on the concentration of cations, anions, and organic molecules. nanocrystals of the required size can be formed in the solution. For example, one of the reactions for formation of zinc sulphide (ZnS) nanocrystals is: zinc acetate + sodium sulphide + thiophenol → zinc sulphide + sodium acetate + zinc-thiophenol

$$Zn(CH_3COO)_2 + Na_2S + C_6H_5SH \rightarrow ZnS + 2Na(CH_3COO) + Zn\text{-}SHC_6H_5$$

Zinc sulphide is insoluble in organic solvents like methanol and hence precipitates out. However, sodium acetate is highly soluble in normal solvents and remains in the solution. Nanocrystal powder thus formed is separated out by using centrifugal machine and is washed several times so that the reaction by-products and unreacted solvents are removed. Different chemical reactions produce nanocrystals with different physical properties.

Nanocrystals can also be obtained by heating the powder of the material with glass, melting the mixture, and re-solidifying the same. Many old churches in Rome, old palaces in India have decorative coloured glasses made from nanocrystals grown in this fashion. We can say that knowledge of growing nanocrystals is very old while they are looked upon with new focus for the last three decades.

The sol-gel process, involves the formation of inorganic networks through the formation of a colloidal suspension (*sol*) and gelation of the sol to form a network in a continuous liquid phase (*gel*). The precursors for synthesising these colloids consist usually of a metal or metalloid element surrounded by various reactive ligands. The starting material is processed to form a dispersible oxide and forms a sol in contact with water or dilute acid. Removal of the liquid from the sol yields the gel, and the sol/gel transition controls the particle size and shape. Calcination of the gel produces the oxide. Various steps involved in the sol-gel process can be summarized as follows:

(i) Formation of different stable solutions of the alkoxide or solvated metal precursor (the sol).

(ii) Gelation is the formation of an oxide- or alcohol-bridged network (the gel) by a polycondensation or polyesterification reaction that results in a dramatic increase in the viscosity of the solution.

(iii) Ageing of the gel (syneresis), during which the polycondensation reactions continue until the gel transforms into a solid mass, accompanied by contraction of the gel network and expulsion of solvent from gel pores. Sometimes the smaller particles merge with the larger particles and form coarse particles. This is called as **Ostwald ripening** or **coarsening**.

(iv) **Drying of the gel**, in this step water and other volatile liquids are removed from the gel network. However, this process is complicated due to fundamental changes in the structure of the gel.

(v) **Dehydration**, in this step the surface-bound M-OH groups are removed and the gel is stabilised against dehydration.

(vi) **Densification and decomposition** of the gels at high temperatures ($T > 800°C$). In this step, the pores of the gel network are collapsed and remaining organic species volatilised.

This synthesis method offers the possibility of synthesising non-metallic inorganic materials *e.g.* glasses, glass ceramics, or ceramic materials at very low temperatures compared to the high temperature process required by melting glass or firing ceramics.

It is very difficult to control the growth of particles and then stop the newly formed particles from mixing together. Moreover, we must ensure that the reactions are complete so that no undesired reactant is left on the product and completely removing any growth aids that may have been used in the process. However, the production rates of nanopowders are very low by this process. The main advantage of any bottom-up approach is that one can obtain same-sized particles.

(iv) Physical methods

Most of the methods of preparing nanoscale materials discussed so far are physical methods of preparing nanoparticles. In addition to these methods, radio frequency (RF) plasma and molecular beam epitaxy (MBE) are the techniques also used to fabricate nanomaterials.

R.F. Plasma

Radio frequency plasma technique is used to synthesise metal nanoparticles. The metal is placed in a specially designed container and kept in a vacuum chamber. By using RF plasma it is evaporated with the help of RF coils surrounding the container. By passing helium gas high-temperature plasma is generated inside the RF coils. The metal vapours condense on to the helium gas atoms and diffuse towards a collector rod, *i.e.* at a lower temperature and form the nanoparticles. These are then passivated by allowing to pass some gas like oxygen through the chamber.

Molecular beam epitaxy

This is an advanced technique used to form nanoparticles. In this technique, a substance is evaporated under ultra-high vacuum (UHV ≈ 10^{-9} torr). The required material is evaporated at a very slow rate (1 μm/h) and a thin film is deposited on a suitable substrate (supporting material) heated to the required temperature. Semiconductor heterojunctions can be grown in a controlled manner by this technique. The substrate material and the material to be deposited must be extremely pure so that any contamination in the device is avoided. This technique offers a very good control over the thickness of the material, it is reasonably reproducible, and the material prepared is also in its purest form. The UHV helps to make it possible to analyse and characterise the material deposited while it is being prepared. This method is highly advanced as well as expensive. However, nanodevices are fabricated by using this method.

8.8.9 Biological Nanomaterials Biomimicking

Biomaterials have received a considerable amount of attention over the last three decades as a means of treating decreases and easing suffering. Biomaterials have found application in about 8000 different kinds of medical devices, which have been used in repairing skeletal systems, returning cardiovascular functionality, replacing organs, and repairing and returning senses. Even though biomaterials have had a pronounced impact in medical treatment a need still exists to be able to design and develop better polymer, ceramic and metal systems.

Nanotechnology is playing an important role to improve biomaterial performance.

Most research focussing on biological micro-electro-mechanical systems (Bio MEMS) is for their use in diagnosis devices and for the detection of DNA, viruses, proteins, and other biologically derived molecules. Nanoscale Bio MEMS could allow for the real-time detection and analysis of signaling pathways, which would further our knowledge and understanding of the basic mechanisms and functions of the cell.

It is now believed that significant evidence exists that highlights the promise nanotechnology has for biological applications.

Many biological materials can be classified as biological nanomaterials *e.g.*, proteins, viruses, and bacteria are some of the familiar examples. The building blocks of proteins are the amino acids. The size of each amino acid is about 0.6 nm. Out of more than 100 odd amino acids that occur naturally, only 20 are involved in the formation of proteins. A protein chain is formed by tying hundreds or sometimes thousands of amino acid molecules together. Such chains resemble nanowires. The basic building blocks for the DNA are four nucleotide molecules that bind together in the form of a double helix. Thus, the DNA in the form of a double helix is also an example of double nanowire.

(i) Micro-electro-mechanical Systems (MEMSs)

Micro-electro-mechanical systems (MEMSs) are small integrated devices or systems that combine electrical and mechanical components. To make MEMSs devices, lithographic techniques are combined with metal deposition processes. MEMSs involve a mechanical response to an applied electrical signal, or an electrical response resulting from a mechanical deformation. The size or MEMSs devices or systems vary in the sub-micrometre (or submicron) level to the millimetre level. A system can have any number of such devices, from a few to millions. The same technology used for fabrication of integrated circuits is extended to fabricate the MEMSs. The MEMSs has mechanical elements such as beams, gears, diaphragms, and springs to devices along with the electrical components on it.

The main advantage of MEMSs devices are miniaturisation, multiplicity, and the ability to directly integrate these devices into micro-electronics. **Multiplicity**, refers to the large number of devices and designs that can be manufactured rapidly, *e.g.* **miniaturisation** has enabled the development of micrometre-sized accelerometers for activating airbags in cars.

MEMSs devices have applications which include inkjet-printer cartridges, accelerometers, miniature robots, microengines, locks, inertial sensors, microtransmissions, micromirrors, microactuators, optical scanners, fluid pumps, transducers, and chemical pressure and flow sensors. New applications or MEMSs devices and systems are emerging as the existing technology is applied to the miniaturisation and integration of conventional devices.

MEMSs systems can sense, control, and activate mechanical processes on the microscale and function individually or in arrays to generate effects on the macroscale. The microfabrication technology enables fabrication of large arrays of devices, which individually perform simple tasks, but in combination can accomplish complicated functions.

Future applications of MEMSs will be driven by processes enabling greater functionality through higher levels of electronic-mechanical integration and also greater numbers of mechanical components working alone or together to enable a complex action. Future MEMSs products will like to have higher levels of electrical mechanical integration and also more intimate interaction with the physical world. Advancing from their success as MEMSs sensors, MEMSs products will be embedded in larger non-MEMS systems, such as printers, automobiles, and biomedical diagnostic equipment, and will also enable new and improved systems. MEMSs devices are manufactured using batch fabrication techniques similar to those used for integrated circuits, unprecedented levels of functionality, reliability, and sophistication can be placed on a small silicon chip at a relatively low cost.

There are several devices and machines that have micrometre-sized elements.

(ii) Nano-electro-mechanical Systems (NEMSs)

NEMSs are similar to MEMS but smaller in size and are related to nanotechnology and nanomechanics. NEMS machines and devices are in the early states of development and many are still in conceptual stages. Numerous computer simulations of possibilities and ideas have been advanced. Because of their nanoscale size, NEMSs could be useful to measure small displacements and forces at a molecular scale. There are two standard methods of fabrication of NEMS machines and devices: the top-down and the bottom-up approach. The top-down approach is accepted as the most standard approach to NEMSs. This approach may be summarised as "a set of tools designed to build a smaller set of tools," *e.g.*, a millimetre-sized factory builds micrometre-sized factories, which in turn can build nanometre-sized devices and machines. The other approach can be thought of as putting together single atoms or molecules until a desired level of complexity and functionality is achieved in a device. Such an approach may utilise molecular self-assembly or mimic molecular biology systems.

The first very large scale integration NEMS device was demonstrated in 2000 by scientists at IBM. Its premise was an array of AFM tips that can heat/sense a deformable substrate in order to function as a memory device. A combination of these approaches may also be used, in which nanoscale molecules are integrated into a top-down framework, *e.g.* the carbon nanotube nanomotor.

More recently, a technique named **nanoimprint lithography** has been developed that may provide a low-cost, high-production rate manufacturing technology. This can produce patterns on a surface having 10 nm resolution at low cost and high rates because it does not require the use of a sophisticated radiation beam generating patterns for the production of each structure.

By exploiting nanoscale effects, NEMS offer a number of unique properties, which pave the way to applications such as force sensors, chemical sensors, and ultrahigh frequency resonators. The most interesting properties of NEMS arise from the behaviour of active parts, which is typically in the form of cantilevers or doubly damped beams with nanoscale dimensions. In NEMS, the charge controls the mechanical motion, and vice versa. The presence of mechanical motion decisions that the moving element to possess high-quality mechanical properties, including high Young's moduli and low dgnsity. While limitations in strength and flexibility compromise the performance of Si-based NEMs actuators. Carbon nanotubes (CNTs) are better candidate to realize the full potential of NEMS, in part due to their one-dimensional structure, high aspect ratio, perfect terminated surfaces, and exceptional electronic and mechanical properties. These properties, now complemented by significant advances in growth and manipulation techniques, make CNTs the most promising building blocks for next generation NEMS. The predicted behaviour of CNT based NEMS switches-basis of NEM devices is favourable, and electronic properties have been shown to be reversible with mechanical deformation by a local probe.

8.9 CARBON NANOTUBES (CNT)

Carbon nanotubes were discovered in 1991 by Sumio Iijima at the NEC laboratories. Although other nanotubes based on boron nitride and molybdenum have also been reported, currently NCTs are by far the most important group. CNTs have occupied a special identity in the field of nanotechnology. Since their discovery, theoretical and experimental studies in different fields, *e.g.* mechanics, optics, electronics, etc. have focussed on both the fundamental physical properties and on the potential applications of nanotubes. In all fields there has been substantial progress over the last two decades. Basically, CNTs are naturally self-organised nanostructure in the form of a tube composed of carbon atoms with completed bonds. The structure of CNTs consists of a single sheet of graphite, rolled into a tube, both ends of which are capped with C_{60} fullerene hemispheres. The tubes can be either open at their ends or capped at one or both ends with half a spheroidal fullerene. Figure 8.13 shows a schematic representation of a carbon nanotube. A nanotube is completely specified by what is referred to as a roll-up vector, which identifies "its helical nature" and diameter. Depending on its roll-up vector, a nanotube can have either metallic (Fig. 8.13 (*a*)), semiconducting (Fig. 8.13 (*b*), (*c*)) properties.

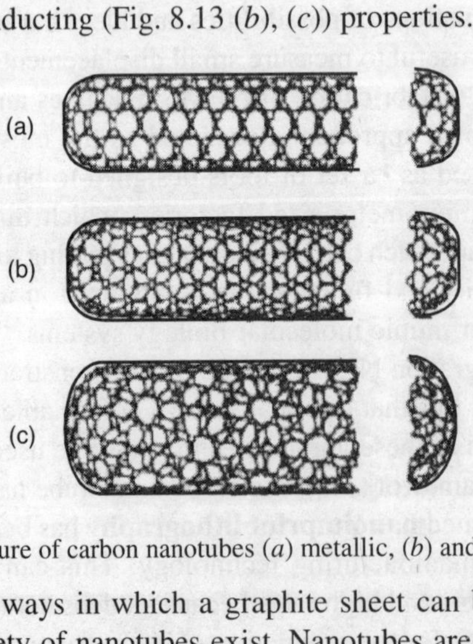

Fig. 8.13 The structure of carbon nanotubes (*a*) metallic, (*b*) and (*c*) semiconducting.

There are a vast number of ways in which a graphite sheet can be rolled up to form a seamless cylinder and hence a wide variety of nanotubes exist. Nanotubes are characterised by their *chiral* (or wrapping) vector, *c* such that $c = na_1 + ma_2$, where a_1 and a_2 are the basis vectors of the graphite (more correctly graphene) lattice and *m*, *n* are integers (Fig. 8.14). The chiral vector spans the circumference

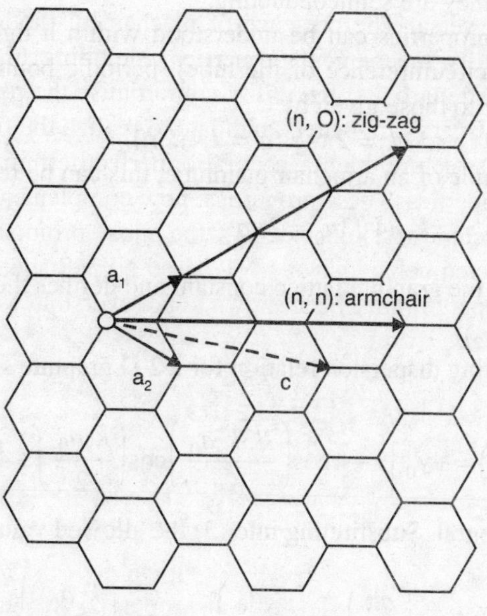

Fig. 8.14 The relationship between the graphite lattice basis vectors a_1, a_2 and the chiral vector $c = na_1 + ma_2$ is used to characterise carbon nanotubes. Two limiting cases are shown: (*n*, O) indices are associated with zig-zag tubes whereas (*n*, *n*) indices are associated with armchair tubes. All other tubes are chiral.

of the tube formed by joining the dotted lines shown in Fig. 8.14. Those tubes with chiral vectors of the form (*n*, *o*) are termed zig-zag tubes (Fig. 8.15 (*a*)), whereas $n = m$ a so-called 'arm chair' tube results (Fig. 8.15 (*b*)). All other values of *n* and *m* produce a chiral tube [Fig. 8.15 (*c*)].

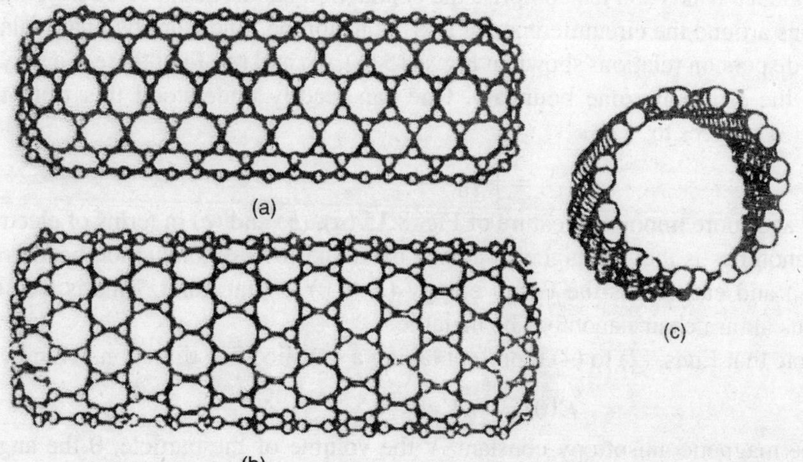

Fig. 8.15 Diagrams of three types of nanotubes: (*a*) armchair, (*b*) zig-zag and (*c*) chiral.

Closely following the synthesis of nanotubes came the remarkable theoretical prediction that their electronic properties could be changed between metallic and semiconducting simply by varying the tube diameter or its helicity, *i.e.* by changing the values of *n* and *m*. A key theoretical result is that armchair nanotubes are metallic, whereas for all other SWNNTs, when $n - m = 3\ 1$ (1 an integer) the tubes are metallic; otherwise they are semiconducting.

These striking electronic properties can be understood within a tight-binding framework. In the direction along c (*i.e.* along circumference of the tube), periodic boundary conditions constrain the values of electron wavevector to those given by

$$c\ .\ k = 2\pi{-}q\ (q = 1, ...,n) \qquad ...(1)$$

Taking the particular example of an armchair nanotube, this can be reduced to

$$k_x n 4\sqrt{3}a_0 = 2\pi q \qquad ...(2)$$

where $a_0 = (\sqrt{3})\,0.142$ nm is the graphite lattice constant and defines the chiral vector for the armchair tube (in the case $c = n\ (a_1 + a_2)$).

The well-known tight binding dispersion relation for a 2 D graphite sheet is

$$E_{2D}\left(k_x, k_y\right) = +\gamma_0\left\{1+4\cos\left(\frac{\sqrt{3}k_x a_0}{2}\right)\cos\left(\frac{k_y a_0}{2}\right)+4\cos^2\left(\frac{k_y a_0}{2}\right)\right\}^{1/2} \qquad ...(3)$$

γ_0 is the C – C overlap integral. Substituting into (3) the allowed values of k_x, given by (2) yields.

$$E_{10}^{A}\left(k_y\right) = \pm\gamma_0\left\{1\pm 4\cos\left(\frac{q\pi}{n}\right)\cos\left(\frac{k_y a_0}{2}\right)+4\cos^2\left(\frac{k_y a_0}{2}\right)\right\}^{1/2} \quad \left(-\pi < k_y a_0 < \pi\right) \qquad ...(4)$$

where the superscript A denotes that (4) is valid only for armchair tubes. It is assumed that the value of the overlap integral for graphite and nanotube are identical. The 1 D dispersion relations given by (4) are shown graphically in Fig. 8.16 for the case of (5, 5) armchair SWNT. There are six dispersion curves for both the valence and conduction bands. In each case, four bands are doubly degenerate (bold continuous lines) and two bands are non-degenerate (thin, dashed lines). Thus ten electronic levels comprise the valence bands and ten comprise the conduction bands, as expected from the fact that there are ten hexagons around the circumference of a (5, 5) nanotube. There are two particularly noteworthy features of the dispersion relations shown in Fig. 8.15 (*a*), (*b*) and (*c*). First there is a very high degree of degeneracy at the Brillouin zone boundary. One can readily understood this from the fact that at $k_y a_0 = \pi$, Eqn. (4) reduces to

$$E_{1D}^{A} = \pm\gamma_0$$

The second and more important feature of Fig. 8.15 (*a*), (*b*) and (*c*) in terms of electronic properties of armchair nanotubes is that the highest valence band and lowest conduction band are degenerate at $k_y = \pm\ 2\pi/(3a_0)$ and each cross the Fermi energy ($E = 0$) at that point. This is true of all armchair nanotubes –thus all armchair nanotubes be metallic.

We must note that Eqns. (2) to (4) represent taking a 1D slice in a direction given by

$$E(\theta) = -KV\cos^2\theta$$

where K is the magnetic anisotropy constant, V the volume of the particle, θ the angle between the magnetisation vector and an easy direction of magnetisation and $E(\theta)$ is the magnetic anisotropy energy for a uniaxial particle, through the 2 D graphite band structure. The crossing at $E = 0$ in Fig. 8.16 occurs because the corresponding 2 D graphite band cross at the K points (*i.e.* the corners) of the 2 D Brillouin Zone. For armchair tubes, there is always an allowed K-vector that goes through K point of the 2 D lattice. However, this alignment of the tube wave vector and K-point of the graphite Brillouin zone does not automatically occur for zig-zag or chiral tubes and hence these can have semiconducting character. Only for $n - m = 31$ there is an alignment of an allowed **k**-vector and a K-point of the 2D Brillouin Zone, giving rise to metallic behaviour.

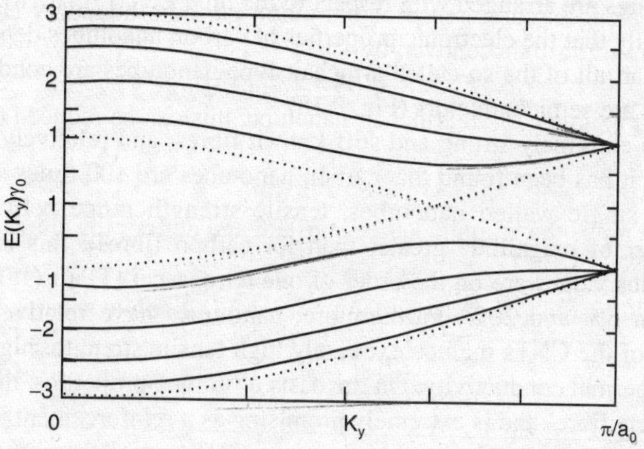

Fig. 8.16 One dimensional dispersion relation for an armchair (5, 5) nanotube.

The *nano* in the nanotube prefix denotes that tube diameters are of the order of a nanometer (*i.e.*, 100 nm or less). Each nanotube is a single molecule composed of millions of atoms; the length of these molecules is much greater (on the order of the thousands of times greater) than its diameter.

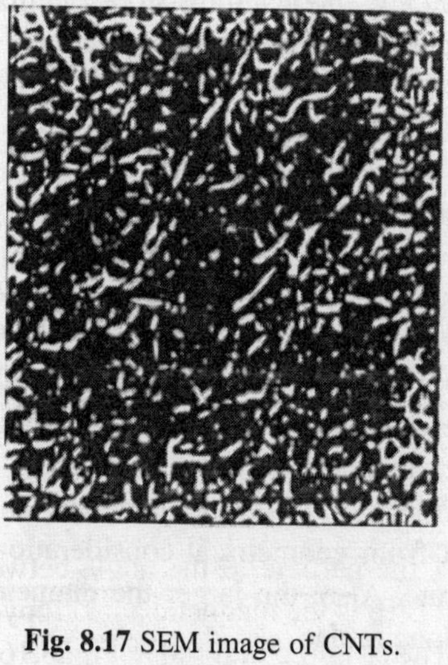

Fig. 8.17 SEM image of CNTs.

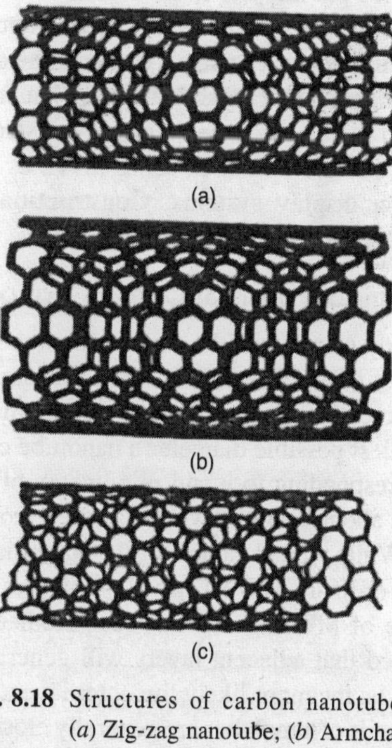

Fig. 8.18 Structures of carbon nanotubes (*a*) Zig-zag nanotube; (*b*) Armchair nanotube; and (*c*) Chiral or helical nanotube.

According to number of layers, carbon nanotubes (CNTs) can be **single-walled** and **multi-walled** (MWNTs). In multi-walled nanotubes, more than one CNTs are coaxially arranged (Fig. 8.17). Iijima realized that graphite could be bent to form MWCNTs with different helicities or chiralities which refer to the way hexagonal rings are arranged with respect to the tube axis (Fig. 8.18).

It is found theoretically that the electronic properties of carbon nanotubes depended on the diameter and helicity: in particular, all of the so-called armchair type nanotubes are conductors (Fig. 8.18) and most zig-zag nanotubes are semiconductors (Fig. 8.18).

These nanotubes are extremely strong and stiff known fibres, and relatively ductile. According to indirect measurements, it has been found that carbon nanotubes are 100 times stronger than steel and six times lighter. For single walled nanotubes, tensile strength range between 50 and 200 GPa (approximately an order of magnitude greater than for carbon fibres); this is the strongest known material. Elastic modulus values are on the order of one tetrapascal [TPa : (1 TPa = 103 GPa)], with fracture strain between 5% and 20%. Furthermore, nanotubes have relatively low densities. The outstanding properties of the CNTs include extremely high tensile strength, high modulus, flexibility, light weight and high thermal conductivity. On the basis of unique properties, the carbon nanotube has been termed the 'ultimate fibre' and is extremely promising as a reinforcement in composite materials. The high stiffness, aspect ratio, and nanoscale tip make CNTs excellent probes for sensing, testing, manipulation, and fabrication on the nanoscale.

Carbon nanotubes also have unique structure sensitive electrical characteristics. Depending on the orientation of the hexagonal units in the graphite plane, *i.e.*, tube well, with the tube axis, the nanotube may behave electrically as either a metal or a semiconductor. Semiconducting and metallic nanotubes recently are put into use in field effect transistors (FET) and single electron transistors. The flat-panel and full-colour displays, *i.e.* TV and computer monitors, have been fabricated using carbon nanotubes as field emitters. These displays should be cheaper to produce and will have lower power requirements than CRT and liquid crystal displays. The mechanical strength of CNT is amazingly 600 times tougher than steel, which finds applications in micro-electro-mechanical systems (MEMSs) and in aerospace. Progress has been made in using CNTs as field emission devices *e.g.* diodes, transistors, etc. for high resolution display systems. Construction of nanoprobes and electrodes in biological and other applications are also being aimed.

8.9.1 Various Forms of Carbon Nanotubes

As already mentioned, the carbon nanotube is a generic term that includes both single and multiwalled nanotubes (SWNTs and MWNTs). The smallest nanotube reported to date has a diameter of only 4 Å. Since increasing the curvature of the graphite wall increases the strain energy, this is also predicted to be the smallest possible diameter a nanotube can have. Typical diameters for SWNTs are between 1 and 2 nm, corresponding to n and m numbers of around 10. In MWNTs, the distance between the adjacent layers is about 0.34 nm, which is much more than the atomic separation (1.42 Å) within the individual layers. While typical values for the outer diameter are 10–20 nm, the inner diameter is typically 3–5 nm. Thus an ordinary MWNT has 10–30 layers, each of which are SWNTs of different diameters. The shell structure of MWNTs has not been studied thoroughly yet, but from geometrical considerations it is concluded that adjacent layers will generally be incommensurate. Also, the larger the diameter of a MWNT is, the more likely the outer layers are disordered and noncircular.

The ends of nanotubes are generally closed in the growth process. While the wall of a carbon nanotube is made up of an exclusively hexagonal pattern, pentagons are needed to cap the tubes. More generally, pentagonal and heptagonal (5– and 7–corpered polygons, respectively) defects enable the graphitic sheet to take up more complicated structures than a simple tube, as we will discuss on the following page.

Unless special set-ups are used for the growth of SWNTs, they are usually assembled into ropes by their mutual van der waals attraction. For example, a rope of a typical diameter of about 10 nm contains ~100 SWNTs. Since multi-walled nanotubes have a much higher bending stiffness, they do not arrange similarly.

Nanotubes grown under suitable conditions have a very low concentration of defects over mm distances, that is, over hundreds or even thousands of interatomic spacings. However, if one were able to control the occurrence of defects, very useful and interesting nanotube structures would emerge. We will next show what kind of modified SWNTs and MWNTs have already been observed due to the occurrence of defects within the nanotubes. A single, stable defect structure in the hexagonal lattice of a graphite sheet is made up of a pentagon-heptagon pair, that is 5- and 7-sided polygons. One such defect can cause a sharp bend in a SWNT. Defective nanotubes are especially interesting for electronic applications where the defect site may act as a tunnelling barrier. In the case of MWNTs, highly defective tubes may be obtained with certain synthesis conditions, for example with relatively low growth temperatures. While high-quality MWNTs are very straight and stiff, very defective ones have a continuous and smooth curvature. On the other hand, the curvature can be highly regular and so result in helices. An SWNT, or a single shell of a MWNT, can have a second nanotube branching out. The beginning point of the branch has similarly to the end sections pentagons dispersed among the hexagons to accommodate the necessary curvature. Such branching has been observed in a few cases to date.

As mentioned earlier, the ends of carbon nanotubes are usually closed. It is, however, possible to open nanotubes by a suitable chemical treatment. This is due to the stronger chemical reactivity of the end section with its larger curvature and pentagon structures. The opened nanotubes can be utilized for creating yet another side branch in the science of nanotubes, namely filled tubes. Both SWNTs and MWNTs have been filled with various materials, such as fullerenes, simple metals, and molecular compounds. Nanotubes with fullerenes inside are called peapods, and are presently intensively investigated. The study of the electronic properties of filled nanotubes is still in its infancy.

8.9.2 Chemistry of Carbon Nanotubes

The chemistry of carbon nanotubes, which encompasses the aspects of purification and functionalisation, is a very important field for tailoring their structural and electronic properties and thus for the development of future applications. Purification aims at removing the nontubular material in raw nanotube samples. Several techniques, more or less complex, have been proposed, including solvent treatment followed by ultrafiltration, flocculation using aqueous surfactants, oxidation and acid washing coupled with centrifugation, resuspension in surfactant solution, and cross-flow filtration steps. These methods however lead to a mixture of nanotubes of differing lengths and still containing potential contamination with nontubular material. Polymer suspensions and chromatographic purification are useful to remove nontubular material. Field-flow fractionation on purified, shortened nanotubes and size-exclusion chromatography on raw nanotubes have demonstrated some success at producing fractions separated by length. Capillary electrophoresis was reported to allow separation based on tube length.

Several methods for attachment of oxygen-containing groups to the nanotube surfaces have been developed. Carboxylic carbonyl and hydroxyl groups were found covering the nanotube walls. The concentration of acid groups on the surfaces of HNO_3 treated nanotubes was higher than that found for graphite treated under the same conditions. Sonication of nanotubes prior to the functionalisation in acids increases the concentration of acid groups. Oxidation of nanotubes with a $H_2SO_4 - HNO_3$ mixture also leads to a higher concentration of functional groups on the surface. Long-term acid treatment of nanotubes also induces modifications. Functional groups can be removed from the surface of nanotubes by healing. It should be noted that SWNTs obtained by different methods exhibit different behaviour in the course of acid treatment and subsequent oxidation. The sensitivity of SWNTs to chemical processing was investigated by TEM.

To enhance the possibilities of atomic force microscopes with nanotube probe tips, the tip ends were functionalized using both solutions and gases. Modification of tips using a discharge in N_2 results in the formation of nitrogen-containing heterocycles at the tip ends. High degree of saturation of nanotube tips with functional groups is achieved by treatment with a low-pressure ammonia plasma followed by oxidation with a $NaClO_3$ solution.

The reactions with silicon of individual nanotubes or nanotube bundles at 970°C in ultrahigh vacuum resulted in the formation of Si-β-SiC (nanorod) nanotube heterojunctions.

8.9.3 Oriented or Localized Growth

The term "oriented" is used here to describe the formation of large quantities of CNTs aligned perpendicularly to the substrate surface, usually MWNTs of controlled dimensions (diameter 20–400 nm, length 0.1–60 μm). The key steps are substrate patterning, catalyst deposition, and the appropriate use of different CVD routes.

The term "localized" describes the self-assembled nanointerconnections made by CNTs between islands on prepatterned substrate. The directed growth of SWNTs parallel to the plane of a silicon substrate was reported. The SWNTs are suspended bridges grown from catalyst material placed on top of regularly patterned silicon tower structures. The key synthetic step in this approach involves developing a series of liquid-phase catalyst precursor materials, allowing for uniform film formation and large-scale catalyst patterning. Contact printing techniques were used to selectively deposit catalyst precursors on top of tac silicon tower arrays. Calcination led to the formation of catalyst particles confined on the tower tops and subsequent CVD (CH_3. 900° C, 20 min) growth yielded SWNTs emanating from the towers. Directed free-standing SWNT networks were formed by nanotubes growing to adjacent towers and becoming suspended above the surface. It is noted that a rational design of tower arrangements could lead to a variety directed SWNT architectures. The liquid phase catalyst precursors consist of three general components: inorganic chloride precursors, a removable triblock copolymer serving as the structure-directing agent for the chlorides, and an appropriate alcohol for dissolution of the inorganic and polymer compounds. The increase in yield was achieved through a so-called conditioning step that uses bulk amounts of catalysts placed near (upstream) the catalyst-patterned substrate in the CVD growth environment.

8.9.4 Electro-mechanical Properties

Along with their electronic properties, carbon nanotubes have been praised for their mechanical properties. The praise is based on high strength, flexibility, and large aspect ratio. A carbon nanotube is a freestanding conductor and its mechanical properties are highly relevant for the electronic properties and the potential technological applications of the tube. For example, nanotube-based mechanical oscillators may have applications in high-frequency electronic circuits, and carbon nanotubes are already used in scanning "probe microscopy both as STM and AFM tips. Thin strong and durable nanotubes come extremely close to an ideal scanning tip.

Mechanical properties of carbon nanotubes are closely related to the properties of a graphite sheet, but the tubular anisotropic form affects the mechanical behaviour. The basis is the graphite sp^2 bond, which is the strongest of chemical bonds. In nanotubes the overall density of defects can be extremely low depending on the synthesising method and prevailing synthesising parameters. This has led to predictions of a very high axial strength.

8.9.5 Basic Mechanical Properties

Theoretical calculations of Young's modulus for individual SWNTs centre around 1 TPa or slightly higher, but values as high as 5.5 TPa have been presented. The spread is due to different interaction models and, also to differing values of nanotube wall thickness (if an isolated nanotube is not a well-defined quantity). Most of the theoretical attention has been on SWNTs because modelling the interlayer interaction in MWNTs is a complicated matter. Young's modulus values for MWNTs as well as SWNTs range from 0.97 TPa to 1.11 TPa with the value increasing slightly with the number of layers.

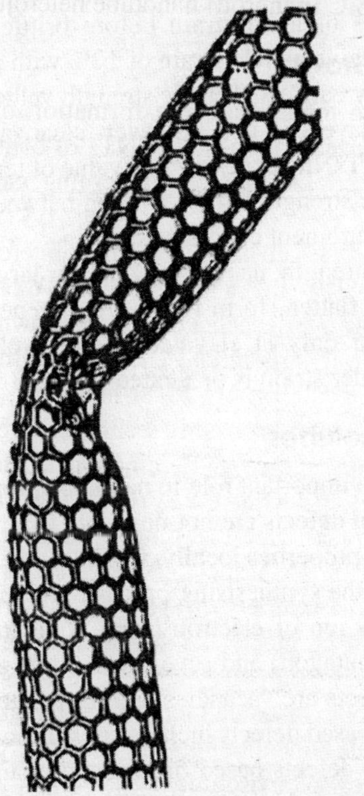

Fig. 8.19 A molecular dynamics simulations based example of nanotube behaviour under strain. The image shows portion of a (14, 0) nanotube of 2632 atoms (tube length is 20 nm) bent 36 degrees with constant bending rate in 300 K temperature and then, to enhance structural reconstruction, annealed in an elevated 3300K temperature for a sequence of three 130 ps annealing and cooling simulation runs. In the bending process the tube buckled but remained hexagonally bonded. The defects visible in the image relax the strain and are a result of the third annealing.

The small size of carbon nanotubes presents challenges also for experimental characterisation. Nevertheless measurements have been performed. The first Young's modulus measurement related thermal vibration amplitudes of MWNTs to their Young's modulus and obtained an average value of 1.8 TPa with a large spread. After that, with AFM techniques, values such as 1.28 ± 0.59 TPa for arc-discharge-produced tubes have been obtained. The latest measuring technique is due to Poncharal group, who induced vibrations on MWNTs by alternating potential and have measured the vibration frequencies. Young's modulus values between 0.7 TPa and 1.3 TPa for tubes with a diameter less than 12 nm and between 0.1 TPa and 0.3 TPa for thicker tubes were reported. The drop is explained by an onset of a wave-like bending mode of the nanotube wall at the inner arc of the bending. Measuring SWNTs is more complicated than measuring MWNTs due to their small diameters and the tendency to bundle. Measurement of individual SWNTs using the thermal vibrations method provide an average value of 1.25 TPa. Although the current measurements suffer from inaccuracies due to vibration amplitude measurement and assumptions made on AFM tip characteristics, the current-agreement is that defect free nanotubes, both SWNTs and MWNTs, have a Young's modulus value around slightly above 1 TPa, which is extremely high and sets nanotubes as the strongest known material albeit challenged by other nanotubular structures such as boron nitride tubes.

Theoretically, nanotube tensile strength is high, and this is supported by calculations in which SWNTs support as high as 30% of axial strain before brittle failure and by more recent kinetic activation based calculations that give a yield strain of 17% with a chirality and temperature-dependent defect formation activation energy barriers. Tensile strength values ranging from 11 GPa to 63 GPa are reported. The role of the inner layers is not, however, clear, and the failure process may be more complicated. For individual SWNTs, the experimental value of tensile strength is still an open question, but for bundles of SWNTs tensile strength values ranging between a few GPa and several tens of GPa depending on the bundle an measurement characteristic.

As a result of the high tensile strength, nanotubes can bear large strain before plastic deformation or brittle failure. The tubes buckle, flatten, form ripples, and generally deform under strain but plastic deformation and relaxation occur only at elevated strain levels and temperatures. An example of simulated nanotube behaviour under strain is presented in Fig. 8.19.

8.9.6 Defects in Carbon Nanotubes

As in any material, defects play an important role in nanotube properties. Structurally, defects make the tubeless strong and thus in general defects are not desirable from the purely mechanical point of view. However, they alter the electronic properties locally, which can be utilized in the creation of single-tube devices. Defects are generated in the synthesising process, and they can also be caused by mechanical manipulation, or, for example, by ion or electron beam bombardment of the tube. The most typical structural defects are five-fold (pentagon) and seven-fold (heptagon) rings in the six-fold (hexagonal) lattice. Other types of typical defects are vacancies and miscellaneous bonding configurations such as amorphous diamond. Noncarbon based defects include substitutional atoms or atom groups. In addition to these, MWNTs exhibit diverse defects based on discontinuous inner layers. Defects may alter the tube form a straight tube to a bulging, kinked, spiral, or even more miscellaneous form. Defects and deformations induce scattering and localisation, which result in conductivity drops, symmetry breaking based gap opening, and conductivity barriers. The scattering effects of isolated point defects are averaged over the circumference, and therefore their influence decreases with an increasing diameter and is small for all but the very thinnest tubes. However, if the defects or deformations accumulate locally, a tunnelling barrier may form. AFM manipulation has been used to make sharp bends in SWNTs, which has resulted in the formation of tunnelling barriers. The created changes are not, however, necessarily permanent. Nanotube defect manipulation offers interesting prospects both for devices with permanent characteristics and for devices with tunable properties, but a breakthrough requires more accurate and flexible mass manipulation of tube properties than is currently available.

8.9.7 Carbon Nanotubes as Mechanical Oscillators

Due to their small size, combined with a low mass density and high Young's modulus, nanotubes are predicted to have characteristic vibration frequencies in the gigahertz regime, which is extremely high for a mechanical vibrator. Such frequencies are not attainable with the current silicon-based micro-mechanical oscillator production techniques with which eigen frequencies in the 10 MHz to 100 MHz regime can be attained. A first approximation for the fundamental modes of a nanotube vibrator can be obtained from elasticity theory and is for a thin cylindrical transversally vibrating rod that is clamped from one end

$$f_1 = 0.28\frac{d_1}{L^2}\sqrt{\frac{Y}{\rho}} \quad ...(5)$$

The values for tube diameter d_1 = 1 nm, Young's modulus Y = 1 TPa, and density ρ = 750 kg/m^3 are used. The mass density of a SWNT is not a well-defined quantity; the value presented here is a rough estimate based on the nanotube atomic spacial 'density and the volume encompassed by a tubular cylinder depicting the tube of 1 nm radius, to compare with graphite mass density 1900–2300 kg/m^3, a frequency of 10 MHz is obtained for a tube length of 1 μm and a frequency of 1 GHz for a length of 100 nm. For a rod clamped from both ends, the constant in front of the equation is altered according to the boundary conditions, but the overall dependency on d_1, L, Y, and ρ remains. Based on the discussion above, nanotubes, and especially SWNTs, are capable of significantly smaller dimensions and higher frequencies than conventional silicon-based vibrators.

Currently, only a few studies of excited mechanical vibrations in nanotubes exist. Pancharal group applied alternating potentials to MWNTs and measured Young's modulus and Roulet group have performed a basic study of acoustoelectric coupling in ropes of SWNTs, and they reported dependence on the excitation power, either electron heating or phase coherence breaking is observed.

There are many things in the mechanical properties that have not been measured, and the measurements that exist suffer from a degree of inaccuracy. Most theory concentrates on SWNTs, and although the SWNTs often show more clearly the nanotube characteristic and desirable electronic properties, the theoretical aspects due to multiple layers still have many questions to answer, as do defectous-structure related matters. However, in nanoscale science and technology, electronic and mechanical behaviour will become more intimately connected, and obviously, the electro-mechanical properties of carbon nanotubes will play an important role.

8.9.8 Electronic Structure and Properties

Tight Binding Model: The carbon atoms in nanotubes are sp^2-hybridised, thus each atom contributes one unpaired π-electron to the tube. Carbon nanotubes are electronically active materials due to these π-electrons. The basic electronic properties of the carbon nanotubes are calculated by extending the tight-binding approach of $2D$ graphite to the geometrical configuration of the $1D$ nanotube system. The crystal momentum vector perpendicular to the nanotube axis is quantised due to the periodic boundary condition imposed by the finite circumference of the nanotube. This relatively simple model has been proven, mainly by low temperature STM spectroscopy, to explain well the basic electronic structure of the carbon nanotube. According to this model, the chirality of the carbon nanotube determines its division into metallic and semiconducting tubes. In the model, the curvature of graphitic wall of the nanotube is ignored. This has lately been shown to bring about important modifications to the model. Still the model survives well enough to continue to serve as a basis for classification of the nanotubes into metallic and semiconducting ones.

A brief presentation of the basic tight binding model of the single shell carbon nanotube is as follows. Due to the tubular $1D$ structure, the nanotube has N separate $1D$ energy bands. In a general form these are derived from the graphite energy levels by the general formula

$$E_\mu(k) = E_{g2D}\left(k\frac{K_2}{|K_2|} + \mu K_1\right)\left(\mu = 0, \dots N-1, \quad -\frac{\pi}{T} < k < \frac{\pi}{T}\right) \quad \dots(6)$$

Where E_{g2D} denotes the energy eigenvalues of a *2D* graphite sheet, K_1 and K_2 are reciprocal lattice vectors corresponding to *C* and *T* respectively. Figure 8.20 shows how the carbon nanotube *k*-vectors are projected to the graphite momentum space lattice. In graphite the conduction and valence bands meet at the so-called *K*-points in the momentum space, which is responsible for its semimetallic behaviour. In the tight-binding picture, the metallicity of a carbon nanotube is determined by whether the *1D* lines of allowed *K*-vectors run through these points.

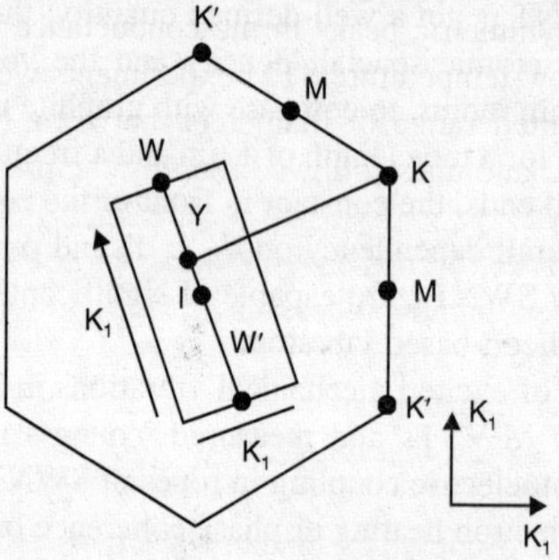

Fig. 8.20 Allowed *K*-vector or carbon nanotubes are formed by the vectors K_1 and K_2 in the reciprocal lattice space of graphite. The nanotube is metallic if the ratio of *YK* to K_1 is an integer.

In an energy diagram, whether the energy bands will cross the Fermi energy or leave an energy gap depends on the *(n, m)* that is the chirality, as follows: The armchair nanotubes, with indices *(n, n)*, have two bands that cross each other and the Fermi energy at the points $k = \pm 2\pi/3a$, they are thus always metallic, while only some of the zig-zag tubes are so. The general condition for the metallicity of a nanotube is that the factor $2n + m$ is a multiple of 3, a condition that indeed is fulfilled for the armchair tubes. The theory also predicts that one-third of all nanotubes are metallic while two-thirds are semiconducting. These rates of occurrence have not been confirmed by the presently performed experiments, which however is not surprising considering doping effects and the fact that almost nothing is known about the possible preference in the growth processes for certain nanotube symmetries over others.

As a general result, in all metallic *1D* systems with half-filled bands, the so-called Peierls distortion will, by an atomic displacement, change the unit cell so that the band becomes filled and a gap opens up at the Fermi energy. The phenomenon is well known among conducting polymers, charge transfer salts, etc. In carbon nanotubes, however, it has been shown theoretically that the effect very quickly diminishes as a function of nanotube radius.

8.9.9 Experimental Studies: STM Imaging and Spectroscopy

The electronic structure of the SWNT that is predicted by the tight-binding calculations described above was put to test at first primarily by scanning tunnelling spectroscopy, done with low temperature STM. In such experiments, it is possible to image the nanotube with atomic resolution along the length of the tube and with a width of a few unit cells. The width of the imaging is of course limited by the curvature of the nanotube. This imaging reveals (in the best cases) the structure and diameter of the nanotube, enabling one to determine the (*n, m*) indices of the nanotube with some confidence. Furthermore, the STM tip be positioned at any point along the nanotube, separated by a tunnelling distance from the nanotube surface, and spectroscopic measurements be performed by taking the current-voltage characteristics. The I-V characteristics reveal the metallic or semiconducting character of the tube since the measured current is proportional to its density of states (DOS). This experimental technique thus addresses the fundamental question of the relation between the structure and the basic electronic properties of the individual nanotube. As seen by STM spectroscopy, in the zero-bias region the metallic tube has a finite conductance while the semiconducting tube has zero, corresponding to the basic Fermi surface properties. At higher voltages asymmetric peaks in the conductance appear. These are the Van Hove peaks due to the 1*D* DOS. The low-temperature STM experiments were carried out with SWNTs, and thus they cover only a very limited range of diameters, between 1 and 2 nm. For this range, the measured E_g for semiconducting nanotubes-followed roughly the predicted $1/d_1$ dependence, being in the range 0.3–0.7 eV.

In recent scanning tunnelling spectroscopy measurements on SWNTs, a curvature-induced "extra" gap was found in zig-zag type metallic nanotubes, but not in tubes of armchair type. This gap is also inversely dependent on the nanotube radius. The measured results were theoretically anticipated in refined calculations. The curvature-induced gap has a size of a few 1 0s of meV for subnanometer tube diameters. From this energy scale we see that for all but the very smallest of possible tubes, the effect is washed out at room temperature.

Other scanning tunnelling spectroscopy experiments have been performed that explore other topics such as individual quantum states in short SWNTs and MWNTs doped nanotubes crossing nanotubes, defects and junctions in SWNTs and peapods (SWNTs filled with C_{60}).

8.9.10 Experimental Studies: Raman Spectroscopy

Another important tool for nanotube characterisation is Raman spectroscopy. Particularly, the strong, so-called radial breathing mode is derived from the tubular structure of the nanotube and is absent from, all other 2*D* graphitic forms of carbon. It is widely utilized to give a measure on the nanotube diameter. Due to the singularities in the DOS of the 1*D* nanotubes and the electron-phonon coupling, very strong resonant Raman effects are observable in studies on carbon nanotubes. This enables the use of Raman spectroscopy to investigate the electronic structure of carbon nanotubes, particularly the occurrence of semiconducting and metallic properties. A possibly very important recent development is the ability to measure the Raman signal from individual nanotubes with the so-called micro-Raman technique. It was claimed to be possible to determine the (*n, m*) indices of the isolated nanotube with this technique. This, of course, opens up the possibility of transport measurements on a nanotube of known symmetry.

8.9.11 Optical Properties

The optical physics of nanotubes has not moved forward in pace with the other developments in the field. This is due to unability to make samples consisting of nanotubes of a single symmetry. Thus all optical signals are a mixture of signals from a very large variety of nanotubes. Furthermore, since nanotubes tend to aggregate together, especially SWNTs, any signal from semiconducting tubes is easily hidden by interaction with neighbouring metallic tubes. Unlike transport measurements, optical measurements on single nanotubes are experimentally less feasible. On the other hand, however, the 1*D* singularities in the optical response of nanotubes give strong signals, just as in the case of Raman spectroscopy. Also, if aligned assemblies, of nanotubes can be produced, the optical anisotropy of such a system is potentially huge. In a recent development, SWNTs were processed by sonication and surface-active chemicals to produce solutions of truly individually solvated tubes, coated with the surface-active hydrocarbon molecules. These SWNTs were rather short (due to the strong sonication), with an average length of only 130 nm. Perhaps for the first time, photoexcited luminescence from nanotube solutions were observed in this case. The luminescence spectrum was closely associated with the measured absorption spectrum, with only a minor red-shift. This work is intensively being pursued and is likely significantly widen research in carbon nanotube systems. Photoconductivity measurements on single nanotubes have so far not been successful.

The research on the optical properties of carbon nanotube may in the near future bring forward a number of potential applications within optoelectroates, thus adding to the list that already exists within other fields of nanotube-based nanotechnology.

8.9.12 Electronic Transport Properties

Within the realm of electronic transport physics, the carbon nanotube offers exciting possibilities both as a testing ground in basic research as well as in nanotechnology. In the case of the former, the near perfection in structure for lengths exceeding 1 μm makes the nanotube the best concrete example of a 1*D* quantum wire. On the other hand, in nanotechnology the prospect of devices made from a single nanotube, that is, intrananotube devices, offer the possibility of truly molecular-level electronics. Even if molecular electronics is eventually pursued with smaller molecules, carbon nanotubes may be used as connecting wires between metal structures of ~100 nm size and molecules of ~1 nm size.

Although the carbon nanotube is mainly "marketed" as a perfect 1*D* object, this is not always the case. In fact, depending on the circumstances one may have to treat the nanotube as anything from a 0*D* to a 3*D* object. Obviously, if a SWNT is short enough, it becomes a 0*D* object, or a quantum dot. If only the outer shell of a MWNT is considered, and the diameter of the tube is large enough with respect to some critical length scale of the particular situation, this outer shell becomes a 2*D* object. Finally, if we again consider a larger MWNT (in diameter), and the charge carrier coupling between the layers is strong enough, then we may have at hand an anisotropic 3*D* object. We can thus see that even without considering nanotubes of different chiralities and more elaborate topologies than the simple tubular one, the subject of electron transport in carbon nanotubes can be rather complicated.

As has been discussed, the energy gap in semiconducting SWNTs is ~ 0.5 eV, which is high enough with a good margin for the nanotube to function at room temperature as the active component *e.g.*, in field effect transistors. Other effects, such as the energy level quantization in short nanotubes (L = 10–100 nm), require cryogenic temperatures for the effects to be observable. Roughly, it can be said that in metallic nanotubes the interesting transport phenomena occur at low temperatures, while in semiconducting tubes much of the work is carried out at room temperature.

8.9.13 Bulk Transport

In this section, we review the work on the transport physics of single SWNTs, SWNT ropes, and MWNTs. This is clearly more significant than the subject of transport in macroscopic amounts of carbon nanotubes. However, the carbon nanotube may offer interesting applications as the conductive component in composites, when mixed together with an insulating host material. In order to have a composite conducting, the volume fraction of the conductive component has to exceed some critical value. Typically, tile conductive material consists μm-sized particles of a more or less rounded shape. For conduction to occur, the particles have to touch each other frequently enough so that the conductive channels are formed over macroscopic distances. As prescribed by percolation theory, this occurs at a volume fraction of 16% whereby the conductivity of the composite rises very sharply (as a function of filling percentage) with several orders of magnitude. However, it has already been demonstrated with conductive material consisting of 1D objects, for example fibrils of polyaniline a conducting polymer, that there is no apparent sharp threshold for the onset of conduction, and more importantly, the composite becomes conducting at a much lower volume fraction of the conductive component. Obviously, this should also be achievable with nanotubes as the conductive components. In fact nanotube-based composite systems are already being considered for commercial applications.

8.9.14 Carbon Nanocones

Theoretical studies predicted the formation of graphitic cones. Subsequently, isolated graphitic cones were produced by carbon condensation on a graphite substrate and by pyrolysis of heavy oils. More recently, single walled aggregates of conical graphitic structures have been prepared by laser ablation of graphite targets. In addition, conical structures consisting of other layered materials such as BN have also been prepared by reacting boron oxide vapours with multi-walled CNTs.

Recently, it is reported that pyrolysis of palladium precursors, always produce conical nanofibres. An important feature of these new nanostructures is that they are held together by van der Waals forces since the fibre is composed of an arrangement of stacked cones which can be opened (lamp-shade), or closed. It is believed that nanocones may be good electron field emitters. Calculations of the electronic properties of nanocones reveal that there is a charge accumulation towards the tip and that there are localized states near the Fermi level. These features make them suitable as field emitters.

8.9.15 Haeckelites

The presence of defects such as pentagons and heptagons in fullerenes change the electronic properties. A new hypothetical type of graphene sheets has been proposed which admit pentagons, heptagons, and hexagons, observing that the number of heptagons and pentagons should be the same in order to compensate the negative curvature of the heptagons and the positive curvature of pentagons (Fig. 8.21). These arrangements have been named as haeckelites. They all exhibit a considerable signal at the Fermi level in the local density of states. Obviously, all these systems should possess a metallic behaviour. Thus, it is possible to roll up haeckelites to form nanotubes, which will be conductors independent of the diameter and the helical arrangement.

Another property of haeckelite tubes is that they keep the stiffness of classical graphite nanotubes composed with just hexagons; the Young's modulus of haeckelite tube is around 1.0 Tera Pa; of the same order as classical carbon nanotubes. Haeckelites also exhibit local rugosity due to the local bending produced by heptagons and pentagons. It might be possible, that electron irradiation experiments can generate heptagon-pentagon pairs (rugosity) in SWCNT.

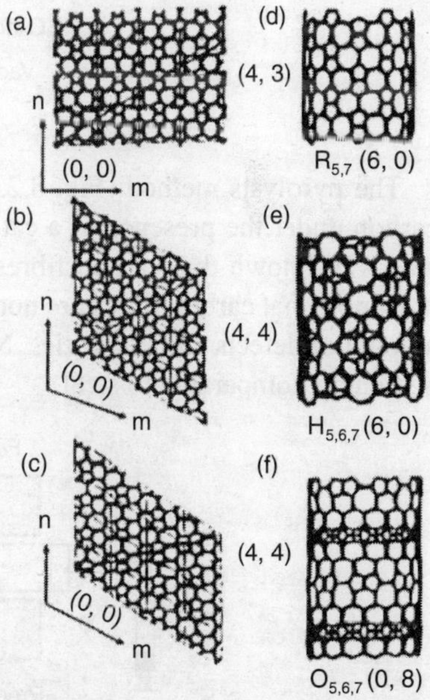

Fig. 8.21 Haeckelites (*a*) Rectangular haeckelite sheet (R); (*b*) Hexagonal haeckelite sheet (H); (*c*) Oblique haecklite sheet (0); (*d*) to (*f*) are the corresponding haeckel tubes. All haeckelites are metallic (sheets and tubes).

8.9.16 Production Methods for CNTs

CNTs can be obtained by using four different methods: (*i*) arc-discharge, (*ii*) pyrolysis, (*iii*) laser vapourisation and (*iv*) electrolysis.

The electric arc-discharge set-up is shown in Fig. 8.22. This method is very similar to the one used for obtaining fullerenes with two main differences: first, the pressure are higher, around 500 torr (for fullerenes the pressure is around 100 torr), and second, the nanotubes are found deposited in the cathode and not in the soot. This method produces well ordered MWCNT with diameters from 2 nm up to 30 nm. The separation between graphite layers is around 3.4 Å, which is very close to that of graphite 3.35 Å and corresponds to turbostratric graphite. The length of these nanotubes can be up to 30μ. The optimum conditions for generating nanotubes with this method consist of using currents of 150 amps, voltages of 25 volts and graphite electrodes between 6 mm and 8 mm in diameter with a separation between them of 1 mm under helium atmosphere. The electric arc reaction is too violent, so it is very difficult to control the formation reaction.

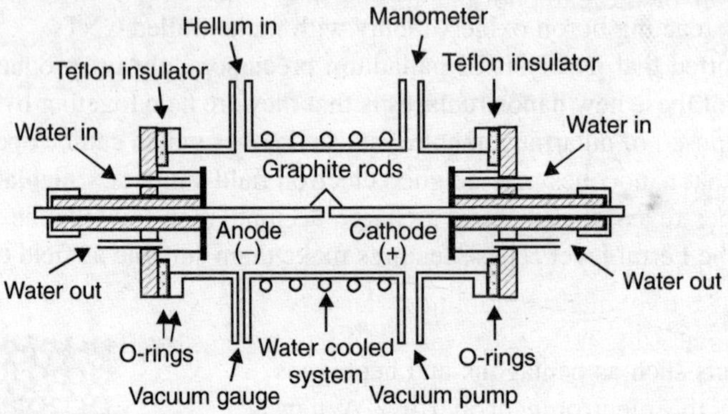

Fig. 8.22 Arc-discharge set-up.

The pyrolysis method (Fig. 8.23) consists of heating a hydrocarbon or a precursor which contains carbon under the presence of a catalyst such as nickel, cobalt or iron. Prior to the discovery of C_{60}, it was well-known that carbon fibres could be obtained by pyrolytic methods using hydrocarbons. We must note that carbon fibres are not nanotubes since the fibres are bigger in diameter and exhibit a great amount of defects and impurities. Nevertheless, the application of carbon fibres is widespread, showing a strength comparable to steel.

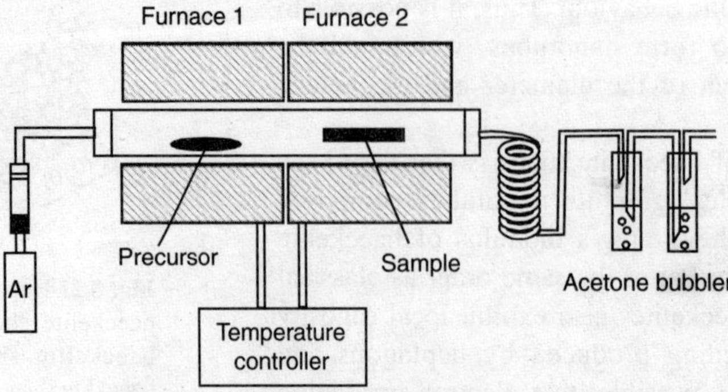

Fig. 8.23 Pyrolysis set-up.

In 1994, Ajayan group found that carbon nanotubes could be aligned by embedding them in a polymer, thus forming a composite material. When the polymer was cut, they observed an aligned arrangement of nanotubes. In this context, a better alternative is to use laser ablation of cobalt thin films and then pyrolyse them with an organic precursor such as 2-amino 4, 6-dichloro-5-tryazine. The nanotubes obtained in this way are aligned in bundles which can grow up to 40μ and diameters between 30 nm and 50 nm.

The laser vapourisation method involves firing a high power laser towards a graphite target inside a furnace at 1200° C. The condensation of material generated by the laser is responsible for the nanotube formation. If we add nickel or cobalt to the graphite target, we obtain single-walled carbon nanotubes (SWCNT). The SWCNT show diameters around 14 Å forming a two-dimensional crystal with a lattice constant of 17 Å. Arrangements of SWCNT can also be obtained by electric arc-discharge with mixtures of carbon, nickel and ytrium.

The electrolysis method is based on using graphite electrodes inside a molten salt such as lithium chloride (700° C) under an argon atmosphere. Depending on conditions 20% to 30% consists of nanotubes. The depth of the cathode and the current (3–10 amps) play an important role in the formation of nanotubes.

8.9.17 Boron Nitride Curved Nanostructures (BNNTs)

Hexagonal boron nitride (h-BN) is another layered material with the structure very similar to graphite having hexagonal ring layers separated by 3.33 Å, in which every boron is connected to three nitrogen atoms and *vice-versa;* and B-N distance is 1.44 Å. Between the layers, every boron interacts with a nitrogen atom through a van der Waals force. Regarding the electronic properties, h-BN is an insulator with a band gap of 5.8 eV, whereas graphite is a semi-metal (band overlap = 0.04 eV). The first BN nanotube were prepared using the electric arc-discharge with a high melting point metal (tantalum and tungsten for example) electrodes and BN powder inside the anode. In 1996, Loiseau and coworkers reported the production of BN nanotubes by acting HfB_2 electrodes in an inert atmosphere. This technique lead to the formation of mainly single- and double-walled BN tubes.

It was reported that in BN nanotubes the tips were squared which means that instead of having pentagonal rings to close the structure there are squares of BN. This makes a sense because a pentagon will exhibit B-B or N-N bonds, which will not be very stable, changing also the proportions of B and N. Therefore, BN cages should possess squares. In terms of curvature, BN admits, with high stability, squares for the positive curvature and octagons for the negative curvature. From functional tight binding calculations, it is found that nitrogen excess cages with B and N are stable. Theoretical studies revealed that BNNTs are semiconductors with a band gap of 5.5 eV, smaller than in bulk BN. This property of BNNTs is independent of tube diameter, chirality, and number of walls. These BNNTs may also be extremely resistant to oxidation and should exhibit high Young's modulii. Therefore, the BNNTs may find important uses in ceramics.

BNNTs bundles can also be produced when carbon nanotubes are exposed to mixture of B_2O_3, V_2O_5, and MoO_3 in a nitrogen atmosphere at elevated temperatures. There are alternative approaches to generate BNNTs and other BN nanostructures: (*a*) thermolysis of borazine ($B_3N_6H_6$) over nickel boride catalyst particles at 1000–1100°C, (*b*) thermal annealing of powdered rhornbohedral boron and h-BN mixtures at 1200°C in lithium vapour, (*c*) plasma jets and (*d*) ball milling of h-BN, followed by thermal annealing at 1300°C.

At this stage, novel composites and nanoscale components need to be fabricated employing BNNTs, which appear to be extremely resistant to oxidation. Experimentally obtained values of Young's modulus of BNNTs (Y or $E = 1.22 \pm 0.24$ TPa) agrees well with the values obtained using theoretical approaches, and reveals that BN tubes are highly crystalline and probably, they may be the strongest insulating nanofibres.

Graphite and h-BN have opened new avenues for novel nanomaterials with different combinations of the type $B_X C_Y N_Z$. Pyrolysis experiments with melamine have been found very helpful in synthesising $C_x N_y$ nanofibres.

Boron doped carbon nanotubes have been synthesised by electric arc discharge with h-BN inside the graphite anode. During this experiment, it is observed that carbon nanotubes are highly graphitic and also very long (–20μ). It was found that B was located at the tips of the nanotubes and that the zig-zag chirality was preferred. It is reported by ab-initio calculations that boron was acting as surfactant avoiding the zig-zag tubes closure. In addition, boron doped nanotubes exhibit conducting properties.

It is reported that nitrogen doped carbon nanotubes can also exhibit interesting features such as pyridine like bonding in the graphitic mesh, which causes peak close to the Fermi level in the LDOS. All these nanotubes, armchair and zig-zag are found to be conductors. Obviously, nitrogen is donating electrons to the nanotube system to produce conductivity.

8.10 KEY ISSUES IN NANOMANUFACTURING

Similar to large-scale fabrication, nanomanufacturing research issues revolve around precursor materials; fabrication processes and characterisation techniques; instrumentation and equipment; theoretical modelling and control; and design and integration of structures into devices and systems.

8.11 NANOSTRUCTURES IN MOTION

Micromechanics, micromachines and moving nanostructures represent a new "micro-revolution". However, the concepts related to micro-mechanics and micro-machining of silicon are not new. A number of commercial products including accelerometers, pressure transducers, thermal print heads and inkjet nozzle arrays have been fabricated for internal industrial use and for the open market.

Since 1982, the field of micro-mechanics has expanded considerably, adding micrometers, movable linkages, gears, micro-turbines, electrical drives and micro-weezers to an ever growing suite of micro-mechanical devices.

8.12 NANO-MATERIAL ADVANTAGE

Nanocrystalline materials are three dimensional solids composed of nanometersised grains, or crystallites. Because of their unique structure, which is characterised by ultrafine grains and a rather high density of crystal lattice defect, these materials have extraordinary fundamental properties that could be exploited to make **'next-generation' superstrong materials**.

Strength and ductility are the central mechanical properties of any material. They are determined by the physical nature of plastic deformation, which in conventional, coarse-grained metals is mainly carried by dislocations-line defects of regular crystal lattice-within individual grains. Study suggest, however, that the mechanism of plastic deformation in nanocrystalline materials may be different and so may lead to novel mechanical properties.

High strength, or hardness has already been observed in many nanomaterials. But in most cases such nanomaterials have very low ductility—they fail when their shape is changed. Some even become brittle when a force or deforming load is applied. Strength and ductility are usually opposing characteristics: the higher the strength, the lower the ductility, and vice versa. This correlation is associated with the nature of plasticity: the more difficult it is for dislocations to appear and to move, the stronger but brittle and less ductile is any crystalline material.

Very recently, Wang and coworkers report their success in engineering an unusual nanocrystalline material that combines two useful properties that are often mutually exclusive strength and ductility. They created a copper nanostructure by rolling the metal at temperature below 77 K and then heating it to around 450 K. The result was a 'bimodel' structure of micrometersized grains (at a volume fraction of around 25%) embedded in a matrix of nanocrystalline grains. The material showed extraordinary high ductility, but also retained its high strength. The reason for this behaviour seems to be that while the nanocrystalline grains provide strength, the embedded larger grains stabilise the tensile deformation of the material.

Processing nanomaterials to achieve both strength and ductility can enhance the material's resistance to fatigue. More than anything else, it is fatigue that at present limits the lifetime, and hence the range of applications, of many advanced materials. Ethics is the study of understanding what is good and what is bad. As nanotechnology steps into centre stage and the bounty it promises begins to become reality, it will raise several issues of ethics, public policy, law and social responsibility. When we try to understand a new technology like the nanotechnology, it is very natural to think about the good and the bad effects of such technology. Any technology can be used for constructive as well as destructive use. This is also true for the nanotechnology. Nanotechnology as of now is still in its infant stage. A variety of applications are being proposed and explored. A majority of these are in the areas related to computer technology, defence, electronics, medicine and consumer products like cosmetics. These applications of nanomaterials can have different impacts on mankind and the world around. It is quite natural to think about the good and the bad effects of nanotechnology. When we think about such effects, we will be in a position to decide whether to take the nanotechnology in one direction or the other. We shall discuss the good and the bad about nanotechnology and the ethical issues it can raise.

8.13 APPLICATIONS OF NANOTECHNOLOGY

We would not like to focus exclusively on potential dangers of nanotechnology. Let us begin by considering some of the good things that the nanotechnology can provide. For example, nanotechnology helps to clean up the environment. Paints based on nanomaterials could convert sunlight into useable energy and remove CO_2 from the atmosphere. There are other materials which could remove other pollutants from the air. Lighter but stronger materials could be developed from designer molecules. Aircraft made of lighter materials with the strength of diamonds would be more fuel-efficient and safer. Clothing made of stronger materials would last longer.

In future, nanobots (nanoscale robots) would travel through blood vessels clearing away plaque and entering cancerous cells to destroy them. Nanotechnology is expected to manufacture food and clean water economically. Computer chips may be made at lower cost from chemical synthesis avoiding toxic by-products. Obviously, nanotechnology is quite promising. Scientists have started understanding nanoscience and would like to exploit it to the benefit of mankind. Human life can be made healthier and safer by the use of nanoscience and nanotechnology. *The potential benefits of nanoscience and nanotechnology are immense but the potential hazards are equally serious.*

Today, nanotech is making its way into our lives, producing spill- and wrinkle-resistant clothing, lighter and stronger sports equipment, self-cleaning windows that breakdown dirt, etc. We will continue to see improvements in everyday products and advances in industries such as electronics. For instance, nanotechnology is helping to develop organic light-emitting diodes that can emit their own light, such as a glow worm. These would be brighter, faster, lighter, more energy efficient and would enable ultra-thin displays viewable from any angle without loss of quality. Nanotechnology will be useful in many other ways in the near and distant future. The following are some of the examples:

- *Transportation:* New materials will make vehicles lighter and stronger; these vehicles will use less fuel and can withstand more damage. Space travel will become inexpensive and no longer limited to the elite few.
- *Computers:* As processing chips become smaller, faster, and more powerful, computers will also shrink in size, allowing them to be truly ubiquitous and embedded in clothing and even in the human body.
- *Military:* Soldiers will wear an exoskeleton that can change its flexibility and become instant armor. Clothes will be able to store energy to deliver superhuman strength when needed, such as for jumping over a 20-ft wall. Weapons can be miniaturised, and smart bullets can be programmed to hit specific targets.
- *Energy:* More powerful and smaller batteries will enable devices to become smaller and operate longer. Efficient molecule-sized solar cells can be mixed into road asphalt to continually harness the sun's energy. Fossil fuels and coal will be replaced with renewable energy sources, even turning garbage into fuel.
- *Environment:* One may be able to rebuild our thinning ozone layer and clean up the environment with nanobots that eat oil spills and other contaminants. More efficient manufacturing processes meanless pollution. We would not need to cut trees to produce paper anymore.
- *Medicine:* Surgeries will be performed with tools more than 1000 times precise than the sharpest scalpel today. Cosmetic surgery won't even require surgery-eyes can perhaps change colour and noses can change shape without cutting. One will able to use medical monitoring devices to detect diseases. New drugs can target specific cancer cells and viruses to destroy them. They can also fix defective genes or otherwise deliver precise treatment. To the extent that getting old is like any other physical disorder, repairing cells can slow, halt or reverse the ageing process.
- *Distant future:* Nanotechnology enables researchers to build materials atom by atom. Scientists therefore predict that they will be able to manufacture anything. Among other things, they can make food from dirt and water, eliminating famine worldwide.

8.14 LIMITATIONS OF NANOTECHNOLOGY

There is no doubt that there are many positive benefits from nanotechnology, but what about the bad effects of nanotechnology? Is nanotechnology ultimately worth the cost, or is the price too high. Let us examine some of the negative consequences predicted for nanotechnology:

- *Medical and health:* Nanoparticles have been shown to be absorbed in the livers of research animals and even cause brain damage in fish exposed to them within 2 days. If nanoparticles find their way into living cells, they can enter the food chain through bacteria and pose a health threat, *e.g.* mercury in fish, pesticides in vegetables, or hormones in meat. Carbon nanotubes look very much like asbestos fibres, but what happens if they get released into air? Being carbon-based, they would not be detected and obviously they are repelled by the usual defence mechanism of our bodies.
- *Environment:* Since nanomaterials are as strong as diamonds, question arises, how decomposable will they be? Will they litter our environment further or pose other disposal problems like nuclear waste and nuclear hazards or space litter? In the distant future, will self-replicating nanobots-necessary to create the trillions of nanoassemblers needed to build any kind of product-run amok, spreading as quickly as virus?

- *Privacy:* With the shrink in size of products, eavesdropping devices too can become invisible to the naked eye and more mobile. This make it easier to invade our privacy. If nanodevices are planted into our bodies, mind controlling nanodevices may be able to affect our thoughts by manipulating brain processes.
- *Terrorism:* Capabilities of terrorists also advances with advances in capabilities of military. Thus weapons become more powerful and portable. Obviously, these devices can also be turned against us. Nanotech may create new, unimaginable forms of torture causing disintegration of a person at the molecular level. Radical groups or terrorists could let loose nanodevices targeted to kill a specific person.
- *Society:* With all these limitations, several expert advocate a strong system to regular and monitor nanotechnology developments. Nanotechnology is a multidisciplinary science. Nanotechnology could be used to restore environment to spread wealth, and to cure more illness. This depends on our action, working within the limits set by the real world.

Nanotechnology has become a very active and vital areas of research, which is fast developing in industrial sectors and spreading to almost every field of science and engineering. Control of matter on a molecular scale is the main aim of all researchers working in this field. Fundamental physical, mechanical, magnetic and biological properties of materials are improved remarkably as the size of their constituent grains decreases to a nanometer scale. Nanostructured materials and associated processing technologies have opened up exciting new possibilities for future applications in aerospace, automotives, electronics, coatings, nonvolatile memories, sensors, acuators, optoelectronics, drug delivery, etc. One of the difficult problems facing the design of nanostructures-based stems is understanding how to interconnect and address them. Nodoubt, the success of nanoscience will depend on the development of new device and manufacturing technologies.

SUMMARY

- *Nanoscience* is, at its simplest, the study of fundamental principles of molecules and structures with at least one dimension roughly between 1 and 100 nanometers. These structures are known as nanostructures.
- *Nanotechnology* is the application of nanostructures into useful nanodevices. Nanotechnology has applications in electronic devices including information storage retrieval, as well as life sciences.
- *Molecular nanotechnology* is the self-assembly of the molecules into an ordered and functional structure.
- *Carbon nanotubes* (CNTS) are seamless graphene tubule structures with nanometer-size diameters and high aspect ratios. This new classes of one-dimensional materials is shown to have exceptional chemical, thermal and novel electronic properties. Nanotubes are categorised as single walled nanotubes (SWNTs) and multi-walled nanotubes (MWNTs). SWNTs has just the single shell of hexagons, whereas in MWNTs, multiple layers are nested like the rings in a tree trunk.
- *Nanotubes* have very high electrical and thermal conductivities, alongwith stability upto very high temperatures.
- *The electronic* properties of nanotubes have been used to fabricate nanoscale transistors.
- Another exciting nanotube application is in flat panel displays for televisions and computer monitors.
- *Quantum dots* are nanometer-sized semiconductor crystals or electrostatically confined electrons. They are fabricated in semiconductor materials and have typical dimensions between nanometers to a few micron (10^{-6} m). These can be easily connected to electrodes and are therefore excellent tools to study atomic like properties.

- *Nanowires* are structures that have a thickness or diamater constrained to tens of nanometers or less and an unconstrained length. Many different types of nanowires exit, including metallic (*e.g.* Ni, Pt, Au), semi-conducting (*e.g.* Si, InP, GaN, etc.), and insulating (*e.g.* SiO_2, TiO_2). Molecular nanowires are composed of repeating molecular units either organic (*e.g.* DNA) or inorganic (*e.g.* $Mo_6S_{9x}I_x$).

SUGGESTED READINGS

1. *Nanostructured Materials and Nanotechnology*, edited by H.S. Nalwa, Academis Press, London (2002).
2. C.P. Poole, and F.J. Owens, *Introduction to Nanotechnology*, Wiley VCH, Weinheim (2003).
3. N. Hosono, et al., *Nanomaterials: From Research to Applications,* Elsevier (2006).
4. K.K. Kar, *Carbon Nanotubes: Synthesis, Characterisation and Applications*, Research Publishing, Singapore (2011).
5. T. Pradeep, *Nano: The Essentials*, Tata McGraw Hill (2008).
6. *Nanotechnology and Nanoelectronics*, edited by W.R. Fahrner, Springer (2005).
7. *Carbon Nanotubes and Related Structures*, edited by D.M. Guldi and N. Martin, Wiley-VCH Verlag GmbH and Co. KGaA (2010).
8. *Carbon Nanotube Electronics*, edited by A. Javey and J. Kong, Springer (2009).
9. S. Reich, et al., *Carbon Nanotubes*, Wiley-VCH Verlag (2004).
10. J.J. Ramsden, *Applied Nanotechnology*, William Andrew (2009).
11. *Nanomaterials Handbook*, edited by Yuri Gogotsi, Taylor and Francis (2006).

REVIEW QUESTIONS

1. What is nanoscience and nanotechnology? What is the difference between these two?
2. Explain the working of SEM and TEM. Compare there two.
3. Discuss the impact of nanoscience in various fields. Give examples of some natural phenomena where nanoscience plays an important role.
4. What are quantum dots? How does the frequency of the light emitted from a quantum dot depend on the size of quantum dot?
5. What are nanowires? In what respects the electrical properties of nanowires are different from conventional wires?
6. Discuss the mechanical properties of nanomaterials. How these can be useful in various applications?
7. Described the sol-gel method of fabrication of nanomaterials.
8. How the nanotechnology will affect the life of a common man?
9. What are carbon nanotubes? Explain their electrical properties.
10. Mention important properties of nanoparticles.
11. How would one make and stabilise quantum dot?
12. What are the different types of quantum dots investigated? What makes quantum dots luminescence attractive?
13. What are the important concerns of materials scientists in the nanoscience area?
14. How many types of nanotubes are there? Mention their characteristics.
15. What are the main allotropic forms of carbon? Compare their important properties.
16. What are the various techniques used to synthesise nanostructured materials/ CNTs? Discuss one of them briefly.

17. "From cars to medicine, nanotubes may be a miracle material". "They are stronger than steel and as flexible as plastic, conduct energy better than almost any material ever discovered and can be made from unexotic raw materials such as methane gas". Comment upon the above statement.

SHORT ANSWER QUESTIONS

1. What are nanomaterials?

Ans. These are the materials containing nanocrystals, *i.e.* their grain size is in the range of 1 to 100 nm. Nanomaterials may be metals, alloys, intermetallics or ceramics.

2. What do you understand by nanotechnology?

Ans. Nanotechnology deals with various structures of matter having dimensions of a billionth of a meter.

3. What is nanoscience and nanotechnology?

Ans. Nanoscience and nanotechnology primarily deal with the synthesis, characterisation, exploration, and exploitation of nanostructured materials characterised by at least one dimension in the nanometer (1 nm = 10^{-9} m) range.

4. How nanoworld is uniquely different?

Ans. The physical and chemical properties of nanostructures are distinctly different from those of a single atom (molecule) and bulk matter with the same chemical composition. These differences between nanomaterials and the molecular and condensed phase materials pertain to the spatial structures and shape, phase changes, energetic, electronic structure, chemical reactivity, and catalytic properties of large, finite systems and their assemblies.

5. What are the most significant applications of nanomaterials?

Ans. Perhaps in nanodevices and nanoelectronics.

6. How nanoscience is interdisciplinary?

Ans. Nanoscience is truly interdisciplinary area covering physics, chemistry, biology, materials and engineering.

7. What is a carbon nanotube?

Ans. Carbon nanotubes are long cylinders of 3-coordinated carbon, slightly pyramidalised by curvature from the pure sp^2 hybridisation of graphene, toward the dimond like sp^3. Infinitely long in principle, a perfect tube is capped at both ends by hemifullerenes, leaving no dangling bonds. A single walled carbon nanotube (SWNT) is one such cylinder, while multi wall tubes (MWNTs) consist of many nested cylinders whose successive radii differ by roughly the interlayer spacing of graphite.

8. What are the advantages of MWNTs over SWNTs?

Ans. Multiwall tubes have two advantages over their SWNTs cousins. The multishell structure is stiffer than the single wall one, especially in compression. Large-scale syntheses by enhanced chemical vapour deposition (CVD) processes are many, while for SWNTs, only the rice HiPco process appears to be scaleable.

9. What are the important concerns of material scientists in the nanoscience area?

Ans. (*i*) Nanoparticles or nanocrystals of metals and semiconductors, nanotubes, nanowires, and nanobiological systems.

(*ii*) Assemblies of nanostructures (*e.g.* nanocrystals and nanowires) and use of biological systems, such as DNA as molecular nanowires and templates for metallic or semiconducting nanostructures.

(*iii*) Theoretical and computational investigations that provide the conceptual framework for structure, dynamics, response and transport in nanostructures.

(*iv*) Applications of nanomaterials in biology, medicine, electronics, chemical processes, high strength material, etc.

10. What are the present goals of the science and technology of nanomaterials?

Ans. To master the synthesis of nanostructures (nano-building units) and their assemblies of desired properties; to explore and establish nanodevices concepts, to generate new classes of high-performance nanomaterials, including biology inspired systems; and to improve techniques for the investigation of nanostructures.

11. What are nanocomposites?

Ans. Nanocomposites are the heterogeneous materials composed of at least one type of nanometer-sized particles.

12. What are van der Waals forces?

Ans. The van der Waals forces represent the weak (secondary) interaction between adjacent molecule or atoms, and arise from their electric dipoles. The van der Waals forces are long-range (even for distances larger than tens of nanometer), and may be attractive or repulsive depending on the distance between interacting atoms or molecules. In particles, the van der Waals forces between nanotubes become attractive when the distance is larger than the equilibrium distance, or repulsive when the distance is smaller than the equilibrium distance. The van der Waals forces between nearest nanotubes play a crucial role in physical properties of multiwall carbon nanotubes and single wall carbon nanotubes ropes.

13. How many kinds of CNTs other than SWCNT and MWCNT have been developed so far?

Ans. (*i*) *Carbon Nanocoil* having unique 3*D* structures.

(*ii*) *Carbon Nanofibre* (CNFS): These are, cylindrical or conical structures having length in the order of a few microns and diameter varying from 10 to 200 nm. The internal structure of CNFS varies and is comprised of different arrangements of modified grapheme sheets.

(*iii*) *Carbon Nanocone* (CNCON): This has a prefect conical structure with symmetry different from other known carbon materials, including nanotubes, nanocoil and bucky balls.

(*iv*) *Carbon Nanohorn* (CNH): This is made of the same graphite structure as CNT. CNH is a single wall carbon cone having structures similar to those of nanotube caps.

(*v*) *Carbon Nanoribbon* (CNR): This is strip of a graphene sheet with morphology different from CNTs. It is highly graphitised and width of ribbon is in the range of 1 to 20 nm and length is in microscopic scale.

14. What is nanoelectronics?

Ans. This is the field of science and engineering dealing with the fabrication, study and application of nano size electronic devices for information technologies. Generally, quantum effects determine the operational principles of these devices.

15. What is Nano-fish memory device?

Ans. This is similar in structure to a MOSFET (MOS field effect transistor) except that it is a three terminal device with two gate electrodes, one on the top of the other. The top electrode forms the control gate, below which a floating gate is capacitively coupled to the control gate. The memory cell operation involves putting charge on the floating gate or removing it, corresponding to two logic levels. These devices utilize single or multiple nanoparticles as the charge storage element.

16. What do you understand by nanophotonics?

Ans. Nanophotonics is the field of science and engineering dealing with the fabrication, study and application of optical phenomena at the nanoscale. It deals with structures and process which are spatially localized in domains smaller than the wavelength of visible radiation. The field includes nanoscale confinement of radiation, nanoscale confinement of matter, nanoscale physical or photochemical transformations.

17. What are quantum wires?

Ans. These are *1D* structures. As compared to quantum films, one more dimension of the structure appears to be so small as provide quantum confinement. Charge carriers are free to move only along the wire. Thus, only one kinetic component along with the confined energy contributes to the total energy. As a consequence, the density of states has $E^{1/2}$ dependence for each of the discrete pairs of states in the confined directions.

18. What are quantum dots?

Ans. These are *0D* structures in which the carriers are confined in all three directions. Their energy state is quantised in all directions and the density of states is represented by series of discrete, sharp picks resembling that of an atom. Quantum dots are also called as **artificial atoms.** These are usually formed by a definite number of atoms. They are typically represented by atomic clusters or **nanocrystallites**. Considering the coupling of quantum dots with external electric circuits, they can be classified as open dots and almost isolated or closed quantum dots. In an open dot, the coupling is strong and the movement of electrons across dot-lead functions is classically allowed. When point contacts, connecting a dot with an external circuit, are pinched off, effective barriers are formed and conductance occurs only by tunnelling. These are almost isolated or closed quantum dots.

19. What do you understand by bucky ball, C_{60}?

Ans. This is a hollow cage carbon molecule consisting of 60 carbon atoms and has resemblance of its molecular structure to geodesic domes.

20. What is density matrix?

Ans. A representation of mixed states generated by right-multiplying the state vector by its adjoint.

21. What is lithography ?

Ans. Literally, lithography means "stone writing". This refers generally to any technique for forming a patterned structure on a solid surface.

22. What is photolithography?

Ans. This is a photochemical technique for etching specific patterns using projected light. It is the most widely used technique today.

23. What is a nanocomputer?

Ans. A computer whose characteristic length scale is between $10^{-7.5}$ to $10^{-10.5}$ m (*i.e.* between ~32 and ~0.32 nm, *i.e.* closer to 1 nm than to 1 micrometer or 1 picometer on logarithmic scale).

24. What are smart materials?

Ans. These are usually defined as those that can sense external stimuli and then respond to those stimuli by some means in real (or near real) time. Such materials are therefore a product that lies somewhere between a material and more complex nanodevice—a sensor, in fact. Some smart materials respond to light, others to pressure and so on. Some are self-repairing. These materials find vide spread applications today. These materials should be recognized along with the familiar metals, plastics, ceramics, composites, powder metals, and speciality type and multifunctional materials (shape metal alloys (SMAs), micro-mechanical systems (MEMSs), functionally graded materials (FGMs), and nanomaterials). By using these materials instead of adding mass, engineers can endow structures with built-in responses to a myriad of contingencies. In their various forms, these materials can perform as actuators, which can adapt to their environments by changing characteristics such as shape and stiffness, as sensors, which provide the actuators with information about structural and environmental changes. Smart materials are beginning to play an important role in civil engineering drawings for dams, bridges, highways, and buildings. There are many more applications of these materials in process.

FILL IN THE BLANKS

1. The prefix nano in the word nanotechnology means a .... (1×10^{-9}). [billionth]
2. Nanostructures constitute a bridge between molecules and infinite .... systems. [bulk]
3. The structure of nanoparticles of Cds, CdSe and such materials is affected by .... [size]
4. The synthesis of nanomaterials includes control of size, shape and .... [structures]
5. The most significant applications of nanomaterials may be in nanodevices and .... [nanoelectronics]
6. The skeleton of C_{60} consists of 20 hexagonal and .... pentagonal rings fused all together. [12]
7. Composite materials reinforced with carbon or graphite are often used in .... goods. [sporting]
8. The discovery and rapid evolution of carbon nanotubes has played a major role in .... triggering the explosive growth of .... in nanotechnology. [research and development]
9. Graphene refers to a nanolayer of sp^2-bonded .... atoms. [carbon]
10. Carbon nanotubes are hollow cylinders of .... sheets. [graphite]
11. A carbon nanotube made of single graphite layer rolled up into a hollow cylinder is called a .... [SWNT]
12. A tube comprising several concentrically arranged cylinders is referred to as .... [MWNT]
13. Multiwall nanotubes have similar lengths to single walled tubes, but much lager .... [diameters]
14. Nanotubes can behave like metals or .... [semiconductors]
15. Graphene is a two-dimensional honeycomb net, made up of .... bounded carbon atoms. [sp^3]
16. Armchair tubes always show .... properties. [metallic]
17. The fundamental band gap in semiconducting nanotubes ranges from .... eV to .... eV being dependent on small variations of the diameter and bonding angle. [0.4, 0.7]

CHAPTER

9

Metamaterials

9.1 INTRODUCTION

Artificial materials, often called *metamaterials* are artificial electromagnetic media whose physical properties are engineered by assembling microscopic and nanoscopic structures in unusual combinations. The study of these materials breaks the traditional frontiers and brings together many disciplines such as physics and microfabrication; electromagnetic theory and computational methods; optics and microwaves; and nanotechnology and nanochemistry. Their possible applications range from electronics and telecommunications to sensing, medical instrumentation and data storage.

The modern history of *'electromagnetics'* metamaterials, usually referred as metamaterials, (MTM) can probably be counted from the seminal paper by Pendry in 2000, where an ambitious goal to create a so-called *perfect lens* was put forward. For creation of such a 'lens'. One needs to design an artificial medium which would possess specific properties, usually not observed in natural materials. The material for the perfect lens should be a medium with both *negative permittivity* and *permeability* in the same frequency range (media with $\varepsilon < 0$ and $\mu < 0$). These necessary conditions ensure a refractive index of n or $\mu = -1$ while guaranteeing to free space. This strange *'left handed'* media did not exist at that time but could be imagined as possible *"composites of the future.* We may note that Veselago in 1968 believed that such media could be found among homogeneous magnetic semiconductors fabricated chemically. However, no magnetic semiconductors with doubly negative material parameters had been obtained. It was emphasised that such media should be strongly dispersive (resonant) composites. It was pointed out that a slab under the condition $\varepsilon = \mu = -1$ will focus a diverging light beam. Under the condition $d = 2D$ where d is the thickness of the left handed slab and D is the distance from the slab interface to source, the light will be focussed at the point distance by D from the back interface without aberrations and reflections. Obviously, these lenses require extremely low-loss characteristics and must be adequately thin. We may note that this is not a lens in its optical meaning since it does not focus a parallel light beam. It focuses only diverging beams.

The phenomenon of free-space subdiffraction imaging, although not using Veselago-Pendry super-lenses has been experimentally demonstrated by many types of metamaterials, *e.g.* plasmonic silver film, the magnetoinductive lens, and the Swiss-roll structure, which recover fine spatial features through evanescent enhancement but cannot be said to possess a negative refractive index.

Subdiffraction imaging phenomena have also been extended to the far field using magnifying hyperlenses that employ a high degree of anisotropy.

The first successful attempt at super lensing of Veselago-Pendry type, and indeed, the first demonstration of subdiffraction imaging using metamaterials employed negative-refractive index transmission line (NRI-TL) metamaterials, which consist of a fine TL grid loaded with inductors and capacitors to control their effective-medium response. The use of periodic lumped reactive loading and properly designed host medium enables the accurate description of the effective permittivity and permeability seen by TEM waves propagating within the NRI-TL network.

In this chapter, we have tried to summarise the state-of-the art in the most recent field of electromagnetic artificial materials, often called *'Metamaterials'*.

9.2 PREHISTORY OF METAMATERIALS

(i) Artificial Dielectrics: The artificial dielectric material is usually defined as a composite reproducing on a much larger scale, processes occurring in the molecules of usual dielectric. This involved arranging metallic elements in a tree-dimensional (3*D*) array or lattice structure to stimulate the crystalline lattices of dielectric materials. Such an array responds to radio waves just as a molecular lattice respond to light waves: Free electrons in the metal elements flow back and forth under the action of the alternating electric field. Metal elements called also as lattice inclusions or lattice particles become oscillating dipoles similar to the oscillating molecular dipoles of a natural dielectric.

Artificial dielectrics are not necessarily regular lattices. They can be random mixtures. Even the nonuniform concentration of particles is sometimes allowed, which offers unusual properties of such composite media. When the concentration of particles exceeds the so-called percolation threshold (particles touch one another and/or the capacitive coupling between adjacent particles is very strong) artificial dielectrics in the same low frequency range become artificial conductors with complex conductivity. Their conductivity can be engineered (*i.e.* controlled by the design parameters) and is in principle tunable magnetically or electrically. Artificial conductors find applications in electro-mechanical devices, fuel cells, and other techniques where controlled heating by electric current is desired.

(ii) Artificial Magnetics: Magnetism without magnetic constituents has been known since 1940s due to works of Schelkunoff and Friis that suggested so-called split-ring resonators (SRRs) (Fig. 9.1).

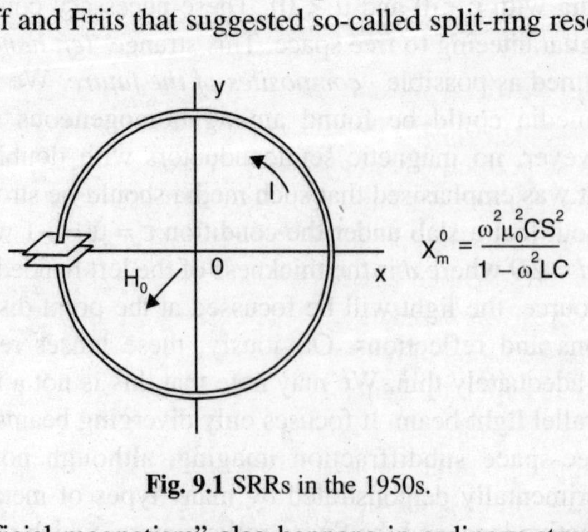

Fig. 9.1 SRRs in the 1950s.

We may note that "artificial magnetism" also happens in ordinary structures, like wet snow. The loop-forming parts of liquid water cause diamagnetic behaviour. However, in lattices of SRRs the artificial magnetism is significantly enhanced in the resonant frequency range (and it is paramagnetic at lower frequencies).

Particles with metal loops of various shapes were studied in the 1980s and in combination with other shapes in 1990s, especially in connection with artificial bianisotropic materials for microwave applications. Polarisabilities of these bianisotropic particles in magnetic and electric fields were studied analytically, numerically and experimentally. Such double SRRs could be used to create artificial magnetics without chirality. In 1997, one finds probably the first experimental demonstration of negative permeability at artificial microwave materials.

The design with strong capacitive coupling between loops suggested by Pendry and coworkers in 1998 seems to be more appropriate for the artificial magnetism.

The strong coupling of two loops allowed one to obtain the magnetic resonance at lower frequencies. This means that the resonant frequency is low enough to consider the lattice of SRRs as a continuous medium. Because of its planar structure SRRs (Fig. 9.2(*a*)) are perhaps very practical ways of creating artificial magnetism at microwaves. Figure 9.2(*b*) shows swiss roll metal scatterers seems to be more efficient as magnetic resonators but work at considerably lower frequencies. The amplitude and frequency bandwidth of magnetic response can be enhanced by using very densely packed stacks of split rings, called metasolenoids.

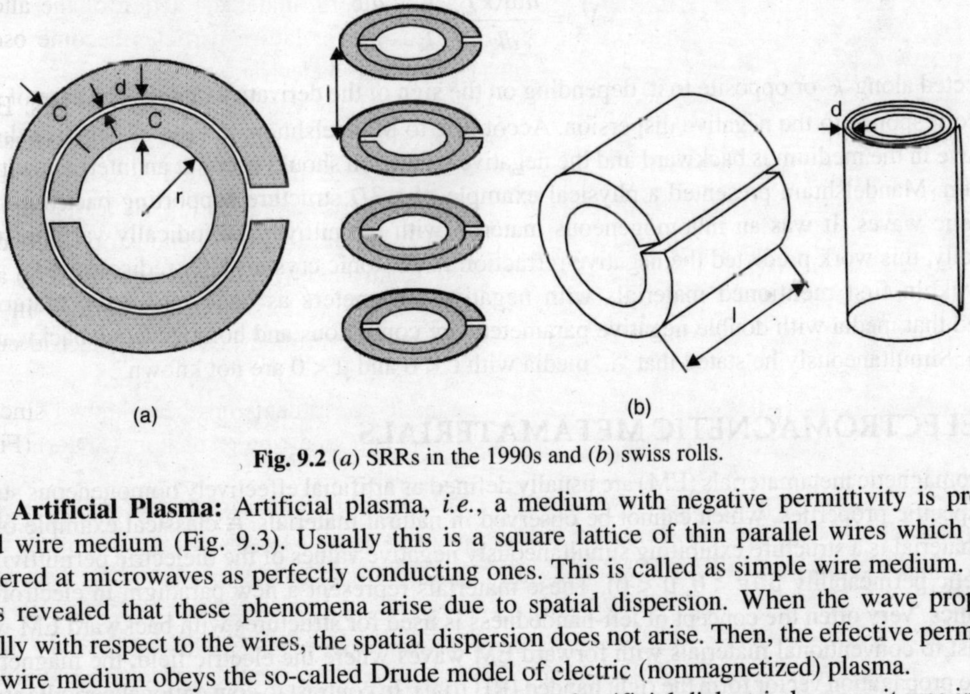

Fig. 9.2 (*a*) SRRs in the 1990s and (*b*) swiss rolls.

(iii) Artificial Plasma: Artificial plasma, *i.e.*, a medium with negative permittivity is presently termed wire medium (Fig. 9.3). Usually this is a square lattice of thin parallel wires which can be considered at microwaves as perfectly conducting ones. This is called as simple wire medium. Recent studies revealed that these phenomena arise due to spatial dispersion. When the wave propagates normally with respect to the wires, the spatial dispersion does not arise. Then, the effective permittivity of the wire medium obeys the so-called Drude model of electric (nonmagnetized) plasma.

The lattice of parallel wires was considered as a kind of artificial dielectric because it was invented and practically used for applications in microwave lenses. In 1970s tunable lattices of wires in which PiN diodes were inserted in order to switch the negative effective permittivity of the lattice to the positive one were created and produced by industry.

The term wire media appeared after theoretically revealing the effects of spatial dispersion in recent work. Experimental verification of some of these effects also appeared in the literature. We may note that the array of parallel swiss rolls mentioned above behaves like a wire medium (of thick wires) with respect to the magnetic field of propagating wave.

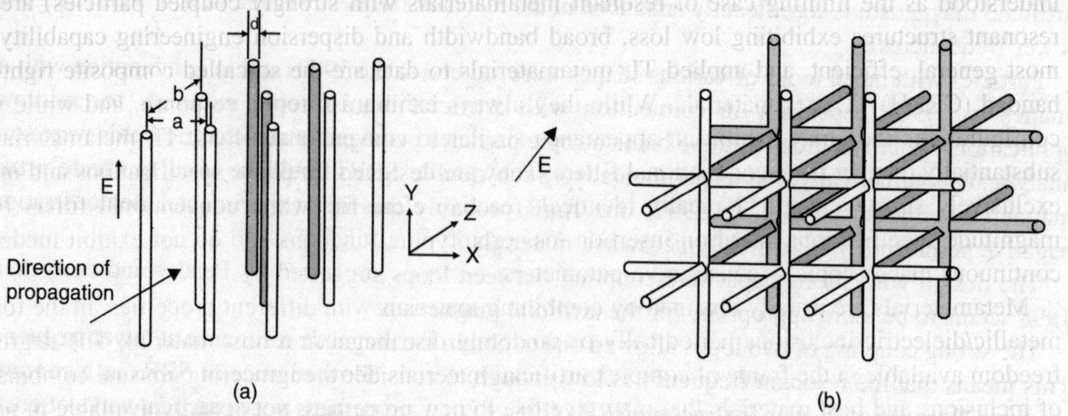

Fig. 9.3 (*a*) Simple wire media in the 1960s; and (*b*) Triple wire media in the 1960s.

(iv) Backward-waves in Bulk Media: In isotropic media, the absolute value of the wave vector is fully determined by the frequency. Therefore the group velocity

$$\vec{V}_g = \frac{d\omega(k)}{d\vec{k}} = \frac{\vec{k}}{k} = \frac{d\omega}{dk} \quad ...(1)$$

is directed along $\vec{k}$ or opposite to it, depending on the sign of the derivative $d\omega/dk$. The case of negative sign corresponds to the negative dispersion. According to Mandelshtam, in case of negative dispersion the wave in the medium is backward and the negative refraction should occur at an interface with such a medium. Mandelshtam presented a physical example of a *3D* structure supporting backward electromagnetic waves. It was an inhomogeneous material with permittivity periodically varying in space. Basically, this work predicted the negative refraction in photonic crystals later rediscovered.

Sivukhin first mentioned materials with negative parameters as backward-wave materials. He noticed that media with double negative parameters are continuous and homogeneous backward-wave media. Simultaneously he stated that "... media with $\varepsilon < 0$ and $\mu < 0$ are not known".

9.3 ELECTROMAGNETIC METAMATERIALS

Electromagnetic metamaterials (EM) are usually defined as artificial effectively homogeneous structures with specific properties, which cannot be observed in natural materials. A classical example of such a metamaterial is a structure exhibiting simultaneously negative values of the dielectric permittivity ε and magnetic permeability μ ($\varepsilon < 0$, $\mu < 0$). These materials represent a new paradigm in electronics and photonics. Very often the concept of left-handedness is used for structures with backward EM waves in contrast to conventional materials with forward EM waves where the electric field, the magnetic field, and the propagation vector form the right handed (RH trial). In contrast to conventional periodic structures that are usually operated in the Bragg regime (period $\approx \lambda_g/2$, where λ_g is the guided wavelength) for filtering, metamaterials are long wavelength structures (average distance between unit elements or atoms $<< \lambda_g$). Since they are not readily available naturally, they are engineered under the form of *1D*, *2D*, or *3D* structures.

Metamaterials may be engineered using resonant particles. However, resonant metamaterials are narroband or lossy. In contrast, transmission line (TL) metamaterials, which are constituted of patterned lumped elements (inductances and capacitances). (We may note that TL metamaterials may be understood as the limiting case of resonant metamaterials with strongly coupled particles) are non-resonant structures exhibiting low loss, broad bandwidth and dispersion engineering capability. The most general, efficient, and applied TL metamaterials to date are the so-called composite right-/left-handed (CRLH) TL metamaterials. While they always exhibit a filtering response, and while in 1*D* configurations they may exhibit an appearance similar to conventional filters. TL metamaterials are substantially distinct from conventional filters. They are designed for phase specifications and include exclusively subwavelength, typically identical, reactive elements, while conventional filters follow magnitude specifications based on insertion loss prototyping functions and do not exhibit media (*i.e.* continuous macroscopic)—constitutive parameters.

Metamaterials are usually obtained by combining materials with different properties, in the form of metallic/dielectric inclusions periodically or randomly distributed in a host material. The degrees of freedom available in the frame of complex artificial materials allow engineering unusual combinations of inclusions and host materials that may give rise to new properties, not actually available in natural materials.

TL approach gives an efficient design tool for microwave applications providing a correct description of physical properties of metamaterials. A conventional TL with a positive phase velocity behaves as a right-handed transmission line (RHTL). An artificial RHTL can be formed as a ladder network of capacitors connected in shunt-and series-connected inductors. Figure 9.4(*a*) shows the unit cell of RHTL. The dual TL can be designed as a ladder network of inductors connected in shunt-and series-connected capacitors as in the unit cell Fig. 9.4(*b*). This line has a negative phase velocity and is

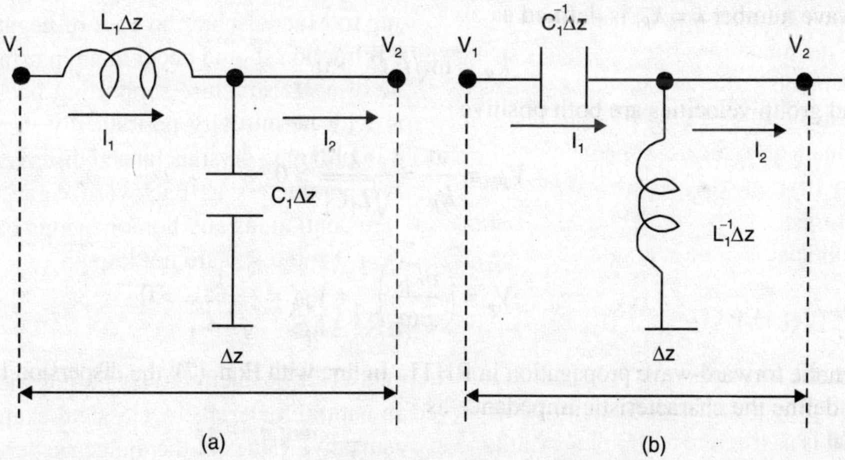

Fig. 9.4 Units cells of (*a*) RH TL and (*b*) LH TL.

referred to as the left handed transmission like. A backward-wave propagates along the LHTL, which can be considered as a 1*D* metamaterial. A more general model of a LHTL is a composite right-/left-handed structure, which includes RH effects. In most practical applications, the influence of RH effects is negligibly small and many interesting features can be observed when a combination of RH and LHLTs is used. The most important feature of the LH and RHTLs is that their dispersion characteristics.

9.4 GENERAL EQUATIONS FOR RH AND TLs

One can describe the homogeneous RHTL presented as cascade connection of unit cells (Fig. 9.4(*a*)) by the telegraph equations

$$\frac{\partial V}{\partial z} = -L_1 \frac{\partial I}{\partial \pi}$$

$$\frac{\partial I}{\partial z} = -C_1 \frac{\partial V}{\partial \pi} \quad ...(2)$$

where L_1 and C_1 are the inductance and capacitance per unit length respectively, V is the voltage and I is the current.

Both V and I are periodical with respect to time t and coordinate z, along which I–V wave propagates.

For sinusoidal wave, $\frac{\partial}{\partial \pi} = i\omega$ and the wave equations look as

$$\frac{\partial^2 V}{\partial z^2} = -\omega^2 L_1 C_1 V \quad ...(3)$$

$$\frac{\partial^2 I}{\partial z^2} = -\omega^2 L_1 C_1 I \quad ...(4)$$

The solutions to Eqns. (3) and (4) are

$$V = V_0 \exp[i(\omega t - kz)] \quad ...(5)$$

$$I = I_0 \exp[i(\omega t - kz)] \quad ...(6)$$

where the wave number $k = k_R$, is defined as

$$k_R = \omega\sqrt{L_1 C_1} > 0 \quad ...(7)$$

Phase and group velocities are both positive

$$V_{Ph} = \frac{\omega}{k_R} = \frac{1}{\sqrt{L_1 C_1}} > 0 \quad ...(8)$$

$$V_g = \left(\frac{\partial k_R}{\partial \omega}\right)^{-1} = V_{Ph} = \frac{1}{\sqrt{L_1 C_1}} > 0 \quad ...(9)$$

Therefore, the forward-wave propagation in RHTL. In line with Eqn. (7), the dispersion law is linear. One can define the characteristic impedance as

$$Z_0 = \frac{V}{I} = \sqrt{\frac{L_1}{C_1}} \quad ...(10)$$

For a homogeneous perfect LHTL formed as a cascaded connection of unit cells (Fig. 9.4(*b*)), the telegraph equations in the case of sinusoidal waves look as

$$\frac{\partial V}{\partial z} = -\frac{1}{i\omega}\left(\frac{1}{C}\right)_1 I \quad ...(11)$$

$$\frac{\partial I}{\partial z} = -\frac{1}{i\omega}\left(\frac{1}{L}\right)_1 V \quad ...(12)$$

where $(1/C)_1$ and $(1/L)_1$ are the inverse capacitance and the inverse inductance per unit length. One can write the wave equations as

$$\frac{\partial^2 V}{\partial z^2} = -\frac{1}{\omega^2}\left(\frac{1}{L}\right)_1\left(\frac{1}{C}\right)_1 V \quad ...(13)$$

$$\frac{\partial^2 I}{\partial z^2} = -\frac{1}{\omega^2}\left(\frac{1}{L}\right)_1\left(\frac{1}{C}\right)_1 I \quad ...(14)$$

The solutions to Eqns. (13) and (14) are the same as in Eqns. (5) and (6), with the wave number defined as

$$\frac{1}{k_L} = -\frac{1}{\omega}\sqrt{\left(\frac{1}{L}\right)_1\left(\frac{1}{C}\right)_1} < 0 \quad ...(15)$$

In this case the phase velocity is negative

$$V_{Ph} = -\omega^2 \frac{1}{\sqrt{\left(\frac{1}{L}\right)_1\left(\frac{1}{C}\right)_1}} < 0 \quad ...(16)$$

whereas the group velocity is positive

$$V_g = \left(\frac{\partial k_L}{\partial \omega}\right)^{-1} = \omega^2 \frac{1}{\sqrt{\left(\frac{1}{L}\right)_1\left(\frac{1}{C}\right)_1}} > 0 \quad ...(17)$$

Obviously, the backward wave propagates in the LHTL. In line with Eqn. (15), the wave number (propagation constant) is inversely proportional to the frequency and the dispersion law is nonlinear.

$$Z_0 = \sqrt{\frac{(1/C)_s}{(1/L)_1}} \quad ...(18)$$

Equation (2) through (18) describe the homogeneous, infinitely long TLs with distributed parameters. One can define a section of such a line of length l by the electrical length θ as

$$\theta_{R,L} = k_{R,L} \cdot l \quad ...(19)$$

In accordance with Eqn. (7), the frequency dependence of the electrical length of a section of RHTL is

$$\theta_R(\omega) = \theta_{OR}\frac{\omega}{\omega_0} \quad ...(20)$$

where $\theta_{OR} = k_R l > 0$ is the electrical length at the freqency ω_0. In the case of section of LHTL, the electrical length can be found from Eqn. (13) and the frequency dependence is expressed as

$$\theta_L(\omega) = \theta_{OL}\frac{\omega_0}{\omega} \quad ...(21)$$

where $\theta_L(\omega) = \theta_{OL}\frac{\omega_0}{\omega}$ is the electrical length at the frequency ω_0.

In practice, one can compose the artificial RH and LHTLs as a periodical structure containing one inductive and one capacitive component in the unit cell of the length l defined by a real length of the lumped components (Fig. 9.5). In this case, we have to use the translation symmetry for a description of the I – V wave propagating along the one-dimensional structure. The telegraph equations for the RHTL can be expressed as

$$V_{n+1} - V_n = -i\,\omega\,L_0\,I_{n+1} \qquad ...(22)$$

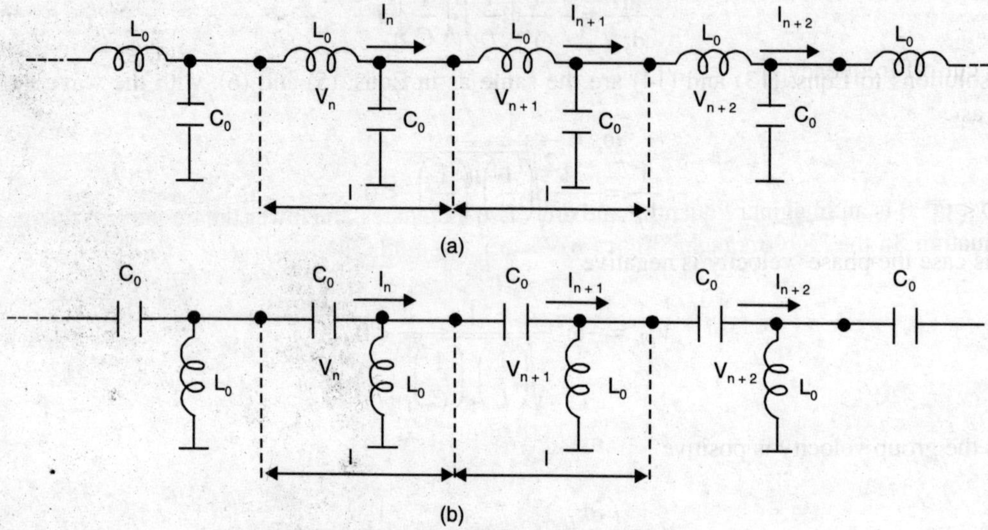

Fig. 9.5 Artificial lumped element TLs: (*a*) RH TL and (*b*) LH TL

$$I_{n+1} - I_n = -i\,\omega\,C_0\,V_n \qquad ...(23)$$

The voltage and current are

$$V_n = V_0\,e^{-in\theta},\; I_n = I_0\,e^{-in\theta} \qquad ...(24)$$

with

$$\theta = kl,\; n = 1, 2, 3, ...$$

Substituting Eqn. (24) into Eqns. (22) and (23), one obtains

$$\sin^2\left(\frac{\theta}{2}\right) = \frac{1}{4}\omega^2 L_0 C_0 = \frac{\omega^2}{\omega_c^2} \qquad ...(25)$$

where the cut-off frequency

$$\omega_c = \frac{2}{\sqrt{L_0 C_0}} \qquad ...(26)$$

For $\omega > \omega_c$, θ is an imaginary quantity and the wave attenuates: the higher the frequency, the more is attenuation. In the low frequency limit, *i.e.* $\omega << \omega_c$, we have

$$\frac{\omega}{\omega_c} = \pm\sin\left(\frac{\theta}{2}\right) \approx \pm\frac{\theta}{2} \qquad ...(27)$$

and

$$k = \frac{2\omega}{l\omega_c} = \omega\sqrt{L_1 C_1} \qquad ...(28)$$

with $L_1 = L_0/l$ and $C_1 = C_0/l$. The artificial lumped element RHTL behaves at $\omega << \omega_c$ as an infinitely long, perfect TL with the linear dispersion law. In a wide frequency range, the artificial periodic RHTL (Fig. 9.5(*a*)) is considered as a low-pass lumped element TL.

The same consideration of the artificial LHTL: (Fig. 9.5(*b*)) gives the dispersion relation.

$$\sin^2\frac{\theta}{2} = \frac{1}{4}\frac{1}{\omega^2 L_0 C_0} = \frac{\omega_c^2}{\omega^2} \quad ...(29)$$

with the cut-off frequency

$$\omega_c = \frac{1}{2\sqrt{L_0 C_0}} \quad ...(30)$$

For $\omega < \omega_c$, θ is an imaginary quantity and the wave attenuates: the lower the frequency, the more is the attenuation. In the high frequency limit ($\omega >> \omega_c$), one finds

$$\frac{\omega_c}{\omega} = \pm\sin\left(\frac{\theta}{2}\right) \approx \pm\frac{\theta}{2} \quad ...(31)$$

and

$$k = \frac{-1}{\omega}\sqrt{\left(\frac{1}{L}\right)_1\left(\frac{1}{C}\right)_1} \quad ...(32)$$

with $\left(\frac{1}{L}\right)_1 = \left(\frac{1}{L_0}\right)\Big/ l$ and $\left(\frac{1}{C}\right)_1 = \left(\frac{1}{C_0}\right)\Big/ l.$

We find that the artificial lumped element LHTL (Fig. 9.5(*b*)) behaves as an infinitely long, perfect TL with *k* inversely proportional to ω. The artificial periodic LHTL is considered as a high-pass lumped element TL. Figure 9.6 shows the dispersion characteristics of RH and LHTLs. In further consideration, the frequency range is limited by the inequalities $\omega << \omega_c$ for the RHTL and $\omega >> \omega_c$ for the LHTL.

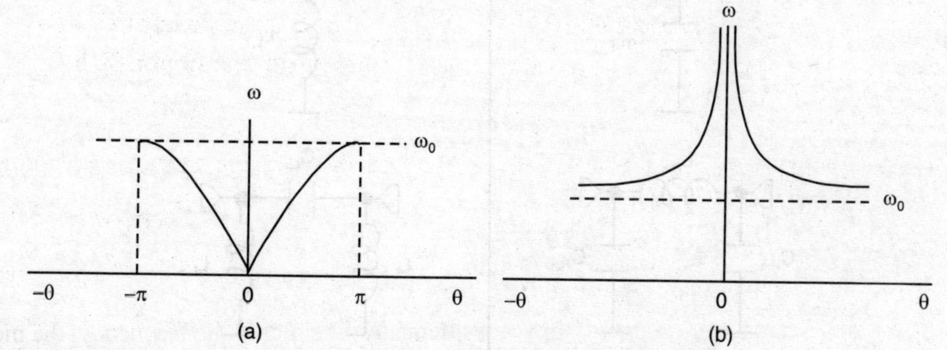

Fig. 9.6 Dispersion characteristics of (*a*) RH lumped element TL and (*b*) LH lumped element TL.

Microwave Phase Shifters based on Switchable LH and RHTLs

The transmission type phase shifter is a lossless two-port providing a change in a phase response of the EM wave under the control signal (current or voltage). The digital phase shifters using switchable channels are well known: the EM waves propagate in turn along two channels formed by TL sections of different electrical lengths.

Using specific dispersion properties of RH and LHTLs makes it possible to combine the benefits of switchable channel phase shifters of both kinds. Using cascade connection of RH and LHTL sections, one can obtain a similar slope of the phase response of two TLs of different electrical lengths. Obviously, in principle, it is possible to design a controllable phase shifter based on switchable metamaterial TLs, providing a flat differential phase response.

Figure 9.7 illustrates the operational principle of a digital phase shifter using switchable RH and LHTL sections.

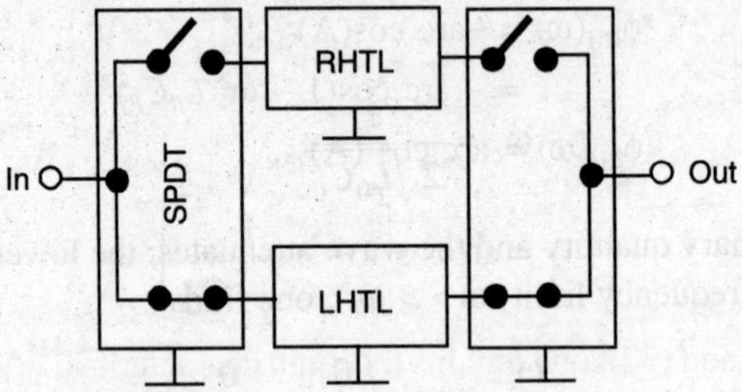

Fig. 9.7 Structure of a digital phase shifter based on switchable TL sections of different electrical lengths.

In one state, the signal goes through the RHTL section with a negative phase response ϕ_1, whereas in another state, it propagates through the LHTL section with a positive phase response ϕ_2. The differential phase response (phase shift), $\Delta\phi = \phi_1 - \phi_2$, is obtained by switching the signal path using two SPDT switches.

It is reasonable to form the phase shifter containing switchable RH and LHTL unit cells as T- or TT networks as shown in Fig. 9.8. For different bits of a digital N-bit phase shifter giving the phase

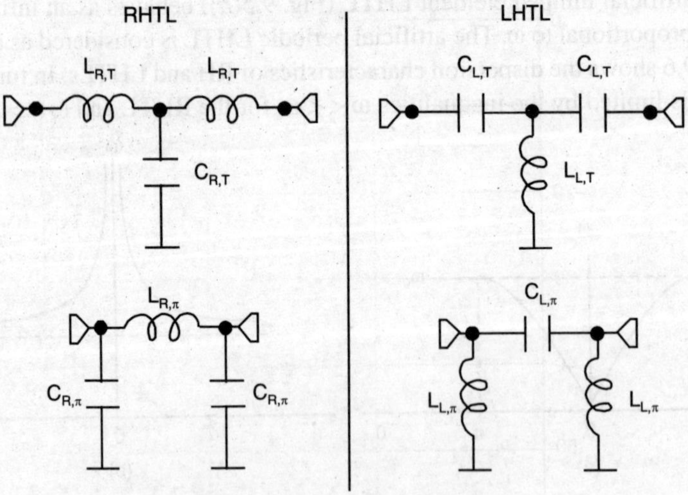

Fig. 9.8 Lumped-element equivalent circuits of the RH and LHTL sections.

shift $\Delta\phi_m$, the equivalent electrical length θ, of the both RHTL and LHTL sections should be chosen as follows:

$$|\phi_m| = \Delta\phi_m/2 \quad ...(33)$$

where

$$\Delta\phi_m = \Delta\phi_{LH} - \Delta\phi_{RH} \quad ...(34)$$

and $m = 1, 2,... N$ is the bit number.

In general, one can describe the TL section by the ABCD matrix

$$\begin{bmatrix} A & B \\ C & D \end{bmatrix}_{TL} = \begin{bmatrix} \cos\theta & iZ_0 \sin\theta \\ i\sin\theta/Z_0 & \cos\theta \end{bmatrix} \quad ...(35)$$

Taking into account the symmetry of the section of a homogeneous TL, it is possible to replace it by the symmetric lumped element T- or π-circuits.

One obtains the dispersion equations as

$$\phi_{RH}(\omega) = -\text{arc}\cos(A)_{T,\pi} = -\text{arc}\cos(1 - \omega^2 L_R C_R) \quad ...(36)$$

$$\phi_{LH}(\omega) = \text{arc}\cos(A)_{T,\pi} = \text{arc}\cos\left[1 - \frac{1}{\omega^2 L_L C_L}\right] \quad ...(37)$$

on differentiation Eqns. (36) and (37), one obtains the main characteristic of interest, *i.e.* slope parameter of phase characteristics:

$$\frac{d\phi_{RH}}{d\omega} = \frac{\sqrt{2L_R C_R}}{\sqrt{1-(\omega^2 L_R C_R/2)}} \quad ...(38)$$

$$\frac{d\phi_{LH}}{d\omega} = \frac{\sqrt{2/L_L C_L}}{\omega^2} \frac{1}{\sqrt{1-(1/2\omega^2 L_L C_L)}} \quad ...(39)$$

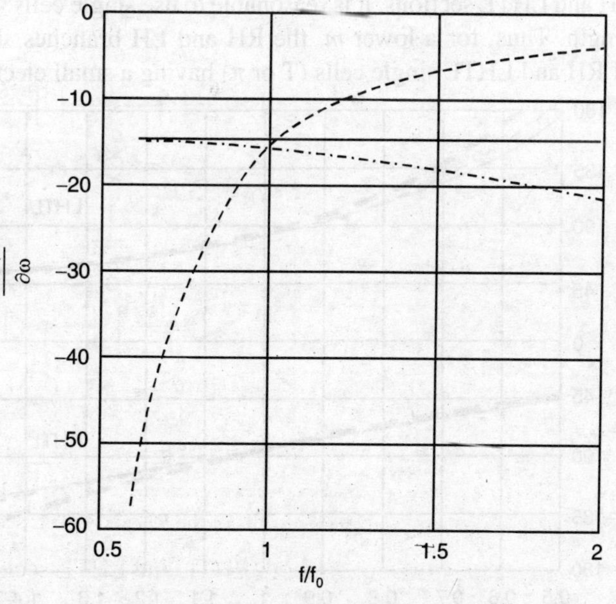

Fig. 9.9 Frequency dependence of derivative $\frac{\partial(\Delta\varphi)}{\partial\omega}$ in [deg] for the natural TL section (solid line), T-section of lumped element RHTL (dash-dotted line), and T-section of lumped element LHTL (dashed line); the electrical length $|\theta|_0 = 45°$ at $f/f_0 = 1$.

The phase response of a TL section relates to the electrical length as $\phi(\omega) = -\theta(\omega)$. For the maximum available digital phase shift, $\Delta\phi = \phi_{LH} - \phi_{RH} = 180°$ ($m = 1$), the electrical lengths of both RH and LH channels are $|\phi_0| = 90°$, and consequently $\omega_0{}^2 L_R C_R = \omega_0{}^2 L_L C_L = 1$. At the central frequency of the operational bandwidth ω_0

$$\left.\frac{d\phi_{RH}}{d\omega}\right|_{\omega=\omega_0} = \left.\frac{d\phi_{LH}}{d\omega}\right|_{\omega=\omega_0} = -2\sqrt{L_{R,L}C_{R,L}} \quad ...(40)$$

The equality is valid at the central frequency only. The frequency dependence of the slope is different for LH and RHTLs.

For $m > 1$, the terms in Eqns. (38) and (39), $\omega_0^2 L_R C_R/2 << 1$ and $\frac{1}{2}\,\omega_0^2 L_L C_L << 1$, both decreases rapidly when m arises. Thus, one can simplify Eqns. (38) and (39) for $m > 1$

$$\frac{d\phi_{RH}}{d\omega} = -\sqrt{2L_R C_R} \quad ...(41)$$

$$\frac{d\phi_{LH}}{d\omega} = -\frac{\sqrt{2L_L C_L}}{\omega^2} \quad ...(42)$$

Figure 9.9 shows the frequency dependencies for the slope parameter of the distributed RHTL section for the natural TL section, lumped element section of RHTL, the lumped element section of LHTL with electrical length $\phi_0 = 45°$ ($m = 2$) at $f/f_0 = 1$. The difference in the slope parameters of the LH and RHTL sections is remarkably pronounced at a lower frequency range ($\omega << \omega_0$) and is less at higher frequencies, though rises at $\omega >> \omega_0$. Concluding, one finds that for a design of the broadband phase shifter on switchable RH and LHTL sections, it is reasonable to use single cells with a small value of the equivalent electrical length. Thus, for a lower m, the RH and LH branches should be designed as a cascaded connection of RH and LHTL single cells (T or π) having a small electrical length.

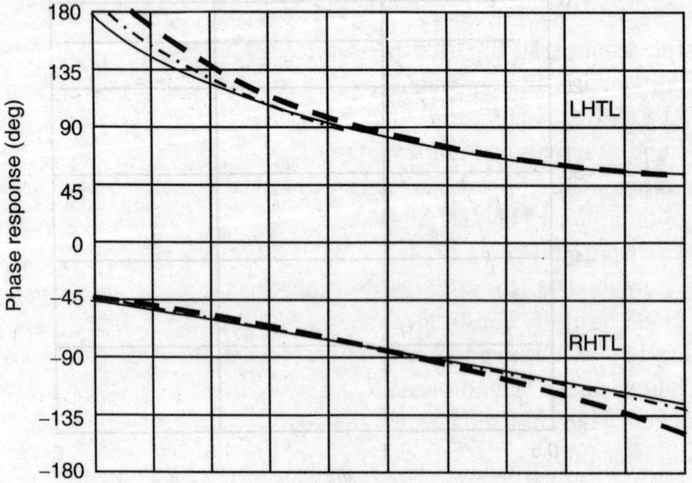

Fig. 9.10 Phase characteristics versus normalised frequency for different numbers of LC section: single LC section, dashed lines; two LC sections, dash-dotted lines; and ideal TL section, solid lines.

Figure 9.10 presents the theoretical phase response of the RH and LHTL sections for different numbers of LC sections (single cells in Fig. 9.8) providing a phase shift of 180° while switching the RH and LH channels.

When the characteristic impedance of both the TLs Z_0 is the same, the perfect matching of the network shown in Fig. 9.7 is provided in any frequency range for both states if Z_0 is equal to the port impedance.

We have seen that combination of the RH and LHTL sections is useful for applications in microwave techniques. A new approach to designing digital phase shifters that are free of spurious response resonators with enlarged functionality resulted in exceptional performance of the devices. The combination of the RH and LHTL sections can also be used for improving the characteristics of directional couplers, power splitters, matching circuits, etc. Modern multilayer technology makes it possible to design fully integrated miniature microwave devices on the RH/LHTLs.

9.5 METAMATERIAL LENSES AND SUPERLENSES

The concept of a perfect lens that is able to restore both the propagating and evanescent spectra of a source at the image is a product of the theoretical work of Veselago, who in 1968 consolidated earlier thoughts about the focussing of rays without aberration through a LH metamaterial slab by way of negative refraction, and Pendry, who in 2000 proposed the conditions that would enable Veselago's slab lens to transmit near-field information by way of the apparent amplification of evanescent waves. Such a lens is called Veselago-Pendry superlens, for it is able to exceed the performance of conventional lenses, which are constrained by the classical diffraction limit to produce images whose resolutions are limited to the order of $\lambda/2$ of illumination. These superlenses have many potential applications in biomedical, microelectronics, and defense related fields in subdiffraction microscopy, lithography, tomography, and sensing, but their design is complicated by the strict requirement for isotropic $\mu = -\mu_0$ and $\varepsilon = -\varepsilon_0$. These necessary conditions ensure a refractive index of $n = -1$ while guaranteeing matching to free space. Furthermore, these lenses require extremely low-loss characteristics and must be adequately thin.

The first successful attempt at superlensing of Veselago-Pendry type, and indeed, the first demonstration of subdiffraction imaging using metamaterials employed negative-refractive index transmission line (NRI-TL) metamaterials, which consist of fine TL grid loaded with inductors and capacitors to control their effective-medium response. This effort began with the realisation of planar NRI-TL lenses based on discrete-lumped loading, followed by the development of a rigorous theory that enabled the design of a planar NRI-TL superlens, which successfully verified the phenomenon of Veselago-Pendry superlensing, albeit in a planar microstrip environment. The need to extend the benefits of the NRI-TL method to free-space realisations led to volumetric topologies that, although polarisation-specific, could be easily constructed, and one of these topologies was shown to exhibit true free-space superlensing of the Veselago-Pendry type. Nowadays, fully isotropic 3*D* metamaterials based on NRI-TL techniques have been proposed, and it seems that topological simplifications, along with ongoing developments in fabrication techniques, should facilitate their realization and suggest practical applications in near future.

9.6 ELECTROMAGNETIC AND PHOTONIC CRYSTALS

Electromagnetic crystals are artificial periodical structures operating at frequencies where the wavelength is comparable with the characteristic period of the structure. In the optical frequency range, such structures are called photonic crystals. The inherent feature of such media is the existence of frequency bands inside which wave propagation is not allowed. These bands are called bandgaps, and therefore these crystals are sometimes also called *electromagnetic bandgap* (EBG) or *photonic bandgap* (PBG) structures. The bandgaps originate physically from the spatial (Bloch) resonances of the crystal and strongly depend on the direction of wave propagation. This reveals that electromagnetic crystals are media with spatial dispersion.

9.7 HYPERLENS

Hyperlens is one of the extraordinary applications of metamaterials: utilizing hyperlens, a metamaterial with permittivities in opposite signs in two orthogonal directions of cylindrical coordinates, to overcome the resolution limit of optical imaging. A hyperlens can be a simple additional part to any existing microscopes that will remarkably improve the resolution.

An EM wave has two distinct properties, the angular frequency (ω) and the wave vector k that is the inverse of wavelength (λ). There is a special relationship between the two and this dispersion in a medium is represented by the following in k-space:

$$\frac{k_r^2}{\varepsilon_\theta}+\frac{k_\theta^2}{\varepsilon_r}=\frac{\omega^2}{c^2} \qquad \text{...(43)}$$

where k_r and k_θ are wave vectors, ε_r and ε_θ are electric permittivities in radial and tangential directions, respectively. ω is the angular frequency and c is the speed of light in vacuum.

Equation (43) is written in cylindrical coordinator for convenience. If a medium exhibits isotropic electric response ($\varepsilon_r = \varepsilon_\theta = \varepsilon$) in all directions (*i.e.*, an isotropic medium), Eqn. (43) can be graphically represented with a circle (Fig. 9.11(*a*) solid). However, for imaging application, this circular dispersion is undesirable because accessible wave vectors (k_θ) are limited to the refractive index of the medium, $n = \sqrt{\varepsilon}$.

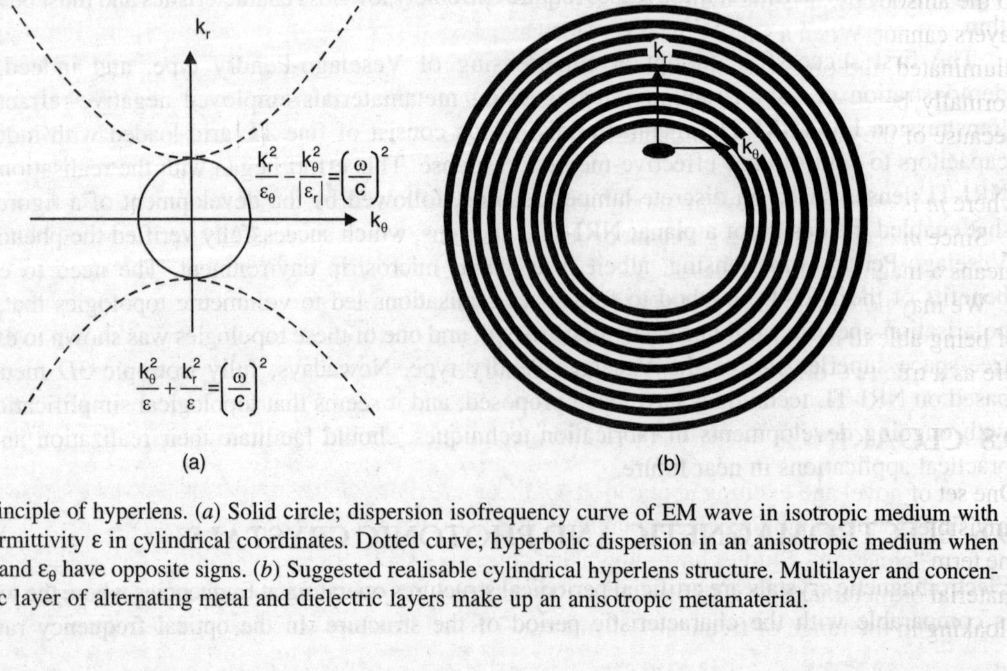

Fig. 9.11 Principle of hyperlens. (*a*) Solid circle; dispersion isofrequency curve of EM wave in isotropic medium with permittivity ε in cylindrical coordinates. Dotted curve; hyperbolic dispersion in an anisotropic medium when ε_r and ε_θ have opposite signs. (*b*) Suggested realisable cylindrical hyperlens structure. Multilayer and concentric layer of alternating metal and dielectric layers make up an anisotropic metamaterial.

This is why higher index materials are sought to be used in immersion microscopy; the larger the k_θ collected, the better the resolution. We know that current immersion methods have their limitations simply because there are not many materials available in nature with high valres of n. However, a material with anisotropic electrical responses ($\varepsilon_r \neq \varepsilon_\theta$) can indeed support the propagation of large wave vectors. This means that if ε_r can be engineered to be negative and ε_θ positive, there is a way to harness arbitrarily k_θ. In such a case, dispersion relation reads as

$$\frac{k_r^2}{\varepsilon_\theta} - \frac{k_\theta^2}{|\varepsilon_r|} = \frac{\omega^2}{c^2} \quad \text{...(44)}$$

which represents a hyberbola. With a hyperbolic dispersion (Fig. 9.10(*a*)), theoretically there is no limit to the maximum k_θ that can be supported by the material. This device was named *'hyperlens'*.

Materials of such characteristics are not found in nature. However, a hyperlens can be engineered as a metamaterial. Figure 9.10(*b*) shows one example of such materials that consists of alternating layers of a metal and a dielectric, whose cylindrical configuration is similar to earlier reported cylindrical superlens. Since in visible frequency, metals tend to have negative and dielectrics have positive permittivities, a metamaterial that contains both metals and dielectrics can be designed to have opposite signs of permittivities in two orthogonal directions using the following equations:

$$\varepsilon_\theta = \frac{\varepsilon_m + \varepsilon_d}{2} \quad \text{...(45)}$$

$$\varepsilon_r = \frac{2\varepsilon_m \varepsilon_d}{\varepsilon_m + \varepsilon_d} \quad \text{...(46)}$$

where ε_m and ε_d are permittivities of the metal and the dielectric layers , respectively. Assuming the layer thickness are much smaller than the workiug wavelength, this cylindrical anisotropy can give rise to a hyperbolic dispersion behaving as one bulk matter, or as an effective matter. We may not that in addition to the anisotropy, it is the cylindrical geometry that provides the imaging capability that a stack of flat layers cannot. When a small object is placed at the centre of the cylindrical hyperlens (Fig. 9.10(*b*)) and illuminated, the object scatters the light and generates a wideband of k_θ. The large k_θ, which would normally be evanescent in common dielectrics, can now propagate along the radial direction, and because of the curvature, the k_θ shrinks due to the conservation of angular momentum:

$$m = k_\theta r \quad \text{...(47)}$$

where m is angular momentum mode number and r is the radius.

Since m is constant for a given mode, the k_θ has to become smaller as r becomes larger, which simply means a magnification process.

We may note that the hyperlens design is not limited to the cylindrical geometry that has a drawback of being able to magnify in only one direction. A spherical hyperlens would be more applicable in real life as a true two-dimensional (2*D*) nanoimaging device.

9.8 CLOAKING

One set of novel and exciting applications of metamaterials deals with making a given object essentially 'invisible' to the impinging electromagnetic wave. This application of metamaterials, associating with it the term "cloaking". Studies have shown that covering a dielectric or conducting object with a plasmonic material or metamaterial with permittivity lower than the one of the background may realize the desired cloaking in the range of frequency of interest. This effect relies on a scattering cancellation, for which the wave scattered from the cloak may cancel the one from the object to be cloaked, leaving an external observer with a very low residual scattering that makes the system practically invisible around the design frequency. Besides this, there are three other cloaking techniques, *e.g.* the possibility of inducing anomalous localised resonances they may isolate and cloak a given object, use of transmission line circuits to match a given network surrounding an object with the background material. It is expected that metamaterial cloaking may become of major interest for various exciting applications in various scenarios, including *camouflaging* and noninvasive probing, and sensing in medical and biological applications.

9.9 METAMATERIAL TRANSMISSION LINES

These are *1D* artificial propagating media that exhibit either a single-negative (SNG) or a double negative (DNG, or left-handed, LH) behaviour. In the first case, signal propagation is not allowed in the frequency region of interest (even though the medium is transparent at other frequencies) since the effective permeability or permittivity are negative in that region. Conversely, LH metamaterial transmission lines exhibit backward-wave propagation in the LH band (due to the coexistence of negative effective permittivity and permeability in that band) and forward-wave propagation in a certain frequency interval at higher frequencies. Purely left-handed (PLH) lines (*i.e.* those lines exhibiting only a backward-wave transmission band) cannot be implemented in practice due to the presence of parasitic elements. This is why, LH lines actually exhibit a composite behaviour, and for this reason these lines are called composite right/lelt-handed (CRLH) lines. Two main approaches can be used for the implementation of metamaterial transmission lines: (*i*) the nonresonant type approach, where a host line is periodically loaded with series capacitors and/or shunt capacitors, and (*ii*) the resonant-type approach, where the host transmission lines are loaded with split ring resonators (SRRs) or complimentary split ring resonators, CSRRs). SRRs have been widely used to design and fabricate left-handed materials (LHMs), because SRRs provide the required μ-negative (MNG) properly, whereas wise medium provides ε negative (ENG) property.

CRLH transmission line (TL) metamaterials which exhibit both a LH frequency band and RH frequency with an infinite wavelength transition frequency when a structure resonances are balanced, represent a new paradigm in electromagnetic theory and engineering. In their metamaterial (effectively homogeneous) range, CRLH structures may be conveniently described as ideal uniform TLs. When a broadband description is required, to account for bandage; anisotropy; stopband; and diffraction effect, CRLH structures may be analyzed using Bloch-Floquet TL analysis (*1D*, *2D* or *3D*). The fundamental parameters of CRLH metamaterials are their dispersion relation and their Bloch impedance, and their operation bandwidth is determined by the lumped LC elements of their unit cell. Useful CRLH metamaterial concepts for antennas include the infinite-wavelength power divider, which may be used as a compact and low-loss series feeding network, multiband capabilities (dual-band for the C-CRLH case, tri-band for the DL-CRLH case, and quad-band for the extended CRLH case). CRLH metamaterials have lead to several breakthrough in both *1D* and *2D* leaky wave antenna and reflector applications.

9.10 CONTROLLABLE METAMATERIALS

The concept of controllability appears with the comparison of the behaviour of two complementary passive metallic bidimensional photonic bandgap (*2D* PBG). The first one, which consists of a square lattice of continuous metallic rods, is well-known for having a wide forbidden band from OHz to its plasmon frequency called plasmon-like band. This last one depends on the geometrical parameters of the structure. In this forbidden band, the effective permittivity is negative. The second one is a square lattice of discontinuous metallic rods. In this case a valence (or transmission) band occurred instead of the plasmon-like band at low frequency. The effective permittivity becomes positive in this allowed band.

Controllable photonic bandgap (CPBG) materials are used to realize planar and conformable antennas for communication at 12 GHz and in the GSM-UMTs. The main advantage of these CPBG materials is their capability to realize multifunction smart antennas. The main drawback of these materials is their thickness, which is of the order of several $\lambda/4$, where λ is the operating wavelength.

9.11 OPTOELECTRONIC CONTROL OF METAMATERIALS

While metamaterials have been utilized to create materials with unprecedented properties such as invisibility cloaks, negative refraction, and perfect lenses, metamaterials alone suffer important drawbacks. The resonant nature of most metamaterial designs to date leads to significant frequency dispersion, resulting in narrow bandwidth may be desirable, while for others a broader bandwidth is required, which present distinct challenges. Truly speaking, the usable range of any given design may be limited to about 5%–10% compared to its centre frequency ω_0. The narrowband resonant response and large metamaterial absorption severely restricts several potential applications. Similarly, the interesting and often desirable physical properties of many natural materials are also subject to limitations. However, composites that result from the combination of natural materials and metamaterials (usually termed hybrid metamaterials) have the potential to address deficiencies of either system, and are relatively unexplored so far.

Hybrid metamaterials form when the properties of a natural material, such as the dynamic photoconductivity of a semiconductor, strongly couples with the resonance of a metamaterial element. The resulting hybrid material will have passive properties (*e.g.*, negative electric response, negative index, or gradient index) that can be selected by the controlled design and patterning of the metamaterial elements, and dynamic properties (photoconductivity, nonlinearity, gain, etc.) that result from strategic incorporation of natural materials. Metamaterials thus not only extend and complement natural material response, but they can also yield tunable properties, enabling them to actively and individually control electric and magnetic fields. Coupled with their ability to yield nearly any EM response from radio frequencies to near optical frequencies underscores their importance and substantiates their use for future state-of-the art devices operating in any technologically relevant spectral range.

Metamaterials hybrid structures that can be controlled optoelectronically may be fabricated on, or constructed with, various semiconductors, *e.g.* semiinsulating gallium arsenide (SI-GaAs), GaAs: ErAs nanoisland superlattices, low temperature grown GaAs, silicon-on sapphire (SOS), silicon-on-insulator, and a thin n-doped GaAs layer, etc.

For optical control of metamaterials, the carrier generation efficiency must be considered. The number of carriers generated per cm^3 is given by

$$n = 6.25 \times 10^{18}\,\frac{I_0\,(I - R)\cdot\tau}{Ad\hbar\omega} \qquad ...(48)$$

where, I_0 is the average power in watts, R is the reflectivity, $\hbar\omega$ is the photon energy in eV, A is the area of the laser excitation and d is either the penetration depth or sample thickness in the appropriate limit.

The expression is valid for continuous (CW) and pulse excitation.

We may note that amplitude or frequency control of the metamaterial resonance may be possible through the dependence of epsilon infinity (of the substrate material) or interband transitions. The frequency dependence of the metamaterial electric response can be approximately modelled by the Brude-Lorentz model:

$$\bar{\varepsilon}(\omega) = \varepsilon_\infty + \frac{\omega_p^2}{\omega_0^2 - \omega^2 - i\gamma\omega} \quad \text{...(49)}$$

where, ε_∞ is the real part of dielectric constant at $\omega = \infty$, ω_0 is the central frequency of the oscillator and γ is damping.

Under optical excitation, one can increase ε_∞ thereby decreasing the frequency at which $\varepsilon_1(\omega) = 0$ on the high-frequency side of the resonance. In such cases, no additional loss is introduced at the metamaterial resonance frequency. In this way, we can extend the range of optical control of metamaterials to the mid-IR and perhaps near-IR regimes; although high powers are essential, significant heating of the substrate is problematic, and relatively small changes are expected.

Active control of metamaterial devices at THz frequencies has been achieved by biasing *n*-GaAs near the gaps of split rings, which effectively tuned the strength of the resonance of the metamaterial. While active control required arrays of conducting wires, it did not affect the oscillator strength of the response, lending potential flexibility to future devices. Dynamical control of metamaterial devices at THz frequencies was realized through photoexcitation of carriers in the substrate of the SRR. This demonstration of optical control of metamaterials allows for fast switching times that are not realizable by electronic means.

9.12 EXTRAORDINARY TRANSMISSION

The experimental discovery of unexpectedly high levels of transmission of light through perforated metal screens initiated interest on the problem of transmission of EM waves through electrically small holes practiced in opaque slabs. This phenomenon was termed extraordinary transmission because classical Bethe's theory for small holes predicted much lower power transmission per hole. Extraordinary transmission phenomenon can be classified under following three different categories:

(*i*) Transmission through periodic (or quasiperiodic) arrays of apertures in opaque screens. These apertures can be either infinitely long slits or finite size holes of various shapes.

(*ii*) Transmission through a single aperture around which the surface of the screen is periodically structured (with corrugations, *e.g.*).

(*iii*) Transmission through one, two, or three holes without periodic patterns around them.

The phenomenon of extraordinary transmission of EM waves through metal screens periodically performated with small holes or narrow slits can be explained in terms of impedance matching concepts. This point of view directly leads to equivalent circuit models that are very useful to make qualitative and quantitative predictions. The circuit models only require the determination of a reduced set of parameters that can be used for obtaining the entire transmission spectrum in the frequency range of interest.

9.13 PHOTONIC CRYSTALS (PHCs)

Although fundamentally photonic crystals are not really subwavelength structures, and are not considered as metamaterials by some negative refraction is possible in these dielectric structures without a negative refractive index or 'backward wave" effect if they contain periodic variations on the scale of wavelength of the electromagnetic radiation. These materials can also be obtained using self-organisation principles. This effect in photonic crystals comes from complex Bragg scattering effects, which differ from effects in left-handed metamaterials.

There are numerous opportunities for PhCs to bring decisive improvements in active systems. In addition to their applications to individual devices, PhCs obiously offer the possibility of revisiting integration in optics. Given the present thrust for integration and for PICs, complex sources combining passive and active devices have a good chance to be the first integrated systems that will exploit PhCs. One advantage of PhCs is simply due to the fact that they impose the polarisation state of emitting devices (typically TE).

It seems that the use of PhCs in practical systems does not rest so much on their bandgap properties. At present, no specific direction of application can be mentioned, future information technologies will undoubtedly benefit from today's research on semiconductor based PhCs. Photonics is a rich and diversified field, where all opportunities to combine light and matter are properly scrutinised.

Photonic metamaterials open up a way to overcome the constraints set by ordinary materials. The basic idea is to create an artificial crystals with deep subwavelength periods. Analogous to an ordinary optical material, such a photonic metamaterial can be treated as an effective medium that is characterised by effective material parameters ε(ω) and μ(ω). However, the proper design of the elementary building blocks ("artificial atoms") of the photonic metamaterial allows for a nonvanishing magnetic response and even $\mu < 0$ at optical frequencies—despite the fact that constituent materials of photonic metamaterials are nonmagnetic.

9.14 FABRICATION OF METAMATERIALS

Lithographic tools have played a pivotal role in the fabrication of metamaterials. As the dimensions and wavelengths have shrunk, the technology of printed circuit boards has been replaced by other lithographic processes, such as optical and electron-beam lithography (EBL). All three technologies share a common base—the use of *resist*. Resist refers to the patterning placed upon a material that can then be transferred into the material by the application of ink or dye. In other words, covering a selected area of material (*e.g.*, the copper in circuit boards, dielectrics, or semiconductor materials) with resist allows it to be processed in a different manner from that used on the rest of the material. Pattern transfer is achieved by first applying resist to the whole surface of a piece of material or substrate and then selectively removing some of the resist. A further processing step, which may be deposition of another material or etching of the existing material, transfers the pattern itself into the material.

The resist is usually a polymer and the selective removal of the resist can be achieved by exposing it to either ultraviolet (UV) light or electrons (X-rays and ions are also possible). Resist comes in either a positive or a negative form: positive organic resists consists of polymers, the chains of which are broken during the exposure process; conversely, in negative organic resists, the exposure to UV light or electrons causes the formation of additional cross-links between the polymer chains. Examples of positive and negative commercially available resists are polymethyl methacrylate (PMMA) and AR-N-a novolak resin-based resist, respectively.

Once the resist has been developed the next stage of processing can begin, which involves either an etching step onto previously defined metallisation or selective deposition of metal through holes developed in the resist layer.

Following methods of fabrication are used:

(A) **Lithography**

A.1 Optical lithography

A.2 Electron-beam lithography

A.3 Lift-off

A.4 Multiple layer structures

A.5 Arbitrary multilayer structures

A.6 Nanoimprint lithography

(B) Focussed ion beam etching.

(C) SRR-based metamaterials in different metallisations.

The approaches include high-precision "top-down" deterministic ones that are typically specified, in the first place, as a set of patterns which are transferred either directly or via a mask to the unstructured material followed by deposition and/or etching.

SUMMARY

1. Metamaterials are artificial engineered materials that can provide unusual elecromagnetic properties not readily available in nature, represent a new paradigm in electronics and photonics.
2. One of the most important properties of metamaterials is the feasibility of left-handed materials (LHMs) from the combination of electric conductors. These LHMs are characterised by their negative values of electric permittivity ($\varepsilon < 0$) and magnetic permeability ($\mu < 0$); and hence, left-handed materials (LHMs) are also referred to as double negative (DNG) materials. Several applications arise from the use of such LHMs, like the negative refraction index, or the feasibility of LH transmission lines.
3. The beauty of metamaterial concept is the simple epiphany that we do not have to rearrange atoms- or subatomic particles therein – in order to engineer a new matter; instead, we fabricate artificial structure with the unit components much smaller than the wavelength of interacting electromagnetic (EM) waves. The collective response of the bulk can result in novel EM properties that are not observed in nature.
4. Metamaterials are beginning to transform optics and microwave technology. The versatile properties of these materials can be tailored according to practical needs and desires.
5. Perhaps the best-known metamaterials are the materials used in the pioneering demonstrations of negative refraction or invisibility cloaking of microwaves, or for negative refraction of near-visible light. These materials consist of metallic cells that are smaller than the relevant electromagnetic wavelength. Each cell acts like an artificial atom that can be tuned by changing the shape and the dimensions of the metallic structure.
6. Different approaches are used for a description of the fundamental EM properties of metamaterials and the practical realization of these materials as well. Among them, the transmission line (TL) approach gives an efficient design tool for microwave applications providing a correct description of physical properties of metamaterials.
7. In contrast to conventional periodic structures that are usually operated in the Bragg regime (period $\approx \lambda_g/2$, where λ_g is the guided wavelength) for filtering, metamaterials are long-wavelength structures (average distance between unit elements or atoms $<< \lambda_g$). Since they are not readily available naturally, they are engineered under the form of *1D*, *2D* or *3D* structures.
8. Composite right/left-handed (CRLH) metamaterials have lead to several breakthroughs in both *1D* and *2D* leaky wave antenna and reflector applications.

9. An interesting aspect of using metamaterials in radiating systems is the possibility to create highly directive beams from simple sources placed inside planar layered structures.
10. Electronically controllable photonic bandgap (CPBG) materials are used to realize planar and conformable antennas for communication at 12 GHz and in the GSM-UMTs band. The main advantage of these CPBG materials is their capability to realize multifunction smart antennas. The main drawback of these materials is their thickness, which is of the order of several $\lambda/4$ where λ is the operating wavelength.

SUGGESTED READINGS

1. J.B. Pendry, *Phys. Rev.* Lett. 85, 3966 (2000).
2. V.G. Veselago, *Sov. Phys.* Uspekhi, 10, 509 (1968).
3. X. Zhu and M. Obtsu (eds), *Near-Field Optics: Principles and Applications*, World Scientific, Singapore (1999).
4. C. Caloz and T. Itoh, *Electromagnetic Metamaterials: Transmission Line Theory and Microwave Applications*, John Wiley (2006).
5. E. Eleftheriades and K.G. Balmain, *Negative-Refraction Metamaterials: Fundamental Principles and Applications*, Wiley, New York (2006).
6. N. Engheta and R. Ziolkowski (eds.) *Metamaterial Physics and Engineering Explorations*, John Wiley, New York (2006).
7. P. Alitala and S. Tretyakov, *Metamaterials*, 1(2), 81–88(2007).
8. F. Capolino (ed.), *Theory and Phenomena of Metamaterials*, CRC Press, Taylor and Francis Group (2009).

REVIEW QUESTIONS

1. What are metamaterials? How they differ from normal materials?
2. What are electromagnetic and photonic crystals? Explain the concept of negative refraction and backward wave effects.
3. Write the properties of materials with extreme anisotropy.
4. What is super resolution effect? Explain the concept of super-resolution imaging with hyperlens.
5. Give an brief account of metamaterials for cloaking applications.
6. What are electromagnetic bandgaps? How they are used to microwave circuit design?
7. Why metamaterials are also referred as double negative (BNG) materials?
8. Give an brief account of microwave phase shifters and filters based on a combination of left-handed and right-handed transmission lines.
9. Give an brief account composite right/left-handed (CRLH) transmission line (TL) metamaterials. How these materials are substantially different from filters?
10. Given an brief account of metamaterial antenna applications.

SHORT ANSWER QUESTIONS

1. What are metamaterials?

Ans. These are artificial engineered materials that can provide unusual electromagnetic properties. One of their most important properties is the feasibility of left-handed materials (LHMs) from the combination of electric conductors. These LHMs are characterised by their negative values of electric permittivity ($\varepsilon < 0$) and magnetic permeability ($\mu < 0$); and hence, LHMs are also referred to as *double negative* materials. Several applications arise from the use of such LHMs, like the negative refraction index, or the feasibility of LH transmission lines.

2. What are electromagnetic and photonic crystals?

Ans. Electromagnetic crystals are artificial periodical structures operating at frequencies where the wavelength is comparable with the characteristic period of the structure. In the optical frequency range, such structures are referred as photonic crystals. The inherent feature of such media is the existence of frequency bands inside which wave propagation is not allowed. These bands are also called band gaps, and therefore these crystals are sometimes also called *electromagnetic band gap* (EBG) or *photonic band gap* (PBG) structures.

3. Mention the properties of materials with extreme anisotropy?

Ans. These materials have remarkable properties. All the extraordinary waves supported by the media are propagating waves. Thus, theoretically, it is possible to transport an arbitrary field distribution with suitable polarisation through such material with no loss of resolution. Lens formed by slabs of such materials provide a unique opportunity to transmit the near-field with super-resolution. At the front interface with free space, the evanescent waves are transformed into propagating waves, which prevents their decay and preserves the subwavelength information. The propagating waves travel through the slab and reproduce the image at the back interface of the lens. This phenomenon is known as *'canalisation'*. This regime is particularly appropriate for the transmission of images with super-resolution over significant distances in terms of wavelength. We may note that this effect cannot be achieved using other available imaging techniques.

4. How the innovation of metamaterials opened wide the door in the field of electromagnetics?

Ans. The beauty of metamaterial concept is the simply epiphany that we do not have to rearrange atoms- or subatomic particles therein—in order to engineer a new matter; instead we fabricate artificial structure with the unit components much smaller than the wavelength of the interacting electromagnetic (EM) waves. The collective response of the bulk can result in novel EM properties that are not observed in nature.

5. What is a hyperlens?

Ans. A hyperlens utilizes unusual characteristic of an anisotropic metamaterial to carry large wave vectors into the far field. Concentric multilayers of metal and dielectrics make up this metamaterial structure. This metamaterial lens was designed such that (*i*) large wave vectors can be collected and (*ii*) to gradually convert them into smaller wave vectors. These small wave vectors are then captured by a conventional microscope that directly delivers the magnified image of nanoscale objects to our eyes. A hyperlens can be a simple additional part to any existing microscopes that will remarkably improve the resolution. Hyperlensing requires no postimaging processes and is a direct far-field imaging technique.

6. What is metamaterial cloaking?

Ans. One set of novel and exciting applications of metamaterials deals with making a given object essentially "invisible" to the impinging electromagnetic wave. This is termed as metamaterial cloaking. Metamaterial cloaking may become of major interest for various exciting applications in various scenarios, including camouflaging and noninvasive probing, and sensing in medical and biological applications.

7. What are metamaterial transmission lines?

Ans. Metamaterial transmission lines are one-dimensional (1*D*) artificial propagating media that exhibit either a single-negative (SNG) or a double-negative (DNG or left-handed, LH) behaviour. In the first case, signal propagation is not allowed in the frequency region of interest (even though the medium is transparent at other frequencies) since the effective permeability and permittivity are negative in that region. Conversely, LH metamaterial transmission lines exhibit backward wave propagation in the LH band (due to the coexistence of negative effective permittivity and permeability in that band) and forward wave propagation in a certain frequency interval at higher frequencies. Purely left-handed (PLH) lines (*i.e.* those lines exhibiting only a backward wave transmission band) cannot be implemented in practice due to the presence of parasitic elements. Hence LH lines actually exhibit a composite behaviour, and for this reason these lines are called composite right/left-handed (CRLH) lines. Two main approaches can be used for the implementation of metamaterial transmission lines: (*i*) the nonresonant-type approach, where a host line is periodically loaded with series capacitors and/or shunt capacitors, and the resonant-type approach, where the host transmission lines are loaded with SRRs or CSRRs.

8. Explain the concept of microwave phase shifters and filters based on a combination of left-handed (LH) and right-handed (RH) transmission lines.

Ans. Different approaches are used for a description of the fundamental FM properties of metamaterials and the practical realization of these materials as well. Among them, the transmission line (TL) approach gives an efficient design tool for microwave applications providing a correct description of physical properties of metamaterials. A conventional TL with a positive phase velocity behaves as right-handed transmission line (RHTL). An artificial RHTL can be formed as a ladder network of capacitors connected in shunt- and series- connected inductors. A combination of the RH and LHTL sections is useful for applications in microwave techniques. A new approach to designing digital phase shifters that are free of spurious response resonators with enlarged functionality resulted in exceptional performance of the devices. One can also use the combination of RH and LHTL sections for improving the characteristics of directional couplers, power splitters, matching circuits, etc. Modern multilayer technology makes it possible to design fully integrated miniature microwave devices on the RH/LHTLs.

9. What is difference between conventional periodic structures and metamaterials?

Ans. Metamaterials are broadly defined as effectively homogeneous structures exhibiting unusual electromagnetic properties not readily available in nature, represent a new paradigm in electronics and photonics. In contrast to conventional periodic structures that are usually operated in the Bragg regime (period ≈ $\lambda_g/2$, where λ_g is the guided wavelength) for filtering, metamaterials are long-wavelength structures (average distance between unit elements or atoms << λ_g). Since they are not readily available naturally, they are engineered under the form of 1*D*, 2*D* or 3*D* structures. The most general, efficient, and applied TL metamaterials to date are the so-called composite right/left-handed (CRLH) TL metamaterials. The fundamental parameters of CRLH metamaterials are their dispersion relation and their Bloch impedance, and their operation bandwidth is determined by the lumped LC elements of their unit cell.

10. What is active and dynamical control of metamaterial devices at THz frequencies?

Ans. Active control is achieved by biasing *n*-GaAs near the gaps of split rings, which effectively tuned the strength of the resonance of the metamaterial. Active control required arrays of conducting wires, but it did not affect the oscillator strength of the response, lending potential flexibility to future devices. Dynamical control is realized through photoexcitation of carriers in the substrate of the split ring resonator (SRR). This demonstration of the optical control allows for fast switching times that are not realizable by electronic means.

11. Is ruby glass a metamaterial?

Ans. Yes, ruby glass contains nanoscale gold colloids that render the glass neither golden nor transparent, but ruby, depending on the size and concentration of the gold droplets. The colour originates from a resonance of the surface plasmons on the metallic droplets. Metamaterials per se are nothing new: what is new is the degree of control over the structures in the material that generate the desired properties.

FILL IN THE BLANKS

1. Electromagnetic crystals are artificial .... structures operating at frequencies where the wavelength is comparable with the .... period of the structure. [periodic, characteristic]
2. When the usual (forward) wave propagates in the isotropic medium, its way vector and its group velocity vector are .... In the backward-wave, these two vectors are by definition .... [parallel, antiparallel]
3. Lens based imaging devices such as conventional microscopes can not provide resolution better than .... , where λ is the wavelength of radiation. This restriction is known as the .... limit and holds irrespectively of the frequency of operation—from microwave frequencies up to the visible range. [$\lambda/2$, diffraction]
4. When a multilayer array of plasmonic nanosphere is placed near an electric dipole, on the other side, the optical near field is .... due to the excitation of .... modes in the layers. [enhanced, resonance]
5. The smallest resolvable feature (Δ) of any optical system can be approximated by the diffraction limit

$$\Delta = \frac{\lambda}{2(NA)}$$

where λ is the wavelength of the light in the free space and NA = $n \sin \theta$ is the ...., where n is the refractive index of the medium and θ is the light collection angle of the optics. [numerical aperture]

6. A hyperlens utilizes unusual characteristic of an .... metamaterial to carry large wave vectors into the far field. [anisotropic]
7. Metamaterials are materials with electromagnetic properties that originate from human-made .... structures. [subwavelength]
8. Perhaps the best known metamaterials are the materials used in the pioneering demonstrations of .... refraction or invisibility .... of microwaves, or for .... refraction of near-visible light. [negative, cloaking, negative]
9. One of the most important properties of metamaterials is the feasibility of .... materials from the combination of electric conductors. [left-handed (LH)]
10. In contrast to conventional periodic structures that are usually operated in the Bragg regime (period $\approx \lambda_g/2$, where λ_g is the guided wavelength) for filtering, metamaterials are....... wavelength structures (average distance between unit elements or atoms << ....). [long, λ_g]

OBJECTIVE QUESTIONS

1. Electromagnetic crystals are media with
(*a*) spatial dispersion (*b*) interference
(*c*) polarisation (*d*) none of these [*a*]
2. Lens-based imaging devices such as conventional microscopes cannot provide resolution better than
(*a*) λ (*b*) $\lambda/2$
(*c*) λ^2 (*d*) λ^3 [*b*]
3. Hyperlen is a metamaterial having signs of permittivities
(*a*) same (*b*) opposite
(*c*) One zero and other positive (*d*) None of these [*b*]
4. Metamaterials are materials with electromagnetic properties that originate from human-made
(*a*) subwavelength structures (*b*) interference
(*c*) diffraction (*d*) polarisation [*a*]
5. Ruby glass is an example of
(*a*) semiconductor (*b*) metal
(*c*) insulator (*d*) metamaterial [*d*]
6. Electromagnetic bandgaps (EBGs) are periodic structures that are able to inhibit signal propagation at certain
(*a*) frequencies (*b*) directions
(*c*) frequencies and directions (*d*) none of these [*c*]
7. Left-handed materials (LHMs) are also referred to as
(*a*) singly negative materials (*b*) double negative materials
(*c*) positive materials (*d*) none of these [*b*]
8. Metamaterials are not readily available naturally, they are engineered under the form of
(*a*) 1*D* structure only (*b*) 2*D* structure only
(*c*) 3*D* structure only (*d*) 1*D*, 2*D* or 3*D* structures [*d*]
9. The proper design of the elementary building blocks (artificial atoms) of the photonic metamaterial allows for a nonvanishing magnetic response and even $\mu = 0$ at optical frequencies–despite the fact that constituent materials of the photonic metamaterial are
(*a*) magnetic (*b*) nonmagnetic
(*c*) superconducting (*d*) none of these [*b*]

CHAPTER

10

Nonlinear Optical Materials

10.1 INTRODUCTION

Nonlinear optics is attracting increasing attention of scientists and engineers due to its application in telecommunications and possibilities for optical information storage and computing. Optical fibre communications has already proven that optics is the method of choice for many purposes, owing to its wide band width and freedom from electromagnetic interference. Nonlinear optics enhances further improvements in the efficiency and versatility of existing advantages of optics, *e.g.* amplification of optical signals. We know that fibre-optic cables have such low absorption that they can transmit signals over many kilometres, but eventually the signals need to be amplified.

At present, this is achieved by converting the weak optical signal to an electronic one, amplifying that electronically, and then converting the strong electronic signal into a strong optical signal again. Obviously, it would be more efficient if the light beam could be amplified directly, say by a laser beam in a suitable medium. Such a process comes into the realm of nonlinear optics.

10.2 NONLINEAR OPTICAL PHENOMENA

Nonlinear optics (NLO) is essentially a material phenomenon. The fundamental laws of optics, as embodied in Maxwell's equations of electromagnetism, apply in all materials. But in any material the solution of the equations depends on *constitutive relations* characteristic of material. In particular, these relate the current density to the electric field through the conductivity, and the electric displacement to the electric field through the permittivity. The latter is of significance here.

The relation between the electric displacement vector $\vec{D}$ and the electric field $\vec{E}$ is

$$\vec{D} = \varepsilon \cdot \vec{E} \quad ...(1)$$

where ε is permittivity tensor. Usually one factors ε is to

$$\vec{\varepsilon} = \varepsilon_0 \vec{\varepsilon}_r \quad ...(2)$$

where ε_0 is the permittivity of free space and $\vec{\varepsilon}_r$ is the *relative permittivity*. (This and subsequent equations will be written in a form assuming a four dimension rationalised system of units) such as

SI units. From definition

$$\vec{D} = \varepsilon_0 \vec{E} + \vec{P} \qquad ...(3)$$

where $\vec{P}$ is the electric polarisation, given by the electric dipole moment per unit volume (plus terms involving gradients of densities of higher electric multipole moments). Combining (1) to (3), one finds

$$P = \varepsilon_0 (\bar{\varepsilon}_r - 1) \cdot \vec{E} = \varepsilon_0 \vec{\chi} \cdot \vec{E} \qquad ...(4)$$

where $\vec{\chi}$ is the electric susceptibility.

In linear materials, the response is always proportional to the stimulus. Then, for example, the induced polarisation is proportional to the field and the susceptibility is independent of the field. In practice this is true at sufficiently low fields. However, at sufficiently high fields the polarisation stops being proportional to the field and hence the susceptibility starts depending on the field. One then has nonlinear response.

Since nonlinear response usually manifests itself as small deviations from linear response, one can express it by writing the induced polarisation as a power series in the field:

$$\vec{P}/\varepsilon_0 = \chi^{(1)} \cdot \vec{E} + \chi^{(2)} : \vec{E}\vec{E} + \chi^{(3)} : \vec{E}\vec{E}\vec{E} + ... \qquad ...(5)$$

Here $\chi^{(1)}$ is the linear susceptibility tensor, $\chi^{(2)}$ the quadratic susceptibility tensor, $\chi^{(3)}$ the cubic susceptibility tensor, and so on. We may note that custom excludes from eqn. (5) the numerical coefficients which would make it a Taylor series. In practice, normally terms beyond cubic are not required.

For static electric fields, eqn. (5) is adequate as it stands: the polarisation simply depends nonlinearly on the field. However, for optics one need to consider the oscillating electric field vectors of electromagnetic waves. In the linear term, the polarisation can oscillate at the same frequency as the electric field, but in the nonlinear terms, the product of two or more oscillating fields gives oscillations at combinations of the frequencies. Further, the linear response to a superposition of fields of different frequencies is the superposition of the separate linear responses, whereas the nonlinear response also includes cross-terms from simultaneous response to the different fields.

Now, a more general form of eqn. (5) with explicit frequency dependence can be written as

$$\vec{P}(-\omega_0)/\varepsilon_0 = \vec{\chi}^{(1)}(-\omega_0;\omega_0) \cdot \vec{E}(\omega_0) + \vec{\chi}^{(2)}(-\omega_0;\omega_1,\omega_2)$$

$$: \vec{E}(\omega_1)\vec{E}_2(\omega_2) + \vec{\chi}^{(3)}(-\omega_0;\omega_1,\omega_2,\omega_3) : \vec{E}(\omega_1)E(\omega_2)E(\omega_2) + ... \qquad ...(6)$$

We note that here the output frequency is ω_0 and is given a negative sign; in each susceptibility the sum of frequencies must be zero, so that $\omega_1 + \omega_2 = \omega_0$ in $\vec{\chi}^{(2)}$, $\omega_1 + \omega_2 + \omega_3 = \omega_0$ in $\vec{\chi}^{(3)}$, and so on. However, the input frequencies may be positive or negative in order to reflect the fact that the product of two fields say $\cos \omega_1 t \cos \omega_2 t$ can be expressed as the sum of two fields at the sum and difference frequencies, $\cos(\omega_1 \pm \omega_2)t$. (Alternatively, one may choose to regard $\vec{E}(\omega)$ as arising from $\cos \omega t$ via either $e^{-i\omega t}$ or $e^{+i\omega t}$ with the same result. Here positive frequencies correspond to absorption and negative ones to emission). As a result there is rich variety of nonlinear phenomena arising from different combination of input fields.

Relative to a set of Cartesian axes, eqn. (6) can be expressed in more detail as

$$\vec{P}_\alpha(-\omega_0)/\varepsilon_0 = \chi_{\alpha\beta}(-\omega_0;\omega_0)E_\beta(\omega_0)$$

$$+ \chi_{\alpha\beta\gamma}(-\omega_0;\omega_1,\omega_2)E_\beta(\omega_1)E_\gamma(\omega_2)$$

$$+\chi_{\alpha\beta\gamma\delta}\left(-\omega_0;\omega_1,\omega_2,\omega_3\right)E_\beta\left(\omega_1\right)$$

$$E_\gamma\left(\omega_2\right)E_\delta\left(\omega_3\right)+\ldots \qquad \ldots(7)$$

Subscripts α, β, γ, δ represent cartesian components for compactness, the Einstein summation convention is employed, in which repeated subscripts are understood to be summed overall three values (typically x, y, z or 1, 2, 3 according to context). Since a susceptibility with $n + 1$ subscripts is necessarily a component of the nth-order tensor $\vec{\chi}^{(n)}$, the subscript n can be omitted in the component form. Now, the nth order susceptibility has 3^{n+1} distinct components.

Now, if some of the frequencies are the same, then not all distinct susceptibility components will be different in value, since they are interchanged when two subscripts are interchanged at the same time as the corresponding frequencies: Obviously, the response of the system to two frequencies cannot depend on which we label ω_1 and which ω_2, so that for example

$$\chi_{\alpha\beta\gamma}(-\omega_0;\ \omega_1,\ \omega_2) = \chi_{\alpha\gamma\beta}(-\omega_0;\ \omega_2,\ \omega_1) \qquad \ldots(8)$$

We note that $\chi^{(1)}$ is always symmetric, *i.e.* $\chi_{\alpha\beta} = \chi_{\beta\alpha}$ for any frequency. (We may note that these results no longer apply in an applied magnetic field).

When all frequencies are zero, then all components $\chi_{\alpha\beta\gamma\ldots}$ are symmetric underall interchanges of the subscripts {αβγ....}. Theoretical treatments reveal that this symmetry will also hold approximately provided all frequencies are remote from the absorption frequencies of the material, and this is called as *Kleinman symmetry.* Experiments confirm the validity of this approximation to about 20% away from absorption bands.

Table 10.1 Symmetry Properties of Nonlinear Optical Susceptibility

Property	*Conditions of validity*	*Transformation that preserves numerical value of* χ	*Example with second-order susceptibility*
Reality of fields	Always true	Change sign of all frequency components and take complex conjugate	$\chi_{ijk}^{(2)}(-\omega_n-\omega_m, -\omega_n, -\omega_m) = \chi_{ijk}^{(2)}(\omega_n+\omega_m, \omega_n, \omega_m)$
Intrinsic permutation symmetry	Always true	Simultaneously interchange last two frequency components and last two Cartesian coordinates	$\chi_{ijk}^{(2)}(\omega_n+\omega_m, \omega_n, \omega_m) = \chi_{ijk}^{(2)}(\omega_n+\omega_m, \omega_m, \omega_n)$
Reality of nonlinear susceptibility tensor	Lossless medium	Take complex conjugate $= \chi_{ijk}^{(2)}(\omega_n+\omega_m, \omega_n, \omega_m)$	$\chi_{ijk}^{(2)}(\omega_n+\omega_m, \omega_n, \omega_m)$
Full permutation symmetry	Lossless medium	Simultaneously interchange any frequency components and the corresponding Cartesian coordinates	$\chi_{ijk}^{(2)}(\omega_n+\omega_m, \omega_n, \omega_m) = \chi_{ijk}^{(2)}(-\omega_n, \omega_m, -\omega_n-\omega_m)$ and, therefore, with the above results, $\chi_{ijk}^{(2)}(\omega_n+\omega_m, \omega_n, \omega_m) = \chi_{ijk}^{(2)}(\omega_n -\omega_m, \omega_n+\omega_m)$
Kleinman's symmetry	χ is independent of frequency and medium is lossless (*e.g.* $\omega_{opt.} \ll \omega_{reson.}$)	Interchange cartesian coordinates without interchanging frequency components	$\chi_{ijk}^{(2)}(\omega_n+\omega_m, \omega_n, \omega_m) = \chi_{ijk}^{(2)}(\omega_n+\omega_m, \omega_n, \omega_m)$
Inversion symmetry	Medium is centrosymmetric	All values of second-order susceptibility are zero	$\chi_{ijk}^{(2)}(\omega_n+\omega_m, \omega_n, \omega_m) = 0$

The symmetry of the material determines the number of independent nonzero components $\chi_{\alpha,\beta,\gamma,...}$. A symmetry operation transforms the material into an indistinguishable orientation but also transforms the components according to standard expressions. However, if the orientations are indistinguishable, then the transformed components must equal the untransformed ones, and this places constraints on thecomponents. According to one general result, in ecosymmetric materials, all components of even-order susceptibilities are zero, notably those of $\vec{\chi}^{(2)}$. This is due to the fact that under inversion a product of an even number of field components remains unchanged, whereas the polarisation changes sign. On the other hand, as a property of a centrosymmetric material, the susceptibility component must not change sign under in version, and these criteria are consistent only if the component is zero. Symmetry properties of nonlinear optical susceptibility is summarized in Table 10.1.

10.2.1 Harmonic Generation

We now consider specific classes of nonlinear optical phenomena. First, suppose the incident light is monochromatic, of frequency ω. Then eqn. (6) permits the generation of higher harmonics of frequency 2ω, 3ω,... . The quadratic susceptibility $\vec{\chi}^{(2)}$ (−2ω; ω, ω) gives rise to *second harmonic generation* (SHG) in non-centrosymmetric materials, the cubic susceptibility $\vec{\chi}^{(3)}$ (−3ω; ω, ω, ω) to *third harmonic generation* (THG), and so on. One may also speak of frequency doubling, tripling, and so on.

As all the input frequencies are equal, the components of the harmonic generation tensors are symmetric under interchange of any subscripts apart from the first. Thus, one finds

$$\chi_{\alpha\beta\gamma}(-2\omega;\omega,\omega)=\chi_{\alpha\gamma\beta}(-2\omega;\omega,\omega) \qquad ...(9)$$

$$\chi_{\alpha\beta\gamma\delta}(3\omega;\omega,\omega,\omega)=\chi_{\alpha\beta\delta\gamma}(-3\omega;\omega,\omega,\omega)=\chi_{\alpha\gamma\beta\delta}(-3\omega;\omega,\omega,\omega),\text{ etc.} \qquad ...(10)$$

For SHG, this symmetry reveals that there are only six independent combinations of subscripts βγ rather than nine. This means that it is usual to replace the double subscripts by a single one according to the prescription 11 → 1, 22 → 2, 33 → 3, 23 → 4, 31 → 5, 12 → 6 (the Voigt abbreviated notation).

Normally this is done not in $\chi_{\alpha\beta\gamma}$ itself but in the SHG coefficient $d_{\alpha\lambda}$, where α = 1, 2 or 3 and λ = 1, ...6; for *e.g.* $d_{123} \to d_{14}$. This is defined to relate the amplitudes of polarisation and the optical electric field, which are the experimentally convenient quantities rather than the complete oscillating quantities in eqn. (7). However, this has the effect of making each coefficient $d_{\alpha\lambda}$ equal to half the corresponding $\chi_{\alpha\lambda}$; if we set $P(\omega) = P_0 \cos 2\,\omega t$ and $E(\omega) = E_0 \cos \omega t$, then from $P(2\omega) = \varepsilon_0\, \chi\, E(\omega)^2$ we obtain $P_0 = \frac{1}{2}\varepsilon_0\chi E_0^2$, since $\cos^2 \omega t = \frac{1}{2}(\cos 2\omega t - 1)$.

The above approach also leads to the idea of a effective SHG coefficients d_{eff} for a particular experimental geometry. One writes each amplitude vector as the product of a scalar amplitude and a unit polarisation vector. Now, if the polarisation vectors are $\vec{e}_p$ for the polarisation and $\vec{e}_1$ and $\vec{e}_2$ for the two fields, then

$$d_{eff}=\frac{1}{2}\vec{e}_p\cdot\vec{\chi}:\vec{e}_1\vec{e}_2 \qquad ...(11)$$

We may note that $\vec{e}_1$ and $\vec{e}_2$ seem the same, as they refer to the same incident light beam, but this is to neglect possible effects of the linear optics of the medium.

SHG occurs inside the medium, and hence the relevant polarisations are those inside the medium. Except in optically isotropic materials, and for special geometries in materials of lower symmetry, plane polarised incident light propagates as two orthogonally polarised waves in the medium (we may remember ordinary and extraordinary rays in the double refraction phenomenon). Thus $\vec{e}_1$ and $\vec{e}_2$ may be the same, corresponding to either of the propagating waves, or they may be different, one corresponding to each wave. We may also note that in anisotropic media the polarisation is determined by the plane in which the electric displacement $\vec{B}$ oscillates, which is not necessarily the same as that in which $\vec{E}$ oscillates: it is $\vec{D}$ which is always transverse to the direction of propagation.

We may note that linear optics also play a role in determining the efficiency of SHG. Inside the medium, a fundamental wave generates second-harmonic waves all along its path, and the second harmonic wave generated at any particular point will have a phase determined by that of the fundamental at that point. At the same point there will be second-harmonic waves which have propagated there with phases determined at other points along the path, and the total second harmonic wave will be the superposition of all these waves. In general, the phases of fundamental and second harmonic waves will vary differently along the path, and as a result the net second-harmonic intensity will vary periodically along the direction of propagation.

One can also look at this to realise that the fundamental and second-harmonic waves will propagate at different speeds, depending on the inverse of their respective refractive indices. Since all the waves will arrive in phase, and hence the intensity will be maximal through constructive interference, when they have each travelled an integral number of cycles at their respective speeds. This distance depends on the mismatch $\Delta\vec{k}$ of their wave vectors where

$$k = 2\pi/\lambda = \omega n/c \qquad ...(12)$$

with λ and n the wavelength and refractive index at frequency ω in the material. Maximum intensity therefore occurs after the *coherence length*

$$l_{\text{coh}} = \pi/\Delta k \qquad ...(13)$$

To maximise SHG efficiency, one seeks geometries which match the wave vectors and hence the phases of the fundamental and harmonic beams. One can distinguish various types of such *phase-matching,* depending on the directions of the beams and on the ordering and dispersion (frequency dependence) of the harmonic and fundamental refractive indices. Typically only a few are explored experimentally.

For collinear harmonic and fundamental beams, type I phase-matching can be achieved when an extraordinary (or ordinary) ray at the harmonic frequency is produced by the combination of two ordinary (or extraordinary) rays at the fundamental frequency such that

$$n_{2\omega} = n_{\omega} \qquad ...(14)$$

the subscript indicating the frequency. This condition is called *index matching,* one can achieve the type-II *phase-matching* when an extraordinary (or ordinary) ray at the harmonic frequency is produced by the combination of one ordinary and one extraordinary ray at the fundamental frequency. The required condition in this case is

$$2n_{2\omega} = n_{\omega}^{0} + n_{\omega}^{c} \qquad ...(15)$$

the subscript indicates the type of ray. Whether the ordinary or extraordinary ray at the harmonic is produced depends on whether n_{ω}^{0} is greater or less than n_{ω}^{c}, *i.e.* on the sign of birefrigence.

We may note that *Walk-off* is also of concern in optically anisotropic media. The energy generated at 2ω propagates according to the Poynting vector $\vec{E} \times \vec{H}$, but in general this is not parallel to the propagation direction (as in general $\vec{D}$ is not parallel to $\vec{E}$). The angle between the *Poynting vector* and the propagation direction is the walk-off angle ϕ. If the wave inside the material is polarised in the direction of unit vector $\hat{e}$ with components (e_1, e_2, e_3) in the principal axes, then ϕ satisfies

$$\cos\phi = \frac{e_1^2/n_1^2 + e_2^2/n_2^2 + e_3^2/n_3^2}{\left(e_1^2/n_1^4 + e_2^2/n_2^4 + e_3^2/n_4^2\right)^{1/2}} \quad ...(16)$$

where the n_i are the principal refractive indices. We may note that walk-off limits the efficiency of SHG, and again its minimisation requires a careful choice of experimental geometry and a detailed knowledge of the linear optical properties of the material.

In the absence of walk-off, the second-harmonic power $W_{2\omega}(z)$ at a distance z along the direction of propagation is related to the fundamental power $W_\omega(O)$ at the surface of the material by

$$W_{2\omega}(z)/W_\omega(O) \propto z^2 d_{\text{eff}}^2 \left[\sin\left(\frac{1}{2}z\Delta k\right)\bigg/\frac{1}{2}z\Delta k\right]^2 \quad ...(17)$$

where it is assumed that $W_{2\omega}(z) << W_\omega(O)$, *i.e.* that there is negligible depletion of pump beam by the SHG. The function is square brackets is usually called $\sin c^2\left(\frac{1}{2}z\Delta k\right)$, and is also met in time dependent perturbation theory (Fig. 10.1). As a function of wave vector mismatch Δk, it equals unity for $\Delta k = 0$,

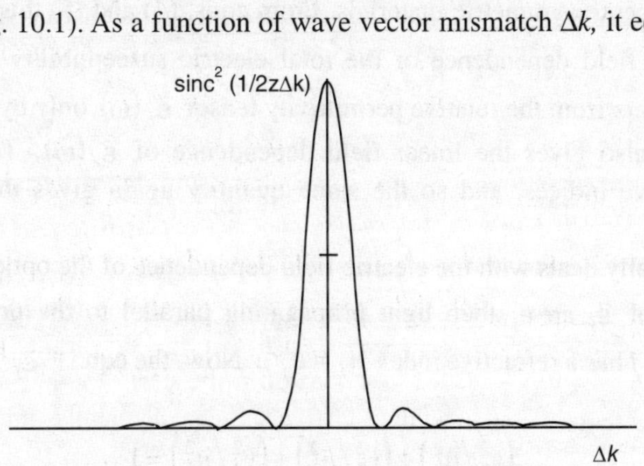

Fig. 10.1 The function $\sin c^2\left(\frac{1}{2}z\Delta k\right)$ as a function of Δk.

in which case the second harmonic power increases quadratically with z. It has zeros at $\Delta k = 2p\pi/z$ for integer p, with maxima of quadratically decreasing magnitude between successive zeros. In order to obtain a reasonable fraction of the maximum intensity for a given z (fixed by the dimension of the sample), one finds $\frac{1}{2}z\Delta k < 1$ or so. Since z may be a few millimeters, this means $\Delta k = 10\text{ cm}^{-1}$, say. In comparison to optical wave vectors of 10000 cm^{-1} or more, one finds that very precise matching is required for reasonable intensity. One may achieve this by very precise orientation of the crystal, perhaps assisted by controlling the crystal temperature to modify the refractive indices slightly.

When precise phase matching cannot be achieved, the net intensity can still be increased by *quasi-phase-matching*. In this technique, active regions of the material of thickness l_{coh} are separated by inactive regions of the same thickness, for example by assembling alternating crystal slices or by spatially varying the treatment of a polymer. The effect is to make inactive the regions where destructive interference would occur. As a result, each active region adds to the net SHG intensity, which therefore increases in a stepwise fashion. One may also be able to arrange the alternating regions with alternating SHG coefficients in such a way that every region contributes to the net SHG intensity. The main problem is to avoid excessive losses at the interfaces.

In the context of harmonic generation one could finally mention *optical rectification.* This is the production of the Zeroth harmonic, *i.e.* a constant field, via the quadratic susceptibility $\chi^{(2)}$ (0; ω, −ω). We may note that it is not often of significance in NLO, though it does help in understanding higher-order effects.

10.2.2 Electro-optic Effects

Let us suppose that the material is subjected to a static electric field as well as monochromatic incident light of frequency ω; in this context it suffices for the 'static' field to oscillate at a frequency well below optical frequencies. The resulting phenomena fall within the scope of eqn. (6): they correspond to electric-field-induced modifications of the optical phenomena, and hence are termed *electro-optic effects.* Since the electric field is at essentially zero frequency, it does not change the output frequency of the optical phenomena it modifies.

The *linear electro-optic* (LEO) or Pockels effect arises from the quadratic susceptibility $\vec{\chi}^{(2)}(-\omega;\omega,0)$ in non-centrosymmetric materials. From eqns. (4) and (5) this quantity can be seen to give the linear electric field dependence of the total electric susceptibility $\vec{\chi}(\omega)$ relating $\vec{P}(\omega)$ to $\vec{E}(\omega)$. Since $\vec{\chi}(\omega)$ differs from the relative permittivity tensor $\vec{\varepsilon}_r(\omega)$ only by the constant unit tensor $\vec{1}$, the same quantity also gives the linear field dependence of $\vec{\varepsilon}_r(\omega)$. The relative permittivity determines the refractive indices, and so the same quantity again gives their linear electric field dependence.

In practice one normally deals with the electric field dependence of the optical indicatrix $\vec{\varepsilon}^{-1}$. If the principal components of $\vec{\varepsilon}_r$ are ε_i, then light propagating parallel to the principal axis i polarised parallel to principal axis j has a refractive index $n_j = \varepsilon_j^{1/2}$. Now, the eqn. $\vec{v}\cdot\vec{\varepsilon}_r^{-1}\cdot\vec{v} = 1$ can be expressed in the principal axes as

$$\left(v_1^2/n_1^2\right)+\left(v_2^2/n_2^2\right)+\left(v_3^2/n_3^2\right)=1 \qquad ...(18)$$

Since away from optical absorptions all $n_i \geq 1$, this equation represents an ellipsoid as the vector $\vec{v}$ varies. Its principal axes are parallel to those of $\vec{\varepsilon}_r$ and its semi-axis lengths are the principal refractive indices; *e.g.*, if we choose $\vec{v}$ parallel to the principal axis 1, then $v_2 = 0 = v_3$ and $v_1 = \pm n_1$. Truly speaking, in any direction the length of the vector $\vec{v}$ from the origin to the surface of the index ellipsoid (18) gives the corresponding refractive index n.

In order to determine the electric field dependence of $\vec{\varepsilon}_r^{-1}$ (for convenience we write this as tensor $\vec{Q}$) and hence of the refractive indices, we need the result for any matrix $\vec{M}$ that

$$\frac{d\vec{M}^{-1}}{dx} = -\vec{M}^{-1}\cdot\frac{d\vec{M}}{dx}\cdot\vec{M}^{-1} \qquad ...(19)$$

which follows at once on differentiating the identity $\vec{M}\cdot\vec{M}^{-1} = 1$ and rearranging. Then we find the *Pockels coefficients* describing the linear electric field dependence of $\vec{Q}$ is

$$r_{\alpha\beta\gamma} = \partial Q_{\alpha\beta}/\partial E_\gamma \qquad ...(20)$$

$$= -Q_{\alpha\lambda}\,\chi_{\lambda\mu\gamma}(-\omega;\omega,0)Q_{\mu\beta} \qquad ...(21)$$

In the principal axes, $Q_{\alpha\beta} = \delta_{\alpha\beta}/n_\alpha^2$ (α not summed), and eqn. (21) reduces to

$$r_{\alpha\beta\gamma} = \chi_{\alpha\beta\gamma}(-\omega; \omega, 0)/n_\alpha^2 n_\beta^2 \qquad ...(22)$$

where the repeated indices α and β are not summed.

We may note that the first-two subscripts of the linear electro-optic coefficient refer to the same frequency, interchanging them leaves the coefficient unchanged. Obviously, there are only six independent combinations, to which the voigt abbreviated notation is again customarily applied; e.g. $r_{123} \rightarrow r_{63}$. One can understand this notation is terms of three-field dependences of six-independent components of the symmetric indicatrix. We have seen that Kleinman symmetry would make $\chi_{\alpha\beta r}$ equal to $\chi_{\alpha\gamma\beta}$; from eqn. (22) we see that this would not make $\chi_{\alpha\beta\gamma}$ equal to $r_{\alpha\gamma\beta}$ unless n_β and n_γ happened to be equal (although if Kleinman symmetry did not apply to $\chi_{\alpha\beta\gamma}$, suitable values of n_β and n_γ could make it apply more closely to $r_{\alpha\beta\gamma}$, as it were by accident).

All materials exhibit a *quadratic electro-optic* or *Kerr effect.* This arises from the coefficient $\vec{\chi}^{(3)}(-\omega;\omega,0,0)$, and corresponds to a quadratic dependence of the total electric susceptibility—and relative permittivity and indicatrix-on electric field. In particular, application of a strong electric field causes an optically isotropic material to acquire a birefringence. This is readily detectable and so the Kerr effect is known for the last about hundred years. Just as the Pockels effect can be characterised by the coefficient $r_{\alpha\beta\gamma}$ giving the linear field dependence of the indicatrix, so the Kerr effect can be characterised by a coefficient $P_{\alpha\beta\gamma\delta}$ giving the quadratic field quadratic field dependence.

We have described the *BC Kerr effect.* One also encounters the *optical* or *AC Kerr effect* in which the optical susceptibility at frequency ω_0 is modified by a strong applied optical field at a frequency ω_1. This is described by the coefficient $\vec{\chi}^{(3)}(-\omega_0;\omega_0,-\omega_1,\omega_1)$: the applied field components at $-\omega_1$ and ω_1 in effect yield a DC field permitting output at ω_0 from input at ω_0.

Like all quadratic effects, the linear electro-optic effect is absent in centro-symmetric materials. One can than think the Kerr effect in such materials as an electric field induced LEO effect, whereby the electric field first distorts the material and destroys its centrosymmetry and then registers an LEO effect in the distorted material. Similarly, there is a version of each $\vec{\chi}^{(2)}$ phenomenon that is induced in centrosymmetric material by an applied static electric field through a suitable $\vec{\chi}^{(3)}$.

Electric field induced second harmonic (EFISH) generation described by the coefficient $\vec{\chi}^{(3)}(-2\omega; \omega, \omega, 0)$ is an important example. It is used to obtain information on molecular first hyper-polarisabilities β (quadratic polarisabilities) in solution. The electric field tends to orient the molecules via their permanent electric dipole moments, which are necessarily nonzero if β is nonzero, since each requires a noncentrosymmetric molecule. The oriented solution is no longer isotropic and exhibits SHG with an intensity dependent on the components of β projected along the dipole moment.

10.2.3 Intensity-dependent Refractive Index

We have seen that AC Kerr effect corresponds to a modification of the linear optical behaviour of the medium at one frequency by an intense beam at another frequency. What happens if the two frequencies coincide? In this case, there is no extra applied beam: the phenomenon arises from a single beam at a single frequency via the coefficient $\vec{\chi}^{(3)}(-\omega; \omega, -\omega, \omega)$. This corresponds to a refractive index which is modified by the BC component of $\vec{E}(\omega)^2$, or equivalently by the intensity of the beam.

In accordance with the latter interpretation, one can write the nonlinear refractive index as

$$n = n_0 + n_2 I \qquad ...(23)$$

where n_0 is the usual low-intensity refractive index, I is the intensity and n_2 is the nonlinear coefficient. Alternatively, one may write n in terms of the electric field amplitude $|\vec{E}|$, so that

$$n = n_0 + n_2' |\vec{E}|^2 \quad ...(24)$$

The intensity is given for weak nonlinearity by

$$I = \frac{1}{2} \varepsilon_0 n_0 c |\vec{E}|^2 \quad ...(25)$$

the two nonlinear coefficients are related by

$$n_2' = \frac{1}{2} \varepsilon_0 n_0 c n_2 \quad ...(26)$$

The coefficient n_2 may be more convenient experimentally, but n_2' is more conventionally related to $\vec{\chi}^{(3)}$. One can write in abbreviated and simplified notation.

$$\chi = \chi^{(1)} + \chi^{(3)} |E|^2 \quad ...(27)$$

then $n^2 = 1 + \chi$ and $n_0^2 = 1 + \chi^{(1)}$, so that

$$n_2' = \chi^{(3)}/2n_0 \quad ...(28)$$

Two important effects arise from the intensity-dependent refractive index. One is an optical device or component such as a waveguide designed in terms of the low-intensity refractive index n_0 will have quantitatively and perhaps qualitatively different properties at high intensities where n differs from n_0, *e.g.* an optical cavity may have a transmission which depends sensitively on the refractive index and thus acts as a switch for the light beam. Under suitable conditions the cavity may also exhibit *bistability*: the transmission changes from a low to high value for a sufficiently high intensity and remains at this level until the intensity is reduced to a finite amount. This hysteresis then allows the nonlinear cavity to act as an optical memory element.

The other effect arising from n_2 is *self-focussing* (or self-defocussing). A light beam caries in intensity across its diameter, often according to Gaussian beam profile. At the centre of the beam the higher intensity produces a refractive index different from that at the edges. If the refractive index is higher at the centre ($n_2 > 0$), the beam is slowed relative to the edges. On the other hand, if $n_2 < 0$, the wavefront becomes concave, corresponding to defocussing. Self-focussing increases the intensity and so increases progressively until it is modified by higher-order processes or (for sufficiently narrow beams) by diffraction. We may note that the enhanced intensity may also damage the material.

We may note that closely related to the intensity dependent refractive index is the *photorefractive index*. In this effect, incident light also changes the refractive index, sometimes at rather low intensity, but the change remains after the light beam is removed. The effects considered hitherto arise from distortion of the electronic and nuclear arrangement in the material. On the other hand, the photo-refractive effect is understood to arise from the separation and migration of charges under the influence of the incident light, *i.e.* through photoconductivity. The charges become trapped, producing high electric fields in their vicinities. These fields then modify the local refractive index by the usual electro-optic effects. Obviously, the refractive index change remains for a time which reflects the rate of carrier recombination, which in turn depends on the temperature and illumination (among other things).

10.2.4 Frequency Mixing

The most general optical phenomenon involves the mixing of several waves to produce output at the various sum and difference frequencies. So far we have considered the phenomena involving special cases where not all frequencies differ.

In noncentrosymmetric materials we encounter *three-wave mixing*. This is the combination of waves at frequencies ω_1 and ω_2 to produce an output wave at frequency ω_0. The reversibility of right paths also reveals the existence of the less obvious inverse process in which a single wave at frequency ω_0 generates outputs at two frequencies ω_1 and ω_2. One may distinguish between variants of these processes according to the relative intensities of the various beams.

Sum-frequency generation corresponds to the process $\omega_0 = \omega_1 + \omega_2$ where the input beams at ω_1 and ω_2 are typically of high intensity to produce a reasonably intense output at ω_0. Obviously, SHG is a special case. Conversely, *difference-frequency generation* corresponds to the process $\omega_0 = \omega_1 - \omega_2$ with high-intensity beams at ω_1 and ω_2. *Parametric up-conversion* is analogous to sum-frequency generation, but now one output beam (say at ω_1) is much more intense than the other. The result is that the output at $\omega_0 = \omega_1 + \omega_2$ is accompanied by a weekened input beam at ω_1 but little at ω_2. *Parametric down conversion* yields output at $\omega_1 - \omega_2$ accompanied mainly by a strengthened input beam at ω_1, and clearly will tend to occur at the same time as up-conversion.

Like SHG, sum frequency generation is useful for producing tunable radiation at high frequencies where individual tunable lasers are not available. Moreover, like SHG, its efficiency is constrained by phase-matching considerations. Optimum efficiency of generation of light at ω_0 from beams at ω_1 and ω_2 requires wave vectors which satisfy $\vec{k}_0 = \vec{k}_1 + \vec{k}_2$. In practice, the efficiency is determined by the smaller of (*i*) the physical interaction length of the beams in the material and (*ii*) the coherence length defined by eqn. (13) in terms of the wave vector mismatch $\Delta\vec{k} = \vec{k}_0 - \vec{k}_1 - \vec{k}_2$. Similarly difference frequency generation is useful for producing tunable infrared radiation of high intensity. In both cases, colinear phase maching requires.

$$\omega_1(n_0 - n_1) + \omega_2(n_0 - n_2) = 0 \qquad ...(29)$$

which clearly can be rather complicated to arrange.

Let us now consider inverse processes. *Optical parametric amplification* uses an intense pump beam at ω_0 interacting with a signal at ω_1 to produce an amplified signal beam, accompanied by a weak additional idler beam at ω_2. As usual, $\omega_0 = \omega_1 + \omega_2$, but in this case a more informative version of this equation is $\omega_1 = \omega_0 - \omega_2$: the beam at ω_1 is strengthened by the difference frequency $\omega_0 - \omega_2$.

Now *image reducing the input intensity* at ω_1. In principle, it can be reduced to zero, until a photon of frequency ω_1 rises only from spontaneous emission. Then as it travels through the medium it generates further photons by optical parametric amplification, so that there is optical gain. Of course, the gain may be small over realistic lengths of material. We may note that this is a standard problem, to which the standard solution is to create an optical cavity within which the beam is reflected back and forth, so increasing the path length and hence the gain. Such an arrangement can yield *optical parametric oscillation*.

In the optical parametric oscillator, one first adjusts the phase matching to select a specific pair of frequencies ω_1 and ω_2 that sum of the chosen pump frequency ω_0. Here ω_1 is the signal frequency of interest and ω_2 the idler. The cavity is bounded by mirrors transparent to ω_0. They are highly reflecting either at just ω_1 (single resonant) or at both ω_1 and ω_2 (doubly resonant). The cavity has resonant frequencies determined by its length and boundary phase shifts at a given frequency. Then the amplitudes of the three beams at ω_0, ω_1 and ω_2 satisfy coupled equations. Which have oscillatory solutions when the pump power exceeds a certain threshold. In practice, single-resonant cavities produce stable output.

Interestingly, all materials exhibit *four-wave mixing.* Obviously, there is a wide variety of possible processes, and although one expects cubic effects to be weaker than quadratic effects (provided the latter are not forbidden by symmetry), there are extra possibilities for resonant enhancement of cubic effects when one relevant frequency approaches an excitation frequency of the material. Thus, one finds phenomena in which three input beams of comparable intensity combine to provide outputs at the sum and difference frequencies, or in which intense pump beams convert or amplify weak signal beams.

Degenerate four-wave mixing has a number of special features. It is degenerate because the four waves all have the same frequency, though they do not all have the same wave vector. One can envisage each pair of waves combining to produce a wave at the second harmonic which then scatters the third-wave to produce the output. Detailed analysis reveals that one can arrange phase-matches output propagation backwards along one of the input beams but with the complex conjugate phase. Clearly, degenerate four-wave mixing can yield phase conjugation.

The phase-conjugate beam mixes with the input beam, but because of its conjugate phase it has the opposite phase distortion to the input beam. This phenomenon provides a means of correcting aberration, using a material configured for degenerate four-wave mixing which as a *phase-conjugate mirror.* Distortion acquired by the input beam, *e.g.* by passage through an optically imperfect medium, is transmitted back reversed in phase. When it passes through the distorting medium, its initial phase distortion exactly cancels that imposed by the medium. Thus one regenerates the information in the original beam prior to distortion.

10.2.5 Units and Dimensions

Two systems of units are commonly used in nonlinear optics. The SI (or *mks*) system is typically used in applied nonlinear optics and photonics and is the primary system of units. The Gaussian system of units is often used in fundamental studies of nonlinear optics. An equation written in SI system can be interpreted as a Gaussian system equation by omitting the prefactors (often ε_0 or $4\pi\varepsilon_0$). For example, equation

$$\vec{P}(t) = \varepsilon_0 \chi^{(1)} \vec{E}(t)$$

is in SI unit and becomes

$$\vec{P}(t) = \chi^{(1)} \vec{E}(t)$$

in the Gaussian system.

Merely conversing units is not enough, though, different systems of units use a different number of base units. Three dimensional systems use only mass, length and time, whereas four-dimensional systems add current (or some other electromagnetic quantity). It is also common to rationalise systems so that the factors of π occur only in equations referring to spherical or cylindrical symmetry. In nonlinear optics one often meet cgs-esu, a three dimensional unrationalised system, and SI units, a four-dimensional rationalized system. Because they are expressed in terms of a different number of base units, corresponding quantities in the different systems have different dimensions. In such a situation, it is necessary to relate the dimensions prior to the conversion of units.

In both system of units, the mechanical quantities have the same dimensions. The force F between two charges Q and Q' a distance r apart is expressed as

$$F = Q_{esu}\, Q'_{esu}/r^2 = Q_{SI}Q'_{SI}/4\pi\varepsilon_0 r^2 \quad ...(30)$$

We deduce

$$Q_{esu} = Q_{SI} \big/ \sqrt{4\pi\varepsilon_0} \quad ...(31)$$

The fields are given by

$$Q_{esu} = Q'_{esu}/r^2;\ E_{SI} = Q'_{SI}/4\pi\varepsilon_0 r^2 \qquad ...(32)$$

whence

$$E_{esu} = \sqrt{4\pi\varepsilon_0}\, E_{SI} \qquad ...(33)$$

The electric dipole moments are given by

$$p_{esu} = q_{esu}\, r;\ p_{SI} = Q_{SI}\, r \qquad ...(34)$$

whence

$$p_{esu} = p_{SI}\big/\sqrt{4\pi\varepsilon_0} \qquad ...(35)$$

With the same equation relating the polarisations P_{esu} and P_{SI}.

One can now relate the susceptibilities. We have

$$P_{esu} = \chi^{(n)}_{esu} E^{n}_{esu};\ P_{SI} = \varepsilon_0\, \chi^{(n)}_{SI}\, E_{SI} \qquad ...(36)$$

or explicitly

$$\chi^{(n)}_{esu} = \chi^{(n)}_{SI}\,\varepsilon_0\Big/\left(\sqrt{4\pi\varepsilon_0}\right)^{n+1} \qquad ...(37)$$

$$\chi^{(1)}_{esu} = \chi^{(1)}_{SI}/4\pi \qquad ...(38)$$

$$\chi^{(2)}_{esu} = \chi^{(2)}_{SI}\Big/4\pi\sqrt{4\pi\varepsilon_0} \qquad ...(39)$$

$$\chi^{(3)}_{esu} = \chi^{(3)}_{SI}/16\,\pi^2\varepsilon_0 \qquad ...(40)$$

The polarisabilities are given by

$$\alpha_{esu} = p_{esu}/E_{esu};\ \alpha_{SI} = p_{SI}/E_{SI} \qquad ...(41)$$

So that

$$\alpha_{esu} = \alpha_{SI}/4\pi\varepsilon_0 \qquad ...(42)$$

The first and second hyperpolarisabilities are given by

$$\beta_{esu} = p_{esu}/E^2_{esu};\ \beta_{SI} = p_{SI}/E^2_{SI} \qquad ...(43)$$

$$\gamma_{esu} = p_{esu}/E^3_{esu};\ \gamma_{SI} = p_{SI}/E^3_{SI} \qquad ...(44)$$

So that

$$\beta_{esu} = \beta_{SI}\Big/\left(\sqrt{4\pi\varepsilon_0}\right)^3 \qquad ...(45)$$

$$\gamma_{esu} = \gamma_{SI}/(4\,\pi\,\varepsilon_0)^2 \qquad ...(46)$$

Now, we can interrelate the measures of these quantities in different units. One obtains

$$\chi^{(1)}_{esu} = \chi^{(1)}_{SI}/12.56637 \qquad ...(47)$$

$$\beta^{(2)}_{esu}/\text{esu} = [\beta^{(2)}_{SI}/\text{mV}^{-1})/4.19169 \times 10^{-4} \qquad ...(48)$$

$$\chi^{(3)}_{esu}/\text{esu} = [\chi^{(3)}_{SI}/(\text{mV}^{-1})^2]/\,1.38920 \times 10^{-16} \qquad ...(49)$$

$$\alpha_{esu}/\text{esu} = (\alpha_{SI}/\text{Fm}^2)/1.11265 \times 10^{-16} \qquad ...(50)$$

$$\beta_{esu}/\text{esu} = (\beta_{SI}/\text{Fm}^3\text{V}^{-1})/3.7140 \times 10^{-21} \qquad ...(51)$$

$$\gamma_{esu}/\text{esu} = (\gamma_{SI}/\text{Fm}^4\text{V}^{-2})/1.23799 \times 10^{-25} \qquad ...(52)$$

The units of quantities in cgs-esu are often marked as 'esu' only. However, α_{esu} is often quoted in cm^3, β_{esu} in cm^5/esu, and γ_{esu} in $cm^7/(esu)^2$. In this case, the esu quoted explicitly is that of charge. However, use of the esu of potential, the 'stat volt' is not common. One also encounters other SI units. If one defines a hybrid quadratic susceptibility $\chi^{(3)}_x$ not by eqn. (36) but by

$$P_{SI} = \chi^{(2)}_X E^2_{SI} \qquad ...(53)$$

Then the extra factor ε_0 disappears from eqn. (37), and one find

$$\chi^2_x/\text{CV}^{-2} = (\chi^{(2)}_{SI}/\text{mV}^{-1})/1.12941 \times 10^{11} \qquad ...(54)$$

10.3 NONLINEAR OPTICAL DEVICES

Nonlinear optical effects become significant at high intensities of light where the photons are observed as interacting. We may remember that photons are regarded as noninteracting and it has been the noninteracting nature of photons that has made them particularly useful in the communication of information. Nonlinear optics offers the intriguing possibility that interacting photons can be employed in several ways including: the generation of coherent light at new frequencies where there are no convenient laser sources; the generation of ultrashort pulses of light; the propagation of ultrashort pulses without distortion from linear dispersion; and is processing of information. We way note that in nonlinear optics the situation is the reverse of superconductivity where a usually highly interacting fermion, the electron is induced to adopt noninteracting photon like qualities.

So far investigations of nonlinear optical effects have produced a wealth of new phenomena but as yet very few devices have made much impact. However, situation is rapidly improving with significant developments in this field. We will restrict to the study of some selected nonlinear devices. The first is *second-harmonic generation for frequency doubling semiconductor diode laser output.* These devices are intended to produce short wavelengths for optical data storage. The main advantages of short wavelengths in optical storage are that the information storage density scales as λ^{-2} and the photochemistry involved is much more sensitive to higher energy photons.

Soliton telecommunication in optical fibres also is developing as an application of third order nonlinear effects. They rely on the intensity-dependent refractive index in very transparent optical fibres and they operate in the negative-dispersion wavelength region around 1.5 μm in these fibres.

All-optical switching devices, which open up the attractive possibility of all-optical logic and all-optical signal processing at speeds unattainable by conventional electronics are attracting a lot of interest from the research community.

10.3.1 Devices Based on Second-order Effects

Second order nonlinearities are commonly employed to generate new frequency components from one or more incident laser beams. For example, frequency-doubling the output of near-infrared semiconductor lasers is a viable option for the generation of compact blue-green sources. Optical parametric oscillators (OPOs) based on difference frequency generation are routinely used to generate frequency-tunable radiation throughout the optical spectrum.

Second-order optical nonlinearities in general give rise to three wave mixing processes. This kind of optical nonlinearity has been used in bulk optics configuration for the generation of coherent light in such devices as the optical parametric oscillator (OPO) and the second harmonic generator (SHG).

Second-harmonic Generation

Figure 10.2 shows a state-of-the-art waveguide second harmonic generation (SHG). It has a waveguide fabricated by an ion-exchange process, and also has a periodic grating structure incorporated in the waveguide for phase matching purposes. The materials employed are $LiNbO_3$, $LiTaO_3$ and KTP (potassium titanyl phosphate). These devices have been shown to efficient, capable of output powers of a few milliwatts CW, and compatible with laser diodes. Their main areas of applications are envisaged as optical data storage and laser printing.

The materials for SHG turn out to have the same symmetry requirements as for electro-optic materials. There has been extensive development of $LiNbO_3$ in particular as a waveguide electro-optic modulator for optical communication applications. However, $LiNbO_3$ suffers from photorefractive damage which is more prevalent at the short wavelengths associated with SHG, and therefore although

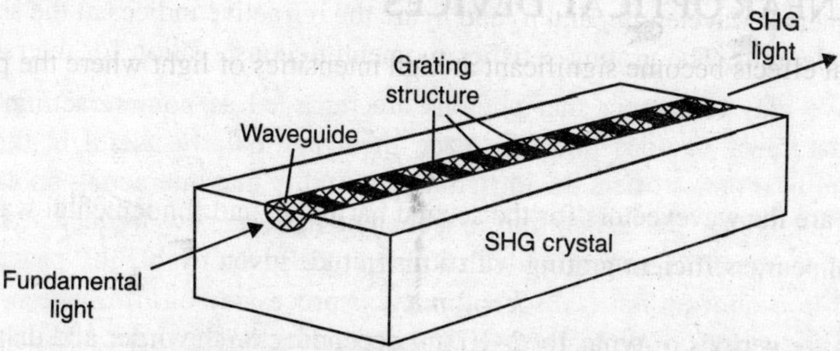

Fig. 10.2 Waveguide second-harmonic generation device.

it has a relative large effective nonlinear coefficient dependent upon crystal systemmetry (d_{eff}) it is unsuitable for laser diode doubling. Photorefractive damage is thought to be due to the creation of F-centre-like defects in the material. These defects involve a misplaced electron in the crystal lattice. They have a large electric field associated with them and this field modifies the local refractive index through the electro-optic effect. There are palliatives for photorefractive damage in $LiNbO_3$; MgO can be added at the melt stage, and proton exchange waveguides seem to have been less prone to damage. Other materials investigated for waveguide QPM-SHG devices, namely $KTiOPO_4$ (KTP) and $LiTaO_3$ (Table 10.2).

Table 10.2 QPM-SHG Waveguide Devices

Material	d_{eff} *(esu)*	*Figure of merit* (d^2/n^3)	*Efficiency* $(W\ cm^2)$
$LiNbO_3$	$d_{13} = 1.2 \times 10^{-8}$	1	
	$d_{33} = 8.5 \times 10^{-8}$	50	40–60%
$LiTaO_3$	$d_{33} = 6.38 \times 10^{-8}$	28.1	29.6%
KTP	$d_{13} \approx 1.0 \times 10^{-8}$	1.5	100%

Waveguide

The role of the waveguide in nonlinear devices is to confine the light over distances that are not limited by diffraction. The devices then become limited by absorption rather than diffraction.

The waveguides are typically manufactured using ion-exchange techniques, *e.g.* in $LiNbO_3$ and $LiTaO_3$ and Li ions can be exchanged for hydrogen ions— this is known as proton exchange. This is done in a bath of dilute benzoic acid at temperatures of around 250–300 C. The sample has a diffusion mask, typically made of aluminum, which allows the exchange process to proceed only through the apertures in the metal diffusion mask. The diffusion mask is made by a photolithographic process and typically the width of a waveguide is around 2–5 μm. For KTP, one can employ Rb for ion exchange. Also in $LiNbO_3$, Ti in-diffusion can be used to produce waveguides.

The grating structure: This is required for *quasi-phase matching* (QPM), and also the first technique suggested for phase matching. The technique relies on spatially modulating the second-order nonlinearity with a period given by 2 ml_{coh}, where $m = 1, 3, 5$, etc. The coherence length is given by

$$l_{coh} = \frac{\lambda}{4(n_2 - n_1)} \qquad ...(55)$$

where λ is fundamental wavelength, and n_2 and n_1 are the refractive indices at the second harmonic and fundamental wavelengths. The grating satisfies momentum conservation for interaction in accordance to

$$\vec{k}_1 - 2\vec{k}_2 = \vec{k} \quad ...(56)$$

where $\vec{k}_2$ and $\vec{k}_1$ are the wavevectors for the second harmonic and fundamental waves. $\vec{k}$ is the wave-vector of the nonlinear coefficient grating with a magnitude given by

$$k = 2\,m\pi/\lambda \quad ...(57)$$

The gratings have periods of typically 2–10 μm depending on the order and dispersion.

The waveguide SHG devices have been specifically designed for frequency doubling AlGaAs/GaAs laser diodes which have an output at around 800–850 nm. The key figure of merit of these devices is the efficiency of operation, and if a comparison is to be made between devices then the efficiency has to be normalised against the length of device. The normalised efficiency is defined as:

$$\eta = \frac{100\left|a^{(2\omega)}(L)\right|^2}{\left|a^{(\omega)}(L)\right|^2 L^2} \quad ...(58)$$

where $a^{(2\omega)}(L)$ is the amplitude of the second harmonic and $a^{(\omega)}(L)$ is the amplitude of the fundamental; L is the length of the device.

10.3.2 Devices Based on Third-order Effects

Third harmonic generation is the process that leads to the creation of a wave at frequency 3ω in response to an applied wave at frequency ω. The source of the third-harmonic wave is the nonlinear polarisation

$$P(3\omega) = \varepsilon_0 \chi^{(3)}(3\omega, \omega, \omega, \omega)\, E(\omega)^3$$

Third harmonic generation is often used as a means of measuring the value of $\chi^{(3)}$ of optical materials.

In this section, we concentrate on devices which utilize the optical *Kerr* effect, where the refractive index is intensity dependent according to eqn. (23). There are several mechanisms which give rise to an intensity-dependent refractive index.

Soliton transmission or propagation

The propagation of pulses of picosecond duration in optical fibres is often governed by the nonlinear Schrodinger equation

$$\frac{\partial A}{\partial z} = -\frac{1}{2}\beta_2 \frac{\partial^2 A}{\partial \tau^2} + i\gamma |A|^2 A \quad ...(59)$$

where A is slowly varying envelope of the electric field, z is the longitudinal coordinate, τ is the local time defined by $\tau = t - zv_g$ where v_g is the group velocity, γ is a nonlinear response parameter given by $\gamma = \{[4\pi\varepsilon_0]/2\pi\}\; n_2 n_0 \omega_0$, and β_2 is the group velocity dispersion parameter defined by

$$a\beta_2 = \left.\left(\frac{\partial^2 \beta}{\partial \omega^2}\right)\right|_{\omega=\omega_0} \quad ...(60)$$

where β is the wave number of the fundamental mode of the fibre. The first term on the r.h.s. of eqn. (59) treats the effects of group velocity dispersion. At the minimum-loss wavelength of standard communication fibres, pulses experience anomalous group-velocity dispersion. In this regime, the higher frequency components of the pulse propagate at a faster group velocity than do lower frequency components, and thus, the pulse broadens and becomes down-chirped. This effect can be balanced against the effects of self-phase modulation which are treated by the second term on the r.h.s. of eqn. (59). Figure 10.3 shows the nonlinear phase shift experienced by a pulse propagating under the influence of self-phase modulation in a medium with a positive nonlinear refractive index. This nonlinear phase modulation induces an instantaneous frequency shift given by

$$\delta\omega = -\frac{\partial \phi_{ne}}{\partial \tau} \qquad \text{...(61)}$$

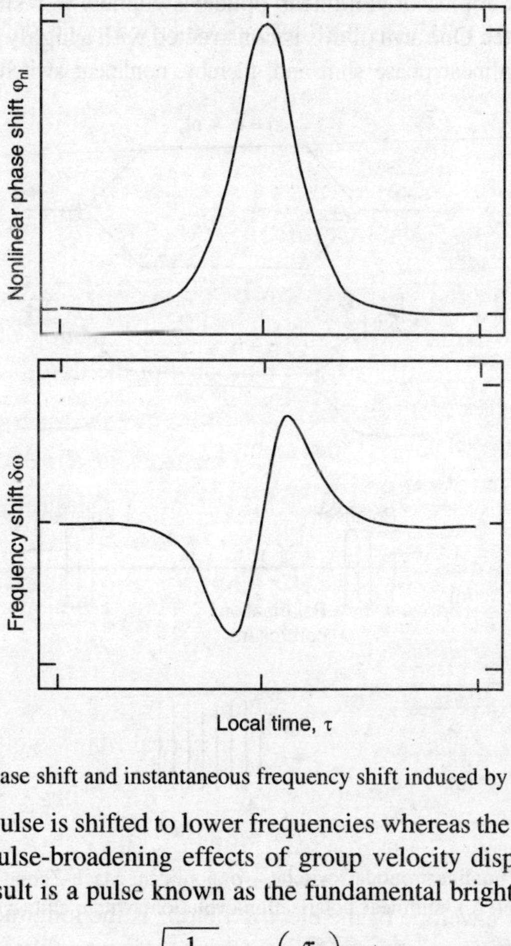

Fig. 10.3 Nonlinear phase shift and instantaneous frequency shift induced by self-phase modulation.

The leading edge of the pulse is shifted to lower frequencies whereas the trailing edge is shifted. This effect can counteract the pulse-broadening effects of group velocity dispersion for the proper pulse shape and intensity. The result is a pulse known as the fundamental bright soliton which has the form

$$A = \sqrt{\frac{1}{L_D \gamma}} \sec h\left(\frac{\tau}{\tau_P}\right) e^{iz/2L_D} \qquad \text{...(62)}$$

where
$$L_D = \frac{\tau_P^2}{|\beta_2|} \qquad \text{...(63)}$$

We may note that this pulse propagates without changing shape and without becoming chirped. In addition, the pulse accumulates a uniform nonlinear phase shift across its profile. This feature makes solitons natural bits of information for the information for the all-optical switching schemes.

Solitons are nonlinear waves which propagate without changing form (or at least change form in a manner that is predictably periodic); they do not interact with each other. This is a nonlinear optical effect and therefore a certain amplitude is required in order to create a solition of a given duration. For a standard fibre with a dispersion of 15 ps nm^{-1} $km^{-1}$and a nonlinearity of $n_2 = 3.2 \times 10^{-16}$ cm^2/W for such pulses (fundamental soliton) of duration τ in picoseconds, we have

$$P(\text{Watts}) = 6.9/\tau^2,\ E(pJ) = 13.6/\tau,$$
$$z(\text{km}) = 0.009\ \tau^2.$$

Prototypical nonlinear optical switches

Figure 10.4 shows four examples of nonlinear optical switches. For simplicity the first example is Mach-Zehnder interferometer. One arm of this is constructed with a highly nonlinear material to produce an intensity-dependent nonlinear phase shift and, thereby, nonlinear switching. Nonlinear common-path

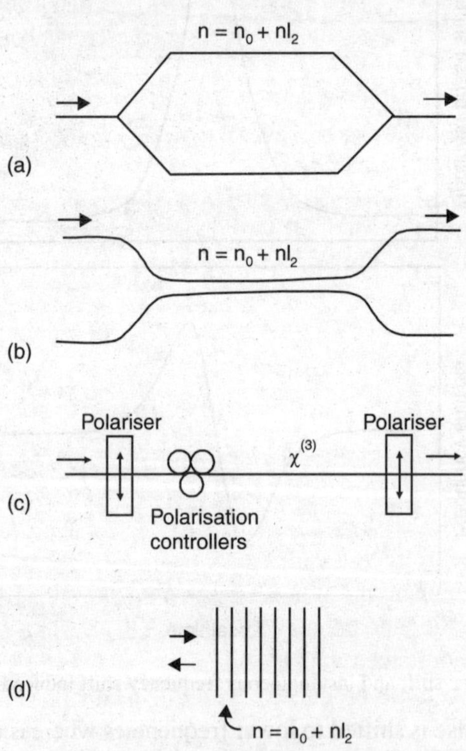

Fig. 10.4 Four examples of nonlinear optical switches: (*a*) nonlinear Mach-Zehnder interferometer, (*b*) nonlinear directional coupler, (*c*) nonlinear polarisation evolution switch, and (*d*) nonlinear Bragg reflector.

interferometers are also being investigated. In particularly, nonlinear loop mirrors (*i.e.*, nonlinear Sagnac interferometers constructed with all-fibre optic components) are being investigated as all-optical switches in communication systems and have been used to mode lock fibre lasers. The nonlinear directional coupler shown in Fig. 10.4(*b*) is another promising waveguide switch. In this device, evanescent coupling occurs between two nonlinear waveguides. Nonlinear directional coupling has been demonstrated in glass fibres, in polymers and in semiconductors. The switch shown in Fig. 10.4(*c*) is based on nonlinear polarisation evolution. The input light is passed through a linear polariser and, then, polarisation controllers to produce elliptically polarised light. Due to the tensor nature of the third-order susceptibility, the light, then undergoes nonlinear polarisation evolution, creating an intensity-dependent transmission through the output polariser. This device has recently been used to mode lock fibre lasers. Figure 10.4(*d*) shows a nonlinear Bragg reflector, constructed with nonlinear material to make the reflectivity intensity-dependent.

Two-photon absorption and nonlinear index saturation limit the performance of the nonlinear switches described above. In order to identify materials suitable for nonlinear switching applications, the following figure of merit have been developed:

$$W = \frac{\Delta n_{max}}{\alpha\lambda},\ T = \frac{2\beta\lambda}{n_2} \qquad \text{...(64)}$$

where α and β are the linear and two-photon absorption coefficients, respectively. α is the usual linear absorption coefficient and β describes processes in which two photons are simultaneously removed from the light beam. Δn_{max} is the limiting value of the nonlinear index change. Useful optical switching demands $W > 2$ and $T > 1$.

We may note that nonlinear optical effects allow photons to interact, and new wavelengths can be generated or light can be used to control light. The large-scale application of nonlinear optical effects requires convenient sources of light, probably semiconductor lasers, and the means (waveguides and fibres) to confine high intensities of light required for efficient nonlinear optical processes.

10.4 OPTICAL NONLINEAR EFFECTS IN SEMICONDUCTORS

The technological importance of semiconductors has resulted in the availability of good-quality material and a mature process and fabrication technology; consequently, nonlinear optical phenomena in semiconductors have been intensively investigated.

The semiconductors which are extensively employed in electronic and opto-electronic applications are covalent inorganic solids. They are composed of elements in groups II to VI of the periodic table combined as II–VI, III–V and IV.

Third order optical nonlinearities arise in semiconductors in some profusion. The third-order effects give rise to an intensity-dependent refractive index and absorption coefficient. These effects are divided into two broad classes called resonant and non-resonant optical nonlinearities. Resonant effects deal with photon energies well below the fundamental absorption edge. With resonant optical nonlinearities, real photogenerated carriers are produced which give rise to the optical nonlinearities. However, the optical nonlinearity associated with the production of real carriers does not fit neatly into the theory developed in this chapter. When considering resonant nonlinearities, in some ways, it is more appropriate to take a broad definition of nonlinear optical effects and to regard them generally as intensity-dependent optical properties. Perhaps the most significant aspect of resonant optical nonlinearities is the insight that is provided into the dynamics of free carriers in semiconductors.

The resonant optical nonlinearity is of larger magnitude but is limited by saturation and thermal effects. There is also a significant relaxation time and absorption associated with the creation of real photogenerated carriers. These effects have limited application of resonant optical nonlinearities. However, resonant optical nonlinearities have proved to be useful as means of studying the dynamics of free carriers. In bulk semiconductors, the nonresonant third-order optical nonlinearity has been shown to be strongly related to two-photon absorption. Resent studies on all-optical switching have shown that it is possible to obtain large nonlinear phase shifts in semiconductors without large absorption only if the photon energies employed are below half the band gap energy of semiconductor.

The significant features of *quantum wells* for third-order optical effects are the stability of the exciton and the two dimensional density of states. Both of these play a role in the resonant optical nonlinearity. The nonresonant AC Stark effect operates on the excitonic absorption and is an ultrafast effect; however, it is limited by nonlinear absorption from the two-photon effect.

The second-order effect in semiconductors can be understood in terms of the asymmetric distribution of charge along the bonds, and, in low dimensional semiconductors, by constructing asymmetric quantum-well structures it has been shown that it is possible to increase the magnitude of the second-order effect. This is an exciting development and opens up the possibility of utilizing the extensive range of semiconductor technology to produce synthetic nonlinear optical materials.

Semiconductors present an abundance of interesting nonlinear optical phenomena, but as yet only a limited number of devices have been produced that utilize optical nonlinearities. However, current semiconductor technology gives a unprecedent degree of flexibility in the design of new materials and devices.

10.5 NONLINEAR GLASSES

These are attractive materials because of their obvious compatibility with existing fibre networks and the sub-picosecond response times of many glass compositions. However, the nonlinearity of silica is very low, $n_2 = 3 \times 10^{-16}$ cm^2 W^{-1}. More efficient signal processing elements could be realized if the nonlinearity was increased; an increase by an order of magnitude in n_2 would imply a similar saving in the power requirements.

Glasses that exhibit permanent or semi-permanent refractive-index changes are becoming an increasingly important area of research. These so-called *photorefractive glasses* offer the potential for many new exciting applications.

Effects based on second-order nonlinearities require non-centrosymmetric materials and hence would not be expected in glasses, due to their isotropic nature. However, experimental results have shown the presence of a second-order nonlinearity in both doped silica fibres and commercial silica samples poled in a DC electric field. The latter effect shows a $\chi^{(2)}$ of the same order of magnitude as that found in $LiNbO_3$.

The origin of nonlinear refractive index in glasses can be divided into two categories, namely the *resonant* and *nonresonant* contributions. The resonant contribution arises when the glass matrix is doped with atoms, or particles, but have an absorbing transition in the spectral region of interest, *e.g.* semiconductor doped glasses. The nonresonant contribution to the nonlinear refractive index has been attributed to the perturbation of the electron clouds surrounding the constituent atoms of the glass. The magnitude of the effect, therefore, depends on the composition of the glass and the types of bond contained within the matrix. Nonresonant effects are smaller than the resonant contribution, though with a much faster response time.

In many respects glass has become the material of choice for the implementation of ultrafast all-optical switching devices. Though it has a relatively low value of n_2, very low loss optical fibres can be pulled, resulting in low-power device operation.

In the anomalous dispersion regime the fibre can support temporal solitons. These pulses, themselves a consequence of fibre nonlinearity, have been shown to be both an effective transmission technique and a means of implementing all-optical switching. It is reported that soliton pulse trains can be transmitted over distances upto one million kilometres.

Other nonlinear glasses may find applications in future communications systems. The power requirements for the system depend on the nonlinear coefficient. By increasing the value of n_2 in a glass the switching power will decrease, leading to more efficient operation. There are reports for increase in magnitude of n_2 for a variety of different glasses. However, we may in mind that when n_2 increases, other detrimental nonlinear effects such as *two-photon absorption* (TPA) and *stimulated Raman scattering* also increase.

Specialised glasses which show photorefractive effects are becoming important areas of research. Optically tuneable filters based on gratings, have obvious applications for the multiplexing/demultiplexing of signals. Read/write holographic memories and direct writing of integrated circuits have also been proposed.

The prospects for the successful application of nonlinear glasses look very promising. Indeed, optical-pulse compressors and mode-locked lasers based on self-phase modulation are commonplace in many research laboratories.

SUMMARY

1. Nonlinear optical phenomena occurs when the response of a material system to an applied optical field depends nonlinearly on the strength of the applied field. The subsequent interaction of the optical field with nonlinear response generates a myriad of interesting physical processes. In some applications, these nonlinear processes are merely limitations to the performance of linear systems, whereas in photonics, nonlinear processes promise to be much more.
2. Applications of nonlinear optics include the frequency doubling of semiconductor lasers, the generation of ultrashort laser pulses, dispersion compensation in communication systems, and all optical switching.
3. Two systems of units are commonly used in nonlinear optics. The SI (or mks) system is typically used in applied nonlinear optics and photonics and is the primary system of units. The Gaussian systems of units is often used in fundamental studies in nonlinear optics.
4. In conventional linear optics, it is assumed that the applied optical field is sufficiently weak that the induced polarisation $\vec{P}(t)$ depends linearly on the applied field strength $\vec{E}(t)$ such that

$$\vec{P}(t)=[\varepsilon_0]\chi^{(1)}\vec{E}(t)$$

where the constant of proportionality $\chi^{(1)}$ is the linear susceptibility. When the applied field is sufficiently strong, one can often accurately describe the induced polarisation as a power series in $\vec{E}(t)$ such that

$$\vec{P}(t)=[\varepsilon_0]\chi^{(1)}\vec{E}(t)+\chi^{(2)}\vec{E}^2(t)+\chi^{(3)}\vec{E}^3(t)+...$$

where $\chi^{(2)}$ and $\chi^{(3)}$ are known, respectively, as the second-and-third-order nonlinear optical susceptibilities. The two lowest order nonlinear terms, which are displayed explicitly in above equation are responsible for most nonlinear optical interactions.
5. Propagation of light in a nonlinear, nonmagnetic medium with no currents and no free charges is governed by the nonlinear wave equation

$$\nabla\times\nabla\times\vec{E}+\frac{1}{[\varepsilon_0]C^2}\frac{\partial^2}{\partial t^2}\vec{D}^{(1)}=-\frac{4\pi}{[4\pi\varepsilon_0]}\frac{1}{C^2}\frac{\partial^2}{\partial t^2}\vec{P}^{NL}$$

where $\vec{P}^{NL}$ is the polarisation due to nonlinear part.

6. Third harmonic generation is the process that leads to the creation of a wave at frequency 3 ω in response to an applied wave at frequency ω. The source of the third-harmonic wave is the nonlinear polarisation

$$P(3\,\omega) = [\varepsilon_0]\chi^{(3)}\,(3\omega, \omega, \omega, \omega)E\,(\omega)^3$$

Third harmonic generation is often used as a means of measuring the value of $\chi^{(3)}$ of optical materials.

7. *Four-wave mixing* is the name given to a process in which input waves at frequencies ω_1, ω_2, and ω_3 interact in a nonlinear material by means of the nonlinear susceptibility $\chi^3(\omega_4, \omega_3, \omega_2, \omega_1)$ to create an output wave at frequency $\omega_4 = \omega_1 + \omega_2 + \omega_3$.
8. The third-order susceptibility is a fourth-rank tensor and, thus, possesses 81 components, each of which is independent in the most general case. However, for degenerate four-wave mixing in an isotropic material, $\chi^{(3)}$ (ω, ω, ω, –ω) possesses only two independent components and the nonlinear polarisation can be expressed as

$$\vec{P} = [\varepsilon_0]\left[A(\vec{E}\cdot\vec{E}^*)\vec{E} + \frac{1}{2}B(\vec{E}\cdot\vec{E})E^*\right]$$

where $A = 6\,\chi^{(3)}_{1122}$ and $B = 6\,\chi^{(3)}_{1221}$

9. The nonlinear susceptibility of many waveguide materials is orders of magnitude smaller than the susceptibility of common bulk nonlinear materials. Nonetheless, nonlinear effects in many waveguides can often be observed at modest power levels due to the small mode size and low loss of the waveguides.
10. In nonlinear optics, photons are regarded as noninteracting and it has been the non-interacting nature of photons that has made them particularly useful in the communication of information.
11. Nonlinear optical effects effects allow photons to interact, and new wavelengths can be generated or light can be used to control light. The large-scale applications of nonlinear optical effects requires sources of light, probably semiconductor lasers, and the means (waveguides and fibres) to confine high intensities of light required for efficient nonlinear optical processes.
12. Semiconductors present an abundance of interesting nonlinear optical phenomena, but only a limited number of devices have been produced so far, which utilize optical nonlinearities. However, current semiconductor technology gives a unprecedented degree of flexibility in the design of new materials and devices, and for this reason, and because semiconductors are widely employed in opto-electronics.
13. Glasses that exhibit permanent or semi-permanent refractive index changes are becoming an increasingly important area of research. These so-called *photorefractive glasses* offer the potential for many new and exciting applications.
14. Nonlinear materials include ABP, KDP, $LiNbO_3$, $LiIO_3$, CdSe, $Ag_3\,AsS_3$, $Ba_2\,NaNb_5\,O_{15}$, urea, β-BaB_2O_4, KTP, $LiB_3\,O_5$, MgO, $LiNbO_3$, $KNbO_3$, MgO: $LiNbO_3$, DAN, D-LAP, POM, MAP, etc. However, the task of growth, characterisation, and fabrication of these crystals remains a challenging one.

SUGGESTED READINGS

1. R.W. Boyd and Grynberg, *Contemporary Nonlinear Optics*, Academic Press (1992).
2. P.N. Butcher and D. Cotter, *The Elements of Nonlinear Optics*, Cambridge University Press (1990).
3. Y.R. Shen, *The Principles of Nonlinear Optics*, Wiley (1984).
4. E. Zernike and J.E. Mid-winter, *Applied Nonlinear Optics*, 2nd edn., Wiley (1973).

5. G. Agrawal, *Nonlinear Fibre Optics*, Academic Press (1989).
6. M.C. Gupta and J. Ballato (eds.), The Handbook of Photonics, 2nd edn. CRC Press (2007).

REVIEW QUESTIONS

1. Explain nonlinear optical phenomena.
2. What are electro-optic effects? Explain linear electro-optic (LEO) or Pockels effect.
3. Explain intensity-dependent refractive index. What effects arise from this?
4. Explain nonlinear optical susceptibility. Give a brief account of symmetry properties of nonlinear optical susceptibility.
5. What do you understand by third order nonlinearities? Explain third harmonic generation and four-wave mixing.
6. Give a brief account of devices based on second-order optical nonlinear effect.
7. Give a brief account of devices which employ third-order optical nonlinearities.
8. Give a brief account of optical nonlinear effects in semiconductors.
9. What are nonlinear glasses? Explain the origins of nonlinearity in glasses.
10. Give a brief account of nonlinear crystals.

SHORT ANSWER QUESTIONS

1. When nonlinear optical phenomena occur?

Ans. This phenomena occur when the response of a material system to an applied optical field depends—nonlinearly on the strength of the applied field.

2. What are the two systems of units used in nonlinear optics?

Ans. The SI (or *mks*) system is typically used in applied nonlinear optics and photonics. The Gaussian systems of units is often used in fundamental studies of nonlinear optics. The transformation between these system of units can be interpreted by writing SI equation or as a Gaussian system equation by omitting the prefactors (ε_0 or $4\pi\varepsilon_0$), *e.g.*

$$\hat{P}(t) = \varepsilon_0 \chi^{(1)} E(t)$$

in SI system becomes

$$\tilde{P}(t) = \chi^{(1)} \tilde{E}(t)$$

in Gaussian system

3. What is nonlinear optical susceptibility?

Ans. In conventional linear optics, it is assumed that the applied field is sufficiently weak that the induced polarisation $\tilde{P}(t)$ depends linearly on the applied electric field strength $\tilde{E}(t)$ such that

$$\tilde{P}(t) = [\varepsilon_0]\chi^{(1)} \tilde{E}(t)$$

where the constant of proportionality $\chi^{(1)}$ is the linear susceptibility.

4. What do you understand by second-order nonlinearities?

Ans. Second-order nonlinearities are commonly employed to generate new frequency components from one or more incident laser beams, *e.g.* frequency-doubling the output of near-infrared semiconductor lasers is a viable option for the generation of compact blue-green laser sources.

Optical Parametric Oscillators (OPOs) based on difference-frequency generation are routinely used to generate frequency-tunable radiation throughout the optical spectrum.

5. What do you understand by third-harmonic generation?

Ans. This is the process that leads to the creation of a wave at frequency 3ω in response to an applied wave at frequency ω. the source of third-harmonic wave is the non-linear polarisation

$$P(3\omega) = [\varepsilon_0]\,\chi^{(3)}\,(3\omega, \omega, \omega, \omega)\,E(\omega)^3$$

Third harmonic generation is often used as a means of measuring the value of $\chi^{(3)}$ of optical materials.

6. What do you understand by four-wave mixing?

Ans. Four-wave mixing is the name given to a process in which input waves at frequencies ω_1, ω_2, and ω_3 interact in a nonlinear material by means of the nonlinear susceptibility $\chi^{(3)}(\omega_4, \omega_3, \omega_2, \omega_1)$ to create an output at frequency $\omega_4 = \omega_1 + \omega_2 + \omega_3$.

7. Write the relation for intensity-dependent refractive index and explain its importance

Ans.
$$n = n_0 + n_2 I \qquad \text{...(1)}$$

where n_0 is usual low-intensity refractive index, I is the intensity and n_2 is the nonlinear coefficient. One can understand the process of degenerate four-wave mixing conceptually as a process in which the refractive index of a material is modified in proportion to the local intensity $I = [4\pi\,\varepsilon_0]\,(nc/2\pi)\,|E|^2$ of light field in accordance to eqn. (1). Closely related to the intensity-dependent refractive index is the *photorefractive effect.*

8. Explain the tensor nature of the third-order susceptibility.

Ans. The third-order susceptibility is a fourth-rank tensor, and thus possesses 81 components, each of which is independent in the most general case. However, for degenerate four-wave mixing in an isotropic material $\chi^{(3)}(\omega, \omega, \omega, -\omega)$ possesses only two independent components and the nonlinear polarisation can be expressed as

$$\vec{\beta} = [\varepsilon_0]\left[A(\vec{E}\cdot\vec{E}^*)\vec{E} + \frac{1}{2}B(\vec{E}\cdot\vec{E})\vec{E}^*\right]$$

where $\quad A = 6\,\chi^{(3)}_{122}$ and $B = 6\,\chi^{(3)}_{121}$.

9. What are solitons or solitary waves?

Ans. Solitons or solitary waves are waves which propagate without dispersion. They rely on the intensity-dependent refractive index in very transparent optical communication fibres and they operate in the negative-dispersion wavelength region around 1.5 μm in these fibres.

10. Give a brief account of second harmonic generation device.

Ans. This has a waveguide fabricated by an ion-exchange process, and also has a periodic grating structure incorporated in the waveguide for phase-matching purposes. The materials employed are $LiNbO_3$, $LiTaO_3$ and KTP (Potassium titanyl phosphate). These devices have been shown to be efficient, capable of output powers of a few milliwatts CW, and compatible with lase diodes. Their main areas of application are envisaged as optical data storage and laser printing.

11. Explain third order optical nonlinear effects in semiconductors.

Ans. Third order optical nonlinearities arise in semiconductors in some profusion. Third order effects give rise to an intensity-dependent refractive index and absorption coefficient. These effects are divided into two broad classes called resonant and nonresonant optical nonlinearities. Resonant effects deal with photon energies close to the fundamental absorption edge in semiconductors and are caused by photogenerated carriers, whereas nonresonant effects are at photon energies were below the fundamental absorption edge. With resonant optical nonlinearities, real photogenerated carriers are produced which give rise to the optical nonlinearities.

12. What is the origin of nonlinearity in glasses?

Ans. The origin of the nonlinear refractive index in glasses can be divided into two categories, namely the resonant and nonresonant contributions. The resonant contribution arises when the glass matrix is doped with atoms, or particles, that have an absorbing transition in the spectral region of interest. The nonresonant contribution to the nonlinear refractive index has been attributed to the perturbation of the electron clouds surrounding the constituent atoms of the glass. The magnitude of the effect, therefore, depends on the composition of the glass and the types of bond contained within the matrix. Nonresonant effects are smaller than the resonant contribution, though with a much faster response time.

OBJECTIVE QUESTIONS

1. In conventional linear optics, it is assumed that the applied optical field is sufficiently weak that the induced polarisation $\vec{P}(t)$ depends linearly on the applied electric field strength $\vec{E}(t)$ such that

   (*a*) $\vec{P} = \varepsilon_0 \vec{\chi}^{(1)} \cdot \vec{E}$ (*b*) $\vec{P} = \varepsilon_0 \vec{E}$

   (*c*) $\vec{P} - \dfrac{\vec{E}}{\varepsilon_0}$ (*d*) $\vec{P} = \mu_0 \varepsilon_0 \vec{E}$ [*a*]

2. The linear electro-optic (LEO) or Pockels effect in non-centro-symmetric materials arises from the

   (*a*) linear susceptibility tensor (*b*) quadratic susceptibility tensor

   (*c*) cubic susceptibility tensor (*d*) none of these [*b*]

3. Intensity-dependent refractive index expression is

   (*a*) $n = n_0 + I$ (*b*) $n = n_0 + n_2$

   (*c*) $n = n_0 + n_2 I$ (*d*) $n = n_0 + \dfrac{I^2}{n_2}$ [*c*]

4. The materials for second-harmonic generation turnout to have the same symmetry requirements as for

   (*a*) electro-optic materials (*b*) non-electro-optic materials [*a*]

Appendices

APPENDIX 1

Units, Conversion Factors, Physical Constants

UNITS

Quantity	Symbol	S.I. Units: Expressed as	S.I. Units: Full form	Other units
1. Activation energy	Q	J/mol	Joule per mole	cal/mole, cal/gm
2. Atomic diameter		nm	Nanometre	Å
3. Bond energy	—	kJ/mole	Kilojoule/mole	eV/atom; kcal/mole
4. Bond length	—	nm	Nnanometre	Å
5. Bragg angle	θ	rad	Radian	0 (degree)
6. Calorific value	—	kJ/m^3	Kilojoule per cubic metre	
7. Coercive field	H_c	A/m	Ampere per metre	Oersted
8. Concentration	C	mol/m^3	Mole per cubic metre	No. of $atoms/cm^3$
9. Contact angle	Φ	rad	Radian	0 (Degree)
10. Current density	J	A/m^2	Ampere per square metre	amp/cm^2
11. Dipole moment	—	C–m	Coulomb metre	debye
12. Dislocation energy	E	J/m	Joule per metre	erg/cm; eV/plane
13. Diffusion co-efficient	D	m^2/s	Metre square per second	—
14. Diffusion constant	D_0	m^2/s	Metre square per second	—
15. Dielectric field strength	E	V/m, N/c	Volt per metre	Newton per Coulomb
16. Electrical conductivity	σ	$Ohm^{-1}\ m^{-1}$	Per ohm per metre	mho/cm
17. Electron affinity	—	kJ/mol	Kilojoule per mole	eV/atom, kcal/ mole
18. Electron energy level		J	Joule	eV (electron volt), erg.
19. Energy of surface imperfections	γ	J/m^2	Joule per square metre	erg/cm^2
20. Enthalpy	H	J/mol	Joule per mole	cal/mol, cal/gm
21. Entropy	S	J/mol/K	Joule per mole/Kelvin	cal/mole/°C cal/gm/°C
22. External energy	—	J/mol	Joule per mole	cal/mole, cal/gm
23. Flux	J	$mol/m^2/s$	Mole per square metre per second	No. of $atoms/cm^2/sec$
24. Force	N	Newton	lbf.	
25. Fracture stress	r_f	MN/m^2	Mega Newton per square metre	kgf/mm^2, PSi

(*Contd.*)

Quantity	*Symbol*	*S.I. Units* *Expressed as*	*Full form*	*Other units*
26. Fracture toughness (Strain energy release rate)	Gc	J/m^2	Joules per square metre	lb-in/in^2
Stress intensity factor	K_c	$MN/m^{3/2}$	Mega Newton per (metre)$^{3/2}$	KSi in$^{1/2}$
27. Frequency of radiation	ν	S^{-1}, Hz	per second; Hertz	
28. Gibbs free energy	G	J/mol	Joule per mole	cal/mole, cal/gm
29. Hysteresis loss		J/m^3	Joule per cubic metre	—
Hall Petch constant	K	MNm^{-32}	Mega Newton/(metre)$^{3/2}$	— Kgf/m$^{3/2}$
30. Interatomic force	F	N	Newton	Kgf, lb, dyne
31. Interfacial energy	Y	J/m^2	Joule per square metre	erg/cm^2
32. Internal energy	E	J/mol	Joule per mole	cal/mole, cal/gm
33. Interplanar spacing	d	nm	Nanometre	Å
34. Kinetic energy	E	J	Joule	erg, eV
35. Lattice parameter	a	nm	Nanometre	Å
36. Magnetic induction	B	W_b/m^2 or T	Weber per square Metre or Tesla	Gauss
37. Magnetic moment	m	Am^2	Ampere square metre	
38. Magnetic permeability	μ	H/m	Henry per metre	
39. Magnetic field strength	H	A·T/m	Ampere turns per metre	Oersted
40. Mass	—	Kg	Kilogram	lb
41. Momentum	—	Kg m/s	Kilogram metre per second	—
42. Power		W	Watt	ft.lbf.
43. Pressure	P	MPa, MN/m^2	Mega pascal; Mega Newton/ Square metre	atmosphere, psi Kg/cm^2, dyne/ cm^2, mm of Hg
44. Resistivity	ρ	Ohm.m	Ohm metre	ohm.cm
45. Shear modulus	u	GN/m^2	Giga Newton per Square metre	Kgf/mm^2; PSi, dyne/cm^2
46. Shear stress	r	N/m^2 P_a	Newton/m^2 pascal	Kgf/mm^2, PSi, dyne/cm^2
Shear strain	γ	—		
47. Specific heat	C_v	J/mol/k	Joule per mole per Kelvin	cal/mol/°C; cal/gm/°c
48. Specific heat	C_p	J/mol/k	Joule per mole per Kelvin	cal/gm/°C
49. Specific volume	—	m^3/Kg	Cubic metre per Kilogram	ft^3/1b, cm^3/gm
50. Specific energy	—	J/Kg	Joule per Kilogram	Btu/lb
51. Strain energy	ε	J/m^3	Joule per cubic metre	erg/cm^3
52. Surface energy	Y	J/m^2	Joule per square metre	erg/cm^2
53. Surface tension	γ	N/M	Newton per metre	dyne/cm
54. Temperature	T	K	Kelvin	°C, °F
55. Temperature co-efficient of resistance	α_0	K^{-1}	per Kelvin	per °F
56. Tensile strength	—	MN/m^2	Mega Newton per square metre	Kgf/mm^2, Psi, dyne/cm^2

(Contd.)

Quantity	Symbol	S.I. Units Expressed as	S.I. Units Full form	Other units
57. Tensile stress	σ	MN/m^2	Mega Newton per square metre	PSi, Kgf/mm^2
58. Thermal energy	*RT*	J/mol	Joule per mole	cal/mole, cal/gm
59. Thermal conductivity	—	W/m/K	Watt per metre per deg Kelvin	Kcal/m/s/°C
60. Ultimate tensile strength	*UTS*	MN/m^2	Mega Newton per Square metre	PSi, Kgf/mm^2
61. Vibration frequency	ν or s^{-1}	S^{-1}	Per second	—
62. Viscosity	η	Pas; N/m^2s	Pascal second Newton per square metre second	Poise
63. Wavelength of radiation	λ	nm	nanometre	Å
64. Young's modulus	*E*	GPa,	Giga pascal;	PSi, Kgf/mm^2
65. Yield strength	σ_y	N/m^2	$Newton/m^2$	Kgf/mm^2

APPENDIX 2

Conversion Factors

1. 1° = 0.01745 rad
2. 1Å $=10^{-10}$ m = 0.1 nm
3. 1 A·hr = 3.6 Kc
4. 1 ampere = 1 C/s
5. 1 atm = 0.101325×10^6 N/m^2 = 760 mm Hg = 10^5 Pascal
6. 1 Btu = 1.05506 kJ
7. 1 Btu/lb = 2326 J/kg
8. 1 Btu/ft^3 = 37.2589 kJ/m^3
9. 1 bar = 10^{-1} MPa
10. 1 Calorie = 4.18 J
11. 1 Coulomb = 1 A-s
12. 1 Debye = 3.33×10^{-30} cm
13. 1 Dyne/cm = 10^{-3} N/m
14. 1 Dyne/cm^2 = 0.1 N/m^2
15. 1 erg = 10^{-7} J
16. 1 erg/cm = 10^{-5} J/m
17. 1 eV = 1.602×10^{-19} J
18. 1 eV/entity = 96.49 kJ/mol (of entities)
19. 1 eV/particle = 96.49 kJ/mol
20. 1 farad = 1 C/V
21. 1 Gauss = 10^{-4} Wb/m^2
22. 1 gram-calorie = 4.185 J
23. 1 Henry = 1 V-s/A
24. 1 Hertz = 1 cycle/s
25. 1 HP = 0.7457 kW
26. 1 Kgf/mm^2 = 9.81 MN/m^2
27. 1 KSi/in = 1.1 MN/m$^{3/2}$
28. 1 lb/Cu.ft = 16.01 kg/m^3
29. 1 lumen = 0.0016 W (at 0.55 μm)

30. 1 Newton	= 1 kg·m/s^2
31. 1 Oersted	= 79.6 A/m
32. 1 Poise	= 0.1 PaS
33. 1 PSi	= 6.89 kN/m^2 10^5
34. T°C	= (T + 273.15) K
35. T°F	= 5/9 (T + 459.67) K
36. 1 torr (mm of Hg)	= 133.3 N/m^2
37. 1 TR	= 3024 kCal
38. 1 Watt	= 1 Joule/sec
39. TR	= 3024 kcal/hg
40. 1 Joule	= 10^7 ergs
41. 1 k-cal	= 4.18 kJ

APPENDIX 3

Physical Constants

Constant	Symbol	Value
1. Acceleration due to gravity	g	$= 9.81\ m/s^2$
2. Atomic mass unit	amu	$= 1.660 \times 10^{-24}$ g
3. Avogadro's number	N	$= 6.023 \times 10^{-23}$g/mol
4. Bohr magneton (magnetic moment)	β	$= 9.273 \times 10^{-24}$ A m^2
5. Boltzmann's constant	k	$= 1.380 \times 10^{-23}$ J/K
		$= 8.620 \times 10^{-5}$ eV/k
6. Electron rest mass	m_e	$= 9.109 \times 10^{-31}$ Kg
7. Electronic charge	e	$= 1.602 \times 10^{-19}$ C
8. Faraday's constant	F	$= 9.649 \times 10^{4}$ C/mol
9. Gas constant	R	= 8.314 J/mol/K
10. Mass of proton	m_P	$= 1.673 \times 10^{-24}$ g
11. Mass of electron	m_e	$= 9.108 \times 10^{-28}$ g
12. Planck's constant	h	$= 6.626 \times 10^{-34}$ Js
13. Permeability of free space	μ_0	$= 1.257 \times 10^{-6}$ H/m
14. Velocity of light in free space	c	$= 2.998 \times 10^{8}$ m/s
15. Permittivity of free space	ε_0	$= 8.854 \times 10^{-12}$ F/m
16. Volume of 1 kg mole of ideal gas at N.T.P.	V	= 22.41 m^3
17. Magnetic constant	μ_i	$= 1.2566 \times 10^{-8}$ H/cm
18. Gravitation constants	G	$= 6.670 \times 10^{-1}$
19. Radius of electron	r_e	$= 2.81777 \times 10^{-15}$ m

APPENDIX 4

Prefix Names, Symbols and Multiplication Factors

Name	Symbol	Factor by which unit is multiplied
1. 1 Deci	d	$= 10^{-1}$
2. 1 exa	E	$= 10^{18}$
3. 1 Giga	G	$= 10^{9}$
4. 1 Kilo	K	$= 10^{3}$
5. 1 Mega	M	$= 10^{6}$
6. 1 micro	μ	$= 10^{-6}$
7. 1 milli	mm	$= 10^{-3}$
8. 1 micron	mie	= 0.001 in = 25 μm
9. 1 Nano m	nm	$= 10^{-9}$ m
10. 1 peta	P	$= 10^{15}$
11. 1 tera	T	$= 10^{12}$

Index

D

E

F

G

H